T0211651

Lecture Notes in Artificial Intelligence 12034

Subseries of Lecture Notes in Computer Science

More information about this series at http://www.springer.com/series/1244

Ngoc Thanh Nguyen · Kietikul Jearanaitanakij ·
Ali Selamat · Bogdan Trawiński ·
Suphamit Chittayasothorn (Eds.)

Intelligent Information and Database Systems

12th Asian Conference, ACIIDS 2020
Phuket, Thailand, March 23–26, 2020
Proceedings, Part II

Editors
Ngoc Thanh Nguyen
Department of Applied Informatics
Wrocław University of Science
and Technology
Wrocław, Poland

Ali Selamat
Faculty of Computer
Science and Information
University Teknologi Malaysia
Kuala Lumpur, Malaysia

Suphamit Chittayasothorn
King Mongkut's Institute
of Technology Ladkrabang
Bangkok, Thailand

Kietikul Jearanaitanakij
King Mongkut's Institute
of Technology Ladkrabang
Bangkok, Thailand

Bogdan Trawiński
Department of Applied Informatics
Wrocław University of Science
and Technology
Wrocław, Poland

ISSN 0302-9743 ISSN 1611-3349 (electronic)
Lecture Notes in Artificial Intelligence
ISBN 978-3-030-42057-4 ISBN 978-3-030-42058-1 (eBook)
https://doi.org/10.1007/978-3-030-42058-1

LNCS Sublibrary: SL7 – Artificial Intelligence

This Springer imprint is published by the registered company Springer Nature Switzerland AG
The registered company address is: Gewerbestrasse 11, 6330 Cham, Switzerland

Preface

ACIIDS 2020 was the 12th event in a series of international scientific conferences on research and applications in the field of intelligent information and database systems. The aim of ACIIDS 2020 was to provide an international forum of research workers with scientific backgrounds on the technology of intelligent information and database systems and its various applications. The ACIIDS 2020 conference was co-organized by King Mongkut's Institute of Technology Ladkrabang (Thailand) and Wrocław University of Science and Technology (Poland) in cooperation with the IEEE SMC Technical Committee on Computational Collective Intelligence, European Research Center for Information Systems (ERCIS), The University of Newcastle (Australia), Yeungnam University (South Korea), Leiden University (The Netherlands), Universiti Teknologi Malaysia (Malaysia), BINUS University (Indonesia), Quang Binh University (Vietnam), and Nguyen Tat Thanh University (Vietnam). It took place in Phuket, Thailand during March 23–26, 2020.

The ACIIDS conference series is already well established. The first two events, ACIIDS 2009 and ACIIDS 2010, took place in Dong Hoi City and Hue City in Vietnam, respectively. The third event, ACIIDS 2011, took place in Daegu (South Korea), followed by the fourth event, ACIIDS 2012, in Kaohsiung (Taiwan). The fifth event, ACIIDS 2013, was held in Kuala Lumpur (Malaysia) while the sixth event, ACIIDS 2014, was held in Bangkok (Thailand). The seventh event, ACIIDS 2015, took place in Bali (Indonesia), followed by the eighth event, ACIIDS 2016, in Da Nang (Vietnam). The ninth event, ACIIDS 2017, was organized in Kanazawa (Japan). The 10th jubilee conference, ACIIDS 2018, was held in Dong Hoi City (Vietnam), followed by the 11th event, ACIIDS 2019, in Yogyakarta (Indonesia).

For this edition of the conference we received 285 papers from 43 countries all over the world. Each paper was peer reviewed by at least two members of the international Program Committee and the international board of reviewers. Only 105 papers with the highest quality were selected for an oral presentation and publication in these two volumes of the ACIIDS 2020 proceedings.

Papers included in these proceedings cover the following topics: knowledge engineering and Semantic Web; natural language processing; decision support and control systems; computer vision techniques; machine learning and data mining; deep learning models; advanced data mining techniques and applications; multiple model approach to machine learning; application of intelligent methods to constrained problems; automated reasoning with applications in intelligent systems; current trends in artificial intelligence; optimization, learning, and decision-making in bioinformatics and bioengineering; computer vision and intelligent systems, data modeling and processing for industry 4.0; intelligent applications of the Internet of Things (IoT) and data analysis technologies; intelligent and contextual systems; intelligent systems and algorithms in information sciences; intelligent supply chains and e-commerce; privacy, security, and trust in artificial intelligence; and interactive analysis of image, video, and motion data in life sciences.

The accepted and presented papers focus on new trends and challenges facing the intelligent information and database systems community. The presenters showed how research work could stimulate novel and innovative applications. We hope that you found these results useful and inspiring for your future research work.

We would like to express our sincere thanks to the honorary chairs for their support: Prof. Suchatvee Suwansawat (President of King Mongkut's Institute of Technology, Ladkrabang, Thailand), Cezary Madryas (Rector of Wrocław University of Science and Technology, Poland), Prof. Moonis Ali (President of the International Society of Applied Intelligence, USA), and Prof. Komsan Maleesee (Dean of Engineering, King Mongkut's Institute of Technology, Ladkrabang, Thailand).

Our special thanks go to the program chairs, special session chairs, organizing chairs, publicity chairs, liaison chairs, and local Organizing Committee for their work towards the conference. We sincerely thank all the members of the international Program Committee for their valuable efforts in the review process, which helped us to guarantee the highest quality of the selected papers for the conference. We cordially thank the organizers and chairs of special sessions who contributed to the success of the conference.

We would like to express our thanks to the keynote speakers for their world-class plenary speeches: Prof. Włodzisław Duch from the Nicolaus Copernicus University (Poland), Prof. Nikola Kasabov from Auckland University of Technology (New Zealand), Prof. Dusit Niyato from Nanyang Technological University (Singapore), and Prof. Geoff Webb from the Monash University Centre for Data Science (Australia).

We cordially thank our main sponsors: King Mongkut's Institute of Technology Ladkrabang (Thailand), Wrocław University of Science and Technology (Poland), IEEE SMC Technical Committee on Computational Collective Intelligence, European Research Center for Information Systems (ERCIS), The University of Newcastle (Australia), Yeungnam University (South Korea), Leiden University (The Netherlands), Universiti Teknologi Malaysia (Malaysia), BINUS University (Indonesia), Quang Binh University (Vietnam), and Nguyen Tat Thanh University (Vietnam). Our special thanks are also due to Springer for publishing the proceedings and sponsoring awards, and to all the other sponsors for their kind support.

We wish to thank the members of the Organizing Committee for their excellent work and the members of the local Organizing Committee for their considerable effort. We cordially thank all the authors, for their valuable contributions, and the other participants of this conference. The conference would not have been possible without their support. Thanks are also due to many experts who contributed to making the event a success.

We would like to extend our heartfelt thanks to Jarosław Gowin, Deputy Prime Minister of the Republic of Poland and Minister of Science and Higher Education, for his support and honorary patronage of the conference.

March 2020

Ngoc Thanh Nguyen
Kietikul Jearanaitanakij
Ali Selamat
Bogdan Trawiński
Suphamit Chittayasothorn

Organization

Honorary Chairs

Suchatvee Suwansawat	President of King Mongkut's Institute of Technology Ladkrabang, Thailand
Cezary Madryas	Rector of Wrocław University of Science and Technology, Poland
Moonis Ali	President of International Society of Applied Intelligence, USA
Komsan Maleesee	Dean of Engineering, King Mongkut's Institute of Technology Ladkrabang, Thailand

General Chairs

Ngoc Thanh Nguyen	Wrocław University of Science and Technology, Poland
Suphamit Chittayasothorn	King Mongkut's Institute of Technology Ladkrabang, Thailand

Program Chairs

Kietikul Jearanaitanakij	King Mongkut's Institute of Technology Ladkrabang, Thailand
Tzung-Pei Hong	National University of Kaohsiung, Taiwan
Ali Selamat	Universiti Teknologi Malaysia, Malaysia
Edward Szczerbicki	The University of Newcastle, Australia
Bogdan Trawiński	Wrocław University of Science and Technology, Poland

Steering Committee

Ngoc Thanh Nguyen (Chair)	Wrocław University of Science and Technology, Poland
Longbing Cao	University of Technology Sydney, Australia
Suphamit Chittayasothorn	King Mongkut's Institute of Technology Ladkrabang, Thailand
Ford Lumban Gaol	Bina Nusantara University, Indonesia
Tu Bao Ho	Japan Advanced Institute of Science and Technology, Japan
Tzung-Pei Hong	National University of Kaohsiung, Taiwan
Dosam Hwang	Yeungnam University, South Korea
Bela Stantic	Griffith University, Australia

Geun-Sik Jo	Inha University, South Korea
Hoai An Le-Thi	University of Lorraine, France
Zygmunt Mazur	Wrocław University of Science and Technology, Poland
Toyoaki Nishida	Kyoto University, Japan
Leszek Rutkowski	Częstochowa University of Technology, Poland
Ali Selamat	Universiti Teknologi Malaysia, Malaysia

Special Session Chairs

Marcin Pietranik	Wroclaw University of Science and Technology, Poland
Chutimet Srinilta	King Mongkut's Institute of Technology Ladkrabang, Thailand
Paweł Sitek	Kielce University of Technology, Poland

Liaison Chairs

Ford Lumban Gaol	Bina Nusantara University, Indonesia
Quang-Thuy Ha	VNU University of Engineering and Technology, Vietnam
Mong-Fong Horng	National Kaohsiung University of Applied Sciences, Taiwan
Dosam Hwang	Yeungnam University, South Korea
Le Minh Nguyen	Japan Advanced Institute of Science and Technology, Japan
Ali Selamat	Universiti Teknologi Malaysia, Malaysia

Organizing Chairs

Wiboon Prompanich	King Mongkut's Institute of Technology Ladkrabang, Thailand
Adrianna Kozierkiewicz	Wrocław University of Science and Technology, Poland
Krystian Wojtkiewicz	Wrocław University of Science and Technology, Poland

Publicity Chairs

Rathachai Chawuthai	King Mongkut's Institute of Technology Ladkrabang, Thailand
Marek Kopel	Wrocław University of Science and Technology, Poland
Marek Krótkiewicz	Wrocław University of Science and Technology, Poland

Webmaster

Marek Kopel	Wroclaw University of Science and Technology, Poland

Local Organizing Committee

Pakorn Watanachaturaporn	King Mongkut's Institute of Technology Ladkrabang, Thailand
Sorayut Glomglome	King Mongkut's Institute of Technology Ladkrabang, Thailand
Watchara Chatwiriya	King Mongkut's Institute of Technology Ladkrabang, Thailand
Sathaporn Promwong	King Mongkut's Institute of Technology Ladkrabang, Thailand
Putsadee Pornphol	Phuket Rajabhat University, Thailand
Maciej Huk	Wrocław University of Science and Technology, Poland
Marcin Jodłowiec	Wrocław University of Science and Technology, Poland

Keynote Speakers

Włodzisław Duch	Nicolaus Copernicus University, Poland
Nikola Kasabov	Auckland University of Technology, New Zealand
Dusit Niyato	Nanyang Technological University, Singapore
Geoff Webb	Monash University Centre for Data Science, Australia

Special Sessions Organizers

1. *CSHAC 2020: Special Session on Cyber-Physical Systems in Healthcare: Applications and Challenges*

Michael Mayo	University of Waikato, New Zealand
Abigail Koay	University of Waikato, New Zealand
Panos Patros	University of Waikato, New Zealand

2. *ADMTA 2020: Special Session on Advanced Data Mining Techniques and Applications*

Chun-Hao Chen	Tamkang University, Taiwan
Bay Vo	Ho Chi Minh City University of Technology, Vietnam
Tzung-Pei Hong	National University of Kaohsiung, Taiwan

3. *CVIS 2020: Special Session on Computer Vision and Intelligent Systems*

Van-Dung Hoang	Quang Binh University, Vietnam
Kang-Hyun Jo	University of Ulsan, South Korea
My-Ha Le	Ho Chi Minh City University of Technology and Education, Vietnam
Van-Huy Pham	Ton Duc Thang University, Vietnam

4. *MMAML 2020: Special Session on Multiple Model Approach to Machine Learning*

Tomasz Kajdanowicz	Wrocław University of Science and Technology, Poland
Edwin Lughofer	Johannes Kepler University Linz, Austria
Bogdan Trawiński	Wrocław University of Science and Technology, Poland

5. *CIV 2020: Special Session on Computational Imaging and Vision*

Manish Khare	Dhirubhai Ambani Institute of Information and Communication Technology, India
Prashant Srivastava	NIIT University, India
Om Prakash	Inferigence Quotient, India
Jeonghwan Gwak	Korea National University of Transportation, South Korea

6. *ISAIS 2020: Special Session on Intelligent Systems and Algorithms in Information Sciences*

Martin Kotyrba	University of Ostrava, Czech Republic
Eva Volna	University of Ostrava, Czech Republic
Ivan Zelinka	VŠB – Technical University of Ostrava, Czech Republic
Pavel Petr	University of Pardubice, Czech Republic

7. *IOTAI 2020: Special Session on Internet of Things and Artificial Intelligence for Energy Efficiency-Recent Advances and Future Trends*

Mohamed Elhoseny	Mansoura University, Egypt
Mohamed Abdel-Basset	Zagazig University, Egypt

8. *DMPI-4.0 2020: Special Session on Data Modelling and Processing for Industry 4.0*

Du Haizhou	Shanghai University of Electric Power, China
Wojciech Hunek	Opole University of Technology, Poland
Marek Krótkiewicz	Wrocław University of Science and Technology, Poland
Krystian Wojtkiewicz	Wrocław University of Science and Technology, Poland

9. *ICxS 2020: Special Session on Intelligent and Contextual Systems*

Maciej Huk	Wroclaw University of Science and Technology, Poland
Keun Ho Ryu	Ton Duc Thang University, Vietnam
Goutam Chakraborty	Iwate Prefectural University, Japan
Qiangfu Zhao	University of Aizu, Japan
Chao-Chun Chen	National Cheng Kung University, Taiwan
Rashmi Dutta Baruah	Indian Institute of Technology Guwahati, India

10. *ARAIS 2020: Special Session on Automated Reasoning with Applications in Intelligent Systems*

Jingde Cheng	Saitama University, Japan

11. *ISCEC 2020: Special Session on Intelligent Supply Chains and e-Commerce*

Arkadiusz Kawa	Łukasiewicz Research Network – The Institute of Logistics and Warehousing, Poland
Justyna Światowiec-Szczepańska	Poznań University of Economics and Business, Poland
Bartłomiej Pierański	Poznań University of Economics and Business, Poland

12. *IAIOTDAT 2020: Special Session on Intelligent Applications of Internet of Things and Data Analysis Technologies*

Rung Ching Chen	Chaoyang University of Technology, Taiwan
Yung-Fa Huang	Chaoyang University of Technology, Taiwan
Yu-Huei Cheng	Chaoyang University of Technology, Taiwan

13. *CTAIOLDMBB 2020: Special Session on Current Trends in Artificial Intelligence, Optimization, Learning, and Decision-Making in Bioinformatics and Bioengineering*

Dominik Vilimek	VŠB – Technical University of Ostrava, Czech Republic
Jan Kubíček	VŠB – Technical University of Ostrava, Czech Republic
Marek Penhaker	VŠB – Technical University of Ostrava, Czech Republic
Muhammad Usman Akram	National University of Sciences and Technology Pakistan, Pakistan
Vladimir Juras	Medical University of Vienna, Austria
Bhabani Shankar Prasad Mishra	KIIT University, India
Ondrej Krejcar	University of Hradec Kralove, Czech Republic

14. *SAILS 2020: Special Session on Interactive Analysis of Image, Video and Motion Data in Life Sciences*

Konrad Wojciechowski	Polish-Japanese Academy of Information Technology, Poland
Marek Kulbacki	Polish-Japanese Academy of Information Technology, Poland
Jakub Segen	Polish-Japanese Academy of Information Technology, Poland
Zenon Chaczko	University of Technology Sydney, Australia
Andrzej Przybyszewski	UMass Medical School, USA
Jerzy Nowacki	Polish-Japanese Academy of Information Technology, Poland

15. *AIMCP 2020: Special Session on Application of Intelligent Methods to Constrained Problems*

Jarosław Wikarek	Kielce University of Technology, Poland
Mukund Janardhanan	University of Leicester, UK

16. *PSTrustAI 2020: Special Session on Privacy, Security and Trust in Artificial Intelligence*

Pascal Bouvry	University of Luxembourg, Luxembourg
Matthias R. Brust	University of Luxembourg, Luxembourg
Grégoire Danoy	University of Luxembourg, Luxembourg
El-ghazali Talbi	University of Lille, France

17. *IMSAGRWS 2020: Intelligent Modeling and Simulation Approaches for Games and Real World Systems*

Doina Logofătu	Frankfurt University of Applied Sciences, Germany
Costin Bădică	University of Craiova, Romania
Florin Leon	Gheorghe Asachi Technical University, Romania

International Program Committee

Muhammad Abulaish	South Asian University, India
Waseem Ahmad	Waiariki Institute of Technology, New Zealand
R. S. Ajin	Idukki District Disaster Management Authority, India
Jesus Alcala-Fdez	University of Granada, Spain
Bashar Al-Shboul	University of Jordan, Jordan
Lionel Amodeo	University of Technology of Troyes, France
Toni Anwar	Universiti Teknologi Petronas, Malaysia
Taha Arbaoui	University of Technology of Troyes, France
Mehmet Emin Aydin	University of the West of England, UK
Ahmad Taher Azar	Prince Sultan University, Saudi Arabia
Thomas Bäck	Leiden University, The Netherlands

Amelia Badica	University of Craiova, Romania
Costin Badica	University of Craiova, Romania
Kambiz Badie	ICT Research Institute, Iran
Hassan Badir	École Nationale des Sciences Appliquées de Tanger, Morocco
Zbigniew Banaszak	Warsaw University of Technology, Poland
Dariusz Barbucha	Gdynia Maritime University, Poland
Ramazan Bayindir	Gazi University, Turkey
Maumita Bhattacharya	Charles Sturt University, Australia
Leon Bobrowski	Białystok University of Technology, Poland
Bülent Bolat	Yildiz Technical University, Turkey
Mariusz Boryczka	University of Silesia, Poland
Urszula Boryczka	University of Silesia, Poland
Zouhaier Brahmia	University of Sfax, Tunisia
Stephane Bressan	National University of Singapore, Singapore
Peter Brida	University of Žilina, Slovakia
Andrej Brodnik	University of Ljubljana, Slovenia
Piotr Bródka	Wroclaw University of Science and Technology, Poland
Grażyna Brzykcy	Poznan University of Technology, Poland
Robert Burduk	Wrocław University of Science and Technology, Poland
Aleksander Byrski	AGH University of Science and Technology, Poland
Tru Cao	Ho Chi Minh City University of Technology, Vietnam
Leopoldo Eduardo Cardenas-Barron	Tecnologico de Monterrey, Mexico
Oscar Castillo	Tijuana Institute of Technology, Mexico
Dariusz Ceglarek	WSB University in Poznań, Poland
Stefano A. Cerri	University of Montpellier, France
Zenon Chaczko	University of Technology Sydney, Australia
Altangerel Chagnaa	National University of Mongolia, Mongolia
Somchai Chatvichienchai	University of Nagasaki, Japan
Chun-Hao Chen	Tamkang University, Taiwan
Rung-Ching Chen	Chaoyang University of Technology, Taiwan
Shyi-Ming Chen	National Taiwan University of Science and Technology, Taiwan
Leszek J. Chmielewski	Warsaw University of Life Sciences, Poland
Sung-Bae Cho	Yonsei University, South Korea
Kazimierz Choroś	Wrocław University of Science and Technology, Poland
Kun-Ta Chuang	National Cheng Kung University, Taiwan
Piotr Chynał	Wrocław University of Science and Technology, Poland
Dorian Cojocaru	University of Craiova, Romania
Jose Alfredo Ferreira Costa	Federal University of Rio Grande do Norte (UFRN), Brazil

Ireneusz Czarnowski	Gdynia Maritime University, Poland
Piotr Czekalski	Silesian University of Technology, Poland
Theophile Dagba	University of Abomey-Calavi, Benin
Tien V. Do	Budapest University of Technology and Economics, Hungary
Grzegorz Dobrowolski	AGH University of Science and Technology, Poland
Rafał Doroz	University of Silesia, Poland
Habiba Drias	University of Science and Technology Houari Boumediene, Algeria
Maciej Drwal	Wrocław University of Science and Technology, Poland
Ewa Dudek-Dyduch	AGH University of Science and Technology, Poland
El-Sayed M. El-Alfy	King Fahd University of Petroleum and Minerals, Saudi Arabia
Keiichi Endo	Ehime University, Japan
Sebastian Ernst	AGH University of Science and Technology, Poland
Nadia Essoussi	University of Carthage, Tunisia
Rim Faiz	University of Carthage, Tunisia
Victor Felea	Universitatea Alexandru Ioan Cuza din Iași, Romania
Simon Fong	University of Macau, Macau SAR
Dariusz Frejlichowski	West Pomeranian University of Technology, Poland
Blanka Frydrychova Klimova	University of Hradec Králové, Czech Republic
Mohamed Gaber	Birmingham City University, UK
Marina L. Gavrilova	University of Calgary, Canada
Janusz Getta	University of Wollongong, Australia
Daniela Gifu	Universitatea Alexandru Ioan Cuza din Iași, Romania
Fethullah Göçer	Galatasaray University, Turkey
Daniela Godoy	ISISTAN Research Institute, Argentina
Gergo Gombos	Eötvös Loránd University, Hungary
Fernando Gomide	University of Campinas, Brazil
Antonio Gonzalez-Pardo	Universidad Autónoma de Madrid, Spain
Janis Grundspenkis	Riga Technical University, Latvia
Claudio Gutierrez	Universidad de Chile, Chile
Quang-Thuy Ha	VNU University of Engineering and Technology, Vietnam
Dawit Haile	Addis Ababa University, Ethiopia
Pei-Yi Hao	National Kaohsiung University of Applied Sciences, Taiwan
Spits Warnars Harco Leslie Hendric	BINUS University, Indonesia
Marcin Hernes	Wrocław University of Economics and Business, Poland
Francisco Herrera	University of Granada, Spain
Koichi Hirata	Kyushu Institute of Technology, Japan

Bogumiła Hnatkowska	Wrocław University of Science and Technology, Poland
Huu Hanh Hoang	Posts and Telecommunications Institute of Technology, Vietnam
Quang Hoang	Hue University of Sciences, Vietnam
Van-Dung Hoang	Quang Binh University, Vietnam
Jaakko Hollmen	Aalto University, Finland
Tzung-Pei Hong	National University of Kaohsiung, Taiwan
Mong-Fong Horng	National Kaohsiung University of Applied Sciences, Taiwan
Yung-Fa Huang	Chaoyang University of Technology, Taiwan
Maciej Huk	Wrocław University of Science and Technology, Poland
Dosam Hwang	Yeungnam University, South Korea
Roliana Ibrahim	Universiti Teknologi Malaysia, Malaysia
Mirjana Ivanovic	University of Novi Sad, Serbia
Sanjay Jain	National University of Singapore, Singapore
Jarosław Jankowski	West Pomeranian University of Technology, Poland
Kietikul Jearanaitanakij	King Mongkut's Institute of Technology Ladkrabang, Thailand
Khalid Jebari	LCS Rabat, Morocco
Janusz Jeżewski	Institute of Medical Technology and Equipment ITAM, Poland
Joanna Jędrzejowicz	University of Gdańsk, Poland
Piotr Jędrzejowicz	Gdynia Maritime University, Poland
Przemysław Juszczuk	University of Economics in Katowice, Poland
Dariusz Kania	Silesian University of Technology, Poland
Nikola Kasabov	Auckland University of Technology, New Zealand
Arkadiusz Kawa	Poznań University of Economics and Business, Poland
Zaheer Khan	University of the West of England, UK
Muhammad Khurram Khan	King Saud University, Saudi Arabia
Marek Kisiel-Dorohinicki	AGH University of Science and Technology, Poland
Attila Kiss	Eötvös Loránd University, Hungary
Jerzy Klamka	Silesian University of Technology, Poland
Frank Klawonn	Ostfalia University of Applied Sciences, Germany
Shinya Kobayashi	Ehime University, Japan
Joanna Kolodziej	Cracow University of Technology, Poland
Grzegorz Kołaczek	Wrocław University of Science and Technology, Poland
Marek Kopel	Wrocław University of Science and Technology, Poland
Józef Korbicz	University of Zielona Gora, Poland
Raymondus Kosala	BINUS University, Indonesia
Leszek Koszałka	Wroclaw University of Science and Technology, Poland
Leszek Kotulski	AGH University of Science and Technology, Poland

Jan Kozak	University of Economics in Katowice, Poland
Adrianna Kozierkiewicz	Wrocław University of Science and Technology, Poland
Ondrej Krejcar	University of Hradec Králové, Czech Republic
Dariusz Król	Wrocław University of Science and Technology, Poland
Marek Krótkiewicz	Wrocław University of Science and Technology, Poland
Marzena Kryszkiewicz	Warsaw University of Technology, Poland
Adam Krzyzak	Concordia University, Canada
Jan Kubíček	VSB – Technical University of Ostrava, Czech Republic
Tetsuji Kuboyama	Gakushuin University, Japan
Elżbieta Kukla	Wrocław University of Science and Technology, Poland
Julita Kulbacka	Wrocław Medical University, Poland
Marek Kulbacki	Polish-Japanese Academy of Information Technology, Poland
Kazuhiro Kuwabara	Ritsumeikan University, Japan
Halina Kwaśnicka	Wroclaw University of Science and Technology, Poland
Annabel Latham	Manchester Metropolitan University, UK
Bac Le	VNU University of Science, Vietnam
Kun Chang Lee	Sungkyunkwan University, South Korea
Yue-Shi Lee	Ming Chuan University, Taiwan
Florin Leon	Gheorghe Asachi Technical University of Iasi, Romania
Horst Lichter	RWTH Aachen University, Germany
Igor Litvinchev	Nuevo Leon State University, Mexico
Rey-Long Liu	Tzu Chi University, Taiwan
Doina Logofatu	Frankfurt University of Applied Sciences, Germany
Edwin Lughofer	Johannes Kepler University Linz, Austria
Lech Madeyski	Wrocław University of Science and Technology, Poland
Nezam Mahdavi-Amiri	Sharif University of Technology, Iran
Bernadetta Maleszka	Wrocław University of Science and Technology, Poland
Marcin Maleszka	Wrocław University of Science and Technology, Poland
Yannis Manolopoulos	Open University of Cyprus, Cyprus
Konstantinos Margaritis	University of Macedonia, Greece
Vukosi Marivate	Council for Scientific and Industrial Research, South Africa
Urszula Markowska-Kaczmar	Wrocław University of Science and Technology, Poland

Takashi Matsuhisa	Karelia Research Centre, Russian Academy of Science, Russia
Tamás Matuszka	Eötvös Loránd University, Hungary
Vladimir Mazalov	Karelia Research Centre, Russian Academy of Sciences, Russia
Héctor Menéndez	University College London, UK
Mercedes Merayo	Universidad Complutense de Madrid, Spain
Jacek Mercik	WSB University in Wrocław, Poland
Radosław Michalski	Wrocław University of Science and Technology, Poland
Peter Mikulecky	University of Hradec Králové, Czech Republic
Miroslava Mikusova	University of Žilina, Slovakia
Marek Milosz	Lublin University of Technology, Poland
Jolanta Mizera-Pietraszko	Opole University, Poland
Nurhizam Safie Mohd Satar	Universiti Kebangsaan Malaysia, Malaysia
Leo Mrsic	IN2data Ltd Data Science Company, Croatia
Agnieszka Mykowiecka	Institute of Computer Science, Polish Academy of Sciences, Poland
Pawel Myszkowski	Wrocław University of Science and Technology, Poland
Grzegorz J. Nalepa	AGH University of Science and Technology, Poland
Fulufhelo Nelwamondo	Council for Scientific and Industrial Research, South Africa
Huu-Tuan Nguyen	Vietnam Maritime University, Vietnam
Le Minh Nguyen	Japan Advanced Institute of Science and Technology, Japan
Loan T. T. Nguyen	International University (VNU-HCMC), Vietnam
Ngoc-Thanh Nguyen	Wrocław University of Science and Technology, Poland
Quang-Vu Nguyen	Korea-Vietnam Friendship Information Technology College, Vietnam
Thai-Nghe Nguyen	Cantho University, Vietnam
Yusuke Nojima	Osaka Prefecture University, Japan
Jerzy Paweł Nowacki	Polish-Japanese Academy of Information Technology, Poland
Agnieszka Nowak-Brzezińska	University of Silesia, Poland
Mariusz Nowostawski	Norwegian University of Science and Technology, Norway
Alberto Núñez	Universidad Complutense de Madrid, Spain
Manuel Núñez	Universidad Complutense de Madrid, Spain
Kouzou Ohara	Aoyama Gakuin University, Japan
Tarkko Oksala	Aalto University, Finland
Marcin Paprzycki	Systems Research Institute, Polish Academy of Sciences, Poland
Jakub Peksiński	West Pomeranian University of Technology, Poland

Danilo Pelusi	University of Teramo, Italy
Bernhard Pfahringer	University of Waikato, New Zealand
Bartłomiej Pierański	Poznan University of Economics and Business, Poland
Dariusz Pierzchała	Military University of Technology, Poland
Marcin Pietranik	Wrocław University of Science and Technology, Poland
Elias Pimenidis	University of the West of England, UK
Jaroslav Pokorný	Charles University in Prague, Czech Republic
Nikolaos Polatidis	University of Brighton, UK
Elvira Popescu	University of Craiova, Romania
Petra Poulova	University of Hradec Králové, Czech Republic
Om Prakash	University of Allahabad, India
Radu-Emil Precup	Politehnica University of Timisoara, Romania
Małgorzata Przybyła-Kasperek	University of Silesia, Poland
Paulo Quaresma	Universidade de Evora, Portugal
David Ramsey	Wrocław University of Science and Technology, Poland
Mohammad Rashedur Rahman	North South University, Bangladesh
Ewa Ratajczak-Ropel	Gdynia Maritime University, Poland
Sebastian A. Rios	University of Chile, Chile
Leszek Rutkowski	Częstochowa University of Technology, Poland
Alexander Ryjov	Lomonosov Moscow State University, Russia
Keun Ho Ryu	Chungbuk National University, South Korea
Virgilijus Sakalauskas	Vilnius University, Lithuania
Daniel Sanchez	University of Granada, Spain
Rafał Scherer	Częstochowa University of Technology, Poland
Juergen Schmidhuber	Swiss AI Lab IDSIA, Switzerland
Ali Selamat	Universiti Teknologi Malaysia, Malaysia
Tegjyot Singh Sethi	University of Louisville, USA
Natalya Shakhovska	Lviv Polytechnic National University, Ukraine
Donghwa Shin	Yeungnam University, South Korea
Andrzej Siemiński	Wrocław University of Science and Technology, Poland
Dragan Simic	University of Novi Sad, Serbia
Bharat Singh	Universiti Teknology PETRONAS, Malaysia
Paweł Sitek	Kielce University of Technology, Poland
Andrzej Skowron	Warsaw University, Poland
Adam Słowik	Koszalin University of Technology, Poland
Vladimir Sobeslav	University of Hradec Králové, Czech Republic
Kamran Soomro	University of the West of England, UK
Zenon A. Sosnowski	Białystok University of Technology, Poland
Chutimet Srinilta	King Mongkut's Institute of Technology Ladkrabang, Thailand
Bela Stantic	Griffith University, Australia

Jerzy Stefanowski	Poznań University of Technology, Poland
Stanimir Stoyanov	University of Plovdiv "Paisii Hilendarski", Bulgaria
Ja-Hwung Su	Cheng Shiu University, Taiwan
Libuse Svobodova	University of Hradec Králové, Czech Republic
Tadeusz Szuba	AGH University of Science and Technology, Poland
Julian Szymański	Gdańsk University of Technology, Poland
Krzysztof Ślot	Łódź University of Technology, Poland
Jerzy Świątek	Wrocław University of Science and Technology, Poland
Andrzej Świerniak	Silesian University of Technology, Poland
Ryszard Tadeusiewicz	AGH University of Science and Technology, Poland
Muhammad Atif Tahir	National University of Computing and Emerging Sciences, Pakistan
Yasufumi Takama	Tokyo Metropolitan University, Japan
Maryam Tayefeh Mahmoudi	ICT Research Institute, Iran
Zbigniew Telec	Wrocław University of Science and Technology, Poland
Dilhan Thilakarathne	Vrije Universiteit Amsterdam, The Netherlands
Satoshi Tojo	Japan Advanced Institute of Science and Technology, Japan
Bogdan Trawiński	Wrocław University of Science and Technology, Poland
Trong Hieu Tran	VNU University of Engineering and Technology, Vietnam
Ualsher Tukeyev	Al-Farabi Kazakh National University, Kazakhstan
Olgierd Unold	Wrocław University of Science and Technology, Poland
Natalie Van Der Wal	Vrije Universiteit Amsterdam, The Netherlands
Jorgen Villadsen	Technical University of Denmark, Denmark
Bay Vo	Ho Chi Minh City University of Technology, Vietnam
Gottfried Vossen	ERCIS Münster, Germany
Wahyono Wahyono	Universitas Gadjah Mada, Indonesia
Lipo Wang	Nanyang Technological University, Singapore
Junzo Watada	Waseda University, Japan
Izabela Wierzbowska	Gdynia Maritime University, Poland
Krystian Wojtkiewicz	Wrocław University of Science and Technology, Poland
Michał Woźniak	Wrocław University of Science and Technology, Poland
Krzysztof Wróbel	University of Silesia, Poland
Marian Wysocki	Rzeszow University of Technology, Poland
Farouk Yalaoui	University of Technology of Troyes, France
Xin-She Yang	Middlesex University, UK
Tulay Yildirim	Yildiz Technical University, Turkey
Piotr Zabawa	Cracow University of Technology, Poland

Sławomir Zadrożny	Systems Research Institute, Polish Academy of Sciences, Poland
Drago Zagar	University of Osijek, Croatia
Danuta Zakrzewska	Łódź University of Technology, Poland
Katerina Zdravkova	Ss. Cyril and Methodius University in Skopje, Macedonia
Vesna Zeljkovic	Lincoln University, USA
Aleksander Zgrzywa	Wroclaw University of Science and Technology, Poland
Jianlei Zhang	Nankai University, China
Zhongwei Zhang	University of Southern Queensland, Australia
Maciej Zięba	Wrocław University of Science and Technology, Poland
Adam Ziębiński	Silesian University of Technology, Poland

Program Committees of Special Sessions

Special Session on Cyber-Physical Systems in Healthcare: Applications and Challenges (CSHAC 2020)

Michael Mayo	University of Waikato, New Zealand
Abigail Koay	University of Waikato, New Zealand
Panos Patros	University of Waikato, New Zealand

Special Session on Advanced Data Mining Techniques and Applications (ADMTA 2020)

Tzung-Pei Hong	National University of Kaohsiung, Taiwan
Tran Minh Quang	Ho Chi Minh City University of Technology, Vietnam
Bac Le	VNU University of Science, Vietnam
Bay Vo	Ho Chi Minh City University of Technology, Vietnam
Chun-Hao Chen	Tamkang University, Taiwan
Chun-Wei Lin	Harbin Institute of Technology, China
Wen-Yang Lin	National University of Kaohsiung, Taiwan
Yeong-Chyi Lee	Cheng Shiu University, Taiwan
Le Hoang Son	VNU University of Science, Vietnam
Vo Thi Ngoc Chau	Ho Chi Minh City University of Technology, Vietnam
Van Vo	Ho Chi Minh University of Industry, Vietnam
Ja-Hwung Su	Cheng Shiu University, Taiwan
Ming-Tai Wu	University of Nevada, Las Vegas, USA
Kawuu W. Lin	National Kaohsiung University of Applied Sciences, Taiwan
Tho Le	Ho Chi Minh City University of Technology, Vietnam
Dang Nguyen	Deakin University, Australia
Hau Le	Thuyloi University, Vietnam
Thien-Hoang Van	Ho Chi Minh City University of Technology, Vietnam
Tho Quan	Ho Chi Minh City University of Technology, Vietnam

Ham Nguyen	University of People's Security, Vietnam
Thiet Pham	Ho Chi Minh University of Industry, Vietnam
Nguyen Thi Thuy Loan	Nguyen Tat Thanh University, Vietnam
Mu-En Wu	National Taipei University of Technology, Taiwan
Eric Hsueh-Chan Lu	National Cheng Kung University, Taiwan
Chao-Chun Chen	National Cheng Kung University, Taiwan
Ju-Chin Chen	National Kaohsiung University of Science and Technology, Taiwan

Special Session on Computer Vision and Intelligent Systems (CVIS 2020)

Yoshinori Kuno	Saitama University, Japan
Nobutaka Shimada	Ritsumeikan University, Japan
Muriel Visani	University of La Rochelle, France
Heejun Kang	University of Ulsan, South Korea
Cheolgeun Ha	University of Ulsan, South Korea
Byeongryong Lee	University of Ulsan, South Korea
Youngsoo Suh	University of Ulsan, South Korea
Kang-Hyun Jo	University of Ulsan, South Korea
Hyun-Deok Kang	Ulsan National Institute of Science and Technology, South Korea
Van Mien	University of Exeter, UK
Chi-Mai Luong	University of Science and Technology of Hanoi, Vietnam
Thi-Lan Le	Hanoi University of Science and Technology, Vietnam
Duc-Dung Nguyen	Institute of Information Technology, Vietnam
Thi-Phuong Nghiem	University of Science and Technology of Hanoi, Vietnam
Giang-Son Tran	University of Science and Technology of Hanoi, Vietnam
Hoang-Thai Le	VNU University of Science, Vietnam
Thanh-Hai Tran	Hanoi University of Science and Technology, Vietnam
Anh-Cuong Le	Ton Duc Thang University, Vietnam
My-Ha Le	Ho Chi Minh City University of Technology and Education, Vietnam
The-Anh Pham	Hong Duc University, Vietnam
Van-Huy Pham	Tong Duc Thang University, Vietnam
Van-Dung Hoang	Quang Binh University, Vietnam
Huafeng Qin	Chongqing Technology and Business University, China
Danilo Caceres Hernandez	Universidad Tecnologica de Panama, Panama
Kaushik Deb	Chittagong University of Engineering and Technology, Bangladesh
Joko Hariyono	Civil Service Agency of Yogyakarta, Indonesia
Ing. Reza Pulungan	Universitas Gadjah Mada, Indonesia
Agus Harjoko	Universitas Gadjah Mada, Indonesia

Sri Hartati	Universitas Gadjah Mada, Indonesia
Afiahayati	Universitas Gadjah Mada, Indonesia
Moh. Edi Wibowo	Universitas Gadjah Mada, Indonesia
Wahyono	Universitas Gadjah Mada, Yogyakarta, Indonesia

Special Session on Multiple Model Approach to Machine Learning (MMAML 2020)

Urszula Boryczka	University of Silesia, Poland
Abdelhamid Bouchachia	Bournemouth University, UK
Robert Burduk	Wrocław University of Science and Technology, Poland
Oscar Castillo	Tijuana Institute of Technology, Mexico
Rung-Ching Chen	Chaoyang University of Technology, Taiwan
Suphamit Chittayasothorn	King Mongkut's Institute of Technology Ladkrabang, Thailand
José Alfredo F. Costa	Federal University (UFRN), Brazil
Ireneusz Czarnowski	Gdynia Maritime University, Poland
Fernando Gomide	State University of Campinas, Brazil
Francisco Herrera	University of Granada, Spain
Tzung-Pei Hong	National University of Kaohsiung, Taiwan
Konrad Jackowski	Wrocław University of Science and Technology, Poland
Piotr Jędrzejowicz	Gdynia Maritime University, Poland
Tomasz Kajdanowicz	Wrocław University of Science and Technology, Poland
Yong Seog Kim	Utah State University, USA
Bartosz Krawczyk	Virginia Commonwealth University, USA
Kun Chang Lee	Sungkyunkwan University, South Korea
Edwin Lughofer	Johannes Kepler University Linz, Austria
Hector Quintian	University of Salamanca, Spain
Andrzej Siemiński	Wrocław University of Science and Technology, Poland
Dragan Simic	University of Novi Sad, Serbia
Adam Słowik	Koszalin University of Technology, Poland
Zbigniew Telec	Wrocław University of Science and Technology, Poland
Bogdan Trawiński	Wrocław University of Science and Technology, Poland
Olgierd Unold	Wrocław University of Science and Technology, Poland
Michał Woźniak	Wrocław University of Science and Technology, Poland
Zhongwei Zhang	University of Southern Queensland, Australia
Zhi-Hua Zhou	Nanjing University, China

Special Session on Computational Imaging and Vision (CIV 2020)

Ishwar Sethi	Oakland University, USA
Moongu Jeon	Gwangju Institute of Science and Technology, South Korea
Jong-In Song	Gwangju Institute of Science and Technology, South Korea
Taek Lyul Song	Hangyang University, South Korea
Ba-Ngu Vo	Curtin University, Australia
Ba-Tuong Vo	Curtin University, Australia
Du Yong Kim	Curtin University, Australia
Benlian Xu	Changshu Institute of Technology, China
Peiyi Zhu	Changshu Institute of Technology, China
Mingli Lu	Changshu Institute of Technology, China
Weifeng Liu	Hangzhou Danzi University, China
Ashish Khare	University of Allahabad, India
Moonsoo Kang	Chosun University, South Korea
Goo-Rak Kwon	Chosun University, South Korea
Sang Woong Lee	Gachon University, South Korea
U. S. Tiwary	IIIT Allahabad, India
Ekkarat Boonchieng	Chiang Mai University, Thailand
Jeong-Seon Park	Chonnam National University, South Korea
Unsang Park	Sogang University, South Korea
R. Z. Khan	Aligarh Muslim University, India
Suman Mitra	DA-IICT, India
Bakul Gohel	DA-IICT, India
Sathya Narayanan	NTU, Singapore

Special Session on Intelligent Systems and Algorithms in Information Sciences (ISAIS 2020)

Martin Kotyrba	University of Ostrava, Czech Republic
Eva Volna	University of Ostrava, Czech Republic
Ivan Zelinka	VŠB – Technical University of Ostrava, Czech Republic
Hashim Habiballa	Institute for Research and Applications of Fuzzy Modeling, Czech Republic
Alexej Kolcun	Institute of Geonics, ASCR, Czech Republic
Roman Senkerik	Tomas Bata University in Zlin, Czech Republic
Zuzana Kominkova Oplatkova	Tomas Bata University in Zlin, Czech Republic
Katerina Kostolanyova	University of Ostrava, Czech Republic
Antonin Jancarik	Charles University in Prague, Czech Republic
Petr Dolezel	University of Pardubice, Czech Republic
Igor Kostal	The University of Economics in Bratislava, Slovakia
Eva Kurekova	Slovak University of Technology in Bratislava, Slovakia

Leszek Cedro	Kielce University of Technology, Poland
Dagmar Janacova	Tomas Bata University in Zlin, Czech Republic
Martin Halaj	Slovak University of Technology in Bratislava, Slovakia
Radomil Matousek	Brno University of Technology, Czech Republic
Roman Jasek	Tomas Bata University in Zlin, Czech Republic
Petr Dostal	Brno University of Technology, Czech Republic
Jiri Pospichal	The University of Ss. Cyril and Methodius (UCM), Slovakia
Vladimir Bradac	University of Ostrava, Czech Republic
Petr Pavel	University of Pardubice, Czech Republic
Jan Capek	University of Pardubice, Czech Republic

Special Session on Internet of Things and Artificial Intelligence for Energy Efficiency-Recent Advances and Future Trends (IOTAI 2020)

Xiaohui Yuan	University of North Texas, USA
Andino Maseleno	Universiti Tenaga Nasional, Malaysia
Amit Kumar Singh	National Institute of Technology Patna, India
Valentina E. Balas	Aurel Vlaicu University of Arad, Romania

Special Session on Data Modelling and Processing for Industry 4.0 (DMPI-4.0 2020)

Jörg Becker	Westfälische Wilhelms-Universität, Germany
Rafał Cupek	Silesian University of Technology, Poland
Helena Dudycz	Wroclaw University of Economics and Business, Poland
Marcin Fojcik	Western Norway University of Applied Sciences, Norway
Du Haizhou	Shanghai University of Electric Power, China
Marcin Hernes	Wroclaw University of Economics and Business, Poland
Wojciech Hunek	Opole University of Technology, Poland
Marek Krótkiewicz	Wrocław University of Science and Technology, Poland
Florin Leon	Technical University Asachi of Iasi, Romania
Jing Li	Shanghai University of Electric Power, China
Jacek Piskorowski	West Pomeranian University of Technology Szczecin, Polska
Khouloud Salameh	American University of Ras Al Khaimah, UAE
Predrag Stanimirović	University of Nis, Serbia
Krystian Wojtkiewicz	Wrocław University of Science and Technology, Poland
Feifei Xu	Shanghai University of Electric Power, China

Special Session on Intelligent and Contextual Systems (ICxS 2020)

Adriana Albu	Polytechnic University of Timisoara, Romania
Basabi Chakraborty	Iwate Prefectural University, Japan
Chao-Chun Chen	National Cheng Kung University, Taiwan
Dariusz Frejlichowski	West Pomeranian University of Technology Szczecin, Poland
Diganta Goswami	Indian Institute of Technology Guwahati, India
Erdenebileg Batbaatar	Chungbuk National University, South Korea
Goutam Chakraborty	Iwate Prefectural University, Japan
Ha Manh Tran	Ho Chi Minh City International University, Vietnam
Hong Vu Nguyen	Ton Duc Thang University, Vietnam
Hideyuki Takahashi	Tohoku Gakuin University, Japan
Intisar Chowdhury	University of Aizu, Japan
Jerzy Świątek	Wroclaw University of Science and Technology, Poland
Józef Korbicz	University of Zielona Gora, Poland
Keun Ho Ryu	Chungbuk National University, South Korea
Khanindra Pathak	Indian Institute of Technology Kharagpur, India
Kilho Shin	Gakashuin University, Japan
Maciej Huk	Wroclaw University of Science and Technology, Poland
Marcin Fojcik	Western Norway University of Applied Sciences, Norway
Masafumi Matsuhara	Iwate Prefectural University, Japan
Min-Hsiung Hung	Chinese Culture University, Taiwan
Miroslava Mikusova	University of Žilina, Slovakia
Musa Ibrahim	Chungbuk National University, South Korea
Nguyen Khang Pham	Can Tho University, Vietnam
Plamen Angelov	Lancaster University, UK
Qiangfu Zhao	University of Aizu, Japan
Quan Thanh Tho	Ho Chi Minh City University of Technology, Vietnam
Rafal Palak	Wroclaw University of Science and Technology, Poland
Rashmi Dutta Baruah	Indian Institute of Technology Guwahati, India
Senthilmurugan Subbiah	Indian Institute of Technology Guwahati, India
Sonali Chouhan	Indian Institute of Technology Guwahati, India
Takako Hashimoto	Chiba University of Commerce, Japan
Tetsuji Kuboyama	Gakushuin University, Japan
Tetsuo Kinoshita	RIEC, Tohoku University, Japan
Thai-Nghe Nguyen	Can Tho University, Vietnam
Zhenni Li	University of Aizu, Japan

Special Session on Automated Reasoning with Applications in Intelligent Systems (ARAIS 2020)

Yuichi Goto	Saitama University, Japan
Shinsuke Nara	Muraoka Design Laboratory, Japan
Hongbiao Gao	North China Electric Power University, China
Kazunori Wagatsuma	CIJ Solutions, Japan
Yuan Zhou	Minjiang Teachers College, China

Special Session on Intelligent Supply Chains and e-Commerce (ISCEC 2020)

Carlos Andres Romano	Polytechnic University of Valencia, Spain
Costin Badica	University of Craiova, Romania
Davor Dujak	University of Osijek, Croatia
Waldemar Koczkodaj	Laurentian University, Canada
Miklós Krész	InnoRenew, Slovenia
Paweł Pawlewski	Poznan University of Technology, Poland
Paulina Golińska-Dawson	Poznan University of Economics and Business, Poland
Adam Koliński	Łukasiewicz Research Network – The Institute of Logistics and Warehousing, Poland
Marcin Anholcer	Poznan University of Economics and Business, Poland

Special Session on Intelligent Applications of Internet of Things and Data Analysis Technologies (IAIOTDAT 2020)

Goutam Chakraborty	Iwate Prefectural University, Japan
Bin Dai	University of Technology Xiamen, China
Qiangfu Zhao	University of Aizu, Japan
David C. Chou	Eastern Michigan University, USA
Chin-Feng Lee	Chaoyang University of Technology, Taiwan
Lijuan Liu	University of Technology Xiamen, China
Kien A. Hua	Central Florida University, USA
Long-Sheng Chen	Chaoyang University of Technology, Taiwan
Xin Zhu	University of Aizu, Japan
David Wei	Fordham University, USA
Qun Jin	Waseda University, Japan
Jacek M. Zurada	University of Louisville, USA
Tsung-Chih Hsiao	Huaoiao University, China
Tzu-Chuen Lu	Chaoyang University of Technology, Taiwan
Nitasha Hasteer	Amity University Uttar Pradesh, India
Chuan-Bi Lin	Chaoyang University of Technology, Taiwan
Cliff Zou	Central Florida University, USA
Hendry	Satya Wacana Christian University, Indonesia

Special Session on Current Trends in Artificial Intelligence, Optimization, Learning, and Decision-Making in Bioinformatics and Bioengineering (CTAIOLDMBB 2020)

Sajid Gul Khawaja — National University of Sciences and Technology, Pakistan
Tehmina Khalil — Mirpur University of Sciences and Technology, Pakistan
Arslan Shaukat — National University of Sciences and Technology, Pakistan
Ani Liza Asmawi — International Islamic University, Malaysia
Martin Augustynek — VŠB – Technical University of Ostrava, Czech Republic
Martin Cerny — VŠB – Technical University of Ostrava, Czech Republic
Klara Fiedorova — VŠB – Technical University of Ostrava, Czech Republic
Habibollah Harun — Universiti Teknologi Malaysia, Malaysia
Lim Kok Cheng — Universiti Tenaga Nasional, Malaysia
Roliana Ibrahim — Universiti Teknologi Malaysia, Malaysia
Jafreezal Jaafar — Universiti Teknologi Petronas, Malaysia
Vladimir Kasik — VŠB – Technical University of Ostrava, Czech Republic
Ondrej Krejcar — University of Hradec Kralove, Czech Republic
Jan Kubíček — VŠB – Technical University of Ostrava, Czech Republic
Kamil Kuca — University of Hradec Kralove, Czech Republic
Petra Maresova — University of Hradec Kralove, Czech Republic
Daniel Barvík — VŠB – Technical University of Ostrava, Czech Republic
David Oczka — VŠB – Technical University of Ostrava, Czech Republic
Dominik Vilimek — VŠB – Technical University of Ostrava, Czech Republic
Sigeru Omatu — Osaka Institute of Technology, Japan
Marek Penhaker — VŠB – Technical University of Ostrava, Czech Republic
Lukas Peter — VŠB – Technical University of Ostrava, Czech Republic
Alice Krestanova — VŠB – Technical University of Ostrava, Czech Republic
Chawalsak Phetchanchai — Suan Dusit University, Thailand
Antonino Proto — VŠB – Technical University of Ostrava, Czech Republic
Naomie Salim — Universiti Teknologi Malaysia, Malaysia
Ali Selamat — Universiti Teknologi Malaysia, Malaysia

Imam Much Subroto	Universiti Islam Sultan Agung, Indonesia
Lau Sian Lun	Sunway University, Malaysia
Takeru Yokoi	Tokyo Metropolitan International Institute of Technology, Japan
Hazli Mohamed Zabil	Universiti Tenaga Nasional, Malaysia
Satchidananda Dehuri	Fakir Mohanh University, India
Pradeep Kumar Mallick	KIIT University, India
Subhashree Mishra	KIIT University, India
Cem Deniz	NYU Langone, USA
P. V. Rao	VBIT Hydrabad, India
Tathagata Bandyopadhyay	KIIT University, India

Special Session on Interactive Analysis of Image, Video and Motion Data in Life Sciences (SAILS 2020)

Artur Bąk	Polish-Japanese Academy of Information Technology, Poland
Grzegorz Borowik	Warsaw University of Technology, Poland
Wayne Brookes	University of Technology Sydney, Australia
Leszek Chmielewski	Warsaw University of Life Sciences, Poland
Zenon Chaczko	University of Technology Sydney, Australia
David Davis	University of Technology Sydney, Australia
Aldona Barbara Drabik	Polish-Japanese Academy of Information Technology, Poland
Marcin Fojcik	Western Norway University of Applied Sciences, Norway
Carlo Giampietro	University of Technology Sydney, Australia
Katarzyna Musial-Gabrys	University of Technology Sydney, Australia
Tomasz Górski	Polish Naval Academy, Poland
Adam Gudyś	Silesian University of Technology, Poland
Doan Hoang	University of Technology Sydney, Australia
Celina Imielińska	Vesalius Technologies LLC, USA
Frank Jiang	University of Technology Sydney, Australia
Henryk Josiński	Silesian University of Technology, Poland
Anup Kale	University of Technology Sydney, Australia
Sunil Mysore Kempegowda	University of Technology Sydney, Australia
Ryszard Klempous	Wroclaw University of Technology, Poland
Ryszard Kozera	The University of Life Sciences - SGGW, Poland
Julita Kulbacka	Wroclaw Medical University, Poland
Marek Kulbacki	Polish-Japanese Academy of Information Technology, Poland
Aleksander Nawrat	Silesian University of Technology, Poland
Jerzy Paweł Nowacki	Polish-Japanese Academy of Information Technology, Poland
Eric Petajan	LiveClips LLC, USA
Andrzej Polański	Silesian University of Technology, Poland

Andrzej Przybyszewski	UMass Medical School Worcester, USA
Joanna Rossowska	Polish Academy of Sciences, Poland
Jakub Segen	Gest3D LLC, USA
Aleksander Sieroń	Medical University of Silesia, Poland
Carmen Paz Suarez Araujo	University of Las Palmas, Spain
José Juan Santana Rodríguez	University of Las Palmas, Spain
Adam Świtoński	Silesian University of Technology, Poland
Agnieszka Szczęsna	Silesian University of Technology, Poland
David Tien	Charles Sturt University, Australia
Konrad Wojciechowski	Polish-Japanese Academy of Information Technology, Poland
Robin Braun	University of Technology Sydney, Australia

Special Session on Application of Intelligent Methods to Constrained Problems (AIMCP 2020)

Peter Nielsen	Aalborg University, Denmark
Paweł Sitek	Kielce University of Technology, Poland
Antoni Ligęza	AGH University of Science and Technology, Poland
Sławomir Kłos	University of Zielona Góra, Poland
Grzegorz Bocewicz	Koszalin University of Technology, Poland
Izabela E. Nielsen	Aalborg University, Denmark
Zbigniew Banaszak	Koszalin University of Technology, Poland
Małgorzata Jasiulewicz-Kaczmarek	Poznan University of Technology, Poland
Robert Wójcik	Wrocław University of Science and Technology, Poland
Arkadiusz Gola	Lublin University of Technology, Poland
Marina Marinelli	University of Leicester, UK
Masood Ashraf	Aligarh Muslim University, India
Ali Turkyilmaz	Nazarbayev University, Kazakhstan
Chandima Ratnayake	University of Stavanger, Norway
Marek Magdziak	Rzeszów University of Technology, Poland

Special Session on Privacy, Security and Trust in Artificial Intelligence (PSTrustAI 2020)

M. Ilhan Akbas	Embry-Riddle Aeronautical University, USA
Christoph Benzmüller	Freie Universität Berlin, Germany
Roland Bouffanais	Singapore University of Technology and Design, Singapore
Bernabe Dorronsoro	University of Cadiz, Spain
Rastko Selmic	Concordia University, Canada
Ronaldo Menezes	University of Exciter, UK
Apivadee Piyatumrong	NECTEC, Thailand
Khurum Nazir Junejo	Ibex CX, Pakistan

Daniel Stolfi	University of Luxembourg, Luxembourg
Juan Luis Jiménez Laredo	Normandy University, France
Kittichai Lavangnananda	King Mongkut's University of Technology Thonburi, Thailand
Jun Pang	University of Luxembourg, Luxembourg
Marco Rocchetto	ALES, United Technologies Research Center, Italy
Jundong Chen	Dickinson State University, USA
Emmanuel Kieffer	University of Luxembourg, Luxembourg
Fang-Jing Wu	Technical University Dortmund, Germany
Hannes Frey	University Koblenz-Landau, Germany
Umer Wasim	University of Luxembourg, Luxembourg
Christian M. Adriano	University of Potsdam, Germany

Intelligent Modeling and Simulation Approaches for Games and Real World Systems (IMSAGRWS 2020)

Alabbas Alhaj Ali	Frankfurt University of Applied Sciences, Germany
Costin Bădică	University of Craiova, Romania
Petru Caşcaval	Gheorghe Asachi Technical University, Romania
Gia Thuan Lam	Vietnamese-German University, Vietnam
Florin Leon	Gheorghe Asachi Technical University, Romania
Doina Logofătu	Frankfurt University of Applied Sciences, Germany
Fitore Muharemi	Frankfurt University of Applied Sciences, Germany
Minh Nguyen	Frankfurt University of Applied Sciences, Germany
Julian Szymański	Gdańsk University of Technology, Poland
Paweł Sitek	Kielce University of Technology, Poland
Daniel Stamate	University of London, UK

Contents – Part II

Privacy, Security and Trust in Artificial Intelligence

Interactive Analysis of Image, Video and Motion Data in Life Sciences

Contents – Part I

Knowledge Engineering and Semantic Web

Natural Language Processing

Decision Support and Control Systems

Computer Vision Techniques

Machine Learning and Data Mining

Advanced Data Mining Techniques and Applications

Multiple Model Approach to Machine Learning

Application of Intelligent Methods to Constrained Problems

Configuration of Employee Competences in IT Projects

Jarosław Wikarek and Paweł Sitek(✉)

Kielce University of Technology, Kielce, Poland
{j.wikarek,sitek}@tu.kielce.pl

Abstract. IT projects are characterized by high complexity, implementation of the latest technological solutions and specialized competence of contractors and high costs. The key element in IT project management is to have employees with appropriate substantive competencies (programmers, designers, analysts, etc.). Before starting an IT project, the IT project manager must answer the following questions: *Do I have the right set of team competences to start the project? If the team lacks competence, then in what area and to what extent*? The article proposes a model of employee competence configuration and a procedure for their verification in the context of IT project management. Also presented is the method of model implementation and numerous computational examples verifying its usefulness and effectiveness of implementation.

Keywords: Project management · Employee competences · Mathematical programming · Constraint Logic Programming · Optimization

1 Introduction

In today's knowledge-based economy, the implementation of new and innovative IT solutions is essential and commonplace. IT solutions ensure greater competitiveness, facilitate management of the organization through effective use of resources, including information, and support the decision-making process. Developing new IT solutions as well as implementing existing ones requires a project approach. What is characteristic for IT projects is primarily the close relationship with the latest technologies, innovation and the requirement for the team implementing a given IT project to have high and advanced substantive and organizational competencies. Very often it is the competences at an appropriate level, and often lack thereof, that decide about the success/failure of an IT project [1, 2]. Therefore, the project manager has to answer a few key questions before starting the project: *Do I have the right set of team competences to start the project? If the team lacks competence, then in what area and to what extent? Is it possible to extend the set of competences of the team?* etc. The article proposes a model of competence configuration for IT projects in the form of CSP (Constraint Satisfaction Problem) [3, 4] supplemented with a set of questions. On the basis of the model, a procedure for verification of the set of competences and its implementation using a hybrid approach was proposed.

N. T. Nguyen et al. (Eds.): ACIIDS 2020, LNAI 12034, pp. 3–12, 2020.
https://doi.org/10.1007/978-3-030-42058-1_1

2 Illustrative Example and Problem Description

The issue of configuration of employee competences will be presented by means of an illustrative example. An example of an IT project is the implementation of an ERP (Enterprise Resource Planning) class system in an enterprise. Within the framework of the implementation, specific tasks should be performed (Table 1 - *Tasks to be carried out*) in the specified order, as illustrated by the graph (Fig. 1).

Figure 2 shows the schedule of tasks, according to the graph. The schedule, which was obtained using the CPM (Critical Path Method) [5, 6] is in many cases imposed before the start of the project. Each task within a project is characterized by a specific time of implementation and requires an employee with specific competences (Table 1 - *Competencies*). The competences of individual employees are presented in Table 1 - *Competences of employees*. For the sake of simplicity, we omit other resources other than employees. The described issue (configuration of employee competences) is to find answers to a few key questions about competences in the context of IT project implementation. Usually, such questions must be answered by every IT project manager, even before the start of the project.

The most important questions are:

- Q1: Is the current set of employee competences sufficient for the realization of a specific set of tasks within the project?
- Q2: How should the set of employee competences be supplemented in order to be able to carry out a specific set of tasks and to keep the cost of competence supplementation as low as possible?
- Q3: How should the set of employee competences be supplemented in order to be able to complete a specific set of tasks, keep the cost (4) of supplementing as low as possible and have each employee work only for the allowed time?
- Q4: How should the set of employee competences be supplemented in order to be able to complete a specific set of tasks, keep the cost of change as low as possible and have the tasks were completed in accordance with the imposed schedule of implementation?

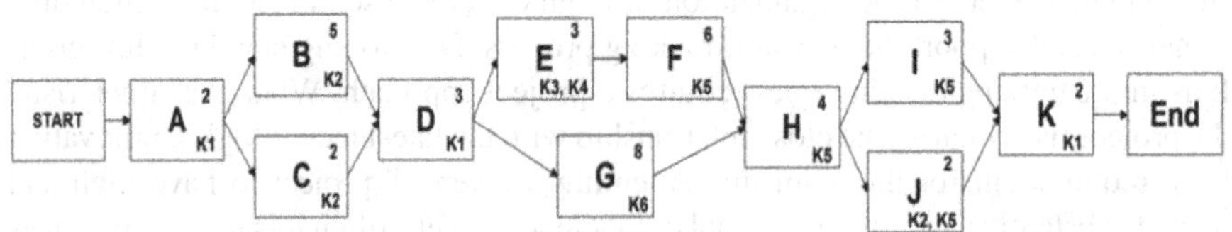

Fig. 1. Graph representing a network of activities for the illustrative example (ERP class system implementation project)

In order to answer the above questions, a mathematical model of competence configuration for IT projects was proposed (Sect. 2.1). The model is formulated as a set of constraints that is supplemented by a Q1..Q4 set of questions. On the basis of this model, a procedure for verification of the set of employee competences was developed, which was implemented with the use of the GUROBI environment [7] (Sect. 2.2).

Table 1. Data for the illustrative example

Tasks to be carried out

N_z	Name	T
A	Setting up a steering committee and defining the budget	2
B	Pre-implementation analysis - processes	5
C	Pre-implementation analysis - infrastructure	2
D	Preparation of the implementation plan	3
E	Installation and configuration of the operating system, the network system and the database	3
F	Installation and configuration of the modules of the implemented system	6
G	Training of employees	8
H	Data migration and test system start	4
I	Debugging	3
J	Configuration of reports and analyses	2
K	Commissioning and production start of the system	2

Sequence of tasks

N_z	N_n	N_z	N_n
A	B	G	H
A	C	F	H
B	D	H	I
C	D	H	J
D	E	I	K
D	G	J	K
E	F		

Competencies

N_k	Name
K1	Project Manager
K2	Analyst
K3	Administrator of databases
K4	Administrator of computer networks
K5	Implementer
K6	Instructor

Employees

N_e	Name	Tp
PA	Employer A	8
PB	Employer B	8
PC	Employer C	4
PD	Employer D	8
PE	Employer E	8
PF	Employer F	8
PG	Employer G	4

Competences of employees

N_e	N_k	N_e	N_z
PA	K1	PE	K3
PB	K2	PE	K4
PB	K5	PE	K5
PC	K2	PF	K5
PC	K5	PG	K5
PD	K5		

Competences required for tasks

Nr_z	Nr_k	Nr_o	Nr_k
A	K1	G	K6
B	K2	H	K5
C	K2	I	K5
D	K1	J	K2
E	K3	J	K5
E	K4	K	K1
F	K5		

N_z	Task index
N_k	Competence index
N_e	Employee index
T	Time of task completion
Tp	Permitted working time
---	cannot acquire the given competence
0	already possesses the given competence
N_n	Successor index (tasks performed after the N_z task)

Cost of change of competence

N_e \ N_k	K1	K2	K3	K4	K5	K6
PA	0	800	900	900	700	700
PB	---	0	900	900	0	700
PC	---	0	900	900	0	700
PD	---	800	900	900	0	400
PE	---	800	0	0	0	700
PF	---	800	900	900	0	700
PG	---	800	900	900	0	200

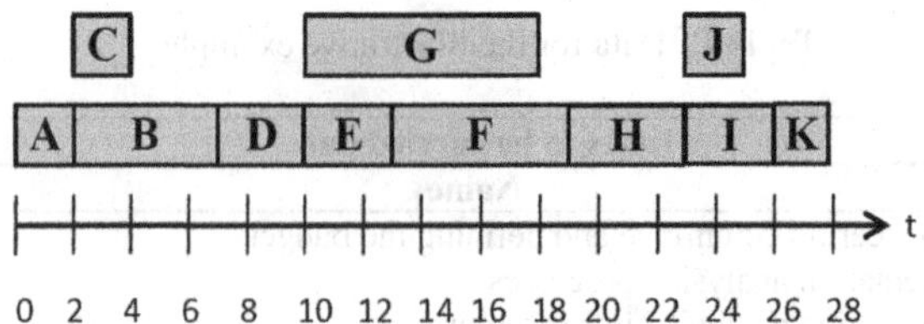

Fig. 2. Example of a schedule (Gantt chart) – for the illustrative example without taking into account resources (competences)

2.1 Formalization of the Problem of Employee Competence Configuration

The model for the discussed issue of employee competence configuration was formulated as a (1)…(9) set of constraints and a Q1…Q4 set of questions. The proposed model can be classified as BIP (Binary Integer Programming) [8] or CSP (Constraint Satisfaction Problem) depending on the question under consideration. Table 3 shows the basic indexes, parameters and decisive variables of the model. Table 2 shows the significance of each model constraint (1)…(9).

Table 2. Description of model constraints

Constraint	Description
(1)	The constraint ensures that an employee with the appropriate competences is assigned to each task
(2)	The constraint states that the worker concerned can only obtain the competences that are permitted for them
(3)	The constraint ensures that if the selected employee is assigned to a specific task, they must have the required competencies
(4)	The constraint determines the cost of acquisition of certain competences by the employees
(5)	The constraint ensures that a given employee works only during the permitted working hours
(6, 7)	Constraints enforce binding of the $Y_{i,k}$ and $X_{i,k,u}$ variables
(8)	Constraint ensures that simultaneous/overlapping tasks cannot be performed by the same employee
(9)	Binarity constraint

$$\sum_{i \in I} X_{i,k,u} = d_{k,u} \; \forall k \in K, u \in U \wedge d_{k,u} = 1 \tag{1}$$

$$Gx_{i,k} \leq go_{i,u} \; \forall i \in I, u \in U \tag{2}$$

$$X_{i,k,u} \leq Gx_{k,u} + g_{k,u} \; \forall i \in I, k \in K, u \in U \tag{3}$$

Table 3. Indices, parameters and decision variables

Symbol	Description
Indices	
I	Set of employees
K	Set of tasks
U	Set of competences
i	Employee index ($i \in I$)
k	Task index a ($k \in K$)
u	Competence index ($u \in U$)
Parameters	
$d_{k,u}$	If an employee with competence u is necessary to perform task k, then $d_{k,u} = 1$, otherwise $d_{k,u} = 0$ ($u \in U, k \in K$)
$g_{i,u}$	If the employee i has the competence u then $g_{i,u} = 1$ otherwise $g_{i,u} = 0$ ($u \in U, i \in I$)
$go_{i,u}$	If the employee i can acquire the competence u then $go_{i,u} = 1$ otherwise $go_{i,u} = 0$ ($u \in U, i \in I$)
$gc_{i,u}$	Cost of acquisition by the employee i of the competence u ($u \in U, i \in I$)
r_k	Task k completion time ($k \in K$)
p_i	Maximum working time of the employee i ($i \in I$)
$m_{k1,k2}$	If the schedule assumes the implementation of task $k1$, which coincides with the implementation of task $k2$ then $m_{k1,k2} = 1$ otherwise $m_{k1,k2} = 0$ ($k1, k2 \in K$)
ta	Arbitrarily large constant
Decision variables	
$X_{i,k,u}$	If the employee i performs task k using the competence u then $X_{i,k,u} = 1$ otherwise $X_{i,k,u} = 0$ ($k \in K, u \in U, i \in I$)
$Gx_{i,u}$	If the tasks require that the employee i to acquire the competence u then $Gx_{i,u} = 1$ otherwise $Gx_{i,u} = 0$ ($u \in U, i \in I$)
$Y_{i,k}$	If the employee i executes task k then $Y_{i,k} = 1$ otherwise $Y_{i,k} = 0$ ($k \in K, i \in I$)

$$\text{Cost} = \sum_{i \in I} \sum_{u \in U} (Gc_{i,u} \cdot Gx_{i,u}) \tag{4}$$

$$\sum_{k \in K} Y_{i,k} \cdot r_k \leq p_i \ \forall i \in I \tag{5}$$

$$\sum_{u \in U} X_{i,k,u} \leq ta \cdot Y_{i,k} \ \forall i \in I, k \in K \tag{6}$$

$$\sum_{u \in U} X_{i,k,u} \geq Y_{i,k} \ \forall i \in I, k \in K \tag{7}$$

$$Y_{i,k1} + Y_{i,k2} \leq 1 \ \forall k1, k2 \in K \wedge d_{k1,k2} = 1, i \in I \tag{8}$$

$$X_{i,k,u} \in \{0, 1\} \ \forall i \in I, k \in K, u \in U$$
$$Gx_{i,u} \in \{0, 1\} \ \forall i \in I, u \in U$$
$$Y_{i,k} \in \{0, 1\} \ \forall i \in I, k \in K, u \in U \quad (9)$$

2.2 Procedure for Verifying the Set of Employee Competences in IT Projects

The employee competence configuration model presented in Sect. 2.1 became the basis for the development of a procedure for verification of such competences for the implemented IT projects. The general scheme of such a procedure is shown in Fig. 3. Based on the constraints and questions and the set of competences and tasks to be performed (optionally, according to a specified schedule), the procedure generates answers to individual questions using the model from Sect. 2.1.

The implementation of the procedure was based on the author's hybrid approach [9, 10], the diagram of which is shown in Fig. 4.

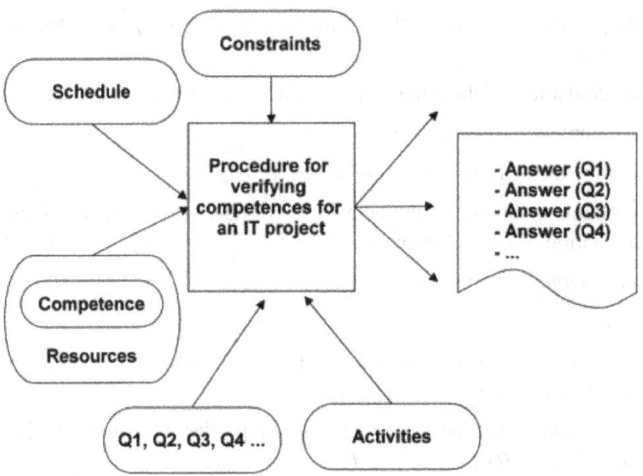

Fig. 3. Concept of a procedure verifying the set of employee competences for IT projects

Two environments were used to implement the procedure using the hybrid approach. To model the problem, its transformation (presolving) and the generation of the model transformed in the MP (Mathematical Programming) solver format, CLP (Constraint Logic Programming), and more precisely the ECLiPSe [11] environment was used. The GUROBI environment was used to solve the transformed model. A detailed description of the transformation of the model that constitutes its presolving is presented in [12].

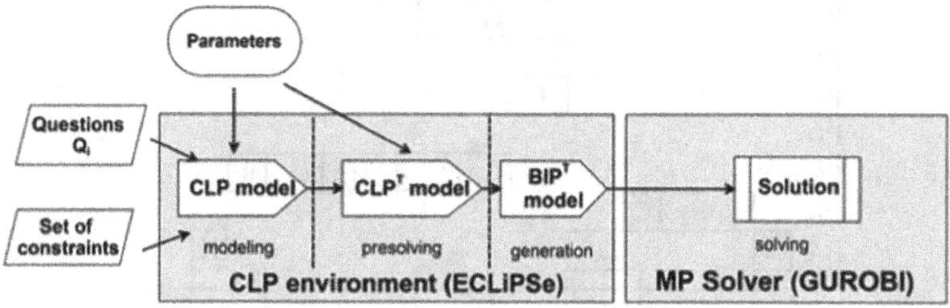

Fig. 4. Diagram of hybrid approach

3 Computational Examples

In order to verify the (1)..(9) model and to evaluate the proposed implementation method, a number of computational experiments were carried out. The experiments were carried out in two stages. The first step involved finding answers to questions Q1..Q4 for an illustrative example (Sect. 2), using the instances of data from Table 1. The results are presented in Table 4, as Answer (Q1)..Answer (Q4) and additionally using Gantt charts for Q2 and Q3 questions (Figs. 5 and 6).

The second stage included a series of experiments aimed at assessing the effectiveness of the proposed implementation in the form of a hybrid approach. The model was implemented for different sets of questions and different data instances. Individual data instances (E1..E20) differed in the number of tasks, employees and competences. In order to assess the effectiveness of the proposed hybrid approach in the context of the implementation of the presented model, it was simultaneously implemented only using mathematical programming for the same data instances. The obtained result is presented in Table 5.

Analysis of the results shows that the hybrid approach enables a significant reduction in the number of decisive variables of the model (up to 20 times) and the computation time (up to 20 times).

Table 4. Results for the illustrative example.

Question	Answer
Q1	No
Q2	The PG employee should be trained to acquire the K6 competence (cost 200)
Q3	The PD employee should be trained to acquire the K6 competence (cost 400)
Q4	The schedule for the implementation of the project is the one shown in Fig. 2. The PD employee should be trained to acquire the K6 competence (cost 400)

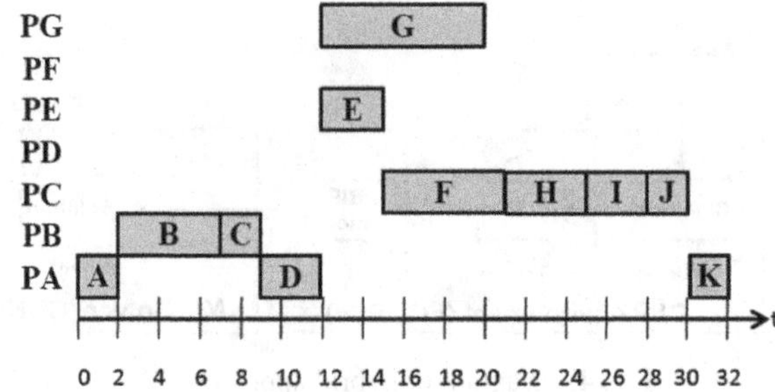

Fig. 5. Example of an implementation schedule – employee allocation (Gantt chart) - Q2

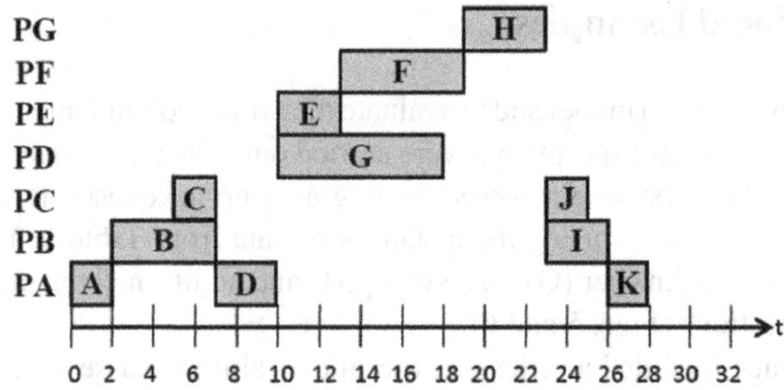

Fig. 6. Example of an implementation schedule – employee allocation (Gantt chart) – Q3

Table 5. Results for hybrid approach and mathematical programming

En	Number of employee	Number of competences	Number of Tasks	Question	Number of variables		Computation time	
					MP	Hyb	MP	Hyb
E1	8	6	12	Q2	624	86	0	0
E2	8	6	12	Q3	720	104	1	1
E3	8	6	12	Q4	720	104	1	1
E4	8	6	12	Q3 + Q4	720	104	1	1
E5	12	8	24	Q2	2 400	121	2	0
E6	12	8	24	Q3	2 688	409	3	1
E7	12	8	24	Q4	2 688	409	3	1
E8	12	8	24	Q3 + Q4	2 688	409	3	1
E9	16	12	36	Q2	7104	412	3	0
E10	16	12	36	Q3	7680	988	63	3
E11	16	12	36	Q4	7680	988	9	2
E12	16	12	36	Q3 + Q4	7680	988	65	9
E13	20	14	42	Q2	12 040	637	7	0
E14	20	14	42	Q3	12 880	1447	103	23

(continued)

Table 5. (*continued*)

En	Number of employee	Number of competences	Number of Tasks	Question	Number of variables		Computation time	
					MP	Hyb	MP	Hyb
E15	20	14	42	Q4	12 880	1447	12	2
E16	20	14	42	Q3 + Q4	12 880	1447	141	33
E17	24	18	60	Q2	26 352	1526	19	1
E18	24	18	60	Q3	31 392	6566	545	34
E19	24	18	60	Q4	31 392	6566	34	4
E20	24	18	60	Q3 + Q4	31 392	6566	643	38

4 Conclusions

The proposed model of employee competence configuration and its implementation in the form of a procedure verifying these competences is an extremely useful tool for any IT project manager. The problem of selection, supplementation and verification of employee competences is a key issue that affects the success of a given IT project.

Ability to assess prior to project commencements whether the competences we possess are sufficient or which are missing and to what extent. The model may also take into account the imposed project implementation schedule/often occurring in practical situations and constrains related to the working time of individual employees. The proposed implementation with the use of the author's hybrid approach of the procedure for verification of employee competences is much more effective than the use of classical mathematical programming. Depending on the question asked, the times needed to obtain answers achieved with the hybrid approach were up to 20 times shorter than those achieved with mathematical programming.

Further research will focus on the development of the model so as to include the situation in which there is a need to assign an employee with more competences to a task, taking into account the absence of employees [13, 14] and the introduction of logical constraints (a given project cannot be implemented by specific employees working together, given tasks within a project cannot be implemented in parallel, etc.) and other types of projects [15]. It is also planned to adapt and apply the proposed model to issues in the area of production [16, 17], logistics [18, 19], transport [20], etc. For larger projects, a multi-agent approach is planned to be implemented [21].

References

1. Hughes, B., Ireland, R., West, B., Smith, N., Shepherd, D.I.: Project Management for IT-Related Projects, 2nd edn. BCS, The Chartered Institute for IT (2012). Hughes, B. (editor)
2. Kuster, J., et al.: Project Management Handbook. MP. Springer, Heidelberg (2015). https://doi.org/10.1007/978-3-662-45373-5
3. Rossi, F., Van Beek, P., Walsh, T.: Handbook of Constraint Programming. Foundations of Artificial Intelligence. Elsevier Science Inc., New York (2006)

4. Apt, K.: Principles of Constraint Programming. Cambridge University Press, Cambridge (2003)
5. Punmia, B.C., Khandelwal, K.K.: Project Planning and Control with PERT & CPM. Firewall Media, New Delhi (2002)
6. Averous, J., Linares, T.: Advanced Scheduling Handbook for Project Managers. Fourth Revolution, Singapore (2015)
7. Gurobi. http://www.gurobi.com/. Accessed 10 Oct 2019
8. Schrijver, A.: Theory of Linear and Integer Programming. Wiley, New York (1998). ISBN 0-471-98232-6
9. Sitek, P., Wikarek, J.: Capacitated vehicle routing problem with pick-up and alternative delivery (CVRPPAD): model and implementation using hybrid approach. Ann. Oper. Res. **273**, 257–277 (2019). https://doi.org/10.1007/s10479-017-2722-x
10. Sitek, P., Wikarek, J., Nielsen, P.: A constraint-driven approach to food supply chain management. Ind. Manage. Data Syst. **117**, 2115–2138 (2017). https://doi.org/10.1108/IMDS-10-2016-0465
11. Eclipse - The Eclipse Foundation open source community website. www.eclipse.org. Accessed 19 Oct 2019
12. Sitek, P., Wikarek, J.: A multi-level approach to ubiquitous modeling and solving constraints in combinatorial optimization problems in production and distribution. Appl. Intell. **48**, 1344–1367 (2018). https://doi.org/10.1007/s10489-017-1107-9
13. Szwarc, E., Bocewicz, G., Banaszak, Z., Wikarek, J.: Competence allocation planning robust to unexpected staff absenteeism. Eksploatacja i Niezawodnosc – Maint. Reliab. **21**, 440–450 (2019). https://doi.org/10.17531/ein.2019.3.10
14. Szwarc, E., Bocewicz, G., Bach-Dąbrowska, I., Banaszak, Z.: Declarative model of competences assessment robust to personnel absence. In: Damaševičius, R., Vasiljevienė, G. (eds.) ICIST 2019. CCIS, vol. 1078, pp. 12–23. Springer, Cham (2019). https://doi.org/10.1007/978-3-030-30275-7_2
15. Relich, M.: Identifying project alternatives with the use of constraint programming. In: Borzemski, L., Grzech, A., Świątek, J., Wilimowska, Z. (eds.) Information Systems Architecture and Technology: Proceedings of 37th International Conference on Information Systems Architecture and Technology – ISAT 2016 – Part I. AISC, vol. 521, pp. 3–13. Springer, Cham (2017). https://doi.org/10.1007/978-3-319-46583-8_1
16. Gola, A.: Reliability analysis of reconfigurable manufacturing system structures using computer simulation methods. Eksploatacja I Niezawodnosc – Maint. Reliab. **21**(1), 90–102 (2019). https://doi.org/10.17531/ein.2019.1.11
17. Janardhanan, M.N., Li, Z., Bocewicz, G., Banaszak, Z., Nielsen, P.: Metaheuristic algorithms for balancing robotic assembly lines with sequence-dependent robot setup times. Appl. Math. Model. **65**, 256–270 (2019)
18. Dang, Q.V., Nielsen, I., Yun, W.Y.: Replenishment policies for empty containers in an inland multi-depot system. J. Marit. Econ. Logist. **15**, 120–149 (2013). ISSN 1479-2931
19. Grzybowska, K., Kovács, G.: Developing agile supply chains – system model, algorithms, applications. In: Jezic, G., Kusek, M., Nguyen, N.-T., Howlett, R.J., Jain, L.C. (eds.) KES-AMSTA 2012. LNCS (LNAI), vol. 7327, pp. 576–585. Springer, Heidelberg (2012). https://doi.org/10.1007/978-3-642-30947-2_62
20. Wirasinghe, S.C.: Modeling and optimization of transportation systems. J. Adv. Transp. **45**, 231–347 (2011)
21. Lasota, T., Telec, Z., Trawiński, B., Trawiński, K.: A multi-agent system to assist with real estate appraisals using bagging ensembles. In: Nguyen, N.T., Kowalczyk, R., Chen, S.-M. (eds.) ICCCI 2009. LNCS (LNAI), vol. 5796, pp. 813–824. Springer, Heidelberg (2009). https://doi.org/10.1007/978-3-642-04441-0_71

Competence-Oriented Recruitment of a Project Team Robust to Disruptions

Eryk Szwarc[1(✉)], Izabela Nielsen[2], Czesław Smutnicki[3], and Grzegorz Bocewicz[1]

[1] Faculty of Electronics and Computer Science, Koszalin University of Technology, Koszalin, Poland
eryk.szwarc@tu.koszalin.pl

[2] Department of Materials and Production, Aalborg University, Aalborg, Denmark
izabela@mp.aau.dk

[3] Faculty of Electronics, Wroclaw University of Technology, Wroclaw, Poland
czeslaw.smutnicki@pwr.edu.pl

Abstract. Selection of competent employees is one of the numerous factors that determine the success of a project. The literature describes many approaches that help decision makers to recruit candidates with the required skills. Only a few of them take into account the disruptions that can occur during the implementation of a project, caused by employee absenteeism, fluctuations in the duration of activities, etc. Collectively, what these approaches amount to is proactive planning of employee teams with redundant competences. Searching for competence frameworks robust to disruptions involves time-consuming calculations, which do not guarantee that an admissible solution will be found. In view of this, in the present study, we propose sufficient conditions, the fulfilment of which guarantees the existence of such a solution. By testing these conditions, one can determine whether there exists an admissible solution, i.e. whether it is at all worth searching for a robust competence framework. The possibilities of practical application of the proposed method are illustrated with an example.

Keywords: Project team · Employee competences · Robust competence framework · Disruptions

1 Introduction

People (a project team) are the key resource that determines successful achievement of a goal (implementation of a project). That is why the right selection of members of a project team is such an important issue already at the stage of defining project resources. It is worth noting that team recruitment boils down to looking for variants of allocation of employees to project tasks/activities [7]. The goal is to determine: when to employ how many contractors with what competences at what cost.

The work was supported by the National Science Centre, Poland, under research project no: 2019/33/N/HS4/00379.

N. T. Nguyen et al. (Eds.): ACIIDS 2020, LNAI 12034, pp. 13–25, 2020.
https://doi.org/10.1007/978-3-030-42058-1_2

In most cases, staffing decisions regarding the appointment of project teams are subject to uncertainty related to temporary unavailability of resources (e.g. employees) during the implementation of a project, delays in task start times, changes in task duration, etc. Moreover, decision-makers are unable to predict the probable, let alone the exact, moment of the occurrence of those disruptions (e.g. which employee will be absent when, which materials will be delivered with what delay, etc.). The effectiveness of reactive approaches [23], which involve modification of the already implemented project schedule (so-called re-scheduling) in the event of a disruption, depends on many factors, including the competences of available employees.

Employee competences are represented collectively as a personnel competence structure or framework (hereinafter referred to as competence framework) [24]. In simplified terms, a personnel competence framework can be identified with the competence matrix [1, 17] commonly used in the construction, production and other industries, which defines current qualifications of staff members and so also their allocation to specific tasks (hereinafter referred to as task assignment).

An alternative to reactive management of disruptions, in cases such as employee absenteeism, is a proactive approach that generates frameworks robust to selected anticipated types of disruptions [4, 12, 14, 18]. The proactive strategy consists in recruiting a team with a competence framework that allows to make changes to the existing project schedule. Competence frameworks that guarantee the possibility of re-scheduling a project are hereinafter referred to as robust to a specific set of disruptions. Put another way, a robust competence framework is one that guarantees the implementation of the assumed plan despite the occurrence of a specific type of disruption. As a consequence, the planning of competence frameworks that can guarantee the execution of the planned tasks is tantamount to seeking (synthesizing) alternative competence frameworks that are robust to the given set of disruptions.

Studies [21] show that due to the high computational complexity of this type of problems ($f(n, m) = O(2^{m \times n})$, where: m – number of employees, n – number of tasks), in special cases, they can only be solved by checking all variants of competence frameworks. This excludes the use of exact methods (e.g. the branch and bound algorithm). This is particularly important from the point of view of the scale of the problems encountered in practice and the fact that there is no guarantee that an admissible solution will be found. An interesting line of inquiry is the search for the so-called sufficient conditions, the satisfaction of which guarantees the existence of an admissible solution. By testing such conditions, as a preliminary step, one can determine whether it is at all worth searching for robust competence frameworks.

Section 2 presents a literature review of assignment planning under uncertainty. In Sect. 3, a reference model is proposed which can be used to search for competence frameworks that allow to develop competence allocation plans robust to a set of anticipated types of disruption, such as absences of individual employees. Based on this model, sufficient condition is proposed. Section 4 reports computational experiments which illustrate the possibilities of applying the proposed sufficient condition. The conclusions are drawn in Sect. 5.

2 Literature Review

Planning of human teams under conditions of uncertainty is a well-researched problem. The solutions proposed in the literature are largely limited to the introduction of time buffers or capacity (resource) buffers. Time buffers (most often understood as additional time windows allowing for the completion of delayed tasks) are used in project management [8], personnel assignment [4, 6, 22] and task planning problems [3]. In turn, the so-called capacity buffers (understood as surplus resources), also called reserve personnel (reserve crew/resources), ensure that surplus employees are present in specific time intervals in case there is a greater than anticipated demand for workers or there is a shortage of personnel. Several authors [4, 13, 22] have studied the use of capacity buffers (reserve crew planning) in the aviation industry. They believe that sufficient back-up staff should be introduced to prevent disruptions, but the number of reserve crew members should not be too high to ensure crew availability for other duties. The size of a capacity buffer is usually determined in advance in a deterministic way. For example, Ingels and Maenhout [11] define capacity buffers by setting a percentage above the expected staffing requirements and taking into account specific time constraints. Another type of buffer is overtime, which increases efficiency in specific time intervals [10]. Ionescu and Kliewer [12] and Shebalov and Klabjan [15] have focused on maximizing the number of crew swapping options for aircraft crew planning [4]. Another solution to disruptions is the exchange of duties between employees, referred to as resource substitution [9, 12, 15]. Reactive strategies can also be improved by extending daily work hours and total work hours in the entire planning horizon [5, 10].

One approach treats competences as capacity buffers [19–21]. The model presented in the studies that follow this approach allows to search for an optimal competence framework. Taking into account the fact that the calculations are time-consuming, and still may not yield a solution, it seems interesting to look for so-called sufficient conditions the fulfilment of which ensures the existence of an admissible solution. Differently put, the knowledge of such conditions allows one to determine whether it is at all worthwhile to start looking for robust competence frameworks.

3 Model

As part of preliminary research [20, 21], a model for synthesizing competence frameworks robust to disruptions was developed. The model is limited to situations in which the disruptions are cases of a priori known (at the stage of planning the allocation of tasks, e.g. at the beginning of the day's shift) absences of single members of a staff assigned to the execution of a given set of tasks (a production order).

Sets:

Z_i: set of tasks indexed by $i = 1, \ldots, n$
P_k: set of employees indexed by $k = 1, \ldots, m$

Parameters

l_i: duration of the i-th task Z_i (in hours)

s_k^j: minimum number of working hours (lower working time limit) of the k-th employee ($s_k \in \mathbb{N}$) when the j-th employee is absent

z_k^j: maximum number of working hours (upper working time limit) of the k-th employee ($z_k \in \mathbb{N}$) when the j-th employee is absent

$w_{a,b}$: a parameter that specifies whether tasks Z_a and Z_b can be performed by the same employee (the tasks are mutually exclusive):

$$w_{a,b} = \begin{cases} 1 & \text{when tasks } Z_a \text{ and } Z_b \text{ are mutually exclusive} \\ 0 & \text{in the remaining cases} \end{cases}$$

R^* expected robustness of a competence framework, $R^* \in [0, 1]$

Decision variables:

G: a competence framework defined as $G = (g_{k,i}|k = 1\ldots m; i = 1\ldots n)$, where $g_{k,i}$ stands for employees' competences to perform tasks; $g_{k,i} \in \{0, 1\}$, $g_{k,i} = 0$ means that the k-th employee has no competences to perform the i-th task, and $g_{k,i} = 1$ means that the k-th employee has the competences to perform the i-th task.

R: a measure of the robustness of competence framework G to the absence of one employee $R \in [0, 1]$. $R = 0$ – stands for lack of robustness, i.e. each absence results in unassigned tasks; $R = 1$ – stands for full robustness, i.e. regardless of which employee is absent, all tasks are assigned to available staff. For example: $R = 0.25$ means that the competence framework ensures allocation of tasks in one-quarter of the possible cases of absence of one employee; $R = 0.5$ means that the competence framework ensures allocation of tasks in half of the possible cases of absence of one employee,

G^j: a competence framework obtained for a situation in which the j-th employee $G^j = (g_{k,i}^j|k = 1\ldots(m-1); i = 1\ldots n)$ is absent from his/her scheduled duty

X^j: task assignment in the situation when the j-th employee is absent, defined as $X^j = (x_{k,i}^j|k = 1\ldots(m-1); i = 1\ldots n)$,, where $x_{k,i}^j \in \{0, 1\}$:

$$x_{k,i}^j = \begin{cases} 1 & \text{when task } Z_i \text{ has been assigned to employee } P_k \\ 0 & \text{in the remaining cases} \end{cases}$$

c^j: an auxiliary variable that specifies whether assignment X^j satisfies the given constraints. The value of variable $c^j \in \{0, 1\}$ depends on variables $c_{1,i}^j$, $c_{2,k}^j$, and $c_{3,k}^j$, which specify whether constraints (3), (4), and (5) are satisfied

Constraints:

1. Construction of a competence framework for situations when the j-th employee is absent from his scheduled duty:

$$g_{k,i}^j = \begin{cases} g_{k,i} & \text{when } k < j \\ g_{(k+1),i} & \text{when } k \geq j \end{cases}. \tag{1}$$

2. Tasks can only be performed by employees who have appropriate competences:

$$x_{k,i}^{j} = 0, \quad when\ g_{k,i}^{j} = 0, \ \text{for}\ k = 1 \ldots (m-1); i = 1 \ldots n; j = 1 \ldots m. \tag{2}$$

3. Task Z_i is assigned to exactly one employee:

$$\left(\sum\nolimits_{k=1}^{m-1} x_{k,i}^{j} = 1\right) \Leftrightarrow \left(c_{1,i}^{j} = 1\right), \ \text{for}\ i = 1 \ldots n;\ j = 1 \ldots m. \tag{3}$$

4. Workload of the k-th employee should be no less than the time limit s_k^j:

$$\left(\sum\nolimits_{i=1}^{n} x_{k,i}^{j} \cdot l_i \geq s_k^j\right) \Leftrightarrow \left(c_{2,k}^{j} = 1\right), \ \text{for}\ k = 1 \ldots (m-1);\ j = 1 \ldots m. \tag{4}$$

5. Workload of the k-th employee should not exceed the upper working time limit z_k^j:

$$\left(\sum\nolimits_{i=1}^{n} x_{k,i}^{j} \cdot l_i \leq z_k^j\right) \Leftrightarrow \left(c_{3,k}^{j} = 1\right), \ \text{for}\ k = 1 \ldots (m-1);\ j = 1 \ldots m. \tag{5}$$

6. Performance of mutually exclusive tasks:

$$x_{k,a}^{j} + x_{k,b}^{j} \leq 1, \ \text{when}\ w_{a,b} = 0, \ \text{for}\ k = 1 \ldots (m-1); i = 1 \ldots n;\ j = 1 \ldots m. \tag{6}$$

7. Robustness of the competence framework:

$$R = \frac{LP}{m}, \tag{7}$$

$$R \geq R^*, \tag{8}$$

$$LP = \sum\nolimits_{j=1}^{m} c^j, \tag{9}$$

$$c^j = \prod\nolimits_{i=1}^{n} c_{1,i}^{j} \prod\nolimits_{k=1}^{m} c_{2,k}^{j} \prod\nolimits_{k=1}^{m} c_{3,k}^{j}. \tag{10}$$

The competence framework and the task assignment are represented in the model by decision variables G, G^j and X^j, respectively. Task assignment X^j which satisfies constraints (2)–(6) is referred to as an admissible assignment in the situation of an absence of the j-th employee. In this context, the question to be considered is the following: *Does there exist a competence framework G that can guarantee robustness $R \geq R^*$ in the event of an absence of a single employee?*

The structure of the proposed model, which includes a set of decision variables and a set of constraints that relate those variables to one another in a natural way, allows to formulate the problem in hand as a CSP and implement it in a constraint programming environment:

$$CS = ((\mathcal{V}, \mathcal{D}), \mathcal{C}), \tag{11}$$

where: $\mathcal{V} = \left\{G, G^1, \ldots, G^m, X^1, \ldots, X^m, R\right\}$ – a set of decision variables which includes: competence framework G, competence sub-frameworks G^j for cases when the j-th employee is absent, the corresponding task assignments X^j, and robustness R. $\mathcal{D}$ – a finite set of decision variable domains $\{G, G^1, \ldots, G^m, X^1, \ldots, X^m, R\}$. $\mathcal{C}$ – a set of constraints specifying the relationships between the competence framework and its robustness (constraints 1–10).

In previous studies [2, 21], the above model and method were used to search for competence frameworks robust to disruptions caused by an absence of one to three employees. The results of those investigations indicate that the search for competence frameworks is effective (competence frameworks were obtained in less than 1500 s) for problems regarding projects with up to 50 tasks and employing up to 10 employees. It is therefore worth identifying sufficient conditions which, when tested, provide an answer to the question whether there exists an admissible solution?

3.1 Illustrative Example

A project is being implemented which consists of 10 tasks: Z_1–Z_{10}, each of which takes one unit of time (t.u.) to complete. The structure of the project is shown in Fig. 1. Tasks Z_1–Z_4 require competence A, tasks Z_5–Z_7 require competence B, tasks Z_8–Z_9 require competence C, and task Z_{10} requires competence D.

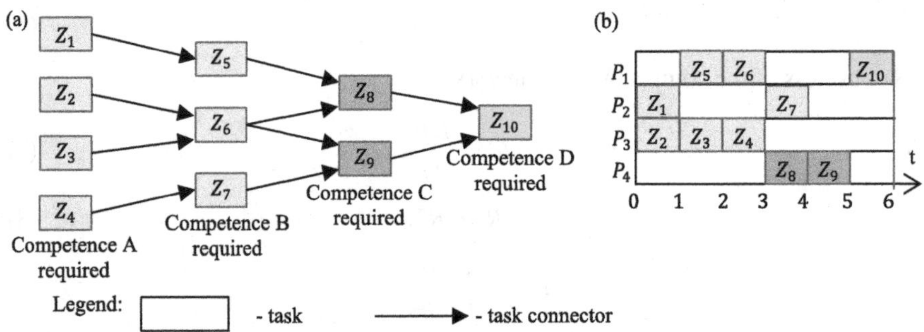

Fig. 1. Project structure (a) and project schedule (b)

The tasks have been assigned to a team of four employees P_1, P_2, P_3, P_4. Their competences make up framework G presented in Table 1, where the values of the cells ($g_{k,i}$) determine whether employee P_k has the competence (value "1") to perform task Z_i. For instance, employee P_1 has competences B and D, indispensable for performing tasks Z_5, Z_6, Z_7 and Z_{10}.

The following task completion rules (constraints) have been adopted:

- a task can only be completed by a competent employee,
- P_1 and P_4 can perform no more than 3 tasks, and P_2 and P_3 can execute no more than 4 tasks,
- all tasks must be assigned to employees.

Competence framework G allows for all planned tasks to be completed, as shown in the schedule in Fig. 1(b). Let us assume, however, that the project is jeopardized by an

Table 1. Competence framework G

	Task									
Employee	Z_1	Z_2	Z_3	Z_4	Z_5	Z_6	Z_7	Z_8	Z_9	Z_{10}
P_1 (B, D)	0	0	0	0	1	1	1	0	0	1
P_2 (A, B)	1	1	1	1	1	1	1	0	0	0
P_3 (A, B, C)	1	1	1	1	1	1	1	1	1	0
P_4 (A, C, D)	1	1	1	1	0	0	0	1	1	1

absence of one employee. *In such a situation, is framework G still sufficient to complete the planned tasks? And if not, does there exist a version of that framework that can guarantee robustness in the event of an absence of one of the employees* ($R = 1$)*?* In order to answer this question, it is necessary to analyze each of the four cases of absence (it is assumed that a team member is absent from the very beginning of the project). As shown in Fig. 2(a, b, and c), in three cases of an absence of a single employee (P_1, P_2, P_3), the competences of the remaining employees and the constraints on the number of tasks they can perform allow to modify the already implemented schedule. A modification is understood here as delegating the tasks of an absent employee to the available personnel (a replacement), without changing the task start/completion dates. In the event of an absence of P_4, the remaining employees are not able to take over his/her duties (Fig. 2(d)) either because they lack the required competences or because they have already reached their limit on the number of tasks they can perform.

It is easy to note that if employee P_2 acquires competence C, he/she will be able to take over the duties of employee P_4. This means that framework G (Table 2), which takes account of this fact, is robust to the absence of any one of the members of the project team.

The same result will be obtained as a solution to problem CS (11). However, attention should be paid to the computational complexity of this problem. For problems of a similar scale (4 employees and 10 tasks), the number of potential competence frameworks that should be considered in order to find a robust variant is 2^{40}. In the general case, the computational complexity is $f(m, n) = 2^{m \times n}$, which limits the use of these types of approaches to small-scale problems only. In order to avoid time-consuming calculations when solving large scale problems normally encountered in everyday practice (e.g. 50 employees, 100 tasks), a sufficient condition has been proposed, the fulfilment of which will guarantee the existence of admissible solutions.

3.2 Sufficient Condition

The proposed recursive sufficient condition, which allows to determine, for a given project structure, whether there exist admissible competence frameworks robust to the absence of a single employee, is as following statement:

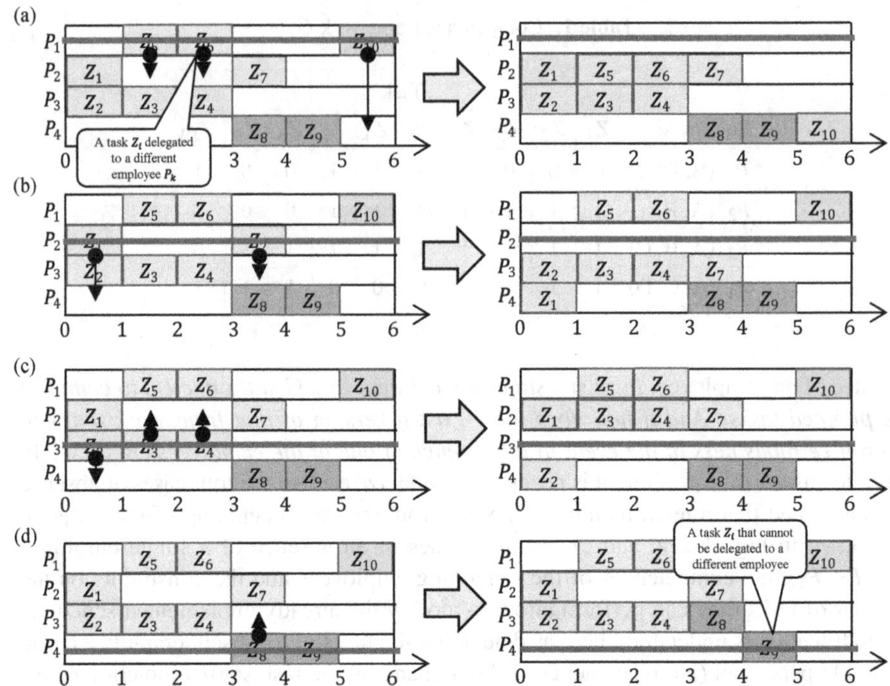

Fig. 2. Replacement scenarios for the individual employees: P_1 (a), P_2 (b), P_3 (c) and P_4 (d).

$\forall z \in Z^*$
$\Theta(z, P)\{$
if $\{P\ is\ empty\}$ then return "no solution"
if $\{\varphi(z) \in \Phi(P_1)$ and $C(\Psi(P_1) \cup \{z\}) = 1\}$ then $\{\Psi(P_1) \leftarrow \Psi(P_1) \cup \{z\}\}$
if $\{\varphi(z) \notin \Phi(P_1)$ and $C(\Psi(P_1) \cup \{z\}) = 1\}$ then $\{\Phi(P_1) \leftarrow \Phi(P_1) \cup \{\varphi(z)\};$
$\Psi(P_1) \leftarrow \Psi(P_1) \cup \{z\}\}$
If $\{C(\Psi(P_1) \cup \{z\}) = 0\}$ then $\{\Theta(z, (P_2, \ldots, P_q))\}$ $\}$

where: $P = \left(P_1, P_2, \ldots, P_q\right)$ – a sequence of an organization's available employees (when employee P^* is absent), $\Phi(P_k)$ – a set of competences of employee P_k, $\Psi(P_k)$ – a set of tasks performed by employee P_k, $C(\mathrm{A})$ – a function that defines whether tasks in set A can be completed in the given organization (if the function satisfies the organization's constraints, it returns 1; if the function does not satisfy the constraints, it returns 0), Z^* – a set of tasks normally performed by employee P^* which need to be delegated to employees P, $\varphi(z)$ – a competence necessary to complete task z.

The above condition is met if the available employees have the competences to take over the tasks of the absent employee or if the acquisition of such competences will allow them to do so. By testing this condition, one can quickly (greedily) determine (computational complexity: $f(n, m) = O(m \times n)$) the existence (or non-existence) of an admissible solution (i.e. a framework robust to the absence of one employee). In the next section, the effectiveness of the proposed condition is shown using a computational experiment.

Table 2. A competence framework G robust to the absence of one employee

	Task									
Employee	Z_1	Z_2	Z_3	Z_4	Z_5	Z_6	Z_7	Z_8	Z_9	Z_{10}
P_1 (B, D)	0	0	0	0	1	1	1	0	0	1
P_2 (A, B)	1	1	1	1	1	1	1	1	1	0
P_3 (A, B, C)	1	1	1	1	1	1	1	1	1	0
P_4 (A, C, D)	1	1	1	1	0	0	0	1	1	1

4 Experiment

Consider a project involving 10 tasks, implemented according to the given schedule, as shown in Fig. 3(a). Each task requires one employee with competences as shown in Table 3. Consider the competence framework presented in Table 4.

Table 3. Required competences.

Task	Z_1	Z_2	Z_3	Z_4	Z_5	Z_6	Z_7	Z_8	Z_9	Z_{10}
Required competences	A	B	D	F	E	C	A	C	F	E

Table 4. Competence framework G.

	Task									
Employee	Z_1	Z_2	Z_3	Z_4	Z_5	Z_6	Z_7	Z_8	Z_9	Z_{10}
P_1 (C)	0	0	0	0	0	1	0	1	0	0
P_2 (A, D)	1	0	1	0	0	0	1	0	0	0
P_3 (C)	0	0	0	0	0	1	0	1	0	0
P_4 (F)	0	0	0	1	0	0	0	0	1	0
P_5 (E)	0	0	0	0	1	0	0	0	0	1
P_6 (A, E)	1	0	0	0	1	0	1	0	0	1
P_7 (B, C)	0	1	0	0	0	1	0	1	0	0
P_8 (B, D, F)	0	1	1	1	0	0	0	0	1	0

Let us consider the employee assignment shown in Fig. 3(b).

It is easy to see that the competence framework G from Table 4 is not robust to an absence of employee P_7, because the only employee who can complete task Z_2 is employee P_8, who is busy at the time of execution of Z_2. This means that one should look for such modifications to the competence framework that will allow to complete

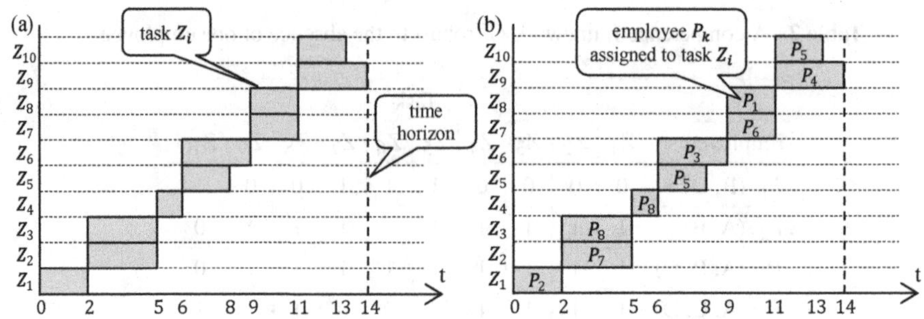

Fig. 3. Project schedule (a), employee assignment (b).

the tasks assigned in all cases of employee absence. The size of the search space for problem CS (11), in the case under consideration, is $f(8, 10) = 280$. This means that to assess whether there exist admissible solutions, in the worst case, one has to search this entire space. To ensure that such a solution exists, testing the sufficient condition can be done. It turns out that a solution exists in the form shown in Table 5. The solution was obtained in less than 1 s. The bold values of the Table indicate the required changes to the existing competences. This does not mean, however, that this is:

- the only admissible solution,
- an optimal solution, i.e. a variant of the competence framework characterized by the smallest number of changes to the existing employee competences (changes from "0" to "1").

Table 5. Competence framework G robust to an absence of one employee.

	Task									
Employee	Z_1	Z_2	Z_3	Z_4	Z_5	Z_6	Z_7	Z_8	Z_9	Z_{10}
P_1 (A, B, C, D, E, F)	**1**	**1**	**1**	**1**	**1**	1	**1**	1	**1**	**1**
P_2 (A, C, D)	1	0	1	0	0	0	1	**1**	0	0
P_3 (C)	0	0	0	0	0	1	0	1	0	0
P_4 (F)	0	0	0	1	0	0	0	0	1	0
P_5 (E)	0	0	0	0	1	0	0	0	0	1
P_6 (A, E)	1	0	0	0	1	0	1	0	0	1
P_7 (B, C)	0	1	0	0	0	1	0	1	0	0
P_8 (B, D, F)	0	1	1	1	0	0	0	0	1	0

As part of further calculations, we searched for the optimal competence framework by repeatedly solving problem CS (11). The calculations were made in the ILOG CPLEX environment using a PC with Intel Pentium i7-4770 and 16 GB RAM. The solution (Table 6) was found in 129 s. As it can be seen, the optimal competence framework requires only one cell to be changed (those marked in italic).

Table 6. Optimal competence framework G_{opt} robust to an absence of one employee.

	Task									
Employee	Z_1	Z_2	Z_3	Z_4	Z_5	Z_6	Z_7	Z_8	Z_9	Z_{10}
P_1 (B, C)	0	*1*	0	0	0	1	0	1	0	0
P_2 (A, D)	1	0	1	0	0	0	1	0	0	0
P_3 (C)	0	0	0	0	0	1	0	1	0	0
P_4 (F)	0	0	0	1	0	0	0	0	1	0
P_5 (E)	0	0	0	0	1	0	0	0	0	1
P_6 (A, E)	1	0	0	0	1	0	1	0	0	1
P_7 (B, C)	0	1	0	0	0	1	0	1	0	0
P_8 (B, D, F)	0	1	1	1	0	0	0	0	1	0

It is worth noting that the proposed sufficient condition can also be effectively used in assessing whether the search space is empty (whether there is no admissible solution). Let's consider the following example with data as given in Tables 3 and 4 and a project schedule as shown in Fig. 4. It is clear that in this structure it is impossible to find a replacement for an absent employee because all staff perform tasks in the same time interval and there is no back-up personnel available. Therefore, it is impossible to make changes to the existing competence framework that would allow to replace the absent employee. To assess whether there exist admissible solutions by solving CS (11), one needs to search a space of size 2^{80}. By contrast, when using the proposed sufficient conditions, one has to perform only 8×10 operations to obtain the information that there does not exist a single variant of the competence framework that is robust to the type of disruption under consideration.

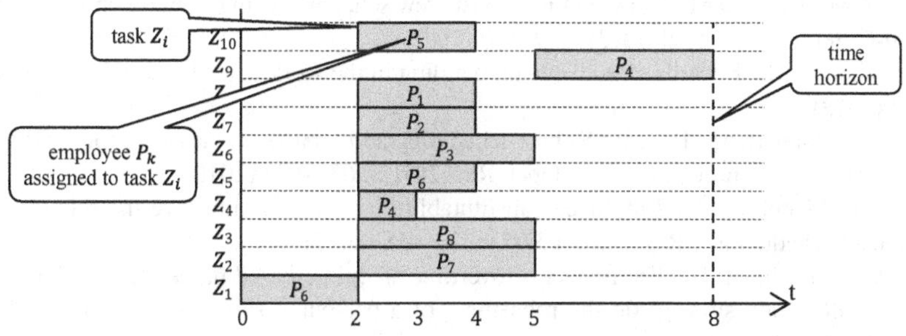

Fig. 4. Project schedule.

5 Conclusions and Future Works

The search for competence frameworks that can guarantee the completion of the planned tasks is tantamount to seeking (synthesizing) alternative competence frameworks that are robust to the given (a priori known) set of disruptions.

Recruiting a project team robust to the selected type of disruptions boils down to seeking (synthesizing) alternative variants of competence frameworks for which it will be possible to complete the planned tasks despite the disruptions. The computational complexity of the problem considered in the present study means that determination of robust competence frameworks in specific cases can be time consuming. To guarantee that it is worth making the tedious calculations, a sufficient condition has been proposed. By testing this condition, one can check whether or not there exist admissible solutions.

In our future work, we plan to focus on the robustness of competence frameworks to other disruptions, such as changing task time durations, etc. It is also planned to use various hybrid approach variants for implementation of the proposed models [16].

References

1. Antosz, K.: Maintenance – identification and analysis of the competency gap. Maint. Reliab. **20**(3), 484–494 (2018)
2. Bocewicz, G., Wikarek, J., Sitek, P., Banaszak, Z.: Robust competence allocation for multi-project scheduling. In: Świątek, J., Borzemski, L., Wilimowska, Z. (eds.) ISAT 2019. AISC, vol. 1051, pp. 16–30. Springer, Cham (2020). https://doi.org/10.1007/978-3-030-30604-5_2
3. Davenport, A., Gefflot, C., Beck, C.: Slack-based techniques for robust schedules. In: Cesta, A., Borrajo, D. (eds.) Proceedings of the Sixth European Conference on Planning, pp. 43–49 (2001)
4. Dück, V., Ionescu, L., Kliewer, N., Suhl, L.: Increasing stability of crew and aircraft schedules. Transp. Res. Part C Emerg. Technol. **20**(1), 47–61 (2012)
5. Easton, F., Rossin, D.: Overtime schedules for full-time service workers. Omega **25**(3), 285–299 (1997)
6. Ehrgott, M., Ryan, D.M.: Constructing robust crew schedules with bicriteria optimization. J. Multi-Criteria Decis. Anal. **11**(3), 139–150 (2003)
7. Fapohunda, T.M.: Towards effective team building in the workplace. Int. J. Educ. Res. **1**(4), 1–12 (2013)
8. Hazir, O., Haouari, M., Erel, E.: Robust scheduling and robustness measures for the discrete time/cost trade-off problem. Eur. J. Oper. Res. **207**(2), 633–643 (2010)
9. Ingels, J., Maenhout, B.: Employee substitutability as a tool to improve the robustness in personnel scheduling. OR Spectrum **37**(3), 623–658 (2017)
10. Ingels, J., Maenhout, B.: The impact of overtime as a time-based proactive scheduling and reactive allocation strategy on the robustness of a personnel shift roster. J. Sched. **21**(2), 143–165 (2018)
11. Ingels, J., Maenhout, B.: The impact of reserve duties on the robustness of a personnel shift roster: an empirical investigation. Comput. Oper. Res. **61**, 153–169 (2015)
12. Ionescu, L., Kliewer, N.: Increasing flexibility of airline crew schedules. Procedia Soc. Behav. Sci. **20**(3), 1019–1028 (2011)
13. Moudani, W., Mora-Camino, F.: Solving crew reserve in airlines using dynamic programming approach. Int. J. Optim. Theory Methods Appl. **2**(4), 302–329 (2010)

14. Nielsen, P., Jiang, L., Rytter, N.G.M., Chen, G.: An investigation of forecast horizon and observation fit's influence on an econometric rate forecast model in the liner shipping industry. Marit. Policy Manag. **41**(7), 667–682 (2014)
15. Shebalov, J., Klabjan, D.: Robust airline crew pairing: move-up crews. Transp. Sci. **40**, 300–312 (2006)
16. Sitek, P., Wikarek, J.: A multi-level approach to ubiquitous modeling and solving constraints in combinatorial optimization problems in production and distribution. Appl. Intell. **48**, 1344–1367 (2018)
17. Smith III, H.H., Smarkusky, D.: Competency matrices for peer assessment of individuals in team projects. In: Proceedings of the 6th Conference on Information Technology Education, SIGITE 2005, 20–22 October 2005, Newark, NJ, USA (2005)
18. Sobaszek, Ł., Gola, A., Kozłowski, E.: Application of survival function in robust scheduling of production jobs. In: Ganzha, M., Maciaszek, M., Paprzycki, M. (eds.) Proceedings of the 2017 Federated Conference on Computer Science and Information Systems (FEDCSIS), pp. 575–578. IEEE, New York (2017)
19. Szwarc, E., Bocewicz, G., Bach-Dąbrowska, I.: Planning of teacher staff competence structure robust to unexpected personnel absence. In: Manufacturing Modelling, Management and Control (MIM), Berlin (2019, in print)
20. Szwarc, E., Bocewicz, G., Bach-Dąbrowska, I., Banaszak, Z.: Declarative model of competences assessment robust to personnel absence. In: Damaševičius, R., Vasiljevienė, G. (eds.) ICIST 2019. CCIS, vol. 1078, pp. 12–23. Springer, Cham (2019). https://doi.org/10.1007/978-3-030-30275-7_2
21. Szwarc, E., Bocewicz, G., Banaszak, Z., Wikarek, J.: Competence allocation planning robust to unexpected staff absenteeism. Maint. Reliab. **21**(3), 440–450 (2019)
22. Tam, B., Ehrgott, M., Ryan, D.M., Zakeri, G.: A comparison of stochastic programming and bi-objective optimisation approaches to robust airline crew scheduling. OR Spectrum **33**(1), 49–75 (2011)
23. Vieira, G.E., Herrmann, J.W., Lin, E.: Rescheduling manufacturing systems: a framework of strategies, policies and methods. J. Sched. **6**(1), 35–58 (2003)
24. Whiddett, S., Hollyforde, S.: The Competency Handbook. Institute of Personnel and Development, London (2000)

Constraint Programming for New Product Development Project Prototyping

Marcin Relich[1(✉)], Izabela Nielsen[2], Grzegorz Bocewicz[3], and Zbigniew Banaszak[3]

[1] Faculty of Economics and Management, University of Zielona Gora, Zielona Gora, Poland
m.relich@wez.uz.zgora.pl

[2] Department of Materials and Production, Aalborg University, Aalborg Øst, Denmark
izabela@m-tech.aau.dk

[3] Faculty of Electronics and Computer Science, Koszalin University of Technology, Koszalin, Poland
bocewicz@ie.tu.koszalin.pl, zbigniew.banaszak@tu.koszalin.pl

Abstract. The paper is concerned with using computational intelligence for identifying the relationships between variables and constraint programming for searching variants of completing a new product development project. The relationships are used to the cost estimation of new product development (NPD) and to the search for possible variants of reaching the desirable NPD cost. The main contribution of this paper is the use of constraint programming to a project prototyping problem in the context of product development. Moreover, the paper presents a method for estimating the NPD cost and searching variants that can ensure the desirable NPD cost. The project prototyping problem is formulated in terms of a constraint satisfaction problem and implemented using constraint programming techniques. These techniques enable declarative description of the considered problem and effective search strategies for finding admissible solutions. An example illustrates the applicability of the proposed approach for solving an NPD project prototyping problem.

Keywords: Project evaluation · Product development cost · Neuro-fuzzy system · Decision support system · Constraint satisfaction problem

1 Introduction

New products development is one of the most important activities in contemporary companies. Increasing competition and changed customer requirements impose more frequent new product launches. Introducing new product into the market before competitors and customer satisfaction are prerequisites for the product success. The company's resources are often limited what causes a need to evaluate an NPD project from the perspective of its potential profitability and searching for possibilities to perform a project within available resources. If the company's resources (e.g. financial, human) are not sufficient to continue an NPD project within a basic project schedule, then there is a need to search possible variants of project performance. Also, the NPD cost may have unacceptable level for the decision makers, and as a result, this leads to a need to verify the

N. T. Nguyen et al. (Eds.): ACIIDS 2020, LNAI 12034, pp. 26–37, 2020.
https://doi.org/10.1007/978-3-030-42058-1_3

possibility of project performance within the preferable cost. The process of identifying possible variants can be divided into two steps. The first step is concerned with selecting variables that impact the NPD cost, identifying the relationships, and estimating the cost. In turn, the second step refers to the use of the identified relationships to search variants of project performance within the assumed constraints.

The formulation of a project prototyping problem using variables and constraints enables its specification in terms of a constraint satisfaction problem (CSP). A CSP that generally belongs to combinatorial problems is solved with the use of constraint programming techniques [1, 2]. Constraint programming includes search strategies that are crucial for improving search efficiency of solving the considered problem [2, 3]. Consequently, constraint programming can be considered as an appropriate framework for designing a decision support system for identifying possible variants of project performance. The use of constraint programming techniques to project selection and scheduling problems has been widely considered in the literature. However, an aspect of identifying variants to improve performance of NPD projects is still neglected.

The paper is organised as follows: Sect. 2 presents problem formulation of an NPD project prototyping in terms of a CSP. A method for searching variants for the desirable NPD cost is shown in Sect. 3. An illustrative example of the proposed approach is presented in Sect. 4. Finally, Sect. 5 concludes this study.

2 Problem Formulation

A project prototyping problem refers to the search for the possibilities to complete an NPD project in an alternative way, taking into account the fulfilment of the assumed constraints. This study is concerned with searching for variants of project completion by the desirable level of the NPD cost. If this cost is unacceptable for the decision maker, then according to a traditional approach to project evaluation, the specific project is no longer considered. However, if the project is important from a strategic point of view, the decision maker can be interested in acquiring the prerequisites that enable the desirable project performance. The proposed approach refers to identification all possibilities (alternatives) to perform a project within constraints that can be related to project objectives, project budget, human resources, machines, etc. Figure 1 presents the traditional approach for project evaluation and the proposed approach for seeking variants within the desirable project performance. The traditional approach may be considered as a project prototyping problem stated in the forward form, whereas the proposed approach – the problem stated in the inverse form.

The proposed approach allows the decision maker to identify prerequisites, for which a project can obtain the desirable project performance within specified constraints, variables, and relationships between these variables. The number of possible variants of project performance depends on constraints, domains related to variables, and granularity of decision variables. Relationships between variables can be identified on the basis of previous experiences with the similar projects and presented, for example, as if-then rules. Then, the identified relationships may be used twofold: to predict the potential of a project (the traditional approach), and to verify the existence of such changes that could reach the desired project completion (the proposed approach).

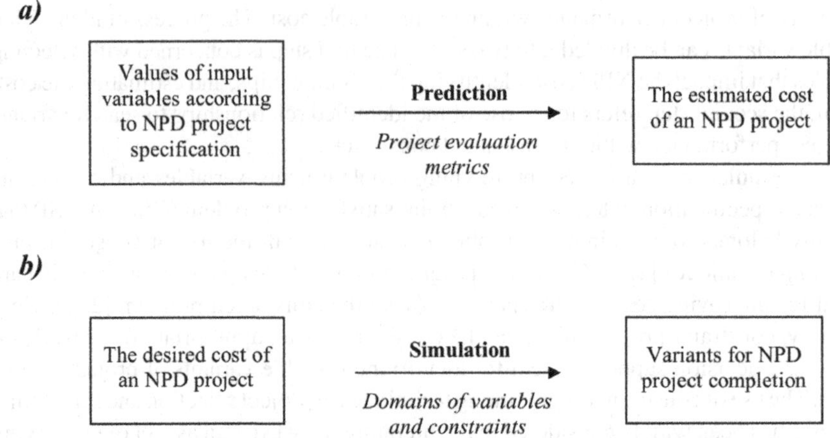

Fig. 1. The traditional (a) vs. proposed (b) approach for project evaluation

The use of the proposed approach requires the specification of variables, their domains, and constraints. This specification enables the identification of all available solutions, if there are any solutions. This approach may be effortlessly formulated in terms of a constraint satisfaction problem (CSP). A CSP may be described in the following form [4]:

$$(V, D, C)$$

where:

V is a finite set of n variables $\{v_1, v_2, \ldots, v_n\}$,
D is a finite and discrete domains $\{d_1, d_2, \ldots, d_n\}$ related to variables V,
C is a finite set of constraints $\{c_1, c_2, \ldots, c_m\}$ that restrict values of variables and link them.

Each constraint is treated as a predicate that may be seen as an n-ary relation defined by a Cartesian product $d_1 \times d_2 \times \ldots \times d_n$. The solution of a CSP is a vector $(d_{1i}, d_{2k}, \ldots, d_{nj})$ that is related to the assessment of a value of each variable that satisfies all constraints C. Generally, constraints may be specified in analytical and/or logical formulas.

Figure 2 presents an example of constraints (depicted as ovals) and variables that impact the cost of new product development, including the product design and prototype tests. Characteristic features of variables that impact on the total cost of a NPD project are their controllability by a company, and the possibility of the simulation to identify a set of the values of variables that satisfy all constraints and ensure the desirable level of costs.

There are the following variables regarding the project prototyping problem:

V_1 – the product development cost
V_2 – the number of employees involved in product design

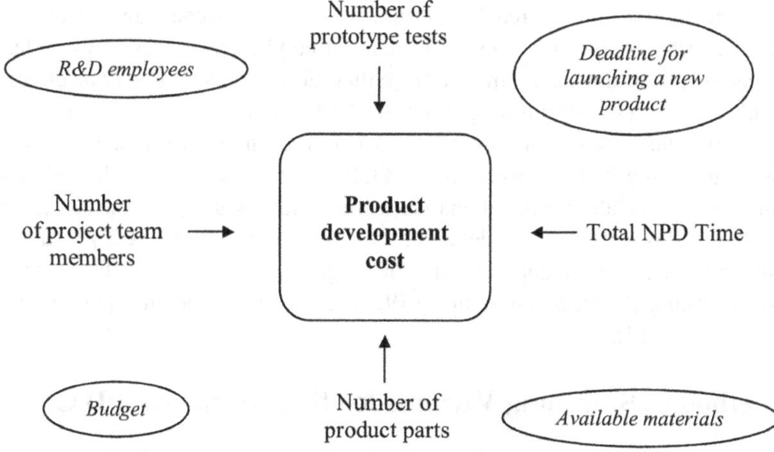

Fig. 2. Variables and constraints for evaluating an NPD project

V_3 – the number of employees involved in prototype tests
V_4 – the duration of product design
V_5 – the duration of prototype tests
V_6 – the number of product parts
V_7 – the number of prototype tests
V_8 – the amount of materials needed to produce a unit of a new product

The set of constraints is as follows:

C_1 – the total number of R&D employees who may be involved in an NPD project
C_2 – the project budget
C_3 – the deadline for launching a new product
$C_4 \geq V_2 + V_3$ (the constraint refers to the number of employees involved in product development)
$C_5 \geq V_4 + V_5$ (the constraint refers to the duration of product design and prototype tests)
C_6 – the minimal number of product parts
C_7 – the minimal number of prototype tests

The model formulation in terms of a CSP integrates technical parameters of a new product, parameters regarding planned project performance, and available resources. The problem solution refers to the search for answers to the following questions:

- what is the NPD cost?
- what values should have the variables to reach the desirable NPD cost?

A project prototyping problem can be formulated in terms of a CSP that in turn can be solved with the use of the specific techniques such as constraint propagation and variable distribution. Constraint propagation applies constraints to prune the search space.

Propagation techniques aim to reach a certain level of consistency, and accelerate the search procedures to reduce the size of the search tree [4]. The values of variables that are excluded by constraints, are removed from their domains. A CSP may be effectively solved with the use of constraint programming (CP) techniques. The declarative nature of a CP is particularly useful for applications where it is enough to state *what* has to be solved without saying *how* to solve it [4]. As CP uses the specific search methods and constraint propagation algorithms, it enables a significant reduction of the search space. Consequently, CP is suitable to model and solve complex problems. A CSP framework and CP techniques have been applied to a wide range of problems, for instance, scheduling [5, 6], planning [7, 8], manufacturing [9, 10], resource allocation [11], or supply chain problem [12, 13].

3 A Method of Searching Variants for the Desirable NPD Cost

The proposed method consists of the following five steps: (1) collecting data from previous projects that are similar to a new project, (2) identifying relationships between variables, (3) modelling a project prototyping problem in terms of a CSP, (4) estimating the NPD cost, and if it is non-acceptable, (5) searching variants for obtaining the desirable NPD cost. Figure 3 illustrates a framework for the proposed decision support system that uses computational intelligence to identify relationships and constraint programming to reduce the search space and verify the possibility of reaching the desirable total cost of a new product.

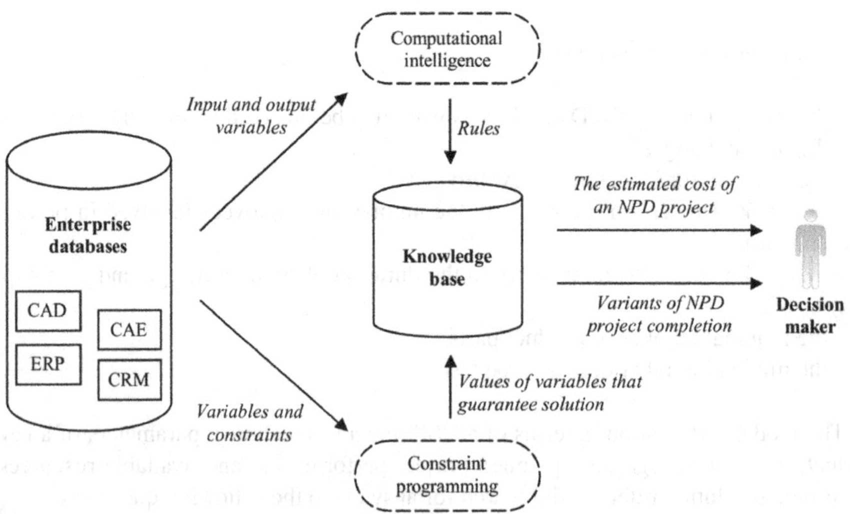

Fig. 3. A framework for the proposed decision support system

The data is collected from databases related to information systems that support project management of NPD. This requires the use of some project management standards, including project performance planning, monitoring, controlling, and using appropriate techniques needed for project planning and executing. Enterprise databases that can be used to information retrieval refer to, for example, computer-aided design (CAD), computer-aided engineering (CAE), enterprise resource planning (ERP), and customer relationship management (CRM) system. The applicability of the proposed method depends on the fulfilment of the following procedures regarding data management: an enterprise adjusts common project management standards to its needs, distinguishes phases in the NPD process, uses standards for specifying tasks in an NPD project and for project portfolio management, measures the success of new products on corporate financial performance, uses the results of a financial performance analysis to improve the effectiveness of NPD projects, registers performance and metrics of NPD projects, uses a primary schedule for monitoring performance in NPD projects, and uses the defined procedure to allocate employees to NPD projects.

The proposed method bases on identification of cause-and-effect relationships that are used to estimate the cost of product development, and search for a desirable outcome of an NPD project. These relationships can be determined using parametric models that may involve econometric models/regression analysis [14, 15], artificial neural networks [16, 17], fuzzy logic systems [18], or hybrid systems such as neuro-fuzzy systems [19, 20]. The variable selection to a parametric model depends on factors such as the significant impact of a variable on the NPD cost and the controllability of a variable by a company. The set of variables includes for example the number of project team members, product components and prototype tests. A set of variables, their domains, and constraints constitutes a CSP that is a framework for obtaining answers to the questions about the value of the cost, and if it is non-acceptable, about the values of variables that enable the desirable total cost of product development.

The last step of the proposed method refers to the search of possible solutions to achieve the desirable total cost of product development. The search space depends on the number of variables chosen to the analysis, a range of domains of decision variables, and constraints that can link variables and limit possible solutions. An exhaustive search always finds a solution, if it exists, but its performance is proportional to the number of admissible solutions. Therefore, an exhaustive search tends to grow very quickly as the size of the problem increases, what limits its usage in many practical problems. Consequently, there is a need to develop more effective methods for searching the space and finding possible solutions. This research proposes constraint programming (CP) to solve efficiently a CSP specified for NPD project prototyping. Figure 4 illustrates a framework for solving this problem in terms of a CSP.

The proposed approach of solving the above-described problem includes three stop/go conditions. The first stop/go condition checks the possibility of existing solution for a set of decision variables within the specified input and output variables, their domains and constraints (1). If there is any solution, then there is verified the possibility to extend domains related to the selected decision variables (2). Finally, if domains of decision variables cannot be extended, then there is verified the possibility to change the set of decision variables (3).

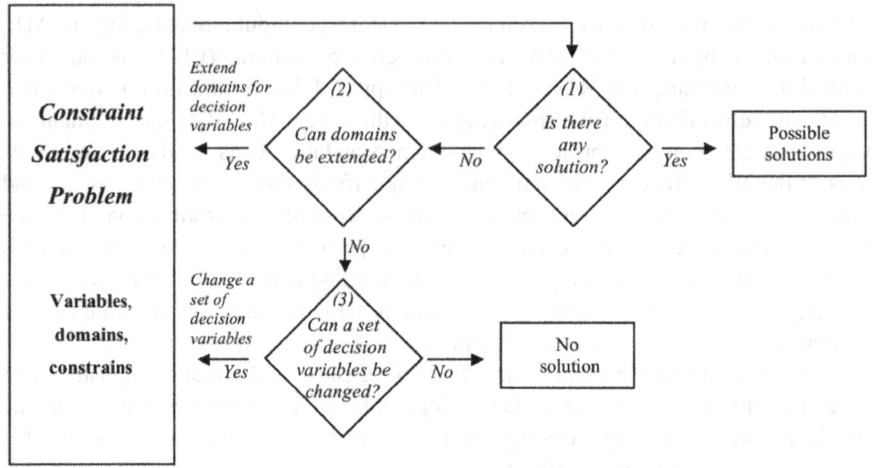

Fig. 4. A framework for solving the problem stated in the inverse form

4 An Example of the Proposed Approach

4.1 Estimating the NPD Cost

The estimation process has been carried out with the use of an adaptive neuro-fuzzy inference system (ANFIS), and compared with multiple regression analysis (MR). The NPD cost (V_1) is estimated using six variables (V_2, …, V_7). The dataset includes 27 completed projects that belong to the same product line as the considered new product projects. The dataset has been divided into training set (21 cases) and testing set (6 cases) to evaluate estimating quality of an estimating model. A neuro-fuzzy system combines the advantages of the artificial neural networks (ability to learning and identifying the complex relations) and fuzzy logic (ability to incorporating expert knowledge and specifying the identified relationships in the form of rules if-then). The learning method of ANFIS and learning parameters have been experimentally adjusted by comparison of errors for methods implemented in Matlab® environment such as grid partition and subtractive clustering. The smallest errors for the considered dataset have been generated with the use of subtractive clustering method with the following parameters: squash factor – 1.25, accept ratio – 0.5, reject ratio – 0.15 and range of influence (RI) from 0.1 to 1.5. The experiments have been performed for various shapes of membership functions. However, the most accurate forecasts have been achieved for a Gaussian membership function.

The ANFIS has generated in the training set less RMSE than the average and linear regression model. However, the RMSE in the testing set for the ANFIS with parameter RI from 0.2 to 0.5 is greater than for the linear regression model. The least RMSE and the small number of rules have been generated by the ANFIS with parameter RI from 0.6 to 1.5. Figure 5 presents the use of the ANFIS (with RI = 1) to estimate the NDP cost by the following values of input variables: $V_2 = 4$, $V_3 = 3$, $V_4 = 5$, $V_5 = 4$, $V_6 = 30$, $V_7 = 9$.

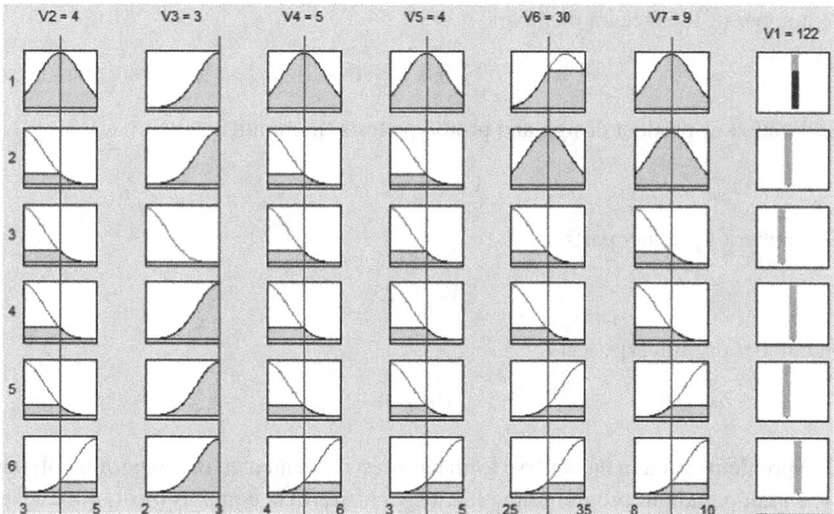

Fig. 5. The NPD cost estimation using ANFIS

After inputting these values to the trained ANFIS, the NPD cost is estimated at 122 thousand €. Let us assume that the NPD cost does not fulfil the decision maker's expectations. To check the possibility of fulfilling these expectations, the problem is reformulated into the inverse form, i.e. there is sought project performance that ensures the desirable NPD cost.

4.2 Searching for the Desirable NPD Cost

An example consists of three cases according to steps presented in Fig. 4: (1) a basic variant for originally declared decision variables, their domains, and constraints; (2) an extension of domains for the selected decision variables; and (3) a change in the set of decision variables.

A Basic Variant. Let us assume that the estimated NPD cost equals 122 thousand € (see Subsect. 4.1), and it does not satisfy the decision maker's expectations. The decision maker is interested in decreasing the cost to 120 thousand €. To verify the possibility of existing solutions, this problem is stated in the inverse form. The solution of the problem stated in the inverse form is sought with the use of constraint programming, and it requires the specification of decision variables, their domains, and constraints, among which are relationships between variables (e.g. their mutual impact on the cost). Domains for considered variables are as follows: $D_2 = \{3, \ldots, 5\}$, $D_3 = \{2, 3\}$, $D_4 = \{4, \ldots, 6\}$, $D_5 = \{3, \ldots, 5\}$, $D_6 = \{25, \ldots, 35\}$, and $D_7 = \{8, \ldots, 10\}$.

There are the following constraints:

– the desirable NDP cost (in thousand €)

$$V_1 \leq 120$$

– the number of R&D team members

$$V_2 + V_3 \leq 6$$

– the duration of product design and prototype tests (in months)

$$V_4 + V_5 \leq 7$$

– the number of product parts

$$V_6 \geq 25$$

– the number of prototype tests

$$V_7 \geq 8$$

The problem stated in the inverse form has been implemented in Mozart/Oz software that is a multiparadigm programming language. Mozart/Oz contains most of the major programming paradigms, including logic, functional, imperative, object-oriented, concurrent, constraint, and distributed programming. The significant strengths of Mozart/Oz are constraint and distributed programming that are able to effectively solve many practical problems, for example, timetabling and scheduling [21].

Table 1 presents 10 possible solutions of the above-described problem in terms of potential changes in variables and their impact on the desirable NPD cost.

Table 1. A set of possible solutions

Case	Values of variables	V_1
1	$V_2 = 3, V_3 = 3, V_4 = 4, V_5 = 3, V_6 = 25, V_7 = 8$	119.8
2	$V_2 = 3, V_3 = 3, V_4 = 4, V_5 = 3, V_6 = 26, V_7 = 8$	119.9
3	$V_2 = 3, V_3 = 3, V_4 = 4, V_5 = 3, V_6 = 27, V_7 = 8$	120.0
4	$V_2 = 3, V_3 = 2, V_4 = 4, V_5 = 3, V_6 = 25, V_7 = 8$	119.7
5	$V_2 = 3, V_3 = 2, V_4 = 4, V_5 = 3, V_6 = 26, V_7 = 8$	119.8
6	$V_2 = 3, V_3 = 2, V_4 = 4, V_5 = 3, V_6 = 27, V_7 = 8$	119.8
7	$V_2 = 3, V_3 = 2, V_4 = 4, V_5 = 3, V_6 = 28, V_7 = 8$	119.9
8	$V_2 = 3, V_3 = 2, V_4 = 4, V_5 = 3, V_6 = 29, V_7 = 8$	119.9
9	$V_2 = 3, V_3 = 2, V_4 = 4, V_5 = 3, V_6 = 30, V_7 = 8$	120.0
10	$V_2 = 3, V_3 = 2, V_4 = 4, V_5 = 3, V_6 = 31, V_7 = 8$	120.0

Let us assume that the decision maker's expectations are related to the decrease of the NDP cost from 122 to 120 thousand €. There is no solution satisfying a new value of the preferable NPD cost, what triggers the next step in the procedure of solving the considered problem, i.e. verification of the possibility to extend domains related to the selected variables (see Fig. 4).

An Extension of Domains for the Selected Variables. This step of the procedure of solving the problem stated in the inverse form requires selecting variables, for which an extension of domains is possible. As a result, the above-described problem is solved again for new domains assigned to some variables, and for other conditions unchanged. For example, domains for the following variables are extended: the number of employees involved in product design (D_2 from {3, …, 5} to {3, …, 6}), the number of employees involved in prototype tests (D_3 from {2, 3} to {2, …, 4}), the duration of product design (D_4 from {4, …, 6} to {4, …, 7}), and the duration of prototype tests (D_5 from {3, …, 5} to {3, …, 6}). Other domains of variables and constraints are the same as in the basic variant.

An extension of domains for four variables causes the increase of choice nodes in the explored search tree. Nevertheless, the fully explored subtrees do not contain any solution nodes. If domains of decision variables cannot be extended or there is no solution (as in the presented example), then there is verified the possibility to change the set of decision variables.

A Modification of the Set of Variables. In the next step of the procedure of solving the problem stated in the inverse form, the decision maker selects another set of variables towards reaching the preferable NPD cost. For example, to the present set of variables is added one variable more, namely the amount of materials needed to produce a unit of a new product (V_8). The planned materials consumption of producing a new product equals 6 units. The decision maker is interested in obtaining information about changes in the NPD costs by materials consumption of producing a new product between 5 and 7 units. Domains for the variables are as follows: V_2 = 3#6, V_3 = 2#4, V_4 = 4#7, V_5 = 3#6, V_6 = 25#35, and V_7 = 8#10, and V_8 = 5#7. The preferable NPD costs should be smaller or equal than 120 thousand €. Other domains of variables and constraints are the same as in the basic variant. There are four admissible solutions for materials consumption of producing a new product equals 5 units.

5 Conclusion

The presented approach supports the decision makers in searching possibilities of alternative project performance within available resources. This approach is especially useful in the case of limited resources (e.g. the project budget), if there is the possibility to continue an NPD project according to new conditions. The limited resources require more effort and attention to manage the NPD projects. Consequently, there is a need to develop a decision support system for identifying variants towards improving project performance. An effective development and updating of a decision support system requires formulating a single knowledge base that reflects a model of NPD project prototyping. The proposed model encompasses the areas related to project management of NPD and company's resources. These areas can be described in terms of a constraint satisfaction problem that includes the sets of decision variables, their domains, and constraints, which link and limit the variables. The project prototyping problem refers to the search of answers to queries about the estimated values of an output variable, and about the prerequisites that ensure the desirable values of an output variable.

Characteristics of the proposed approach include the use of constraint programming to implement a constraint satisfaction problem in the context of NPD project prototyping. The results show that the use of constraint programming improves search efficiency of solving the considered problem, especially for a larger number of admissible solutions. Moreover, this study presents the use of a neuro-fuzzy system to determine the relationships for estimating the cost of an NPD project. The identified relationships are stored in a knowledge base in the form of if-then rules and used to generate a set of possible variants to improve project performance. If project performance within original specification is unacceptable for the decision makers, then the sought variants can support them in identifying the impact of one variable on other variables within the specified constraints. Drawbacks of the proposed approach can be considered from the perspective of collecting enough amounts of data of the past similar NPD projects, and specifying several parameters to build and learn a neuro-fuzzy network. The future work is related to the verification of the proposed approach in project-oriented companies belonged to different business sectors. Moreover, the future research will extend the application of the presented approach towards incorporating the warranty cost in the total NPD cost.

References

1. Frühwirth, T., Abdennadher, S.: Essentials of Constraint Programming. Cognitive Technologies. Springer, Heidelberg (2003). https://doi.org/10.1007/978-3-662-05138-2
2. Liu, S.S., Wang, C.J.: Optimizing project selection and scheduling problems with time-dependent resource constraints. Autom. Constr. **20**, 1110–1119 (2011)
3. Apt, K.R.: Principles of Constraint Programming. Cambridge University Press, Cambridge (2003)
4. Banaszak, Z., Zaremba, M., Muszyński, W.: Constraint programming for project-driven manufacturing. Int. J. Prod. Econ. **120**, 463–475 (2009)
5. Baptiste, P., Le Pape, C., Nuijten, W.: Constraint-Based Scheduling: Applying Constraint Programming to Scheduling Problems. Kluwer Academic Publishers, Norwell (2001)
6. Bocewicz, G., Nielsen, I.E., Banaszak, Z.: Production flows scheduling subject to fuzzy processing time constraints. Int. J. Comput. Integr. Manuf. **29**, 1105–1127 (2016)
7. Do, M., Kambhampati, S.: Planning as constraint satisfaction: solving the planning graph by compiling it into CSP. Artif. Intell. **132**, 151–182 (2001)
8. Relich, M.: Identifying project alternatives with the use of constraint programming. In: Borzemski, L., Grzech, A., Świątek, J., Wilimowska, Z. (eds.) Information Systems Architecture and Technology: Proceedings of 37th International Conference on Information Systems Architecture and Technology – ISAT 2016 – Part I. AISC, vol. 521, pp. 3–13. Springer, Cham (2017). https://doi.org/10.1007/978-3-319-46583-8_1
9. Banaszak, Z.A.: CP-based decision support for project driven manufacturing. In: Józefowska, J., Weglarz, J. (eds.) Perspectives in Modern Project Scheduling. ISOR, vol. 92, pp. 409–437. Springer, Boston (2006). https://doi.org/10.1007/978-0-387-33768-5_16
10. Soto, R., Kjellerstrand, H., Gutiérrez, J., López, A., Crawford, B., Monfroy, E.: Solving manufacturing cell design problems using constraint programming. In: Jiang, H., Ding, W., Ali, M., Wu, X. (eds.) IEA/AIE 2012. LNCS (LNAI), vol. 7345, pp. 400–406. Springer, Heidelberg (2012). https://doi.org/10.1007/978-3-642-31087-4_42
11. Modi, P.J., Jung, H., Tambe, M., Shen, W.-M., Kulkarni, S.: A dynamic distributed constraint satisfaction approach to resource allocation. In: Walsh, T. (ed.) CP 2001. LNCS, vol. 2239, pp. 685–700. Springer, Heidelberg (2001). https://doi.org/10.1007/3-540-45578-7_56

12. Sitek, P., Wikarek, J.: A multi-level approach to ubiquitous modeling and solving constraints in combinatorial optimization problems in production and distribution. Appl. Intell. **48**(5), 1344–1367 (2018)
13. Grzybowska, K., Kovács, G.: The modelling and design process of coordination mechanisms in the supply chain. J. Appl. Logic **24**, 25–38 (2017)
14. Liu, H., Gopalkrishnan, V., Quynh, K.T., Ng, W.K.: Regression models for estimating product life cycle cost. J. Intell. Manuf. **20**(4), 401–408 (2009)
15. Nielsen, P., Jiang, L., Rytter, N.G., Chen, G.: An investigation of forecast horizon and observation fit's influence on an econometric rate forecast model in the liner shipping industry. Marit. Policy Manag. **41**(7), 667–682 (2014)
16. Seo, K.K., Park, J.H., Jang, D.S., Wallace, D.: Approximate estimation of the product life cycle cost using artificial neural networks in conceptual design. Int. J. Adv. Manuf. Technol. **19**(6), 461–471 (2002)
17. Relich, M.: A knowledge-based system for new product portfolio selection. In: Różewski, P., Novikov, D., Bakhtadze, N., Zaikin, O. (eds.) New Frontiers in Information and Production Systems Modelling and Analysis. ISRL, vol. 98, pp. 169–187. Springer, Cham (2016). https://doi.org/10.1007/978-3-319-23338-3_8
18. Kłosowski, G., Gola, A.: Risk-based estimation of manufacturing order costs with artificial intelligence. In: Federated Conference on Computer Science and Information Systems, pp. 729–732 (2016)
19. Efendigil, T., Önüt, S., Kahraman, C.: A decision support system for demand forecasting with artificial neural networks and neuro-fuzzy models: a comparative analysis. Expert Syst. Appl. **36**(3), 6697–6707 (2009)
20. Relich, M., Bzdyra, K.: Knowledge discovery in enterprise databases for forecasting new product success. In: Jackowski, K., Burduk, R., Walkowiak, K., Woźniak, M., Yin, H. (eds.) IDEAL 2015. LNCS, vol. 9375, pp. 121–129. Springer, Cham (2015). https://doi.org/10.1007/978-3-319-24834-9_15
21. Van Roy, P.: Multiparadigm Programming in Mozart/Oz. LNCS, vol. 3389. Springer, Heidelberg (2005). https://doi.org/10.1007/b106627

Automated Reasoning with Applications in Intelligent Systems

A Knowledge Base for Industrial Control Network Security Analysis and Decision-Making with Reasoning Method

Hongbiao Gao[1], Jiaming Liu[1], Jianbin Li[1(✉)], and Jingde Cheng[2]

[1] School of Control and Computer Engineering, North China Electric Power University, Beijing, China
{gaoh,liujiaming,lijb87}@ncepu.edu.cn
[2] Department of Information and Computer Sciences, Saitama University, Saitama, Japan
cheng@aise.ics.saitama-u.ac.jp

Abstract. Because more and more industrial control systems are connected to networks and more and more information technologies are used in the industrial control systems, the security of industrial control network is a problem. To solve the problem, we can use strong relevant logic as basic logical system to perform reasoning to make security analysis and provide decisions for industrial control network before the actual fault occurred. Although the method has been proposed in our previous work, the method needs a knowledge base to save the collected and reasoned out empirical knowledge. This paper proposes a knowledge base for the method.

Keywords: Industrial control network · Security analysis · Decision-making · Reasoning method · Knowledge base

1 Introduction

At present, information technology are widely used in the field of industry, besides, industrialization and information technology have been integrated continuously. Industrial control system has been widely used in various industrial fields, such as electric power, transportation, aerospace and so on, therefore as an important part of the key infrastructure in any country, industrial control system security is vital to national economic and social development [11].

Because lots of hardware and software are widely used in industrial control system, and enterprise management information systems are integrated, industrial control systems become open with various network data exchange [12]. Therefore, the terrorists, commercial spies, and illegal personnel have more opportunities to attack the open industrial control network, which are industrial control systems in the physical environment of the relative closure in the past. Improving the security of industrial control network is necessary and how to

N. T. Nguyen et al. (Eds.): ACIIDS 2020, LNAI 12034, pp. 41–52, 2020.
https://doi.org/10.1007/978-3-030-42058-1_4

improve the security of industrial control network is a future research direction [10,14]. Industrial communication and advanced manufacture need industrial control network. To predict security risks which are high-temperature radiation, electromagnetic interference, malicious attacks, and system failures is an important research trend [9]. In order to improve the security of industrial control network, we can predict risks and make the corresponding decisions before the terrorists, commercial spies, and illegal personnel attack the open industrial control network to make it occurs actual fault [4,8,13].

Various methods and technologies of artificial intelligence have been widely applied for improving the security of industrial control network, in which expert system is a kind of common methods and technologies to be used to predict risks and make the corresponding decisions to improve the security of industrial control network. It is sure that expert system holds some benefits. Because expert system encapsulate related empirical knowledge of one certain field and logic knowledge as reasoning mechanism, it can express and imitate thinking and idea of security analysist and decision makers in a certain way, and use interactive dialogue to guide them. However, the traditional methods of expert system hold some problem when we use them to predict risks and make the corresponding decisions for industrial control network. First, empirical knowledge is embedded into the expert system, so the scope of the knowledge and reasoning rules is limited. Once the needed empirical knowledge is beyond the scope, the system cannot do anything. The second problem is that traditional methods of expert system only respond to users, so the traditional methods cannot predict risks, make the corresponding decisions for industrial control network before the accident or attack happen. According to the above reason, traditional methods are usually passive, and defense ability of the accident and attack is not enough. The third problem is that traditional methods of expert system use classical mathematical logic or its various extensions to underlie reasoning. However, they are not suitable logical systems for reasoning method [2].

To solve the problem of traditional methods, we can use strong relevant logic as basic logical system to perform reasoning to make security analysis and provide decisions for industrial control network before the actual fault occurred. Although the method has been proposed in our previous work [7], the method needs a knowledge base to save the collected and reasoned out empirical knowledge. This paper proposes a knowledge base for the method. The rest of the paper is organized as follows: Sect. 2 introduces the security analysis and decision-making method for industrial control network with reasoning based on strong relevant logic. Section 3 shows the requirement analysis of the knowledge base. Section 4 shows the design of the knowledge base. Section 5 shows the implementation of the knowledge base. Finally, some concluding remarks are given in Sect. 6.

2 The Proposed Method

A formal logic system L is an ordered pair $(F(L), \vdash_L)$ where $F(L)$ is the set of well formed formulas of L, and $\vdash_L$ is the consequence relation of L such that for a set P of formulas and a formula C, $P \vdash_L C$ means that within the framework of L taking P as premises we can obtain C as a valid conclusion. $Th(L)$ is the set of logical theorems of L such that $\phi \vdash_L T$ holds for any $T \in Th(L)$. According to the representation of the consequence relation of a logic, the logic can be represented as a Hilbert style system, a Gentzen sequent calculus system, a Gentzen natural deduction system, and so on [1,2].

Let $(F(L), \vdash_L)$ be a formal logic system and $P \subseteq F(L)$ be a non-empty set of sentences. *A* formal theory with premises P based on L, called a L-theory with premises P and denoted by $T_L(P)$, is defined as $T_L(P) =_{df} Th(L) \cup Th^e_L(P)$ where $Th^e_L(P) =_{df} \{A|P \vdash_L A \; and \; A \notin Th(L)\}$, $Th(L)$ and $Th^e_L(P)$ are called the logical part and the empirical part of the formal theory, respectively, and any element of $Th^e_L(P)$ is called an empirical theorem of the formal theory [1,2].

Our proposed method [7] can be summarized as "for any given premises P, we construct a meaningful formal theory $T_L(P)$ and then find new and important empirical theorems in $Th^e_L(P)$ automatically?".

The method chooses strong relevant logic as the logic part, and chooses empirical knowledge of industrial control network as empirical part. By using the method, we want to find important empirical theorems, which are important rules, risks, decisions, facts and so on. We have defined several factors to find the important empirical theorems, which are the abstract level of empirical theorems, deduction distance of empirical theorems, degree of logical connectives, and propositional schema of empirical theorems [7]. We may get lots of conclusions and hope to find important ones from those conclusions. We consider that the method used in the case studies to find new and interesting mathematical theorems in our previous work can be referred [5,6].

In detail, there are five phases [7] in the method as shown in Fig. 1. We prepare logical fragments of strong relevant logic in Phase 1. To prepare the logical fragments of strong relevant logic automatically, we can use general reasoning engine FreeEnCal [3] as the tool. We should input the axioms and inference rules of strong relevant logic, and input the degree of each logical connectives. We collect empirical knowledge of Industry control systems as empirical premises in Phase 2. The collected empirical knowledge are rules and data in industrial control network. We define a semi-lattice according to the abstract level of collected empirical knowledge of industrial control network [7]. According to the defined semi-lattice, we can systematically perform security analysis and decision-making from low abstract level to high abstract level. Besides, new predicate and functions can also be defined from the empirical theorems and added into the semi-lattice.

We use general reasoning engine FreeEnCal [3] as the tool to perform reasoning in Phase 3. Then we perform abstract in Phase 4, and find important empirical theorems in Phase 5. We perform Phase 3 to Phase 5 not in one time, but perform them loop by loop until all of prepared empirical premises have been used.

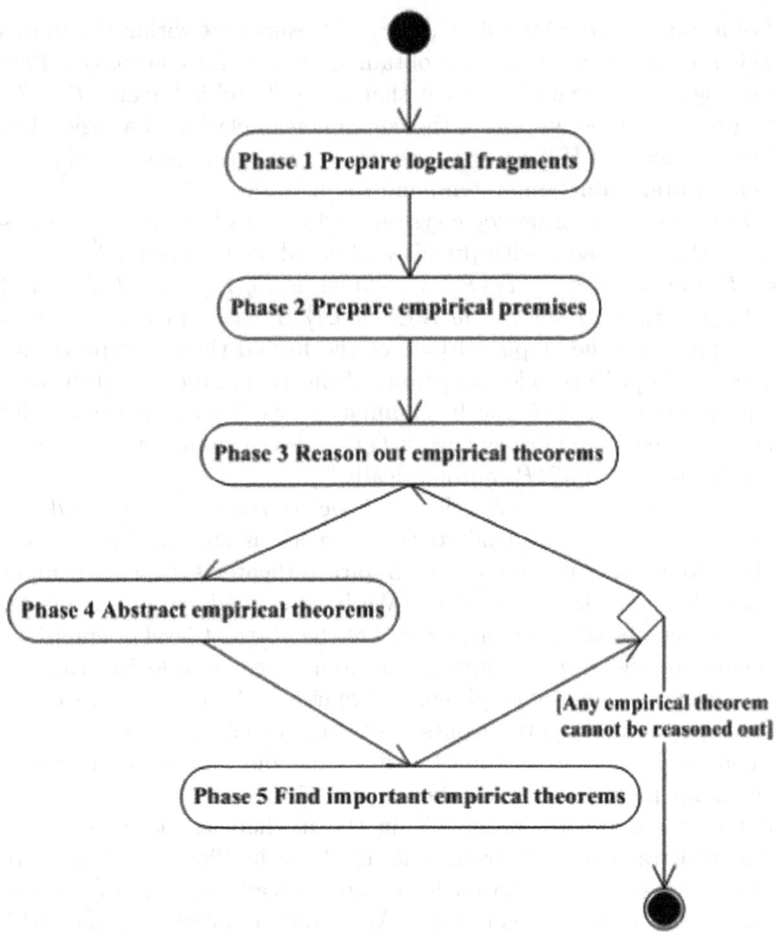

Fig. 1. The activity diagram of the method.

3 Requirement Analysis of Knowledge Base

According to the proposed security analysis and decision-making method, the usage of the knowledge base can be described as following three parts. First, users can use the knowledge base to save collected empirical rules and knowledge of industrial control network and reasoned out new empirical rules and knowledge. As new empirical rules and knowledge is reasoned out, maybe some new predicates and function are abstracted, they should be also saved into the knowledge base. All of the empirical rules and knowledge should be saved as natural language, logical formula, and category language which is used by FreeEnCal. Second, users can use the knowledge base to save reasoned out conclusions based on real-time data of industrial control network. Maybe some new predicates and function are abstracted, they should be also saved into the knowledge base. Third, users can use the knowledge base to perform security analysis and decision-making based on the information provided by the knowledge base. The information includes those defined several factors to find the important empirical theorems, which are the abstract level of empirical theorems, deduction distance of empirical theorems, degree of logical connectives, and propositional schema of empirical theorems and so on.

Based on the usage of the knowledge base, we analyzed requirements of the knowledge base and list up requirements of the knowledge base as below.

R1: Users can use the knowledge base to save, display, delete, revise and search the collected empirical knowledge of industrial control network.
R2: Users can use the knowledge base to find and output the needed empirical knowledge as inputted empirical premises of FreeEnCal.
R3: Users can use the knowledge base to save, display, delete, revise and search the reasoned out empirical knowledge of industrial control network by FreeEnCal.
R4: Users can use the knowledge base to analyze the degree of logical connectives, propositional schema, abstract level, and deduction distance of empirical theorems.
R5: Users can use the knowledge base to save, display, delete, revise and search the logical formula of strong relevant logic and category language of FreeEnCal of each empirical knowledge.
R6: Users can use the knowledge base to save, display, delete, revise and search the consequence relation of each empirical knowledge.
R7: Users can use the knowledge base to display and search all of individual constants, individual variable, predicates, functions and their abstract levels, corresponding logical symbols, expression of the category language in the empirical knowledges.

4 Design of Knowledge Base

Based on the requirement analysis of the knowledge base, we design Entity-Relationship model for the knowledge base as shown in Figs. 2, 3, 4, and 5. In the model, there are nine entities, which are predicate entity, function entity, individual constant/variable entity, propositional schema entity, abstract level entity, degree entity, deduction distance entity, empirical knowledge of industrial control system (ICS) entity, and identification entity. Besides, there are 5 relations between those entities. The relation between the empirical knowledge of ICS entity and propositional schema entity is called "represent". The relation between the empirical knowledge of ICS entity and deduction distance entity, degree entity, abstract level entity are called "possess". The relation between the empirical knowledge of ICS entity and identification entity is "attach". We also

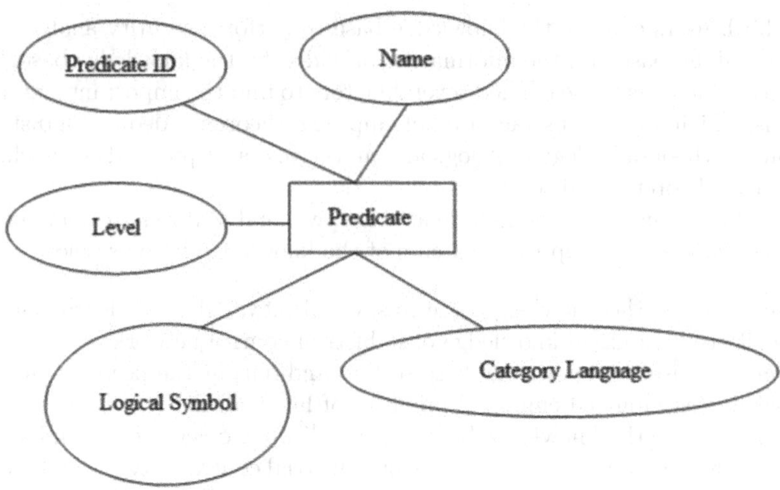

Fig. 2. The predicate entity

defined properties for each entity. In detail, we defined four properties for empirical knowledge of ICS entity, which are ID, natural language, logical formula and category language. Category language means a kind of language used by FreeEnCal. We defined two properties for propositional schema entity, which are ID and schema. We defined two properties for deduction distance entity, which are ID and number. Here, number means the number of deduction distance for one formula. We defined two properties for abstract level entity, which are ID and level. Level means the abstract level of one formula. We defined four properties for degree entity, which are ID, degree of logical connective "entailment", degree of logical connective "conjunction", and degree of logical connective "not". We have defined five properties for predicate entity, which are predicate ID, name, level, logical symbol, and category language. Here, name means the name of the predicate, level means the abstract level of the predicate, logical symbol means the logical symbol of the predicate, and category language means the category language of the predicate. Similarly, we defined five properties for function entity, and defined four properties for individual constant/variable entity. Finally, we defined five properties for identification entity, which are ID, individual constant/variable ID, predicate ID, Function ID, and Premises ID. Premises ID means the ID defined in empirical knowledge of ICS entity.

We also designed nine tables in the knowledge base according to the designed Entity-Relationship model, which are predicate table, function table, individual constant/variable table, propositional schema table, abstract level table, degree table, deduction distance table, empirical knowledge of industrial control system table, and identification table. Then, we designed each row of the tables which correspond to each property of each entity.

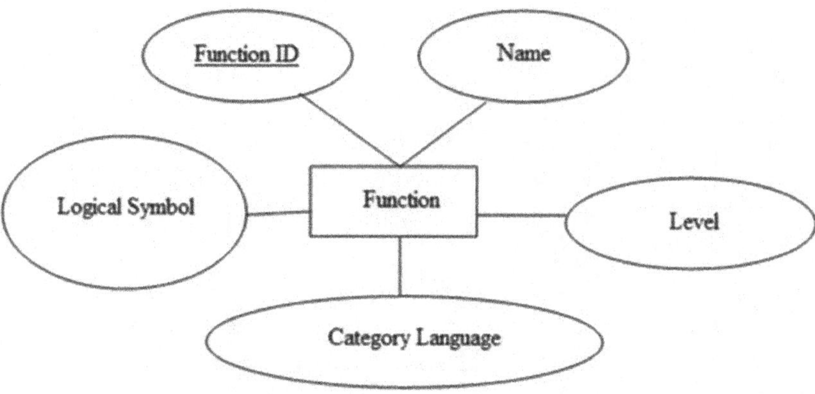

Fig. 3. The function entity

5 Implementation of Knowledge Base

From many database management systems, we choose MySQL to realize the knowledge base because of its excellent performance, easy maintenance, supporting many operating systems, stable usage, and the corresponding visual interface. In detail, we installed and configured MySQL and we chose Navicat 8 as visualization software. We established a database and made 9 tables according to the design of the knowledge base. Then, we made each row of the tables and input the property of each row, and set the primary key of each table. After the tables have been created, we inputted the collected empirical knowledge of industrial control networks into the database. Then, we also inputted the logical

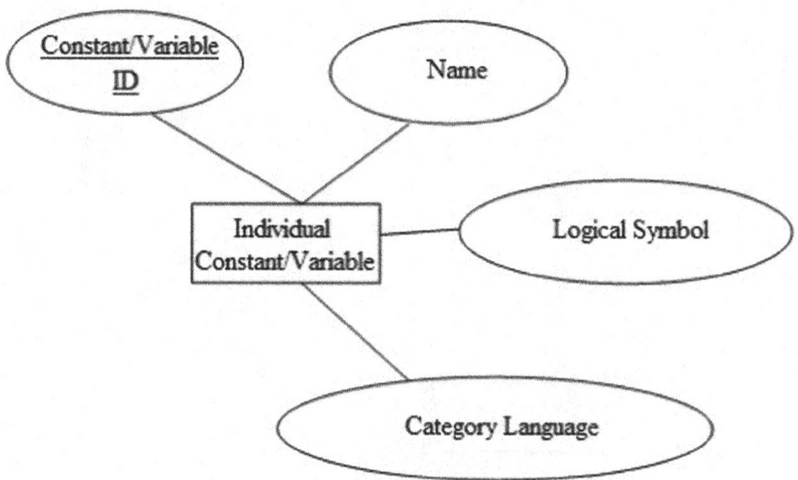

Fig. 4. The individual constant/variable entity

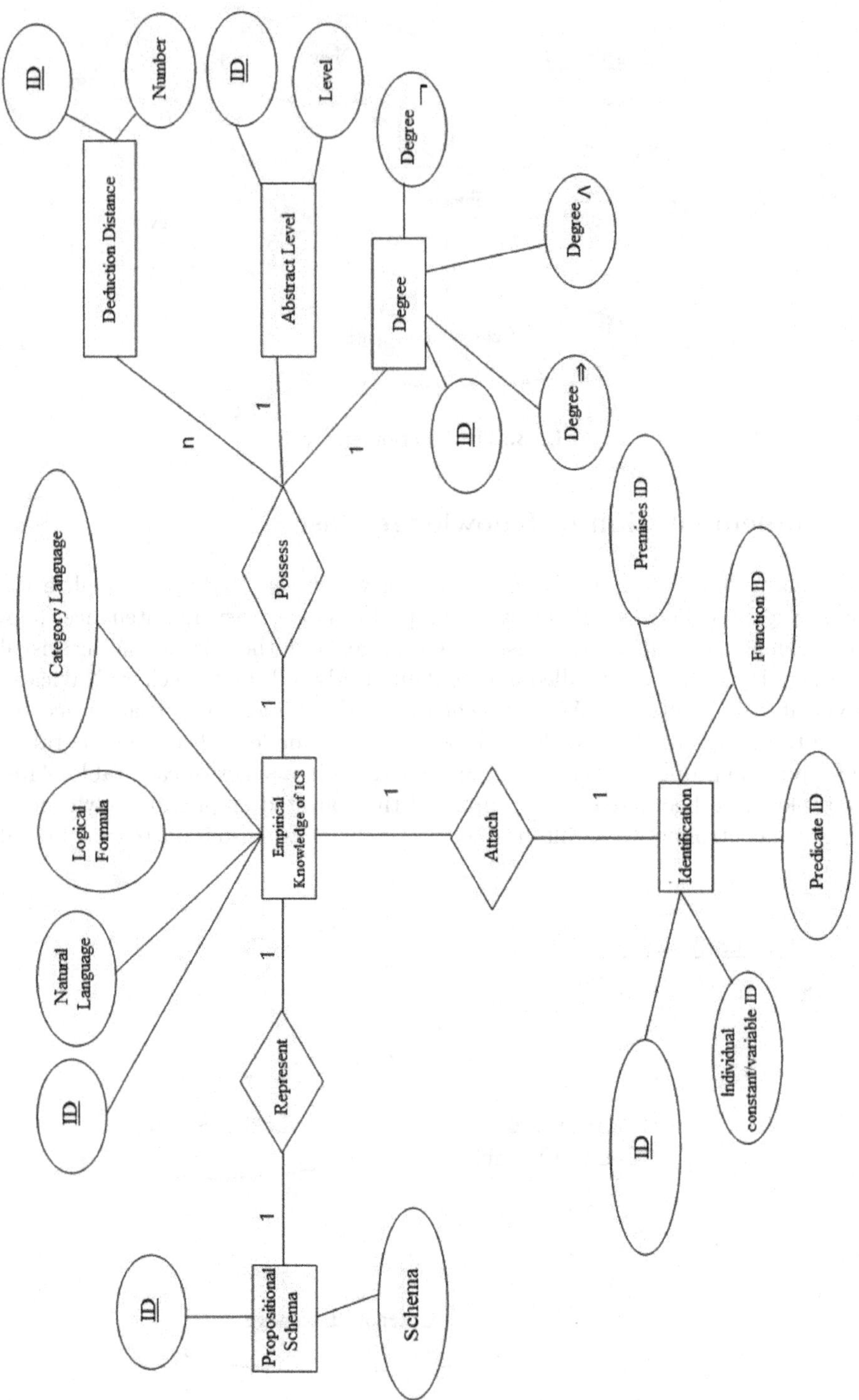

Fig. 5. The Entity-Relationship model of knowledge base.

formula, category language, degree, abstract level, propositional schema of each empirical knowledge into the database. We also inputted the predicates, functions, and their abstract level, the individual constants and variables into the database.

After all of information is added, users can save, display, delete, revise and search the collected and reasoned out empirical knowledge of industrial control network by using natural language, logical formula and category language. Besides, users can perform security analysis and decision-making of industrial control network by analyzing the information provided by the knowledge base, such as degree of logical connectives, propositional schema, abstract level, and deduction distance of empirical theorems, and consequence relation of each empirical knowledge. We implemented the knowledge base and showed some implemented results as examples in Figs. 6, 7, 8, 9, 10 and 11. In detail, Fig. 6 shows the implemented empirical knowledge of industrial control system table. Figure 7 shows the implemented predicate table of the knowledge base. Figure 8 shows the implemented function table of the knowledge base. Figure 9 shows the implemented propositional schema table of the knowledge base. Figure 10 shows how to use the knowledge base to search empirical knowledge. Figure 11 shows how to use the knowledge base to search the abstract level, deduction distance, degree of logical connectives, and propositional schema. Besides, all of the other requirements introduced in Sect. 3 have been implemented. Because of the limitation of pages, we cannot show all of the functions of the knowledge base at here.

ID	自然语言	逻辑式
1	当断路器与隔离开关串联使用时，合闸时先合隔离开关，后合断路器。	∀a∀b(SC(a,b)∧CLOSE({a,b})⇒CLOSE(b)∧CLOSE(a))
2	当断路器与隔离开关串联使用时，拉闸时先拉开断路器，后拉开隔离开关。	∀a∀b(SC(a,b)∧OPEN({a,b})⇒OPEN(a)∧OPEN(b))
3	母线停电时，先断开母线上各出线及其他元件断路器，再拉线路侧隔离开关、	∀a∀b∀c∀d(PF(c)⇒OPEN(a)∧OPEN(b@d)∧OPEN(b@c))
4	母线送电时，先合上母线侧隔离开关、线路侧隔离开关，再合上母线上各出线	∀a∀b∀c∀d(PD(c)⇒CLOSE(b@c)∧CLOSE(b@d)∧CLOSE(a))
5	母线故障时倒闸母线，用母线"冷倒"。	∀c(BD(c)∧DZ(c)⇒LD(c))
6	母线正常时倒闸母线，用母线"热倒"。	∀c(ZC(c)∧DZ(c)⇒RD(c))
7	母线"冷倒"时，先断开断路器，再拉开故障母线侧隔离开关，再合上运行母线	∀c(LD(c))⇒∀a∀b∀e∀f(OPEN(a)∧OPEN(b@e)∧CLOSE(b@f)∧CLOSE(a))
8	当带发电机运行的线路跳闸后，将导致发电机解列。	∀d∀t(TZ(d^t)⇒JL(d))
9	单电源变压器停电时，先断开负荷侧断路器，再断开电源侧断路器，最后拉开	∀g(PF(g))⇒∀a∀j(OPEN(a@j))∧∀a∀k(OPEN(a@k))∧∀b(OPEN(b))
10	单电源变压器送电时，先合上隔离开关，再合上电源侧断路器，最后合上负荷	∀g(PD(g))⇒∀b(CLOSE(b))∧∀a∀k(CLOSE(a@k))∧∀a∀j(CLOSE(a@j))
11	三电源变压器停电时，先断开低压侧断路器，再断开中压侧断路器，再断开高	∀i(PF(i))⇒∀a∀l(OPEN(a@l))∧∀a∀m(OPEN(a@m))∧∀a∀n(OPEN(a@n))∧∀b
12	三电源变压器送电时，先合上隔离开关，再合上高压侧断路器，再合上中压侧	∀i(PD(i))⇒∀b(CLOSE(b))∧∀a∀n(CLOSE(a@n))∧∀a∀m(CLOSE(a@m))∧∀a
13	双电源变压器送电时，先合上隔离开关，再合上高压侧断路器，最后合上低压	∀h(PD(h))⇒∀b(CLOSE(b))∧∀a∀n(CLOSE(a@n))∧∀a∀l(CLOSE(a@l))
14	双电源变压器停电时，先断开低压侧断路器，再断开高压侧断路器，最后拉开	∀h(PF(h))⇒∀a∀l(OPEN(a@l))∧∀a∀n(OPEN(a@n))∧∀b(OPEN(b))
15	电压过高会击穿。	∀p(HI(p))⇒∀u(JC(u))
16	击穿使得金属层汽化。	∀u(JC(u)⇒QH(u))
17	金属层汽化后，会自愈。	∀u(QH(u)⇒ZY(u))
18	自愈会剔除一批电弱点，使得耐压等级提高。	∀u(ZY(u))⇒∃v(TC(v))∧∀w(UP(w))
19	在输电线路中，发生短路、雷击等会使电流上升并且电压下降。	∀d(SD(d)∧DL(d))⇒∀o(UP(o))∧∀p(DW(p))
20	在输电线路中，过负荷会使电流上升而电压保持不变。	∀d(SD(d)∧GFH(d))⇒∀o(UP(o))∧∀p(ST(p))
21	电流上升且电压下降会导致延时。	∀o(UP(o))∧∀p(DW(p))⇒∀d(YS(d))
22	延时且电流电压不再改变时，断路器QF会跳开。	∀d(YS(d)))∧∀p(ST(p))⇒∀a(OPEN(a))
23	电流上升会导致温度上升。	∀o(UP(o))⇒∀q(UP(q))
24	温度上升会产生CH4。	∀q(UP(q))⇒∃x(CS(x))
25	CH4到达一定浓度时，会采取瓦斯保护动作。	∀x(HI(x))⇒∀d(BH(d))
26	变压器低压侧有电源，变压器中性点必须直接接地。	∀k∀r∀l(CZ(k@(r*l)))⇒∀r∀s(JD(r*s))
27	变压器跳闸后，最直接的后果就是造成负荷转移，使相关的并联变压器负荷	∀r(TZ(r))⇒∃r∃j(ZY(j)∧UP(r*j))
28	电网频率升高会导致电网的损耗增加。	∀y(UP(y))⇒∀z(UP(z))

Fig. 6. Empirical knowledge of industrial control system table

谓词编号	自然语言	符号	抽象等级	FreeEnCal语言
1	串联	SC	1	C_1_0
2	关闭	CLOSE	1	C_1_1
3	打开	OPEN	1	C_1_2
4	停电	PF	1	C_1_3
5	送电	PD	1	C_1_4
6	故障	BD	1	C_1_5
7	倒闸	DZ	1	C_1_6
8	冷倒	LD	1	C_1_7
9	热倒	RD	1	C_1_8
10	正常运行	ZC	1	C_1_9
11	短路	DL	1	C_1_10
12	输电	SD	1	C_1_11
13	上升,提高	UP	1	C_1_12
14	下降	DW	1	C_1_13
15	过负荷	GFH	1	C_1_14
16	保持不变	ST	1	C_1_15
17	延时	YS	1	C_1_16
18	存在	CZ	1	C_1_17
19	接地	JD	1	C_1_18
20	增加出力	ZJCL	1	C_1_19
21	跳闸	TZ	1	C_1_20
22	解列	JL	1	C_1_21
23	击穿	JC	1	C_1_22
24	过高	HI	1	C_1_23
25	汽化	QH	1	C_1_24
26	自愈	ZY	1	C_1_25
27	剔除	TC	1	C_1_26
28	产生	CS	1	C_1_27
29	瓦斯保护	BH	1	C_1_28
30	转移	ZY	1	C_1_29

Fig. 7. Predicate table.

函数编号	自然语言	符号	抽象等级	FreeEnCal语言
1	a和b都	{a,b}	1	C_0_2
2	a侧上的b	b@a	1	C_0_31
3	a的b	a*b	1	C_0_32
4	带有a的b	b^a	1	C_0_33

Fig. 8. Function table.

ID	图示
1	$A \wedge B \Rightarrow C \wedge D$
2	$A \wedge B \Rightarrow C \wedge D$
3	$A \Rightarrow B \wedge C \wedge D$
4	$A \Rightarrow B \wedge C \wedge D$
5	$A \wedge B \Rightarrow C$
6	$A \wedge B \Rightarrow C$
7	$A \Rightarrow B \wedge C \wedge D \wedge E$
8	$A \Rightarrow B$
9	$A \Rightarrow B \wedge C \wedge D$
10	$A \Rightarrow B \wedge C \wedge D$
11	$A \Rightarrow B \wedge C \wedge D \wedge E$
12	$A \Rightarrow B \wedge C \wedge D \wedge E$
13	$A \Rightarrow B \wedge C \wedge D$
14	$A \Rightarrow B \wedge C \wedge D$
15	$A \Rightarrow B$
16	$A \Rightarrow B$

Fig. 9. Propositional schema table.

```
SELECT
`1`.ID,
`1`.`自然语言`,
`1`.`逻辑式`
FROM
`1`
WHERE
`1`.ID = '1'
```

所得结果如图 6-10 查询结果 1 为：

ID	自然语言	逻辑式
	当断路器与隔离开关串联使用时，合闸时先合隔离开关，后合断路器。	∀a∀b(SC(a,b)∧CLOSE((a,b))→CLOSE(b)∧CLOSE(a))

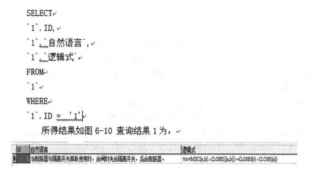

Fig. 10. Search empirical knowledge

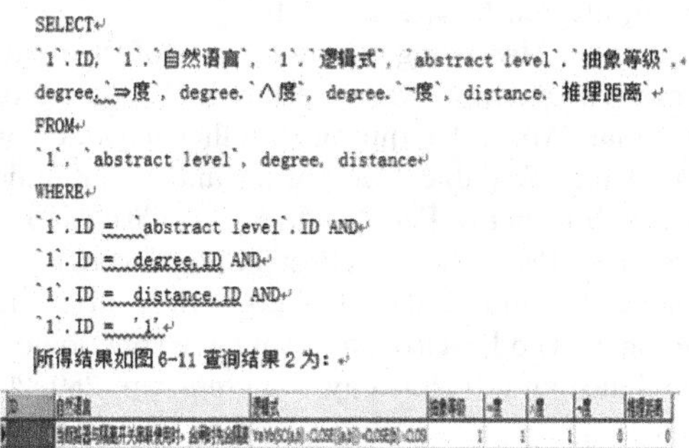

```
SELECT
`1`.ID, `1`.`自然语言`, `1`.`逻辑式`, `abstract level`.`抽象等级`,
degree.`⇒度`, degree.`∧度`, degree.`¬度`, distance.`推理距离`
FROM
`1`, `abstract level`, degree, distance
WHERE
`1`.ID = `abstract level`.ID AND
`1`.ID = degree.ID AND
`1`.ID = distance.ID AND
`1`.ID = '1'
```

所得结果如图 6-11 查询结果 2 为：

Fig. 11. Search factors of importance of the empirical knowledge

6 Concluding Remarks

We have designed a knowledge base for industrial control network security analysis and decision-making with reasoning method. We also showed how we implemented the knowledge base.

We also present some challenging research problems as future works. First, we will expand the knowledge base to a multilingual knowledge base for industrial control network security analysis and decision-making with reasoning method. Currently, the empirical knowledge of the knowledge base is limited in Chinese. Second, we will establish a method to use knowledge graph to represent the empirical knowledge in knowledge base to perform industrial control network security analysis and decision-making.

References

1. Cheng, J.: Entailment calculus as the logical basis of automated theorem finding in scientific discovery. Systematic Methods of Scientific Discovery: Papers from the 1995 Spring Symposium, pp. 105–110. AAAI Press - American Association for Artificial Intelligence (1995)
2. Cheng, J.: A strong relevant logic model of epistemic processes in scientific discovery. Front. Artif. Intell. Appl. **61**, 136–159 (2000)
3. Cheng, J., Nara, S., Goto, Y.: FreeEnCal: a forward reasoning engine with general-purpose. In: Apolloni, B., Howlett, R.J., Jain, L. (eds.) KES 2007. LNCS (LNAI), vol. 4693, pp. 444–452. Springer, Heidelberg (2007). https://doi.org/10.1007/978-3-540-74827-4_56
4. Figueira, P., Bravo, C., Lopez, J.: Improving information security risk analysis by including threat-occurrence predictive models. Comput. Secur. **88**, 101609 (2020)
5. Gao, H., Goto, Y., Cheng, J.: A systematic methodology for automated theorem finding. Theor. Comput. Sci. **554**, 2–21 (2014)
6. Gao, H., Li, J., Cheng, J.: Measuring interestingness of theorems in automated theorem finding by forward reasoning: a case study in Tarski's geometry. In: Proceedings of 2018 IEEE SmartWorld, Ubiquitous Intelligence and Computing, Advanced and Trusted Computing, Scalable Computing and Communications, Cloud and Big Data Computing, Internet of People and Smart City Innovations, pp. 168–173. IEEE Computer Society Press (2018). (Best Paper Award)
7. Gao, H., Li, J., Cheng, J.: Industrial control network security analysis and decision-making by reasoning method based on strong relevant logic. In: Proceedings of 4th IEEE Cyber Science and Technology Congress, pp. 289–294. IEEE Computer Society Press (2019)
8. Huang, K., Zhou, C., Qin, Y., Tu, W.: A game-theoretic approach to cross-layer security decision-making in industrial cyber-physical systems. IEEE Trans. Ind. Electron. **67**(3), 2371–2379 (2020)
9. Karimireddy, T., Zhang, S.: A hybrid method for secure and reliable transmission on industrial automation and control networks in Industry 4.0. In: Proceedings of the 2019 International Conference on Automation and Computing, pp. 1–6 (2019)
10. Lin, C., Wu, S., Lee, M.: Cyber attack and defense on industry control systems. In: Proceedings of the 2017 International Conference on Dependable and Secure Computing, pp. 524–526 (2017)
11. Line, M., Zand, A., Stringhini, G., Kemmerer, R.: Targeted attacks against industrial control systems: is the power industry prepared. In: Proceedings of the 2nd Workshop on Smart Energy Grid Security, pp. 13–22 (2014)
12. Qin, Y., Zhang, Y., Feng, W.: TICS: trusted industry control system based on hardware security module. In: Wen, S., Wu, W., Castiglione, A. (eds.) CSS 2017. LNCS, vol. 10581, pp. 485–493. Springer, Cham (2017). https://doi.org/10.1007/978-3-319-69471-9_37
13. Schumann, M., Drusinsky, D., Michael, J., Wijesekera, D.: Modeling human-in-the-loop security analysis and decision-making processes. IEEE Trans. Softw. Eng. **40**(2), 154–166 (2014)
14. Wood, A., He, Y., Maglaras, L., Janicke, H.: A security architectural pattern for risk management of industry control systems within critical national infrastructure. Int. J. Crit. Infrastruct. **13**, 113–132 (2017)

An Extension of Formal Analysis Method with Reasoning for Anonymity

Yating Wang and Yuichi Goto(✉)

Department of Information and Computer Sciences, Saitama University, Saitama 338-8570, Japan
{wangyating,gotoh}@aise.ics.saitama-u.ac.jp

Abstract. Formal analysis method with reasoning has been proposed as an alternative formal analysis method for cryptographic protocols. In the method, at first, analysts formalize the participant's and attacker's behaviors in order to carry out forward reasoning, then analysts check whether the logic formulas that represents security flaws of the target protocol exist or not in deduced logical formulas. However, the current method can deal with security flaws related to authentication, confidentiality, fairness, and non-repudiation, but not anonymity. This paper proposes an extension of formal analysis method with reasoning for dealing with security flaws related to anonymity. The paper also gives a case study with the proposed method in the Bolignano protocol. The result shows that the extension method is useful to detect security flaws related to anonymity.

Keywords: Formal analysis method with reasoning · Cryptographic protocols · Anonymity

1 Introduction

In the rapid development of informative society, cryptographic protocols are necessary techniques to ensure the security of many applications in cyberspace. Cryptographic protocols are information exchange protocols that provide security-related functions in networks. Flaws of the cryptographic protocols will cause serious security problems to the cyberspace application. Therefore, security analysis of the cryptographic protocols become an indispensable process.

Formal analysis method is used for security analysis of cryptographic protocols [6,10,11]. In the formal analysis of cryptographic protocols, the theorem proving method and model checking method [1,3] are widely used. Those methods are proving methods because analysts should enumerate the concrete security specifications in advance, then prove or check whether the target cryptographic protocol satisfies the enumerated security specifications or not, thereby verifying whether flaws exist or not. Therefore, all concrete security specifications must be enumerated when using the formal analysis method with proving to verify the security of a cryptographic protocol. However, analysts usually find it difficult

N. T. Nguyen et al. (Eds.): ACIIDS 2020, LNAI 12034, pp. 53–64, 2020.
https://doi.org/10.1007/978-3-030-42058-1_5

to thoroughly enumerate the security specifications and lead to some security-related flaws cannot be detected.

Formal analysis method with reasoning [4,18,20] has been proposed and improved to solve the limitation of the proving methods. The essential difference between traditional proving methods and the proposed reasoning method is whether analysts should enumerate the verification targets before doing formal analysis or not. By contrast with traditional proving methods, in the reasoning method, analysts enumerate only the protocol itself, and participants' and attackers' behaviors as premises of forward reasoning. Then they get all possible executed result under the protocol by forward reasoning. Finally, they check whether the deduced results are related to security flaws of the protocol. Thus, theoretically, analysts do not have to know the verification targets of the formal analysis under the reasoning method. However, current formal analysis method with reasoning has some limitations. The method only can deal with the flaws related to authentication, confidentiality [20], fairness and non-repudiation [22] but is unable to detect to the flaws related to anonymity.

To detect more types of cryptographic protocol flaws, this paper puts forward a new method to perform forward reasoning to detect the flaws related to the security of anonymity. The paper also takes the protocol of Bolignano as a case for carrying out security analysis. By the result of analyzing, it is proved that the improved formal analysis method with reasoning is valid to detect the security flaws related to the security property of anonymity.

The remainder of the paper is structured as follows. The basic notions of participant, identity, anonymity are explained in Sect. 2. Section 3 presents a new method to perform forward reasoning to detect the flaws related to anonymity. Section 4 describes the Bolignano protocol as a case for security analysis. Finally, we make some concluding remarks in Sect. 5.

2 Basic Notions

A protocol is a series of steps designed to accomplish a task involving two or more entities. "Series of steps" is a protocol sequence from beginning to end. Each step must be carried out in turn, and no step can be taken before the corresponding step has been completed. "Two or more entities involved" means the completion of the protocol requires at least two people. Cryptographic protocols are protocols that use cryptography to perform such security-related functions. Security-related functions include the prevention or security of an unauthorized entity access to information or the intentional destruction of the information.

Participant is an entity authorized to execute cryptographic protocols. There are two types of participants. One is an honest participant sending or receiving information strictly in line with the cryptographic protocol's steps. The other is a dishonest participant who might be lying in executing a cryptographic protocol or not at all executing a cryptographic protocol, attempting to impersonate or mislead the other participant for various unlawful purposes. A third party is a party that participates and is someone other than the directly involved entities. *TTP* (a trusted third party) is a third party that is not responsible for misconducting or colluding with any party to the detriment of the other [13]. We supposed that *TTP* (a trusted third party) and *CS* (currency server) are honest participants at all times.

Identity is the true participant's personal privacy information that does not want to reveal to others [13]. In the highly informative society, there are problems relating to identity theft, whereby the participant's personal information is stolen and used for illegal purposes. Adequate measures need to ensure that the data is encrypted to avoid revealing the identity information. Nowadays commonly encrypt identity information by digital signatures, etc. We based on the understanding in this paper to determine whether the identity is protected or not.

Anonymity [9,13,15] is one of the security properties that cryptographic protocol should ensure. Anonymity implies that the true identity of a participant is not revealed to any entity [16]. Anonymity could be defined as a condition in which an individual's true identity is unknown. On the other hand, privacy is a person's right to control access to his or her private data. From the above definitions, it can be clearly understood that anonymity and privacy are not the same and are in fact entirely different things. Anonymity, in the scope of this research, means that the true identity of the participant is hidden. While entering into a protocol, the participant does not have any obligation to reveal his or her true identity to other participant, and there is no way other participant has the ability to track the true identity of the participant. For ensuring the privacy of data, various cryptographic mechanisms such as encryption are used. Anonymity is thus a mechanism that hides the identity of the participant and keeps it secret during a cryptographic protocol. It helps protect the participant's privacy.

Participants can observe a series of occurring events but they can't know who made them. The following two forms distinguish anonymity:

1. Sender Anonymity
   Anonymity provides anonymous protection for the sender and information about the identity of the sender should be hidden.
2. Receiver Anonymity
   Similarly, anonymity should provide anonymous protection for the receiver and information about the identity of the receiver should be hidden.

In this paper, we concentrated on realizing anonymity that has been acquired which protects the identity.

3 An Extension of Formal Analysis Method with Reasoning

3.1 Overview of Extension Method

Analysts formalize "participant behaviors", "attacker behaviors" and "common behaviors between participants and an attacker" as targets of formalization with reasoning in the previous formal analysis method [20]. The result of the previous formal method of analysis is not clear in which information is protected, and it is impossible to analyze whether the participant's personal information is revealed. Anonymity can not be detected using the previous method of formal analysis.

In the extended method, we have added "anonymous behaviors". The "anonymous behaviors" is a set of rules for judging information that are protected. We followed premises about "anonymous behaviors" and know which information are protected. Thus able to detect anonymity using the extended method for reasoning. Following we provide a formal overview of behaviors for flawed anonymity detection. Analysts formalize "behaviors of participants", "common behaviors" and "anonymous behaviors" as the formalization targets. "Behavior of participants" is a collection of participants' rules. "Common behaviors" rules in addition to participants' behaviors of sending and receiving information. "Anonymous behaviors" is a set of rules that determine the information that is protected by each participant.

3.2 Formalization of Cryptographic Protocols

Overview of Formalization. Analysts have formalized "behaviors of participants", "common behaviors" and "anonymous behaviors" as formalization targets in the extended method. In order to formalize the above behaviors, we defined predicates, functions, and individual constants in the cryptographic protocols that represent participants' behavior or data.

We defined predicates as follows:

- $Start(p_1, p_2)$: Participant p_1 and participant p_2 start a communication process.
- $Parti(p)$: p is a participant.
- $Send(p_1, p_2, x)$: Participant p_1 sends x to participant p_2.
- $Recv(p, x)$: Participant p receives x.
- $Anoy(p, x)$: x is protected from participant p.
- $Eq(x_1,x_2)$: x_1 and x_2 are equal.

We defined functions as follows:

- $id(p)$: Identifier of a participant p.
- $data(x_1, \cdots, x_n)$ $(n \in N)$: A set of data that is sent and received x_1, and x_n.
- $nonce(p)$: Nonce of a participant p.
- $enc(k, x_1, \cdots, x_n)$: A data set made up of encrypted data x_1, and x_1 by k.
- $sk(p)$: Secret key of a participant p.
- $pk(p)$: Public key of a participant p.
- $old(x)$: Old data of x.
- $symk(p_1, p_2)$: Symmetric key of participant p_1 and participant p_2.
- $sig(p, x_1, \cdots, x_n)$: A compilation of data consisting of $x_1, \cdots,$ and x_n with participant p's signature.
- $plus(x)$: Incremented data of x.

- $h(x)$: A one-way hash function of message x.
- $tstamp(p)$: Timestamp of a participant p.

We defined individual constants as follows:

- a, b, a_1, $\cdots$, a_n ($n \in N$): Persons (If the number of person is more than 8, it shall be represented as a_1, $\cdots$, a_n).
- ttp: A trusted third sever.
- i: An attacker.

Furthermore, individual constants are unique in definition that are assigned to uniquely defined data in a protocol.

Behaviors of Participants. Behaviors of Participants is a set of rules between actions in each step as "If a participant receives such information, another data will be sent by the participant". In step M, sending and receiving are defined as M in the specification of cryptographic protocols. $X_1 \Rightarrow X_2$: Y_1, Y_2, $\cdots$, Y_z, it means that X_1 sends data Y_1, Y_2, $\cdots$, Y_z ($z \in N$) to X_2 in step M.

1. Represent the participants' behaviors in each step of a cryptographic protocol through formulas. p_i and x_i are independent variables and the number of information sent or received is n and m.
   a. If step 1 of the protocol or all previous senders of information is similar in step 1, use the following formula (1), meaning "if p_1 and p_2 start the communication process, p_1 sends data x_1, $\cdots$, x_n to p_2".

$$Start(p_1, p_2) \Rightarrow Send(p_1, p_2, data(x_1, ..., x_n)) \tag{1}$$

   b. When step 2 and beyond the protocol are not similar in step 1, use the following formula (2), meaning "if the data is received by the participant p_1, p_1 sends next data to p_2". But p_i, x_i are individual variables, as well as n, m are the amount of sent or received information.

$$Recv(p_1, data(x_1, ..., x_m)) \Rightarrow (Parti(p_1) \Rightarrow Send(p_1, p_2, data(x_1, ..., x_n))) \tag{2}$$

2. Replace individual variables p_1 and p_2 with ttp or p_i in the previous task in formulas (1) and (2).
   a. The individual variable p_1 will be replaced by the individual constant ttp if the sender is ttp (a third trusted party).
   b. The individual variable p_2 will be replaced by the individual constant ttp if the sender is ttp (a third trusted party).
3. Replace individual variables x_1, $\cdots$, x_n with terms in formulas (1) and (2) in accordance with the following rules.
   a. Replace a function $f(p_i)$ or $f(ttp)$ or an individual variable that is uniquely defined if Y_i is not encrypted.

b. Replacement $plus(x'_i)$. x'_i and previous task 3-a are replaced if Y_i is incremented data.
c. Replace $enc(k, x'_i, \cdots, x'_n)$ and replace k if Y_i is encrypted data, depending on key types such as the public key or symmetric key. In addition, k is not replaced if k is other kinds of the key.
d. Substitute $sig(p_i, x'_i, \cdots, x'_n)$ and replace p_i with p_1, p_2, ..., p_n or ttp if Y_i is signed data.

4. Define a constant individually and replace the variable with the constant if a variable is included only in A_1 or A_2 in part of formulas $A_1 \Rightarrow A_2$ (A_1, A_2 is formulas).
5. In those formulas add quantifier $\forall$ corresponding to the individual variables k, x_i, and p_i.
6. Generate a Start(p_1, p_2) formula to substitute a participant's individual constant with p_1 and p_2.

Common Behaviors. Common behaviors mostly describe veiled behaviors. For example, the decryption of encrypted data is included in this target.

1. Generate the formula which implies participant p receives original information if participant p receives information encrypted by the symmetric key of participant p.
$\forall p_1 \forall p_2 \forall x_1 ... \forall x_n\ (Recv(p_1, enc(symk(p_1, p_2), x_1, \cdots, x_n)) \Rightarrow$
$Recv(p_1,\ data(x_1,\ \cdots,\ x_n)))$
2. Generate the formula which implies participant p receives information encrypted with a session key known to participant p, participant p receives original information.
$\forall p_1 \forall x_1 \cdots \forall x_n (Recv(p_1, enc(sesk(p_1), x_1, \cdots, x_n)) \Rightarrow Recv(p_1,\ data(x_1,\ \cdots$
$x_n)))$
3. Generate the formula which implies participant p original information if participant p receives information encrypted by the public key of participant p.
$\forall p \forall x_1 \cdots \forall x_n (Recv(p, enc(pk(p), x_1, \cdots, x_n)) \Rightarrow Recv(p,\ data(x_1,\ \cdots,\ x_n)))$
4. Generate the formula which implies participant p receives additional information if participant p receives multiple information.
$\forall p((Recv(p, data_1) \wedge \cdots \wedge Recv(p, data_n) \Rightarrow Recv(p,\ data')))$
5. $symk(p_1,\ p_2)$ and $symk(p_2,\ p_1)$ are equal. (This formula needs to carry out symmetrical key reasoning in the method.)
$\forall p_1 \forall p_2 (Eq(symk(p_1, p_2), symk(p_2, p_1)))$
6. A appears as a target specification participant.
$Parti(\alpha)$ in which α is an individual constant that represents a participant or a third trusted server.
7. The attacker is not a participant at all.
$\neg\ Parti(i)$

Anonymous Behaviors. Anonymous Behaviors provide the participants with the basis for judging whether the identity is protected or not.

1. Generate the formula which implies participant p_1 receives the data x signed by $p_2(\{x\}_{sig_{p_2}})$, then p_2 can protect the data of x.
$\forall p_1 \forall p_2 \forall x(Recv(p_1, sig(p_2, x)) \Rightarrow Anoy(p_2, x))$
2. Generate the formula which implies participant p_1 receives the data encrypted by participant p_2's secret key $(\{x\}_{sk_{p_2}})$, but participant p_1 cannot decrypt the private key of participant p_2, then participant p_2 can protect the data of x.
$\forall p_1 \forall p_2 \forall x(Recv(p_1, sk(p_2, x)) \Rightarrow Anoy(p_2, x))$
3. Generate the formula which implies participant p_1 receives the data of x not encrypted by anything, then participant p_1 cannot protect the data of x.
$\forall p_1 \forall x(Recv(p_1, x) \Rightarrow \neg\, Anoy(p_1, x))$
4. If the data set is protected by the participant p_1, participant p_1 can also protect per data, even if the other way around. They are described as $\forall p_1 \forall x_1 \cdots \forall x_n(Anoy(p_1, data\{x_1 \cdots x_n\}) \Rightarrow Anoy(p_1, x_1) \wedge \cdots \wedge Anoy(p_1, x_n))$ and $\forall p_1 \forall x_1 \cdots \forall x_n(Anoy(p_1, x_1) \wedge \cdots \wedge Anoy(p_1, x_n) \Rightarrow Anoy(p_1, data\{x_1 \cdots x_n\}))$
5. If participant p_1 protects the data encrypted by the key $(\{x\}_k)$ and k is the key of participant p_1, then participant p_1 also can protect the data of x.
$\forall p_1 \forall x(Anoy(p_1, enc(k, x)) \wedge Anoy(p_1, k) \Rightarrow Anoy(p_1, x))$
6. If participant p_1 protects the data encrypted by hash function $(h(x))$, then participant p_1 also can protect the data of x.
$\forall p_1 \forall x(Anoy(p_1, h(x)) \Rightarrow Anoy(p_1, x))$
7. *TTP* and *CS* keep honest at all times. If participants p_1 and participant p_2 transfer data by *TTP* or *CS* and participant p_1 protect the data of x, participant p_2 protects the data of x as well.
$\forall p_1 \forall p_2 \forall x(Anoy(ttp, x) \wedge Anoy(p_1, x) \Rightarrow Anoy(p_2, x))$

3.3 Forward Reasoning

In the proposed method, a forward reasoning program is used to do forward reasoning automatically. The logic system underlying forward reasoning is strong relevant logics [4]. Forward reasoning underlying strong relevant logics ensures that all deduced formulas are related to given premises. The number of forward reasoning executions in the extended method depends on the steps to be detected in the cryptographic protocol.

Analysts use FreeEnCal [5] to automatically carry out forward reasoning in the extension of the formal analysis method with reasoning. Analysts put the first step of the "behaviors of participants" into FreeEnCal for the first execution of forward reasoning, and put generated logical formulas of "common behaviors" and "anonymous behaviors" into FreeEncal, and then add the formula to the result of the first execution of the second forward reasoning for the second step of the protocol. Effectively add the formula representing each protocol step. If the added logical formula is interpreted as the cryptographic protocol's last step, forward reasoning will be completed.

3.4 Analysis of the Result

Analysts verify the deducted formulas after forward reasoning has been completed in each execution result. All information obtained by the participants are collected in the final execution document. According to the anonymous behaviors, if a participant protects the identity of himself/herself in the protocol, the other part does not protect the data of this identity, then this participant keeps anonymity. In contrast the data of identity is also protected by other participants, prove that the identity is revealed, so this participant leaves out anonymity.

The Results Are Analyzed as Follows

- As a result, we can determine which sender identifiable information is protected. If the sender identifiable information is protected by a participant other than itself, proves that the sender personal identifiable information is known by other participants, and the sender identifiable information is revealed, that is, exists the flaws related to sender anonymity.
- Similarly, in the result, we can determine which receiver identifiable information is protected. If the receiver identifiable information is protected by a participant other than itself, proves that the receiver personal identifiable information is known by other participants, and the receiver identifiable information is revealed, that is, exists the flaws related to receiver anonymity.

4 Case Study in Bolignano Protocol

4.1 Bolignano Protocol

Regarding the development of mobile technology, replaced cash payments with online payments. Payment protocols provide the required security features, among them, anonymity plays an important role. Customers would really like to keep identity unrevealed at times. So, as a case study, we used the Bolignano protocol to analyze the security to confirm the usefulness with reasoning of extending the formal method of analysis. The Bolignano protocol is a payment protocol put forward by Bolignano in 1997 [2], and it only concerns customer A finishing commodity browsing processes, selecting the later purchase and payment accreditation. Customer A has held an intact indent before the protocol is executed and he/she agrees with its content and item [17]. There are three participants in the Bolignano protocol, they are customer A, merchant B and currency server CS.

First, the definition of some symbols used in Bolignano's description is as follows:

- A, B: The logo of customer or merchant, and does not include personal identity.
- *order*: Payment order details.
- *payment*: Payment bill.
- *result*: Payment result.

The protocol specification is the following. In step 1, the message is sent by the customer A to the merchant B, and ask a unique merchant identity $id(b)$ from the merchant B.

Then the customer A generates a unique set of transaction identifier ($id(a)$, $id(b)$) after it received $id(b)$ sent from the merchant B.

In step 3, the customer A encrypts the transaction identifier with its secret key and sends it to merchant B along with public key of the merchant B to encrypt order and public key of the currency server CS to encrypt payment.

In step 4, the merchant B encrypts the same transaction identifier with its own secret key and then the merchant B sends it together with the public key of the currency server CS to encrypt payment to the currency server CS.

In step 5, the currency server CS decrypts to get payment. Then the currency server CS encrypts the payment result together with the hash value of the transaction identifier with the secret key and sends to the merchant B.

In step 6, the merchant B forwards the result to customer A. Similarly, customer A also checks if the results and the transaction identifier are consistent with the former purchase requests. In summary, all involved participants may receive the necessary information after the protocol is normally implemented.

The Basic Cryptographic Specification of Bolignano Protocol

1. $A \Rightarrow B$: A, B
2. $B \Rightarrow A$: $id(B)$
3. $A \Rightarrow B$: $\{id(A), id(B)\}_{sk_a}$, $\{order\}_{pk_b}$, $\{payment\}_{pk_{cs}}$
4. $B \Rightarrow CS$: $\{id(A), id(B)\}_{sk_a}$, $\{id(A), id(B)\}_{sk_b}$, $\{payment\}_{pk_{cs}}$
5. $CS \Rightarrow B$: $\{result, h(id(A), id(B))\}_{sk_{cs}}$
6. $B \Rightarrow A$: $\{result, h(id(A), id(B))\}_{sk_{cs}}$

We can realize that this protocol does not satisfy the anonymity of the sender and receiver, according to the definition of anonymous property. In the second step, A receives the identity of B, and then A knows the identity of B. So the participant of B is revealed in the cryptographic protocol therefore this protocol leaves out receiver anonymity. In the third step, B decrypts to get a unique set of transaction identifier, and the transaction identifier contains the identity of A. So the participant of A is revealed in the cryptographic protocol therefore this protocol leaves out sender anonymity. This protocol has the anonymous flaw according to the definition of anonymity.

4.2 Analysis of the Case Study

3781 logical formulae were deduced by the extended method as a result of the formal analysis. After executing the second step,

- $\neg$ $Anoy(a, id(b))$

was deducted, which means the identity of B did not be protected, that is, the identity information of B was known to participant A and the identity of B is revealed that did not satisfy receiver anonymity. After executing the third step,

- $Anoy(a, id(b))$

was deducted, which means the identity of B was protected from participant A, that is, the identity information of B was known to participant A and the identity of B was revealed that did not satisfy receiver anonymity. After executing the fourth step,

- $Anoy(b, id(a))$

was deduced, which means the identity of A was protected from participant B, that is, the identity information of A was known to participant B and the identity of A was revealed that did not satisfy sender anonymity.

Bolignano's anonymous property is listed in Table 1. Analysis based on the results mentioned above, the Bolignano protocol has left out the sender anonymity and receiver anonymity, that is, the Bolignano protocol has the security flaw of anonymity.

Table 1. The anonymity of Bolignano protocol.

Protocol	Sender anonymity	Receiver anonymity
Bolignano protocol	–	–

In summary, we proposed a new method to perform forward reasoning related to the flaws of anonymity to detect more types of cryptographic protocol flaws. We extended the "anonymous behaviors" in the forward reasoning to determine which personally identifiable information is protected. In the cryptographic protocol, if the personally identifiable information is protected by a participant other than itself, proves that the personal identifiable information is known by other participants, and the personal information is revealed, that is, exists the flaws related to anonymity. We also took the protocol of Bolignano as a case for carrying out security analysis. The analysis has shown that the method of extension is valid to detect anonymous flaw. My extension method compares with the previous method is listed in Table 2. Obviously, my extension method is more efficient than the previous method in detecting anonymous flaw.

Table 2. My extension method compares with the previous method.

Security property	The previous method	My extension method
Anonymity	–	√

Although we successfully detected the anonymous flaw through the case of the Bolignano protocol, but a case analysis is too lacking. In the future, we have to analyze more encrypted forms of the protocol to verify our extension method that it is valid to detect the anonymous security of the cryptographic protocol.

5 Concluding Remarks

We have extended the formal method of analysis with reasoning for cryptographic protocols to detect the flaws related to anonymous security. We also analyzed the Bolignano protocol using the method proposed and successfully detected the flaw. Furthermore, the extended method can detect the security properties flaws of authentication, confidentiality, fairness, non-repudiation and anonymity on the basis of the ability to detect flaws in authentication, confidentiality, fairness and non-repudiation.

For the future, more cryptographic protocols will be analyzed to verify the validity of the extension method. We will analyze the anonymity analysis when adding the attacker behavior in the cryptographic protocols. And we will consider several ways to automatically narrow the scope of deduced formulas and analyze formulas.

References

1. Avalle, M., Alfredo, P., Bogdan, W.: Formal verification of security protocol implementations: a survey. Formal Aspects Comput. **26**(1), 99–123 (2014)
2. Bolignano, D.: Towards the formal verification of electronic commerce protocols. In: The 10th IEEE Computer Security Foundation Workshop, Rockport, USA, pp. 133–146 (1997)
3. Bau, J., Mitchell, J.C.: Security modeling and analysis. IEEE Secur. Priv. **9**(3), 18–25 (2011)
4. Cheng, J., Miura, J.: Deontic relevant logic as the logical basis for specifying, verifying, and reasoning about information security and information assurance. In: The 1st International Conference on Availability, Reliability and Security (ARES 2006), pp. 601–608. IEEE Computer Society, Vienna (2006)
5. Cheng, J., Nara, S., Goto, Y.: FreeEnCal: a forward reasoning engine with general-purpose. In: Apolloni, B., Howlett, R.J., Jain, L. (eds.) KES 2007. LNCS (LNAI), vol. 4693, pp. 444–452. Springer, Heidelberg (2007). https://doi.org/10.1007/978-3-540-74827-4_56
6. Cortier, V., Kremer, S., Warinschi, B.: A survey of symbolic methods in computational analysis of cryptographic systems. J. Autom. Reason. **46**(3–4), 225–259 (2011)
7. Dreier, J., Kassem, A., Lafourcade, P.: Formal analysis of e-cash protocols. In: The 12th International Conference on Security and Cryptography (SECRYPT 2015), Colmar, France, pp. 65–75 (2015)
8. Jiang, Y., Gong, H.: Modeling and formal analysis of communication protocols based on game. In: 2010 International Conference on Computer Application and System Modeling (ICCASM 2010), vol. 12, no. 3, pp. 470–473 (2013)
9. Javan, S., Bafghi, A.: An anonymous mobile payment protocol based on SWPP. Electron. Commer. Res. **14**(4), 635–660 (2014)
10. Meadows, C.A., Meadows, C.A.: Formal verification of cryptographic protocols: a survey. In: Pieprzyk, J., Safavi-Naini, R. (eds.) ASIACRYPT 1994. LNCS, vol. 917, pp. 133–150. Springer, Heidelberg (1995). https://doi.org/10.1007/BFb0000430
11. Meadows, C.: Formal methods for cryptographic protocol analysis: emerging issues and trends. IEEE J. Sel. Areas Commun. **21**(1), 44–54 (2003)

12. Pfitzmann, A., Waidner, M.: Networks without user observability. Comput. Secur. **6**(2), 158–166 (1987)
13. Polychronis, A.: Fair exchange protocols with anonymity and nonrepudiation for payments. Technical Report RHULMA20132, University of London, Royal Holloway (2013)
14. Schneier, B.: Applied Cryptography: Protocols, Algorithms, and Source Code in C. Wiley, Hoboken (1996)
15. Tellez, J., Zeadally, S.: An anonymous secure payment protocol in a payment gateway centric model. In: The 9th International Conference on Mobile Web Information Systems (MobiWIS 2012), Niagara Falls, Ontario, Canada, vol. 10, pp. 758–765 (2012)
16. Walker, J., Li, J.: Key exchange with anonymous authentication using DAA-SIGMA protocol. In: Chen, L., Yung, M. (eds.) INTRUST 2010. LNCS, vol. 6802, pp. 108–127. Springer, Heidelberg (2011). https://doi.org/10.1007/978-3-642-25283-9_8
17. Wen, J., Zhao, L., Jiang, H.: Formal analysis of electronic payment protocols based on game theory. In: 2010 International Conference on Computer Application and System Modeling (ICCASM 2010), Taiyuan, China, vol. 6, pp. 319–323 (2010)
18. Wagatsuma, K., Goto, Y., Cheng, J.: A formal analysis method with reasoning for key exchange protocols. IPSJ J. **56**(3), 903–910 (2015). (in Japanese)
19. Wang, C., Shu, N., Wang, H.: Formal analysis of a model for electronic payment systems. In: International Conference on Communication and Electronic Information Engineering (CEIE 2016), vol. 116, pp. 613–620 (2016)
20. Yan, J., Wagatsuma, K., Gao, H., Cheng, J.: A formal analysis method with reasoning for cryptographic protocols. In: 12th International Conference on Computational Intelligence and Security, pp. 566–570. IEEE Computer Society, Wuxi (2016)
21. Yan, J., Ishibashi, S., Goto, Y., Cheng, J.: A study on fine-grained security properties of cryptographic protocols for formal analysis method with reasoning. In: 2018 IEEE SmartWorld, Ubiquitous Intelligence, Computing, Advanced, Trusted Computing, Scalable Computing, Communications, Cloud, Big Data Computing, Internet of People and Smart City Innovations, pp. 210–215. IEEE-CS, Guangzhou (2018)
22. Yan, J., Wang, Y., Goto, Y., Cheng, J.: An extension of formal analysis method with reasoning: a case study of flaw detection for non-repudiation and fairness. In: Carlet, C., Guilley, S., Nitaj, A., Souidi, E.M. (eds.) C2SI 2019. LNCS, vol. 11445, pp. 399–408. Springer, Cham (2019). https://doi.org/10.1007/978-3-030-16458-4_23
23. Zhang, J.: Research on secure e-payment protocols. In: 2011 International Conference on Information Management, Innovation Management and Industrial Engineering, Shenzhen, China, pp. 121–123 (2011)

An Extension of Reciprocal Logics for Trust Reasoning

Sameera Basit and Yuichi Goto(✉)

Department of Information and Computer Science, Saitama University, Saitama 338-8570, Japan
{sameera,gotoh}@aise.ics.saitama-u.ac.jp

Abstract. Trust reasoning is an indispensable process to establish trustworthy and secure communication under open and decentralized systems that include multi-agents and humans. Trust reasoning is the ability to reason about trust targets and their relationships whether these targets are trusted or not. Reciprocal logics is an expectable candidate for a logic system underlying trust reasoning. However, current reciprocal logics does not cover various trust properties of trust relationship. This paper presents an extends reciprocal logics to deal with various trust properties for trust reasoning. The paper also shows an example of usage the extension logic. The extended logics illustrates the general properties of trust that can facilitate the engineering of trustworthy systems.

Keywords: Reasoning about trust · Strong relevant logics · Trust properties

1 Introduction

Trust relationship is one of important reciprocal relationships in our society and cyber space. There are many reciprocal relationships that must concern two parties, e.g., parent-child relationship, relative relationship, friendship, cooperative relationship, complementary relationship, trade relationship, buying and selling relationship, and so on [7]. Especially, trust relationship is basis of communications between human to human, human to system, and system to system, and basis of decision making of human and/or system. For example, we trust the data that we use for a decision to be reliable; when we cross an intersection of streets, we trust the cars in other directions to follow the traffic signals [13].

Trust reasoning is an indispensable process to establish trustworthy and secure communication under open and decentralized systems that include multi-agents and humans. Trust reasoning is the ability to reason about trust targets and their relationships whether these targets are trusted or not. In open and decentralized systems, although it is difficult to know whether a system, an agent, or a human that require to connect with our system can be trusted or not before communication with it, we want to know that to establish trustworthy and secure communication. Thus, we should calculate the degree of trust of

N. T. Nguyen et al. (Eds.): ACIIDS 2020, LNAI 12034, pp. 65–75, 2020.
https://doi.org/10.1007/978-3-030-42058-1_6

target system, agent, or human by using already known fact, hypotheses, and observed data. Trust reasoning is a process to calculate the degree of trust or decide which target can be regarded as trust one.

Reciprocal logics [7] is an expectable candidate for a logic system underlying trust reasoning. Classical mathematical logics and its various conservative extensions are not suitable for logic systems underlying reasoning, because those have paradoxes of implication [1,2]. Strong relevant logics have rejected those paradoxes of implication, and are considered as the universal basis of various applied logics for knowledge representation and reasoning [6]. Thus, strong relevant logics and its conservative extensions are candidates for logic systems underlying reasoning. Reciprocal logics is a conservative extension of strong relevant logics to deal with various reciprocal relationships. The reciprocal logics provide several predicates to describe propositions about trust relationship [7].

On the other hand, there are various trust properties for trust relationships. Basically, trust is established in interaction between two entities: trustor and trustee. Trustee provides trustworthy data to make a trustor trust in a trustee. For example, home appliance devices (trustee) provide energy-related data for users (trustor) to control these devices in use case of home energy management [16]. Trust is affected by several subjective such as social status and physical properties; and objective factors such as competence and reputation and classified the properties influencing trust into [12,23]. Trust is a trustors belief about some trustees property [9]. Many trust properties has been identified in literature, trust in reliability, honesty, credibility [15,22,25]. Most authors focused on only one dimension such as trust in the reliability [14,18], trust in the sincerity [19] and other researchers dealt with trust and cooperation. Demolombe [8] provided a formal definition for 6 trust properties based on modal logics. Current reciprocal logics does not cover such trust properties.

This paper aims to extends reciprocal logics to deal with various trust properties for trust reasoning. At first, we have surveyed and identified properties that are relevant to the trustworthy relationship, and then extends reciprocal logics by introducing new predicates for the identified trust properties. The paper also shows an example of usage the extension logic. The extended logics illustrates the general properties of trust that can facilitate the engineering of trustworthy systems.

The rest of the paper is organized as follows: Sect. 2 shows previous works about trust properties. Section 3 gives overview of reciprocal logics and its limitation. Section 4 presents an extension of reciprocal logics and an example of usage the extension logic. Finally, some concluding remarks are given in Sect. 5.

2 Trust Properties in Previous Works

In the domain of trust reasoning, several work show the interest to clearly differentiate between trust in reliability and trust in honesty which is later called credibility [15,22,25]. However, a large set of works focused only on trust in the reliability [14,18] and trust in the sincerity. Leturc et al. [17] defined trust as a subjective probability just refers to one dimension of trust while ignoring the other dimension which is objective probability. The subjective probability assembles too many important parameters and beliefs, that are very relevant in social reasoning [19] while there is a need to define other parameters and beliefs related to trust objectives such as competence. Yan et al. [23,24] only discusses trust properties and its classification but its formal representation is not presented.

Demolombe [8] provided a formal definition for trust that distinguishes between different properties an agent may have trust in. Other study also shows that having high (reliability) trust in a person in general is not necessarily enough to decide to enter a situation of dependence on that person [21]. There is a need to consider other factors that influence trustworthy relationship. Yan et al. [23] presented suggestions to evaluate trust with regard to competence, benevolence, integrity, and predictability and also targeted trust at different context and technology areas.

Trust properties presented in [8] are based on modal logics. These properties are defined in terms of material implications which leads to a well-known paradox. Other review shows that [12,19] are also interested to model trust with modal logics, graph-based approach [20], Second order propositional logic [10] based on classical mathematical logics.

3 Reciprocal Logics and Its Limitation

Relevant logics were constructed during the 1950s in order to find a mathematically satisfactory way of grasping the elusive notion of relevance of antecedent to consequent in conditionals, and to obtain a notion of implication which is free from the so-called 'paradoxes' of material and strict implication [10,20].

Strong relevant logics [6] were proposed in order to find a satisfactory logic to underlie relevant reasoning. These logic requires that the premises of a argument represented by an entailment include no unnecessary and needless conjuncts and the conclusion of that argument includes no unnecessary and needless disjuncts, and rejected those conjunction-implicational paradoxes and disjunction-implicational paradoxes [6].

Reciprocal logics [7] was established for specifying, verifying and reasoning about reciprocal relationships. These reciprocal logics underlie relevant reasoning as well as truth-preserving reasoning in the sense of conditional, ampliative reasoning, paracompletes reasoning, and paraconsistent reasoning. Moreover, the logics can be used for reasoning about relative relations among points as well as regions. Cheng [5] shows that various reciprocal logics can be obtained by introducing predicates and related axioms about reciprocal relationships into strong relevant logics.

Reciprocal relationship such as trust relationships may be symmetrical or unsymmetrical transitive or non transitive. Althought there are many definition based on the concept of trust. Trust Relationships have usually something in common. In general a trust relationship must concern two entities say trustor and trustee such that trustor trust trustee to do something. Trust relationships is not necessarily symmetrical and not necessarily transitive. In many cases, a trust relationship is conditional in the form that A trust B to do something if the condition is true [7]. Predicates which are already defined in [7] for one of the reciprocal logics, i.e., trust relationships are as follows.

The limitation of reciprocal logics is that the logic does not provide enough predicates to describe sentences about trust easily. This papers aims to describe various predicates for trust properties. Further, they can be used to represent and specify various reciprocal relationships and then to reason about the trustworthy decision and actions [7].

- Defined predicates by Cheng [7]:

- $TR(pe_1, pe_2)$ means pe_1 trusts pe_2.

- $NTR(pe_1, pe_2) =_{df} \neg(TR(pe_1, pe_2))$, $NTR(pe_1, pe_2)$ means pe_1 doesnot trust pe_2.

- $TREO(pe_1, pe_2) =_{df} TR(pe_1, pe_2) \wedge (TR(pe_2, pe_1))$, $TREO(pe_1, pe_2)$ means pe_1 and pe_2 trust each other.

- $ITR(pe_1, pe_2, pe_3) =_{df} \neg(TR(pe_1, pe_2) \wedge (TR(pe_1, pe_3)$, $ITR(pe_1, pe_2, pe_3)$ means pe_1 doesnot trust both pe_2 and pe_2 (Incompatibility).

- $XTR(pe_1, pe_2, pe_3) =_{df} (TR(pe_1, pe_2) \vee (TR(pe_1, pe_3)) \wedge (NTR(pe_1, pe_2) \vee (NTR(pe_1, pe_3))$ $XTR(pe_1, pe_2, pe_3)$ means pe_1 trust either pe_2 or pe_3 but not both (exclusive disjunction).

- $JTR(pe_1, pe_2, pe_3) =_{df} \neg(TR(pe_1, pe_2) \vee (TR(pe_1, pe_3))$, $JTR(pe_1, pe_2, pe_3)$ means pe_1 trust either pe_2 or pe_3 (joint denial).

- $TTR(pe_1, pe_2, pe_3) =_{df} (TR(pe_1, pe_2) \wedge (TR(pe_2, pe_3)) \Rightarrow (TR(pe_1, pe_3)$, $TTR(pe_1, pe_2, pe_3)$ means pe_1 trust pe_3 if pe_1 trusts pe_2 and pe_2 trust pe_3.

- $CTR(pe_1, pe_2, pe_3) =_{df} (TR(pe_1, pe_3) \Rightarrow (TR(pe_2, pe_3))$, $CTR(pe_1, pe_2, pe_3)$ means pe_2 trusts pe_3 if pe_1 trusts pe_3.

- $NCTR(pe_1, pe_2, pe_3) =_{df} (TR(pe_1, pe_3) \Rightarrow (TR(pe_2, pe_3))$, $NCTR(pe_1, pe_2, pe_3)$ means pe_2 trusts pe_3 if pe_1 doesnot trusts pe_3.

- $CNTR(pe_1, pe_2, pe_3) =_{df} \neg(TR(pe_1, pe_3) \Rightarrow \neg(TR(pe_2, pe_3))$, $CNTR(pe_1, pe_2, pe_3)$ means pe_2 does not trusts pe_3 if pe_1 doesnot trusts pe_3.

- $TRpo(pe_1, o_1) =_{df} \forall\ pe_2(B(pe_2, o_1) \wedge (TR(pe_1, pe_2))$, $TRpo(pe_1, o_1)$ means pe_1 trusts o_1.

- $NTRpo(pe_1, o_1) =_{d_f} \forall\ pe_2(B(pe_2, o_1) \wedge (NTR(pe_1, pe_2)),\ NTRpo(pe_1, o_1)$ means pe_1 doesnot trusts o_1.

- $NTRpo(o_1, pe_1) =_{d_f} \forall\ pe_2(B(pe_2, o_1) \wedge (NTR(pe_2, pe_1)),\ NTRpo(o_1, pe_1)$ means o_1 doesnot trusts pe_1.

- $TRoo(o_1, o_2) =_{d_f} \forall\ pe_1\ \forall\ pe_2\ (B(pe_1, o_1) \wedge (B(pe_2, o_2)) \wedge (TR(pe_1, pe_2),$ $TRoo(o_1, o_2)$ means o_1 trusts o_2.

- $NTRoo(o_1, o_2) =_{d_f} \forall\ pe_1\ \forall\ pe_2\ (B(pe_1, o_1) \wedge (B(pe_2, o_2)) \wedge (NTR(pe_1, pe_2),$ $NTRoo(o_1, o_2)$ means o_1 doesnot trusts o_2.

4 An Extension of Reciprocal Logics for Trust Properties

4.1 Description of Trust Properties

Trust is an essential element of any coherent trustworthy relationship. Trust relationships are more tractable with the aid of trust properties. So, we need to extend reciprocal logics with trust properties that supports trustworthy decisions. We studied various trust properties and classified them according to trustee and trustor objectives and subjectivities respectively. According to our survey we have identified trust properties which are regularly assigned to trust and influence the concept of computational trust. We have also adopted some ideas of trust properties by Demolombe [8]. We further presents a formal representations of identified trust properties and applies them on context based cases.

Current reciprocal logics are not enough to represent sincerity, validity, vigilance, obedience, reliability, credibility, cooperativity, completeness, willingness because at present it only deals with the basic trust relationships which doesn't includes a piece of information exchange between two entities.

- Brief definitions of trust properties are as follows:
- Sincerity: Sincerity is the relationship between what the trustee says and what he believes [3].
- Validity: Validity is the relationship between what the trustee says and what is true [3].
- Completeness: Completeness is the relationship between what is true and what the trustee says [3].
- Cooperativity: Cooperativity is the relationship between what the trustee believes and what he says [3]. The number of interactions between entities that have been held in positive manner [16].
- Credibility: Credibility indicates the degree to which the trustor believes that trustee will participate in the collaboration [16]. This information might be measured based on the level of uncertainty.

- Vigilance: Vigilance is the relationship between what is true and what the trustee believes [3].
- Reliability: Reliability means that any cyber object might imply that it fulfils the required quality of service and it can be measured as probability that an entity correctly performs a required job in a specified period of time under stated conditions [16].
- Obedience: Trustor is said to be obedient if it behaves according to the trustees standards.
- Willingness: Trustor is said to be willing if it is relying on the actions of another party. Trustee decide and intends to do what the trustor have propose to do.

4.2 Formal Representation of Extended Predicates

This section shows trust properties and its formal representation. If any trust relationship between trustor and trustee in a particular context does not meet any of the trust properties, we consider that the communication is not trustworthy. Targets behave different in centralized and distributed environment. A target in a distributed environment may not have direct knowledge of other target so there is a need for mechanisms to support establishment of trust relationships between distributed targets. For this we analyzed various trust properties which will also support the trust relationship among targets involved in a distributed environment.

Predicates

- C is the content
- $Trustor$ is represented with pe_1.
- $Trustee$ is represented with pe_2.
- $T(c)$ C is true
- $PER(pe_1)$ pe_1 performs according to c.
- $DE(pe_2, c)$ pe_2 decides according to c.
- $BEH(pe_2, c)$ pe_2 behaves according to c.
- $I(pe_1, pe_2, c) pe_1$ has informed pe_2 about c.
- $BEL(pe_2, c)$ pe_2 believe in c.
- $STR(pe_1, pe_2, c)$ pe_1 trust pe_2 in his sincerity about c.
- $ValTR(pe_1, pe_2, c)$ pe_1 trustpe_2 in his validity about c.
- $VigTR(pe_1, pe_2, c)$ pe_1 trust pe_2 in his vigilance about c.
- $ObTR(pe_1, pe_2, c)$ pe_1trust pe_2 in his obedience about c.
- $ReTR(pe_1, pe_2, c)$ pe_1 trust pe_2in his Reliability about c.
- $WTR(pe_1, pe_2, c)$ pe_1 trust pe_2in his willingness about c.
- $CreTR(pe_1, pe_2, c)$ pe_1 trust pe_2 in his Credibility about c.
- $CoTR(pe_1, pe_2, c) pe_1$ trust pe_2in his Cooperativity about c.
- $CptTR(pe_1, pe_2, c)$ pe_1 trust pe_2 in his Completeness about c.

Defined Predicates:

1. $STR(pe_1, pe_2, c) =_{d_f} BEL(pe_1$,I($pe_1$,$pe_2$,c)$\Rightarrow$ $BEL(pe_2$,T(c))), $STR(pe_1, pe_2, c)$ means trustor believes that if it is informed by the trustee about some content,then the trustee believes that this content is true.
2. $ValTR(pe_1, pe_2, c) =_{d_f} BEL((pe_1$,I($pe_2$,$pe_1$,c)$\Rightarrow$T(c)), $ValTR(pe_1, pe_2, c)$ means truster believes that if it is informed by the trustee about some content, then this content is true.
3. $CptTR(pe_1, pe_2, c) =_{d_f} BEL(pe_1$,T(c)$\Rightarrow$I($pe_2$,$pe_1$,c)), $CptTR(pe_1, pe_2, c)$ means that truster believes that if some content is true, then the truster is informed by the trustee about this content.
4. $CoTR(pe_1, pe_2, c) =_{d_f} BEL(pe_1$,BEL($pe_2$,T(c)))$\Rightarrow I(pe_2, pe_1, c)), CoTR(pe_1, pe_2, c)$ means truster believes that if the trustee believes that some content is true, then the truster is informed by the trustee about this content.
5. $CreTR(pe_1, pe_2, c) =_{d_f} BEL(pe_1$, (BEL($pe_2$,T(c))) $\Rightarrow c$), $CreTR(pe_1, pe_2, c)$ means truster believes that if the trustee believes that some content is true, then this content is true.
6. $VigTR(pe_1, pe_2, c) =_{d_f} BEL(pe_1, T(c) \Rightarrow$ (BEL(pe_2,c))) , $VigTR(pe_1, pe_2, c)$ means truster believes that if some content is true, then the trustee believes that this content is true.
7. $ObTR(pe_1, pe_2, c) =_{d_f} BEL(pe_1$,S($pe_2$,c)$\Rightarrow I(pe_1, pe_2, c))$, $ObTR(pe_1, pe_2, c)$ means trustor believe that trustee would satisfies the informed content C.
8. $ReTR(pe_1, pe_2, c) =_{d_f} BEL(pe_1$,PER($pe_2$,c)$\Rightarrow I(pe_1, pe_2, c))$, $ReTR(pe_1, pe_2, c)$ means trustor believes that trustee will perform according to content.
9. $WTR(pe_1, pe_2, c) =_{d_f} BEL(pe_1$,DE($pe_2$,c)) , $WTR(pe_1, pe_2, c)$ means trustee believes that trustor has decided to do what it has been informed by trustor.

A summary of our classification is presented in Table 1. It shows identified trust properties classified according to trustee and trustor objectives and subjectivities respectively. This table is used as a frame of reference in context based cases to identify whether targets (Person, Agent or Service) and trust relationships among them are trustworthy or not.

Context Based Cases

Context refers to the circumstances and associations of the target in a trust relationship decision. Context based case means the case which has some context. For example, an agent providing a description for an item, where the agent may be a vendor selling that item, or as a consumer advocate reporting on that item [4].

Secondly context based cases involves some target. A target is an entity which is being evaluated or given trust varies with the perspective of the problem [4]. A target could be one from any domain. Human user from social domain, Web services from web domain, agent from network domain, systems and objects from cyber and physical domain respectively.

Trust is usually specified in terms of a relationship between a trustor, the subject that trusts a target entity, which is known as the trustee, i.e., the entity that is trusted and it forms the basis for allowing a trustee to use or manipulate resources owned by a trustor or may influence a trustor's decision to use a

Table 1. Classification of trust properties w.r.t objective's and subjectivities.

Properties	Trustee objective	Trustee subjective	Trustor objective	Trustor subjective
Sincerity		√		
Validity	√	√		
Completeness	√	√		
Cooperativity		√		
Credibility	√			
Vigilance		√		
Reliability	√			√
Obedience			√	
Willingness				√

service provided by a trustee [11]. A target can behave as a trustor or a trustee depending on the context. For example, a service is only trusted if it response less than 0.2 ms or a system protecting files from the accidental deletion [11].

This paper considers a case from a distributed enviorment which involves a user and a server. A user u wants to know about the weather. The user has found a web service *s1* holds by a server s that presents weather information, but he doesn't know whether it delivers up to date information, or whether the information is correct at all. User asks the server about web service which is said to be content in this case. If the server do not provide the correct information then the user concludes that the server is not trustworthy.

Table 2. Shows trust predicates used to define relationship between two targets.

From\To	Trust	Obedience	Sincerity	Reliability	Competence
User u	$TR(u, s)$	$ObTR(s, u, c)$	$STR(s, u, c)$	–	–
Server s	$TR(s, u)$	–	–	$ReTR(u, s, c)$	$CptTR(u, s, c)$

These trust predicates are used to reason about a case, a user requests a service from a server. We can see that server is sincere about user that it will provide valid details in order to access a web service from server s. User trusts server regarding its reliability and believe that a valid service will be provided by a server (Table 2).

Identification of targets:

- User
- Server

A target may behaves as a trustor or a trustee. In this case if a user behaves as a trustor and server behaves as a trustee than a user should access the required service through proper channel following trustor's standards. For this predicate $ObTR(pe_1, pe_2, c)$ has been defined in Sect. 4 and $ReTR(pe_1, pe_2, c)$ predicate shows server should be reliable and provide services as required. If a user behaves as a trustee and server behaves as a trustor than a user should provide valid login details. If the login details are correct then server should complete the request.

1. (u, s): the user requests a service.
2. $ObTR(s, u, c)$: the user follows the rules (which is c in the predicate) while accessing server.
3. $STR(s, u, c)$: the user provides valid login details.
4. $TR(s, u)$: the server trusts the user.
5. $ReTR(u, s, c)$: the server is reliable and provides requested service.
6. $CptTR(u, s, c)$: the server is capable to provide a service and perform the function as expected.
7. $TR(u, s)$: the user trusts the server.
8. $TREO(u, s)$: at this point both the server and the user trusts each other.

From the above case, we obtain two beliefs that user believes that service is valid and the server is reliable. It can also be said that using predicates defined for trust properties using reciprocal logics provides us with a criterion of logical validity of reasoning.

5 Conclusion

This paper discussed a significant concept of reciprocal relationships. We have extended reciprocal logics with various trust properties for trust reasoning. We have also provided definitions of trust properties and introduced new predicates to reciprocal logics.

As future work, we plan to investigate and develop actual framework for reciprocal relationships which could be said as a complete and comprehensive set of basis for trustworthy systems. Studying trust dynamics in reputation is another direction for further investigation.

References

1. Anderson, A.R., Belnap Jr., N.D.: Entailment: The Logic of Relevance and Necessity, vol. I. Princeton University Press, Princeton (1975)
2. Anderson, A.R., Belnap Jr., N.D., Dunn, J.M.: Entailment: The Logic of Relevance and Necessity, vol. II. Princeton University Press, Princeton (1992)
3. Amgoud, L., Demolombe, R.: An argumentation-based approach for reasoning about trust in information sources. Argument Comput. **5**(2–3), 191–215 (2014). https://doi.org/10.1080/19462166.2014.881417
4. Artz, D., Gil, Y.: A survey of trust in computer science and the semantic web. Web Semant.: Sci. Serv. Agents World Wide Web **5**(2), 58–71 (2007). https://doi.org/10.1016/j.websem.2007.03.002
5. Cheng, J.: The fundamental role of entailment in knowledge representation and reasoning. J. Comput. Inform. **2**(1), 853–873 (1996)

6. Cheng, J.: A strong relevant logic model of epistemic processes in scientific discovery. In: Kawaguchi, E., et al. (eds.) Information Modeling and Knowledge Bases XI. Frontiers in Artificial Intelligence and Applications, vol. 61, pp. 136–159. IOS Press, Amsterdam (2000)
7. Cheng, J.: Reciprocal logic: logics for specifying, verifying, and reasoning about reciprocal relationships. In: Khosla, R., Howlett, R.J., Jain, L.C. (eds.) KES 2005. LNCS, vol. 3682, pp. 437–445. Springer, Heidelberg (2005). https://doi.org/10.1007/11552451_58
8. Demolombe, R.: Reasoning about trust: a formal logical framework. In: Jensen, C., Poslad, S., Dimitrakos, T. (eds.) iTrust 2004. LNCS, vol. 2995, pp. 291–303. Springer, Heidelberg (2004). https://doi.org/10.1007/978-3-540-24747-0_22
9. Demolombe, R.: Reasoning about trust and aboutness in the context of communication. J. Appl. Non-Class. Log. **27**(3–4), 292–303 (2017). https://doi.org/10.1080/11663081.2017.1420316
10. Fan, T., Liau, C.: A logic for reasoning about evidence and belief. In: Proceedings of the International Conference on Web Intelligence, Leipzig, Germany, pp. 509–516. ACM, New York (2017). https://doi.org/10.1145/3106426.3106519
11. Grandison, T., Sloman, M.: A survey of trust in internet applications. Commun. Surv. Tutor. **3**(4), 2–16 (2000). https://doi.org/10.1109/COMST.2000.5340804
12. Herzig, A., Lorini, E., Hubner, J., Vercouter, L.: A logic of trust and reputation. Log. J. IGPL **18**(1), 214–244 (2010). https://doi.org/10.1093/jigpal/jzp077
13. Huang, J., Fox, M.: An ontology of trust. In: ICEC 2006 Proceedings of the 8th International Conference on Electronic Commerce: The new e-commerce: Innovations for Conquering Current Barriers, Obstacles and Limitations to Conducting Successful Business on the Internet, New Brunswick, Canada, pp. 259–270. ACM (2006). https://doi.org/10.1145/1151454.1151499
14. Josang, A., Ismail, R., Boyd, C.: A survey of trust and reputation systems for online service provision. Decis. Support Syst. **43**(2), 618–644 (2007). https://doi.org/10.1016/j.dss.2005.05.019
15. Koutrouli, E., Tsalgatidou, A.: Credibility enhanced reputation mechanism for distributed e-communities, In: Proceedings of the 2011 19th International Euromicro Conference on Parallel, Distributed and Network-Based Processing, pp. 627–634. IEEE Computer Society, Washington (2011). https://doi.org/10.1109/PDP.2011.68
16. Lee, G., Lee, H.: Standardization of trust provisioning study. Technical report, International Telecommunication Union (2005)
17. Leturc, C., Bonnet, G.: A normal modal logic for trust in the sincerity. In: Proceedings of the 17th International Conference on Autonomous Agents and MultiAgent Systems, Stockholm, Sweden, pp175–183. International Foundation for Autonomous Agents and Multiagent Systems, Richland (2018)
18. Liau, C.: Belief, information acquisition, and trust in multi-agent systems: a modal logic formulation. Artif. Intell. **149**(1), 31–60 (2003). https://doi.org/10.1016/S0004-3702(03)00063-8
19. Munindar, P.S.: Trust as dependence: a logical approach. In: 10th International Conference on Autonomous Agents and Multiagent Systems, Taipei, Taiwan, pp. 863–870. International Foundation for Autonomous Agents and Multiagent Systems, Richland (2011)
20. Parsons, S., Sklar, E., McBurney, P.: A simple logical approach to reasoning with and about trust. In: Proceedings of AAAI Spring Symposium on Logical Formalizations of Commonsense Reasoning, Stanford, California (2011)

21. Tan, Y., Castelfranchi, C.: Trust and deception in virtual societie. Kluwer Academic Publishers, USA (2001)
22. Vallee, T., Bonnet, G.: Using KL divergence for credibility assessment. In: Proceedings of the 2015 International Conference on Autonomous Agents and Multiagent Systems, Istanbul, Turkey, pp. 1797–1798. International Foundation for Autonomous Agents and Multiagent Systems, Richland (2015)
23. Yan, Z., Holtmanns, S.: Trust modeling and management: from social trust to digital trust. In: Subramanian, R. (ed.) Computer Security, Privacy and Politics: Current Issues, Challenges and Solutions, pp. 290–300. IGI Global (2007)
24. Yan, Z., Zhang, P., Vasilakos, A.: A survey on trust management for internet of things. J. Netw. Comput. Appl. **42**, 120–134 (2014)
25. Zhao, H., Li, X.: A group trust management system for peer-to-peer desktop grid. J. Comput. Sci. Technol. **24**(5), 833–843 (2009). https://doi.org/10.1007/s11390-009-9275-7

Current Trends in Artificial Intelligence, Optimization, Learning, and Decision-Making in Bioinformatics and Bioengineering

Linear Classifiers with the L_1 Margin from a Small Number of High-Dimensional Vectors

Leon Bobrowski[1,2] and Tomasz Łukaszuk[1(✉)]

[1] Faculty of Computer Science, Bialystok University of Technology, Wiejska 45A, Bialystok, Poland
{l.bobrowski,t.lukaszuk}@pb.edu.pl

[2] Institute of Biocybernetics and Biomedical Engineering, PAS, Warsaw, Poland

Abstract. Designing linear classifiers from a small number of high-dimensional feature vectors is linked to a special principle of the maximal margin. This situation occurs, among others during building linear classifiers from genetic data sets. Maximizing margins is the basic principle of classifiers designing in the support vector machines (*SVM*) method. The Euclidean (L_2) margins are used in the *SVM* method. An alternative approach to linear classifiers designing can be based on maximizing L_1 margins.

Keywords: Linear classifiers · Small data sets · Regularized *CPL* criterion functions · Maximizing L_1 norm margins

1 Introduction

Classification rules are designed in order to reach a good assignment of individual objects to appropriate categories (classes). Classification rules can be designed on the based on data sets within various pattern recognition methods [1]. In accordance with the pattern recognition methodology, objects are represented by feature vectors in order to achieve this goal. Feature vectors assigned to a particular category are collected as one learning set.

In accordance with statistical inference principles, each learning set should be representative for a given category [2]. It means that the number of feature vectors in each learning set should be sufficiently large. This postulate cannot always be realized in practice. The number of feature vectors m in a given data set may be a low in comparison with the number n of features. Obvious examples of this type of data sets can be found in genetics area, where feature vectors have thousands of components (genes), while numbers of objects (patients) oscillate around several dozen.

A small number m of feature vectors compared to a large number n of their components (features) is an obstacle, for designing linear classifiers and assessment of their quality [3].

Several features n in a given data set can be reduced by using some feature selection methods, for example *LASSSO* [4] or *RLS* [5]. Dimensionality n of feature vectors can

N. T. Nguyen et al. (Eds.): ACIIDS 2020, LNAI 12034, pp. 79–89, 2020.
https://doi.org/10.1007/978-3-030-42058-1_7

also be reduced by using one of feature extraction methods. Fundamental role is played here by linear transformations of the principal component analysis (*PCA*) [6]. The quality of classifiers is typically evaluated by using the cross validation technique [7].

The basic method of designing linear classifiers used in machine learning is the *Support Vector Machine* (*SVM*) [8]. In accordance with the *SVM* approach, the optimal hyperplane should separate two learning sets with the maximal margin. We are considering another method of designing the optimal separating hyperplanes with a margin. This method is based on minimization of the *convex and piecewise linear* (*CPL*) criterion functions. The *SVM* optimal separating hyperplanes have the margins determined by the Euclidean (L_2) norm [8]. The *CPL* separating hyperplanes are linked to the margins determined by the maximizing margins with the L_1 norm [9, 10]. The properties of the *CPL* solutions in the case of long feature vectors are analysed in the presented paper. The procedure of designing the separating hyperplanes with the L_1 margins is initially tested.

2 Linear Separability of the Learning Sets

We assume that m objects (patients) O_j have been represented by feature vectors $\mathbf{x}_j = [x_{j,1}, \ldots, x_{j,n}]^T (j = 1, \ldots, m)$ constituting points in the n-dimensional feature space $\boldsymbol{F}[n]$ ($\mathbf{x}_j \in \boldsymbol{F}[n]$). The i-th component $x_{j,i}(x_{j,i} \in R^1)$ of the vector $\mathbf{x}_j$ is the numerical value of the i-th feature $X_i(i = 1, \ldots, n)$ of the j-th object O_j.

We consider two learning sets $\boldsymbol{G}^+$ and $G^-(\boldsymbol{G}^+ \cap \boldsymbol{G}^- = \varnothing)$:

$$\boldsymbol{G}^+ = \{\mathbf{x}_j : j \in J^+\} \; and \; \boldsymbol{G}^- = \{\mathbf{x}_j : j \in J^-\} \tag{1}$$

where J^+ and J^- are non-empty sets of m indices $j(J^+ \cap J^- = \varnothing)$.

The *positive* set $\boldsymbol{G}^+$ contains m^+ examples of feature vectors $\mathbf{x}_j$. Similarly, the *negative* set $\boldsymbol{G}^-$ contains m^- examples of another feature vectors $\mathbf{x}_j$ $(m = m^+ + m^-)$.

Definition 1: The learning sets $\boldsymbol{G}^+$ and $\boldsymbol{G}^-$ (1) are formed by *long* feature vectors $\mathbf{x}_j$, if and only if the number m of vectors $\mathbf{x}_j$ is less than their dimensionality n $(m < n)$.

A possibility of separation of the learning sets G^+ and $\boldsymbol{G}^-$ (1) by the below hyperplane $\boldsymbol{H}(\mathbf{w}, \theta)$ in a feature space is examined:

$$\boldsymbol{H}(\mathbf{w}, \theta) = \{\mathbf{x} : \mathbf{w}^T\mathbf{x} = \theta\} \tag{2}$$

where $\mathbf{w} \in \boldsymbol{R}^n$ is the *weight vector*, $\theta \in R^1$ is the *threshold*, and $\mathbf{w}^T\mathbf{x} = \sum_{,i} w_i \, x_{j,i}$ is the inner product.

Definition 2: The learning sets G^+ and G^- (1) are *linearly separable* in the feature space $\boldsymbol{F}[n]$, if and only if there exists such a *weight vector* $\mathbf{w} = [w_1, \ldots, w_n] \in \boldsymbol{R}^n$, and a threshold $\theta \in R^1$ that these sets can be separated by the hyperplane $\boldsymbol{H}(\mathbf{w}, \theta)$ (2):

$$\begin{aligned} (\exists \mathbf{w}, \theta) \quad & (\forall \mathbf{x}_j \in G^+) : \; \mathbf{w}^T\mathbf{x}_j \geq \theta + 1 \; and \\ & (\forall \mathbf{x}_j \in G^-) : \; \mathbf{w}^T\mathbf{x}_j \leq \theta - 1 \end{aligned} \tag{3}$$

If the above inequalities are fulfilled, then all m^+ feature vectors $\mathbf{x}_j$ from the set G^+ are located on the positive side of the hyperplane $H(\mathbf{w}, \theta)$ (3) and all m^- vectors $\mathbf{x}_j$ from the set G^- are located on the negative side of this hyperplane.

Let us introduce the *m augmented* feature vectors $\mathbf{y}_j$ ($\mathbf{y}_j \in \boldsymbol{F}[n+1]$) [1]:

$$\begin{aligned}(\forall \mathbf{x}_j \in G^+)\ \mathbf{y}_j = [\mathbf{x}_j^T, 1]^T \ \textit{and} \\ (\forall \mathbf{x}_j \in G^-)\ \mathbf{y}_j = -[\mathbf{x}_j^T, 1]^T\end{aligned} \tag{4}$$

and the *augmented* weight vector $\mathbf{v} = [\mathbf{w}^T, -\theta]^T$.

The linear separability inequalities (3) now take the below form:

$$(\exists \mathbf{v})\ (\forall \mathbf{y}_j \in G^+ \cup G^-)\ \ \mathbf{v}^T\mathbf{y}_j \geq 1 \tag{5}$$

The above inequalities are used in the below definition of the perceptron function.

3 The Perceptron Penalty Functions and the Criterion Function

The penalty function $\varphi_j(\mathbf{v})$ can be defined for each feature vector $\mathbf{x}_j$ (1) (augmented vector $\mathbf{y}_j$ (4)):

$$(\forall \mathbf{x}_j \in G^+ \cup G^-)$$

$$\varphi_j(\mathbf{v}) = \begin{cases} 1 - \mathbf{y}_j^T\mathbf{v} & if\ \ \mathbf{y}_j^T\mathbf{v} < 1 \\ 0 & if\ \ \mathbf{y}_j^T\mathbf{v} \geq 1 \end{cases} \tag{6}$$

The convex and piecewise-linear (*CPL*) penalty functions $\varphi_j(\mathbf{v})$ are aimed at reinforcing the linear separability inequalities (5) (Fig. 1).

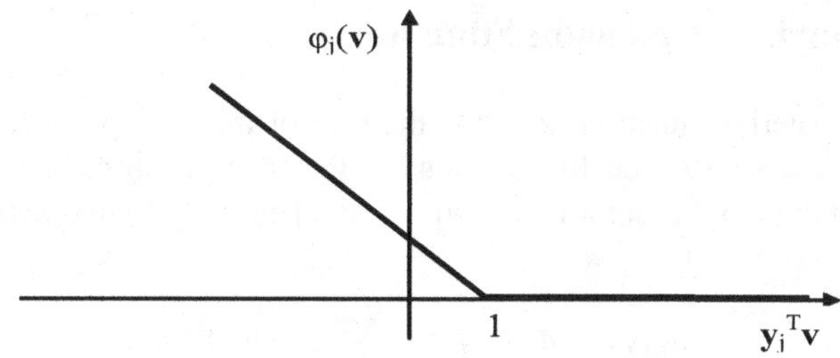

Fig. 1. The perceptron penalty functions $\varphi_j(\mathbf{v})$ (6).

The *perceptron* criterion function $\Phi(\mathbf{v})$ is defined as the weighted sum of the penalty functions $\varphi_j(\mathbf{v})$ (6) [3, 10]:

$$\Phi(\mathbf{v}) = \sum_{j \in J} \alpha_j\, \varphi_j(\mathbf{v}) \tag{7}$$

where the positive parameters $\alpha_j (\alpha_j > 0)$ determine the *prices* of particular feature vectors $\mathbf{x}_j$, and $J = J^+ \cup J^- = \{1, \ldots, m\}$ is the set of the indices j (1). The default (standard) values of the parameters α_j are specified below:

$$\begin{aligned} &if\ \mathbf{y}_j \in G^+,\ then\ \alpha_j = 1/(2\,m^+),\ and \\ &if\ \mathbf{y}_j \in G^-,\ then\ \alpha_j = 1/(2\,m^-) \end{aligned} \tag{8}$$

It can be seen that the non-negative criterion function $\Phi(\mathbf{v})$ (7) is convex and piecewise-linear (*CPL*). The minimum value Φ^* of the *CPL* criterion function $\Phi(\mathbf{v})$ (7) determines the optimal vector $\mathbf{v}^*$:

$$(\exists \mathbf{v}^*)\ (\forall \mathbf{v} \in R^{n+1})\ \Phi(\mathbf{v}) \geq \Phi(\mathbf{v}^*) = \Phi^* \geq 0 \tag{9}$$

The below theorem has been proved [9]:

Theorem 1: The minimum value $\Phi^* = \Phi(\mathbf{v}^*)$ (9) of the perceptron criterion function $\Phi(\mathbf{v})$ (8) is equal to zero ($\Phi^* = 0$) if and only if the learning sets G^+ and G^- (1) are linearly separable (3).

It can also be proved, that the minimum value $\Phi^* = \Phi(\mathbf{v}^*)$ (9) of the criterion function $\Phi(\mathbf{v})$ (7) with the parameters α_j specified by (8) is near to one ($\Phi^* \approx 1.0$) if the learning sets G^+ and G^- (1) are completely overlap ($G^+ \approx G^-$). The below standardization inequalities reflect these properties [9]:

$$0 \leq \Phi(\mathbf{v}^*) \leq 1.0 \tag{10}$$

The minimum value $\Phi(\mathbf{v}^*)$ (9) of the perceptron criterion function $\Phi(\mathbf{v})$ (7) is used as a measure of the lack of linear separability of two learning sets G^+ and G^- (1) in a given feature space $\boldsymbol{F}[n]$.

4 The Regularized Criterion Function

The regularized criterion function $\Psi_\lambda(\mathbf{v})$ is the sum of the perceptron function $\Phi(\mathbf{v})$ (7) and the additional *CPL* penalty functions in the form of absolute values $|w_i|$ of components w_i of the weight vector $\mathbf{w} = [w_1, \ldots, w_n]$ multiply by the *costs* $\gamma_i (\gamma_i > 0)$ of particular features X_i [8]:

$$\Psi_\lambda(\mathbf{v}) = \Phi(\mathbf{v}) + \lambda \sum_{i \in \{1,\ldots,n\}} \gamma_i |w_i| \tag{11}$$

where $\lambda \geq 0$ is the *cost level*. The values of the cost parameters γ_i may be equal to one:

$$(\forall\, i = 1, \ldots, n)\ \ \gamma_i = 1.0 \tag{12}$$

The *relaxed linear separability* (*RLS*) method of feature subset selection is based on the minimization of the criterion function $\Psi_\lambda(\mathbf{v})$ (11) [5]. The regularization component $\lambda \sum \gamma_i |w_i|$ used in the function $\Psi_\lambda(\mathbf{v})$ (11) is similar to that used in the *Lasso* method developed as a part of the regression analysis for the model selection [4].

The main difference between the *Lasso* and the *RLS* methods is in the types of the basic criterion functions. The basic criterion function used in the *Lasso* method is the *residual sum of squares*, whereas the perceptron criterion function $\Phi(\mathbf{v})$ (7) is the basic function used in the *RLS* method. This difference affects, among others, the computational techniques the criterion functions minimization. Quadratic programming is used both in the *Lasso* feature selection and in the *Support Vector Machines* (*SVM*) method of classifiers or regression models designing [11, 12]. The basis exchange algorithm, similar to linear programming, is used in the *RLS* method [5].

The regularized criterion function $\Psi_\lambda(\mathbf{v})$ (11) is convex and piecewise-linear (*CPL*), similarly to the perceptron criterion function $\Phi(\mathbf{v})$ (7). The basis exchange algorithm allows us to find the minimum $\Psi_\lambda(\mathbf{v}_\lambda^*)$ efficiently [9]:

$$(\exists \mathbf{v}_\lambda^*)\ (\forall \mathbf{v})\quad \Psi_\lambda(\mathbf{v}) \geq \Psi_\lambda(\mathbf{v}_\lambda^*) \tag{13}$$

where (6):

$$\mathbf{v}_\lambda^* = [(\mathbf{w}_\lambda^*)^T, -\theta_\lambda^*]^T = [w_{\lambda 1}^*, \ldots, w_{\lambda n}^*, -\theta_\lambda^*]^T \tag{14}$$

The optimal parameters $w_{\lambda i}^*$ are used in the below feature *reduction rule* [5]:

$$(w_{\lambda i}^* = 0) \Rightarrow (\text{the } i - \text{th feature } X_i \text{ is omitted}) \tag{15}$$

The reduction of this feature X_i, which is related to the weight $w_{\lambda i}^*$ equal to zero ($w_{\lambda i}^* = 0$), does not change the value of the inner product $(\mathbf{w}_\lambda^*)^T\mathbf{x}_j$. The positive ($(\mathbf{w}_\lambda^*)^T\mathbf{x}_j > \theta_\lambda^*$) or the negative location ($(\mathbf{w}_\lambda^*)^T\mathbf{x}_j < \theta_\lambda^*$) of all feature vectors $\mathbf{x}_j$ concerning the optimal separating hyperplane $H(\mathbf{w}_\lambda^*, \theta_\lambda^*)$ (2) remains unchanged.

5 Vertices in the Parameter Space

Each of the augmented feature vectors $\mathbf{y}_j$ (4) defines the below hyperplane h_j^1 in the parameter space:

$$(\forall j = 1, \ldots, m)\quad h_j^1 = \left\{\mathbf{v}: \mathbf{y}_j^T\mathbf{v} = 1\right\} \tag{16}$$

Each of the unit vectors $\mathbf{e}_i$ defines the hyperplane h_i^0 in the parameter space:

$$(\forall i = 1, \ldots, n)\quad h_i^0 = \left\{\mathbf{v}: \mathbf{e}_i^T\mathbf{v} = 0\right\} \tag{17}$$

Definition 3: The *m - vertex* $\mathbf{v}_k$ in the parameter space $\boldsymbol{R}^{n+1}$ ($\mathbf{v} \in \boldsymbol{R}^{n+1}$) is located at the intersection point of m hyperplanes h_j^1 (16) and $n + 1 - m$ hyperplanes h_i^0 (18) ($i \in I_k$).

If feature vectors $\mathbf{y}_j$ (4) are linearly independent, then any m - vertex $\mathbf{v}_k$ in the parameter space $\boldsymbol{R}^{n+1}$ ($\mathbf{v} \in \boldsymbol{R}^{n+1}$) can be defined by the below system of (n + 1) linear equations [9]:

$$\begin{gathered}(\forall j = 1, \ldots, m)\ \mathbf{v}_k^T\mathbf{y}_j = 1 \\ \textit{and}\ (\forall i \in I_k)\quad \mathbf{v}_k^T\mathbf{e}_i = 0\end{gathered} \tag{18}$$

where I_k is the k-th subset of indices i of $n + 1 - m$ base unit vectors $\mathbf{e}_i$ in the $(n + 1)$ – dimensional feature space.

The linear Eq. (18) can also be represented by the below matrix formula:

$$\mathbf{B}_k \mathbf{v}_k = \mathbf{1}_k \tag{19}$$

where $\mathbf{B}_k$ is the squared matrix with the first m rows defined by the vectors $\mathbf{y}_j (j = 1, \ldots, m)$ and the last $n + 1 - m$ rows defined by the unit vectors $\mathbf{e}_i (i \in I_k)$:

$$\boldsymbol{B}_k = \left[\mathbf{y}_1, \ldots, \mathbf{y}_m, \mathbf{e}_{i(1)}, \ldots, \mathbf{e}_{i(n+1-m)}\right]^T \tag{20}$$

The vector $\mathbf{1}_m = [1, \ldots, 1, 0, \ldots, 0]^T$ has the first m components equal to 1 and the last $n + 1 - m$ components equal to 0.

If the matrix $\mathbf{B}_k$ (20) is nonsingular, then it is called the k-th *basis* and the vertex $\mathbf{v}_k$ linked to this basis can be computed by the below equation [9]:

$$\mathbf{v}_k = \mathbf{B}_k^{-1} \mathbf{1}_k = \mathbf{r}_{k,1} + \ldots + \mathbf{r}_{k,m} \tag{21}$$

where $\mathbf{B}_k^{-1} = \left[\mathbf{r}_{k,1}, \ldots, \mathbf{r}_{k,n+1}\right]$ is the inverse matrix with the columns $\mathbf{r}_{k,i} (i = 1, \ldots, n + 1)$.

Remark 1: If the basis $\boldsymbol{B}_k$ (20) contains the i-th unit vector $\mathbf{e}_i$, then the i-th component $v_{k,i}$ of the vector $\mathbf{v}_k = [v_{k,1}, \ldots, v_{k,n+1}]^T$ (21) is equal to zero ($v_{k,i} = 0$).

The above property of the m - vertex $\mathbf{v}_k$ (21) is linked to passing of the i-th dual hyperplane h_i^0 (18) through this vertex. As a result, the vector $\mathbf{v}_k$ (21) has the below structure (Definition 2):

$$\mathbf{v}_k = \left[v_{k,1}, \ldots, v_{k,n+1}\right]^T = \left[v_{k,1}, \ldots, v_{k,m}, 0, \ldots, 0\right]^T \tag{22}$$

Lemma 1: If the matrix $\mathbf{B}_k$ (20) is non-singular, then the learning sets G^+ and G^- (1) are linearly separable (6) in the feature space $\boldsymbol{F}[n]$.

Proof: If the matrix $\mathbf{B}_k$ (20) is non-singular, then the m - vertex $\mathbf{v}_k$ can be computed in accordance with the Eq. (21). It means that the m dual hyperplanes h_j^1 (16) pass through this vertex. The following m equations are fulfilled in this case:

$$(\forall j = 1, \ldots, m) \quad \mathbf{v}_k^T \mathbf{y}_j = 1 \tag{23}$$

The equalities (24) mean that the linear separability inequalities (6) hold. □

Lemma 2: The minimum value $\Phi(\mathbf{v}_k)$ (9) of the perceptron criterion function $\Phi(\mathbf{v})$ (7) defined on m feature vectors $\mathbf{y}_j$ (4) is equal to zero ($\Phi(\mathbf{v}_k) = 0$) in each m - vertex $\mathbf{v}_k$ (21).

Proof: All m dual hyperplanes h_j^1 (16) pass through each m - vertex $\mathbf{v}_k$ (Definition 3). In s consequence, the all m penalty functions are equal to zero in such vertex $\mathbf{v}_k$,

$$(\forall j = 1, \ldots, m) \quad \varphi_j(\mathbf{v}_k) = 0 \tag{24}$$

In a consequence, the minimum value $\Phi(\mathbf{v}_k)$ (9) of the perceptron criterion function $\Phi(\mathbf{v})$ (7) is equal to zero in the m - vertex $\mathbf{v}_k$ ($\Phi(\mathbf{v}_k) = 0$). □

6 Properties of the Optimal Vertex

If the learning sets G^+ and G^- (1) are linearly separable (3), then the minimum value Φ^* $= \Phi(\mathbf{v}^*)$ (9) of the perceptron criterion function $\Phi(\mathbf{v})$ (7) is equal to zero in the optimal point $\mathbf{v}^*$ (9). (Theorem 1). It means, that all the penalty function $\varphi_j(\mathbf{v})$ (6) are equal to zero in the point $\mathbf{v}^*$ (9).

$$(\forall j = 1, \ldots, m) \quad \varphi_j(\mathbf{v}^*) = 0 \tag{25}$$

All the penalty functions $\varphi_j(\mathbf{v})$ (8) are equal to zero in the point $\mathbf{v}$ if this point is located on the positive side each of the hyperplane h_j^1 (16):

$$(\forall j = 1, \ldots, m) \quad \mathbf{y}_j^T \mathbf{v} \geq 1 \tag{26}$$

The parameter vectors $\mathbf{v}$ located on the positive side of each hyperplane h_j^1 (16) form the convex polyhedron $\boldsymbol{R}$ in the $(n + 1)$ - dimensional parameter space [9]:

$$\boldsymbol{R} = \{\mathbf{v} \colon (\forall j = 1, \ldots, m)\ \mathbf{y}_j^T \mathbf{v} \geq 1\} \tag{27}$$

Remark 2: The convex polyhedron $\boldsymbol{R}$ (27) is a non-empty set if and only if the learning sets G^+ and G^- (1) are linearly separable (5) [9].

Remark 3: Each vertex $\mathbf{v}_k$ (21) of the convex polyhedron $\boldsymbol{R}$ (27) is the *m - vertex* $\mathbf{v}_k$ in the parameter space $\boldsymbol{R}^{n+1}$ (Definition 3).

It can be proved based on the fundamental theorem of linear programming, that the minimum value $\Phi^* = \Phi(\mathbf{v}_k^*)$ (9) of the perceptron criterion function $\Phi(\mathbf{v})$ (7) can be located in one of the vertices $\mathbf{v}_k$ (22) [13]. Similarly, the minimum value $\Psi_{k,\lambda}(\mathbf{v}_{k,\lambda}^*)$ (13) of the regularized criterion function $\Psi_{k,\lambda}(\mathbf{w})$ (11) can also be located in the optimal vertex $\mathbf{v}_{k,\lambda}^*$ (13) linked to some basis $\mathbf{B}_k$ (20).

If the learning sets G^+ and G^- (1) are linearly separable (5), then the minimum value of the criterion function $\Phi(\mathbf{v})$ (7) is equal to zero in each point $\mathbf{v}$ of the set $\boldsymbol{R}$ (27):

$$(\forall \mathbf{v} \in \boldsymbol{R}(27))\ \Phi(\mathbf{v}) = 0 \tag{28}$$

In particularly, the value $\Phi(\mathbf{v}_k)$ (9) of the perceptron criterion function $\Phi(\mathbf{v})$ (7) in each *m - vertex* $\mathbf{v}_k$ (21) is equal to zero (Definition 3).

Let us consider the regularized criterion function $\Psi_\lambda(\mathrm{v})$ (11) with the below values of the parameters λ and γ_i:

$$\lambda = 1 \ and\ (\forall i = 1, \ldots, n)\ \gamma_i = 1 \tag{29}$$

The value $\Psi_\lambda(\mathbf{v}_k)$ (29) of the regularized criterion function $\Psi_\lambda(\mathbf{v})$ (11) with (29) in each vertex $\mathbf{v}_k = [\mathrm{v}_{k,1}, \ldots, \mathrm{v}_{k,n+1}]^T$ is equal to the L_1 norm $||\mathbf{v}_k||_{L1}$ of this vector $\mathbf{v}_k$ (22):

$$(\forall \mathbf{v}_k \in \boldsymbol{R}(27)) \quad \Psi_\lambda(\mathbf{v}_k) = \sum_{i=1,\ldots,m} |\mathrm{v}_{k,I}| = ||\mathbf{v}_k||_{L1} \tag{30}$$

The values $\Psi_\lambda(\mathbf{v}_k)$ (30) can be computed in each vertex $\mathbf{v}_k$ of the polyhedron $\boldsymbol{R}$ (27).

The problem of the regularized criterion function $\Psi_\lambda(\mathrm{v})$ (11) minimization (13) can be formulated in the below constrained manner:

$$min\{\sum_{i\in\{1,\ldots,n\}} |w_i| : \mathbf{v} \in \boldsymbol{R}(27)\} = \sum_{i\in\{1,\ldots,n\}} |w_i^*| \tag{31}$$

where $\mathbf{v}_k^* = [(\mathbf{w}_k^*)^T, -\theta_k^*]^T = [w_1^*, \ldots, w_n^*, -\theta^*]^T$ (13).

The relation (31) means, that the best vertex $\mathbf{v}_k^*$ (13) of the convex polyhedron $\boldsymbol{R}$ (27) is characterized by the lowest L_1 length of the weight vector $\mathbf{w}_k$, where $\mathbf{v}_k = [\mathbf{w}_k^T, -\theta_k]^T$ (14).

The minimization problem considered in the *support vector machines* (*SVM*) method can be formulated in the below manner [2]:

$$min\{\mathbf{w}^T\mathbf{w}: \mathbf{v} \in \boldsymbol{R}\} = (\mathbf{w}_{SVM}^*)^T \mathbf{w}_{SVM}^* \tag{32}$$

The solution $\mathbf{w}_{SVM}^*$ of the *SVM* optimization problem (32) is characterized by the lowest Euclidean L_2 norm $\|\mathbf{w}_{SVM}^*\|$, in contrast to the L_1 norm used in the *CPL* solution $\mathbf{v}_k^*$ (31).

7 Minimization of the Regularized Criterion Function

The minimization (13) of the regularized criterion function $\Psi_\lambda(\mathrm{v})$ (11) in the case of the long feature vectors $\mathbf{x}_j (m < n)$ can be carried out in two stages:

Stage i: Designing the basis $\boldsymbol{B}_k$ (21) from the m augmented feature vectors $\mathbf{y}_j$ (4) and the $n + 1 - m$ unit vectors $\mathbf{e}_i (i \in I_k)$, where $m < n$.
Stage ii: Finding the solution $\mathbf{v}_k^*$ of the constrained minimization problem (31).

The *Stage i* can be realized in accordance with the Gauss – Jordan transformation through the replacement of selected m unit vectors $\mathbf{e}_i (i \in I_k^C)$ by the augmented feature vectors $\mathbf{y}_j$ (4) [3]. The symbol I_k means the k-th subset (18) of indices i of $n + 1 - m$ base unit vectors $\mathbf{e}_i$. The set I_k^C is formed by m indices i of such unit vectors $\mathbf{e}_i$, which are not contained in the basis $\boldsymbol{B}_k$ (20):

$$I_k \cup I_k^C = \{1, \ldots, n\} \tag{33}$$

Finding the optimal solution $\mathbf{v}_k^*$ of the constrained minimization problem (32) can be realized during the successive steps of *Stage ii* by using the basis exchange algorithm based on the Gauss - Jordan transformation [9]. During the k-th step of this algorithm, one unit vector $\mathbf{e}_{i(k)}$ is removed from the basis $\boldsymbol{B}_k$ (20) and placed in the set I_k^C (33). During this step the $i(k)$ unit vector $\mathbf{e}_{i(k)}$ in the basis $\boldsymbol{B}_k$ (20) is replaced by the $i(k + 1)$ unit vector $\mathbf{e}_{i(k+1)} (i(k + 1) \in I_k^C$ (33)). Each permissible replacement of the unit vector $\mathbf{e}_{i(k)}$ by the vector $\mathbf{e}_{i(k+1)}$ should result in decreasing of the minimum value $\Psi_\lambda(\mathbf{v}_k^*)$ (13) of the regularized criterion function $\Psi_\lambda(\mathrm{v})$ (11):

$$\Psi_\lambda(\mathbf{v}_k^*) > \Psi_\lambda(\mathbf{v}_{k+1}^*) \tag{34}$$

The process of unit vectors $\mathbf{e}_i$ exchange in the basis $\boldsymbol{B}_k$ (20) is stopped after finite number k^* of steps when the condition (34) can not be further fulfilled. After finite number k^* of reducing steps, the optimal feature subspace $\boldsymbol{F}_k^*[m]$ is formed. The reduced feature subspace $\boldsymbol{F}_k^*[m]$ is formed by m basic features X_i with indices i belonging to the optimal subset $I_k^*(i \in I_k^*)$ (33).

$$(\forall j = 1, \ldots, m)\ \mathbf{z}_j \in \boldsymbol{F}_k^*[m] \tag{35}$$

The m-dimensional reduced vectors $\mathbf{z}_j$ (36) are obtained from the n-dimensional ($m < n$) feature vectors $\mathbf{x}_j = [x_{j,1}, \ldots, x_{j,n}]^T (\mathbf{x}_j \in \boldsymbol{F}[n])$ by neglecting components $x_{j,i}$ linked to $n - m$ non-basic features $X_i (i \notin I_k^*)$ (34). The positive learning set $\boldsymbol{G}_k^+$ and the *negative* learning set G_k^- (1) are formed by m reduced vectors $\mathbf{z}_j$:

$$\boldsymbol{G}_k^+ = \{\mathbf{z}_j: j \in J^+\}\ and\ \boldsymbol{G}_k^- = \{\mathbf{z}_j: j \in J^-\} \tag{36}$$

Remark 4: If the learning sets G^+ and G^- (1) are linearly separable (6) in the feature space $\boldsymbol{F}[n]$, then the sets G_k^+ and G_k^- (37) of reduced vectors $\mathbf{z}_j$ are also linearly separable (6) in the m-dimensional ($m < n$) feature subspace $\boldsymbol{F}_k^*[m]$ (35).

Following the support vector machines (*SVM*) approach, the optimal separating hyperplane $\boldsymbol{H}(\mathbf{w}, \theta)$ (2) is characterized by the weight vector $\mathbf{w}_{SVM}^*$ (32) with the minimum Euclidean L_2 length $\|\mathbf{w}_{SVM}^*\| = ((\mathbf{w}_{SVM}^*)^T \mathbf{w}_{SVM}^*)^{1/2}$ [8]. In contrast, the minimization (13) of the regularized criterion function $\Psi_\lambda(\mathrm{v})$ (11) can lead to the optimal vertex $\mathbf{v}_k^* = [(\mathbf{w}_k^*)^T, -\theta_k^*]^T$ (31) characterized by the weight vector $\mathbf{w}_k^*$ with the smallest L_1 length $\sum_{i=1,\ldots,n} |w_i^*|$ [3].

8 Experimental Results

The data set *Breast cancer* [14] was used in the experiment. The set contains 78 objects representing patients diagnosed with breast cancer, each object is described by 24481 features, expressions of human genes. 34 patients are labeled "less than 5 years disease-free survival" (class 1) and 44 other patients are labeled "greater than 5 years disease-free survival" (class 0).

For the purpose of the experiment, the data set was randomly divided into 2 subsets, one for training and one for testing. The training subset consisted of 14 objects, 6 of class 1 and 8 of class 0. The remaining 64 objects were the testing subset. The sampling of the training and testing subsets was repeated 10 times. In this way 10 pairs of subsets were obtained.

Fourteen objects from the training subset were the basis for the determination of the optimal, in terms of the value of the criteria function (11), hyperplane (2) separating objects of class 0 from objects of class 1. Then, the obtained decision rule was applied to 64 objects from the testing subset. The quality of the model was assessed based on the number of misclassified objects from the test subset.

Table 1. The results of the experimental evaluation of 10 linear classifiers optimized through the minimization of the *CPL* regularized criterion function $\Psi_\lambda(\mathbf{v})$ (12)

Sample number	TN	FP	FN	TP	Error rate
1	23	13	16	12	0,453
2	26	10	10	18	0,313
3	21	15	10	18	0,391
4	24	12	17	11	0,453
5	36	0	16	12	0,250
6	31	5	16	12	0,328
7	18	18	3	25	0,328
8	23	13	11	17	0,375
9	27	9	9	19	0,281
10	30	6	25	3	0,484
Average	25,9	10,1	13,3	14,7	0,366

The results shown in the above table characterize 10 linear classifiers by relatively high error rates. Such results may be caused by an extremely small number m of feature vectors $\mathbf{y}_j$ ($m = 14$) in comparison to a large number n of features (genes) X_i ($n \approx 25$ 000) (Table 1).

9 Concluding Remarks

Linear classifiers play a particularly important role when feature vectors are *"long"*, which means that the number m of feature vectors $\mathbf{x}_j$ in a data set is much smaller than the number n ($m < n$) of their components $x_{j,i}$ (features $X_{,i}$). The great importance of the linear classifiers, in this case, is because, under this condition, two learning sets are usually linearly separable [3]. Lemma 1 clarifies the new rules in this area. In accordance with Lemma 1, the linear separability (3) of the learning sets G^+ and G^- (1) is assured if the matrix $\boldsymbol{B}_k$ (21) is non-singular.

The basic method of designing linear classifiers used in machine learning is the *Support Vector Machine* (*SVM*) [8]. Following the *SVM* approach, the optimal hyperplane should separate two learning sets with the maximal margin. The *SVM* optimal separating hyperplanes are aimed at maximization of the margins determined by the Euclidean (L_2) norm [8].

We are examining another method of designing the optimal separating hyperplanes through the minimization of the *convex and piecewise linear* (*CPL*) criterion functions The separating hyperplanes obtained through *CPL* criterion function maximization can be linked to the maximal margins determined by the L_1 norm [9]. Properties of the *CPL* solutions in the case of long feature vectors are analysed in the presented paper. The procedure of designing the separating hyperplanes with the L_1 margins maximization is proposed in this paper. Some computational experiment has been done in order to evaluate error rate of the *CPL* linear classifiers with the maximized L_1 margin.

From a numerical point of view, the *SVM* method is based on quadratic programmings [12]. The minimization of the convex and piecewise linear criterion functions is performed with the computational techniques, which is called the *basis exchange algorithms* [9]. The basis exchange algorithm are based on the Gauss-Jordan transformation and, for this reason, are similar to the *Simplex* algorithm from linear programming [13]. The basis exchange algorithms allow for efficient and precise minimization of the *CPL* criterion functions even in the case of large, multidimensional data sets.

Acknowledgments. The presented study was supported by the grant S/WI/2/2018 from Bialystok University of Technology and funded from the resources for research by Polish Ministry of Science and Higher Education.

References

1. Duda, O.R., Hart, P.E., Stork, D.G.: Pattern Classification. Wiley, New York (2001)
2. Bishop, C.M.: Pattern Recognition and Machine Learning. Springer, New York (2006)
3. Bobrowski, L.: Data Exploration and Linear Separability, pp. 1–172. Lambert Academic Publishing, Saarbrücken (2019)
4. Tibshirani, R.: Regression shrinkage and selection via the Lasso. J. R. Stat. Soc. Ser. B **58**(1), 267–288 (1996)
5. Bobrowski, L., Łukaszuk, T.: Relaxed linear separability (RLS) approach to feature (Gene) subset selection. In: Xia, X. (eds.) Selected Works in Bioinformatics, pp. 103–118. INTECH (2011)
6. Jolliffe, I.T.: Principal Component Analysis. Springer, New York (2002). https://doi.org/10.1007/b98835
7. Lachenbruch, P.A.: Discriminant Analysis. Hafner Press, New York (1975)
8. Vapnik, V.N.: Statistical Learning Theory. Wiley, New York (1998)
9. Bobrowski, L.: Data mining based on convex and piecewise linear criterion functions. Białystok University of Technology (2005). (in Polish)
10. Bobrowski, L.: Design of piecewise linear classifiers from formal neurons by some basis exchange technique. Pattern Recognit. **24**(9), 863–870 (1991)
11. Aliferis, C.F., Statnikov, A., Tsamardinos, I., Mani, S., Koutsoukos, X.D.: Local causal and Markov blanket induction for causal discovery and feature selection for classification. Part II: analysis and extensions. J. Mach. Learn. Res. **11**, 235–284 (2010)
12. Best, M.J.: Quadratic Programming with Computer Programs. CRC Press, Taylor Francis Group, New York (2017)
13. Vanderbei, R.J.: Linear Programming: Foundations and Extensions. International Series in Operations Research & Management Science, vol. 114, 3rd edn. Springer, New York (2008). https://doi.org/10.1007/978-0-387-74388-2
14. Van't Veer, L.J., et al.: Gene expression profiling predicts clinical outcome of breast cancer. Nature **415**(6871), 530 (2002)

A Quantitative and Comparative Analysis of Edge Detectors for Biomedical Image Identification Within Dynamical Noise Effect

Dominik Vilimek(✉), Kristyna Kubikova, Jan Kubíček, Daniel Barvik, Marek Penhaker, Martin Cerny, Martin Augustynek, David Oczka, and Jaroslav Vondrak

VSB-Technical University of Ostrava, FEECS, K450, 17. Listopadu 15, Ostrava-Poruba, Czech Republic
{dominik.vilimek,kristina.kubikova.st,jan.kubicek,daniel.barvik, marek.penhaker,martin.cerny,martin.augustynek,david.oczka, jaroslav.vondrak}@vsb.cz

Abstract. Image processing plays a key role in many medical imaging applications, by automation and making delineation of regions of interest more simple. The paper describes image processing such as image properties, noise generators and edge detectors. The work deals with methods of edge detection in biomedical images using real data sets. The aim of this work are experiments providing information about the detector noise resistance. Another aim is own implementation of selected edge detection operators and an application on different types of data created by magnetic resonance imaging and computed tomography. Theoretical and experimental comparisons of edge detectors are presented.

Keywords: Image processing · Edge detection · Edge detectors · MATLAB

1 Introduction

Vision is one of the basic senses of humans and is an important tool for the perception of the outside world. Through our eyes, our brain receives a lot of information. People have always tried to keep this information in a form that is portable. With the advent of digitalization and the advancement of technology, the field of image processing has experienced a huge boom. Nowadays, the image is created mostly using techniques such as cameras, but also industrial or medical devices.

Image processing is widely used in the field of biomedicine. There are many devices that store information in the form of an image (i.e. X-rays, CT, microscopes, thermovision and many others). The quality and readability of such images may be critical to get correct diagnosis, which may have a direct impact

N. T. Nguyen et al. (Eds.): ACIIDS 2020, LNAI 12034, pp. 90–101, 2020.
https://doi.org/10.1007/978-3-030-42058-1_8

on patient health. The quality of these images is determined by a large number of aspects, from imperfections of the acquisition device to environmental influences and so on [2,10,18].

Edge detection is one of the methods that can be applied to and image. It is one of the most basic techniques for detecting objects in an image. It's widely used i.e. in image analysis, pattern finding, and computer vision. Edge detection is used to enhance the edges of an image, through separating important elements in the image from the background and noise. Edge detection highlights important elements in the image. There are many methods for edge detection called as edge detectors. In purpose of this work, detectors approximating the first and second derivatives are described as Sobel, Roberts, Prewitt, Canny, Laplacian and Zerocross detector. A method using fuzzy logic is described as a representative of unconventional methods [2,20].

The aim of this paper is the implementation of experiments, which will allow us to determine the most suitable method of edge detection for various noise in the image. In the individual chapters the theoretical introduction to edge detectors will be gradually described. Then the own implementation of algorithms and the evaluation of results will be represented.

2 Biomedical Applications of Edge Detectors

The edge can be defined as a place where the brightness function of adjacent pixels changes significantly. It's given by the properties of the image element and its surroundings. It describes the rate of change and direction of the largest growth of the image funcion. The edge detection algorithm can be used for object detection and classification. Edge detection, along with noise reduction, brightness and geometric transformations, is one of the basic image operations. An edge is a place in a image that exhibits a high spatial frequency such as rapid change in luminance function. However, each real image contains non-zero amount of noise, which behaves locally same as edge. Partial derivatives are used to calculate the function change [14,17].

There are four basic models of edges called step, roof, ramp and thin lines, these models can be described as ideal (see Fig. 1). In real images only noisy edges can be found. The reason is that in real images the brightness of the image changes gradually, not in a step. There are many edge detectors that use different edge detection methods, as was mentioned in the previous section. Basically, edge detectors can be divided into detectors using the first derivation or second derivation of the luminance function.

Image segmentation, which also includes edge detection, plays a key role in many medical imaging applications. It automates or facilitates the definition of anatomical structures and other areas of interest. Edge detection can be used for example in panoramic dental X-ray where it is actually widely used by dentists because it simplifies diagnosis of the jaw, teeth and soft tissues. A panoramic image requires image processing that could produce a sharper and clearer image or improve the quality of the information contained in the image. Teeth segmentation from panoramic X-ray images can be used for tooth extraction and can

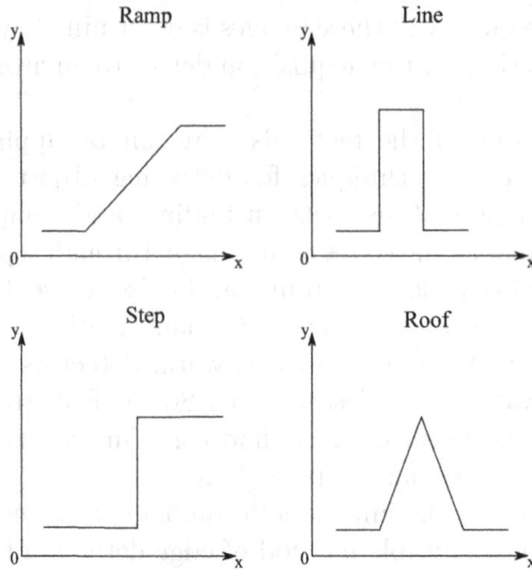

Fig. 1. One-dimensional edge profiles.

play a key role in early diagnosis and treatment that could help the dentist make better decisions. [12,15,21].

Another use of edge detection in biomedical images may be tumor detection at an early stage. There are many types of different tumors that can occur in the human body. One of possibilities to detect tumors can be by positron emission tomography (PET), computed tomography (CT) and magnetic resonance imaging (MRI) [1,19]. For example authors in their study [1,9] used a proposed algorithm for seven MRI images. There were also evaluated results for Sobel based method and an improved version of Sobel edge detection for brain tumor segmentation of MR images. The authors mentioned the need of improvement in enlargement of the region of interest (ROI) and reduce the thickness of the area border lines. Authors specified in the study [7] that edges are of critical importance to the visual appearance of images. It is desirable to preserve important features during denoising process. This study analyze the effectiveness of Sobel, Roberts, Prewitt, Canny, Laplacian and Zerocross methods. These methods will be compared with each other on different datasets. Paper contributes with a comparasion of all methods to chose the most resistant one for a dynamic noise change.

3 Used Methods of Edge Detectors

To analyze the effectiveness of individual methods, several steps are needed to do for allowing us to compare them. Tested methods include: Roberts, Prewitt, Sobel, Laplacian, Canny, and Zerocross edge detectors. These methods will be compared with each other on different types of datasets with different types of implemented noise. The main goal of the testing is used to compare these methods and their resistance to noise.

In this section the implemented tested detectors will be described.

3.1 Roberts Edge Detector

It is one of the oldest edge operators, designed by Lawrence Roberts in 1965. It is also one of the simplest operators. Roberts edge detector uses a matrix of 2×2 adjacent pixels to determine differences in two orthogonal diagonal directions. The main disadvantage is the high sensitivity to noise as it uses a small number of pixels to estimate the gradient [21,22]. The edge size at (x, y) is calculated by following formula [21]:

$$f(x,y) = \sqrt{(f(x,y) - f(x+1,y+1)^2) + (f(x+1,y) - f(x,y+1))^2} \tag{1}$$

where (x, y) indicates the coordination of the input data.

The form of convolution mask could be defined as:

$$k_1 = \begin{bmatrix} 1 & 0 \\ 0 & -1 \end{bmatrix} \qquad k_2 = \begin{bmatrix} 0 & 1 \\ -1 & 0 \end{bmatrix} \tag{2}$$

3.2 Prewitt Edge Detector

The gradient is estimated for eight directions and uses a 3×3 pixel convolution mask or it can also have larger dimensions, for example 5×5. Regularly the mask with the largest gradient module is selected. The derivative is calculated by the mean in the x, y direction, which contributes positively to the noise reduction. For calculation of the edge size at point (x, y) is used image function values in all surrounding pixels, which for brevity we mark as [5,22]:

$$k_1 = \begin{bmatrix} A & B & C \\ D & f(x,y) & F \\ G & H & I \end{bmatrix} \tag{3}$$

The edge size can be calculated by using formula:

$$f_x(x,y) = \frac{1}{3}((C-A) + (F-D) + (I-G)) \tag{4}$$

$$f_y(x,y) = \frac{1}{3}((A-G) + (B-H) + (C-I) \tag{5}$$

where $f_x(x,y)$ and $f_y(x,y)$ are final results of edge size in x and y direction. Convolutional masks have the following form:

$$k_1 = \begin{bmatrix} 1 & 1 & 1 \\ 0 & 0 & 0 \\ -1 & -1 & -1 \end{bmatrix} \qquad k_2 = \begin{bmatrix} 0 & 1 & 1 \\ -1 & 0 & 1 \\ -1 & -1 & 0 \end{bmatrix} \qquad k_3 = \begin{bmatrix} -1 & 0 & 1 \\ -1 & 0 & 1 \\ -1 & 0 & 1 \end{bmatrix} \tag{6}$$

3.3 Sobel Edge Detector

This operator is similar to the Prewitt edge detector, except that it uses a weighted average to calculate derivatives. Generally is used for the detection of horizontal and vertical edges. The edge size at (x, y) could be calculated as [5]:

$$f_x(x,y) = \frac{1}{4}((C-A) + 2(F-D) + (I-G)) \tag{7}$$

$$f_y(x,y) = \frac{1}{4}((A-G) + 2(B-H) + (C-I) \tag{8}$$

where $f_x(x,y)$ and $f_y(x,y)$ are final results of edge size in x and y direction. Convolutional masks have the following form:

$$k_1 = \begin{bmatrix} 1 & 2 & 1 \\ 0 & 0 & 0 \\ -1 & -2 & -1 \end{bmatrix} \quad k_2 = \begin{bmatrix} 0 & 1 & 2 \\ -1 & 0 & 1 \\ -2 & -1 & 0 \end{bmatrix} \quad k_3 = \begin{bmatrix} -1 & 0 & 1 \\ -2 & 0 & 2 \\ -1 & 0 & 1 \end{bmatrix} \tag{9}$$

3.4 Laplacian Operator

The operator approximates the second derivative. The operator approximates the second derivative and can only indicates the size of the edges, but not its direction, so its suitable only when we can detect the edge. The disadvantages include high noise sensitivity and double responses to thin lines in the image. The Laplace operator could be defined by the relation [8,12]

$$\nabla^2 f(x,y) = \frac{\partial^2 f(x,y)}{\partial x^2} + \frac{\partial^2 f(x,y)}{\partial y^2} \tag{10}$$

If the image function is discrete, we use the difference instead of the derivatives:

$$\nabla^2 f(x,y) = f(x+1,y) + f(x-1,y) + f(x,y-1) - 4f(x,y) \tag{11}$$

Convolutional masks have the following form:

$$k_1 = \begin{bmatrix} 0 & 1 & 0 \\ 1 & -4 & 1 \\ 0 & 1 & 0 \end{bmatrix} \quad k_2 = \begin{bmatrix} 1 & 1 & 1 \\ -1 & -8 & 1 \\ 1 & 1 & 1 \end{bmatrix} \tag{12}$$

3.5 Canny Edge Detector

It was published in 1986 by Canny [4]. It can be referred to as an ideal edge detector. It was designed to meet three basic criteria: Minimum error rate, localization with minimal difference between actual and detected edge position and evident response. Edge detection using Canny operator is realized by noise elimination, usually using Gaussian filter, gradient determination with suitable convolution mask and searching for local maximum. Another important parameter is elimination of insignificant edges cause the last step is thresholding by hysteresis.

Hysteresis is used to track remaining pixels that have not been suppressed. Hysteresis uses two thresholds, if the gradient size is below the first threshold, it is set to zero, and if it is above the second, higher threshold, it creates an edge automatically [13,16].

4 Design of Edge Detection Efficiency on the Intensity of Noise

The experiments monitor the efficiency of edge detection on the dynamic change of the noise intensity. The data set was acquired from hospital Podlesi Trinec and contain 2812 images. Data set was devided into four parts. The first dataset cointains total of 126 tested images created by CT. These images shows a calcification of the blood vessels. The second dataset contain only 31 images of the liver created by CT scanner. Another dataset include 18 images of articular cartilage created by MRI and the biggest dateset contain total of 2636 images with blood vessels created again by MRI. As the datasets contained images in RGB format for the further processing, conversion to monochrome format was necessary. The image conversion was done in MATLAB using the rgb2gray function. This function removes hue and saturation information while preserving brightness. The function converts RGB values to grayscale values by creating a weighted sum of R, G, B components by following formula:

$$x = 0.299 \cdot R + 0.587 \cdot G + 0.114 \cdot B \tag{13}$$

The varied type of noise is applied on converted images. In this study a Gaussian, Salt and Pepper, Speckle and Localvar type of noise is chosen. Description of the noise can be found i.e. in studies [3,6,11].

The Fig. 2 shows an example of one experiment.

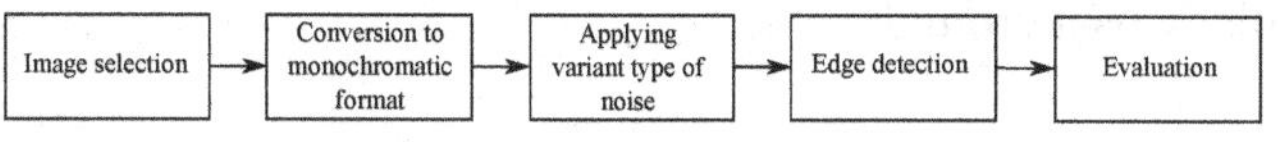

Fig. 2. Flowchart of the experiment.

4.1 Implementation of Edge Operators

In order to compare the efficiency of the edge detection function, the impleman-tion of Sobel, Roberts and Prewitt edge operators was created. The implementation was performed using the native MATLAB convolution function, *conv*2, and also with own implementation of this function. The own implementation of the convolution is shown in the Fig. 3. The input for the convolution is the image A in the form of a matrix and the convolutionary core. Subsequently, the size of the convolution core is determined and the application come in two cycles

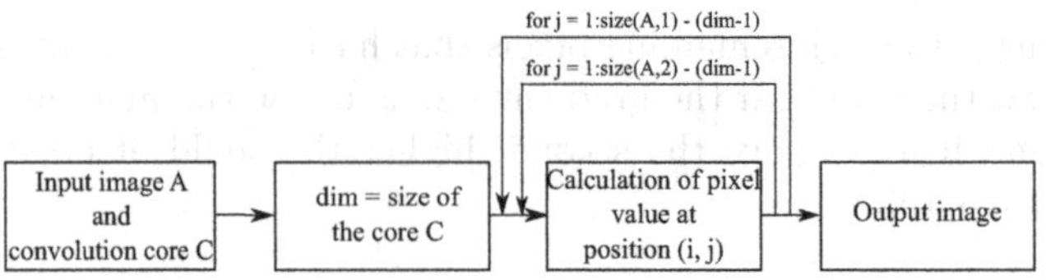

Fig. 3. Flowchart for own convolution.

that pass through the image matrix. Inside the cycle, the value is calculated at coordinates i and j.

If the native MATLAB function is used for convolution, the G_x and G_y values are calculated using the *conv*2 function, it is necessary to define convolution masks. In the case of the own implementation of the convolution, the size of the first derivation for the x and y axis for the Sobel method is calculated according to the following formula:

$$G_x = ((I(x+2,y)+2*I(x+2,y+1)+I(x+2,y+2))-(I(x,y)+2*I(x,y+1)+I(x,y+2)) \tag{14}$$

$$G_y = ((I(x,y+2)+2*I(x+1,y+2)+I(x+2,y+2))-(I(x,y)+2*I(x+1,y)+I(x+2,y)) \tag{15}$$

Where I is input image. Then the gradient size is calculated by formula:

$$|\nabla f(x,y)| = \sqrt{G_x^2 + G_y^2} \tag{16}$$

The result is a binary matrix where 1 corresponds to edge points and 0 corresponds to a black background.

5 Results and Evaluation

In this section the results and methods for evalution of effectivity and robutness are presented. Figure 4 shows a example of the tested dataset. Different types of edge detectors were applied to these data, where we observed the efficiency of individual edge operators with dynamic increasing intensity of noise intensity. To compare the similarity between the output noisy image with detected edges and input image we choose two evaluation techniques. Mean squared error and the correlation.

Mean Squared Error (MSE)
MSE is the mean value of the squared differences between two variables, in this case the difference between the values of the matched pixels of a noisy image

with detected edges and input image with detected edges without noise. Both images must be the same size for calculation. The formula could be describe as:

$$MSE = \frac{1}{n}\sum_{i=1}^{n}(y_i - y_i') \tag{17}$$

where n is the number of pixels in the image, y_i is the original image with detected edges and y_i' is image with detected edges and corrupted by noise.

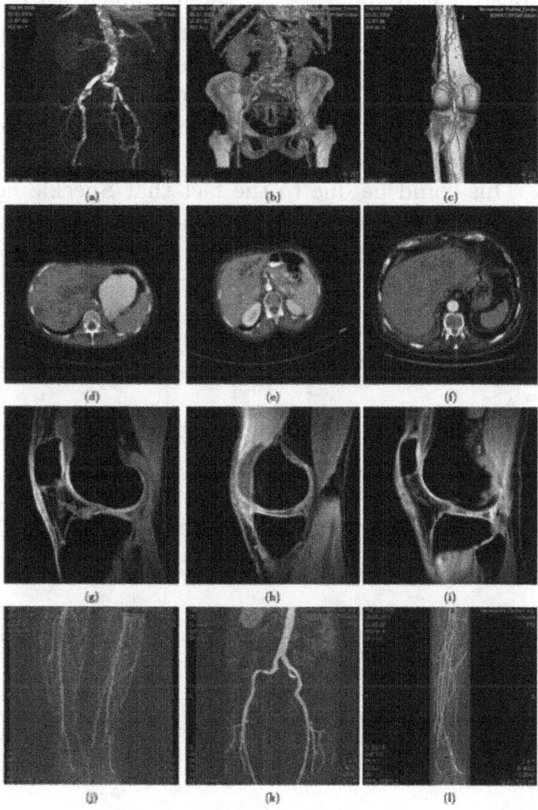

Fig. 4. Example of tested dataset (a)–(c) CT images of vascular calcification, (d)–(f) CT images of liver, (g)–(i) MR images of articular cartilage, (j)–(l) MR images of blood vessels.

Correlation (Corr)

Correlation is statistical technique that shows relationship between two variables. The main correlation results is called the correlation coefficient and its range is −1 to 1. If the correlation coefficient approaches closer to 1 or −1 than the similarity of two variables will be greater but if will be near to zero than the similarity will be lower. Correlation can be calculated as:

$$corr(A,B) = \frac{\sum_m \sum_n (A_{mn} - \bar{A}) \cdot (B_{mn} - \bar{B})}{\sqrt{(\sum_m \sum_n (A_{mn} - \bar{A}))^2 \cdot (\sum_m \sum_n (B_{mn} - \bar{B}))^2}} \tag{18}$$

where A and B are the compared images, $\bar{A}$, $\bar{B}$ are the mean values of these images and m nad n are the row and column numbers of the individual matrices.

The results below were obtained by averaging the resulting values of individual images for one or more datasets, a reference image set containing 17–30 images was selected from each dataset, according to the dataset's capabilities. Figure 5 shows comparison of Sobel operator behavior on different types of data. As can be seen in Fig. 5 for Localvar, Salt and Pepper and Gaus noise the MSE does not change significantly on different types of data and trend is almost the same. Conversely, using Speckle noise, the MSE changes significantly on different types of data. This could be due to the fact that Speckle noise adds noise only to the image objects, not to the background, so it depends on how much background area the object obscures, which changes as the dataset changes.

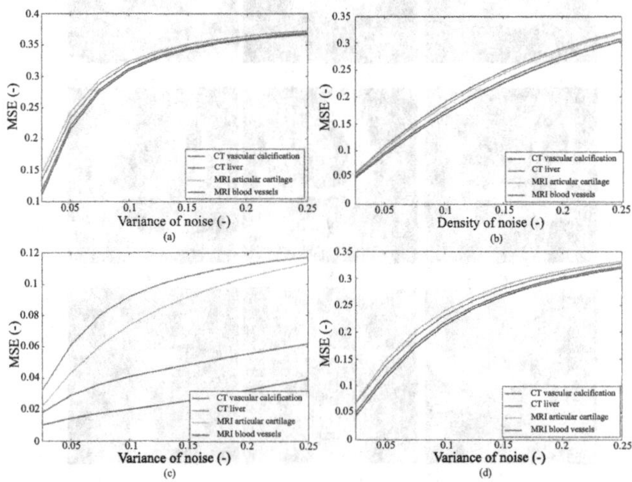

Fig. 5. Progress of MSE on different types of data for the Sobel method with threshold 0.1 corrupted by noise (a) Gaussian, (b) Salt and Pepper, (c) Speckle, (d) Localvar.

Graphs below Fig. 6 show the correlation change of images with the edges detected by the Laplacian method when changing the data set with different type of noise. The smallest correlation change occur with changing the data set of blood vessels images. These are very similar datasets, which could explain the similarity of the development of their correlation curves.

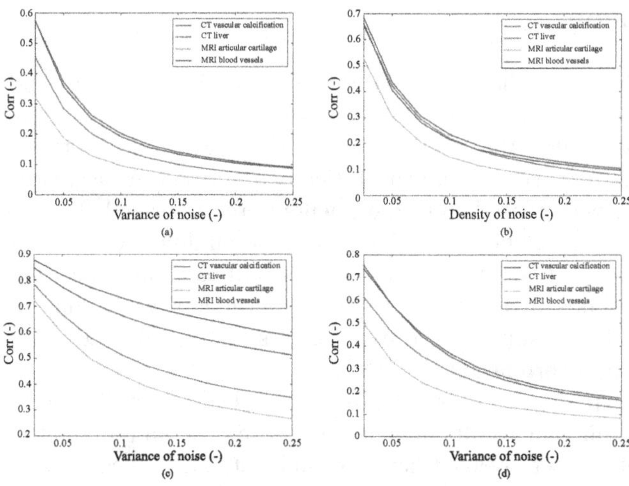

Fig. 6. Progress of Correlation on different types of data for the Laplacian method with threshold 0.01 corrupted by noise (a) Gaussian, (b) Salt and Pepper, (c) Speckle, (d) Localvar.

To determine the best operator, the results for each method were averaged across the threshold and parameter intervals of all noise generators used. The correlation results are shown in the following Table 1.

Table 1. Table of average correlation values.

Operator	CT vessels	CT liver	MRI a. cartilage	MRI vessels	overall average
Sobel	0.2184	0.1774	0.1133	0.2085	0.1794
Roberts	0.2205	0.1484	0.1020	0.2121	0.1707
Prewitt	0.2128	0.1737	0.1100	0.2042	0.1752
Canny	0.2401	0.2268	0.1704	0.2288	0.2165
Laplacian	0.3962	0.3088	0.2436	0.4028	0.3379
Zero Cross	0.3962	0.3086	0.2437	0.4028	0.3378

The Zero Cross and Laplacian methods achieved the best results, but these methods use a different threshold interval. Therefore, we evaluate the best method using the approximation of the first derivative for which the same threshold interval was used. Of these methods, based on average correlation, the Canny method gives the best results.

6 Discussion

This paper describes the basic introduction to edge detectors and important terms in this topic. It also describes the basic principles of image segmentation, which further include the main theme of this work, edge detection. Selected edge detectors were studied and theoretically described. A database of MRI and CT image data was created and experiments were performed. These experiments tracking changes in the behavior of each method on different datasets, sensitivity of each method to different type of noise as well as comparing the efficiency of individual methods in noisy image. All experiments were done in MATLAB environment. The results of the experiments were evaluated using the mean squared error and correlation.

Based on experiments It is not possible to clearly identify the best method. Methods give different results depending on the type and amount of noise. Different types of dataset does not have a significant effect on edge detection results if we will be considering a speckle noise. Gaussian noise has the greatest consequence on edge detection quality, and Speckle has the least effect.

One of the potential improvements or future enhancements would be implementation of other advanced methods such as Multiple Morphological Gradient (MMG), Otsu, or wavelet transform. Experiments could easily be applied to other datasets in the future and help for segmentation of ROI more precisely.

Acknowledgment. The work and the contributions were supported by the project SV450994/2101Biomedical Engineering Systems XV'. This study was also supported by the research project The Czech Science Foundation (GACR) 2017 No. 17-03037S Investment evaluation of medical device development run at the Faculty of Informatics and Management, University of Hradec Kralove, Czech Republic. This study was supported by the research project The Czech Science Foundation (TACR) ETA No. TL01000302 Medical Devices development as an effective investment for public and private entities.

References

1. Aslam, A., Khan, E., Beg, M.S.: Improved edge detection algorithm for brain tumor segmentation. Proced. Comput. Sci. **58**, 430–437 (2015)
2. Berezsky, O., Pitsun, O., Verbovyy, S., Datsko, T., Bodnar, A.: Computer diagnostic tools based on biomedical image analysis. In: 2017 14th International Conference The Experience of Designing and Application of CAD Systems in Microelectronics (CADSM), pp. 388–391. IEEE (2017)
3. Boyat, A.K., Joshi, B.K.: A review paper: noise models in digital image processing. arXiv preprint arXiv:1505.03489 (2015)
4. Canny, J.: A computational approach to edge detection. IEEE Trans. Pattern Anal. Mach. Intell. **6**, 679–698 (1986)
5. Chaple, G.N., Daruwala, R.D., Gofane, M.S.: Comparisions of robert, prewitt, sobel operator based edge detection methods for real time uses on FPGA. In: 2015 International Conference on Technologies for Sustainable Development (ICTSD), pp. 1–4. IEEE, Mumbai, February 2015. https://doi.org/10.1109/ICTSD.2015.7095920

6. Farooque, M.A., Rohankar, J.S.: Survey on various noises and techniques for denoising the color image. Int. J. Appl. Innov. Eng. Manage. (IJAIEM) **2**(11), 217–221 (2013)
7. Jain, P., Tyagi, V.: A survey of edge-preserving image denoising methods. Inform. Syst. Front. **18**(1), 159–170 (2016). https://doi.org/10.1007/s10796-014-9527-0
8. Kovalevsky, V.: Edge detection. In: Modern Algorithms for Image Processing, pp. 87–99. Apress, Berkeley (2019)
9. Kubicek, J., Penhaker, M., Augustynek, M., Cerny, M., Oczka, D.: Segmentation of articular cartilage and early osteoarthritis based on the fuzzy soft thresholding approach driven by modified evolutionary abc optimization and local statistical aggregation. Symmetry **11**(7), 861 (2019)
10. Kubicek, J., Tomanec, F., Cerny, M., Vilimek, D., Kalova, M., Oczka, D.: Recent trends, technical concepts and components of computer-assisted orthopedic surgery systems: a comprehensive review. Sensors **19**(23), 5199 (2019)
11. Kubicek, J., et al.: Prediction model of alcohol intoxication from facial temperature dynamics based on k-means clustering driven by evolutionary computing. Symmetry **11**(8), 995 (2019)
12. Lakhani, K., Minocha, B., Gugnani, N.: Analyzing edge detection techniques for feature extraction in dental radiographs. Perspect. Sci. **8**, 395–398 (2016). https://doi.org/10.1016/j.pisc.2016.04.087
13. Maini, R., Aggarwal, H.: Study and comparison of various image edge detection techniques. Int. J. Image Process. (IJIP) **3**(1), 1–11 (2008)
14. Marr, D., Hildreth, E.: Theory of edge detection. Proc. Roy. Soc. London Ser. B Biol. Sci. **207**(1167), 187–217 (1980)
15. Naam, J., Harlan, J., Madenda, S., Wibowo, E.P.: The algorithm of image edge detection on panoramic dental X-Ray using multiple morphological gradient (MMG) method. Int. J. Adv. Sci. Eng. Inform. Technol. **6**(6), 1012–1018 (2016)
16. Nikolic, M., Tuba, E., Tuba, M.: Edge detection in medical ultrasound images using adjusted Canny edge detection algorithm. In: 2016 24th Telecommunications Forum (TELFOR), pp. 1–4. IEEE, Belgrade, November 2016
17. Pham, D.L., Xu, C., Prince, J.L.: Current Methods in Medical Image Segmentation. Annu. Rev. Biomed. Eng. **2**(1), 315–337 (2000). https://doi.org/10.1146/annurev.bioeng.2.1.315
18. Rangayyan, R.M.: Biomedical Image Analysis. CRC Press, Boca Raton (2004)
19. Saini, P.K., Singh, M.: Brain tumor detection in medical imaging using MATLAB. Int. Res. J. Eng. Technol. **2**(02), 191–196 (2015)
20. Shrivakshan, G.T., Chandrasekar, C.: A comparison of various edge detection techniques used in image processing. Int. J. Comput. Sci. Issues (IJCSI) **9**(5), 269 (2012)
21. Sonka, M., Hlavac, V., Boyle, R.: Image processing, analysis, and machine vision, 3rd edn. Thompson Learning, Toronto (2008). oCLC: ocn123776599
22. Spontón, H., Cardelino, J.: A review of classic edge detectors. Image Process. Line **5**, 90–123 (2015). https://doi.org/10.5201/ipol.2015.35

Monitoring and Statistical Evaluation of the Effect of Outpatient Pleoptic-Orthoptic Exercises

Pavla Monsportova[1], Martin Augustynek[1], Jaroslav Vondrak[1(✉)], Jan Kubíček[1], Dominik Vilimek[1], Jaromír Konecny[1], and Juraj Timkovic[2]

[1] FEECS, VSB-Technical University of Ostrava, K450, 17. Listopadu 15, Ostrava-Poruba, Czech Republic
{pavla.monsportova,martin.augustynek,jaroslav.vondrak, jan.kubicek,dominik.vilimek,jaromir.konecny}@vsb.cz

[2] Clinic of Ophthalmology, University Hospital Ostrava, Ostrava, Czech Republic
juraj.timkovic@vsb.cz

Abstract. The subject of this work is to create an electronic database and its graphical interface with subsequent evaluation of the effect of outpatient pleoptic-orthoptic exercises. The resulting software stores data in a database, works with already stored data and is able to generate time trends for statistical evaluation. The database was created using the relational database system Firebird, which was subsequently connected to the Microsoft Visual Studio 2015 development environment through the C# programming language. The aim of this work is to facilitate the work of orthoptic nurses during outpatient pleoptic-orthoptic exercises in the Center for Children with Visual Impairment at the University Hospital in Ostrava. There is an analysis of pleptic-orthoptic exercises and binocular vision disorders, the treatment of which is supported by these exercises. Furthermore, the creation of electronic database and its graphical interface is introduced.

Keywords: Electronic database · Ophthalmology · Pleoptic-orthoptic exercises · Squint · Blindness

1 Introduction

Many specialist articles and books deal with the issue of diagnosis and treatment of myopia and squinting. Work on this issue expresses the view that the performance of pleoptico-orthoptic exercises contributes to a faster improvement of visual deficits and thus more effective treatment of blindness by occlusive method, atropine and spectacle correction.

The fact that preschool children with visual impairment is more likely to attend nursery schools for the visually impaired than regular nursery schools, because in these specialized facilities more attention is given to children and they are significantly improved by the use of simple exercises [1].

In particular, early diagnosis and appropriate follow-up treatment are extremely important for treatment success [2, 3]. It has also been shown that parents play an

N. T. Nguyen et al. (Eds.): ACIIDS 2020, LNAI 12034, pp. 102–113, 2020.
https://doi.org/10.1007/978-3-030-42058-1_9

important role during treatment, as they have a significant influence on their children and thus on adherence to treatment. The motivation to treat both children and their parents is also important.

1.1 Simple Binocular Vision and Its Disorders

Simple binocular vision (SBV) means seeing simultaneously with both eyes. It is the ability to see a fixed object in duplicate. It is the coordinated sensorimotor activity of both eyes. SBV is not congenital and develops together with the retina of the eye until the child is one year of age and consolidates within about six years [4, 5].

Binocular vision is divided into three levels: superposition, fusion, spatial vision (stereopsis). Superposition is the possibility to simultaneously see the retina of both eyes. This vision is examined and then trained on a troposcope, where each eye is presented with two different images, for example a lion and a cage. The patient should see the lion in the cage. There is an overlap, superposition of two adjacent images and a single impression. If the patient sees only one picture, it is a decrease. The essence of fusion is to create one perception by combining the same images from the right and left eye. Stereopsis, or sensory fusion, is the creation of a deep perception by combining multiple images. Their fusion creates a three-dimensional perception [5–7].

1.2 Pleoptics

Pleoptical therapy is based on training the vision of the amblyopic eye by eliminating the healthy eye by occlusion, which is mediated by a patch, occluder on glasses, contact lens, or pharmacological dripping of the pupil. With the increasing age of the child, the restitution of visual functions slows down and thus the probability of their complete recovery decreases. An occluder is a label made of an opaque material (plastic, cloth) that attaches to the glasses in front of a healthy eye, forcing the amblyopic eye to work, leading to its correction [8].

To enhance the success of the treatment of amblyopia, i.e. to create the right monocular vision, pleoptical exercises are aimed at active exercises of the blind eye, in which the healthy eye is completely obscured by occlusion. Exercises are based on close proximity using touch, hearing and memory. The speed of improvement in visual acuity depends on the degree of eye strain [4].

1.3 Orthoptics

Orthopedic therapy is based on the practice of eye cooperation and spatial vision. Orthopedics is usually performed at specialized workplaces using orthoptic devices. Its essence is the treatment of squint, which consists in correcting and improving SBV in parallel position of both eyes without the use of occlusal. So, both eyes are trained [4, 9, 10].

Spectacular correction, surgery, pleoptic and orthoptic exercises are usually necessary to achieve the goals of strabismus treatment [8].

2 Web Interface Design for Evaluation of Pleoptic-Orthoptic Exercises

The requirement to create a theoretical design of the web interface came from the medical staff of the Center for Children with Visual Impairment at the University Hospital in Ostrava. The web interface design is designed to serve as a basis for its future implementation. The web interface will provide information to the third person on the progress of each session the patient has undergone. The "third" person is the child's legal representative. The most important part of the whole interface will be a statistical evaluation of the effect of pleoptic-orthoptic exercise in the form of time trends, which will show whether during the treatment distance vision, near vision and squint deviations improved or worsened [5].

2.1 Web Interface Design

The web interface will be used to access the "third" person to the patient and session information. It will be important to differentiate the approach of the patient's legal guardian and the medical staff, as each person will receive different powers in the system. The healthcare professional (doctor, nurse) will be able to intervene unlimitedly in all blocks of the system, while the patient will have been able only to read data and make requests to change data or change the session plan. The web interface design is shown in Fig. 1.

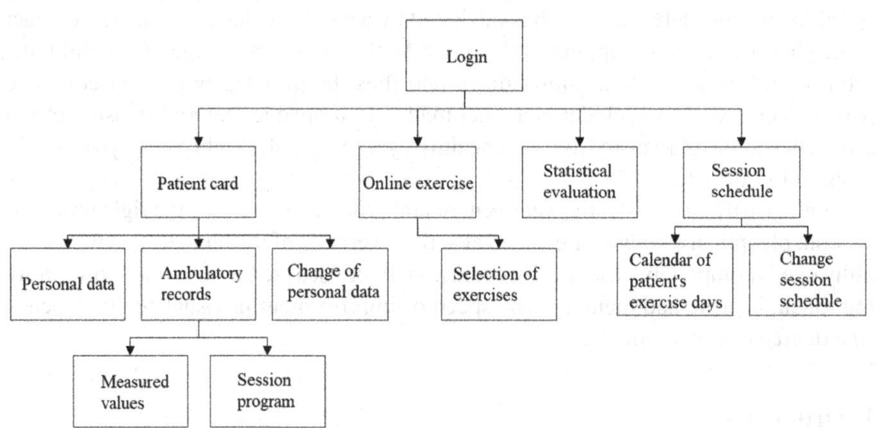

Fig. 1. Web interface block diagram.

The *Patient Card* option will contain patient information. This data will be the same as the information in our patient database, which will be used in the practice room during each session, so it would be appropriate to link the database to this web system. Therefore, the *Patient Card* will include the following options: *Personal Data, Outpatient Records* and *Change of Personal Data.* The *personal data* will be filled in with the name, surname, health insurance company and personal identification number. *Ambulatory records* will include: *Measured values* (Squint Deviation, Distance Vision, Near Vision) required for statistical evaluation of the effect of pleoptic-orthoptic exercises and *session program* in which the nurse will record the date of the session, the start time of the exercises, and the names of the exercises the patient completed on that day. The *change of personal data* will be used to send a request for some modification, e.g. change of the patient's health insurance company. These requests will be processed by the administrator, who will be medical staff.

The *online exercise* will offer a selection of 12 exercises that the patient can perform from home. This option is crucial, as this will allow more frequent eye exercises, which should result in better treatment efficacy. The *Statistical Evaluation* option will include tables with measured values for squint, distant vision, and near vision. These values, measured on the troposcope, the Snellen and Jaegr tables, will be recorded by orthopedic nurses during pleoptical-orthoptical exercises. An integral part of the Statistical Evaluation will be graphs showing the time trends of the measured values, which will show whether there has been an improvement or deterioration of long-distance, near-vision and squint deviations.

The last option, *Session Schedule*, will include *Calendar* and *Change Session Schedule*. In the *Calendar*, the patient will record the days and times of their sessions at the Center for Children with Visual Impairment at the University Hospital in Ostrava. After consultation with the patient, the nurse writes the date and time of the session in the *Calendar*. By *changing the session schedule*, the patient will be able to change the date and time of the session [5, 11].

3 Database Design and Implementation

A database, or data base, is a place to gather and organize a variety of information, such as information about people, products, or orders. Data stored in the database is accessed via the Data Base Management System (DBMS). Commonly, database designation refers to both stored data and software (DBMS).

3.1 Database Requirements

The database requirements came from nurses and doctors of the Center for Children with Visual Impairment at the University Hospital in Ostrava. The database will be operated by orthopedic nurses present in outpatient pleoptico-orthoptic exercises. The aim is to assist orthopists in writing patient cards and subsequent evaluation of the effect of pleoptic-orthoptic exercises. The requirements were as follows:

• login to the database system	• differentiation of treatment cycles
• Search for a patient in the list by birth number and surname	• adding sessions to the treatment cycle
• adding a new patient to the database	• printing of individual sessions
• removing the patient from the list	• statistical evaluation of distance vision in the form of graphs
• printing of patient's personal data	

3.2 Implementation of the Database

From many existing database systems, only three were selected to create the database - Oracle, PostgreSQL, Firebird. The properties of these database systems were investigated, in particular the possibility of connecting the database to the selected Microsoft Visual Studio 2015 development environment.

Based on the requirements in Sect. 3.1, a logical schema of the database was created, consisting of three tables - Patient, Cycle and Session. Subsequently, individual tables were created in the relational database system Firebird.

The Patient table contains 10 attributes: patient_id, RC, firstname, surname, ZP, note, orthoptic_nurse, correction_RE, correction_LE and status. The primary key of this table is the patient_id attribute. This is a number that is automatically assigned to each patient when it is stored in the database. Each patient has a different patient_id and it is not possible that two patients have the same ID. The orthoptic_nurse attribute records the name and surname of the orthoptist who has saved the patient in the database, thus becoming responsible for the patient's card. The status attribute, which determines whether a patient is displayed in the Patient List in the resulting software or not, plays a special role. The status is only numeric 1 and 0. The status is automatically set to 1 when the patient is stored in the database. This leads to the patient being displayed in the Patient List. However, if the status value is 0, the patient does not appear in the Patient List, but remains in the database (Table 1).

Table 1. Patient table.

The name of the attribute	Data type	Key	Note
patient_id	integer	PK	Patient identification number
RC	varchar		Patient's birth number
Firstname	varchar		Patient's name
Surename	varchar		Patient's last name
ZP	integer		The patient's health insurance number
Note	varchar		Note to the patient
orthoptic_nurse	varchar		The name of the orthopedic nurse responsible for the patient card
correction_RE	varchar		Correction value on the right eye
correction_LE	varchar		Correction value on the left eye
Status	integer		The attribute is only 1 and 0

The Cycle table includes only 4 attributes: cycle_id, patient_id, queue_cycle, and date. The primary key is the cycle_id attribute, and it has the same rules as the patient_id in the Patient table. It only takes numerical values and each cycle has a different cycle ID. In this table, the Patient_id foreign key that links the Cycle table to the Patient table. The attribute, queue_cycle takes numerical values and is used to distinguish individual treatment cycles that a patient undergoes, since one patient can have multiple treatment cycles.

Table 2. Cycle table.

The name of the attribute	Data type	Key	Note
Cycle_id	integer	PK	Treatment cycle identification number
Patient_id	integer	FK	Foreign key from the Patient table
Queue_cycle	integer		Treatment cycle order
date	varchar		Date of the first treatment cycle session

The Session table contains 46 attributes. The primary key is Session_ID, which again has only numeric values and each session has its session_ID. The foreign key is the ID_cycle attribute from the Cycle table, which guarantees a link between these tables. The appropriate attributes are used to describe the session described for each patient.

4 Design and Implementation of Graphical User Interface

A graphical user interface (GUI) is a collection of windows, buttons, icons, forms, sliders, and text fields that the user controls a computer to obtain the necessary output from programs. However, the GUI is not only used in computers, but generally in all devices with a display or monitor.

4.1 Graphical User Interface Requirements

Requirements for the GUI were also made by orthopedic nurses and doctors of the Center for Children with Visual Impairment at the University Hospital in Ostrava. Emphasis was placed on:

• Simplicity, clarity and logical arrangement of windows	• Possibility to select the name of individual exercises from the drop-down menu
• Possibility to select patient and treatment cycle by double-clicking on the row in the table	• Possibility to print patient information and individual sessions
• Display the surname, first name and birth identification number of the selected patient in the title bar of each window	• Statistical evaluation of the effect of pleoptic-orthoptic exercise in the form of time trends
• Show graphs for the right eye in red and for the left green in graphs	

4.2 Realization of GUI

First, a block diagram of the graphical user interface (Fig. 2) was created that captures all the tasks that can be performed in the resulting software. Subsequently, the individual forms were gradually designed and linked to the created database in the relational database system Firebird. The resulting software was called OrtopEye. The current version of the product is the first version. Over time, it is expected that the software will be improved and developed in the next releases, as user comments and suggestions are expected from long-term use of OrtopEye 1.0.

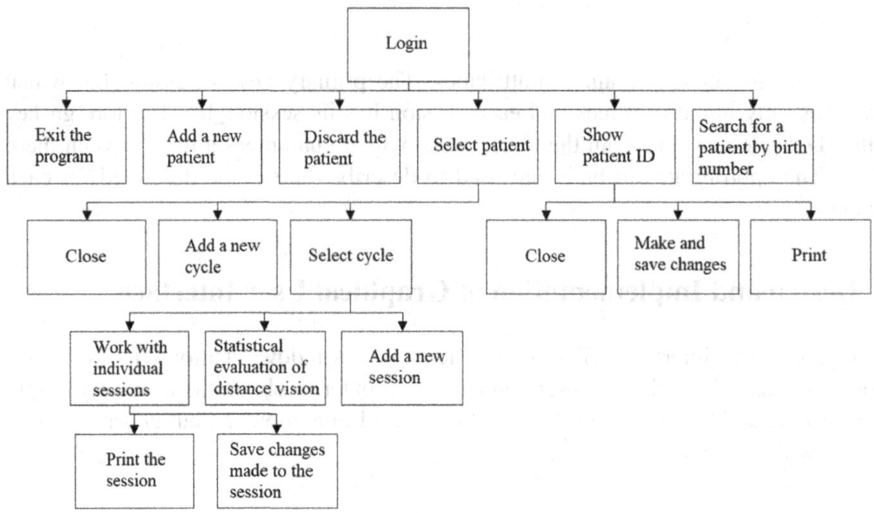

Fig. 2. GUI block diagram.

When OrtopEye 1.0 starts, the Login window is displayed. In this window the user enters Login or user name and Password. Login is based on a comparison of two character strings - a user-entered Login with a fixed Login and a user-entered Password with a fixed Password. If the user enters an incorrect Login or Password, a window is displayed informing you of this fact. If the credentials are correctly entered, the user is given access to the OrtopEye program and the Patient List window is displayed, which is the main program window. From here it is possible to:

- Terminate the program,
- Search for a patient by birth number or surname,
- Display patient identification data,
- Choose a patient,
- Add a new patient,
- Remove the patient from the list.

The majority of the window (Fig. 3) occupies the Patient List, which is retrieved from the database from the Patient table. Patient search is performed by filtering the displayed Patient List by birth number or surname that the user enters in the text box above the Patient List.

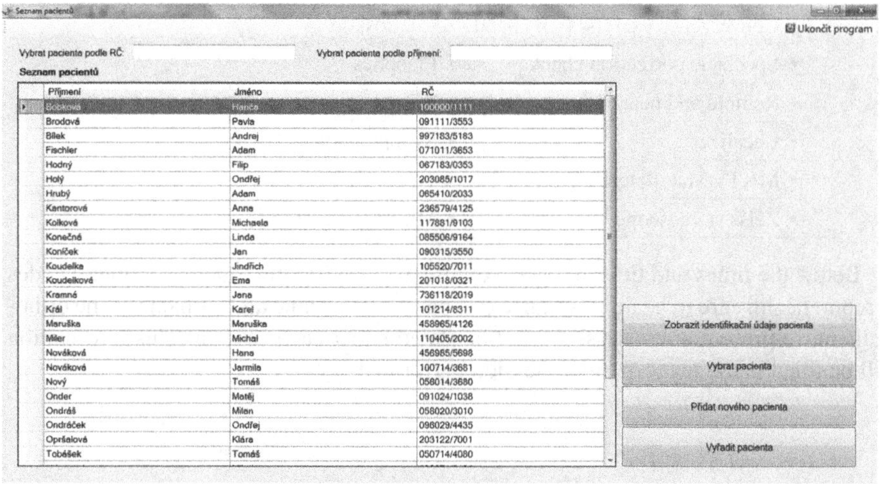

Fig. 3. The main window of the program.

It is also possible to add new patients to the database. Clicking the Add New Patient button opens the New Patient window where you need to fill in all necessary information such as surname, first name, social security number, health insurance number, right and left eye correction, pupillary distance, occluded eye and the name of the orthopedic nurse who established the patient and is therefore responsible for their patient card. After filling in the data, it is necessary to press the Save button to save all data entered in this window into the Patient table.

Button Show patient ID is used to load personal information about the patient and new window shows. This window allows you to make changes to your personal information, print this information, and return to the Patient List window. Changes can be made by overwriting the information and then saving using the Save Changes button, which makes adjustments to the values of each attribute in the database in the Patient table. Pressing the Print button raises an event that opens a Microsoft Excel spreadsheet window that loads patient data. The user can thus check the correctness of the data again.

The Select Cycle button in the Treatment Cycle window opens the Session window. The treatment cycle can also be selected by double-clicking on the row in the table. In the header, the patient's basic data - last name, first name and birth number - are retrieved from the Patient table again. The largest part of the window is occupied by information about individual sessions in a given cycle, which are sorted in ascending order. This information is retrieved from the database from the Session table. Each session has its own number or order and the date it was held. During each session, the patient undergoes 6 exercises, whose names and the exact start times are recorded on the left side of the Session window. The user selects the names of individual exercises from the drop-down menu, which contains the following options:

• AMT - Amblyrainer	• Convergence exercises - machine
• CAM - Campbell Stimulator	• Motility and convergence exercises
• CRS - Central resolution	• Holmes' stereoscope
• Convergence exercises - by hand	• Cheiroskop
• Spectacle correction check	• Pleoptics
• Remote and near vision control	• Prisms
• Localizer	• Synoptophor
• MKT - Makulotest	• Troposkop
• Zeiss stereoscope	• Mirror exercise

Below the times and titles of each exercise is a text box for entering numeric codes to report health care to health insurance companies. The name and surname of the orthopedic nurse present at the session must be stated under each session. This information will be filled in the text box titled Session performed.

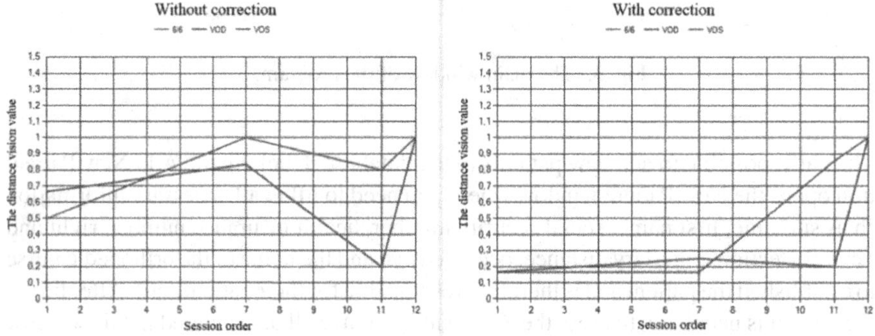

Fig. 4. Statistical evaluation of distance vision. (Color figure online)

In the right part of the Session window, values related to the current patient status are read. These values are determined using a troposcope, Snellen and Jaegr tables. The horizontal and vertical aberrations are detected on the troposcope, the value of which is positive and negative integers. When assessing spatial vision, the user enters information in the text fields for superposition, fusion, and stereopsis. Long-range vision shall be considered separately, for the right eye with and without correction, and separately for the left eye with and without correction. These values are obtained using Snellen tables and then processed into time trends. Jaeger's tables are used to assess near vision, whose values take positive real numbers. If measurements are made on a troposcope during a given session, the measured box under the troposcope heading must be checked. If distance vision is detected, the check box is checked under the heading 'vision to the far' and also similarly for detecting near vision. The Measured fields are essential for collecting data for statistical evaluation of the effect of the pleoptic-orthoptic exercise. Because only the data from the session where the Measured field for the given evaluation area is checked is included in the statistical evaluation.

The button Statistical evaluation of distant vision opens the window of the same name, where basic information about the patient - surname, first name and birth number - is retrieved from the database from the Patient table into the header. The majority of the window is occupied by two graphs (Fig. 4) showing the time trends of distance vision values measured over one treatment cycle. The graph on the left show changes in the right and left eyes when measured without correction. The graph to the right shows the changes in the right and left eyes when corrected. The constant curve at 1 is blue and shows normal vision. The red curve shows the far-sighted vision for right eye (VOD) and the green curve the far-sighted vision for left eye (VOS) [12].

5 Testing of the Resulting Software

After the final software OrtopEye 1.0 was completed, testing was carried out in the exercise room of the Center for Children with Visual Impairment at the University Hospital in Ostrava. A total of 35 patients were tested. A new patient was added, his/her personal details edited and printed. At least one treatment cycle was created for each patient and at least one session was added to each patient. The values of the squint deviation measured on the troposcope, distant vision and near vision were recorded. The training room of the Center for Children with Visual Impairment at the University Hospital in Ostrava also received training of staff, who proposed minor adjustments to the graphical user interface during testing. It was filtering the Patient List. At first, filtration was possible only by birth number. After editing, the ability to filter by last name was added. Another suggestion for improvement was to select the patient by double-clicking on the row in the Patient List. Previously, it was only possible to select a patient by using the Select Patient button, first selecting the appropriate patient in the Patient List and then pressing the Select Patient button.

Testing and training of medical staff was very important as it pointed out possible program margins that would not have been lost without testing in practice, and thus led to the improvement of the final OrtopEye 1.0 software.

Date	dd/mm/yyyy		Cycle: 1		Session:	12	Instructor: MB			
Deviation						**Spatial vision**		**Distant vision**		
Horizontal			Vertical		OD/OS	Superpozitior	s útl. OS	with/without correction		
	obj.	subj.	obj.	subj.	-	fusion	I, II útl. OS	VOD	6/6	6/6
without corr	-3	-3	0	0		sterepse	j. prok.	VOS	6/24	6/12
with corr	-2	-2	0	0						
note	0									
Session:								**Near vision**		
1. 14:00	ATM							VOD	1	
2. 14:10	Zeiss stretoscope							VOS	1	
3. 14:20	xxx									
Code:	065 13 2x, 065 17 2x									

Fig. 5. Output of resulting Excel file

Practical verification of the program at the Center for Children with Visual Impairment at the University Hospital in Ostrava also confirmed all expectations. The resulting software leads to better orientation during the individual sessions and throughout the treatment cycles as printed patient cards are more legible and clearer than nurses' handwritten patient cards. Now the patient card will no longer have individually invented names for individual exercises, because in the program the names of the exercises are selected from the scroll menu and cannot be manually edited. All information has its place on the printed patient card. All sessions are written in the same form. Another advantage of the software is an immediate graphical evaluation of changes in distance vision, which is mediated by graphs showing the time trends of the quantity. These graphs then clearly show the success and effectiveness of pleoptico-orthoptic exercises in a given patient. The resulting patient examination demonstration sheet is shown in Fig. 5.

6 Conclusion

The aim of this work was to create software that will facilitate and facilitate the work of orthoptic nurses in the Center for Children with Visual Impairment at the University Hospital in Ostrava. The resulting OrtopEye 1.0 software will be used to record the patient's personal data, record the progress of each session a patient undergoes, and statistically evaluate the effect of pleoptic-orthoptic exercises for distance vision. The statistical evaluation itself is carried out by means of graphs showing changes in distance vision over the entire treatment cycle.

The database is based on three tables - Patient, Cycle and Session. The tables are interconnected and store data for the entire course of treatment. The resulting solution facilitated the work of nurses at the University Hospital in Ostrava in the form of digital processing of records instead of manual entry. The resulting database was linked to a graphical user interface that is simple, clear, logical and intuitive. The resulting software functionality meets the requirements of medical staff. However, there is no shortage of calls for further work. This would include, for example, color coding of ongoing and completed cycles, better clarity of the printed card (block boundaries, addition of hospital logo, change of fonts of text), extension of statistical evaluation by adding time trends for near vision and squint deviations.

Acknowledgment. The work and the contributions were supported by the project SV450994/2101Biomedical Engineering Systems XV'. This study was also supported by the research project The Czech Science Foundation (GACR) 2017 No. 17-03037S Investment evaluation of medical device development run at the Faculty of Informatics and Management, University of Hradec Kralove, Czech Republic. This study was supported by the research project The Czech Science Foundation (TACR) ETA No. TL01000302 Medical Devices development as an effective investment for public and private entities.

References

1. Teichmannova, B., Grellova, G.: Orthopedic Exercise in Kindergartens - PRACTICE. FLORENCE – Specialized magazine for nursing and other health professions, p. 32 (2010)
2. Birch, E.E.: Amblyopia and binocular vision. Prog. Retin. Eye Res. **33**, 67–84 (2013)

3. Wang, J.: Compliance and patching and atropine amblyopia treatments. Vis. Res. **114**, 31–40 (2015)
4. Dwyer, P., Wick, B.: The influence of refractive correction upon disorders of vergence and accommodation. Optom. Vis. Sci. Off. Publ. Am. Acad. Optom. **72**, 224–232 (1995)
5. Rowe, F.J.: Clinical Orthoptics. Wiley, Wiley (2012)
6. Smith, A., Latter, S., Blenkinsopp, A.: Safety and quality of nurse independent prescribing: a national study of experiences of education, continuing professional development clinical governance. J. Adv. Nurs. **70**, 2506–2517 (2014)
7. Harley, R.D.: Paralytic strabismus in children: etiologic incidence and management of the third, fourth, and sixth nerve palsies. Ophthalmology **87**, 24–43 (1980)
8. Hycl, J.: Squinting and Blindness: Patient Information. Triton, Prague (2000). ISBN 80-725-4088-2
9. Novohradska, H.: Selected Chapters from Ophthalmopaedia. University of Ostrava, Faculty of Education, Ostrava (2009). ISBN 978-807-3687-311
10. Yu-Wai-Man, P.: Chronic progressive external ophthalmoplegia. Acta Ophthalmologica, 95 (2017)
11. Hussaundeen, J.R., et al.: Efficacy of vision therapy in children with learning disability and associated binocular vision anomalies. J. Optom. **11**, 40–48 (2018)
12. Miyagawa, H., et al.: One-year results of voluntary-based supervised exercise or treatment at orthopedic clinic for radiographic severe knee osteoarthritis. J. Phys. Ther. Sci. **28**, 906–910 (2016)

Quantitative Analysis and Objective Comparison of Clustering Algorithms for Medical Image Segmentation

Alice Krestanova, Jan Kubíček(✉), Jiri Skandera, Dominik Vilimek, David Oczka, Marek Penhaker, Martin Augustynek, and Martin Cerny

FEECS, VSB-Technical University of Ostrava, K450, 17. Listopadu 15, Ostrava-Poruba, Czech Republic
{alice.krestanova,jan.kubicek,jiri.skandera.st,dominik.vilimek, david.oczka,marek.penhaker,martin.augustynek, martin.cerny}@vsb.cz

Abstract. The paper describes the implementation of non-hierarchical methods k-means and fuzzy c-means on nosily images from different medical modalities as computed tomography and magnetic resonance. Modern devices are created on the basis of advanced technology, both during the actual acquisition of the image and subsequently during its processing. The problem is caused by the unexpected disturbance of the image by parasitic noise, which may already occur in the electronics of the device or in dependence on the phenomena caused by the external environment. The testing was carried out on 3 datasets of medical images and the evaluation per individual images was determined based on the correlation factor and the mean quadratic error. The result is evaluation of non-hierarchical clustering techniques for the creation of mathematical models of tissue depending on the noise intensity.

Keywords: Cluster analysis · k-means · Segmentation · Unsupervised learning · Fuzzy c-means

1 Introduction

Cluster analysis is used to divide or classify a set of unlabeled elements, so that the features of elements show significant similarity belong to the same group, as opposed to elements outside this group, where the similarity is minimal. In our case, these elements are pixels in the image [1, 2].

Measures of similarity, dissimilarity or distance metrics are most often used to determine the similarity of objects. The distance of two points can be recorded in a symmetrical square matrix with zeros on the main diagonal. In clustering the minimum distance of two objects, their maximum distance and average distance can be used to determine the distance of two clusters. The decision on the final number of clusters is based on both the theoretical distance of the clusters and the decision, if we are clustering individual objects of a variable or data category [3].

N. T. Nguyen et al. (Eds.): ACIIDS 2020, LNAI 12034, pp. 114–125, 2020.
https://doi.org/10.1007/978-3-030-42058-1_10

The image may be unexpectedly disturbed by parasitic noise, which may already occur in the electronics of the apparatus or depending on the effects caused by the external environment. Therefore, using a suitable design for evaluation of clustering based on nonhierarchical methods, it can be achieved by testing data in the presence of additive noise and its intensity.

2 Related Work

The clustering analysis and methods based on this idea can be used for image and signal processing. Cluster analysis is a method of grouping a set of diverse unlabelled elements in such a way that elements in the same group are maximally similar to each other, unlike elements outside the group, where the similarity is minimal. The accuracy of most these algorithms will depend not only on their shape and frequency, but especially on the spatial relationships and the distance of the clusters between themselves [4, 5].

Hierarchical methods create a sequence of partitions with a single main cluster containing all of the given objects placed on one side and separate single element clusters placed on the other. Each of the intermediate levels can be considered as a combination of two clusters from the lower level [3, 6]. Agglomerative approach (starting with individual elements and their aggregation into clusters) represents the closest neighbor, the farthest neighbor, the average bond, the Ward method, the centroid method [7, 8]. Divisional clustering is based on the division from the largest cluster, assuming that all objects initially form a single cluster, which they then split into two clusters with minimal similarity [5, 9, 10].

The nonhierarchical methods include the k-means method, which was named according to the number of clusters k into which the data will be grouped, the number must be before the analysis specified together with an initial estimate of the membership of all clusters. Fuzzy C-means is used for data mining, pattern recognition, classification and image segmentation [11]. The Clustering Large Applications (CLARA) algorithm is an extension of the PAM method for large data sets, where objects reach thousands of thousands [12]. The CLARANS method starts with a randomly selected node and examines neighboring nodes for a preselected scan distance. If a better matching node is found, it selects it and continues from it. If it does not find such a node, mark the existing node as a local minimum. The final state is if a predetermined number of iterations are reached. Bagged clustering is a newer method that combines a hierarchical and a non-hierarchical approach [8].

3 Materials and Methods

Within the design of algorithm for adding additive noise to images and subsequently a evaluation of nonhierarchical clustering techniques for generating mathematical tissue models depending on the noise intensity. First dataset contains images from angiography of vascular calcification and includes 10 images with resolution 1024×1024. Images are color, they are in RGB color model. Second dataset contains images cut of liver, which were taken by CT. Dataset includes 10 color images with resolution 630×630 and 442×442. Third dataset contains 10 images of articular cartilage taken by MRI. Images are color with resolution 512×512 pixels.

3.1 Implementation of Additive Noise

Each image contains certain kinds of noise of various origins. Noise represents a parasitic part of the image that deteriorates its quality. Noise can already be generated in the electronics of the recording device or by external influences such as temperature, humidity or dustiness of the environment. The deterioration and noise of the image also occurs during its subsequent adjustment or compression. Removing noise from image data is an important part of preprocessing after further image analysis. An important step in eliminating noise is to preserve its original properties, so there should be no distortion or blur in the edges of the image.

Firstly, is uploaded image and after it was implemented noise Gaussian, Localvar, Speckle, Salt and Pepper (see Fig. 1).

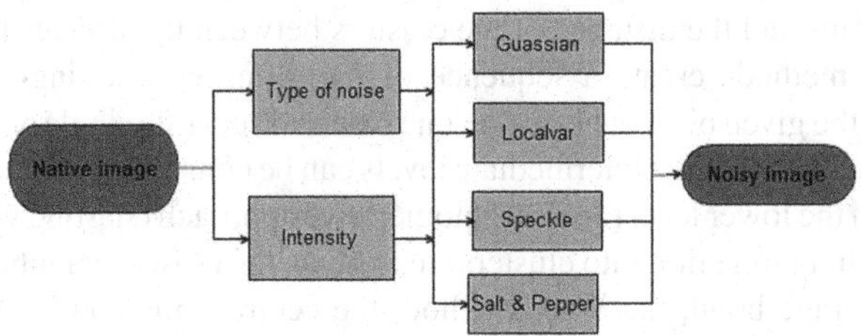

Fig. 1. Block diagram of application noise

3.2 Implementation of Algorithm

In the next step the algorithm was divided into first part for segmentation of native images, when the segmented output images work as gold standard and second part for segmentation of images with noise. Gold standard is compared with segmented noisy images. In algorithm, they were used two nonhierarchical cluster analysis methods k-means and fuzzy c-means (see Fig. 2).

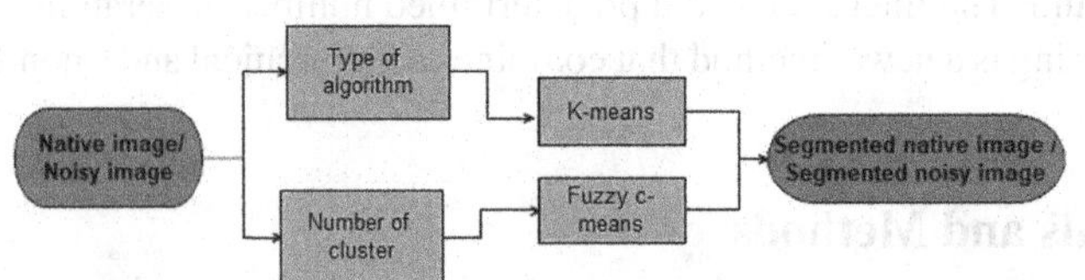

Fig. 2. Implementation of non-hierarchical cluster analysis methods

3.3 Implementation of Noise

Noise with different variability of noise intensity was implemented to the images to monitor noise dynamics in context degradation of medical images. Gaussian noise, which is a statistical noise having a probability of distribution density equal to the normal distribution of the Gaussian distribution, was implemented first. For noise, the input parameter is the mean and dispersion. It is described by the formula:

$$G(x) = \frac{1}{\sigma\sqrt{2\pi}} e^{\frac{(x-\mu)^2}{2\sigma^2}}, \tag{1}$$

where *x* is brightness intensity of noise σ is variance, μ is mean value. Another applied noise called salt & pepper, known as pulse noise. It occurs as sparsely spaced black and white pixels. For noise, the input parameter is density (see Fig. 3) [13].

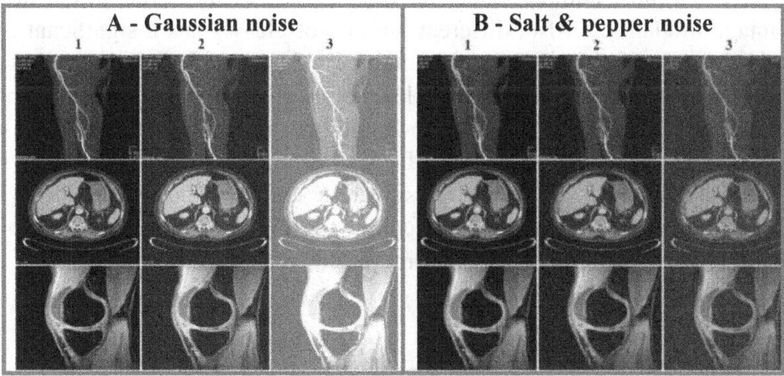

Fig. 3. A – Gaussian noise: application to CT images: 1 – Native image; 2 – $\mu = 0, \sigma = 0.1$, 3 – $\mu = 0, \sigma = 0.5$, B – Salt & Pepper noise: application to CT images: 1 – Native image, 2 – $d = 0.1$, 3 – $d = 0.3$

Speckle noise is most common in coherent imaging systems such as ultrasonic systems, lasers or magnetic resonance imaging. For noise, the input parameter is the mean and variance (see Fig. 4A) [14].

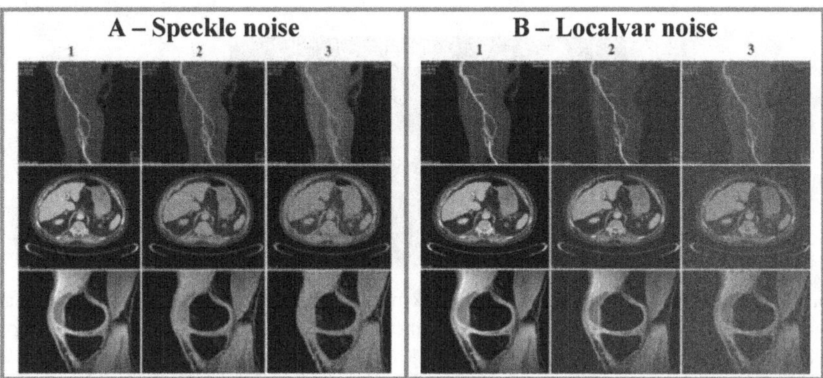

Fig. 4. A – Speckle noise: application to CT images: 1 – Native images, 2 – $\mu = 0, \sigma = 3$, 3 – $\mu = 0, \sigma = 5$; **B – Localvar noise:** Application to CT images: 1 – Native images, 2 – $\mu = 0, \sigma = 0.1$, 3 – $\mu = 0, \sigma = 0.3$

Localvar noise is a type of Gaussian white noise with local variance in the image. For noise, the input parameter is the mean and variance (see Fig. 4B).

3.4 Implementation of Cluster Analysis: k-Means, Fuzzy c-Means

In this study, we are aimed on the non-hierarchical clustering algorithms which are able to classify image pixels based on its homogeneity to individual segmentation classes (clusters). We are particularly aimed on the centroid algorithms where classification is done based on the distance measurement between pixels and centroid of each cluster. In our study we compared different number of the segmentation classes. This comparative analysis points out how number of classes reflects individual tissues within different medical image modalities. Also, different number of classes has a significant impact on the time consumption. We selected the number of classes: 3, 5 and 8 to do testing of the clustering performance for different level of the classification. The segmentation model works differently in the case of a lower number of classes (5), and for instance in the case of a higher classes (8). These facts consist a good basis for testing of the segmentation robustness for different classification settings. The first method chosen for image segmentation is the k-means method (see Fig. 5). When assigning objects to individual clusters, we can only determine whether the given object belongs to the cluster or not at logical level 1 or 0.

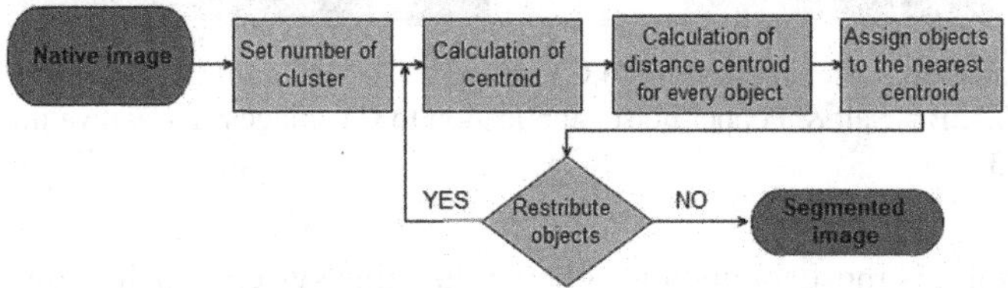

Fig. 5. Algorithm of method k-means

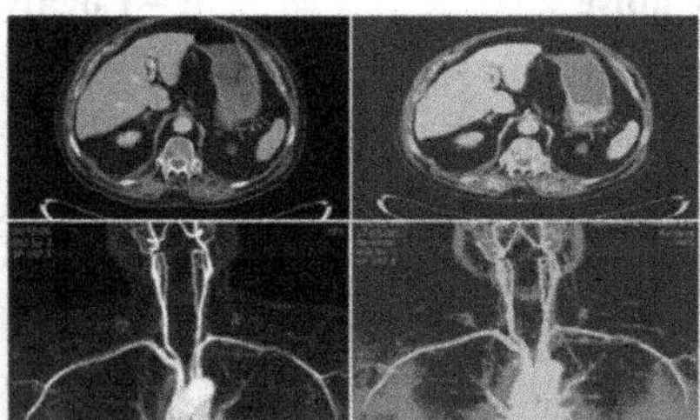

Fig. 6. Regional image segmentation based on k-means method for 3 clusters with native and segmented image: part of abdomen with liver (above), blood vessels and heart (below) (Color figure online)

In Fig. 6 shows an implementation of the k-means method for regional segmentation with the setting of 3 initial clusters. Objects in the image with lower brightness were suppressed in the background. Visible are mainly brightly areas in the tissue, which are represented by a cluster of yellow colours. In Fig. 6, in the upper part is the area of the liver, which is particularly visible, in down part is blood circulation, as is the heart, aorta and main arteries in the image below.

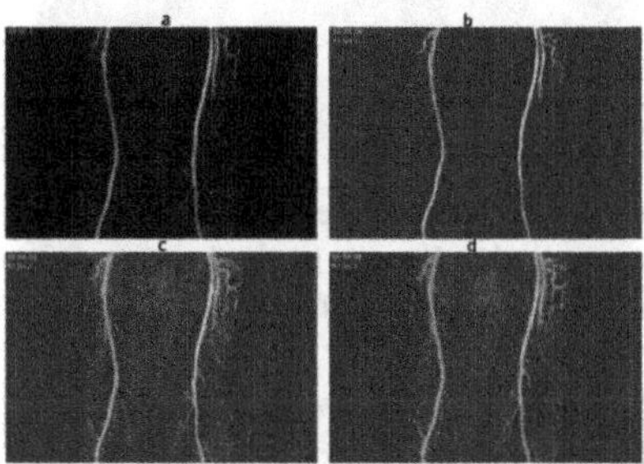

Fig. 7. (a) Native image, (b) Segmented image by method k-means for 3 clusters, (c) Segmented image by method k-means for 5 clusters, (d) Segmented image by method k-means for 8 clusters

In Fig. 7(a) shows blood vessels of the legs, magnetic resonance imaging images, and angiography. The native image of Fig. 7(a) has brightly different portions, and only in the area of the vessels where the contrasting agent has been applied. The background and surrounding of the vessel is unnecessary, so when 3 clusters are set, it is filtered out, see Fig. 7(b). In the settings of 5 and 8 clusters in Fig. 7(c, d) the background is again visible, but the representation of the cv remains critical, so setting up 3 clusters appears to be the best solution.

The fuzzy c-means method is based on fuzzy logic. The input number of clusters for fuzzy c-means was specified in the same number as the k-means method. The block diagram of algorithm for segmentation of images by fuzzy c – means is below Fig. 8.

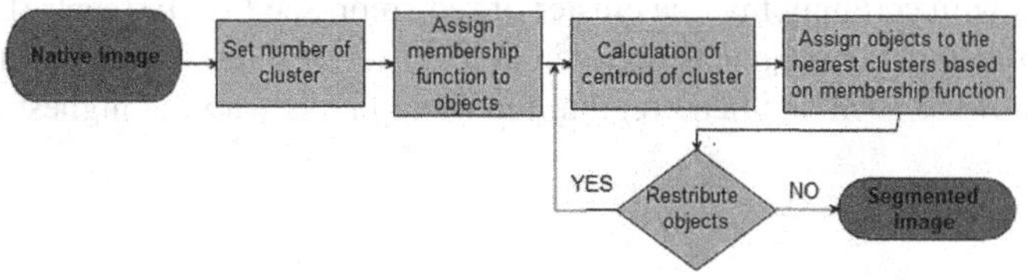

Fig. 8. Algorithm of method fuzzy c-means

In Fig. 9, a selection of the kidney-capturing image portion is shown. Segmentation was performed using the fuzzy c-means method. Individual images from a to f show an increasing number of clustering classes. In Fig. 9(c) the kidneys are difficult to recognize. The best segmented image is shown in Fig. 9(d) at 3 clusters, as the parasitic portions of the background have been suppressed and only important portions of the image remain.

In method Fuzzy c-means the image matrix is divided into pixels from 0 to 255, where 0 is black and 255 is white. FCM algorithm utilizes a membership function which is based on measuring of the Euclidean distance between individual pixels and centroids.

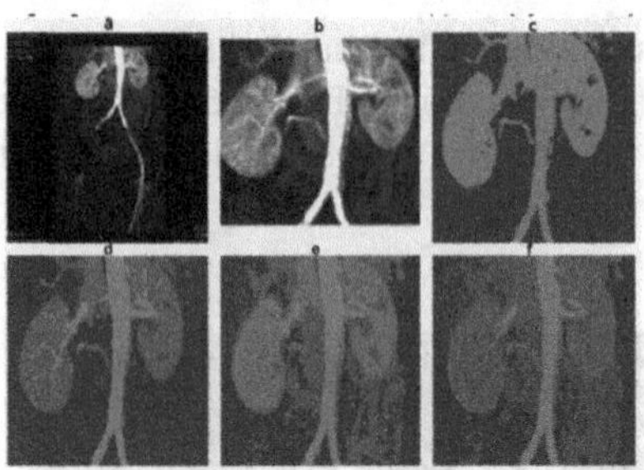

Fig. 9. (a) Native image, (b) Native image of selected area, (c) Segmented image by method c-means for 2 clusters, (d) Segmented image by method c-means for 3 clusters, (e) Segmented image by method c-means for 4 clusters, Segmented image by method c-means for 5 clusters,

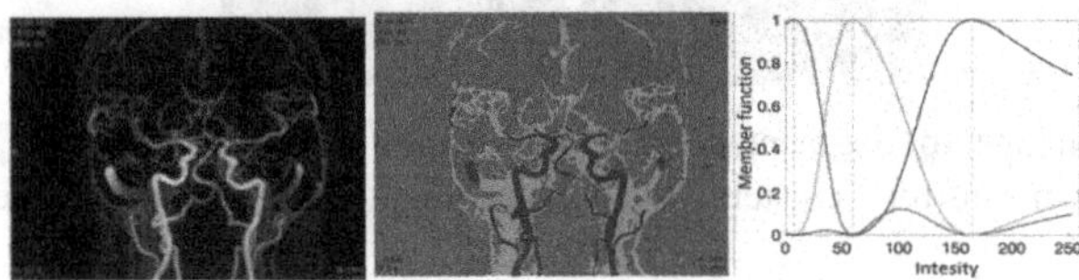

Fig. 10. Native image (left), segmented image with method fuzzy c-means (in the middle), graph membership function of each pixels belonging to clusters (right) (Color figure online)

By this way, pixels are grouped into predetermined number of segmentation classes based on the similarity with the centroid. At Fig. 10 (right) shows the dependence of brightness intensities of pixels on their belonging to clusters. Each of the colour curves represents exactly one cluster. In the FCM algorithm, we are using the Euclidean distance, measuring distance between individual pixels and cluster's centroids. The x-axis is the brightness intensity value of each pixel. On the y-axis are the values representing the pixel density of the definite intensity to the clusters. Pixels with membership function 1 absolutely do not belong to the given cluster, on the contrary 0 absolutely do not belong. It can be said with certainty that the cluster of red color, see Fig. 10 (centre) represents the darkest pixels in the image, as opposed to a cluster of blue colours that depicts the vascular bed of the neck and head region, featuring pixels with the highest luminance intensity.

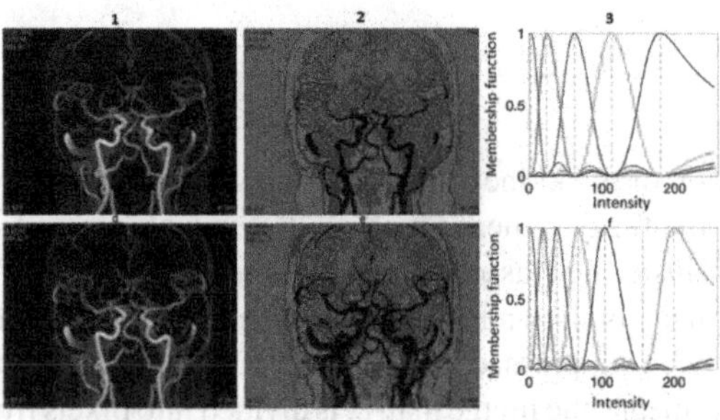

Fig. 11. Native image (left), segmented image with method fuzzy c-means (in the middle), graph membership function of each pixels belonging to clusters (right)

Figure 11 represents angiographically the magnetic resonance imaging of the neck and head. The left side shows the native image in individual shades of gray. The middle image is fuzzy c-means after segmentation and corresponds to the individual graphs on the right-hand side of Fig. 11. These graphs represent the degree of belonging of individual pixels to clusters. In this case, for the number of 5 clusters (top) and 6 clusters (bottom) of the above-mentioned Fig. 11.

4 Testing and Evaluation

In our study, we are aimed on the analysis dealing with the impact of different models of the image noise on the segmentation model based on the centroid clustering algorithms. We present the results of this analysis in the form of evolution characteristics, describing image segmentation robustness for various noise intensity. Regarding statistic evaluation, we compare the performance of different noise settings for various settings of K-means and FCM. Such evaluation points out the sensitivity of particular method.

4.1 Evaluation Metrics

Segmented output from noisy image was compared with segmented native image as gold standard. For evaluation were used mean square error and correlation coefficient.

Mean squared error (MSE) is the easiest method for objective evaluation of quality between images. The parameters determine the square of the average difference between these images. As the only of the analyzed parameters is not standardized. A lower value indicates a better registration process.

The equation for MSE is:

$$MSE = \frac{1}{n}\sum\nolimits_{i=1}^{n} \left(\hat{y}_1 - y_i\right)^2, \tag{2}$$

where i is sequence of number from i to last number n, y is coordinate and $\hat{y}_1$ is coordinate value which is situated on the created line.

Correlation coefficient represents the measure of linear relationship between two quantities. correlation coefficient takes values from -1, determines total negative match by 1, which determines total positive match between images. The correlation coefficient can be defined using the following mathematical formula:

$$r = \frac{\sum_m \sum_n \left(A_{mn} - \bar{A}\right)\left(B_{mn} - \bar{B}\right)}{\sqrt{\left(\sum_m \sum_n \left(A_{mn} - \bar{A}\right)^2\right)\left(\sum_m \sum_n \left(B_{mn} - \bar{B}\right)^2\right)}}, \tag{3}$$

where $\bar{A}$ is the mean of all of the first coordinates of data A_{mn} and $\bar{B}$ is the mean of all of the second coordinates of the data B_{mn}.

4.2 Evaluation of Tested Data

The input for the overall analysis in both methods was necessary to choose the number of clusters. This parameter was chosen equally for both methods. Testing was performed with Gaussian noise set for the following graphs. On the left, Fig. 12 shows a graph for the k-means method and a graph for the fuzzy c-means method on the right. For both methods, the number are set to 3, 5 and 8 clusters. From graphs follows, that the curve for 8 clusters has the highest possible correlation coefficient values, with a 10% difference to the curve for the 3 set clusters.

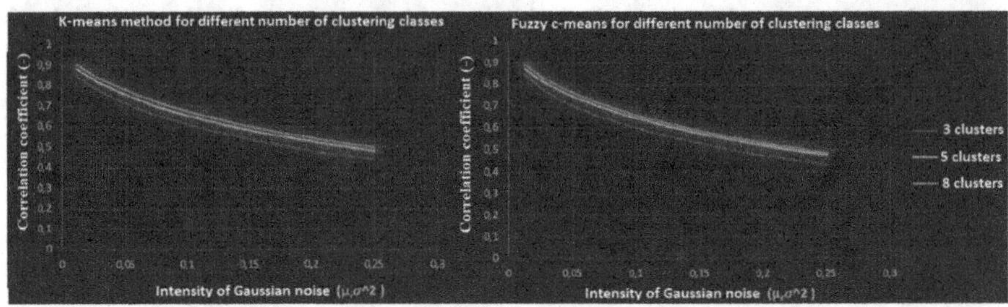

Fig. 12. Dependency number of clusters to quality of segmentation image

4.3 Analysis Methods k-Means and Fuzzy c-Means for Every Type of Noise

In this analysis, it was compared the different results between the k-means method and fuzzy c-means for each of the applied noise. In the individual graphs of Fig. 13, curves representing correlation coefficients of methods depending on the noise intensity can be seen. Although the fuzzy c-means method is more robust and its calculation is mathematically more complex, in this analysis the graphs for both methods are almost duplicate. As for the better success of one method over the other, for no noise there is no noticeable difference. A more detailed analysis of the difference can be seen in Fig. 14.

4.4 Analysis of Image Distortion Based on Noise Type Selection

The resulting graphs (see Fig. 15) represent all types of implemented noise depending on their intensity. The analysis shows which of the noises aggressively affect medical images. Speckle noise, which shows a decrease in image quality at a total intensity of only 10% within the correlation coefficient, was least affected. This is due to the distribution of its noise, which is primarily located on the brightness intensity of the image. The most influential is Gaussian noise, which shows about 40% distortion of the image. This is because this noise affects all pixels in the image regardless of their brightness intensity. Salt & pepper and localvar noises show distortion of approximately 30%.

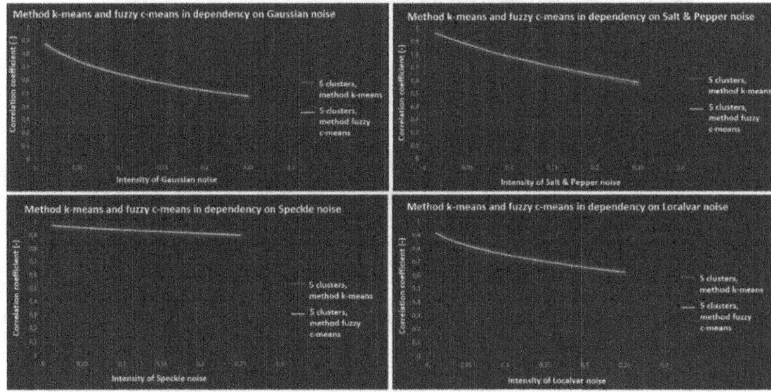

Fig. 13. Comparing of method k-means and fuzzy c-means in dependency on each noise

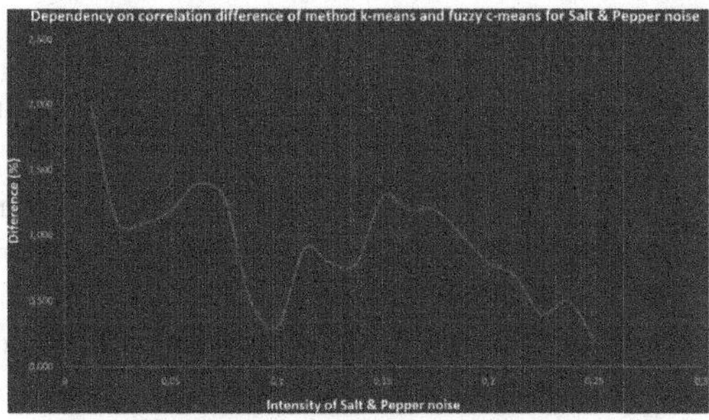

Fig. 14. Graph of degree of difference between methods k-means and fuzzy c-means

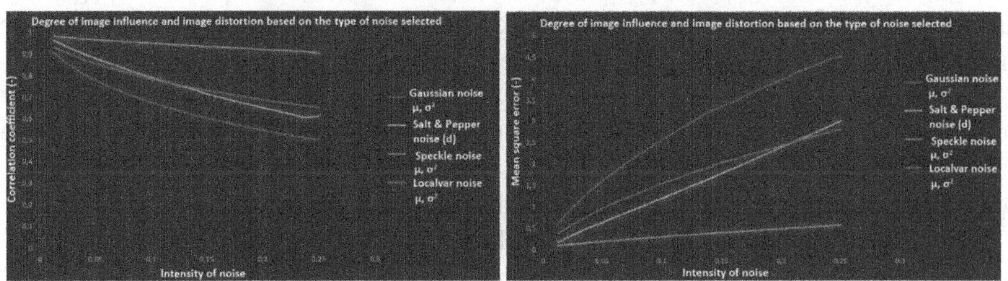

Fig. 15. Graph: degree of image influence and image distortion based on the type of noise selected

4.5 Difference in the Effect of Noise on the Image Depending on the Dataset Type

Here, the difference in mean quadratic error was analyzed depending on the intensity Gaussian noise for each of the datasets. CT scan of the liver and MRI of articular cartilage are images that are least affected by noise. This is mainly due to the similarity of the individual images in the dataset as opposed to angiography of the vessels, where the type of images varies significantly, especially in angle and sensing portions of the individual vessel areas. The result of vascular angiography is thus about 15% lower than other datasets (see Fig. 16).

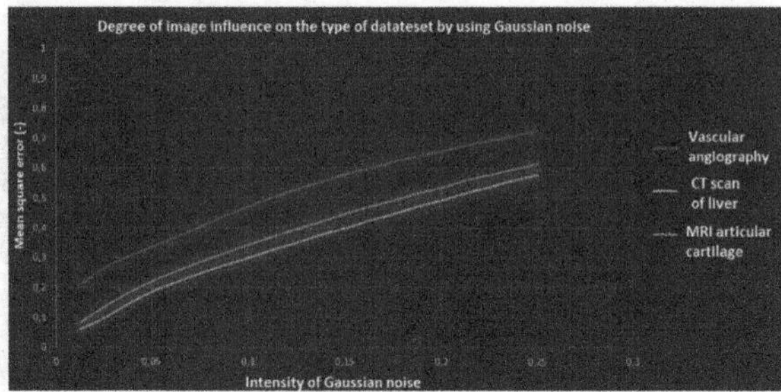

Fig. 16. Comparison of Gaussian noise influence rate according to individual datasets

5 Conclusion

In this paper, we compare the segmentation performance, effectivity and robustness of the centroid clustering algorithms. We omitted comparison with different regional segmentation techniques due to different mathematical principles, like is statistical descriptors or histogram analysis. Therefore, such analysis would have been unobjective regarding the principle of K-means and FCM.

The main implementation of the k-means and fuzzy c-means method on image data was done by 3 datasets obtained mainly from magnetic resonance and computed tomography. Additive noise was applied to all images. This was followed by testing both methods to set the number of primary clusters 3, 5 and 8. Each noise and the native segmented image were then compared with each other.

The comparison was made using the correlation coefficient and the mean square error. All resulting values were then averaged for each image and a certain intensity value. Based on testing, a comparison of the two methods revealed that although the fuzzy c-means method is more robust than the k-means method, the overall result is that the methods are almost equal in performance to all types of noise.

Variance is in percent units. For an analysis involving image-to-noise variation by noise type, the largest effect is set when applying Gaussian noise, with up to 40% distortion relative to the native image. The least distortion, 10%, was observed for speckle noise. For salt & pepper and localvar noise, distortion was up to 30%.

The analysis for selecting the number of clusters showed the best results in 8 clusters, but also a longer computation time. In comparison with the setting of clusters 3 there was about 15% difference. Thus, the analysis shows that a higher number of clusters is more robust to additive image noise. Most affected by noise, the angiographic dataset was captured by magnetic resonance imaging, due to the sensing of various parts of the patient's body, in contrast to articular cartilage images and CT sections of the liver, where the images were obtained for the most part at the same angle.

Acknowledgement. The work and the contributions were supported by the project SV450994/2101 Biomedical Engineering Systems XV'. This study was also supported by the research project The Czech Science Foundation (GACR) 2017 No. 17-03037S Investment evaluation of medical device development run at the Faculty of Informatics and Management, University of Hradec Kralove, Czech Republic. This study was supported by the research project The Czech

Science Foundation (TACR) ETA No. TL01000302 Medical Devices development as an effective investment for public and private entities.

References

1. Garreta, R., Moncecchi, G.: Learning Scikit-Learn: Machine Learning in Python. Packt Publishing Limited, Birmingham (2013). 118 pages. ISBN 978-1-78328-193-0
2. Mitchell, T.M.: Machine Learning. McGraw-Hill, New York (1997). ISBN 978-0070428072
3. Kumar, S., Toshniwal D.: A data mining framework to analyze road accident data. J. Big Data **2**(1) (2015). https://doi.org/10.1186/s40537-015-0035-y. ISSN 2196-1115
4. Aggarwal, C.C., Reddy, C.K. (eds.): Data Clustering. Chapman and Hall/CRC, Boca Raton (2018). https://doi.org/10.1201/9781315373515. ISBN 9781315373515
5. Olson, D.L., Lauhoff, G.: Descriptive Data Mining. Computational Risk Management, pp. 129–130. Springer, Singapore (2019). https://doi.org/10.1007/978-981-13-7181-3_8. ISBN 978-981-13-7180-6
6. Cohen-Addad, V., Kanade, V., Mallmann-Trenn, F., Mathieu, C.: Hierarchical clustering. J. ACM **66**(4), 1–42 (2019). https://doi.org/10.1145/3321386. ISSN 00045411
7. Siddique, M.A., Arif, R.B., Khan, M.M., Ashrafi, Z.: Implementation of fuzzy c-means and possibilistic c-means clustering algorithms, cluster tendency analysis and cluster validation. arXiv, abs/1809.08417 (2018)
8. Huang, H., Meng, F., Zhou, S., Jiang, F., Manogaran, G.: Brain image segmentation based on FCM clustering algorithm and rough set. IEEE Access **7**, 12386–12396 (2019). https://doi.org/10.1109/ACCESS.2019.2893063. ISSN 2169-3536
9. Dudarin, P., Samokhvalov, M., Yarushkina, N.: An approach to feature space construction from clustering feature tree. In: Kuznetsov, S.O., Osipov, G.S., Stefanuk, V.L. (eds.) RCAI 2018. CCIS, vol. 934, pp. 176–189. Springer, Cham (2018). https://doi.org/10.1007/978-3-030-00617-4_17. ISBN 978-3-030-00616-7
10. Berenguer, R., et al.: Radiomics of CT features may be nonreproducible and redundant: influence of CT acquisition parameters. Radiology **288**(2), 407–415 (2018). https://doi.org/10.1148/radiol.2018172361. ISSN 0033-8419
11. Schubert, E., Rousseeuw, P.J.: Faster *k*-medoids clustering: improving the PAM, CLARA, and CLARANS algorithms. In: Amato, G., Gennaro, C., Oria, V., Radovanović, M. (eds.) SISAP 2019. LNCS, vol. 11807, pp. 171–187. Springer, Cham (2019). https://doi.org/10.1007/978-3-030-32047-8_16. ISBN 978-3-030-32046-1
12. Fuentes-Penailillo, F., Ortega-Farias, S., Rivera, M., Bardeen, M., Moreno, M.: Using clustering algorithms to segment UAV-based RGB images. In: 2018 IEEE International Conference on Automation/XXIII Congress of the Chilean Association of Automatic Control (ICA-ACCA), pp. 1–5. IEEE (2018). https://doi.org/10.1109/ica-acca.2018.8609822. ISBN 978-1-5386 5586-3
13. Singh, V., Dev, R., Dhar, N.K., Agrawal, P., Verma, N.K.: Adaptive type-2 fuzzy approach for filtering salt and pepper noise in grayscale images. IEEE Trans. Fuzzy Syst. **26**(5), 3170–3176 (2018). https://doi.org/10.1109/TFUZZ.2018.2805289. ISSN 1063-6706
14. Khwairakpam, A., Kandar, D., Paul, B.: Noise reduction in synthetic aperture radar images using fuzzy logic and genetic algorithm. Microsyst. Technol. **25**(5), 1743–1752 (2019). https://doi.org/10.1007/s00542-017-3474-x. ISSN 0946-7076

Computer Vision and Intelligent Systems

Locally Adaptive Regression Kernels and Support Vector Machines for the Detection of Pneumonia in Chest X-Ray Images

Ara Abigail E. Ambita(✉), Eujene Nikka V. Boquio, and Prospero C. Naval Jr.

Computer Vision and Machine Intelligence Group, Department of Computer Science, University of the Philippines Diliman, Quezon City, Philippines
{aeambita,evboquio,pcnaval}@up.edu.ph

Abstract. In this study, the problem of detecting pneumonia in chest x-ray images is addressed. The method used is based on the computation of locally adaptive regression kernel (LARK) descriptors, which, when used in conjunction with the novel Support Vector Machine (SVM) classifier, obtained a 98% precision and 98% recall tested in only 400 images and a 96% precision and 95% recall tested in 1000 images in a chest x-ray images dataset. Different sample sizes were also tested to show that the method is robust even with a small sample size. This method was also shown to obtain better performance compared to two other classifiers, namely Random Forest (RF) and decision trees (DT).

Keywords: Computer vision · Kernel regression · X-ray images · Image classification · Locally adaptive kernel regression · Support vector machine

1 Introduction

Pneumonia is the leading infectious cause of death among children worldwide, killing approximately 808, 694 children under the age of five (5) in 2017. It is an inflammation in one or both lungs caused by bacteria, viruses, fungi, or other pathogens which require different diagnosis based on its classification [19]. The disease is easily detected through the analysis of chest x-ray images. More so, x-rays are low cost and affordable even for underdeveloped areas. However, pneumonia is prevalent mostly in impoverished developing areas (e.g. some countries in South Asia) where there's a lack of trained personnels and medical resources [4]. Such socio-economic factors impact the correctness of classification and treatment of the disease. One possible solution to address the problem is to employ CAD (Computer-Aided Diagnostic) tools to assist medical personnels in decision-making. CAD tools combine the power of computer vision and artificial intelligence with radiological image processing and therefore open

N. T. Nguyen et al. (Eds.): ACIIDS 2020, LNAI 12034, pp. 129–140, 2020.
https://doi.org/10.1007/978-3-030-42058-1_11

up new opportunities for building CAD systems in medical applications [14]. Over decades, CAD systems has been used to detect lung diseases in chest x-ray images including pneumonia [6].

Pneumonia classification in chest x-rays pose a difficulty due to the variability in the input data thus the demand for a method robust enough to learn features from a large collection of complex data [15]. Also, x-rays are shadow-like images with noise such as Gaussian noise or salt-and-pepper noise. Therefore, preprocessing of the images to remove the noises is crucial.

While image preprocessing and the choice of classifier can help determine the classification performance in x-ray images, image representation is just as crucial. One way to represent the features in an image is to use locally adaptive regression kernel (LARK) descriptors, which capture the local structure of the data effectively.

In this study, an x-ray image classification problem is addressed. In particular, we present a method to detect pneumonia in x-ray images. This method is based on the use of locally adaptive regression kernel (LARK) descriptors, which is "a self-similarity measure based on signal-induced distance between a center pixel and surrounding pixels in a local neighborhood" [17]. To improve the quality of the x-ray images, image processing methods such as power law transformation, Gaussian filters, and Otsu thresholding were used. To reduce the dimensionality of the LARK descriptors, Principal Components Analysis (PCA) was used. A Support Vector Machine (SVM) was then used to classify images with or without pneumonia.

This paper is organized as follows. In Sect. 2, previous works and concepts used are discussed. Section 3 presents the work flow of this study, as well as details of the experiment, dataset used, and the metrics for evaluation of the classification accuracy. The demonstration of the proposed method's performance with the results of the experiment are discussed in Sect. 4. Finally, the paper is concluded in Sect. 5.

2 Related Literature

2.1 Image Processing

Image processing plays a key role in improving the appearance and visualization of images [16]. It also helps in the extraction of features, structure, and other information they reveal, by making these features more prominent. Over the years, image processing proved to be vital to the various computer vision and machine learning problems such as object detection, image stitching, face analysis, and classification problems. Methods for image processing such as power law transformation (Gamma correction), Gaussian filters, and Otsu thresholding are described below.

Power Law Transformation. An image can be represented as a function $f : R^2 \longrightarrow R$ where $f(x, y)$ gives the intensity at position (x, y). As with other functions, transformations can be applied to them. One type of gray level transformation function is the power law transformation. Its transformation function, which is also referred to as Gamma correction, is described as $s = cr^{\gamma}$ where r is the input gray level, s is the output gray level, and c and γ are positive constants. This transformation process compensates for the nonlinear response of the image [8]. Varying the value of γ would obtain a family of possible transformations.

Gaussian Filter. X-ray images contain a lot of noise such as Gaussian noise and salt-and-pepper noise. In [3], several noise reduction techniques to address these problems are discussed. One technique is the Gaussian filter, to reduce the Gaussian noise. The two-dimensional Gaussian filter is given by Eq. 1.

$$G(x, y) = \frac{1}{\sqrt{2\pi}\sigma} \exp\left(-(x^2 + y^2)/2\sigma^2\right) \tag{1}$$

where σ^2 is the variance of the Gaussian filter [5]. Here, a larger value of variance smooths out the noise better, however, it can also result to distortion of parts with big changes in brightness.

Otsu Thresholding. Image segmentation, or the process of separating an image into various segments, can be make images easier to analyze. One way to effectively do so is through thresholding, which separates an object from the background [20].

The Otsu Thresholding method in [12] selects an optimal threshold that maximizes the separability of the classes in gray levels. This reduces the grayscale image into a binary image, for easier image analysis.

2.2 Locally Adaptive Regression Kernels

Regression functions in kernel regression are used to represent images. Here, spatial differences to weigh the input values are used. In locally adaptive kernel regression, pixel gradients are used in addition to the spatial differences [1]. LARK features, by measuring the similarity of two pixels based on gradients, can robustly obtain the local structure of images even with variations in the pixel intensities, noise, blur, and orientations [21]. They have been shown to have better performance as object descriptors as compared to Haar-like, SIFT, and HOG features [2]. This makes them ideal especially in the presence of noise and blur. In [1], locally adaptive kernel regression, particularly steering kernel regression, is used to detect people in videos. In [17], LARK features were used for face verification and it achieved state-of-the art performance.

Let K denote a LARK. It is defined as function that measures the similarity between a center and the surrounding pixels as shown in Eq. 2.

$$K(C_l, \Delta x_l) = \exp\left(-ds^2\right) = \exp\left(-\Delta x_l^{\top} C_l \Delta x_l\right) \tag{2}$$

where $\Delta x = [dx_1, dx_2]^{\top}$, C is the local gradient covariance matrix, $l = 1, 2, \ldots, P$ and P is the total number of samples in a local analysis window around a sample position at the pixel of interest x. ds^2 is the differential arclength on the image surface surface $S(x_1, x_2) = x_1, x_2, z(x_1, x_2)$ given by $ds^2 = dx_1^2 + dx_2^2 + dz^2$ [17].

The covariance matrix $C_l \in \mathbb{R}^{2\times 2}$, which is based on gradients (z_{x_1}, z_{x_2}), is given by:

$$C_l = \sum_{m\Omega_1} \begin{bmatrix} z_{x1}^2(m) & z_{x_1}(m)z_{x_2}(m) \\ z_{x_1}(m)z_{x_2}(m) & z_{x2}^2(m) \end{bmatrix} \tag{3}$$

where z_l is a visual signal containing values of a patch Ω_1 of pixels whose center is located at l along spatial (x_1, x_2) axes.

The LARKs computed from Eq. 2 are normalized into unit vectors to achieve illumination invariance. In this study, we denote the collection of n LARKs from an image by $K = [k_1, \ldots, k_n] \in \mathbb{R}^{P\times n}$ where k is a vectorized version of K.

2.3 Principal Components Analysis

With high-dimensional data sets becoming more common, it is necessary to reduce data dimensionality while still retaining most of the information in the data set in order to improve overall efficiency, space and time complexity. One of the most popular and simplest dimensionality reduction technique is the Principal Components Analysis (PCA). PCA reduces the dimensionality of data without much loss of information by emphasizing the variation and drawing out strong patterns in the data set [7].

PCA reduces the dimensionality of a large data by transforming a dataset into a new dimensional space using linear combinations of the original features, called principal components (PCs). PCs are obtained by solving for the eigenvectors of the covariance matrix of a normalized dataset.

2.4 Support Vector Machines

Support vector machines (SVMs) are based on the statistical learning theory concept of decision planes that define decision boundaries. Given a labeled training data, an SVM is constructed by finding *support vectors* that attempt to accurately describe a hyperplane separating the data points [9]. A hyperplane is optimal if a plane has the maximum margin, i.e, the maximum distance between training points of different classes. This can be expressed as Eq. 4:

$$y = b + \sum \alpha_i y_i x(i) x \tag{4}$$

where y_i is the class of training example $x(i)$, x represents a test example and $x(i)$ refer to the support vectors. b and α_i are parameters that determine the hyperplane. For a nonlinear case, SVM maps the points to a sufficiently high dimension so that they will be separable by a hyperplane. The high-dimensional version of Eq. 4 is represented as:

$$y = b + \sum \alpha_i y_i K(x(i), x) \tag{5}$$

The function $K(x(i), x)$ is defined as the kernel function. With kernels, the kernel data is transformed so that a non-linear decision surface is transformed to a linear equation in a higher number of dimensions. Popular kernel functions include linear, polynomial, and radial basis functions (RBF) [11].

3 Methodology

This section presents the details of the proposed method, such as the work flow, dataset used, and the metrics for evaluation.

3.1 Work Flow

The methodology flowchart is shown in Fig. 1. The proposed method involves three main stages: preprocessing, feature extraction, and classification process.

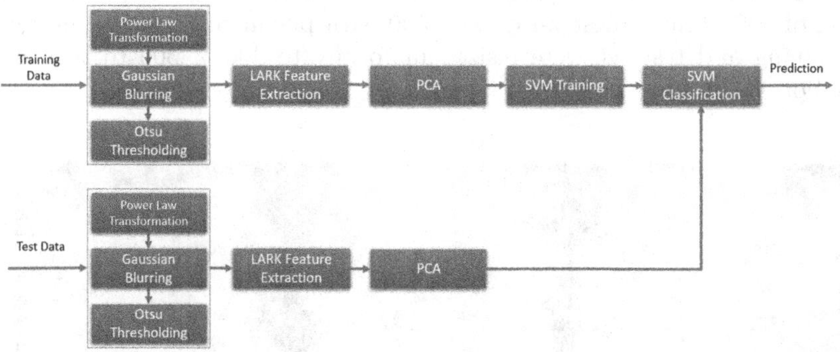

Fig. 1. Flowchart of methodology

Medical x-ray images have a varying intensity of gray levels. To reduce the level gray noise in the images in both the training and test sets, preprocessing is necesary. Power law transformation, Gaussian blurring, and Otsu thresholding were performed.

Once the noise in the images have been reduced, the feature descriptors were extracted. This was done by computing the LARK descriptors, normalizing them, and applying PCA to reduce their dimensionality. Then, an SVM classifier was used to train the images in the training set using the extracted LARK descriptors as input and predict the labels of the images in the test set.

3.2 Dataset

The proposed method was tested on a public data set containing chest x-ray images [10] which contain 2 classes: normal, labeled as 0, and pneumonia, labeled as 1. The images labeled 0 are the x-ray images that without pneumonia, while those labeled 1 are images with pneumonia.

3.3 Metrics

To evaluate the classification performance of the method, the precision and recall whose equations are given by Eqs. 6 and 7 respectively were considered:

$$\text{Precision} = \frac{TP}{TP + FP} \tag{6}$$

$$\text{Recall} = \frac{TP}{TP + FN} \tag{7}$$

where TP stands for True Positive, TN stands for True Negative, FP stands for False Positive and FN stands for False Negative.

4 Experimental Results

In this section, we carry out experiments using a subset of the dataset comprising of 500 normal chest x-rays and 500 with pneumonia. All the images in the training and test set were resized uniformly to 400 × 300 to reduce the dimension.

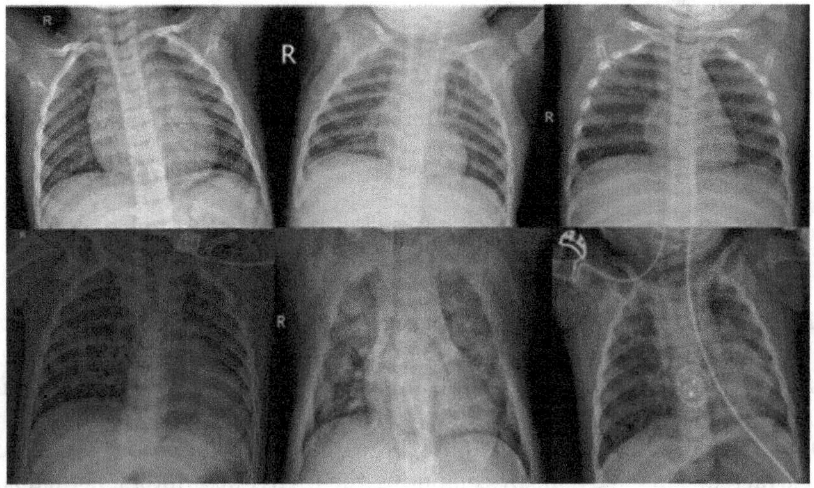

Fig. 2. Example of images with varying orientations and gray pixel intensities

The gray level intensity of medical x-ray images varies considerably which adds up to the complexity of the classification. There are variations in the size and orientation of the images which make it difficult to generalize the intensity, contrast, scale, and angle. In Fig. 2, images with varying orientations and gray level pixel intensities are shown.

4.1 Image Preprocessing

Figures 3 and 4 show the output images of the preprocessing method applied to a normal x-ray image and an x-ray image with pneumonia respectively.

If images are not gamma corrected, they allocate too many bits for the bright tones that humans cannot differentiate and too few bits for the dark tones. By the gamma correction, or power law transformation, we remove this artifact. Figures 3 and 4 show the normalized image (first image from the left) and the result of applying the power law transformation operator at gamma = 0.4 (second image from the left). It can be observed that the details of the gamma corrected image are more visible as compared to the unprocessed image. For denoising, we apply a Gaussian blurring method with a 3 × 3 size kernel to the gamma corrected (power law transformed) image. The high frequency content (e.g., noises, edges) were removed and the pixel values unrepresentative of their surroundings were eliminated.

The last preprocessing step is the application of Otsu thresholding to the Gaussian-smoothed image to segment the image into distinctive subsets such that the overlap between the two subsets are minimized. We can see in the rightmost images of both figures the distinct shape of the white pixels. In a normal chest x-ray, the shape is generally smaller than the shape of the x-ray with pneumonia.

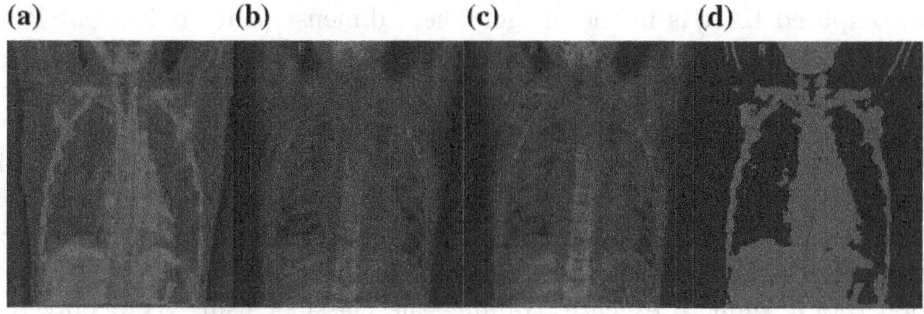

Fig. 3. Image preprocessing on normal chest x-ray (a) Input image (b) Gamma transformation (c) Gaussian blurring (d) Otsu thresholding

4.2 Feature Extraction

After applying Otsu thresholding, the image was converted into grayscale and normalized between 0 to 255. The parameters used in to compute the LARK feature descriptors are displayed in Table 1. Window size refers to the patch size of the LARK, interval defines the covariance matrix to be computed interval-pixels apart, and α is a sensitivity parameter between 0 and 1.

The results of the computation are the vectorized LARKs that act as key points and descriptors for the image. The overlapping patches are then removed

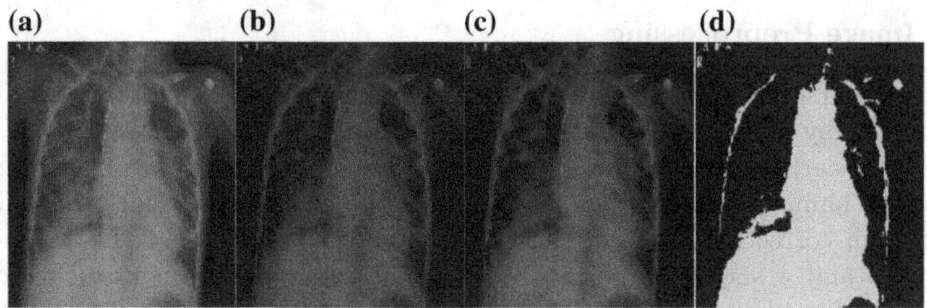

Fig. 4. Image preprocessing on chest x-ray with pneumonia (a) Input image (b) Gamma transformation (c) Gaussian blurring (d) Otsu thresholding

Table 1. Parameters for Lark Feature Extraction

Parameter	Value
Window size	5
Smoothing parameter	3
Interval	2
α	0.13

to give a visual impression of the generated LARKs. Figure 5 displays the plot of the computed LARKs in the image. Then, dimensionality reduction (PCA) was applied to the generated larks to retain only the salient characteristics of the local steering kernels and filter out the non-informative features. We choose the largest eight (8) principal components.

The overall feature extraction using LARK including the dimensionality reduction took an average of 24 min to complete iterating through all the images. The slow execution resulted from several linear filters and matrix multiplications necessary to compute the local steering kernels (LSK). While the LARK computation took 24 min to execute, training the classifier using SVM only took 1.4 min.

The overall feature extraction using LARK including the dimensionality reduction took an average of 24 min to complete iterating through all the images. The slow execution resulted from several linear filters and matrix multiplications necessary to compute the local steering kernels (LSK).

4.3 Classification Results

Eighty (80) percent of the dataset was randomly selected for training and twenty (20) percent for testing.

SVM was used for feature training and classification. The choice of a kernel and the penalty factor influence the performance of SVM [13]. We obtained a maximum performance by training the classifier using the radial basis function (RBF) kernel, the spread of the kernel (γ) into 0.01, and the penalty for misclassifying a

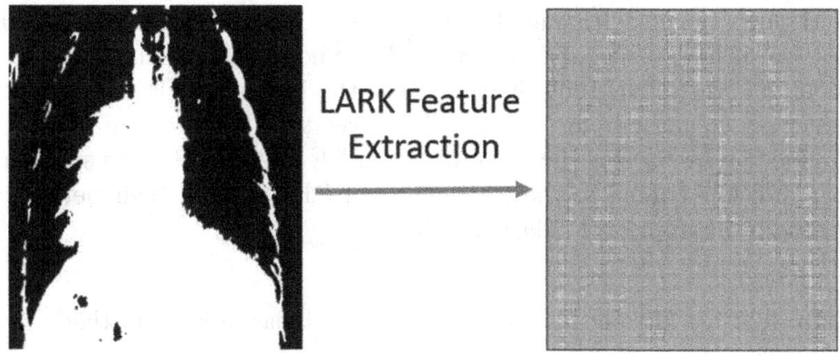

Fig. 5. Visualization of LARK feature descriptors

Table 2. Parameters for SVM classifier

Parameter	Value
Kernel	Radial basis function
C	10
γ	0.01

data point (C) into 10. These parameters are shown in Table 2. We chose the RBF kernel because it showed a better accuracy as compared to linear and polynomial kernel. The kernel and C and γ values were held constant for all the experiments.

Table 3 shows the performance of SVM in classifying the x-ray image on different levels of image preprocessing. The performance increased by 4% when power law transformation and Gaussian blurring were applied to the raw input images. A slight increase of 1% was observed when Otsu thresholding was utilized.

Table 3. SVM prediction performance for different preprocessing techniques (sample size = 1000)

Preprocessing	Precision (%)	Recall (%)
No preprocessing	92	92
Power law transformation, Gamma correction	95	94
Power law transformation, Gamma correction, Otsu thresholding	96	95

Next, we compared our SVM model to two state-of-the-art machine learning methods, Random Forest (RF) and Decision Tree (DT) classifiers. Table 4 shows that SVM outperformed RF and DT methods. RF and SVM are comparable, both with promising results and only a 1% difference on the recall. Note that the experiment was tested on 1000 images. Since the input models did not include transformation to account for the non-linear relationship between parameters, SVM produced a better classification rate than RF and DT. Given a set of transformation and fine-tuning to the parameters of RF and DT, their performance can potentially be improved (Fig. 6).

Table 4. Performance comparison of different classification methods

Method	Precision (%)	Recall (%)
RF	95	94
SVM	**95**	**95**
DT	83	83

Table 5. SVM prediction performance on varying sample sizes (n)

Sample size	Precision (%)	Recall (%)
400	98	98
800	96	95
1000	96	95

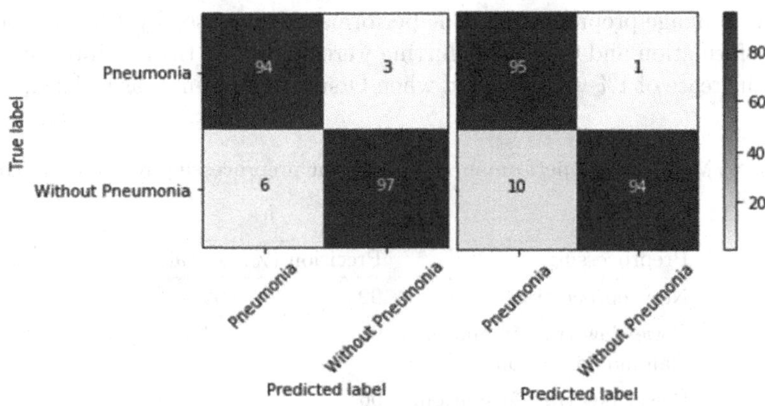

Fig. 6. Confusion matrix of classification with and without Otsu Thresholding

Large number of training data and multi-classification cases causes the SVM to perform undesirably [13]. Hence, the classification model was tested on varying sample sizes to improve the accuracy of the predictions. In Table 5, the increase in performance have increased by 2% when the sample size was downsampled into 400 images (200 normal and 200 with pneumonia). This shows that LARK and SVM combined is robust with just a small sample size.

The performance improved as we decreased the sample size because SVM is robust to large number of variables and small samples. The decision function depends only on a small subset of the training data, i.e., the support vectors. Thus, if patterns corresponding the support vectors are known in advance, the same solution is applicable in solving a much smaller problem that involves only the support vectors [18].

5 Conclusion and Future Work

Local adaptive regression kernels are effective in computing feature descriptors. Combined with PCA dimensionality reduction, it was robust in identifying the salient features regardless of the varying gray pixel level intensities, position, and orientation of the chest x-ray images even with just a small sample size. Support vector machines also proved to be superior in the classification method as compared to other state-of-the-art methods such as Random forest and Decision Trees.

The study was constrained to a binary classification (normal or with pneumonia). Future research directions could include the demonstration of the general applicability of our model in classifying whether a pneumonia is viral or bacterial in order to provide the appropriate diagnosis, therefore improving the clinical outcomes.

Further scope include testing on different image preprocessing techniques to enhance the detection of prominent features and include pairing LARK descriptors with other machine learning techniques to improve the classification. Moreover, using GPU to speed up the calculations, especially the matrix-matrix and LSK computations, may also be considered.

References

1. Alaniz II, A.L., Mantaring, C.M.G.: Using local steering kernels to detect people in videos
2. Boiman, O., Shechtman, E., Irani, M.: In defense of nearest-neighbor based image classification. In: 2008 IEEE Conference on Computer Vision and Pattern Recognition, pp. 1–8. IEEE (2008)
3. Chikhalekar, A.: Analysis of image processing for digital x-ray. Int. Res. J. Eng. Technol. (IRJET) e-ISSN 2395–0056 (2016)
4. Correa, M., et al.: Automatic classification of pediatric pneumonia based on lung ultrasound pattern recognition. PLoS One 13(12) (2018). https://doi.org/10.1371/journal.pone.0206410
5. Deng, G., Cahill, L.: An adaptive Gaussian filter for noise reduction and edge detection. In: 1993 IEEE Conference Record Nuclear Science Symposium and Medical Imaging Conference, pp. 1615–1619. IEEE (1993)

6. Gu, X., Pan, L., Liang, H., Yang, R.: Classification of bacterial and viral childhood pneumonia using deep learning in chest radiography. In: Proceedings of the 3rd International Conference on Multimedia and Image Processing, pp. 88–93 (2018). https://doi.org/10.1145/3195588.3195597. https://dl.acm.org/citation.cfm?id=3195597
7. Jolliffe, I.: Principal component analysis. In: International Encyclopedia of Statistical Science, pp. 1094–1096. Springer, Heidelberg (2011)
8. Maini, R., Aggarwal, H.: A comprehensive review of image enhancement techniques. arXiv preprint arXiv:1003.4053 (2010)
9. Mohanty, N., John, A.L.S., Manmatha, R., Rath, T.: Shape-based image classification and retrieval, vol. 31. Elsevier (2013). https://doi.org/10.1016/B978-0-444-53859-8.00010-2
10. Mooney, P.: Chest x-ray images (pneumonia), March 2018. https://www.kaggle.com/paultimothymooney/chest-xray-pneumonia/version/2
11. Nisbet, R., Miner, G., Yale, K.: Advanced algorithms for data mining handbook of statistical analysis and data mining applications (2018). https://doi.org/10.1016/B978-0-12-416632-5.00008-6
12. Otsu, N.: A threshold selection method from gray-level histograms. IEEE Trans. Syst. Man Cybern. **9**(1), 62–66 (1979)
13. Yu, P., Xu, H., Zhu, Y., Yang, C., Sun, X., Zhao, J.: An automatic computer-aided detection scheme for pneumoconiosis on digital chest radiographs. J. Digit. Imaging **24**(3), 382–393 (2011)
14. Qin, C., Yao, D., Shi, Y., Song, Z.: Computer-aided detection in chest radiography based on artificial intelligence: a survey. BioMed. Eng. OnLine **17**(1), 113 (2018). https://doi.org/10.1186/s12938-018-0544-y
15. Rajaraman, S., Candemir, S., Kim, I., Thoma, G., Antani, S.: Visualization and interpretation of convolutional neural network predictions in detecting pneumonia in pediatric chest radiographs. Appl. Sci. **8**(10) (2018). https://doi.org/10.3390/app8101715. https://www.mdpi.com/2076-3417/8/10/1715
16. Russ, J.C.: The Image Processing Handbook. CRC Press, Boca Raton (2016)
17. Seo, H.J., Milanfar, P.: Face verification using the lark representation. IEEE Trans. Inf. Forensics Secur. **6**(4), 1275–1286 (2011)
18. Wang, J., Neskovic, P., Cooper, L.N.: Training data selection for support vector machines. In: ICNC 2005: Advances in Natural Computation, pp. 554–564 (2005)
19. World Health Organization: Pneumonia (2019). https://www.who.int/news-room/fact-sheets/detail/pneumonia
20. Xu, X., Xu, S., Jin, L., Song, E.: Characteristic analysis of otsu threshold and its applications. Pattern Recogn. Lett. **32**(7), 956–961 (2011)
21. Zhou, H., Wei, L., Lim, C.P., Creighton, D., Nahavandi, S.: Robust vehicle detection in aerial images using bag-of-words and orientation aware scanning. IEEE Trans. Geosci. Remote Sens. **99**, 1–12 (2018)

Deep Feature Extraction for Panoramic Image Stitching

Van-Dung Hoang[1(✉)], Diem-Phuc Tran[2], Nguyen Gia Nhu[2], The-Anh Pham[3], and Van-Huy Pham[4]

[1] Intelligent Systems Laboratory, Quang Binh University, Dong Hoi City, Vietnam
zunghv@gmail.com
[2] Duy Tan University, Da Nang City, Vietnam
phuctd@gmail.com, nguyengianhu@duytan.edu.vn
[3] Hong Duc University, Thanh Hoa City, Vietnam
phamtheanh@hdu.edu.vn
[4] Faculty of Information Technology, Ton Duc Thang University, Ho Chi Minh City, Vietnam
phamvanhuy@tdtu.edu.vn

Abstract. Image stitching is an important task in image processing and computer vision. Image stitching is the process of combining multiple photographic images with overlapping fields of view to produce a segmented panorama, resolution image. It is widely used in object reconstruction, panoramic creating. In this paper, we present an approach based on deep learning for image stitching, which are applied to generate high resolution panoramic image supporting for virtual tour interaction. Different from most existing image matching methods, the proposed method extracts image features using deep learning approach. Our approach directly estimates locations of features between pairwise constraint of images by maximizing an image- patch similarity metric between images. A large dataset high resolution images and videos from natural tourism scenes were collected for training and evaluation. Experimental results illustrated that the deep feature approach outperforms.

Keywords: Image stitching · Feature extraction · Deep learning · Feature matching

1 Introduction

Image stitching has a wide variety of applications. The essential task is handing the required comparing multiple images of the same scene. This problem is very common in the field of object reconstruction, structure from motion (SfM) as well as for optical flow, medical imagery processing, satellite image analysis. Since the 2010s, the image stitching processing has mostly used traditional approaches based on apparent features. These methods are almost based on three sub-tasks: Detection and description of features (such as key-points, key-lines, patches), feature matching, and image warping. Most of popular approaches are using the selected points of interest (PoI) in set of images, which associated PoIs in the referenced images to their equivalent in the

N. T. Nguyen et al. (Eds.): ACIIDS 2020, LNAI 12034, pp. 141–151, 2020.
https://doi.org/10.1007/978-3-030-42058-1_12

sensed image and the transformed images, which are aligned images. In the field of feature detection and description, there are some approaches, such as keypoint feature and patch learning based features. The keypoint approach is related to the important and distinctive points such known as set of major points of interest, e.g. corners, edges, etc. Each keypoint descriptor is represented by a feature vector containing essential characteristics. There are many algorithms for keypoint detection and feature description such as SIFT (Scale-invariant feature transform) [1], SURF (Speeded Up Robust Features) [2], ORB (Oriented FAST and rotated BRIEF) [3], … Among of that, the SIFT method is a robust approach for image registration. The SIFT transforms an image into a large collection of feature vectors, which are feature descriptors stable with invariant to rigid translation, scaling, rotation and robust to local geometric affine distortion, and partially invariant to illumination, brightness changing. SURF is a speeded up and robust for feature extraction and matching. The SURF is a extractor and descriptor, which is the basic improving based in the SIFT approach with the advantage faster. The camera rotation can be computed according to graph-based sampling scheme [4]. The approach in [5] uses the sphere model, the error is minimized in the angle value, on contrary, the error is the distance of re-projection 3D points and feature points in case of perspective camera.

The feature detection and description method is an important step in computer vision pipes. Some groups of researchers were proposed based on machine learning for estimation corresponding features. The convolutional neural network used unlabeled data [6] for training. The feature descriptor enables for robust transformation estimation and outperformed SIFT descriptor for matching task. The learning approach for keypoint detect and matching based on deep learning architecture was presented in [7]. The method contributed new approach for learning to detect and describe keypoints from images deep architecture. The learning model was trained from a large dataset to detect and describe feature keypoints. In contrast, a LF-Net model based on CNNs was proposed to learn from multiple mages for extracting robust features [8]. The method based on inspiration deep semantic feature matching is a novel approach in the field of applied the semantic matching based on convolutional feature pyramids and activation of feature selection. This approach based on pre-trained CNN model for salient feature extraction. The matching task uses a sparse graph matching approach with robust features are selected from a small subset of nearest neighbors of keypoint position [9]. The idea deep convolutional features for non-rigid registration was developed [10]. They used layers of a pre-trained Resnet network to predict feature descriptions which extracts both convolutional features and position capabilities. Experimental results illustrated that outperforms SIFT detector.

2 Multiple View Constraint for Image Matching

The scene modeling task is an important issue in various applications of virtual environment, panorama high-resolution generation, scene planning and navigation of autonomous mobile robot. Although, some progress has been made in the field of virtual 3D scene during the last few decades, still there are limitation of that methods, which are satisfied the requirement of high accurate as well as stable for different kind of dataset.

Some these methods require a large amount of work done manually or apparatus, airborne light detection and ranging. They are usually expensive and require much more time for data acquisition.

Multiple view geometry is the effective method to extract the corresponding information of the real scenes from the camera positions. The general idea is that the camera captures the sequence images of the world from different positions or angle of view, then computer will compute the panoramic structure of world and recover the camera poses, such as in [11]. An example of the constraint of triple views of three observed image is shown in Fig. 1.

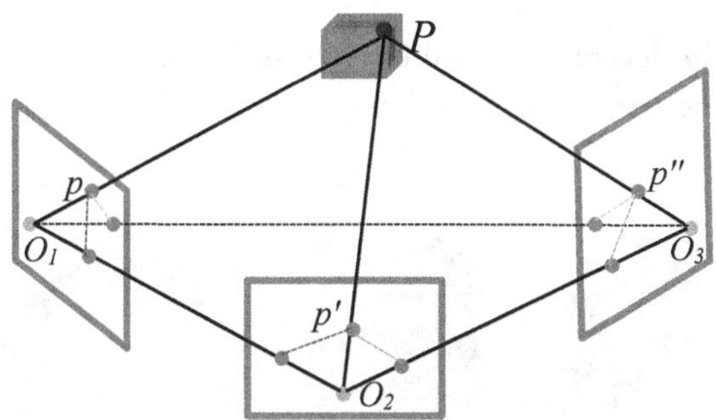

Fig. 1. Multiple view constraints

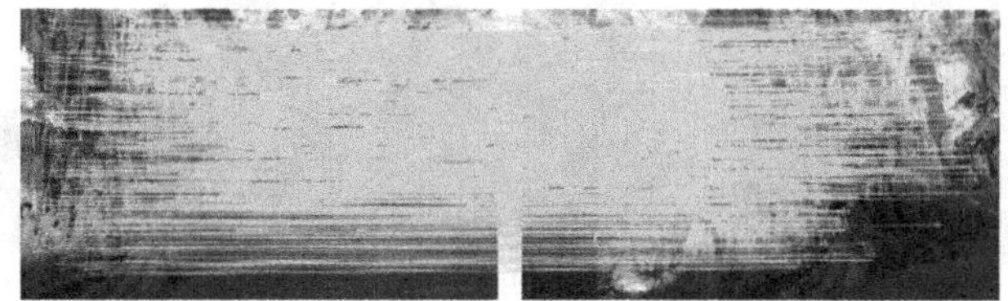

Fig. 2. Keypoint detection and matching of two images for transformation parameters extracting

There are many approaches for extracting the corresponding features (points, lines, planes) of each pair of images, which supports to compute the structure and camera positions. In the field of corresponding point extraction, SIFT and SURF are known as traditional blob detection, supervised machine learning on the patches for extracting dense pixels for training high level of features. Let consider this canonical problem in the mathematic equation. Given the correspondences p and p', the problem is how to find the 3D point P and camera poses with the smallest error. That requirement leads to optimization problem [12] (Fig. 2):

$$\min_{\hat{p}}\left[d(p,\hat{p})^2 + d(p',\hat{p}')^2\right] \tag{1}$$

In the stitching processing, the alignment task needs to transform an image to match the viewpoint of the image. It is simple changed in the coordinates system to a new coordinate system which outputs image matching the required viewpoint of another image. The transformation is mathematically described as follows:

$$p'^T Hp = 0 \tag{2}$$

where the set of (p and p') is corresponding points from pair of images, that is equivalented set p is points in the benchmark coordinate system and p' is the corresponding points in the transformed image, and H is a homography matrix.

The task of image warping, we applied the method in [12]. The image warping processing is following: given source image I and the correspondence between the original position $p_i = (x_i, y_i)$ of a point in image I and its desired new position $p_i' = \left(x_i', y_i'\right)$ with $i = 1, \ldots, n$. *The* generated warped image I′ such that $I\prime\left(p_i'\right) = I(p_i)$ for each point i in image. The warped image uses from 4 images for example is illustrated in Fig. 3.

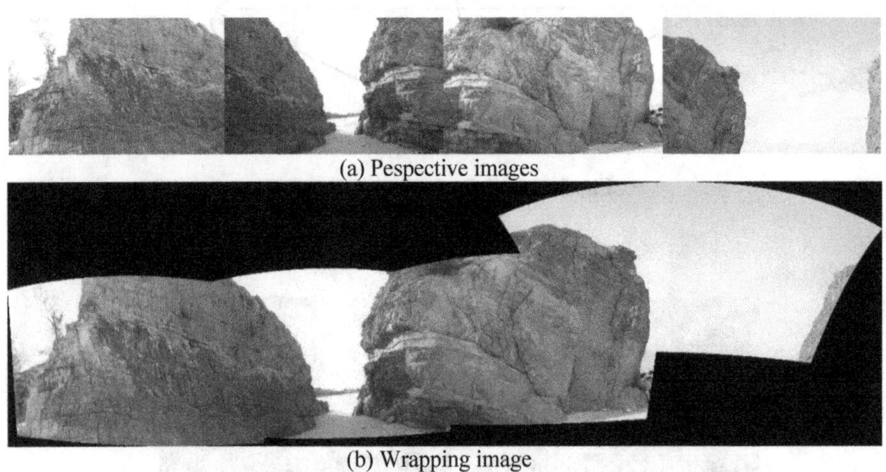

(a) Pespective images

(b) Wrapping image

Fig. 3. Part of panoramic image based image stitching using deep feature extraction

3 Feature Extraction for Image Stitching

3.1 Overview Flowchart of Stitching

Image registration is a process for matching multiple images, which are usual overlapped each other, of the sample scene into the same coordinate system. The spatial relationships between these images can be rigid transformation, affine, homographies, or complex deformations models. In this research, we focus on improving the accuracy in the situation of rigid transformation (translation and rotation), which supports generating panoramic high-resolution images. The general architecture for stitching multiple images is presented in Fig. 3.

The fundamental stitching based on traditional approach is following algorithm:

Step 1: Detecting keypoints (DoG, Harris, etc.) and extracting local invariant descriptors (SIFT, SURF, ORBFASTBRIEF, etc.) from multiple input images
Step 2: Matching the feature descriptors between the multiple images
Step 3: Estimating the homography matrixes using the set of matched features of each pair images
Step 4: The warping transformation image using the homography matrix of each pair images, which are obtained from previous step.

In this study, machine learning approach is proposed for deep feature selection for keypoint extraction and description, as illustrated in Fig. 4. The feature matching is processed for resulting strong corresponding points between each pair of images. The CNN model is reform and retrained multiscale keypoint detections. The learning model is retrained our training dataset, which collected outdoor images and videos from natural scenic spots. Then the essential approach of multiple view constraint is applied for estimating homography matrix from corresponding points.

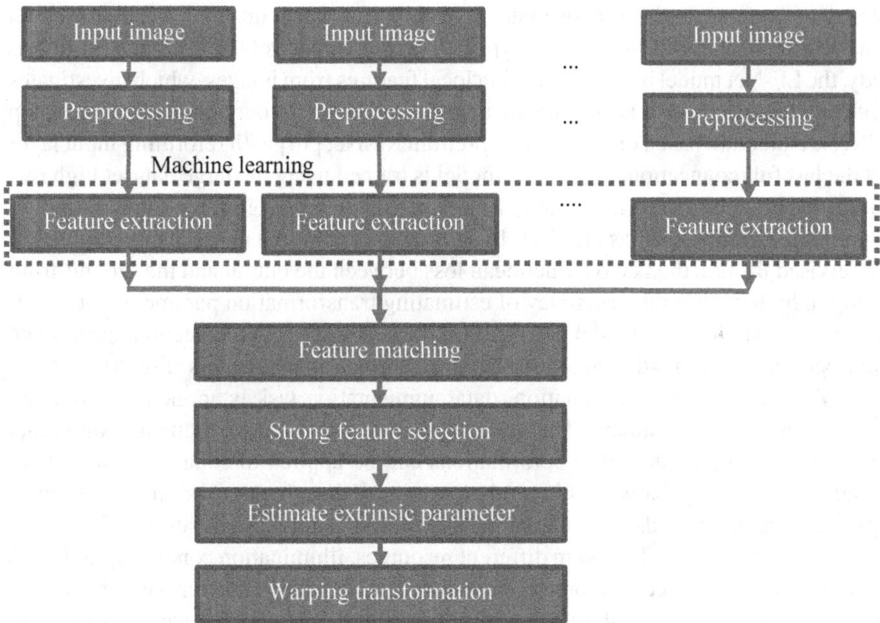

Fig. 4. General flowchart of stitching images

3.2 Learning Approach-Based Feature Extraction

Samples are chosen from training image by selecting a rectangular region of the scene. Color features of interest regions can be calculated based on RGB, HSL and Lab color space approximated human perception. Color features represented for blob regions by its mean values and robustness in HSL space.

Fig. 5. Extracting image samples for learning process

Recently, there are some approach for homography estimation based on deep learning. Among of these methods is the learning-based CNN VGG approach trained on patches of large-scale dataset for matching multiscale keypoints [7]. Another method using learning model to find good keypoint correspondences [13]. The closest to this study, the LF-Net model based on learning local features from images, which investigates depth and relative camera pose to create a virtual target [8]. In our approach, we develop a CNN architecture based on the basic of pretrain ResNet [14] with reforming input layer and the last full connection layers. The model is trained using the large dataset high resolution images and videos from natural tourism scenes. In general, we can use any DL for training a model, such as [15–17]. In another approach, the model can be trained in a supervised fashion thanks to a Euclidean loss between the output and the ground-truth homography for improving accuracy of estimating transformation parameters of multiple views constraint. The model performs on the task of keypoint detection appropriate supported for feature matching to estimate transformation parameters (Fig. 6).

To deal with the natural situation, data augmentation task is applied for covering all situations of posed camera. The underlying concept of image augmentation is that the rotation, flipping, and other deformations can be applied to enhance data without changing and loosing characteristics and properties of data. In our experiments, we have applied several kinds of data augmentation as follows. Color normalization and balance: the facial images were collected in different resources, illumination conditions and from different types of devices. Geometry transformations: affine transformations such as shearing, distorting and scaling randomly warp stroke data for image classification. Thus, affine transformations are very well suited to augment data for improving the overall performance and mitigating overfitting of the training task. We used various techniques for resampling such as rotation, stretching, shearing: random rotation with angle between $-10°$ and $10°$, random shearing with angle between $-10°$ and $10°$; random shearing stretching with stretch factor between $5°$ and $15°$; flipping is also applied in this situation because almost facial shapes are mirror.

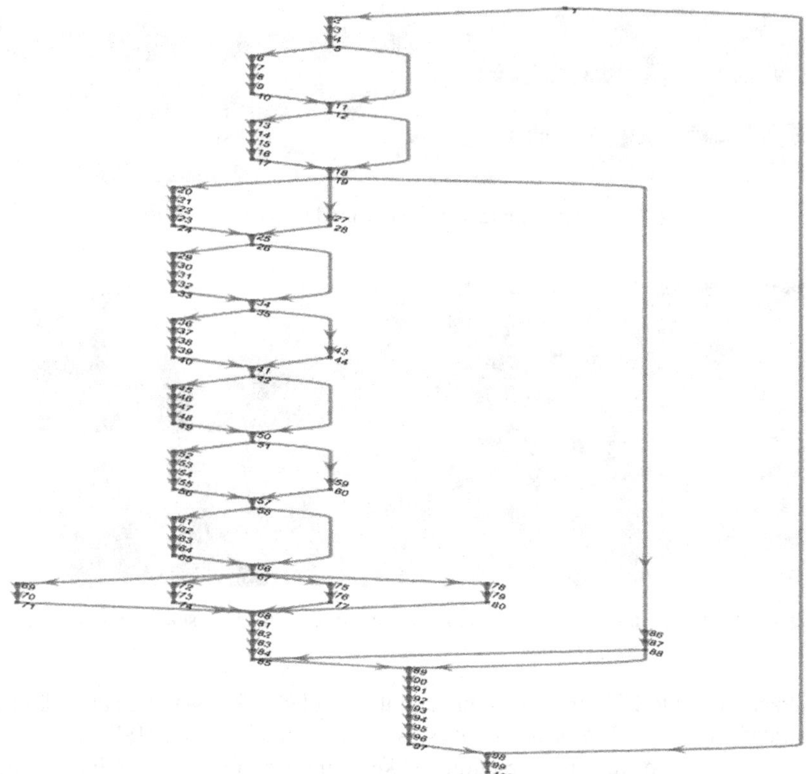

Fig. 6. Deep learning architecture based on pretrain ResNet model

4 Experimental Results

The experiment data was collected from outdoors images about scenic spots, which related to attractions tourism. The image data was acquisited from 15 scenic spots of famous tourist places. For example, Phong Nha-Ke Bang National Park, which is composed of 300 caves and grottos, Paradise Cave, Dark Grotto, Chay rive, and so on. In this objective, image stitching based on registration and wrap to create high resolution panoramic images. It requires professional camera and specialized capturing images with horizontally elongated fields of view. The term has also been applied transformation to a relatively wide aspect ratio of image and video. This dataset is large images, which consists of 46,000 high resolution images (5,148 × 3,456) and 150 high resolution videos (1,920 × 1,088pixels), some image samples are illustrated in Fig. 4. The dataset is splitted to provide training and validation data. The roughly half of the available individual images are used for training. In the panoramic photography, images are captured with more transformation and small rotation (Figs. 7 and 8).

In this experiment, we use pretrain ResNet [14] with reforming input and output layers for retraining in our training data. The input image of the first layer is [1920,1088, 3]. Some sepecific parameters of layers is used 7x7 convolutional blocks with Batch

Fig. 7. Some example datasets captured from scenic spots

Fig. 8. Activation dense orientation and scale-space score map for selecting keypoints

Normalization and ReLUs and are architecturally similar to ResNet, the network graph is illustrated in Fig. 5. The network takes as input a three channel HSL image sized 1920 × 1088 × 3. For feature extraction and keypoint matching, the two input images are stacked channel-wise and fed into the network for extracted the set of corresponding points to estimate homography transformation. For post-processing matching, the RANSAC technique is applied for removing outliners. Particular attention is given to the RANSAC-based fitting method [5]. Without using any additional scene assumptions, except that the scene is represented point-clouds or patches, the RANSAC-based approach in corresponding points matching and removing outliners are like the canonical line fitting problem.

First, a point in the set of points is randomly chosen. Then, the *n* closet points within a certain distance are also selected. This can be considered as the error associated with this method. The points inside the upper and lower plane are inliers. The RANSAC algorithm is used to reject the outliers and fit points to this subset. The process of randomly selecting the first points is repeated until the maximum number of iterations is reached or the number of remaining points is smaller than *n*. We see that the size of the points is dependent on the number of points *n*. In this problem, the homography matrix requires 8-points correspondence, for more details referred to [12] (Figs. 9 and 10).

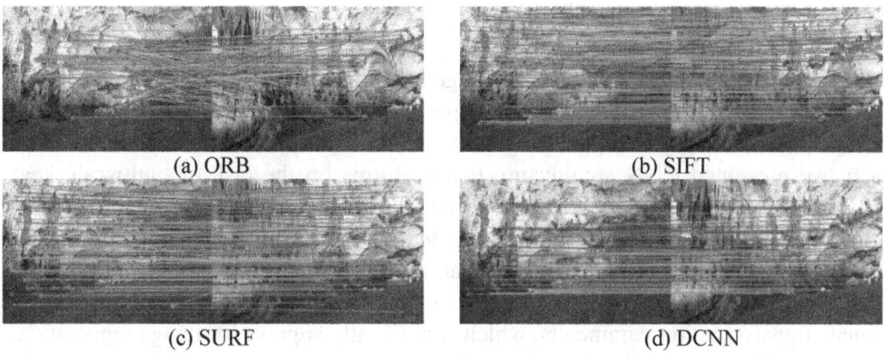

(a) ORB (b) SIFT

(c) SURF (d) DCNN

Fig. 9. Comparison results of ORB, SIFT, SURF, DCNN for keypoint extraction and matching

(a) Set of individual images

(b) Stitching image using 8 images from (a)

(c) Full image stitching using 76 images

Fig. 10. Panoramic image of the Phong-Nha cave based on image stitching technique.

5 Conclusion

This paper presents a stitching approach for generating high resolution panoramic image of scenic spots. The system is essential approach in computer vision for estimating homography matrix of each pair images based on deep feature selection. In this approach, we presented CNN architecture that performs on the task of multiscale keypoint detection and feature description for matching and selecting corresponding points between each pair of images. To retraining detection model, we created large dataset high resolution images and videos from natural tourism scenes. Experimental results illustrated that the deep feature extraction is appropriate for extracting keypoints to estimate transformation parameters, which significantly support for image registration, wrapping, and stitching. The plan further focuses to investigate the deeper architecture and train on larger benchmark dataset for outperformance and improve for multiscale feature detection and description.

References

1. Lowe, D.: Distinctive image features from scale-invariant keypoints. Int. J. Comput. Vision **60**, 91–110 (2004)
2. Bay, H., Tuytelaars, T., Van Gool, L.: SURF: speeded up robust features. In: Leonardis, A., Bischof, H., Pinz, A. (eds.) ECCV 2006. LNCS, vol. 3951, pp. 404–417. Springer, Heidelberg (2006). https://doi.org/10.1007/11744023_32
3. Rublee, E., Rabaud, V., Konolige, K., Bradski, G.R.: ORB: an efficient alternative to SIFT or SURF. In: International Conference on Computer Vision, vol. 11, pp. 2–10 (2011)
4. Govindu, V.: Robustness in motion averaging. In: European Conference Computer Vision (2006)
5. Le, M.-H., Trinh, H.-H., Hoang, V.-D., Jo, K.-H.: Automated architectural reconstruction using reference planes under convex optimization. Int. J. Control Autom. Syst. **14**, 814–826 (2016)
6. Fischer, P., Dosovitskiy, A., Brox, T.: Descriptor matching with convolutional neural networks: a comparison to sift. arXiv preprint arXiv:1405.5769 (2014)
7. Altwaijry, H., Veit, A., Belongie, S.J., Tech, C.: Learning to detect and match keypoints with deep architectures. In: BMVC (2016)
8. Ono, Y., Trulls, E., Fua, P., Yi, K.M.: LF-Net: learning local features from images. In: Advances in Neural Information Processing Systems, pp. 6234–6244 (2018)
9. Ufer, N., Ommer, B.: Deep semantic feature matching. In: Proceedings of the IEEE Conference on Computer Vision and Pattern Recognition, pp. 6914–6923 (2017)
10. Yang, Z., Dan, T., Yang, Y.: Multi-temporal remote sensing image registration using deep convolutional features. IEEE Access **6**, 38544–38555 (2018)
11. Hoang, V.-D., Le, M.-H., Jo, K.-H.: Motion estimation based on two corresponding points and angular deviation optimization. IEEE Trans. Industr. Electron. **64**, 8598–8606 (2017)
12. Hartley, R., Zisserman, A.: Multiple View Geometry in Computer Vision. Cambridge University Press, Cambridge (2004)
13. Moo Yi, K., Trulls, E., Ono, Y., Lepetit, V., Salzmann, M., Fua, P.: Learning to find good correspondences. In: Proceedings of the IEEE Conference on Computer Vision and Pattern Recognition, pp. 2666–2674 (2018)
14. He, K., Zhang, X., Ren, S., Sun, J.: Deep residual learning for image recognition. In: Proceedings of the IEEE Conference on Computer Vision and Pattern Recognition, pp. 770–778 (2016)

15. Tran, D.-P., Hoang, V.-D.: Adaptive learning based on tracking and ReIdentifying objects using convolutional neural network. Neural Process. Lett. **50**, 263–282 (2019)
16. Hoang, V.-D., Le, M.-H., Tran, T.T., Pham, V.-H.: Improving traffic signs recognition based region proposal and deep neural networks. In: Nguyen, N.T., Hoang, D.H., Hong, T.-P., Pham, H., Trawiński, B. (eds.) ACIIDS 2018. LNCS (LNAI), vol. 10752, pp. 604–613. Springer, Cham (2018). https://doi.org/10.1007/978-3-319-75420-8_57
17. Tran, D.-P., Hoang, V.-D., Pham, T.-C., Luong, C.-M.: Pedestrian activity prediction based on semantic segmentation and hybrid of machines. J. Comput. Sci. Cybern. **34**, 113–125 (2018)

Video-Based Traffic Flow Analysis for Turning Volume Estimation at Signalized Intersections

Khac-Hoai Nam Bui, Hongsuk Yi(✉), Heejin Jung, and Jiho Cho

Korea Institue of Science and Technology Information, Daejeon 34141, Korea
hoainam.bk2012@gmail.com, {hsyi,bigbear,jhcho}@kisti.re.kr

Abstract. Traffic flow analysis in complex areas (e.g., intersections and roundabouts) plays an important part in the development of intelligent transportation systems. Among several methods for analyzing traffic flow, image and video processing has emerged as a potential approach to extract the movements of vehicles in urban areas. In this regard, this study develops a traffic flow analysis method, which focuses on extracting traffic information based on Video Surveillance (CCTV) for turning volume estimation at complex intersections, using advanced computer vision technologies. Specifically, state-of-the-art techniques such as Yolo and DeepSORT for the detection, tracking, and counting of vehicles have enveloped to estimate the road traffic density. Regarding the experiment, we collected data from CCTV in an urban area during one day to evaluate our method. The evaluation shows the proposing results in terms of detecting, tracking and counting vehicles with monocular videos.

Keywords: Traffic flow analysis · Computer vision · Deep learning · Detection and tracking · Turning movement counts · Traffic operation

1 Introduction

Accurate information on traffic flow is an important step in the development of various traffic control applications in Intelligent Transportation Systems (ITS). Specifically, vehicle counting is the key technique to evaluate traffic status for conducting traffic flow estimation and analysis. For instance, results from the vehicle counting technique can be used to analyze the operational performance of signalized intersections under different traffic conditions (e.g., peak and non-peak hours) [19]. Specifically, several well-known methods have been introduced for capturing traffic situations such as manual counts, pneumatic road tube sensors, and video detection using computer vision [6]. Among vehicle counting approaches, with the recent rapid development of Deep Learning (DL), vehicle detection using computer vision methods have several advantages than others due to the following reasons:

- Video footage can be used to verify and improve the evaluation systems.

N. T. Nguyen et al. (Eds.): ACIIDS 2020, LNAI 12034, pp. 152–162, 2020.
https://doi.org/10.1007/978-3-030-42058-1_13

- Various applications can be applied for traffic monitoring such as automatic plate number recognition, vehicle type, and speed detection.
- Video-based systems can track and count multiple vehicles moving in different directions across several lanes in terms of sustainability and scalability.

Fig. 1. Considered scenario.

In this regard, this paper focuses on implementing a video-based traffic flow analysis from CCTV for turning volume estimation at complex intersections of an urban area. Specifically, several advanced techniques in video and image processing are enclosed to improve the performance of the system. Generally, the main issues of this study are described as follows:

- We take an investigation on how to apply advanced technologies of computer vision for analyzing the video-based traffic flow which is the emergent issue in this research field.
- We focus on complex scenarios in which the input videos from a single view of the CCTV which is normally used for traffic surveillance at intersections. For instance, Fig. 1 depicts the scenario that we take into account in this study. Therefore, several processing techniques are introduced (e.g, tracking method and virtual line setting) to improve the accuracy.

The rest of this paper is organized as follows: Sect. 2 presents the literature review of traffic monitoring using computer vision for the development of ITS and recent state-of-the-art approaches of object detection and tracking using deep learning. In Sect. 3, we propose a video-based traffic flow analysis approach by applying Yolov3 and DeepSORT methods for vehicle detection and tracking. Section 4 demonstrates the results of our implementation in which the data have collected at a complex intersection of an urban area. Discussions and future works are concluded in Sect. 5.

2 Literature Review

2.1 Traffic Monitoring Using Computer Vision

In recent years, ITS have paid attention to detailed monitoring of road traffic, especially in complex areas such as roundabout and intersections, in order to provide intelligent traffic control [1,2]. The conventional methods using sensors at each of the entrances/exits areas for counting the traffic flow currently have a limitation in the sensing problem. Moreover, sensory systems require a large number of sensors since the complex areas involve conflict situations.

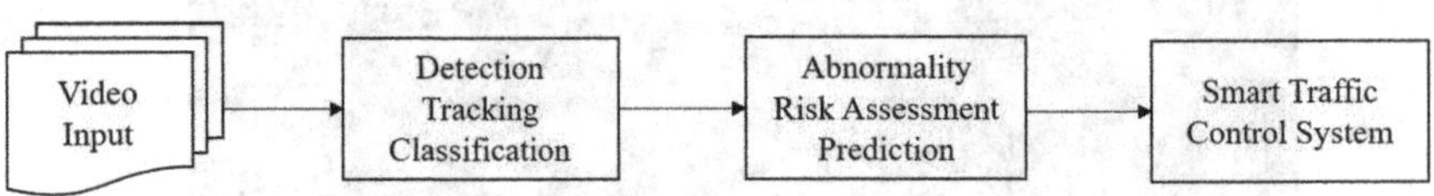

Fig. 2. Classic steps in video monitoring.

Recently, with the rapid growth of advanced techniques in computer vision, many studies focus on detecting and tracking various vehicles by analyzing a camera capture with the help of computer vision. Specifically, Vehicle monitoring based on video analysis consists of two steps as shown in Fig. 2 [5]. The first step includes vehicle detection and tracking, and the output of this level provides vehicle information such as positions, speeds, and classes. The second step, based on the output of the previous level, performs tasks such as predictions of specific behaviors or abnormal detection of the traffic flow [20,21].

2.2 Object Detection and Tracking Using Deep Learning

Object detection and tracking from video sequences has been widely applied in people's life and currently are the main challenges in computer vision. In particular, with the rapid development of deep learning networks for detection tasks, especially, Convolution Neural Network (CNN) based methods have recently achieved great success in object detection [9]. Specifically, many promising approaches have been introduced for detecting moving objects which can be classified into two categories such as single-stage and two-stage methods. Single-stage approaches perform in a single pass feature extraction and propose regions by a final regression layer which are able to achieve high inference speed. Several well-known approaches of this method are SSD [12], YOLO [14], and RetinaNet [10]. On the other hand, two-stage approaches such as Fast R-CNN [7], Mask R-CNN [8], and R-FCN [4] are able to achieve more accurate results by optimizing regressions on each individual region. In the case of object tracking, the recent approaches are classified into categories such as single object trackers (SOT) and multiple object trackers (MOT) which depends on the objective of the tracking process [23]. Specifically, SOT methods track a certain object by specifying from the beginning which does not rely on the detection process. In contrast, MOT is the tracking by detection paradigm in which the tracking algorithms rely on localization results from the object detection methods.

3 Traffic Flow Analysis Based on Video Resources

3.1 The Overview System

Figure 3 depicts the overview of our proposed method. Specifically, the video stream is captured by the monocular cameras that are located in complex areas (e.g., intersections or roundabouts).

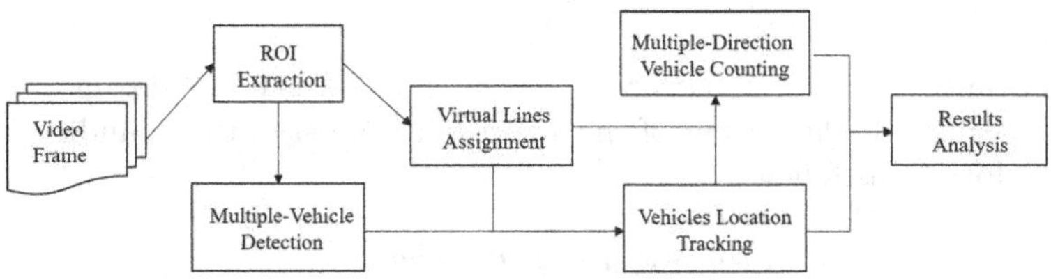

Fig. 3. The overview of video-based traffic flow analysis.

Consequently, a Region of Interest (ROI) is processed to remove noises and correct perspective distortion. Then, the vehicle detection and tracking process identifies the presence, location, and type of numerous vehicles in each frame and tracks them across the entire video, respectively. Finally, vehicles in the predefined virtual lines, which are set in different directions of the considered area, are recorded and counted.

3.2 Methodology

3.2.1 Vehicle Detection

Vehicle Detection is an essential process for extracting traffic information from video resources. Technically, as mentioned above, the generic vehicle detection can be categorized into two types such as the region proposal based methods and the regression/classification based methods [22]. In this study, we apply Yolo for the detecting vehicles due to the following reasons:

- The method is able to achieve high performance with a single end-to-end model that can perform vehicle detection in real-time on video or streaming input from the camera.
- The new version of Yolo (refer as Yolov3) includes 53 convolutional layers (Darknet-53) which are able to provide the better detection process of multiple vehicles with various dimensions in each frame [15].

Fig. 4. Vehicle detection using Yolov3.

For instance, Fig. 4 demonstrates the results of the considered scenario using Yolov3. Specifically, the output of the algorithm is a list of the bounding box in which the format as follows:

$$< class, x, y, w, h, confidence > \tag{1}$$

where class denotes the classification of the detected object, (x, y, w, h) represents the parameters of the bounding box, and confidence is the score of the detected object. In order to maximize the utility of the vehicle detection, in this study, we try to filter the output in which the detected objects involved in several cases as follows:

- The confidence of the detected object is less than a given threshold (α).
- The detected object belongs to non-vehicle classes such as pedestrians, traffic lights, and so on.
- The bounding boxes, that have overlap within the same frame, is measured by the Intersection-over-Union (IoU) score with a given threshold (β).

3.2.2 Vehicle Location Tracking

Technically, bounding boxes have been used to count the number of obstacles in the same class (e.g., crowd, drones, surveillance cameras, and autonomous robots). However, the limitation of object detection is that with the same class, obstacles from the same color cannot be set independently, especially in the complex videos which include various objects [13]. In this regard, with the input bounding box from vehicle detection, the vehicle tracking process is able to assign a unique ID to the particular vehicle, track the vehicle moving around the considered area, and predicting the new location in the next frame based on various attributes of the frame (e.g., gradient) [3].

Currently, two state-of-the-art methods, which follow the tracking by detection paradigm for vehicle tracking, are DeepSORT and TC [16]. DeepSORT is an online method that combines deep learning features with Kalman-filter and

Hungarian algorithm for data association [18], and TC is an offline method that proposes the fusion of visual and semantic features for both single-camera tracking (SCT) and inter-camera tracking (ICT) [17]. In this study, since our main issue focuses on extracting information on traffic flow at a certain area, we apply DeepSORT for tracking the location of vehicles across different frames. Specifically, the state of a target vehicle at a certain point is represented as follows:

$$< x, y, h, r, \dot{x}, \dot{y}, \dot{h}, \dot{r} > \tag{2}$$

where (x, y), r and h denote the centre, aspect ratio, and height of the bounding boxes, respectively, and the other variables are the respective velocities of the variables. In particular, there are two main approaches in DeepSORT in order to track the moving objects in video sequences:

- Hungarian algorithm determines whether a vehicle in the current frame is the one in the previous frame or not based on the score threshold (e.g., IoU). In particular, the result is applied to the association and attribution of the identification number.
- Kalman Filter algorithm predicts future positions based on the current position. This result is used to update the target state.

3.2.3 Vehicle Counting

There are two main methods for the existing vehicle counting systems from videos such as setting baselines and using virtual loops [11] in which the line detection is used since the high speed of vehicles. In this study, since we consider the traffic flow in multiple directions, multiple virtual lines are set in different positions for counting the vehicles.

Consequently, in order to count the number of vehicles pass through in each position, tracking list from vehicle location tracking, which contains the centroids of bounding boxes, is defined. Specifically, supporting (x, y, w, h) and (x',y', w', h') denote the parameters of the current and previous of a certain bounding box, respectively, the positions of the current and previous centroids (C_{cur} and C_{pre}) of the bounding box are calculated as follows:

$$\begin{aligned} C_{cur} &= (x + \frac{w - x}{2}; y + \frac{h - y}{2}) \\ C_{pre} &= (x' + \frac{w' - x'}{2}; y' + \frac{h' - y'}{2}) \end{aligned} \tag{3}$$

Therefore, the vehicle will be counted in a certain route if it passes the virtual line of that direction. Specifically, the Algorithm 1 demonstrates the counting algorithm in our study. Furthermore, since the overlap view of multiple directions from the camera capture, the compared position between current and previous centroids is also taken into consideration to check the status IN/OUT of vehicles in each virtual line.

Algorithm 1. Vehicle counting with multiple directions

```
Data: Current Bounding Box T_cur; Previous Bounding Box T_pre; Set
      L(x,y)={l_1, l_2, ..., l_n} of virtual line positions in all directions
Result: Number of vehicles passing in each direction
Set (x, y, w, h) = (T_cur[0]), T_cur[1],T_cur[2], T_cur[3])
Set (x', y', w', h') = (T_pre[0]), T_pre[1],T_pre[2], T_pre[3])
C_cur = (x + (w-x)/2; y + (h-y)/2)
C_pre = (x' + (w'-x')/2; y' + (h'-y')/2)
while i ∈ L do
    if intersect(C_cur, C_pre, i[x], i[y] = True) then
        count_i += 1
    end
end
Function intersect(A,B,C,D)
return check(A,C,D) != check(B,C,D) and check(A,B,C) != check(A,B,D)
Function check(A,B,C)
return (C[y] − A[y]) ∗ (B[x] − A[x]) > (B[y] − A[y]) ∗ (C[x] − A[x])
```

4 Implementation

4.1 Data Description and Implemented Setup

For the implementation, we first collect data from the intersection of an urban area. Then, we use *Keras*[1], a python deep learning library to develop our method for the evaluation. In order to collect the video data, a digital camera was set-up at an intersection of an urban area. There are 30 input videos which are recorded from 6:00 AM. to 8:00 PM. The spatial resolution of the collected

Fig. 5. The location of the considered scenario.

[1] https://keras.io/.

videos is 1920×1080 pixels, with a temporal resolution of 30 Frames Per Second (FPS). Figure 5 depicts the location of our scenario for the implementation[2]. Specifically, our work based on the Keras implementation of Yolov3 for object detection[3] and DeepSORT for object tracking in [18]. The experiments work well by a PC with Core i7 16-GB CPU and 32 GB GPU memories in which the GPU is used for the acceleration. Particularly, Table 1 depicts the parameters that we use for the implementation based on the input videos.

Table 1. Parameter setting

Parameter	Values
Resolution	1920×1080 pixels
Time duration	3–5 min/input video
Temporal resolution	30 FPS
Score detection threshold (α)	0.6
IoU detection threshold (β)	0.5
Number of virtual lines	4 (routes)

Fig. 6. Considered scenario using proposed method.

4.2 Results Analysis

Regarding the running time of our implementation, the executed time is from 12 to 13 FPS when only using Yolov3 for the detection and from 11 to 12 FPS after adding DeepSORT for the tracking. Figure 6 depicts the traffic analysis from videos using the proposed method. As shown in the figure, there are 4 routes (4 directions in the intersection) that we take into account for analyzing traffic

[2] http://www.skymaps.co.kr/.
[3] https://github.com/qqwweee/keras-yolo3.

flow. Furthermore, from our observation, the thresholds of the detected score and IoU are 0.6 and 0.5, respectively, that can give the best results for vehicle detection.

In order to measure the performance, we compare our results with the manual count. Specifically, there are three input videos with different environments for evaluations such as normal and congestion of traffic flow during the day and night times which are shown in Table 2. The accuracy of vehicle counting of our approach achieves 93,88 % and 87,88 % for the normal and congestion of traffic flow, respectively, and 82,1 % at the night time.

Table 2. Results of the proposed method

Video	Time duration	Frames	Environment	Ground truth	Detected vehicles	Accuracy (%)
vdo. 1	08:07:21-08:11:21	7210	Day, Congestion	396	348	87,88 %
vdo. 2	11:57:23-12:02:23	9018	Day, Normal	343	322	93,88 %
vdo. 3	18:57:28-19:00:28	5388	Night, Normal	313	257	82,1 %

In this study, we consider the scenario with input videos from one view of the CCTV, therefore, the view of several directions will be limited. In this regard, the results are acceptable in this regard. Specifically, Table 3 explains in more detail the number of vehicles in each direction. Particularly, in case of the routes 3 and 4 (Fig. 6), since the vehicles move in the opposite directions of the CCTV, the accuracies are lower than the vehicle counting in route 1 and route 2.

Table 3. Number of vehicles in each direction

Video	Route 1/Ground Truth	Route 2/Ground Truth	Route 3/Ground Truth	Route 4/Ground Truth
vdo. 1	63/70	175 /185	50/62	60/79
vdo. 2	81/82	101/106	34/38	106/117
vdo. 3	55/64	63/73	33/39	106/137

5 Conclusion and Future Work

With the rapid growth of Deep learning, the field of computer vision is shifting from statistical methods to deep learning neural network methods that are able to provide various applications such as the retail stores, automotive, healthcare, and transportation system. In this study, we propose a traffic flow analysis methods using advanced techniques methods in computer vision for improving traffic counting systems from CCTV. Specifically, object detection using Yolov3 has been employed for detecting moving vehicles at a certain intersection. Furthermore, since the complexity of vehicle detection in which the vehicle moves fast and there are many vehicles in the same frame, vehicle tracking using DeepSORT has been applied to improve the performance. The implementation of the specific data that we collect at an intersection of a certain area with different environments has shown the promising results of our method. However, since the vehicles move fast and sometimes look similar (e.g., similar color and shape), the switch ID problem occur seriously. Hence, research on switch ID problem is required for improving the traffic analysis based on Videos in which we take the issue as our future work of this study.

Acknowledgment. This work was partly supported by Institute for Information & communications Technology Promotion (IITP) grant funded by the Korea government (MSIT) (No. 2018-0-00494, Development of deep learning-based urban traffic congestion prediction and signal control solution system) and Korea Institute of Science and Technology Information (KISTI) grant funded by the Korea government (MSIT) (K-19-L02-C07-S01).

References

1. Bui, K.H.N., Jung, J.J.: Cooperative game-theoretic approach to traffic flow optimization for multiple intersections. Comput. Electr. Eng. **71**, 1012–1024 (2018). https://doi.org/10.1016/j.compeleceng.2017.10.016
2. Bui, K.H.N., Jung, J.J.: Computational negotiation-based edge analytics for smart objects. Inf. Sci. **480**, 222–236 (2019). https://doi.org/10.1016/j.ins.2018.12.046
3. Ciaparrone, G., Sánchez, F.L., Tabik, S., Troiano, L., Tagliaferri, R., Herrera, F.: Deep learning in video multi-object tracking: a survey. CoRR abs/1907.12740 (2019). http://arxiv.org/abs/1907.12740
4. Dai, J., Li, Y., He, K., Sun, J.: R-FCN: object detection via region-based fully convolutional networks. In: Proceedings of the 30th Annual Conference on Neural Information Processing Systems (NIPS 2016), pp. 379–387. Curran Associates Inc. (2016)
5. Datondji, S.R.E., Dupuis, Y., Subirats, P., Vasseur, P.: A survey of vision-based traffic monitoring of road intersections. IEEE Trans. Intell. Transp. Syst. **17**(10), 2681–2698 (2016). https://doi.org/10.1109/TITS.2016.2530146
6. Ghanim, M., Shaaban, K.: Estimating turning movements at signalized intersections using artificial neural networks. IEEE Trans. Intell. Transp. Syst. **20**(5), 1828–1836 (2019). https://doi.org/10.1109/TITS.2018.2842147
7. Girshick, R.B.: Fast R-CNN. In: Proceedings of the 2015 IEEE International Conference on Computer Vision(ICCV 2015), pp. 1440–1448. IEEE Computer Society (2015). https://doi.org/10.1109/ICCV.2015.169
8. He, K., Gkioxari, G., Dollár, P., Girshick, R.B.: Mask R-CNN. In: Proceedings of the 2017 IEEE International Conference on Computer Vision (ICCV 2017), pp. 2980–2988. IEEE Computer Society (2017). https://doi.org/10.1109/ICCV.2017.322
9. Jiao, L., et al.: A survey of deep learning-based object detection. IEEE Access **7**, 128837–128868 (2019). https://doi.org/10.1109/ACCESS.2019.2939201

10. Lin, T., Goyal, P., Girshick, R.B., He, K., Dollár, P.: Focal loss for dense object detection. In: Proceedings of the 2017 IEEE International Conference on Computer Vision (ICCV 2017), pp. 2999–3007. IEEE Computer Society (2017). https://doi.org/10.1109/ICCV.2017.324
11. Liu, F., Zeng, Z., Jiang, R.: A video-based real-time adaptive vehicle-counting system for urban roads. PLoS ONE **12**(221), e0186098 (2017). https://doi.org/10.1371/journal.pone.0186098
12. Liu, W., Anguelov, D., Erhan, D., Szegedy, C., Reed, S., Fu, C.-Y., Berg, A.C.: SSD: single shot multibox detector. In: Leibe, B., Matas, J., Sebe, N., Welling, M. (eds.) ECCV 2016. LNCS, vol. 9905, pp. 21–37. Springer, Cham (2016). https://doi.org/10.1007/978-3-319-46448-0_2
13. Luo, H., Xie, W., Wang, X., Zeng, W.: Detect or track: towards cost-effective video object detection/tracking. In: Proceedings of the 33th AAAI Conference on Artificial Intelligence (AAAI 2019), pp. 8803–8810. AAAI Press (2019). https://doi.org/10.1609/aaai.v33i01.33018803
14. Redmon, J., Divvala, S.K., Girshick, R.B., Farhadi, A.: You only look once: unified, real-time object detection. In: Proceedings of the 26th IEEE Conference on Computer Vision and Pattern Recognition (CVPR 2016), pp. 779–788. IEEE Computer Society (2016). https://doi.org/10.1109/CVPR.2016.91
15. Redmon, J., Farhadi, A.: Yolov3: an incremental improvement. CoRR abs/1804.02767 (2018). http://arxiv.org/abs/1804.02767
16. Tang, Z., et al.: Cityflow: a city-scale benchmark for multi-target multi-camera vehicle tracking and re-identification. In: Proceedings of the 29th IEEE Conference on Computer Vision and Pattern Recognition (CVPR 2019), pp. 8797–8806. IEEE Computer Society (2019)
17. Tang, Z., Wang, G., Xiao, H., Zheng, A., Hwang, J.: Single-camera and inter-camera vehicle tracking and 3d speed estimation based on fusion of visual and semantic features. In: Proceedings of the 28th IEEE Conference on Computer Vision and Pattern Recognition (CVPR 2018), pp. 108–115. IEEE Computer Society (2018). https://doi.org/10.1109/CVPRW.2018.00022
18. Wojke, N., Bewley, A., Paulus, D.: Simple online and realtime tracking with a deep association metric. In: Proceedings of the 24th International Conference on Image Processing (ICIP 2017), pp. 3645–3649. IEEE (2017). https://doi.org/10.1109/ICIP.2017.8296962
19. Xia, Y., Shi, X., Song, G., Geng, Q., Liu, Y.: Towards improving quality of video-based vehicle counting method for traffic flow estimation. Sig. Process. **120**, 672–681 (2016). https://doi.org/10.1016/j.sigpro.2014.10.035
20. Yi, H., Bui, K.-H.N.: VDS data-based deep learning approach for traffic forecasting using LSTM network. In: Moura Oliveira, P., Novais, P., Reis, L.P. (eds.) EPIA 2019. LNCS (LNAI), vol. 11804, pp. 547–558. Springer, Cham (2019). https://doi.org/10.1007/978-3-030-30241-2_46
21. Yi, H., Bui, K.H.N., Jung, H.: Implementing a deep learning framework for short term traffic flow prediction. In: Proceedings of the 9th International Conference on Web Intelligence, Mining and Semantics (WIMS 2019), pp. 7:1–7:8. ACM (2019). https://doi.org/10.1145/3326467.3326492
22. Zhao, Z., Zheng, P., Xu, S., Wu, X.: Object detection with deep learning: a review. CoRR abs/1807.05511 (2018). http://arxiv.org/abs/1807.05511
23. Zhong, Z., Yang, Z., Feng, W., Wu, W., Hu, Y., Liu, C.: Decision controller for object tracking with deep reinforcement learning. IEEE Access **7**, 28069–28079 (2019). https://doi.org/10.1109/ACCESS.2019.2900476

Slice Operator for Efficient Convolutional Neural Network Architecture

Van-Thanh Hoang and Kang-Hyun Jo[(✉)]

School of Electrical Engineering, University of Ulsan, Ulsan, Korea
thanhhv@islab.ulsan.ac.kr, acejo@ulsan.ac.kr

Abstract. Convolutional neural networks (CNNs) have shown remarkable performance in various computer vision tasks in recent years. However, the increasing model size has raised challenges in adopting them in real-time applications as well as mobile and embedded vision applications. A number of efficient architectures have been proposed in recent years, for example, MobileNet, ShuffleNet, MobileNetV2, and ShuffleNetV2. This paper describes an improved version of ShuffleNetV2, which uses the Channel Slice operator with slice-step parameters to make information interaction between two channels, instead of using Channel Shuffle and Channel Split operators. Because the Channel Slice and Channel Split operators are similar and the proposed architecture does not have Channel Shuffle operator, it has lower memory access cost than ShuffleNetV2. The experiments on ImageNet demonstrate that the proposed network is faster than ShuffleNetV2 while still achieves similar accuracy.

1 Introduction

Deep convolutional neural networks (CNNs) have shown significant performance in many computer vision tasks in recent years. The primary trend for solving major tasks is building deeper and larger CNNs [5,18]. The most accurate CNNs usually have hundreds of layers and thousands of channels [5,10,19,22]. Many real-world applications need to be performed in real-time and/or on limited-resource mobile devices. Thereby, the model should be compact and low computational cost. The model compression work is actually investigating the trade-off between efficiency and accuracy.

Recently, many research works focus on the field of designing efficient architecture. Inspired by the architecture proposed in [12], the Inception module is proposed in GoogLeNet [18] to build deeper networks without increase model size and computational cost. Then it is further improved in [19] through factorizing convolution. The Depthwise Separable Convolution (DWConvolution) generalized the factorization idea and decomposed the standard Convolution into a depthwise convolution followed by a pointwise 1×1 convolution. MobileNets [8,17], ShuffleNet [13,24] and other networks [3,20] have designed CNNs for limited-resource applications based on DWConvolution and shown that this

N. T. Nguyen et al. (Eds.): ACIIDS 2020, LNAI 12034, pp. 163–173, 2020.
https://doi.org/10.1007/978-3-030-42058-1_14

operation to be able to achieve comparable results with a fewer number of parameters.

The model size (number of parameters) and computation complexity (FLOPs - the number of multiply-adds) is widely used to measure the computational cost. However, they are indirect metrics. They are approximation of, but usually not equivalent to the direct metric that we really care about, such as the inference speed test on real devices, considering that there are many other factors that may influence the actual time cost, like caching, memory access, I/O, hardware/software optimization, etc. [13]. Therefore, using model size and FLOPs as the only metrics for computation complexity is insufficient and could lead to sub-optimal design.

Additionally, operations with the same FLOPs could have different running time, depending on the platform. The main reason is maybe because of the different hardware/software optimization. For example, the latest CUDNN [2] library is specially optimized for 3×3 Convolution. So it is not certain that 3×3 Convolution is 9 times slower than 1×1 Convolution.

With these observations, there are two principles should be considered for effective network architecture design. First, the direct metric (e.g., speed or latency) should be used instead of the indirect ones (e.g., FLOPs and/or model size). Second, this metric should be evaluated on the target platform.

This paper proposes an improved version of ShuffleNetV2 [13]. To make information interaction between two channels, the ShuffleNetV2 uses Channel Shuffle and Channel Split operators, these two operators have no floating-point operation but high memory access cost (MAC). For the same purpose, the proposed network uses the Channel Slice operator with slice-step parameters. Because the Channel Slice and Channel Split operators are similar and the proposed network does not have Channel Shuffle operator, it has lower MAC than ShuffleNetV2. Experiments on ImageNet datasets and COCO datasets demonstrate that the proposed models can achieve similar accuracy with less inference time on different platforms: GPU and x86 CPU.

2 Related Work and Background

2.1 Related Work

Recently, there are many studies focus on the designing efficient architectures approach [8,10,17,24,25]. They have explored efficient CNNs that can be trained end-to-end. Three well-known applicants of this kind of approach that are sufficiently efficient to be deployed on mobile devices are MobileNet [8,17], ShuffleNet [13,24], and Neural Architecture Search networks (NASNet) [25]. All these networks use DWConvolutions, which greatly reduce computational requirements without significantly reducing accuracy. A practical downside of these networks is DWConvolution are not (yet) efficiently implemented in most prominent deep-learning platforms. Therefore, some studies use the well-supported group convolution operation [25], such as CondenseNet [9] and Res-NeXt [21], leading to better computational efficiency in practice.

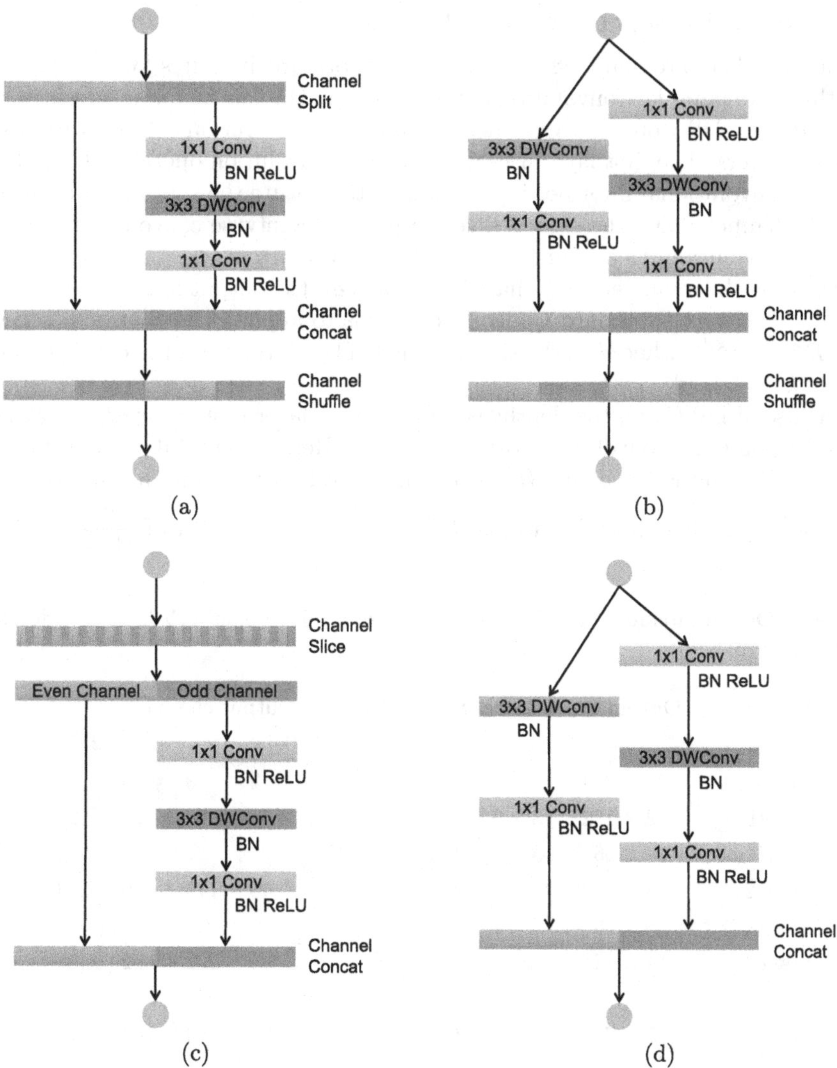

Fig. 1. Architecture of building blocks of ShuffleNetV2 [13] and proposed network. (a) The basic ShuffleNetV2 unit; (b) The ShuffleNetV2 unit for spatial down sampling (2×). (c) The basic proposed unit; (d) The proposed unit for spatial down sampling when output channel is double input. The down sampling will be in the DWConv layer. **DWCon**: depthwise separable convolution.

For designing an efficient CNN architecture, the ShuffleNetV2 [13] introduced some practical guidelines. They are: (1) Equal channel width minimizes memory access cost (MAC); (2) G2 Excessive group convolution increases MAC; (3) Network fragmentation reduces degree of parallelism; and (4) Element-wise operations are non-negligible.

2.2 Depthwise Separable Convolutions

Nowadays, there are many efficient neural network architectures [3,8,17,24] use Depthwise Separable Convolutions (DWConv) as the key building block. The basic idea of DWConv is to replace a standard convolutional layer with two separate layers. The first layer uses a depthwise convolution operator. It applies a single convolutional filter per input channel to capture the spatial information in each channel. Then the second layer employs a pointwise convolution, means a 1×1 convolution, to capture the cross-channel information.

Suppose the input tensor L_i has size $h \times w \times d_i$, the output tensor L_j has size $h \times w \times d_j$. So, the standard Convolution needs to apply a convolutional kernel $K \in \mathcal{R}^{k \times k \times d_i \times d_j}$, where k is the size of kernel. Therefore, it has the computation cost of $h \cdot w \cdot d_i \cdot d_j \cdot k \cdot k$.

In case of DWConv, the depthwise convolution layer costs $h \cdot w \cdot d_i \cdot k \cdot k$ and the 1×1 pointwise convolution costs $h \cdot w \cdot d_i \cdot d_j$. Hence, the total computational cost of DWConv is $h \cdot w \cdot d_i \cdot (k^2 + d_j)$. Effectively, the computational cost of DWConv is smaller than the standard Convolution by a factor of $\frac{k^2 \cdot d_j}{(k^2 + d_j)}$.

Table 1. Overall architecture of proposed network and ShuffleNetV2, for 4 different levels of complexities.

Layer	Output size	KSize	Stride	Repeat	Output channels			
					0.5×	1×	1.5×	2×
Image	224 × 224				3	3	3	3
Conv1	112 × 112	3 × 3	2	1	24	24	24	24
MaxPool	56 × 56	3 × 3	2					
Stage2	28 × 28		2	1	48	116	176	244
	28 × 28		1	3				
Stage3	14 × 14		2	1	96	232	352	488
	14 × 14		1	7				
Stage4	7 × 7		2	1	192	464	704	976
	7 × 7		1	3				
Conv5	7 × 7	1 × 1	1	1	1024	1024	1024	2048
GlobalPool	1 × 1	7 × 7						
FC					1000	1000	1000	1000

3 Network Architecture

As mentioned above, the proposed network replaces Channel Split and Channel Shuffle operators in ShuffleNetV2 with Channel Slice operator to make information interaction between two channels. The architecture of building blocks of ShuffleNetV2 and the proposed network are shown in Fig. 1.

As can be seen, at the beginning of each basic ShuffleNetV2 unit (as illustrated in Fig. 1a), the input of c feature channels are split into two branches with $c-c'$ and c' channels, respectively (in their experiments, $c' = c/2$, means the two branches have same number of channels). One branch remains as the identity. The other branch consists of three convolutions with the same input and output number of channels. After convolution, the two branches are concatenated. So, the number of channels of input and output of the unit are the same. The Channel Shuffle operation is then used to enable information communication between the two branches. After the shuffling, the next unit begins.

For spatial downsampling, the unit is slightly modified and illustrated in Fig. 1b. The Channel Split operator is removed. The two 1×1 is used to control output channels of two branches. The two branches have the same number of output channels.

In case of the proposed network, because the two branches have same channels, the Channel Split can be replaced with Channel Slice operator with step = 2 (as in Fig. 1c). It means the identity branch has the even-index features of input (please note that in programming, the index starts from 0). The other branch has odd-index features of the input. Similar to ShuffleNetV2, these odd-index features are passed through three convolutions with the same input and output channels. After convolution, the two branches are concatenated. Because the information communication between the two branches is enabled in Channel Slice, there is no Channel Shuffle operator in the proposed network. For spatial downsampling, the proposed network can use similar architecture to ShuffleNetV2 (without Channel Shuffle), as illustrated in Fig. 1d.

Table 1 shows the overall architecture of ShuffleNetV2 and the proposed network. Their architectures are similar. The number of channels in each block is scaled to generate networks of different complexities, marked as $0.5\times$, $1\times$, $1.5\times$ and $2\times$. The network uses a 3×3 convolution layer with stride = 2 at the beginning followed by a Max Pooling layer with kernel size = 3 and stride = 2. This kind of design can reduce the size of features quickly to save the FLOPs. Then the building blocks are repeatedly stacked to construct the stages. The first unit of each stage has stride = 2, other units have stride = 1. There is an additional 1×1 convolution layer is added right before global averaged pooling to mix up features.

4 Experiments

4.1 Experimental Configuration

This paper implements networks on Gluon module of MXNet open source deep learning framework [1]. The speed of networks is tested on Ubuntu 16.04 OS with the libraries: MXNet 1.4.1; CUDA 10.1; and CUDNN 7.5.1. They are compared together with the following platforms:

- *GPU-tuned*: run networks on the GPU NVIDIA RTX Titan device. The MXNet framework runs performance tests to find the best algorithm to execute the networks before actually running. It takes longer at the startup, but the network will run faster.

- *GPU*: run networks on the GPU NVIDIA RTX Titan device, without tuning network.
- *CPU-MKL*: run networks on the CPU Core i7-8700K 3.40 GHz device with tuning by Intel Math Kernel Library (Intel MKL) which is optimized to run deep learning model as well as other math routines.
- *CPU*: run networks on the CPU Core i7-8700K 3.40 GHz device without tuning network with Intel MKL.

4.2 Image Classification

For image classification task, this paper evaluates own implementation of MobileNet, MobileNetV2, ShuffleNetV2, and the proposed network on the ImageNet dataset [16] and compares with state-of-the-art architectures such as Xception [3], DenseNet [10], CondenseNet [9], and PeleeNet [20].

Dataset. The ImageNet ILSVRC 2012 classification dataset [16] consists 1.2 million images for training, and 50,000 for validation, from 1,000 classes. This paper adopts the same data augmentation scheme for training images as in [5,6], and apply a single-crop with size 224 × 224 at test time. Following [5,6], this paper reports the Top-1 and Top-5 classification errors on the validation set.

Implementation Details. This paper implementes MobileNet, MobieNetV2, ShuffleNetV2, and proposed networks with training procedure follows the schema described in [23]. Follow the settings of GluonCV package[1], all models are trained using back-propagation [11] by Stochastic Gradient Descent [15] with Nesterov momentum [14] (NAG) optimizer implemented by MXNet for 120 epochs. At the first 5 epochs, the learning rate increases linearly from 0 to 0.4. Then, its value will be decayed with a cosine shape (the learning rate of epoch $t \leq 120$ is set to $0.5 \times \text{lr} \times (\cos(\pi \times t/120) + 1)$). The parameters are initialized by Xavier's initializer [4]. The other settings are: weight decay of 0.0001, momentum of 0.9, and batch size of 512 (the 2× complexity models just have batch size of 256 due to the out-of-memory error).

Results on ImageNet ILSVRC 2012. Following the common practice [8,13, 17,24], all networks in comparison have 4 levels of computational complexity, i.e. about 40, 140, 300 and 500+ MFLOPs. Table 2 shows the comparison between own implemented models mentioned above and also with other state-of-the-art models.

The Top-1 errors (orig.) of networks are obtained from [13] and original papers. The Top-1 errors (impl.) are obtained by running own reimplemented-models. Top-5 errors are also obtained by ourselves, except PeleeNet.

[1] https://gluon-cv.mxnet.io/.

Table 2. Top-1 and Top-5 error rates (%) on ImageNet datasets. In case of MobileNets' name, the number indicates the value of width multiplier parameter α. In case of ShuffleNetV2s' and the proposed networks' name, the number indicates the complexity of the network (as in Table 1).

Model	#Params	FLOPs	Top-1 err (%) (orig.)	Top-1 err (%) (impl.)	Top-5 err (%)
MobileNet-0.25 [8]	0.47M	41M	49.40	47.07	23.04
MobileNetV2-0.25 [17]	1.52M	34M	-	48.87	25.24
DenseNet 0.5× [10]	-	42M	58.60	-	-
Xception 0.5× [3]	-	40M	44.90	-	-
ShuffleNetV1 0.5× (g = 3) [24]	-	38M	43.20	-	-
ShuffleNetV2 0.5× [13]	1.37M	41M	39.70	42.30	19.58
Proposed network 0.5×	1.37M	41M	-	41.53	19.21
MobileNet-0.5 [8]	1.33M	149M	36.30	34.79	13.65
MobileNetV2-0.5 [17]	1.96M	100M	-	35.50	14.64
DenseNet 1× [10]	-	142M	45.20	-	-
Xception 1× [3]	-	145M	34.10	-	-
ShuffleNetV1 1× (g = 3) [24]	-	140M	32.60	-	-
ShuffleNetV2 1× [13]	2.28M	146M	30.60	32.00	12.01
Proposed network 1×	2.28M	146M	-	32.22	12.02
MobileNet-0.75 [8]	2.59M	325M	31.60	29.73	10.51
MobileNetV2-0.75 [17]	2.63M	197M	-	30.81	11.26
DenseNet 1.5× [10]	-	295M	39.90	-	-
Xception 1.5× [3]	-	305M	29.40	-	-
CondenseNet (G = C = 8) [9]	-	274M	29.00	-	-
ShuffleNetV1 1.5× (g = 3) [24]	-	292M	28.50	-	-
ShuffleNetV2 1.5× [13]	3.50M	299M	27.40	28.57	9.68
Proposed network 1.5×	3.50M	299M	-	28.63	9.68
MobileNet-1.0 [8]	4.23M	569M	29.40	26.71	8.71
MobileNetV2-1.0 [17]	3.50M	328M	28.00	28.50	9.90
DenseNet 2× [10]	-	519M	34.60	-	-
Xception 2× [3]	-	525M	27.60	-	-
CondenseNet (G = C = 8) [9]	-	529M	26.20	-	-
ShuffleNetV1 2× (g = 3) [24]	-	524M	26.30	-	-
PeleeNet [20]	2.8M	508M	27.40	-	9.40
ShuffleNetV2 2× [13]	7.39M	591M	25.10	25.70	8.14
Proposed network 2×	7.39M	591M	-	25.76	8.19

As can be seen, the proposed network and re-implemented ShuffleNetV2 have similar errors due to the alike overall and building block architectures for all 4 levels of complexity. The proposed network and ShuffleNetV2 outperform other state-of-the-art models like PeleeNet, DenseNet, and CondenseNet in term of Top-1 and Top-5 errors in case of same level of complexity.

The re-implemented MobileNets can get lower errors than the original, and even lower than MobileNetV2 (both original and re-implemented results). However, unfortunately, this paper cannot reproduce ShuffleNetV2 results. The re-implemented models get higher error than the original because of the different framework and/or training schema. But theoretically, the proposed network can achieve similar accuracy to ShuffleNetV2 if they are implemented on the same framework and training schema.

Table 3. Speed of variants of MobileNet, MobileNetV2, ShuffleNetV2, and Proposed network (images/second).

Model	#Params	FLOPs	GPU (Imgs/s)		CPU (Imgs/s)	
			Tuned	-	MKL	-
MobileNet-0.25 [8]	0.47M	41M	4431.31	3680.86	448.83	156.52
MobileNetV2-0.25 [17]	1.52M	34M	2507.89	2361.65	126.61	119.25
ShuffleNetV2 0.5× [13]	1.37M	41M	2026.41	1991.57	128.39	184.55
Proposed network 0.5×	1.37M	41M	2313.37	2223.60	171.93	190.07
MobileNet-0.5 [8]	1.33M	149M	3319.99	2880.54	297.47	81.23
MobileNetV2-0.5 [17]	1.96M	100M	2070.37	1948.28	103.12	64.42
ShuffleNetV2 1× [13]	2.28M	146M	1759.31	1700.56	133.51	107.14
Proposed network 1×	2.28M	146M	2020.50	1906.16	148.16	107.58
MobileNet-0.75 [8]	2.59M	325M	2560.31	2499.72	208.83	53.36
MobileNetV2-0.75 [17]	2.63M	197M	1706.80	1738.71	84.54	44.39
ShuffleNetV2 1.5× [13]	3.50M	299M	1522.39	1470.93	94.96	75.51
Proposed network 1.5×	3.50M	299M	1745.11	1665.59	120.29	75.33
MobileNet-1.0 [8]	4.23M	569M	2104.56	2070.80	165.30	38.86
MobileNetV2-1.0 [17]	3.50M	328M	1439.67	1513.36	68.82	32.91
ShuffleNetV2 2× [13]	7.39M	591M	1255.78	1236.03	89.31	52.94
Proposed network 2×	7.39M	591M	1461.07	1416.76	93.78	52.68

Speed on Real Devices. This section evaluates the real speed of own implemented models on the 5 platforms mentioned in Sect. 4.1. The speed is calculated by the average time of processing 100 pictures with 8 batch size for GPU and 1 batch size for CPU and ARM. This paper runs 100 picture processing for 10 times separately and averages the time for each model.

As can be seen in Table 3, the MobileNet is the fastest model on all platforms. Although MobileNetV2s has lower complexity, the actual speed of the model is slower than that of MobileNet and others. The main ground is maybe its architecture conflicts with the 4 practical guild-lines introduces in [13] while others follow them.

In comparison between ShuffleNetV2, the proposed network at 4 levels of complexity, it is clear to see that the the proposed network is the fastest. Please note that the number of units which have stride = 2 in the proposed network is just 3, means there are just 3 replacement for parallel 1×1 convolutions, but it is easy to see that the speed increases, especially for small models. So, other network which have parallel branches like Xception [3], PeleeNet [20], and PydMobileNet [7] can use group convolution instead of parallel convolutions to become faster.

For deeper analysis, the difference in speeds on GPU-tuned and GPU is not much because the CUDA, CUDNN and MXNet libraries have not well optimized yet for the new hardware architecture on the GPU Titan RTX. Usually, the speed of GPU-tuned can be 5–15 times faster on older NVIDIA GPU devices.

The speeds on CPU-MKL are 2–5 times faster than CPU except ShuffleNetV2 0.5× and proposed network 0.5×. The main reason for this may be because of the not well optimization of the MKL library. This is another strong evidence for the suggestion that it is necessary to optimize the deep learning network on the targeted platforms with specific hardware and software.

5 Conclusion

This paper introduced an improved version of ShuffleNetV2 which replaces the Channel Shuffle and Channel Split operators with the Channel Slice operator with slice-step parameters (step = 2) to make information interaction between two channels. Because the Channel Slice and Channel Split operators are similar and the proposed network does not have Channel Shuffle operator, it has lower memory access cost than ShuffleNetV2. The experiments on ImageNet and COCO datasets show that the proposed network is faster than ShuffleNetV2 while still achieve similar accuracy.

In the future, it is necessary to test the speed of models on ARM device like iOS/Android mobile-phone and another edge devices like Raspberry Pi, Google TPU or NVIDIA Jetson. Additionally, the atrous Convolution should be considered because it is an efficient way to capture difference spatial information without increasing computational cost much.

Acknowledgments. This work was supported by the National Research Foundation of Korea (NRF) grant funded by the Korea government (MSIP, Ministry of Science, ICT & Future Planning) (No. 2019R1F1A1061659).

References

1. Chen, T., et al.: Mxnet: a flexible and efficient machine learning library for heterogeneous distributed systems. In: Proceedings of the Neural Information Processing Systems, Workshop on Machine Learning Systems (2015)
2. Chetlur, S., Woolley, C., Vandermersch, P., Cohen, J., Tran, J.: cuDNN: efficient primitives for deep learning. arXiv preprint arXiv:1410.0759 (2014)

3. Chollet, F.: Xception: Deep learning with depthwise separable convolutions. In: Proceedings of the IEEE Conference on Computer Vision and Pattern Recognition, pp. 1251–1258 (2017)
4. Glorot, X., Bengio, Y.: Understanding the difficulty of training deep feedforward neural networks. In: Proceedings of the International Conference on Artificial Intelligence and Statistics, pp. 249–256 (2010)
5. He, K., Zhang, X., Ren, S., Sun, J.: Deep residual learning for image recognition. In: Proceedings of the IEEE Conference on Computer Vision and Pattern Recognition, pp. 770–778 (2016)
6. He, K., Zhang, X., Ren, S., Sun, J.: Identity mappings in deep residual networks. In: Leibe, B., Matas, J., Sebe, N., Welling, M. (eds.) ECCV 2016. LNCS, vol. 9908, pp. 630–645. Springer, Cham (2016). https://doi.org/10.1007/978-3-319-46493-0_38
7. Hoang, V.T., Jo, K.H.: Pydmobilenet: improved version of mobilenets with pyramid depthwise separable convolution. arXiv preprint arXiv:1811.07083 (2018)
8. Howard, A.G., et al.: Mobilenets: efficient convolutional neural networks for mobile vision applications. arXiv preprint arXiv:1704.04861 (2017)
9. Huang, G., Liu, S., van der Maaten, L., Weinberger, K.Q.: Condensenet: an efficient densenet using learned group convolutions. In: Proceedings of the IEEE Conference on Computer Vision and Pattern Recognition, pp. 2752–2761 (2018)
10. Huang, G., Liu, Z., Maaten, L.V.D., Weinberger, K.Q.: Densely connected convolutional networks. In: Proceedings of the IEEE Conference on Computer Vision and Pattern Recognition, pp. 4700–4708 (2017)
11. LeCun, Y., et al.: Backpropagation applied to handwritten zip code recognition. Neural Comput. **1**(4), 541–551 (1989)
12. Lin, M., Chen, Q., Yan, S.: Network in network. In: Proceedings of the International Conference on Learning Representations (2014)
13. Ma, N., Zhang, X., Zheng, H.-T., Sun, J.: ShuffleNet V2: practical guidelines for efficient CNN architecture design. In: Ferrari, V., Hebert, M., Sminchisescu, C., Weiss, Y. (eds.) Computer Vision – ECCV 2018. LNCS, vol. 11218, pp. 122–138. Springer, Cham (2018). https://doi.org/10.1007/978-3-030-01264-9_8
14. Nesterov, Y.E.: A method for solving the convex programming problem with convergence rate o $(1/k^2)$. In: Dokl. Akad. Nauk SSSR, vol. 269, pp. 543–547 (1983)
15. Robbins, H., Monro, S.: A stochastic approximation method. Ann. Math. Stat. **22**, 400–407 (1951)
16. Russakovsky, O., et al.: Imagenet large scale visual recognition challenge. Int. J. Comput. Vision **115**(3), 211–252 (2015)
17. Sandler, M., Howard, A., Zhu, M., Zhmoginov, A., Chen, L.C.: Mobilenetv 2: inverted residuals and linear bottlenecks. In: Proceedings of the IEEE Conference on Computer Vision and Pattern Recognition, pp. 4510–4520 (2018)
18. Szegedy, C., et al.: Going deeper with convolutions. In: Proceedings of the IEEE Conference on Computer Vision and Pattern Recognition, pp. 1–9 (2015)
19. Szegedy, C., Vanhoucke, V., Ioffe, S., Shlens, J., Wojna, Z.: Rethinking the inception architecture for computer vision. In: Proceedings of the IEEE Conference on Computer Vision and Pattern Recognition, pp. 2818–2826 (2016)
20. Wang, R.J., Li, X., Ling, C.X.: Pelee: A real-time object detection system on mobile devices. In: Proceedings of the Advances in Neural Information Processing Systems, pp. 1967–1976 (2018)
21. Xie, S., Girshick, R., Dollár, P., Tu, Z., He, K.: Aggregated residual transformations for deep neural networks. In: Proceedings of the IEEE Conference on Computer Vision and Pattern Recognition, pp. 1492–1500 (2017)

22. Zagoruyko, S., Komodakis, N.: Wide residual networks. In: Proceedings of the British Machine Vision Conference (2016)
23. Zhang, H., Cisse, M., Dauphin, Y.N., Lopez-Paz, D.: Mixup: beyond empirical risk minimization. In: Proceedings of the International Conference on Learning Representations (2018)
24. Zhang, X., Zhou, X., Lin, M., Sun, J.: Shufflenet: an extremely efficient convolutional neural network for mobile devices. In: Proceedings of the IEEE Conference on Computer Vision and Pattern Recognition, pp. 6848–6856 (2018)
25. Zoph, B., Vasudevan, V., Shlens, J., Le, Q.V.: Learning transferable architectures for scalable image recognition. In: Proceedings of the IEEE Conference on Computer Vision and Pattern Recognition, pp. 8697–8710 (2018)

Weighted Stable Matching Algorithm as an Approximated Method for Assignment Problems

Duc Duong Lam(✉), Van Tuan Nguyen, Manh Ha Le, Manh Tiem Nguyen, Quang Bang Nguyen, and Tran Su Le

Command and Control Center, VHT, Viettel Group, Hanoi 10000, Vietnam
duongld11@viettel.com.vn

Abstract. The Hungarian algorithm is widely known as a method that is capable of solving the linear assignment problem. This algorithm has a computational complexity of O(n^3). We find this complexity unsuitable for our problems and our further studies, therefore, we propose another approach in this report. We will take Gale-Shapley (Deferred Acceptance) algorithm which provides a stable matching solution into scrutinization. The total weight score of this solution is found to be close to the optimal solution of the Hungarian method. Interestingly, the score difference between the two algorithms decreases as the data size increases ($n \to \infty$). Moreover, the stable matching solution is unique. Though, the Gale-Shapley algorithm proposes significantly faster performance as in finding this unique solution. All in all, our goal is to show that the Deferred Acceptance principle could be used as a potential approximation method for the linear assignment problem and its variations, especially on a large scale dataset. Besides, we will introduce a simple greedy algorithm that can provide a nearer optimal solution for the problem.

Keywords: Stable matching · Hungarian algorithm · Assignment problem

1 Introduction

The Hungarian – Munkres algorithm solves assignment problems in polynomial time with the computational complexity of O(n^3) [1]. Typically, there is a two-sided market with two sets of n elements disjoint together. The goal of this algorithm is to minimize or maximize the total cost of all the assignments. We consider a linear assignment problem with n objects and n agents. One object will be matched with one agent. The relation between elements in the two sets is represented by a square matrix W of size n, which contains $n \times n$ weighted parameters w_{ij}. The optimal assignment problem can be formulated as a linear program:

$$Minimize \sum_{j=1}^{n} \sum_{i=1}^{n} w_{ij} x_{ij}, \quad (1)$$

subject to

$$\sum_{i=1}^{n} x_{ij} = 1, j = 1, \ldots, n, \quad (2)$$

N. T. Nguyen et al. (Eds.): ACIIDS 2020, LNAI 12034, pp. 174–185, 2020.
https://doi.org/10.1007/978-3-030-42058-1_15

$$\sum_{j=1}^{n} x_{ij} = 1, i = 1, \ldots, n, \quad (3)$$

$$x_{ij} \in \{0, 1\}, i, j = 1, \ldots, n, \quad (4)$$

where $\{x_{ij}\}_{n \times n}$ is the binary matrix, $x_{ij} = 1$ if and only if the ith object is assigned to the jth agent. The Hungarian algorithm is based on this key observation: adding or subtracting a number from any row or column of the weight matrix does not affect the optimal assignment of the original matrix. Some variations of assignment problems have already been investigated such as the unbalanced assignment problem with unequal number of objects and agents [2], k-cardinality problems with m objects and n agents but only k assignments [3] ($k < m, n$), just to name a few.

The most challenging factor in this assignment problem is that the running time increases when the size of the dataset increases. Recently, many applications possess a bigger dataset and quick decision making is required, even real-time analysis. Therefore, we come to the question of whether there is another approach to a big problem with expected shorter solving time and an acceptable nearly-optimal solution?

To overcome this problem of responding speed, we have decided to apply the idea of the Gale-Shapley's Deferred Acceptance algorithm [4]. The target of this method is completely different from the Hungarian algorithm. Instead of finding the optimal solution, this approach focuses on finding a solution in which, all matching pairs are *stable*. For your understanding, a pair of a man and a woman is *unstable* if they both prefer each other than their current partner. There are some aspects of this solution to be well-scrutinized: (i) what is the criterion of maximizing (or minimizing) the total weighted score of the stable matching solution?; (ii) how near is the solution to the exactly optimal one when varying the size of problems?; (iii) how promising is the Deferred Acceptance approach when solving a combinatorial optimization?

The complexity of this algorithm is found to be $O(n^2)$ in the worst-case, which is better than the performance of the Hungarian algorithm. As a result, we conclude that while the Stable Matching solution provides a nearly optimal solution with lower computational complexity. Hereby, we call one side in the market is *man* and the other one is *woman* for both methods. Although the weighted score between *man* and *woman* is uncommon, it is possible to replace them by job – machine, object – agent, facility – customer, etc., in real-life applications.

Related Works

In the literature, there have already existed several algorithms that can achieve near-optimal solutions under tight time constraints. Some of them are heuristic methods such as the greedy randomized adaptive [5] and the deep greedy switching [6] algorithms. Recently, by a machine learning approach, Lee *et al.* [7] used Deep Neural Networks to solve this optimization problem. To date, the stable matching approach we used in this work has not been considered for this purpose but for finding stable points, which often has an impact on game theory and economics. The Weighted Stable Matching has been studied as a variation of stable matching problem when it is useful to consider the weighted score (such as profit or cost) rather than a qualitative preference ordering. Pini *et al.* [8] defined new notions of stability and optimality and solved by adapting existing algorithms for the classical stable marriage problem. Amira *et al.* [9] studied the Weighted Stable Matching problem with the billboard and fully distributed models. They provided the other algorithms to find the stable matching solution and demonstrated the lower bound of computational complexities. From our best of our knowledge, none of the above studies and other investigations have worked on finding the optimal (or nearly optimal) solutions for assignment problems.

In this article, we will consider the given assignment problem as the Weighted Stable Matching to transform it into the classical Stable Matching problem by ranking its weighted scores. Our end goal is examining the optimality of this solution and the computational time in comparison to the exact optimal answer obtained by the Hungarian method. Furthermore, we will analyze the structure of the ranking distribution of stable matching solutions and tentatively explain the efficiency of the Gale-Shapley in this special case. A simple greedy algorithm is proposed in Sect. 2 and its results are presented along with the two methods. Also, we have applied this technique to solve the unbalanced assignment problem (a variation) and discuss its future perspectives for mathematical combinatorial optimization problems.

2 Methods

A series of data size *n*, varying from 5 to 500 has been studied systematically. For each size, we created a set of problems that are randomly generated as square matrices, which represented the weighted scores. The result gathered from each of these divisions will then be calculated by getting the averaging value to eliminate the random variation.

The optimal solution of the Hungarian method is calculated by the Munkres algorithm - the typical algorithm that can find the solution for minimum total weighted. However, in our approach, to find the maximum one, we will replace each weight with the large constant number subtracted by the weight.

For the stable matching solutions, we will form two preference lists by sorting the weighted scores. In the traditional stable matching, there is a set of solutions that satisfies the stable condition which is provided by The Gale-Shapley algorithm. In our simulation, there is a unique stable solution, so for each dataset, we calculate only one result.

Based on the principle of the Deferred Acceptance algorithm, the solution of the Gale-Shapley algorithm satisfies the stable condition. However, in this context, we want to find the optimal solution to get the maximal total weighted score. It motivated us to modify the Gale-Shapley algorithm, in which, it gives priority to improving the score. Inspired by the Deferred Acceptance principle, we search and temporarily engage the pair of a free *man* and a free *woman* (with the highest-rank possible). And then, at the comparison step, we will decide to choose between two options: confirm this pair or swap it with another pair to obtain an improvement for the total weighted score. The pseudo-code is described below:

ALGORITHM 1: Greedy weighted matching Algorithm

```
Sorting the weighted score for each man → man's preference
Sorting the weighted score for each woman → woman's preference
Initialize all men and women to be free
while there exists a free man m, do:
    w = m's highest-ranked such woman to whom he has not yet visited and w is free
    For w' in the list of woman the man m has visited (from the highest-ranked), do:
    if weight(m,w') + weight(m',w) > weight(m',w') + weight(m,w):
        swap two pairs
        break
    else:
        (m,w) become engage
end
```

Although the solution of this algorithm is not identical to the optimal solution, the average result is found to better than that of the classical Gale-Shapley algorithm. We analyze and compare its solution and computational time with the Stable Matching and the Hungarian methods in the next section.

3 Simulation and Experiment Results

3.1 The Score of the Stable Matching Solution

At first, the two preference lists are created by sorting the weighted scores. In this context, the matching problem satisfies the *symmetric utilities hypothesis* [9–11] with an asymmetric function $w : N^2 \to R$, that is $w(i, j) = w(j, i)$ for all *i, j* in *(1... n)*. The consequence is the uniqueness of the stable matching solution [9, 12]. We have checked the *man-propose* and *woman-propose* stable matching solutions and these two solutions are identical [13].

From the experiment process, the total weighted score of the unique stable matching solution is estimated and compared to the optimal solution. To understand how close are those solutions, we calculate the relative difference of the total weighted score between the Stable Matching (SM) – Hungarian. That relative difference value of the greedy - Hungarian method is plotted in the same graph. Figure 1 describes the data size *n* dependence of the relative difference to the optimal score for the two algorithms:

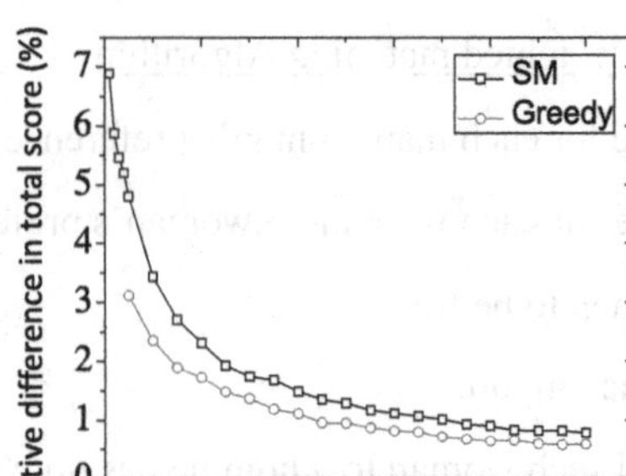

Fig. 1. Data size dependence of the relative difference in the total score between the stable matching (black) and the greedy (red) vs the Hungarian methods. (Color figure online)

The relative difference is acceptably small in our practice problem: less than *7%* at the small dataset ($n < 15$), and at $n = 500$, it is less than *1%*. We have performed this experiment with a larger size at $n = 5000$ (not shown in the Fig. 1), and the relative difference is *~0.1%*. Such a small difference demonstrates that the stable matching method's solution at large size problems is nearly optimal. According to Lee e*t al.* [7], the Convolutional Neural Network (CNN) can reduce the computational complexity to $O(n^2)$, but the loss of accuracy is higher than ours. And the accuracy of their method will drop when the difficulty of the problem increases. While in our result, we see the opposite tendency: the larger the problem is, the higher the accuracy we obtain (see Table 1).

Table 1. Comparison of stable matching with CNN [7]

Size	CNN	Stable matching
8	77.80%	94.26%
16	65.70%	94.65%
5000		99.90%

To test the time efficiency, we have measured the running time of the three algorithms with various sizes of data (in Sect. 3.3). This result proves that the stable matching approach can be an effective way to solve assignment problems within a limited time in real applications. Also based on the Deferred Acceptance principle, the greedy method (red line) even shows a smaller "distance" to the optimal solution but its running time is higher (see Sect. 3.3) as a trade-off.

At our current progress, it is difficult to establish a theory to explain the above observations, however, we can have a better understanding of the results by analyzing the structure of the solutions of the three algorithms in the next section.

3.2 Ranking Structure

To understand how good the optimal total score in Fig. 1, we consider the average rank of partners in all pairs of stable matching. If a man is matched with a woman, who is at i^{th} order $\left(1^{st}, 2^{nd}, 3^{rd}, \ldots, n^{th}\right)$ in his preference list, the partner's rank of this man is i. In this unique matching solution, the roles of man and woman are symmetric, so the average rankings of them are identical in principle. In Fig. 2, we use random data with the size $n = 200$ to present the typical distribution of the partner's rank for the three methods. While the ranks in the Hungarian method are distributed in densely at low region (<10), the ranks in the stable matching locate widely. It shows the principle of the Deferred Acceptance algorithm. The number of the lowest rank (1) in the stable matching method is more than that in the Hungarian method. However, some of the ranks are very large (up to ~*200*). In other tests, we also found that the largest rank could reach to n. But with the symmetric utility function in this context, the ranking difference of *man* and *woman* is small ($<$~*10*) as shown in Fig. 2(d). It means, there are some pairs of *man* and *woman* with a similar and low level in their preference lists. The ranking difference of the Hungarian method is even smaller since its ranking values itself is small (Fig. 2(c)).

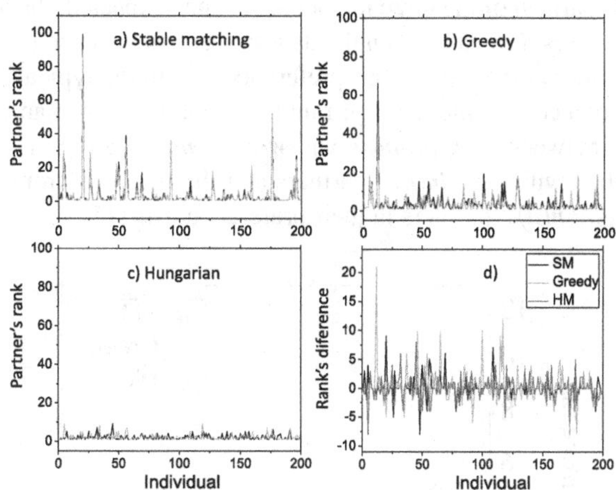

Fig. 2. The partner's rank for each man (black) and woman (red) in the market of size n = *200* for the stable matching algorithm (a), greedy (b), Hungarian method, and the rank's difference of three methods (d) (Color figure online)

By counting the frequency of the rank, we present the distribution of ranking for three methods in Fig. 3. Although, there are some matched pairs with high ranks the stable matching algorithm has many individuals with rank = 1, the best choice. For the Hungarian method, rank 1st is lower but the higher ranks at 2nd, 3rd are larger than that of the stable matching method. Understanding of this structure might help us to effectively apply this approach in various practice situations. In Sect. 3.4, we show an application of this method to a variation of the assignment problem.

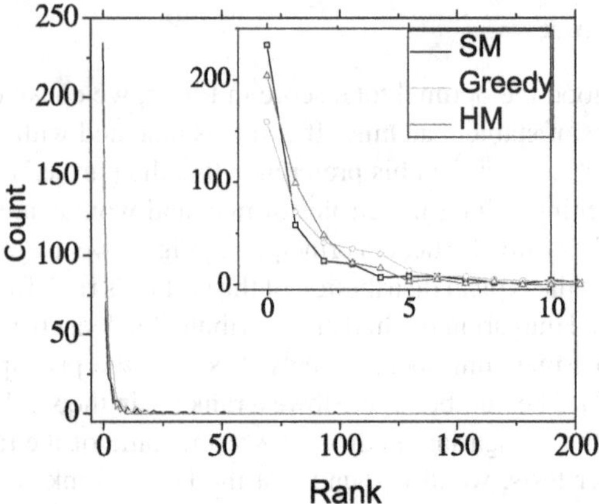

Fig. 3. The distribution of ranking spectra for stable matching, greedy and Hungarian methods. The insert graph is the zoom part of the low rank region.

We plot the average rankings of two sides with varying the data size n in Fig. 4 for a systematical comparison of the three methods. As once expected, the average ranking decreases as *n* increases. It means when the data size grows up, each individual tends to match with the partner at the top of their preference list. In the typical stable matching with random preference list, the average ranking of partners of man and woman are strongly different between *man-propose* and *woman-propose* regime. The proposing side will be matched with the $\sim ln(n)^{th}$ partners and the received-proposal side will be matched with the $\sim n/ln(n)^{th}$ partners in their preference lists [15].

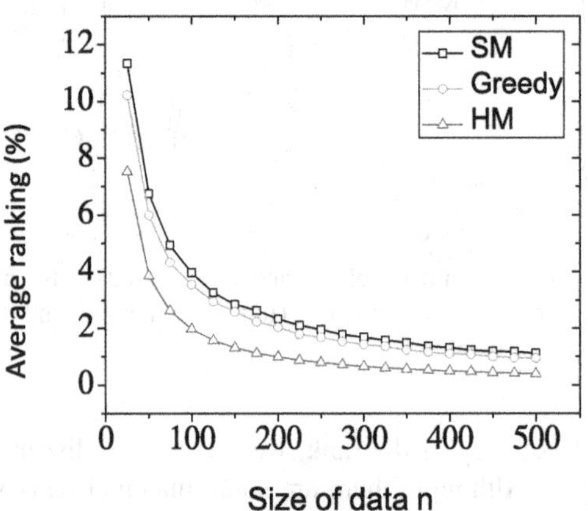

Fig. 4. The data size dependence of the average ranking in all matched pairs between the stable matching and Hungarian methods.

The graph shows the reduction of the ranks for all methods as n increases. This tendency indicates the fact that each individual will have a larger change to find a proper partner in their most preferable list (in percentage) in a big community. The ranking percentage values of the stable matching algorithm are larger than those of the Hungarian method. Moreover, the size n dependent of the average rank in the stable matching method well fits the line *2log(n)*, which is under the *ln(n)* curve, the asymptotic value of optimal-side in the typical stable matching [14, 15]. This could be an intuitive explanation for the nearly optimal solution of this approach which is designed for different purposes.

3.3 Computational Time Comparison

In this section, we compare the computational time between the stable matching, greedy and the Hungarian algorithms. Figure 5 plots the running times of each method with various sizes of data. Since the complexity of the Hungarian algorithm is $O(n^3)$, it has the highest running time. While the stable matching algorithm has the complexity of $O(n^2)$ in its worse-case. The larger the data size gets, the more effective this method is. We can see that at point $n = 500$, the stable matching method (including the sorting time to make preference lists) solves the problem with ~ two order lower measured time than the Hungarian method does.

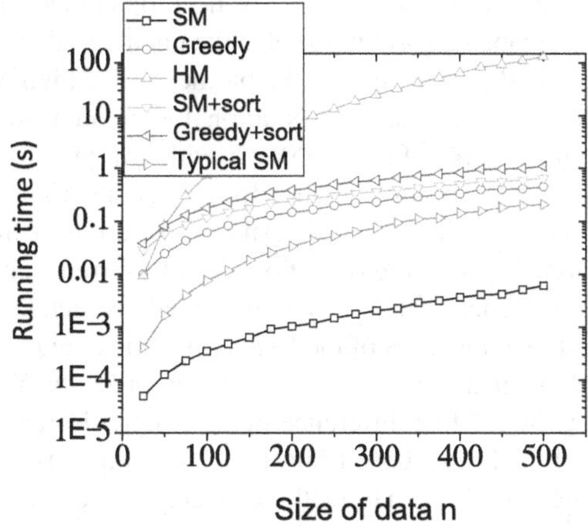

Fig. 5. Size of data dependence of the computational time for the stable matching (black), greedy (red), Hungarian (blue), stable matching with sorting time (green), greedy with sorting time (pink) and the typical stable matching (yellow). (Color figure online)

In the above comparison, we take into account the time used to create the two preference lists from the weighted matrix for the stable matching method. We also measure the time for running only the Gale-Shapley algorithm as shown by the black-rectangular curve. It has ~ two order lower running-time compared to the stable matching method including the sorting time for preference lists. It indicates that this stable matching and the greedy method spend a major time just for building their preference matrices.

Considering the efficiency of the stable matching algorithm in this context, we also plot the measured running time of the typical stable matching with two randomly generated preferences (the yellow-triangular curve). Surprisingly, there is a drastically difference (>one order) in computing time between the two cases, meanwhile, they use the exactly same Gale-Shapley algorithm. The typical stable matching will find one (man-optimal or woman-optimal) solution in its set of stable solutions. However, the stable matching algorithm in our situation can provide only one solution, which is unique. In Sect. 3.1, we have discussed the uniqueness of the stable matching solution in this circumstance. This uniqueness property comes from the two perfectly intercorrelated preference lists, which satisfy the symmetric utility function. In consequence, the process that can find the unique stable solution must have a distinguishing structure. A tentative explanation of these observations will be discussed in Sect. 4.

3.4 Application to the Unbalanced Assignment Problem

A popular variation of the classic assignment is the unbalanced problem in which there are m objects and n agents (m and n differ) [2]. We can easily convert into a balanced problem by adding a sufficient number of "dummy" objects or agents. On the other hand, the recent work of Ashlagi *et al.* [15] on the unbalanced random matching markets (m men and n women) could give an idea to solve this variation of assignment problems. They experimentally proved even the slightest imbalance can help each agent on the short side of the market is matched with one of his (or her) top choices. It means that if we apply the Deferred Acceptance algorithm to solve the unbalanced assignment problems, agents on the short side will preferably find the partner with a high weighted score.

We have performed the additional computational simulation for the problem with 200 women and a varying number of men: $200 + k$, with k varying from 1 to 200. For the Hungarian method, we add k "dummy" women to convert the problem into the classical assignment problem with the size of $n + k$. And the solution of the stable matching method is found by woman-proposing (the short side) Deferred Acceptance algorithm. For each woman, her preference list has $n + k$ man ranked in order. Despite how large the imbalance magnitude, the uniqueness of the Stable Matching solution is conserved. All of the analysis procedures mentioned above are shown in Fig. 6. When increasing the imbalance number (k), the relative difference in total score decreases drastically from ~1.6% (with k = 0 in Fig. 1) to 0.05% (with k = 200) (Fig. 6a), the average ranking of both stable matching and Hungarian methods decrease and get closer to each other (Fig. 6b). The running time of the Hungarian method is not only larger than that of the Stable Matching including its sorting time but also increases more quickly with k. Similarly to the results in Sect. 3.3, the stable matching method spends a major amount of time in sorting its preference lists, meanwhile the Deferred Acceptance process is less than 1 ms and decreases when k increases. This reduction of time comes from the higher chance of finding a top preferable partner for each woman when there are more men join into the market [15] as we will discuss in Sect. 4.

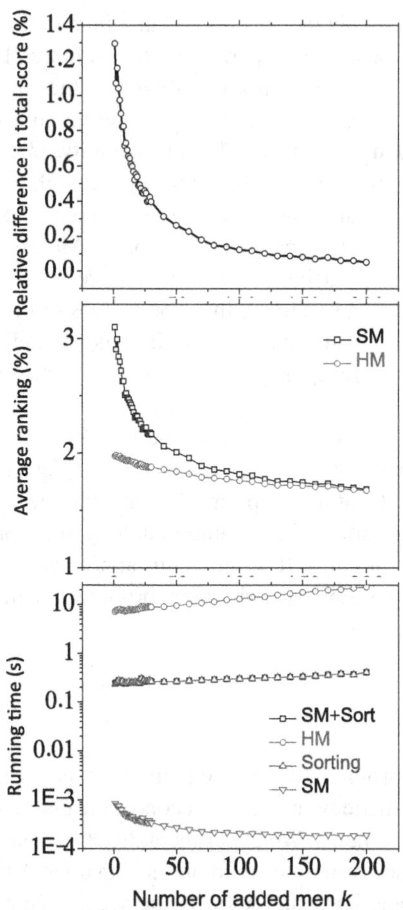

Fig. 6. The effect of imbalanced problem: the number of added man dependence of the relative difference in total score (a), the average ranking of solutions by stable matching (back) and Hungarian (red) methods (b), the running time of Hungarian method (red), the stable matching with sorting time (black), the sorting time (blue) and the stable matching (green). (Color figure online)

4 Discussion

The significant shorter computational time of the Weighted Stable Matching observed in Figs. 5 and 6c has not been investigated in previously published studies. There is an important distinguishing point at the comparison procedure of the "getting-proposal" or "chosen" side. With highly intercorrelated preferences in this circumstance, the Gale-Shapley algorithm has a very fast comparison process. In the man-proposed regime, when a woman receives two proposals, she will compare them by her preference list, which is an unsorted list. Typically, the Gale-Shapley algorithm will do an O(n) search each time to know the index of her partner. She must go through all values starting from the first position in the list and looking for the man who is proposing to her. As soon as she finds out, she will obtain his index (ranking) and finishes her searching. In classical stable matching problems with random preferences and mutually independent, the rank of the man in her list will be randomly distributed in the range from one through n with an equal probability. Meanwhile, in our case, the ranking values of the proposing man locate at a low indexing region with high probability. In common words, the man who highly prefers the woman will probably also get a high priority in her preference list. In consequence, when searching from her top preference, the woman quickly finds the better man between the two men who proposed. Moreover, we even can speed up the comparison process by using the weighted score matrix. Instead of getting the indexes of her partners, she can compare the score values of the two men directly (with O(1) complexity) to find the better one. The further understanding of the combinatorial structure and the Deferred Acceptance processes, we might find on these theoretical works [16, 17].

For a future perspective, we would raise the benefit of optimality that the stable matching solution gains when the market is unbalanced [2] (even very slight) in Sect. 3.4. And, looking back the distribution of partner's rank (Fig. 3, Sect. 3.2), we can obtain a higher average ranking value of the stable matching solution by removing a small number of pairs with low rankings. It suggests this approach could provide a good near-optimal solution for the k-cardinality assignment problem [3], the variation that we have mentioned at the beginning.

5 Conclusions

In our specific matching problem, where the number of individuals could be very large and the decision must be made within a few seconds, the classical Hungarian algorithm does not meet these requirements. We experimentally show that the two methods based on the Deferred Acceptance principle could provide an approximated approach to solve the assignment problem and its variations with the acceptable computational time. The stable matching solution is unique and the average ranking of the stable matching solution is larger than the one of the Hungarian algorithm. When the size n increases, the average ranking decreases, the difference ratio of the total weighted score between two methods decreases. A simple greedy algorithm that is modified from the original Gale-Shapley method can even provide a better assignment. Further, this initial study suggests that the Deferred Acceptance approach has a promising potential in solving combinatorial optimization problems, especially when the size of the problem is getting larger. If we accept an amount of certainty of the answer, we are able to achieve much faster speed. Also, in regard to big problems, we foresee the cost of getting a faster answer to be reduced.

References

1. Jonker, R., Volgenant, A.: A shortest augmenting path algorithm for dense and sparse linear assignment problems. Computing **38**, 325–340 (1987)
2. Caron, G., Hansen, P.: The assignment problem with seniority and job priority constraints. Oper. Res. **47**, 449–453 (1999)

3. Dell'Amico, M., Martello, S.: The k-cardinality assignment problem. Discret. Appl. Math. **76**, 103–121 (1997)
4. Gale, D., Shapley, L.S.: College admissions and the stability of marriage. Am. Math. Monthly **69**, 1 (1962)
5. Feo, T.A., Resende, M.G.C.: Greedy randomized adaptive search procedures. J. Glob. Optim. **6**, 109–133 (1995)
6. Naiem, A., El-Beltagy, M.: Deep greedy switching: a fast and simple approach for linear assignment problems. In: 7th International Conference of Numerical Analysis and Applied Mathematics (2009)
7. Lee, M., Xiong, Y., Yu, G., Li, G.Y.: Deep neural networks for linear sum assignment problems. IEEE Wirel. Commun. Lett. **7**, 962–965 (2018)
8. Pini, M.S., Rossi, F., Venable, K.B., Walsh, T.: Stability, optimality and manipulation in matching problems with weighted preferences. Algorithms **6**, 782–804 (2013)
9. Amira, N., Giladi, R., Lotker, Z. Distributed weighted stable marriage problem. In: 17th International Colloquium on Structural Information and Communication Complexity (SIROCCO), Sirince, Turkey (2010)
10. Rodrigues-Neto, J.A.: Acyclic roommates. Econ. Lett. **118**, 304–306 (2013)
11. Alvaro Rodrigues-Neto, J.: Representing roommates' preferences with symmetric utilities. J. Econ. Theory **135**, 545–550 (2007)
12. Park, J.: Competitive equilibrium and singleton cores in generalized matching problems. Int. J. Game Theory **46**, 487–509 (2017)
13. McVitie, D.G., Wilson, L.B.: The stable marriage problem. Commun. ACM **14**, 486–490 (1971)
14. Pittel, B.: The average number of stable matchings. SIAM J. Discrete Math. **2**(4), 530–549 (1988)
15. Ashlagi, I., Kanoria, Y., Leshno, J.D.: Unbalanced random matching markets: the stark effect of competition. J. Polit. Econ. **125**, 69–98 (2016)
16. Knuth, D.E., Motwani, R., Pittel, B.: Stable husbands. Random Struct. Algorithms **1**, 1–14 (1990)
17. Pittel, B.: On likely solutions of the stable matching problem with unequal numbers of men and women. Math. Oper. Res. **44**, 1 (2018)

Data Modelling and Processing for Industry 4.0

User-Friendly MES Interfaces: Recommendations for an AI-Based Chatbot Assistance in Industry 4.0 Shop Floors

Soujanya Mantravadi[1](✉), Andreas Dyrøy Jansson[2], and Charles Møller[1]

[1] Department of Materials and Production, Aalborg University, Aalborg, Denmark
{sm,charles}@mp.aau.dk

[2] Department of Computer Science and Computational Engineering, UiT – Arctic University of Norway, Narvik, Norway
andreas.d.jansson@uit.no

Abstract. The purpose of this paper is to study an Industry 4.0 scenario of 'technical assistance' and use manufacturing execution systems (MES) to address the need for easy information extraction on the shop floor. We identify specific requirements for a user-friendly MES interface to develop (and test) an approach for technical assistance and introduce a chatbot with a prediction system as an interface layer for MES. The chatbot is aimed at production coordination by assisting the shop floor workforce and learn from their inputs, thus acting as an intelligent assistant. We programmed a prototype chatbot as a proof of concept, where the new interface layer provided live updates related to production in natural language and added predictive power to MES. The results indicate that the chatbot interface for MES is beneficial to the shop floor workforce and provides easy information extraction, compared to the traditional search techniques. The paper contributes to the manufacturing information systems field and demonstrates a human-AI collaboration system in a factory. In particular, this paper recommends the manner in which MES based technical assistance systems can be developed for the purpose of easy information retrieval.

Keywords: AI applications · Manufacturing · Chatbot

1 Introduction

Although manufacturing execution systems have been critical information systems for production planning and control among manufacturing practitioners, their interaction with humans in Industry 4.0 needs further understanding in theory and in practice. MES as an information system has come a long way. It was first introduced in the 70s to assist the online management of production execution. Later in the 90s, it emerged as a powerful software tool to replace paper-based activities of manufacturing operations management (MOM) [1]. Over the years, it became a critical tool for a manufacturing enterprise as it acquired additional functionalities for automation of information exchange due to computer advancements and shop floor systems integration projects [2].

N. T. Nguyen et al. (Eds.): ACIIDS 2020, LNAI 12034, pp. 189–201, 2020.
https://doi.org/10.1007/978-3-030-42058-1_16

For future factories, studying the enhancements for MES in the light of Industry 4.0 context can bring new opportunities that were not identified before. The Industry 4.0 paradigm is predominantly information-centric and guides manufacturing enterprises to assess their future needs and acquire digital capabilities. Industry 4.0 is an IT-driven enabler of smart factories [3] with the design principles of [4]: (a) Interconnection (b) Information transparency (c) Decentralized decision-making (d) Technical assistance. In this paper, we focus on the design principle of 'technical assistance' for the factory workforce, which is about humans receiving support from assistant systems for production-related tasks [5, 4].

Artificial intelligence capability is an asset for manufacturing because the digital age promises hardware with high processing power combined with the vast amounts of available real-time production data generated by MES. Analogous to natural intelligence, an AI system is able to learn from the experience, where data equals experience. A larger set of training production data makes for a robust decision-making system for operations management. AI applications in the manufacturing field are not new and they are widely studied to manage uncertainty, complexity and dynamic changes in the manufacturing systems [6, 7]. On this premise, combining MES with AI can drastically improve automated decision-making capabilities and the workforce can benefit from this situation.

Furthermore, intuitive assistant systems designed based on AI techniques can help the workforce make informed decisions in enterprises [8, 9]. Motivated by the need to study their potential industrial engineering applications, we emphasize on user aspects and design a chatbot for the information system (MES). We apply techniques such as natural language processing and artificial neural networks, which are key parts in making a chatbot 'intelligent' to collaborate with the MES user (intelligence means predictive power in this paper). Hence, our research objective is to use MES to accommodate AI-based technical assistance.

Section 2 presents the related theoretical work and Sect. 3 describes the research methodology. Findings are presented in Sect. 4 and discussed in Sect. 5. Finally, conclusions are drawn in Sect. 6.

2 Related Work

2.1 MES Interfaces

Even though MES is meant to interact with humans on the shop floor for smoother production management, this aspect has not been studied extensively in the academic literature. The design purpose of MES is to support human decision making for activities related to manufacturing operations management. MES does this by making the production data accessible in real-time. Due to this, it is deemed a 'manufacturing cockpit' [10]. There have been studies to improve processes by analyzing real-time production data from MES [2], but not many of them focus on enhancing the efficiency of shop floor operations by deploying conversational assistants (with predictive power) as collaborating agents.

Production data acquisition was formerly needed to calculate machine utilization, whereas, current day's factories need it for real-time process adjustments. Production data is valuable for manufacturing enterprises to meet future market demands for product variability and faster deliveries. Timely availability of production data will boost the existing operational procedures related to work scheduling, customer interaction, order fulfillment, toolmaking, costing and supplier interaction, etc.

MES can have different user groups with customized interfaces and modules for each group. As an example, the management personnel can have access to the monitoring module and might be able to calculate key performance indicators (KPIs) from the acquired data, whereas the machine operators might have access to its planning module. The table below has a broad classification of MES users and their possible role if MES is combined with AI-based prediction systems for information exchange.

Table 1. MES user on the shop floor.

Category	Existing role	Possible engagement
Management personnel	Monitoring the production process	foresee inventory shortage and manage orders*
Operator personnel	Production planning, scheduling, dispatching, tracking	Anomaly detection and take corrective action on the production line [11]

* Focus of this study

User-friendly MES interfaces provide easy access to data, thus supporting human decisions in manufacturing operations. There are standards such as ISA 101 to suggest the best practices of human-machine interfacing in manufacturing [12]. ISA 101 covers menu hierarchies, screen navigation conventions, security methods and electronic signature attributes, pop-up conventions, configuration interfaces to databases, servers, etc.

In this regard, modern MES also comes with dashboards that display production, process insights using data visualization tools. They enhance the human-machine interface and particularly benefit management personnel that performs process data mining and monitoring. Some MES vendors also offer the dashboards as apps. However, these data access tools are oftentimes expensive and non-conversational.

2.2 Idiosyncrasies of Shop Floor

Manufacturing control and management had a variety of applications on the shop floor. Industry practices around it were mostly about shop floor operators manually modifying the processes using ad hoc methods. There was a minimal need for re-planning or on-line measurement and real-time feedback control [13]. However, with changing manufacturing requirements (such as demand for product variability, shorter time to market, engineer-to-order production strategies and real-time factory scheduling) the future demands on in-process management measurement and feedback control cannot be met with traditional practices.

Over the years, MES successfully served as a factory database [14] but faced resistance due to its inflexibility, monolithic architecture and for being an expensive investment. With the ever-increasing volumes of logged production data, manufacturing enterprises might reconsider MES for building a foundation for their future data management initiatives. Considering the importance of data visibility, we argue that intelligent planning and control concepts must be infused in MES to predict and coordinate production activities. Due to this, the MES user can engage in dynamic manufacturing operations with rapid responsiveness.

2.3 Potentials in Chatbots

Context-sensitive user interfaces and real-time learning assistants are some forms of interactions in the cyber-physical world. These technologies simplify the complexity of workers with information and interaction possibilities [5]. Chatbots can be used as conversational information systems [15] and are promising for filtering and processing information. However, chatbots are a nascent technology and their application in enterprises is still not well understood [15]. Chatbots use artificial intelligence and combining them with MES for human-machine interaction can complement the production workforce by demanding flexibility and creativity [16].

A chatbot for MES provides a more natural interaction platform compared to a traditional MES human interface. Being an automated system for enhancing human-computer interaction, chatbots can parse MES user input in the form of natural language and generate an appropriate response text. This may be done in several ways. The simplest is by providing the bot with a database of questions and answers. This approach works well for knowledge bases and static information but falls short in a dynamic environment [17]. Chatbot as an interface layer can also serve as a replacement for MES dashboards or can complement it. A chatbot for MES caters to the digitalization of MES, which is needed for building Industry 4.0 capabilities in a factory [18]. Owing to these principles, we hypothesize:

(H) A chatbot interface is a user-centric design enhancement for MES and it serves to monitor manufacturing operations by easy retrieval of information on demand.

3 Approach

We used a combination of empirical study and an experiment for this paper. First, we conducted a selective literature review using the databases Google Scholar and Scopus, with an emphasis on the publications from 2015-2019. The keywords 'MES', 'manufacturing operations management', 'technical assistance in Industry 4.0', 'chatbots' etc. were searched and Mendeley was used to manage the references. The review results provided knowledge on MES functionalities and its usefulness for coordinating manufacturing operations when combined with conversational virtual assistants. Based on the results, we identified applicable AI techniques. The literature review also helped in driving the empirical study.

Second, to ensure the relevance aspect of the MES design problem and to determine the requirements of the manufacturing industry, empirical evidence on the existing state of operations management technologies were needed. Therefore, we studied four companies to understand business and technological drivers to design technical assistance using MES. Due to commercial confidentiality, they are called Alpha, Beta, Gamma, and Delta. We conducted empirical research, where data triangulation was done using semi-structured interviews, field studies, archival documents, and industry reports.

In the following table, Finding 1 concern the company's drive for choosing MES and Finding 2 concerns the assistance requirements for MES users:

Table 2. Summary of the collected qualitative data.

Company	Data type	Description
Alpha	Size	>10,000 employees
	Industry	Dairy
	Finding 1	MES to develop the Industrial IoT ecosystem in the company using enterprise systems
	Finding 2	The requirement to provide flexibility and visibility to the shop-floor personnel using dashboards based on real-time production data from MES
Beta	Size	>10,000 employees
	Industry	Meat processing
	Finding 1	MES to streamline processes and gain competitiveness
	Finding 2	The requirement to react to supply/demand problems using timely available production data
Gamma	Size	>10,000 employees
	Industry	Electrical equipment
	Finding 1	MES to secure manufacturing intelligence
	Finding 2	The requirement for real-time monitoring and control of cross-business-unit coordination
Delta	Size	>10,000 employees
	Industry	Energy equipment
	Finding 1	Primarily use MES for order execution, but also learn from the acquired production data
	Finding 2	The requirement to use technologies that can give insight into data and enhance the human-machine interface (HMI). Currently, the production line controller is the HMI for shop floor personnel

Data provided insights into the phenomenon of technical assistance for manufacturing planning and control, its dynamics and the importance of information flows in a factory. Based on the gathered requirements a representative use case of information flow for order management is identified. It falls under the category of management personnel (see Table 1). Consequently, the design requirements were documented (see Sect. 4.1).

Lastly, as an experiment, we programmed a prototype chatbot and connected it to a simple web service endpoint, simulating a web-based MES database (we used Odoo, a cloud-based enterprise software offered as a service) of AAU Smart Production lab [19]. To create a simple demonstrator, we used AIML. AIML is a free and open-source artificial intelligence markup language and is the framework of choice for major chatbot platforms, including Pandorabots. The language is XML-based and can be used to create both simple and complex, state-aware chatbots. Using basic tags combined with pattern recognition, the chatbot was able to provide a rational response from its knowledge base [20]. Thus, we had a proof of concept for MES based technical assistance (see Sect. 4.2). The collected qualitative data also helped in evaluating the benefits of such chatbots in the manufacturing context. The following section presents the findings on the requirements of a User-friendly MES interface.

4 Findings

4.1 High-Level Goals

The literature and empirical evidence suggest that enterprises are re-considering MES to launch their future Industry 4.0 data management initiatives. Furthermore, we also noted that the role of humans shifts from machine operator to strategic decision-maker and a flexible problem-solver due to technical assistance in Industry 4.0 [4]. Therefore, we studied if the assistance systems are compatible with a typical MES system, which has the factory data access in real-time. MES focusses on ‘production data collection’, which is defined by IEC 62264-3 standard as an activity of gathering and managing information on work processes and production orders [2]. For this purpose, the multiple requirements of MES users are identified and their developments are presented in the figure below:

Fig. 1. Requirements breakdown for an MES based technical assistance.

Bussmann & McFarlane summarize the future manufacturing demands as increasing complexity and continual change, where anticipated control system properties will not suffice. Futuristic operations need intelligent control and data is key to increase the responsiveness of computer control systems and to aid human operations. To achieve such intelligence, the first step for the system is to have context-awareness, meaning fewer assumptions are to be made on process and component behavior [21]. MES supports control actions being taken in real-time. Live access to production status can improve operational procedures of the workforce and for this purpose; the MES system can be developed with added intelligence in a manufacturing enterprise. We identified that MES can be extended with a chatbot prediction system to enhance user experience and to support human decision making.

In the changing manufacturing times, the strategy to improve customer responsiveness and reduce time to market is extremely crucial. Hence, we studied order management practices to design a system that can proactively suggest relevant information based on the previous input. We focused on the system's capability to update the state of order and anticipate delays.

4.2 Low-Level Goals

In a factory environment, chatbot for MES can be made 'intelligent' using machine learning techniques [22] for it to learn from the repeated queries of users. The relevant approaches are NLP and ANN (discussed further in Sect. 5.2). The proposed MES design enhancement to accommodate AI-based technical assistance consists of:

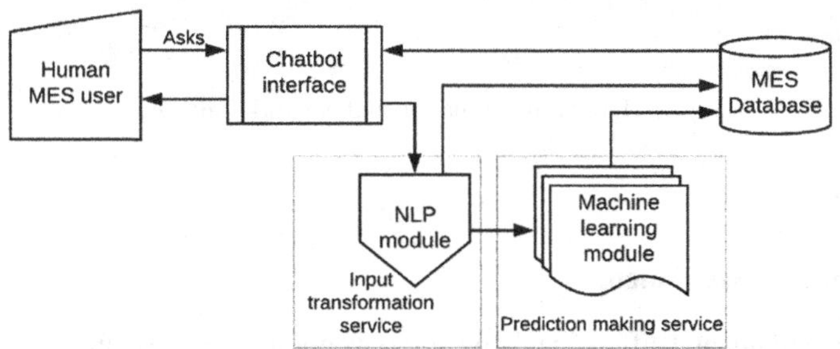

Fig. 2. MES based Technical assistance system.

In Fig. 2 above, the relationship between various parts of the system is shown. Here, a human user interacts with a chatbot interface and may ask about the current state of

```
Input: Natural language query from the user
Output: Requested data value from MES on natural language form, chat log
Algorithm: User interaction
while user has a question
    user enters query
    detected and extract keywords
    for each keyword k in query
        lookup k in MES data fields
        save k for Algorithm 2
        return MES data value v
    end for
    concatenate v with predefined reply string
end while
```

```
Input: Chat log D where D_size > 1
Output: Training set for prediction system
Algorithm: Creating training set for prediction system
for each query q in D
    for each keyword k in q
        map k to numerical value (0, 1]
        add k to training set – input vector
    end for
    for each keyword k' in q+1
        map k' to numerical value (0, 1]
        add k' to training set – output vector
    end for
end for
((q+1) is the successive query)
```

Fig. 3. Chatbot interaction and training algorithms.

affairs using natural language. This interaction may be represented using the following algorithm (see Fig. 3).

A graphical representation of Algorithm 2 is presented in Fig. 4 below. The first known word in the list is assigned 0.1, the last 1.0. The input vector size is set to 4. If there are fewer keywords than this, the value 0.0 is used. If the number of keywords is greater than 4, the most significant words according to the inverse document frequency are used.

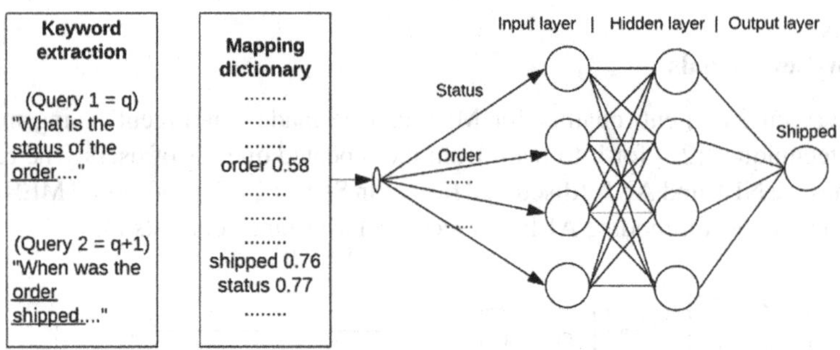

Fig. 4. Input transformation and network training.

4.3 Output and Assessment

To study the fundamental differences between conventional MES user interface and MES with a chatbot user interface, the prototype is assessed based on the chatbots' ability to retrieve order information. However, the chatbot is able to retrieve any data from MES given that they are exposed through a web service. Figure 5 shows examples of returned output. The input from the chatbot client is returned along with a message from the server.

```
Human: please contact the server
Robot: What is your request?
Human: what is the status of order 00055?
Robot: WH/OUT/000555 scheduled for 04/08/2019 13:39:47
Robot: Related information:
Human:
```

Fig. 5. Assessment of conversational MES interface.

We assess the success of MES based technical assistance system based on the ease with which user received a response without having to enter complete order ID. Our qualitative assessment suggests that this approach is easier than the regular database lookup.

The output (see Fig. 5), as well as the assessment, supports the hypothesis that it is easy to retrieve production information using a chatbot interface. The collected empirical data (see Table 2) also suggests that such retrieval process is beneficial to manufacturing. Challenges regarding the training of the system are presented in the following discussion section.

5 Discussion

Previous studies on enterprise systems attempted to close the gap of their integration with shop floor devices in a service-oriented way [23], but they did not focus on the MES user empowerment in the age of industrial internet. In response to the growing interest in practitioners to digitize manufacturing operations and to pursue Industry 4.0 [4], we proposed a concept that provides assistance to the production personnel (team) to make planning and control decisions by reacting quickly. The rationale for choosing MES instead of any other information system in the factory is also explained. The next iteration of this paper is expected to investigate its impact on the existing operational procedures for process improvement. Such a study can expose the technical and organizational challenges in implementing MES based technical assistance systems.

5.1 Shop Floors with Virtual Intelligent-Assistant

The concept of 'Internet of people' connects the workforce to the internet using interfaces. It promotes people-centric design enhancements with the principles of being social, personal and proactive [24]. In a factory's dynamic environment, a chatbot can be connected to the sensors of the physical assets on the shop floor or to external data providers to generate answers on the fly. A chatbot with a prediction system can also be connected to a database supported knowledge base [25]. Since MES is a factory database with real-time production data access and a unified functional platform of IT, we designed a 'human-AI collaboration system' around it.

A chatbot could, in theory, be connected to any data providing service. For the shop floor workforce, a chatbot creates an alternative to looking through menus and tables of MES to find the correct information and aids smoother information access [26]. Since MES is a database application with multiple user groups in a factory, designing a collaborative system (a combination of processes and software to support a group activity) around it is possible. Vision for the MES collaborative system needs further inquiry on the processes that involve group activity on the shop floor. Such an effort also aligns with the concept of 'digital workplace' that promotes virtual assistance [27].

The system may be extended to generate complex answers from the bot and to create a more helpful interface for MES. This prediction system may also be connected to other external data sources to provide additional responses relevant to shop floor activities such as materials inventory, shipping, traffic, exchange rates, etc.

5.2 Real-Time AI Techniques for MES

Contrary to the popular belief that the application of artificial intelligence to automate workplace tasks might lead to the unemployment crisis [28], we present a situation where the advances in AI techniques are utilized to assist the workforce (see Fig. 1). We used

NLP that is crucial for creating a chatbot as it enables a computer system to interpret and find the intent of the MES user's input. In this context, NLP can be implemented using the inverse document frequency algorithm to detect keywords, along with Levenshtein-distance [29] to compensate for spelling errors. This approach approximates MES user input with known chatbot vocabulary as per the dictionary mapping discussed earlier.

Enabling interaction with MES using natural language instead of navigating through complex or even unintuitive user interfaces is imagined to increase user-friendliness, especially for novice users. The user only gets the information they are requesting and do not have to be distracted by the information irrelevant to their current task. This is evident from Fig. 4, where the user was not required to enter the complete order ID in order to get a response. The reply from the bot contains full information as retrieved from MES. This functionality is akin to a sophisticated search function, although it comes with the added benefit of flexible queries in the form of natural language. However, this also enables traditional keyword-based searches in the same interface (The NLP system we presented is available only in the English language).

Training the System. Artificial neural networks are often used for prediction systems [30, 31] and the ANN in our design is trained on the transformed MES user input to suggest relevant information, which turns the bot into an intelligent assistant for MES users. Being able to get related information automatically is expected to be a helpful feature. However, the suggestions given must be evaluated for their usefulness. If the system is not properly trained, offering irrelevant or even wrong information will serve as a source of frustration. Giving feedback will help the system learn and provide appropriate suggestions. However, this should not affect the normal routines of workers. Using the time to train the system would defeat its purpose. Ideally, some person could be assigned to train and improve the accuracy of the chatbot. Training an ANN is an iterative process and is performed until the produced output is within a set error limit of the expected output. Over time, accuracy is expected to improve and less time will be needed to monitor the chatbot.

5.3 In-House Chatbot Solution

Deploying an MES based chatbot and its services are expected to be a major advantage (also cost-effective) compared to existing virtual assistants. Examples of this technology include Google's Assistant and Apple's Siri. Nonetheless, as they are proprietary systems by external vendors, the information must pass through their systems. For sensitive production data, this can be a major drawback. They also come packed with features that a shop floor might not require, where the focus is to get the required information as swiftly as possible. A domain-specific expert chatbot solution like the one presented in this paper will require less initial setup, as it is only connected to the internal services. Furthermore, the chatbot and its accompanying web service may further tailored to fit a specific purpose, as new and unexpected needs might arise in production. In order to realize the full potentials of the chatbot interface layer, we recommend using web-based MES because it makes it easy to scale-up, add clients and functionalities. Its architecture can be presented as (Fig. 6):

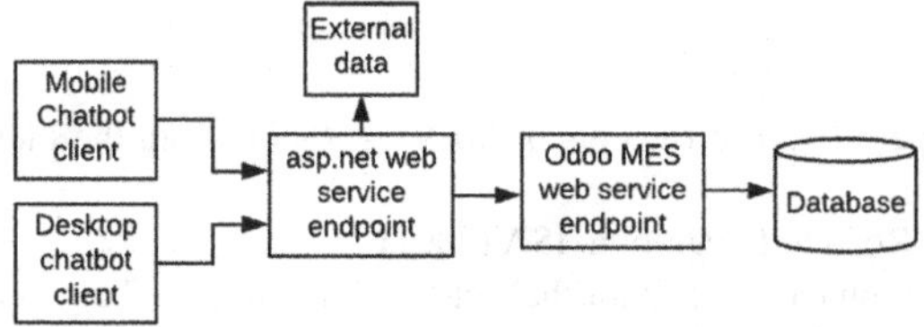

Fig. 6. Web service architecture for MES.

Such design eliminates the traditional monolithic MES architecture and makes it flexible to develop next-generation infrastructural capabilities for the shop floor. Based on a service-oriented architecture method, a digitally enriched MES will be able to facilitate the easy integration of applications as well as enable itself to connect to the network services like the one presented in this paper.

Finally, the data gathered by the chatbot in the form of user queries may also be analyzed to get an insight into the state of production and worker satisfaction. The presented user-friendly MES interface layer demonstrated such behavior, even though the efficacy and accuracy of chatbots for data extraction remains questionable. The current study gave an example of 'order management' but the future work is expected to analyze its impact on a wider range of processes related to manufacturing operations management, based on the aforementioned design.

6 Conclusions

Manufacturing enterprises are reconsidering MES for implementing their future data management initiatives (see Sect. 4.1). Therefore, we attempted to digitally enrich the MES platform for technical assistance on data retrieval. The findings suggest that the intelligent chatbot for MES can assist the shop floor personnel and can aid the operator's decision making (see Sect. 4.3). Because of this design criterion, MES users could engage in dynamic operations with enhanced responsiveness. The study demonstrated a human-AI collaboration in a factory, where MES fostered unstructured collaboration as it feeds the users' responses (data) into the system for further learning. The collected empirical data also helps us in concluding that the chatbot interface for MES is highly beneficial in the manufacturing context. As a result, recommendations for AI-based chatbot assistance in Industry 4.0 shop floors were presented (see Sect. 5).

As far as we know, this is the first time an intelligent chatbot is used for MES to enable a factory as a digital workplace and to scientifically approach the Industry 4.0 design principle of 'Technical assistance'. Future work intends to add more complexities to the proposed design to test its benefits and advance the theory of manufacturing digitalization.

Acknowledgment. This research work is partially funded by the Manufacturing Academy of Denmark. The authors would like to thank the informants from the companies for sharing their knowledge.

References

1. MESA White Paper #01: The Benefits of MES: A Report from the Field. Presented at the (1997)
2. Scholten, B.: The Road to Integration. ISA (2007)
3. Sauer, O.: Information technology for the factory of the future - State of the art and need for action. Procedia CIRP **25**, 293–296 (2014). https://doi.org/10.1016/j.procir.2014.10.041
4. Hermann, M., Pentek, T., Otto, B.: Design principles for industrie 4.0 scenarios. In: Proceedings of the Annual Hawaii International Conference on System Sciences, March 2016, pp. 3928–3937 (2016)
5. Gorecky, D., Schmitt, M., Loskyll, M., Zühlke, D.: Human-machine-interaction in the industry 4.0 era. In: Proceedings - 2014 12th IEEE International Conference on Industrial Informatics, INDIN 2014, pp. 289–294 (2014)
6. Monostori, L., Váncza, J., Kumara, S.R.T.: Agent-based systems for manufacturing. CIRP Ann. - Manuf. Technol. **55**, 697–720 (2006). https://doi.org/10.1016/j.cirp.2006.10.004
7. Hatvany, J., Nemes, L.: Intelligent manufacturing systems—a tentative forecast. IFAC Proc. **11**, 895–899 (1978). https://doi.org/10.1016/S1474-6670(17)66031-2
8. Maedche, A., Morana, S., Schacht, S., Werth, D., Krumeich, J.: Advanced user assistance systems. Bus. Inf. Syst. Eng. **58**, 367–370 (2016). https://doi.org/10.1007/s12599-016-0444-2
9. Stoeckli, E., Uebernickel, F., Brenner, W.: Exploring affordances of slack integrations and their actualization within enterprises - towards an understanding of how chatbots create value. In: Proceedings of the 51st Hawaii International Conference on Systems Science, pp. 2016–2025 (2018). https://doi.org/10.24251/hicss.2018.255
10. Kletti, J.: Manufacturing Execution Systems – MES. Springer, Heidelberg (2007). https://doi.org/10.1007/978-3-540-49744-8
11. Mantravadi, S., Li, C., Møller, C.: Multi-agent manufacturing execution system (MES): concept, architecture & ML algorithm for a smart factory case. In: Scitepress, pp. 477–482 (2019). https://doi.org/10.5220/0007768904770482
12. ISA101: Human-Machine Interfaces. https://www.isa.org/
13. Albus, J.S.: An intelligent systems architecture for manufacturing. In: International Conference Intelligent Systems: A Semiotic Perspective, pp. 20–23 (1996)
14. Younus, M., Hu, L., Yuqing, F., Yong, C.P.: Manufacturing execution system for a subsidiary of aerospace manufacturing industry. In: Proceedings - 2009 International Conference on Computer and Automation Engineering, ICCAE 2009, pp. 208–212 (2009). https://doi.org/10.1109/ICCAE.2009.12
15. Meyer von Wolff, R., Hobert, S., Schumann, M.: How may I help you? – State of the art and open research questions for chatbots at the digital workplace. In: Proceedings of the Hawaii International Conference on Systems Science, vol. 6, pp. 95–104 (2019)
16. Autor, D.H., Levy, F., Murnane, R.J.: The skill content of recent technological change: an empirical exploration. Q. J. Econ. **118**, 1279–1333 (2003)
17. Augello, A., Pilato, G., Machi, A., Gaglio, S.: An approach to enhance chatbot semantic power and maintainability: experiences within the FRASI project. In: Proceedings - IEEE 6th International Conference Semantic Computing, ICSC 2012, pp. 186–193 (2012). https://doi.org/10.1109/ICSC.2012.26
18. Demartini, M., Tonelli, F., Damiani, L., Revetria, R., Cassettari, L.: Digitalization of manufacturing execution systems: the core technology for realizing future smart factories. In: Proceedings of the Summer School Francesco Turco, September 2017, pp. 326–333 (2017)
19. Madsen, O., Møller, C.: The AAU smart production laboratory for teaching and research in emerging digital manufacturing technologies. Procedia Manuf. **9**, 106–112 (2017)

20. Marietto, M., et al.: Artificial Intelligence Markup Language: a brief tutorial. Int. J. Comput. Sci. Eng. Surv. (IJCSES) **4**(3), 1–20 (2013)
21. Bussmann, S., McFarlane, D.: Rationales for holonic manufacturing control. In: Proceedings of the 2nd Intelligent Agent Based Framework for Manufacturing Systems, pp. 1–8 (1999)
22. Wu, Y., Wang, G., Li, W., Li, Z.: Automatic chatbot knowledge acquisition from online forum via rough set and ensemble learning. In: Proceedings - 2008 IFIP International Conference on Network and Parallel Computing, NPC 2008, pp. 242–246 (2008). https://doi.org/10.1109/NPC.2008.24
23. Karnouskos, S., Baecker, O., De Souza, L.M.S., Spieß, P.: Integration of SOA-ready networked embedded devices in enterprise systems via a cross-layered web service infrastructure. In: IEEE International Conference on Emerging Technologies Factory Automation ETFA, pp. 293–300 (2007)
24. Miranda, J., et al.: From the Internet of Things to the Internet of People. IEEE Internet Comput. **19**, 40–47 (2015)
25. Reshmi, S., Balakrishnan, K.: Implementation of an inquisitive chatbot for database supported knowledge bases. Sadhana - Indian Acad. Sci. **41**, 1173–1178 (2016)
26. Carayannopoulos, S.: Using chatbots to aid transition. Int. J. Inf. Learn. Technol. **35**, 118–129 (2018)
27. Lestarini, D., Raflesia, S.P., Surendro, K.: A conceptual framework of engaged digital workplace diffusion. In: Proceeding 2015 9th International Conference on Telecommunication System, Services, and Application, TSSA 2015 (2016)
28. Levy, F.: Computers and populism: artificial intelligence, jobs, and politics in the near term. Oxford Rev. Econ. Policy. **34**, 393–417 (2018). https://doi.org/10.1093/oxrep/gry004
29. Gilleland, M.: Levenshtein Distance, in Three Flavors. https://people.cs.pitt.edu/~kirk/cs1501/Pruhs/Spring2006/assignments/editdistance/LevenshteinDistance.htm
30. Kimoto, T., Asakawa, K., Yoda, M., Takeoka, M.: Stock market prediction system with modular neural networks, vol. 1, pp. 1–6 (2002). https://doi.org/10.1109/ijcnn.1990.137535
31. Sheth, A., et al.: A sensor network based landslide prediction system. In: SenSys, pp. 280–281 (2005)

Modelling Shipping 4.0: A Reference Architecture for the Cyber-Enabled Ship

Georgios Kavallieratos[1](✉), Sokratis Katsikas[1,2], and Vasileios Gkioulos[1]

[1] Department of Information Security and Communications Technology, Norwegian University of Science and Technology, Gjøvik, Norway
{georgios.kavallieratos,sokratis.katsikas,vasileios.gkioulos}@ntnu.no

[2] School of Pure and Applied Sciences, Open University of Cyprus, Latsia, Nicosia, Cyprus
sokratis.katsikas@ouc.ac.cy

Abstract. There is intense activity of the maritime industry towards making remotely controlled and autonomous ships sail in the near future; this activity constitutes the instantiation of the Industry 4.0 process in the maritime industry. Yet, a reference model of the architecture of such vessels that will facilitate the "shipping 4.0" process has not yet been defined. In this paper we extend the existing Maritime Architectural Framework to allow the description of the cyber-enabled ships (C-ESs), and we demonstrate the use of the extended framework by developing descriptions of the architecture of variants of the Cyber-enabled ship. The results can be used not only to systematically describe the architecture of Cyber-enabled ships in a harmonized manner, but also to identify standardization gaps, and to elicit the cybersecurity requirements of the C-ES ecosystem.

Keywords: Autonomous ships · Reference architecture · Cyber-physical systems

1 Introduction

Industry 4.0 describes the trend towards increasing automation and connectivity, by leveraging technologies such as the Internet of Things (IoT), Artificial Intelligence (AI), and Big Data Analytics. In the maritime sector, despite the fact that nowadays almost all ships are automated in some way, the shipping industry is coming to alignment with Industry 4.0 with the emergence of autonomous vessels [9]. However, this is not a direct process towards a fully autonomous system, but rather a gradual shift towards the digital transformation of maritime operations both ship- and shore-side [19]. In this "Shipping 4.0" process, the interaction and dynamics between ship/land, ship/authorities and ship/ship are expected to change fundamentally [32].

In modern systems engineering, specifications are created by means of employing some requirements engineering process. Such a process is used for eliciting the information needed to create a solution architecture, and subsequently

N. T. Nguyen et al. (Eds.): ACIIDS 2020, LNAI 12034, pp. 202–217, 2020.
https://doi.org/10.1007/978-3-030-42058-1_17

to implement and operate it. Thus, the system architecture is a key element of the process of implementing and deploying a system according to the specifications. For simple systems, this process can be carried out semi-formally, by direct communication among the different teams involved in the process. However, this approach does not work in the case of complex systems-of-systems, where a large number of engineering teams are responsible for different components and parts of the system, and the knowledge and work is much more fragmented. This situation calls for a formalized and governed process, where communication is done in a formal and knowledge-intensive manner and where standards are needed at a certain point. One part of the solution to this problem is to use a method which has proven to be useful, namely the development of a *Reference Architecture* [33].

A Reference Architecture describes the structure of a system, with its element types and their structures, as well as their interaction types, among each other and with their environment. By describing these, a Reference Architecture defines restrictions for an instantiation (concrete architecture). Through abstraction from individual details, a Reference Architecture is universally valid within a specific domain. Further architectures with the same functional requirements can be constructed based on the reference architecture [3,17].

A Maritime Architecture Framework (MAF) was proposed in [36], to facilitate the development and adoption of new systems and technologies in the maritime domain. The development process of the MAF followed that of the Smart Grid Architectural Model (SGAM) [6]; accordingly, the MAF has been developed taking into consideration existing maritime architectures, including the Common Shore Based System Architecture [4] and the International Maritime Organization's (IMO) e-Navigation architecture [2].

The IMO uses the term *MASS (Maritime Autonomous Surface Ship)* for the autonomous ship [14]. Cyber-Enabled ships (C-ES) are ships that integrate Cyber Physical Systems (CPSs) within their architectures, and whose operations may be fully or partially carried out autonomously. Thus, a C-ES may be a conventional, remotely controlled or autonomous ship. Further, a C-ES can be manned or unmanned, depending on its operational procedures and its infrastructure. According to the IMO, the levels of autonomy for a MASS are defined as follows:

- *AL0: Ship with automated processes and decision support:* Seafarers are on board to operate and control shipboard systems and functions. Some operations may be automated.
- *AL1: Remotely controlled ship (with seafarers on board):* The ship is controlled and operated from another location, but seafarers are on board.
- *AL2: Remotely controlled ship (without seafarers on board):* The ship is controlled and operated from another location. There are no seafarers on board.
- *AL3: Fully autonomous ship:* The operating system of the ship is able to make decisions and determine actions by itself.

AL0 describes the conventional ship, where the C-ES's operations are the same with those of traditional vessels. Although many contemporary ICT systems can be on board in order to support processes related with navigation and engine control, human operators maintain the central role. For the remotely controlled variants (AL1, AL2), most of the ship's systems are capable of performing predefined actions without human intervention. The ship's operations depend on the communication with the Shore Control Center (SCC) and at the same time are influenced from on-board crew and CPSs. The human operator at these levels gives directions and controls the vessel's systems either locally (AL1) or remotely (AL2), whilst operations such as mooring, navigating, cargo loading and unloading are performed entirely by remote control. At the last level of autonomy (AL3), most of the ship's operations rely on the on-board CPSs, although some of the operations may be supervised by a SCC. Furthermore, the ship is equipped with contemporary navigation, engine and control systems, such as collision avoidance systems. At this level, the human vector does not exist and advanced systems are responsible for the availability, maintainability and reliability of the operations.

In this paper we extend the MAF to include autonomous vessels. We then demonstrate the use of the MAF to create architectural instances of autonomous vessels with varying level of autonomy, and we identify their functional and operational requirements. Finally, we identify and analyze the interdependencies and interconnections among the CPSs that are components of the C-ES. The contribution of this work is as follows:

- The development of an extended Maritime Architectural Framework that can accommodate autonomous vessels;
- The instantiation of this reference architectural model to classes of autonomous vessels, with varying degree of autonomy;
- The identification of functional and operational requirements for autonomous vessels within the architectural model;
- The identification and analysis of the interdependencies and interconnections of the cyber-physical components of the C-ES.

The remainder of the paper is structured as follows: In Sect. 2 the related work is briefly reviewed. Section 3 briefly reviews the MAF and presents the proposed extension. In Sect. 4 we demonstrate the use of the extended MAF to create architectural instances for variants of the C-ES, by identifying the functional and operational requirements of the C-ES; the CPSs comprising the C-ES; and the interdependencies and interconnections among them. Finally, Sect. 5 summarizes our conclusions and indicates directions for future research.

2 Related Work

Reference architectures have been developed for the smart grid [6,21]; service oriented architectures [26]; Industries 4.0 (RAMI4.0) [38]. In the maritime domain, the MITS [29] architecture describes the ICT components in the maritime industry and it has been used in [30] to describe the architecture of the unmanned merchant ship. A limitation of this model is the use of the OASIS [26] reference model to identify the vessel's Operational Technology (OT) infrastructure. The IMO has proposed its e-navigation architecture, covering mostly ship to ship communications, the relationships and the sharing of information between stakeholders [37]. However, this architecture pertains exclusively to conventional ships; hence it cannot be directly applied to the autonomous ship case. ARK-TRANS [24] is a reference architecture framework which captures responsibilities, relations, and dependencies in the transport sector. The European project Maritime Navigation Information Services (MarNIS) [23] has adopted the aforementioned framework. This captures the overall conceptual, logical, and technical aspects of the maritime sector. Yet, the framework is inappropriate for the C-ES case, as it is unable to capture technical operations which are crucial to understanding the operational objectives of the C-ES.

Little published work on the architecture of the C-ES exists. The MUNIN project developed a reference model of the architecture of the unmanned merchant ship [31]. The developed architecture is based on the MITS [29] architecture and on the OASIS [26] reference model. Further, [12] describes an architecture of the autonomous ship that considers only the connectivity of systems, ending up with a communication architecture. In [13] an autonomous ship architecture is proposed, based on IT components, without however taking into account the OT infrastructure.

3 The Extended Maritime Architectural Framework

The MAF is a domain specific architectural methodology that was developed to overcome the challenge to coordinate the development of new systems between technology issues, governance aspects and users between existing architectures in the maritime domain. As such, the MAF establishes clear relationships between technical systems, users and related governance aspects. Similarly with other approaches, the MAF is divided into two parts, namely the multidimensional cube that provides a graphical representation of the underlying maritime domain and the examined system architecture; and a methodology to structure the examined system including the system requirements and (possible) use cases in a consistent way. The methodology is composed of three main steps leading to enable an easy mapping of system architectures to the MAF-Cube. The scope of this process is to structure the system engineering phases starting from planning over the identification of requirements to the use case development in a harmonized and formal way. This allows the user to map the results, to visualize them in the MAF-Cube, to explore interoperability issues, and to identify spots which need to be standardized [36].

The main element of the MAF is the multidimensional cube, that combines different viewpoints to provide a graphical representation of the underlying maritime domain and the examined system architecture. The cube captures three dimensions, alias *axes*, namely interoperability; hierarchical; and topological. The topological axis represents the logical location where a technology component is located. The interoperability axis addresses communication, data and information, usage and context of a maritime system. The hierarchical axis substructures management and control systems of the maritime domain, for example for maritime transportation systems from the traffic management of a coastal area down

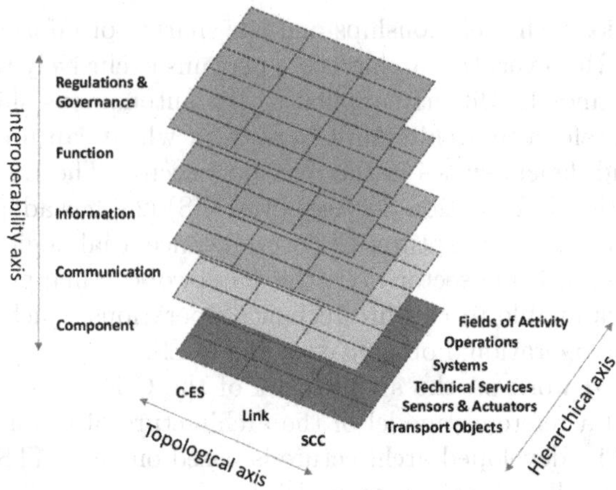

Fig. 1. Extended MAF

to the radar echo of a vessel [36]. Each axis breaks down to a number of *layers*. The layers of the topological axis (ships; link; shore) are derived from IMO's breakdown of the maritime domain [25]. The layers of the interoperability axis (Regulation Governance; function; information; communication; component) cover organizational, informational and technical aspects and include the different levels of interaction (operational, functional, technical and physical) as stated in IMO's e-Navigation vision [16]. Finally, the layers of the hierarchical axis (Fields of activity; operations; systems; technical services; sensors actuators; transport objects) cover economic, technical and physical issues of a maritime system.

Information, technology, and people are crucial elements of the C-ES ecosystem [11], and the MAF is able to capture these elements. Therefore, the MAF can in principle be used for representing and analyzing the C-ES ecosystem. However, the MAF in its current form cannot capture specific characteristics of autonomous vessels; some modifications are required in order to describe the new concepts and technologies. Specifically, the topological axis should be extended to include the C-ES, the SCC, and the Link between them. This should reflect the integration of new concepts, technologies, and operational models across all the components of the interoperability and hierarchical axes. The aforementioned extensions are described below:

- **C-ES layer:** Representing the ship-side entities such as the vessel's infrastructure, operational and functional goals, processes, and systems.
- **SCC layer:** Entities of the shore side infrastructure are represented along with processes, and systems which are vital for the C-ES's operation and facilitate the interaction with in/out of maritime sector entities.
- **Link layer:** Represents the telecommunication methods and protocols between C-ES and SCC.

This conceptual extension includes important aspects of the C-ES's ecosystem. The resulting extended MAF is shown in Fig. 1.

Table 1. Functional requirements

Functions	Description
System functions	The necessary system functions to facilitate the ship's voyage (e.g. engine functions)
Collision avoidance	The avoidance of collision with manned objects, physical obstacles, and marine animals
Search and rescue	The provision of the necessary assistance to other ships or persons which are in danger at sea
Technical reliability	The assurance of the operations, functions, and maintenance of the C-ES's systems
Voyage planning	The C-ES conducts route planning, determines its position, course, and speed and follows a predefined route
Keep general lookout	The C-ES promotes its situational awareness of the area surrounding the vessel
Cyber-security	The C-ES's infrastructure is protected against cyber-attacks
Physical security	The C-ES protects its infrastructure, cargo and humans from physical attacks

4 Putting the Extended MAF in Action

4.1 C-ES Functional and Operational Requirements

Identifying functional and operational requirements is the first step towards modelling the C-ES ecosystem. Functional requirements support the actions of the vessel systems, whilst operational requirements support the business and organizational requirements of the C-ES ecosystem.

The operational and functional requirements of autonomous vessels have been examined in the literature. DNV-GL in [35] clarified the main functional requirements for the conventional ship. [5] has identified the functional requirements for the remote and autonomous ships, focusing on the system specifications. Further, [31] analyzed the functional requirements of a merchant ship. The functional requirements of six main autonomous vessel systems have been analyzed in [34]. Additionally, the navigational, vessel engineering, and communication functions of autonomous vessels have been described in [10]. Although functions and operations described in [35] are included in [31], a set of the functions described in [5] could not be included in this classification. Considering these works and by leveraging the MAF, we identify the functional and operational requirements for the C-ES as depicted in Tables 1 and 2 respectively.

4.2 Cyber-Physical Systems of the C-ES

In order to use the extended MAF to analyze the variants of the C-ES deriving from the four autonomy levels, and in particular in order to analyze the system and component layers of the Hierarchical and Interoperability axis, we need to identify and classify the C-ES's CPSs. In [18] we identified the CPSs of the C-ES; these are shown in Fig. 2. Specifically, the C-ES ecosystem comprises three different classes following the MAF classification, namely the Vessel, the SCC, and the Link.

Table 2. Operational requirements

Operations	Description
Navigation	Ensuring ship navigation during the voyage
Control	The SCC is able to intervene at any time in order to control various ship's operations and functions
Weather	The ship must be capable to operate under harsh weather conditions
Mooring	The C-ES should be able to secure its location in permanent anchor location in the water
Enter a port	The C-ES should be able to secure its location in the port's infrastructure
Fail to safe	In case of emergency, the ship must stop its operations
Rendezvous	Under specific circumstances, the crew should proceed onboard the vessel
Transport cargo	The C-ES should have the appropriate infrastructure in order to transport cargo securely and safely
Load/unload cargo	The C-ES should have the appropriate infrastructure in order to load/unload securely and safely cargo
Transport people	The C-ES should have the appropriate infrastructure in order to transport passengers securely and safely
Communication	The C-ES must establish powerful communication networks within its infrastructure and with external actors
Passenger utilities	The C-ES must have adequate infrastructure to serve passenger's needs
Environmental observations	The C-ES has to use contemporary sensors in order to increase its situational awareness
Anchoring	The C-ES has to anchor in ports or anywhere under the supervision of the SCC
Ensure seaworthiness	The C-ES must comply to the corresponding legal framework for its operations
Maintain personnel and environmental safety	The C-ES should identify potential risks related to the safety of the crew and its environment
Maintain preparedness	The C-ES should develop and maintain resilience-aware activities to mitigate risks and to increase its situational awareness
Human resources	The C-ES should ensure the management of the connected human resources systems
Third parties relationships	The C-ES should manage relationships with suppliers, vendors, and other entities that influence its operational environment
Cyber security	The C-ES should follow cybersecurity standards and procedures and enforce the necessary security policies

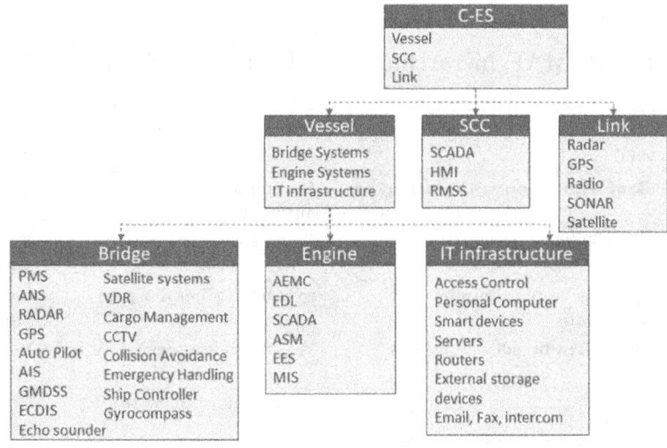

Fig. 2. C-ES's cyber-physical systems

4.3 C-ES Architectural Instances

We used the extended MAF to analyze the variants of the C-ES deriving from the four autonomy levels; the result of this analysis for the hierarchical axis is presented in Table 3. Note that AL1 and AL2 are merged in this table, as both represent a remotely controlled vessel.

By examining Table 3 we conclude that AL1-AL2 and AL3 share all fields of activity among them and with the AL0, with the exception of the communication with a SCC, which is not relevant for AL0 vessels. The operations of the vessels remain the same for all ALs. The *systems* row captures the integrated systems. Although vessels belonging to AL1-AL3 inherit the systems of AL0, advanced decision support and remote control systems are introduced. These are depicted in Fig. 2.

Additional technical services of the remotely controlled and of the autonomous ship respectively have been identified. CPSs identified in the previous layer reflect the technical services of each ship variant and therefore services are increasing as more CPSs are included in the infrastructure. The sensors and actuators installed in the conventional ship's infrastructure accommodate simple vessel functions, such as AutoPilot and weather observations. On the other hand, remotely controlled and autonomous ships will be equipped with advanced sensors systems able to facilitate functions such as docking, mooring, and engine maintenance. Finally, the transport objects (e.g. cargo and humans) for all vessel variants remain unaltered.

Table 4 contains the result of the analysis of the C-ES along the interoperability axis of the MAF cube. Many regulations and guidelines have been established for the AL0 ships depending on their type and fields of activity. Regulations such as [8,15,22] are applicable to different types of cargo ships (e.g., container, ferries, and tanker). The analysis of the regulations regarding the AL1-AL3 reveals that there is no established legal framework which marks boundaries of their operations, functions, and fields of activity [20]. Nevertheless, a lot of effort has

Table 3. MAF hierarchical axis for the C-ES ecosystem

C-ES: Functions			
	AL0	AL1-AL2	AL3
Fields of activity	Communication with authorities	Communication with authorities	Communication with authorities
	Ensure seaworthiness	Ensure seaworthiness	Ensure seaworthiness
	Systems to handle port operations	Systems to handle port operations	Systems to handle port operations
	Vessel Traffic service (VTS)	Vessel Traffic service (VTS)	Vessel Traffic service (VTS)
	–	Communication with SCC	Communication with SCC
Operations	Navigation	Navigation	Navigation
	Docking	Docking	Docking
	Mooring	Mooring	Mooring
Systems	Automatic Identification System (AIS)	Automatic Identification System (AIS)	Automatic Identification System (AIS)
	Electronic Chart Display and Information System (ECDIS)	Electronic Chart Display and Information System (ECDIS)	Electronic Chart Display and Information System (ECDIS)
	Global Maritime Distress and Safety System (GMDSS)	Global Maritime Distress and Safety System (GMDSS)	Global Maritime Distress and Safety System (GMDSS)
	Personnel safety systems	Personnel safety systems	–
	–	Remote maneuvering System	Remote maneuvering System
	–	Collision avoidance system	Collision avoidance system
	–	Autonomous Navigation System (ANS)	Autonomous Navigation System (ANS)
Technical services	Broadcast AIS data	Broadcast AIS data	Broadcast AIS data
	Fire protection	Fire protection	Fire protection
	Power generation	Power generation	Power generation
	Load/unload cargo	Load/unload cargo	Load/unload cargo
	–	Broadcast control commands	Broadcast control commands
	–	Sensors data fusion	Sensor data fusion
	–	–	AEMC
	–	–	Decision making
Sensors/Actuators	Auto Pilot	Auto Pilot	Auto Pilot
	Weather sensors	Weather sensors	Weather sensors
	Traffic sensors	Traffic sensors	Traffic sensors
	–	Docking actuators	Docking actuators
	–	Engine sensors/actuators	Engine sensors/actuators
Transport object	Humans	Humans	Humans
	Container	Container	Container

been put on the development of guidelines from classification societies such as DNV-GL [10], Lloyd's Register [27], China classification society [7], and Beureu Veritas [34]. The functions of the sensors and actuators between different autonomy levels are differentiated, since the complexity of the sensor infrastructure of AL1-AL3 vessels is increased. The information exchange between the ship variants differs with the autonomy level. Specifically, AL1-AL3 ships rely heavily on the information of sensors and actuators since advanced systems such as collision avoidance and decision making demand high information accuracy. The *communication* plane of the MAF can capture different protocols between sensors and actuators. AL0 ships usually employ protocols such as Modbus and radio signals, while AL1-AL3 ship communications may be established by leveraging contemporary communication protocols such as ZigBee, WiFi and Satellite connections. The *components* plane exhibits diversity in different autonomy levels. In particular, the autopilot, weather sensors and other environmental analysis sensors are crucial for AL0 ships, whilst contemporary engine actuators, navigation and docking sensors are vital for AL1-AL3 vessels.

4.4 Interconnections, Dependencies and Interdependencies Among CPS

To complete the architectural description of the C-ES, the interconnections, dependencies and interdependencies among the CPSs need to be also identified. Two CPSs are interconnected when there exists information exchange between them; when two CPSs are connected and the state of one system influences the state of the other, the systems are dependent. Two systems are interdependent when there exists bilateral dependency between them.

Table 5 depicts the interconnections along with the control flows within the CPSs of the C-ES. In particular, the data and control flows for each system are represented with blue arrows and red arrows respectively.

The C-ES's CPSs are all complex components in which changes may occur as a result of operational and/or functional processes. This complexity derives from the combination of IT and OT systems, the size of the C-ES ecosystem, the diversity of the installed components, and the different fields of activity. According to [28] an effective way to examine complex systems is to view them as a group of interacting systems. Accordingly, we examine the dependencies and interdependencies of the C-ES's systems considering three main groups of systems; the Bridge Automation System (BAS), the Engine Automation System (EAS), and the SCC. Furthermore, the dependencies and interdependencies of the three critical onboard components [18], namely the AIS; the ECDIS; and the GMDSS, all subsystems of the BAS are depicted in Figs. 4, 5, and 6.

Figure 3 represents the systems that can directly or indirectly be affected by potential systems state's changes. Figure 4 depicts the dependencies and interdependencies of the AIS, Fig. 5 those of the ECDIS, and Fig. 6 those of the GMDSS. With an eye towards identifying the most critical CPSs, and understanding the impact propagation among them, by way of considering their interconnections, dependencies and interdependencies, we first map the information in Table 5

Table 4. MAF interoperability axis for the C-ES ecosystem

C-ES: Sensors & actuators			
	AL0	AL1-AL2	AL3
Regulations	COLREGs	Could be adopted from conventional	–
	NMEA 2000		–
	Directive 2010/65/EU		–
Functions	Navigation	Navigation	Navigation
	Environment monitoring	Environment monitoring	Environment monitoring
	Temperature, speed and vibration measurements	Temperature, speed and vibration measurements	Temperature, speed and vibration measurements
	–	Mooring	Mooring
	–	Berthing	Berthing
Information	State/value of collision avoidance sensors	State/value of collision avoidance sensors	State/value of collision avoidance sensors
	State/value of steering sensors	State/value of steering sensors	State/value of steering sensors
	State/value of engine room sensors	State/value of engine room sensors	State/value of engine room sensors
	Distance from the port	Distance from the port	Distance from the port
	Depth of sea	Depth of sea	Depth of sea
	–	Objects at sea	Objects at sea
Communication	Modbus	Modbus	Modbus
	Satellite Com	Satellite Com	Satellite Com
	Radio (VHF)	Radio (VHF)	Radio (VHF)
	–	WiFi	WiFi
Components	Auto Pilot	Auto Pilot	Auto Pilot
	Weather sensors	Weather sensors	Weather sensors
	Temperature, speed and vibration sensors	Temperature, speed and vibration sensors	Temperature, speed and vibration sensors
	–	–	Docking actuators
	–	Engine actuators	Engine actuators
	–	Depth sounders	Depth sounders
	–	–	Navigation sensors and actuators

onto two graphs, whose nodes represent CPSs and edges represent connections. We then employ certain graph analysis metrics, that were calculated by leveraging the CASOS ORA tool from Carnegie Mellon University [1], to analyze the systems' criticality.

The exponential ranking centrality (ERC) defines the centrality of each system as its trustworthiness; it is based on the degree of trust that other systems have in it; the AIS and the ECDIS have the highest ERC value (1 and 0,985 respectively). The Betweenness Centrality (BC) metric allows the identification of the systems which hold the most critical position considering the connections and interconnections. The higher the value of the BC of a system, the more sys-

Table 5. Interconnections among C-ES CPSs

	AEMC	MIS	EES	AEC	EmH	ANS	Radar	Gyro Compass	Satellite	GPS	AIS	ECDIS	GMDSS	ASC	EDL	VDR	ASM	Cargo Management	Access Control	PMS	Auto Pilot	Collision Avoidance	HMI	RMSS
AEMC		↕↔	↕	↕	↕↔	↕←			↕					↕←	↕→	↕	↕						↕←	↕←
MIS	↕↔		↕	↕↔	↕↔				↕					↕←	↕		↕						↕←	
EES	↕	↕		↕	↕	↕								↕←	↕	↕	↕							↕←
AEC	↕	↕↔	↕		↕↔	↕↔								↕←	↕		↕→					↕		
EmH	↕↔	↕↔	↕	↕↔		↕↔			↕				↕→	↕↔	↕	↕	↕→				↕	↕	↕	
ANS	↕→		↕	↕↔	↕↔		↓↔	↕→	↕→	↕→	↕→	↕→	↕→			↕→	↕→				↕→	↕→	↕←	↕←
Radar						↑↔		↓→	↕	↕	↕	↕	↕		↑	↕				↑	↕	↕		
Gyro Compass						↕←	↑←			↕	↕	↓	↕	↕		↕					↕←	↕		
Satellite	↕	↕			↕	↕←	↕				↕	↕	↕								↕	↕		↕
GPS						↕←	↕	↕			↕	↕	↕	↕		↕	↕				↕	↕		
AIS						↕←	↕	↕	↕	↕		↕	↕	↕		↕	↕	↕			↕	↕←		
ECDIS						↕←	↕	↕	↕	↕	↕			↕		↕	↕				↕	↕	↕←	
GMDSS					↕←	↕←	↕	↕	↕	↕	↕			↕←				↕	↕→	↕	↕	↕		
ASC	↕→	↕→	↕→	↕→	↕↔	↕↔		↕		↕	↕	↕	↕→		↕→	↕	↕→	↕←	↕←	↕←	↕→	↕→	↕←	↕←
EDL	↕←	↕	↕	↕	↕		↓							↕←										
VDR	↕		↕		↕	↕←	↕	↕		↕	↕	↕		↕				↕				↕		
ASM	↕	↕	↕	↕←	↕←	↕←				↕	↕	↕		↕←				↕		↕	↕	↕→		
Cargo Management											↕		↕	↕→		↕	↕						↕←	
Access Control													↕←	↕→									↕←	
PMS							↓						↕	↕→			↕						↕←	
Auto Pilot						↕←	↕	↕→	↕	↕	↕	↕		↕←			↕					↕	↕←	↕←
Collision Avoidance				↕	↕	↕→	↕	↕	↕	↕	↕→	↕	↕	←		↕	↕←				↕			
HMI	↕→	↕→			↕→	↕→						↕→		↕→				↕→	↕→	↕→	↕→			↕→
RMSS	↕→		↕→			↕→			↕					↕→							↕→		↕←	

tems are connected to it and therefore a potential failure would affect the whole system. Finally, The degree centrality estimates the number of connections a system has. In particular, the Total Degree (TD) is the sum of the links in and from the systems. A system with high TD is a well connected node; therefore its operations and functionalities can influence other systems and, in case of failure, may provoke bigger damage to the infrastructure.

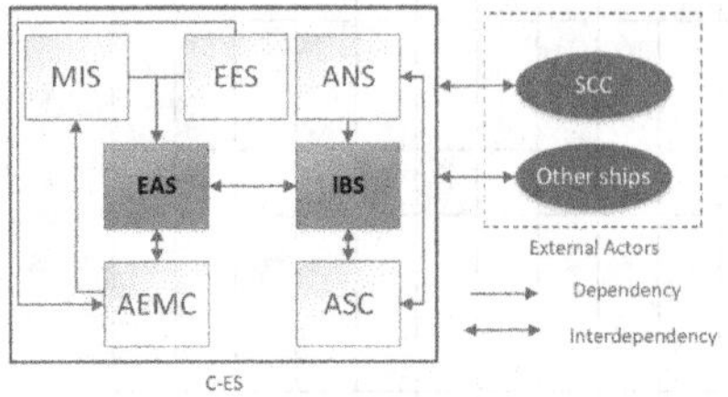

Fig. 3. General Ecosystem

Fig. 4. AIS

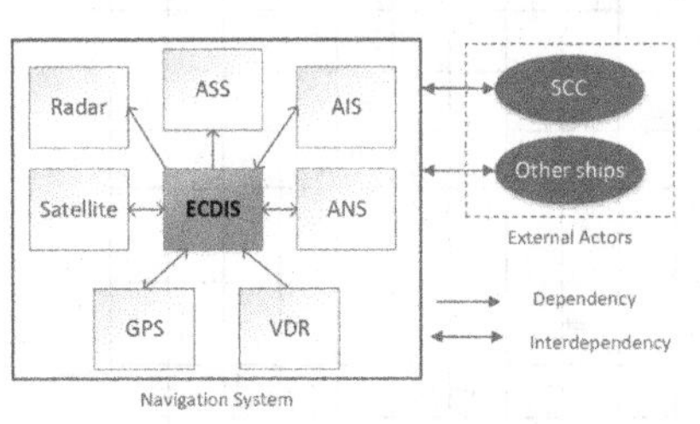

Fig. 5. ECDIS

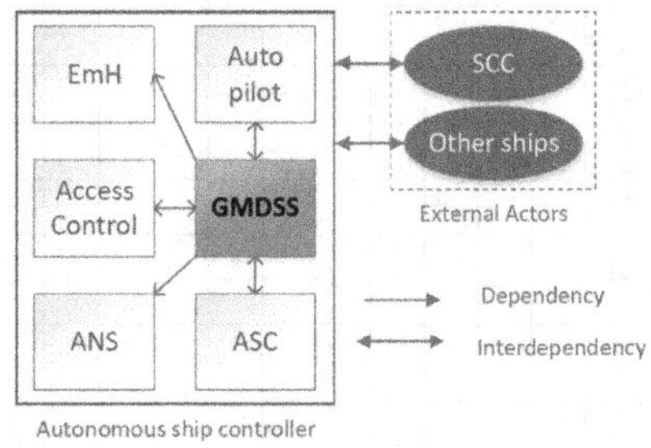

Fig. 6. GMDSS

The aforementioned analysis identified the most critical CPSs of the C-ES, taking into account the trustworthiness (ERC), the percentage of the paths that pass through each system (BC), and the number of the connections that each CPS has (TD). The analysis focused on both connections/interconnections and dependencies/interdependencies of the systems. The ANS and the Autonomous Ship Controller (ASC) have the highest values as it can be seen in Table 6. This denotes that a potential failure of such systems could provoke a sequence of failures among CPSs and therefore increase the impact to the ship. Additionally,

Table 6. Graph analysis results

CPS	ERC
AIS	1
ECDIS	0.985
GPS	0.933

CPS	BC
ASC	0.161
ANS	0.051
GMDSS	0.049

CPS	Total Degree
ASC	0.891
ANS	0.739
ASM	0.609

according to Table 6, the AIS and the ECDIS are the most trustworthy. Thus, a malfunction of these can lead to cascading effects on other ship's systems.

5 Conclusions

A central trend within the digital transformation of the maritime industry is increased autonomy of vessels. This needs to be supported by engineering specifications, regulations, standards, etc. This, in turn, necessitates the existence of an architectural framework that will facilitate the specification, implementation, and operation of such vessels. In this paper we extended the MAF to allow the representation of autonomous vessels; we used this reference architecture to define instances of cyber-enabled ships; we mapped functional and operational requirements of such systems on the reference architecture; and we identified and anlyzed the dependencies and interconnections of cyber-physical systems that comprise a cyber-enable ship. We intend to use this reference architecture and the results obtained herein, along with an appropriate requirements engineering method, to elicit cybersecurity requirements for the cyber-enabled ship.

References

1. CASOS. http://www.casos.cs.cmu.edu/index.php. Accessed 10 Sept 2019
2. e-Navigation. http://www.imo.org/en/OurWork/Safety/Navigation/Pages/eNavigation.aspx. Accessed 10 Oct 2019
3. Smart grid reference architecture. Technical report, CEN-CENELEC-ETSI Smart Grid Coordination Group (2012)
4. A technical specification for the common shore-based system architecture (cssa). Technical report, International Association of Marine Aids to Navigation and Lighthouse Authorities (2015)
5. Bergström, M., Hirdaris, S., Banda, O.V., Kujala, P., Sormunen, O., Lappalainen, A.: Towards the unmanned ship code, June 2018
6. CEN-CENELEC-ETSI Smart Grid Coordination Group: Smart grid reference architecture. Technical report (2012)
7. China classification society: Guidelines for autonomous cargo ships 2018. Technical report (2018)
8. Council of European Union: Regulation (EC) No 725/2004 (2004)
9. Cross, J., Meadow, G.: Autonomous ships 101. J. Technol. **12**, 23–27 (2017)
10. DNVGL: Autonomous and remotely operated ships, class guideline. Technical report (2018)
11. Fitton, O., Prince, D., Germond, B., Lacy, M.: The future of maritime cyber security. Technical report (2015)
12. Höyhtyä, M., Huusko, J., Kiviranta, M., Solberg, K., Rokka, J.: Connectivity for autonomous ships: architecture, use cases, and research challenges. In: International Conference on Information and Communication Technology Convergence (ICTC), 2017, pp. 345–350. IEEE (2017)
13. Im, I., Shin, D., Jeong, J.: Components for smart autonomous ship architecture based on intelligent information technology. Proc. Comput. Sci. **134**, 91–98 (2018)

14. International Maritime Organization : IMO takes first steps to address autonomous ships (2018). http://www.imo.org/en/mediacentre/pressbriefings/pages/08-msc-99-mass-scoping.aspx. Accessed 24 May 2019
15. International Maritime Organization: Convention on the international regulations for preventing collisions at sea (1972)
16. International Maritime Organization: Msc 85/26/add.1, annex 20 strategy for the development and implementation of e-navigation. Technical report (2009)
17. Systems and software engineering–Architecture description. Standard, International Organization for Standardization (2011)
18. Kavallieratos, G., Katsikas, S., Gkioulos, V.: Cyber-attacks against the autonomous ship. In: Katsikas, S.K., et al. (eds.) SECPRE/CyberICPS -2018. LNCS, vol. 11387, pp. 20–36. Springer, Cham (2019). https://doi.org/10.1007/978-3-030-12786-2_2
19. Kitada, M., et al.: Command of vessels in the era of digitalization. In: Kantola, J.I., Nazir, S., Barath, T. (eds.) AHFE 2018. AISC, vol. 783, pp. 339–350. Springer, Cham (2019). https://doi.org/10.1007/978-3-319-94709-9_32
20. Komianos, A.: The autonomous shipping era. operational, regulatory, and quality challenges. TransNav Int. J. Marine Navig. Safety Sea Transp. **12**, 335–348 (2018). https://doi.org/10.12716/1001.12.02.15
21. National Institute of Standards and Technology : Introduction to nistir 7628 guidelines for smart grid cyber security. Technical report (2010)
22. National Marine Electronics Association (NMEA): Nmea 2000 standard (1972)
23. Natvig, M.: Final report on the marnis e-maritime architecture. Technical report (2008)
24. Natvig, M., Westerheim, H., Christiansen, I.: Arktrans the Norwegian system framework architecture for multimodal transport systems supporting freight and passenger transport, June 2019
25. NCSR 1–28: Report to the maritime safety committee, international maritime organization, sub-committee on navigation communications and search and rescue. Technical report (2014)
26. OASIS: Reference architecture foundation for service oriented architecture. Technical report (2009)
27. Register, L.: Cyber-enabled ships: deploying information and communications technology in shipping-lloyds register's approach to assurance. London: Lloyds Register (2016). http://www.marinelog.com/index.php
28. Rinaldi, S.M., Peerenboom, J.P., Kelly, T.K.: Identifying, understanding, and analyzing critical infrastructure interdependencies. IEEE Control Syst. Mag. **21**(6), 11–25 (2001). https://doi.org/10.1109/37.969131
29. Rødseth, Ø,J.: e-maritime standardisation requirements and strategies (2009). http://www.mits-forum.org/architecture.html
30. Rødseth, Ø.J., Tjora, Å.: A system architecture for an unmanned ship. In: Proceedings of the 13th International Conference on Computer and IT Applications in the Maritime Industries (COMPIT) (2014)
31. Rødseth, O.J., Tjora, A., Baltzersen, P.: Munin d4.5: Architecture specification. Technical report (2013)
32. Tran, T.M.N.: Integrating requirements of industry 4.0 into maritime education and training: case study of vietnam (2018)
33. Uslar, M., et al.: Applying the smart grid architecture model for designing and validating system-of-systems in the power and energy domain: a European perspective. Energies **12**(2), 258 (2019)
34. Veritas, Bureau: Guidelines for autonomous shipping. Technical report (2017)

35. Vindøy, V.: A functionally oriented vessel data model used as basis for classification. In: 7th International Conference on Computer and IT Applications in the Maritime Industries, COMPIT, vol. 8 (2008)
36. Weinert, B., Hahn, A., Norkus, O.: A domain-specific architecture framework for the maritime domain. Informatik 2016 (2016)
37. Weintrit, A.: Development of the IMO e-Navigation concept – common maritime data structure. In: Mikulski, J. (ed.) TST 2011. CCIS, vol. 239, pp. 151–163. Springer, Heidelberg (2011). https://doi.org/10.1007/978-3-642-24660-9_18
38. ZVEI Die Elektroindustrie: Reference architecture model industrie 4.0. Technical report (2015)

Effective Data Redistribution Based on User Queries in a Distributed Graph Database

Lucie Svitáková[1(✉)], Michał Valenta[1], and Jaroslav Pokorný[2]

[1] Faculty of Information Technology, Czech Technical University in Prague, Prague, Czech Republic
{svitaluc,michal.valenta}@fit.cvut.cz

[2] Škoda Auto University, Mladá Boleslav, Czech Republic
jaroslav.pokorny@savs.cz

Abstract. The problem of data distribution in NoSQL databases is particularly difficult in the case of graph databases since the data often represent a large, highly connected graph. We face this task with monitoring of user queries, for which we created a logging module providing information serving as an input to a redistribution algorithm which bases on a lightweight method of Adaptive Partitioning but incorporates our enhancements overcoming its present drawbacks (local optima, balancing, edge weights). The results of our experiments show 70% – 80% reduction of communication between cluster nodes which is a comparable result to other methods, which, however, are more computationally demanding or suffer from other shortcomings.

Keywords: Graph databases · Graph partitioning · Redistribution · NoSQL · Pregel

1 Introduction

In recent times, fast-growing amount of data of various types has driven attention to NoSQL databases. The high volume of data forces the NoSQL systems to distribute the data across several cluster nodes. The problem of data distribution is a nontrivial task that requires dedicated management. Particularly difficult is the distribution of data in graph databases since the data often represent a large, highly connected graph. Its division into sub-graphs leaves an immense number of edges abstractly connecting elements between individual cluster nodes, which introduces high communication overhead when working with the data.

In our current work, we focus on so-called property graphs, which are directed labeled multi-graphs. Nevertheless, we generalize the task for undirected graphs which are sufficient for solving the problem of graph data redistribution. Current systems for property graph storage, such as JanusGraph, OrientDB, ArangoDB or others, determine a storage destination for a record mainly with a random

N. T. Nguyen et al. (Eds.): ACIIDS 2020, LNAI 12034, pp. 218–229, 2020.
https://doi.org/10.1007/978-3-030-42058-1_18

selection. Subsequent user queries can, however, reveal traffic patterns for which there could exist a different, more suitable data storage grouping.

We face the task of graph data redistribution first with monitoring of user queries, for which we created a logging module of the TinkerPop framework – a computing framework for graph databases, supported by the main current systems [12]. Then, we use this information in a redistribution algorithm which bases on a lightweight method of Adaptive Partitioning but incorporates our enhancements overcoming its present drawbacks. The results of our experiments show a significant reduction in cluster intra-communication, reaching 70% – 80% improvement.

After a short introduction, we provide the background and related works of the redistribution problem in graph databases in Sect. 2, with a focus on Pregel framework and Adaptive Partitioning method. Section 3 discusses analysis of user queries and introduces our enhancements for the redistribution algorithm of Adaptive Partitioning. Subsequently, Sect. 4 presents the implementation and our experiments with results. Finally, we provide a conclusion, including our vision of future work.

2 Background and Related Works

In terms of graph databases, the problem of data redistribution can be rephrased as a graph partitioning problem. First, let us define a *partition*. According to Bichot and Siarry [2] (modified):

Let $G = (V, E)$ be a graph and $P = P^1, ..., P^k$ a set of k subsets of V. The P is said to be a *partition* of G if

- no element of P is empty: $\forall i \in \{1, ...k\}, P^i \neq \emptyset$
- the elements of P are pairwise disjoint: $\forall (i,j) \in \{1, ..., k\}^2, i \neq j, P^i \cap P^j = \emptyset$
- the union of all the elements of P is equal to V: $\bigcup_{i=1}^{k} P^i = V$.

The graph partitioning problem is then a task of finding a partition P while minimizing edge cuts between the different parts of the partition P. There exist a lot of various algorithms solving this NP-complete problem. However, a significant portion of them, such as spectrum-based, graph-growing, or flow algorithms, need to access any part of the entire graph while computing their solution [3]. Methods based on the Kernighan-Lin algorithm or multilevel partitioning methods require cheap random access to any vertex of the graph as well [5]. In the case of a distributed graph, the requirement of working with a whole graph loaded in memory cannot be met. An algorithm operating only with local scope is thus desired.

This is the reason why we wanted to utilize the computational model Pregel. It offers an elegant solution to graph problems with a local approach. First, we shortly explain the Pregel model. Then, we describe one of the Pregel-compliant methods – Adaptive Partitioning. Eventually, we provide a brief survey of other Pregel-compliant methods, with no intention of exhaustiveness or full detail.

2.1 Pregel

The computational model Pregel [7] serves for computation over large distributed graph datasets. It is composed of so-called *supersteps*. Within each superstep (among others)

- each vertex of the graph reads messages that were sent to it in the previous superstep
- each vertex performs a user-defined function
- each vertex can send a message to a neighboring vertex
- each vertex and its incident edges can change its values.

Not only is it possible to perform the computations separately on individual cluster nodes, but the "think like a vertex" approach also enables rich parallelism on each host. This is very convenient; therefore we studied Pregel-compliant methods, one of which we are going to introduce more specifically later.

2.2 Adaptive Partitioning

Vaquero et al. [14] presented their Adaptive Partitioning of Large-Scale Dynamic Graphs, which takes advantage of the Label Propagation method, originally presented in [10]. Vaquero et al. bases on Ugander and Backstrom [13], who presented adjustments for balanced label propagation. In favor of better clarity, we first present four main steps of Balanced Label Propagation, and then we proceed with a brief explanation of the Adaptive Partitioning.

Steps of the Balanced Label Propagation ([10,13], modified):

1. Label of each vertex is initialized with a unique value. So, at a time 0, the function L assigning a label to a vertex v is equal to some unique id. $L_v(0) = uid(v)$.
2. All the vertices are sorted in a random order $v_1, ..., v_m < v < v_{m+1}, ...v_k$.
3. In a time t, $t \in \{1, 2, ...\}$ a vertex label is determined according to labels of its neighbors as follows

$$L_v(t) = f(L_{v_1}(t), ..., L_{v_m}(t), L_{v_{(m+1)}}(t-1), ..., L_{v_k}(t-1)), \tag{1}$$

   where $k \in \{1, ..., |D_v|\}$ with D_v being a set of neighbors of the vertex v and f returns the most frequent label among the neighbors with ties being broken uniformly randomly.
   The adoption of labels from the same or previous iteration based on the sorting from the second step is incorporated to avoid oscillation of labels back and forth between two vertices.
4. If there is no other label adoption possible, the algorithm finishes. Otherwise, the next iteration, starting from the second step, is started, incrementing the time variable.

The algorithm of Vaquero et al. then mainly comprises the following steps:

1. The graph is initially partitioned, regardless of a partitioning method.

2. The label propagation method is utilized with three main differences:
   - Instead of initializing labels with a unique id, the labels acquire a value of the part of the partition on which the vertex resides.
   - Instead of sorting the vertices in each iteration, it confronts the oscillation issue with a probability factor s with which the label is or is not adopted.
   - The label is adopted only if the constraints, described below, are met.
3. A new-coming vertex, if any, is stored on a cluster node according to the same logic - the one which is prevailing among its neighbors.

Constraints. Vaquero et al. do not mention any lower bound. Regarding the upper bound, they define the maximum number of vertices allowed to change a partition part P^i to a part P^j, $i \neq j$ in an iteration t as

$$Q^{i,j}(t) = \frac{C^j(t)}{|P(t)| - 1}, \tag{2}$$

where $C^j(t)$ is a capacity constraint on a partition part P^j such that $|P^j(t)| \leq C^j$; $P(t) = \{P^1(t), ..., P^n(t)\}$, so $|P(t)| = n$. In other words, the remaining free capacity of the part P^j is equally divided among the remaining parts, and that fraction is the maximum number of vertices that can migrate from a particular part to P^j.

Drawbacks. The described algorithm is a convenient, lightweight solution for graph partitioning. However, it also suffers from a number of drawbacks.

First, the method does not confront the problem of local optima. In each iteration, the prevailing label is always adopted, which introduces a threat of reaching a local optimum, and never finding the best global solution.

Secondly, the constraints do not consider balancing in practice. If we have, for example, a cluster of five hosts, we want a solution with a balanced distribution over all nodes. Or we may want to set a percentual imbalance allowed. The described algorithm does not incorporate a lower bound, so some hosts can even get entirely depleted.

Thirdly, the notion of weighted graphs is not discussed at all. Often, some edges have higher importance, expressed with weight, than others. Label adoption should count with that possibility as well.

2.3 Other Pregel-Compliant Methods

Apart from the Adaptive Partitioning method just described, there are other related works trying to solve the task of graph partitioning, and compatible with or directly using the Pregel platform. The Ja-be-Ja method [11] approaches the problem with label swapping, meaning it is always able to retain precise balance among partition parts. On the other hand, it does not allow for any balance elasticity, and mainly, its computation requires more steps and communication between cluster nodes than the mentioned algorithm.

The algorithm Spinner [8] also bases on Label Propagation and the following work of [13]. However, the authors built balancing constraints on "encouraging a similar number of edges across the different partitions". They do not actually solve the problem of reducing communication as their solution can easily end up with equally spread edges among cluster nodes with still high communication demands in between them. In other words, they focus on balancing but not communication minimalization. Furthermore, they neither handle local optima.

The mechanism of Dong et al. [4] includes a pre-partition phase followed by a superblock construction phase, which is significantly computationally demanding (mainly the first phase) and does not bring any other added value in comparison with our proposed method below, based on lightweight Adaptive Partitioning.

To our knowledge, there are no other contributions regarding graph partitioning on Pregel framework; at least not such bringing any improvements over our work.

3 Query Analysis and Redistribution

In order to properly decide about data redistribution, we need sufficient information about past queries. Each query triggers a traversal which goes through the graph to obtain a correct solution. It is necessary to know about all the paths that the traversal performs as each step can cause intra-host communication.

Based on the logged traversal paths, we can collect the necessary information for data redistribution. Note that we do not consider replicas of records in our current work as that would introduce another degree of freedom in our target problem. We do, however, count on their incorporation in our subsequent work. When talking about balancing the data on hosts of the cluster, we consider all of the vertices to have equal size. The introduction of unequal vertex sizes is a simple adjustment that we omit in favor of clarity. With the mentioned simplifications, the sufficient information for the decision about redistribution is a consolidated number of passed traversals on each edge. We can represent that number as a weight of the edge.

3.1 Redistribution Algorithm

The shortcomings listed in Sect. 2.2 let to our adjustments of the Adaptive Partitioning method, which tend to overcome the individual drawbacks.

Local Optima. The method of Adaptive Partitioning does not handle the problem of local optima. While adopting a label, the prevailing one withing neighbors is always selected. If we allowed acceptance of worsening solutions (not only a prevailing label) in the first several iterations, and then we decreased the probability of potentially degrading adoptions, we could increase the probability of overcoming a local optimum.

Therefore, we incorporate the method of simulated annealing. We propose Eq. 3 for the label selection.

$$\left\lfloor r \cdot \frac{T}{T_{init}} \cdot \left(\left|l\left(D_v\right)\right| - 1\right)\right\rfloor, \tag{3}$$

where r is a random number $r \in [0;1]$, T is a current temperature, T_{init} is an initial temperature, D_v is a set of neighbors of a vertex v, and l is a function returning a set of labels in the provided set of vertices.

We anneal the temperature T from the initial value of T_{init} down to zero. Note that the fraction T/T_{init} always falls in the interval $[0;1]$. Therefore, the initial value of T_{init} does not matter; it is the cooling factor, with which we decrease the T value, that influences the result.

The resulting number of the equation is a label position in an ordered list of the labels from the D_v set, sorted according to the number of occurrences, from the most to the least frequent.

This approach can also influence the oscillation problem discussed before. It can be assumed that with a higher temperature, the endless swapping is less likely to happen. For that reason, we also propose an adoption factor equal to

$$S + \frac{T}{T_{init}} \cdot (1 - S), \tag{4}$$

where S is a resulting probability of label adoption when the temperature already reaches 0. Based on the [14], the S is set to 0.5.

Balancing. The paper of Balanced Label Propagation [13] presents an upper and a lower bound on each part of the partition, equal to $\left\lfloor (1-f)\frac{|V|}{n} \right\rfloor$ and $\left\lceil (1+f)\frac{|V|}{n} \right\rceil$, respectively. f is a fraction $f \in [0;1]$, and n is the number of partitions.

Our proposition of lower bound bases on the original paper, but it takes into account a possibility of different capacities of individual cluster nodes. Therefore, we set the lower bound for a part P^i as

$$\left\lfloor (1-f) \cdot |V| \cdot \frac{C_i}{\sum\limits_{j \in J} C_j} \right\rfloor, \tag{5}$$

where f is a fraction, $f \in [0;1]$, C_i is a capacity of the partition P^i, and $J = \{1, ..., n\}$, n equals the number of parts in the partition.

If $f < 1$, the condition ensures that a cluster host will never be vacant (unless the host is already vacant in the initial state). Incorporating both the upper and the lower bound serves for better balancing of the final result. In order to enable some vertex shuffling, the fraction f must be higher than zero. This also means that there is always a slight imbalance allowed. So, the solution serves for better balancing and for some elasticity of natural grouping as well.

Weighted Graphs. Incorporation of weighted graphs to the proposed enhancements is rather straightforward. The label selection uses an ordered list of neighboring labels. We adjust the list ordering according to a sum of weights of edges connecting a given vertex with neighbors of a particular label.

4 Implementation and Experiments

To the best of our knowledge, a utility providing logging of traversal paths does not exist. For that reason, we implemented a module to the TinkerPop framework, watching each traversal, and logging all its passed walks into a log file.

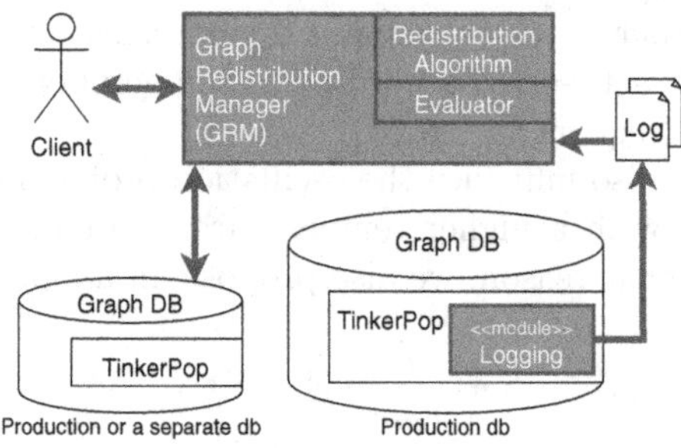

Fig. 1. Architecture of our solution for data redistribution in a graph database

Then we created a Graph Redistribution Manager (GRM). It is an application that can read the logged files, store appropriate information from the logs to a graph database, run our redistribution algorithm, and evaluate each redistribution based on the logged processed queries. The graph database, into which the GRM loads its data, can be configured to a source production database or to any other instance in order to avoid a computation overhead in a production environment.

The implemented solution is currently considered as a proof of concept. Its architecture can be seen in the Fig. 1.

4.1 Experiments

The experiments were run on the JanusGraph, a graph database with a separate storage backend for which we selected Cassandra, a very popular wide column store. The choice of a particular product does, however, not influence our results as we monitor following system-independent measurements

- intra-host communication improvement within a cluster counted as

$$1 - \frac{\textit{number of crossings after the redistribution}}{\textit{number of crossings before the redistribution}}.$$

- number of iterations of the algorithm
- capacity of individual cluster nodes.

Two datasets were utilized, one representing data fraction of Twitter [9], and one containing a road map of Pennsylvania [6]. Their statistical characteristics are stated in the Table 1. The Twitter data form a so-called *scale-free network* in which the number of vertices grows exponentially with a lower degree [1]. There are 3,769 different vertex degrees in the Twitter dataset. On the other hand, the Pennsylvania dataset constitutes of vertices with only nine different degrees.

Table 1. Twitter and Pennsylvania dataset statistics [6,9]

Property	Twitter	Pennsylvania
Nodes	76,268	1,088,092
Edges	1,768,149	1,541,898
Average clustering coefficient	0.5653	0.0465
Diameter (longest shortest path)	7	786

For each dataset we created about 10,000 typical user queries which processing was later logged. We mimic a Twitter users as logging into the application and loading tweets from their followed users. The user can also search for new people to follow, which we express as a search for people who are followed by people being followed by the user.

In the Pennsylvania dataset, we search for the shortest path between two cities. To avoid logging of the whole graph, we first pick a random city, and we find its counterpart with a random walk of ten steps. When these cities are selected, we search for the shortest path with a maximum distance of ten.

4.2 Results

The results are averaged from ten measurements on the Twitter dataset and four measurements on the Pennsylvania dataset; Pennsylvania results evincing much lower standard deviation of resulting values, so more results did not bring any additional information. The results are presented individually for each mentioned factor – the cooling factor of simulated annealing, the imbalance factor of our lower bound, the adoption factor. Eventually, results on a different number of cluster nodes are monitored as well.

Cooling Factor. We examined a cooling factor of the simulated annealing method. Figure 2 shows the behavior of individual runs with different settings. It reveals that the values 0.98 and 0.99 are the most promising ones.

Change in the improvement of the Pennsylvania dataset in Fig. 3 is rather expected. The more balanced topology allows for a higher number of suitable distributions. Even a slow cooling is beneficial and can find some better results.

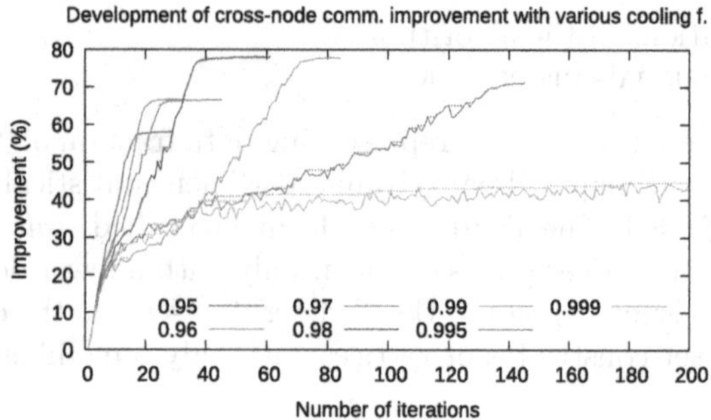

Fig. 2. Cross-node communication improvement with a cooling factor. Dots represent the maximal value of the process so far. Measured on the Twitter dataset.

On the contrary, the Twitter scale-free network does not have so many relevant solutions, and a too-high factor gets already stuck in a local optimum. In the scope of our two datasets, we could conclude that a cooling factor of 0.99 can be set. Nevertheless, different values of the cooling factor should be checked first, when working with a new dataset, because the factor is data-sensitive.

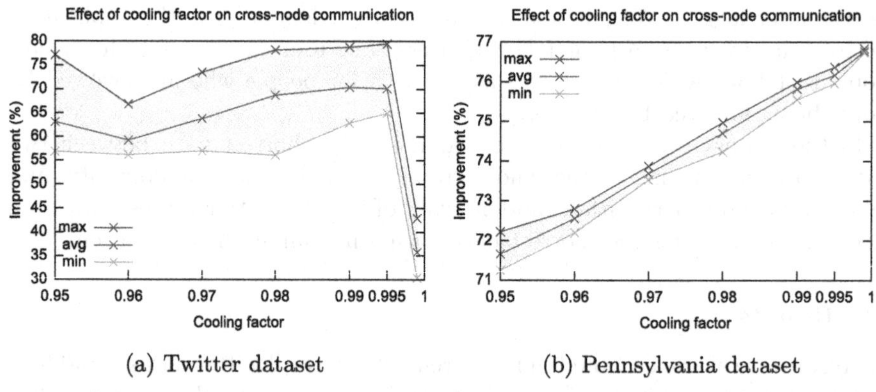

(a) Twitter dataset (b) Pennsylvania dataset

Fig. 3. Effect of cooling factor on cross-node communication.

Imbalance Factor. Figure 4 shows results of the Pennsylvania dataset (Twitter dataset evincing analog results). It is very appealing that even with a very small imbalance factor of 0.01–0.1, the average improvement is still very similar, within a range of [77.38–75.82]. This means that even an imbalance of 1% can produce a high-quality redistribution. With a further growing imbalance over 10%, the improvement further rises as the vertices tend to group on one or more cluster nodes, leaving some nodes with just a small portion of the graph – the maximum allowed imbalance tends to be always fully utilized.

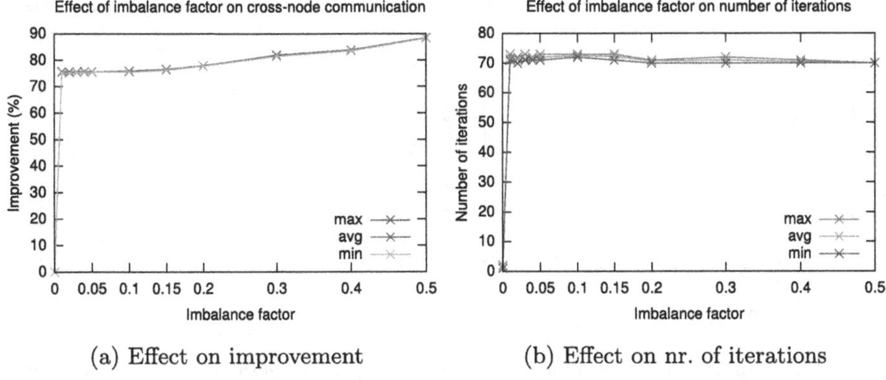

(a) Effect on improvement

(b) Effect on nr. of iterations

Fig. 4. Effect of imbalance factor, measured on the Pennsylvania dataset.

Adoption Factor. The adoption factor proposed in Eq. 4 does not seem to have a significant effect on the results. It behaves differently on both the datasets and with an insignificant influence on the resulting values. Therefore, we suggest to use the original value of 0.5 within the whole computational process to save few computational steps.

Number of Nodes in Cluster. As seen in Figs. 5 and 6, when the number of nodes in a cluster is higher, the number of necessary iterations grows as the task is more difficult and can produce more possible outcomes.

In the case of the Pennsylvania dataset, the higher the number of cluster nodes, the higher the necessary communication among them. However, this behavior is sensitive to a given topology. The scale-free network of Twitter enables suitable organization of vertices also on a higher number of nodes. This behavior is caused by a few ego vertices containing a tremendous number of relationships, followed by a high number of low-degree vertices. Additionally, the behavior is affected by user queries, which are more local than in the map dataset.

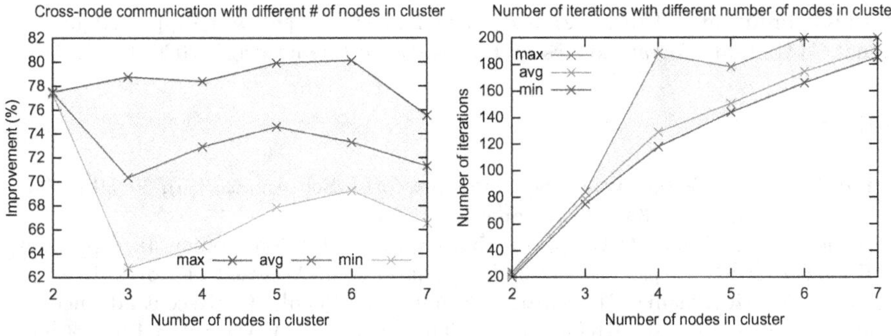

Fig. 5. Effect of the number of nodes in cluster on cross-node communication and number of iterations. Measured on the Twitter dataset.

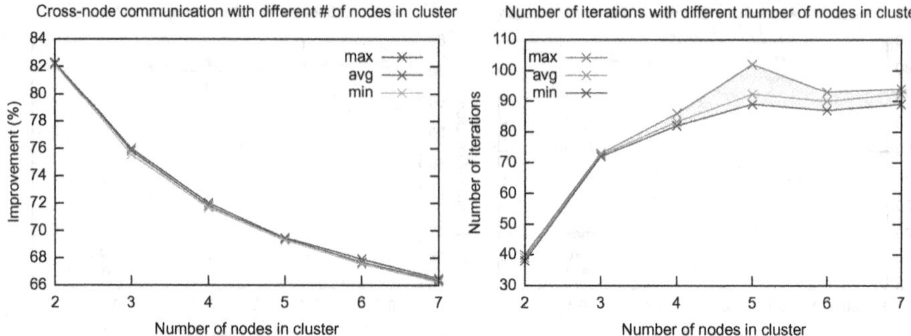

Fig. 6. Effect of the number of nodes in cluster on cross-node communication and number of iterations. Pennsylvania dataset.

5 Conclusion

We created a module for the TinkerPop framework, which logs necessary information about user queries. This information serves as an input to a redistribution algorithm for which we used the lightweight Adaptive Partitioning method with significant enhancements regarding overcoming of local optima, balancing individual cluster nodes, and incorporating the notion of weighted graphs.

Our experiments evince eminent results of 70%–80% reduction of necessary communication between cluster nodes. This improvement is comparable with another mentioned method Ja-be-Ja, which proclaims to acquire similar results. However, Ja-be-Ja has higher computational demands. The other mentioned graph partitioning algorithms compliant with the Pregel platform then suffered from other imperfections. Therefore, we consider our outcomes as very satisfying, and we want to continue in our commenced work.

Besides some enhancements regarding preprocessing of the logged data and other optimizations of the Graph Redistribution Manager, we would like to incorporate, for example, the notion of replicas or information fading, in order to better match the real settings in practice.

Acknowledgements. The research described in this paper was supported by the internal CTU grant "Advanced Research in Software Engineering", 2020.

References

1. Barabasi, A.L., Bonabeau, E.: Scale-free networks. Sci. Am. **288**(5), 50–59 (2003). https://doi.org/10.1038/scientificamerican0503-60
2. Bichot, C.E., Siarry, P.: Graph Partitioning, 1st edn. John Wiley, Incorporated, Hoboken (2013). https://ebookcentral.proquest.com. Accessed 21 Nov 2018
3. Buluç, A., Meyerhenke, H., Safro, I., Sanders, P., Schulz, C.: Recent advances in graph partitioning. In: Kliemann, L., Sanders, P. (eds.) Algorithm Engineering. LNCS, vol. 9220, pp. 117–158. Springer, Cham (2016). https://doi.org/10.1007/978-3-319-49487-6_4

4. Dong, F., Zhang, J., Luo, J., Shen, D., Jin, J.: Enabling application-aware flexible graph partition mechanism for parallel graph processing systems: superblock an application-aware graph partition mechanism. Concurr. Comput.: Pract. Exp. **29**(6), e3849 (2016). https://doi.org/10.1002/cpe.3849
5. Karypis, G., Kumar, V.: A fast and high quality multilevel scheme for partitioning irregular graphs. SIAM J. Sci. Comput. **20**(1), 359–392 (1998). 10.1.1.39.3415
6. Leskovec, J., Lang, K., Dasgupta, A., Mahoney, M.: Community structure in large networks: natural cluster sizes and the absence of large well-defined clusters. Internet Math. **6**(1), 29–123 (2009). https://snap.stanford.edu/data/roadNet-PA.html. Accessed 11 Oct 2019
7. Malewicz, G., et al.: Pregel: a system for large-scale graph processing. In: Proceedings of the 2010 ACM SIGMOD International Conference on Management of data, pp. 135–146. Indianapolis, Indiana, USA (2010). https://doi.org/10.1145/1807167.1807184
8. Martella, C., Logothetis, D., Loukas, A., Siganos, G.: Spinner: Scalable graph partitioning in the cloud. In: 2017 IEEE 33rd International Conference on Data Engineering, pp. 1083–1094 (04 2017)
9. McAuley, J., Leskovec, J.: Learning to discover social circles in ego networks. In: Advances in Neural Information Processing Systems (2012). https://snap.stanford.edu/data/ego-Twitter.html. Accessed 11 Oct 2019
10. Raghavan, U.N., Albert, R., Kumara, S.: Near linear time algorithm to detect community structures in large-scale networks. Phys. Rev. E **76**(3), 036106 (2007). https://doi.org/10.1103/PhysRevE.76.036106
11. Rahimian, F., Payberah, A.H., Girdzijauskas, S., Jelasity, M., Haridi, S.: A distributed algorithm for large-scale graph partitioning. ACM Trans. Auton. Adapt. Syst. **10**(2), 1–24 (2015). https://doi.org/10.1145/2714568
12. The Apache Software Foundation: Apache tinkerpop, August 2019. http://tinkerpop.apache.org/. Accessed 11 Oct 2019
13. Ugander, J., Backstrom, L.: Balanced label propagation for partitioning massive graphs. In: Proceedings of the Sixth ACM International Conference on Web Search and Data Mining, Rome, Italy, pp. 507–516 (2013). https://doi.org/10.1145/2433396.2433461
14. Vaquero, L.M., Cuadrado, F., Logothetis, D., Martella, C.: Adaptive partitioning for large-scale dynamic graphs. In: 2014 IEEE 34th International Conference on Distributed Computing Systems, Madrid, Spain, pp. 144–153 (2014). https://doi.org/10.1109/ICDCS.2014.23

The Functionalities of Cognitive Technology in Management Control System

Andrzej Bytniewski, Kamal Matouk, Anna Chojnacka-Komorowska, Marcin Hernes(✉), Adam Zawadzki, and Agata Kozina

Wroclaw University of Economics and Business, Wroclaw, Poland
{andrzej.bytniewski,kamal.matouk,anna.chojnacka-komorowska, marcin.hernes,adam.zawadzki,agata.kozina}@ue.wroc.pl

Abstract. Cognitive technologies are an important factor accelerating the development of business and society. They are an artificial intelligence tool with cognitive skills enabling learning through empirical experience acquired through direct interaction with the environment. The application of these technologies in the management control system enriches it with the functions of automatic analysis of real data mapping business processes that are implemented at all levels of management. The aim of this paper is to analyze the functionalities of cognitive technology for the intelligent knowledge processing process by the management control system. To achieve this goal, the following research methods were used: literature analysis, observation of phenomena, deduction, and induction. The main results of researches are conclusions that cognitive technology should be used, among other, for realize following functions: automatic creation of a new strategic and operating budget plan, automatic real-time conversion of budgets in case of a change in the organizational structure of the enterprise, automatic deviations control, interpretation and identification the causes of deviations.

Keywords: Management control system · Artificial intelligence · Cognitive technology · Integrated management information system · Enterprise resource planning · Industry 4.0

1 Introduction

Management control system functioning as part of the integrated management information system (IMS) plays a crucial role in strategic and operational management of enterprises, constituting one of the most important solutions in gaining a competitive advantage. It allows the collection, processing and transmission of large amounts of information as well as inference based on this information, which contributes to the organization's knowledge. However, these options are no longer sufficient. There is an increasing need to make decisions based not only on knowledge, but also on the basis of experience, which until now was treated only as the human domain. Due to the high turbulence of the business environment and the need to make decisions in near-real time, the management control system should possess the function of automatic analysis the meaning of real business data (operational, tactical and strategic data). The results of

N. T. Nguyen et al. (Eds.): ACIIDS 2020, LNAI 12034, pp. 230–240, 2020.
https://doi.org/10.1007/978-3-030-42058-1_19

this analysis should lead to automatic decision making including: automatic creation of a new strategic and operating budget plan [1], automatic real-time conversion of budgets in case of a change in the organizational structure of the enterprise, automatic deviations control, interpretation and identification the causes of deviations [2]. In the recent related works, the analysis the meaning of real business data and decision taking is performed separately by different IT tools (often decisions are made manually by human). Therefore the main research problem is to develop IT tools for performing analysis and making decision at the same time (integration the automatic analysis and decision making). The knowledge stored in the management control system should be represented and processed using semantic methods understood as methods that reflect the ontology (semantics) of knowledge in this purpose [3].

The aim of the paper is to analyze the concept of using cognitive technology in the implementation of intelligent processing of knowledge contained in the management control subsystem functioning as part of an integrated management information system. The main contribution is the analysis of possibility of cognitive technologies for performing the automatic analysis the meaning of real business data and automatic decision-making.

The remaining part of the paper is as follows: the next part presents the related works in considered fields and basic notions. Next, the cognitive technologies are characterized. The last part presents the functionalities of cognitive technologies in management control system.

2 Related Works

The use of artificial intelligence has created new opportunities to support decision-making, in particular in the context of risk management. An overview of the possibilities of using artificial intelligence in this area is presented in [7]. The use of artificial intelligence in the Lean Management concept can, in the case of construction projects, lead to improved performance by improving planning, reducing costs, reducing time, and introducing a more efficient use of resources [8]. Item [9] presents examples of current solutions supporting information resource management, in particular focused on the technology of intelligent agents (cognitive agents). This item indicates examples of effective use of cognitive technologies in a modern enterprise and presents methodologies for the operation of cognitive agents working in an uncertain environment. Artificial intelligence and cognitive technologies should be particularly related to decision support systems [10]. The introduction and use of artificial intelligence and the transition from traditional industry-to-industry 4.0 is determined by specific aspects [11]: technological, organizational and environmental. The aim of the study contained in the work was to check the degree of readiness of Swedish small and medium-sized enterprises for the use of artificial intelligence and what factors influence such readiness. A review of the possibilities of using artificial intelligence and cognitive technologies in the area of marketing [12] indicates that Intelligent Systems possess special capabilities and grounds for supporting decision-making processes also in the strategic area. The use of business analysis technology and analysis in management accounting processes [13] supports reporting and decision making. A review of the literature in this area indicates a research gap. Automation of business processes using cognitive agents [14] is an innovative activity that contributes to the growth in productivity. The use of multi-agent systems for managing an ontology-based corporate network [15] was presented for two applications: support for technology monitoring, and assistance in the integration of a new employee. For the purposes of the work, the focus was on: designing multi-agent architecture, interaction between agents; ontology structure and corporate memory structure using semantic web technologies; design and implementation of subgroups of agent associations dealing with annotation and ontology management, and protocols at the root of these agent groups, in particular techniques for distributing annotations and queries between agents. Enterprise resource planning (ERP) and supply chain management (SCM) as well as customer relationship management (CRM) contain tools supporting the decision-making process. The work [16] presents concepts of support for decision-making systems (DSS) in these areas in the use of artificial intelligence (IDSS). This concept represents a multi-purpose conceptual structure (ERP) – (IDSS) which includes SCM and CRM. The result is the integration of decision support in these areas and the provision of appropriate guidance in process planning. Multi-agent systems (MAS) provide a new paradigm for the development of Integrated Business Information Systems (IBIS). Item [17] indicates that they can also be excellent in modelling and implementing IBIS systems. The paper proposes a conceptual framework for Integrated Business Information Systems based on numerous agents (MIBIS) and a unified system of eight orthogonal ontological constructions. Trends [18] in the application of artificial intelligence in a business environment until 2030 show the wide possibilities of using artificial intelligence as a special and fundamental innovation for solutions of Integrated Knowledge Management Systems, automation of business processes, including decision support processes.

Since the beginning of the nineties of the last century, the practical use of ICT has contributed to the acceleration of the quantitative and qualitative development of integrated management information systems, and in them the extensive management control system. Since then, the ERP system has been constantly developing and it covers various additional areas of business activity creating new quality, as well as a new class referred to as ERP IV class [4]. Importantly, there is a growing awareness among users of the benefits of these systems. These systems are being implemented not only in large enterprises, but also in small and medium-sized enterprises.

The 2017 report of the CEM Institute of Market and Public Opinion Research [4] shows that 67% of companies surveyed in Poland believe that an integrated ERP management information system should be characterized by: optimization and automation of daily tasks, practical usefulness, as well as be associated with handling business processes. 40% highly value the possibilities of adjusting the system to the company's structure. 36% of respondents emphasize that an integrated system should provide comprehensive support for all areas in the enterprise.

On the other hand, in its report on creating information resources, KPMG showed that companies that are able to quickly and effectively analyze information, especially that of a managerial nature, are better at managing risk as well as creating long-term strategies and introducing changes. For 44% of the surveyed companies, data analysis is important from the point of view of monitoring information needs, improving financial reports, and analyzing customer information needs [4].

However, in our best knowledge, there is not works related to use cognitive technologies in management control system. Therefore, in this paper, the attention will be directed focused on this research problem.

3 Traditional Functioning of a Management Control System Within the Integrated Management Information System

The management control subsystem, as previously indicated, is an integral part of IMS; it should be emphasized that it possesses mutual information links. The place of the management control subsystem within the integrated management system together with information connections is shown in Fig. 1. From the point of view of the issues under consideration, it occupies a central place, as it is the recipient and donor of data and information from other systems [19].

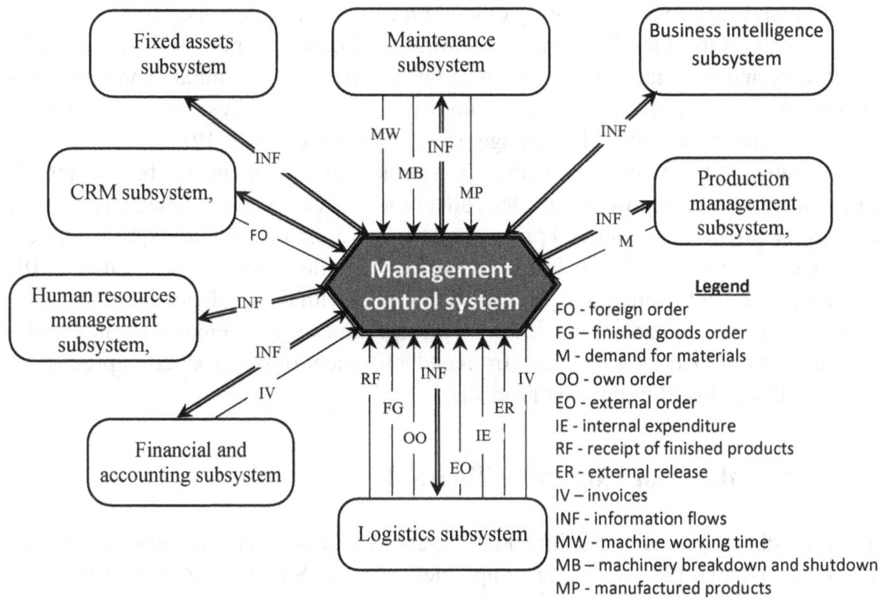

Fig. 1. Information links of the controlling subsystem with other subsystems

As it results from Fig. 1, the management control subsystem has links with all subsystems belonging to IMS. It generates strategic and operational plans that derive data from various subsystems, and in addition, it itself downloads data on real processes mapped in other subsystems located within IMS.

The management control subsystem possesses the most information links with the logistics subsystem. These connections relate to information related to the registration of own orders (sent to suppliers) and the receipt and issue of raw materials, other materials (for production purposes), finished products (for sales purposes). In addition, there are flows of information related to invoices and the orders for products.

From the financial and accounting subsystem, it collects information on the generic system of costs, receivables, turnover and account balances, for detailed analysis in the context of the parameters of the detailed budgets and the budget for the entire enterprise [17].

From the human resources management subsystem, it collects, in particular, data on remuneration (basic wages, bonuses, special supplements, etc.) according to various sections, e.g. cost centers, profit centers, cost carriers), the number of employees (by professional groups, education, etc.) [5].

The management control subsystem also has links with the CRM subsystem, from which it retrieves data on external orders, while providing feedback on cooperation with clients, e.g. turnover, discounts granted to clients, participation in marketing campaigns, etc. [6].

The subsystem of fixed assets provides data on the value of purchased and sold fixed assets, their depreciation according to different classification groups.

The management control subsystem also retrieves information from the maintenance subsystem, and they relate to, among others, machine and device operating time data, data that includes machine failure and downtime, and data illustrating the performance of machinery and equipment. On the basis of this information, the unitary and total costs of machinery and equipment, costs resulting from machine failures, costs of removing failures, etc. are calculated in the management control subsystem [19].

The relationship between the management control subsystem and the business intelligence subsystem is manifested in the collection of data from the latter, taken from the environment, e.g. raw material prices, materials present on the market, loan interest rates, exchange rates, etc. These data are necessary to build budgets as verify them [20].

The management control subsystem also has links with the production management subsystem. From this subsystem, it draws information on the demand for materials, the volume of production in progress, or the advancement of the production process of various details included in finished products.

4 Functionalities of Cognitive Technology

Cognitive technologies are an artificial intelligence tool based on computer models that integrate knowledge [21], which are implemented in IMS and the way they function resembles the action of the human mind. Their task is to implement specific patterns of cognitive functions enabling testing of these functions on a wide range of issues [22]. They do not only allow quick access to information and more efficient search for necessary information, its analysis and drawing conclusions. They also, in addition to responding to stimuli from the environment, possess cognitive abilities enabling learning through empirical experience acquired through direct interaction with the environment, which, as a consequence, allows for automatic generation of decision variants. They allow even for making and implementing decisions, in many cases [23]. By implementing the above-mentioned activities, cognitive technology simultaneously learns by assimilating real events occurring in the enterprise and its environment. Examples of such solutions may be: interpretation of previously unknown management control indicators, proposing corrective actions when sales decrease in an enterprise and the competition introduces a new product, and identifying significant events occurring in the enterprise's environment. It is worth emphasizing that cognitive technologies have the function of priming symbols, i.e. assigning specific symbols of the natural language to relevant real-world objects [24]. It is necessary to properly process unstructured knowledge, written mainly using natural language, for example customer opinions about products. Currently, this type of knowledge is becoming more and more important for an enterprise, because it can affect its level of competitiveness, e.g. by analyzing customer opinions on a given product, one can predict (of course with a certain level of probability) the volume of sales of this product in the future, and thus, at the same time, plan production.

Currently, cognitive technologies are becoming a key area of development connecting the needs of business and society. According to Gartner analysts, by 2020 cognitive technologies will already be an inseparable part of management information systems, which will contribute to modernizing up to 80% of business processes [25]. Meanwhile, experts from Deloitte, in their report "Technology, Media and Telecommunication Predictions, 2016" also predict that by the end of 2020, 95% of the largest global companies creating management information systems for enterprises will integrate cognitive technologies with their products [26].

The use of cognitive architecture in knowledge processing allows for recording information on ontology and taxonomy of data, semantically ordered (large amounts of information resources on phenomena are ordered according to a given semantic dimension). This allows, for example, multidimensional visualization of information describing economic events in any system determined ad hoc by the user, using, for example, the managerial cockpit, which includes detailed interpretation of deviations of the main performance indicators in the enterprise (KPI).

Therefore, the advantages of using cognitive architecture in knowledge processing include:

- flexibility of knowledge representation methods,
- the ability to modify the knowledge structure while the system is running (in runtime mode) or even when it is being used by users,
- effective processing of queries on semantically complex data,
- high speed of searching for semantically complex knowledge,
- the ability to represent rich semantics of knowledge in graphic form.

Cognitive technologies enable the processing of large amounts of data, information and knowledge and can perform innovative functions, although at the same time the complexity of architecture requires the use of better in terms of equipment performance than other technologies. Cognitive technologies can learn and predict what information will potentially interest the recipient [27, 28]. The use of cognitive technology used in integrated management information systems is described in the paper [29]. In particular, the implementation of cognitive technologies allows to introduce new solutions in the field of:

- semantic representation and processing of knowledge,
- automatic conclusions based on information processed by the system,
- automatic generation of decision proposals,

- assessment (evaluation) of the quality of knowledge accumulated in the system,
- automatic monitoring of phenomena occurring in the company's environment,
- automatic implementation of the controlling function, with particular emphasis on the automatic interpretation of deviations and updating budgets based on changes in the structure of budget cells and changes in the operating conditions of the entity,
- implementation of the system's permanent learning process,
- analysis of the current legal status and, consequently, the construction of rules related to the implementation of certain functions,
- intelligent personalization of the system to help users use individual functionalities.

The issues of semantic representation and processing of knowledge are raised in [30].

Cognitive technology is characterized by specific possibilities in the flexible composition of budget plans as well as determination and interpretation of deviations, and it affects the efficiency of management control processes. The problem of efficiency of management control processes is raised in [30]. These possibilities include:

1. Automatic creation of a new strategic budget plan based on budgets from previous years, but unlike the classical approach, cognitive technology can analyze completed or future changes in the implementation of business processes, based on information from other subsystems, and include them in the new budget. For example, cognitive technologies analyze completed and unrealized foreign orders (due to the lack of production capacity) received from the CRM subsystem, further adjusted by the increase in sales resulting from the assumed level of enterprise development, expressed in quantity and value. On this basis, a strategic budget plan is created, and proposals are made to increase or decrease production capacity, and possibly to increase or decrease employment. In addition, production costs are projected on the basis of actual costs from previous years, together with an update resulting from previously proposed changes (resulting from an increase in production capacity and possibly employment). Cognitive technologies explore cyberspace by analyzing changes in the prices of raw materials, other materials, products sold, and rates of remuneration. They use this data to verify the budget plan of revenues and cost items in the created budget cost plans. These plans can be created according to various scenarios, and the specific scenario is automatically suggested by the system, but ultimately the company management chooses the specific one.
2. Automatic creation of the operating budget (usually for a year) based on a slice of the strategic budget plan, adjusted by resources recorded in other subsystems. Cognitive technology analyses the multi-attribute and multi-valued data structures of these resources, updates them with current data obtained from the CRM subsystem, and containing foreign orders for products, corrected for the demand indicator not covered by orders. This indicator is determined on the basis of actual data from previous years. Costs are similarly budgeted, although they take into account their actual level from the last period. Subsequently, this plan is adjusted for updated assumptions regarding the company's development regarding the current planning period.

3. Automatic real-time conversion of budgets in the event of a change in the organisational structure of the enterprise. Due to the fact that budget plans are created based on tangible, intangible and human resources, when the organisational structure changes, cognitive technology transfers resources from existing structures to new structures. On this basis, a new layout of the budget plan is created for the new structures. Cognitive technology collects data from all subsystems and on this basis carries out analyses and draws conclusions. Subsequently, the newly created plan is presented to decision makers for approval. Policy makers have the option of accepting the budget plan without any changes or adjusting it.
4. Automatic deviation control and interpretation. It launches the analysis of significant deviations and automatic determination of deviation limit values, while determining the symmetry of deviations and financial effects, and identifying the causes of deviations. These functions are implemented as a result of learning by cognitive technology based on past events that had an impact on the occurrence of deviations and the automatic extraction of the values of significant semantic attributes (e.g. data from structural and technological files, files of completed orders, working time of people and machines, etc.) by drilling down and deeply learning about the underlying deviations.
5. Automatic identification of the causes of deviations. Based on information from other subsystems, cognitive technology analyses the relationships between deviations and recorded information regarding the implementation of business processes. For example, if a failure and downtime of machines were recorded in the maintenance subsystem, this fact is considered in the process of determining the causes of cost deviations by multi-faceted cognitive technologies (e.g. the effect of this failure is a decrease in energy and fuel consumption, an increase in repair costs, a reduction in the number of products manufactured, etc.) in the scheduled time period.

The examples of application of cognitive technologies in the management control subsystem show their great possibilities of streamlining the process of creating strategic, tactical and operational plans as well as determining the reasons for deviations from plans and their automatic interpretation.

5 Conclusions

Integrated management information systems are recognized as the highest level in the development of computer applications. They support all areas of the company's operations, including management processes implemented by the management control subsystem. They offer numerous possibilities, such as: automatic creation of strategic and operational plans, determination of deviations from these plans, and automatic interpretation of their causes. However, the use of cognitive technology may allow the processing of unusual (unstructured) data contained in various documents obtained from cyberspace. These can be sound files, video clips and images. Cognitive programs are able to process them for decision-making. It should be emphasized that this type of information was not previously included in the analyses made by this subsystem. The use of cognitive technologies belonging to artificial intelligence in the management control subsystem will also allow the integrated management IT system to process all types of data, information and knowledge faster and efficiently, without the need for complex analyses using traditional time-consuming and resource-consuming methods of data mining. The system will be able to automatically monitor internal processes and phenomena occurring in the company's environment, automatically draw conclusions, generate decision proposals, and automatically perform the management function.

The main limitation of use the cognitive technology in management control system is the complexity of its architecture and high computational resource consuming.

The further works will be related to implementation the cognitive technology in Cognitive Information Management System [3] and to perform research experiments related to its performance.

Acknowledgment. "The project is financed by the Ministry of Science and Higher Education in Poland under the programme "Regional Initiative of Excellence" 2019–2022 project number 015/RID/2018/19 total funding amount 10 721 040,00 PLN".

References

1. Wildavsky, A.: Budgeting and Governing. Routledge, Abingdon (2017)
2. Stilley, K.M., Inman, J.J., Wakefield, K.L.: Planning to make unplanned purchases? The role of in-store slack in budget deviation. J. Consum. Res. **37**(2), 264–278 (2010)
3. Hernes, M.: A cognitive integrated management support system for enterprises. In: Hwang, D., Jung, J.J., Nguyen, N.-T. (eds.) ICCCI 2014. LNCS (LNAI), vol. 8733, pp. 252–261. Springer, Cham (2014). https://doi.org/10.1007/978-3-319-11289-3_26
4. https://www.panorama-consulting.com/what-does-our-2019-erp-report-reveal-about-the-erp-industry/. Accessed 12 Oct 2019
5. Hernes, M., Bytniewski, A.: Knowledge representation of cognitive agents processing the economy events. In: Nguyen, N.T., Hoang, D.H., Hong, T.-P., Pham, H., Trawiński, B. (eds.) ACIIDS 2018. LNCS (LNAI), vol. 10751, pp. 392–401. Springer, Cham (2018). https://doi.org/10.1007/978-3-319-75417-8_37
6. Hernes, M., Bytniewski, A.: Knowledge integration in a manufacturing planning module of a cognitive integrated management information system. In: Nguyen, N.T., Papadopoulos, G.A., Jędrzejowicz, P., Trawiński, B., Vossen, G. (eds.) ICCCI 2017. LNCS (LNAI), vol. 10448, pp. 34–43. Springer, Cham (2017). https://doi.org/10.1007/978-3-319-67074-4_4
7. Wu, D.D., Chen, S.H., Olson, D.: Business intelligence in risk management: some recent progresses. Inf. Sci. **256**, 1–7 (2014)
8. Velagapalli, V., Bommareddy, S.S.R., Premalatha, V.: Application of lean techniques, enterprise, resource planning and artificial intelligence in construction project management. In: International Conference on Advances in Civil Engineering (ICACE 2019), 21–23 March 2019. University Vijayawada, A.P., India, vol. 7 (2019). International Journal of Recent Technology and Engineering (IJRTE). ISSN 2277-3878
9. Ryan, J., Snyder, Ch.: Intelligent agents and information resource management. In: Proceedings of the Tenth Americas Conference on Information Systems, New York, pp. 4627–4630 (2004)
10. Gupta, J.N.D., Forgione, G.A., Mora T., M. (eds.): Intelligent Decision-Making Support Systems. Foundations, Applications and Challenges. Springer, London (2006). https://doi.org/10.1007/1-84628-231-4

11. Ryfors, D., Wallin, M., Truve, T.: Swedish manufacturing SMEs readiness for Industry 4.0. What factors influence an implementation of Artificial Intelligence and how ready are manufacturing SMEs in Sweden? Jonkoping University (2019)
12. Martinez-Lopez, F.J., Casillas, J.: Artifical intelligence-based systems applied in industrial marketing: An historical overview, current and future insights. Ind. Mark. Manage. **42**(4), 489–495 (2018)
13. Rikhardson, P., Yigitbasioglu, O.: Business intelligence & analytics in management accounting research: status and future focus. Int. J. Acc. Inf. Syst. **29**, 37–58 (2018)
14. Zebec, A.: Cognitive BPM: Business Process Automation nad Innovation with Artificial Intelligence. http://ceur-ws.org/Vol-2420/paperDC1.pdf. Accessed 10 Nov 2019
15. Gandon, F.: Distributed Artificial Intelligence and Knowledge Management: Ontologies and Multi Agent Systems for a Corporate Semantic Web, Universite Nice Sophia Antipolis, 2002 (tel-00378201). https://tel.archives-ouvertes.fr/tel-00378201. Accessed 10 Nov 2019
16. Lee, M.C., Cheng, J.F.: Development multi-enterprise collaborative enterprise intelligent decision support system. J. Converg. Inf. Technol. **2**(2), 64 (2007)
17. Kishore, R., Zhang, H., Ramesh, R.: Enterprise integration using the agent paradigm: foundations of multi-agent-based integrative business information systems. Decis. Support Syst. **42**(1), 48–78 (2006)
18. Roland Berger Trend Compedium 2030, Megatrend 5 Dynamic technology & innovation (2017)
19. Bytniewski, A. (ed.): An Architecture of Integrated Management System. Wydawnictwo Uniwersytetu Ekonomicznego we Wrocławiu, Wrocław (2015). (in Polish)
20. Chojnacka-Komorowska, A., Hernes, M.: Knowledge representation in controlling subsystem. In: Ganzha, M., Maciaszek, L., Paprzycki, M. (eds.) Position Papers of the 2015 Federated Conference on Computer Science and Information Systems, vol. 6, pp. 187–193. Polskie Towarzystwo Informatyczne (2015)
21. Goertzel, B.: OpenCogBot – achieving generally intelligent virtual agent control and humanoid robotics via cognitive synergy. In: Proceedings of ICAI 2010, Beijing (2010)
22. Kollmann, S., Siafara, L.C., Schaat, S., Wendt, A.: Towards a cognitive multi-agent system for building control. Procedia Comput. Sci. **88**, 191–197 (2016)
23. Snaider, J., McCall, R., Franklin, S.: The LIDA framework as a general tool for AGI. In: The Fourth Conference on Artificial General Intelligence (2011)
24. Lasota, T., Telec, Z., Trawiński, B., Trawiński, K.: a multi-agent system to assist with real estate appraisals using bagging ensembles. In: Nguyen, N.T., Kowalczyk, R., Chen, S.-M. (eds.) ICCCI 2009. LNCS (LNAI), vol. 5796, pp. 813–824. Springer, Heidelberg (2009). https://doi.org/10.1007/978-3-642-04441-0_71
25. Deloitte: Technology, Media & Telecommunications Predictions 2016 (2016). https://www2.deloitte.com/content/dam/Deloitte/au/Documents/technology-media-telecommunications/deloitte-au-tmt-predictions-2016-report-050218.pdf
26. http://www.insoftconsulting.pl/raport-idc-2014/. Accessed 15 Nov 2019
27. Hernes, M., Bytniewski, A.: Towards big management. In: Król, D., Nguyen, N.T., Shirai, K. (eds.) ACIIDS 2017. SCI, vol. 710, pp. 197–209. Springer, Cham (2017). https://doi.org/10.1007/978-3-319-56660-3_18
28. Park, N.: Secure UHF/HF dual-band RFID: strategic framework approaches and application solutions. In: Jędrzejowicz, P., Nguyen, N.T., Hoang, K. (eds.) ICCCI 2011. LNCS (LNAI), vol. 6922, pp. 488–496. Springer, Heidelberg (2011). https://doi.org/10.1007/978-3-642-23935-9_48

29. Hernes, M.: Performance evaluation of the customer relationship management agent's in a cognitive integrated management support system. In: Nguyen, N.T. (ed.) Transactions on Computational Collective Intelligence XVIII. LNCS, vol. 9240, pp. 86–104. Springer, Heidelberg (2015). https://doi.org/10.1007/978-3-662-48145-5_5
30. Nowosielski, K.: Performance improvement of controlling processes: results of theoretical and empirical researches. Przegląd Organizacji **5**, 50–58 (2014)

Towards the Continuous Processes Modeling of Grain Handling in Storage Facility Using Unified Process Metamodel

Krystian Wojtkiewicz[1], Rafał Palak[1], Marek Krótkiewicz[1], Marcin Jodłowiec[1](✉), Wiktoria Wojtkiewicz[2], and Katarzyna Szwedziak[3]

[1] Faculty of Computer Science and Management, Wrocław University of Science and Technology, Wybrzeże Stanisława Wyspiańskiego 27, 50-370 Wrocław, Poland
{krystian.wojtkiewicz,rafal.palak, marek.krotkiewicz,marcin.jodlowiec}@pwr.edu.pl

[2] Knowledge and Information Engineering Group, Wrocław, Poland
wiktoria.wojtkiewicz@kieg.science

[3] Department of Biosystem Engineering, Opole University of Technology, Prószkowska 76, 45-758 Opole, Poland
katarzyna.szwedziak@po.edu.pl

Abstract. The paper presents a case study of modelling real-case industrial scenario of a grain-trading Polish company with Unified Process Metamodel (UPM). UPM is a novel approach for continuous process modelling. Many industrial processes, such as grain storage, are hard to describe by the use of conventional tools. With this example the authors emphasize the benefits of continuous process modeling paradigm over the token-based methodology and show the facilities of storage modeling and behavioral annotation in the UPM.

Keywords: Unified process modelling · Grain assessment · Industry 4.0 · Storage process · Continuous process · Flowing resource

1 Introduction

The newest revolution in technology applies to all fields of industry, including agriculture. Smart manufacturing is one of the aspects of Industry 4.0 [3]. In most basic form, this process relies on the quality of input (resources), and throughout the technology-enhanced production processes, it provides a high-quality output (products). In the field of agriculture, the key element for many manufacturers is the usage of appropriate quality seeds [10]. Research shows that harvesting technology, transport and drying conditions, as well as storage, have an impact on the quality of seeds, determining their suitability for industry. The smallest irregularities can cause irreversible changes and significantly reduce the technological value of seeds and their processing products. Official reports

N. T. Nguyen et al. (Eds.): ACIIDS 2020, LNAI 12034, pp. 241–252, 2020.
https://doi.org/10.1007/978-3-030-42058-1_20

show that 20% of the world's grain crops are damaged due to improper storage conditions. The main causes of losses are seed metabolism, cereal pest activity, improper storage conditions, and transport [7].

Right now, the whole process is complicated and time-consuming. This makes it almost impossible to make changes and to optimize it. The solution that might help to solve such a situation could be a general model of the whole process. The process begins with the sample being taken by a pneumatic probe, then the material collected by the probe is transported by means of a pneumatic system for measurement in the laboratory. In the laboratory, the material is analyzed for the required quality parameters. Bulk density, the cumulative distribution of average grain size, humidity, and falling number of particles of pre-ground flour are tested in succession. The protein and gluten content is also assessed. Cereals are tested for pests. Field pests do not pose a danger to the elevator, because small insects quickly die in the process of inside-elevator transport. However, the occurrence of even one warehouse pest in the sample disqualifies the entire delivery of material to the elevator. Grain infected with warehouse pests is dangerous for the entire elevator. Assessment of the measured quality parameters is made based on standards using conventional and modern methods of analysis [13,14]. Thus, after receiving the collected sample, the bulk density is determined by the means of a brass vessel with a specific volume into which ungraded quarry grain is poured, the cumulative distribution system is determined using a pile of sieves. Gluten and protein content is assessed using a modern spectrometer, the falling time is determined after milling the grain by a laboratory grinder, while the type of flour and cereal pests should be reminded by a specially placed table. The process is not only complicated and time-consuming but also costly. Many people are involved in this process and the equipment is expensive to buy and maintain. However, the precondition check of the grain is only one of the steps in the general algorithm of grain handling. Next one is the storage of grain in a special storage facility. Unlike any other material, grain requires constant monitoring of storage condition and material state. While the whole grain handling process can be abstracted to three stages, namely grain acquisition, storage and distribution, the authors decided to focus on only one phase - storage.

The main contribution of this paper is the case study aiming at the implementation of continuous process modelling in the field of agriculture process modelling with Unified Process Metamodel (UPM). The UPM is a novel process modelling tool, that prior to this paper has been presented in few places [4–6,15,16]. In this paper, authors aimed at proving its capability to model complex storage process as the continuous abstraction. It is worth noting that this approach to process description is a highly unique property of UPM. Most of nowadays used process modelling methods focus on process' discrete aspects paying no attention to its continuous nature [1,2,11,12].

The paper is organized as follows: the next section describes the technological process of grain handling. Section 3 provides an overview of the UPM process modelling, while Sect. 4 includes the UPM model of the continuous grain assessment process. The last section provides a summary of the paper.

2 Technological Process of Grain Handling in Trading Company

One of the basic goals of modeling is to provide a high level view of the phenomena under consideration, which enables analysis. In particular, process modeling creates a unique opportunity to analyze the performance of process components and test various optimization options. In this study, the grain handling process by AGROPOL Ltd. is analyzed. It is the main process of the company business and thus its performance is crucial for the company existence. The process itself has continuous nature, i.e., the time in the process does not have a discrete nature. All inputs and outputs used by the process' nodes relate to one type of material, i.e., grain. Once the grain (of the same type) is mixed, it cannot be distinguished from one other. Thus it can be assumed that material (*flowing resource*[1]) has also continuous nature.

The process consists of three main phases. First, grain acquisition relates to obtaining new grain by the company. The crucial part of this process is grain categorization based on its contamination, that is briefly described in Subsect. 2.1. Once the grain is accepted by the company, it is stored in a storage facility and handled to keep the highest possible quality. The idea of storage is presented in Subsect. 2.2. The last phase presented in Subsect. 2.3 relates to grain redistribution, i.e., logistics process that aims at the transportation of grain from storage to the customer.

Due to the limitation of the paper and the primary objective of the article, authors have decided that the in depth analysis and modelling of the process will be focused on the storage aspect.

2.1 Categorization of Grain Contamination

Assessment of grain quality is crucial for the company. The acceptance of material that doesn't meet quality requirements might lead to lowering the quality grade of all material in the storage where the wrongly assessed grain would be sent. Thus, this task is highly formalized and need to be performed in accordance with nationwide regulations.

The basic grade grain might contain various impurities. They are considered *useful* or *useless*. Useful impurities are: buttocks - lean grains, underdeveloped, sprouted grain, moldy, decayed, mechanically or pests damaged, as well as grains darkened by the result of improper storage, burnt during drying and grains of other species. The second type (useless) contaminants are all mineral impurities such as sand, lumps of earth, small stones, pieces of glass and metal parts as well as organic impurities.

If the assessed grain meets the quality requirements, it is accepted for acquisition and is directed to storage.

[1] *Flowing Resource* (FR) is one the main UPM concept.

2.2 Grain Storage

The storage aims at low term and long term deposition of grain. AGROPOL Ltd. operates on six flat storage facilities and two silos. Each of them can be used for storage of a different kind of grain; however, once the type of grain is assigned to the storage unit, it cannot be changed unless the unit is clear from all grain. While the grain is stored in the unit, the most important aspect is the constant monitoring of the storage condition and changes of material properties, namely temperature and moisture. The detailed description if this phase is subject of further sections, thus will not be elaborated here.

2.3 Grain Redistribution

The last phase in the grain handling process is the redistribution. There are two types, namely internal and external. The first one relates to the transportation of the grain to other company facilities both for storage or as sowing material to the field. The other type, in turn, is the direct result of the selling process.

3 Principles of UPM Process Modelling

Unified Process Metamodel is a novel approach towards modelling of continuous processes in structural and behavioural aspects. A complete definition of the metamodel was included in Wojtkiewicz's dissertation [16], while partial descriptions of both the approach and specific solutions can be found in [4–6,15]. For the sake of ease of understanding the UPM example presented in the next section, below the idea of the metamodel will be presented, along with the basic syntax elements.

One of the basic concepts that were used in the construction of UPM was the idea of scale. Modeling in UPM can, therefore, be carried out at different levels (on a different scale), and in each of them with different detail adapted to the needs of the field. The highest of them is the map of processes. The process map consists of functional layers highlighted (Fig. 1). The next level of abstraction

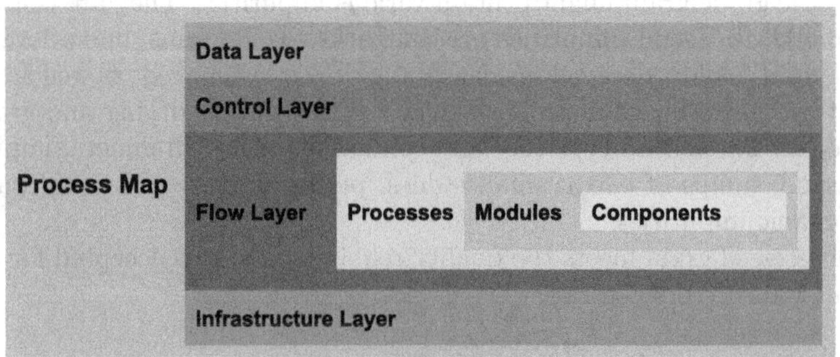

Fig. 1. UPM process layers

is the process. However, this level is not a classic single-layer level, because processes in UPM can be nested. This level should, therefore, be considered as a complex one. The next levels are the component and module levels in turn (Fig. 2). The components are made of modules that constitute the lowest level of abstraction assumed in UPM. These levels of presentation can intertwine each other as needed. In a particular view (diagram), there may be elements typical for each of the specified modeling levels. For better model presentation, it is possible to create dedicated diagrams presenting the internal structure of individual components (including embedded processes).

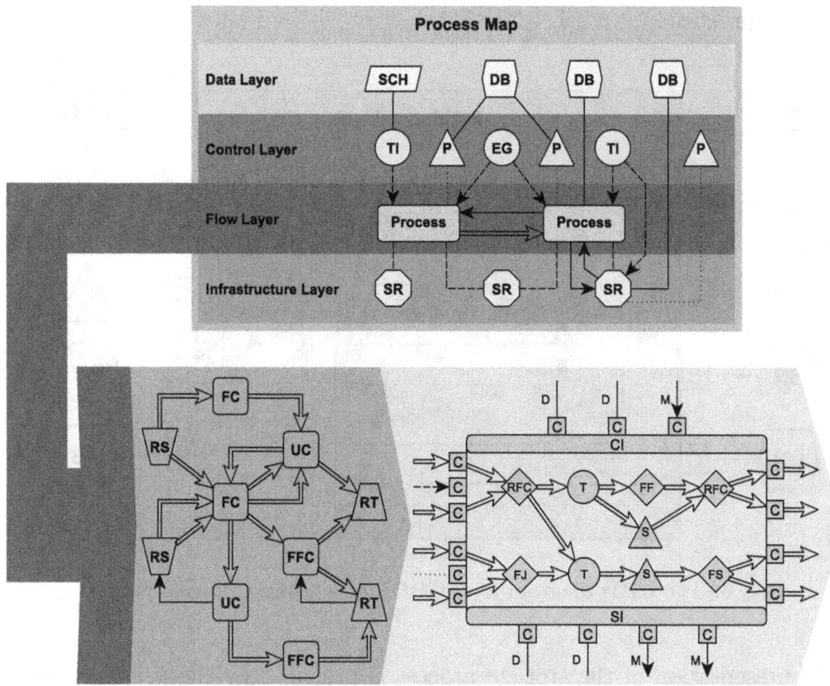

Fig. 2. UPM process layers, components and modules

Another idea that found application in UPM was its universality. It consists in the fact that the language syntax has not been overloaded with too many primitives. The trade-off involves the need to define specialized components for the needs of a specific model. This solution allows perfect adjustment of used elements to the specificity of the modelled fragments of reality. On the other hand, each time it requires the recipient to become familiar with a new set of defined, specific components.

4 UPM Model of Continuous Grain Assessment Process

The given example of a process modeled in UPM is a presentation of the actual grain storage process at AGROPOL Ltd. (a company in Poland trading grain). The goal was to present a whole process by describing the operation of a warehouse complex, storing different types of grains. As part of this process, continuous quality testing of grain storage conditions is undertaken and possible actions are invoked to maintain optimal conditions. The example itself is a simplification of the real system, in which the reference to the exact features of individual elements present in it was omitted, i.e., specific technological solutions were not referred. The given example is to present the general possibilities of modeling continuous processes.

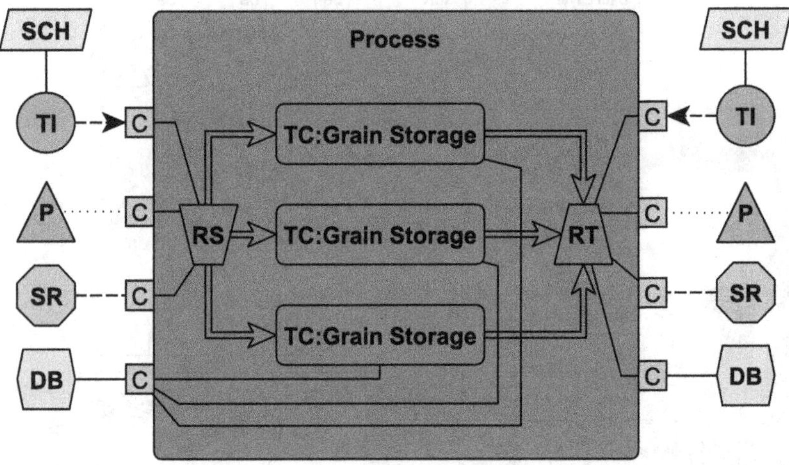

Fig. 3. The UPM diagram for AGROPOL's storage process model

The presentation of the storage process is prepared by the use of two-scale points. The first one relates to the high level of the process overview, that presents the combination of multiple storage units[2]. This scale is presented in the Fig. 3, while the lower level of abstraction is showed in the Fig. 4. The presented model was simplified by omitting obvious elements such as the electrical installation necessary for the proper functioning of individual system components. This approach allows focusing on the presented semantics of individual components without additional processing description of the elements that are not key to show the principles of modeling in UPM. As part of the analysis of the presented example, it should be noted that the following flowing resources [5] were defined for the needs of the process: `Grain` (green color), `Electricity` (red color), `Water` (blue color).

[2] The actual number of storage units in AGROPOL Ltd. is higher than shown; however it was limited in the paper for the sake of simplification.

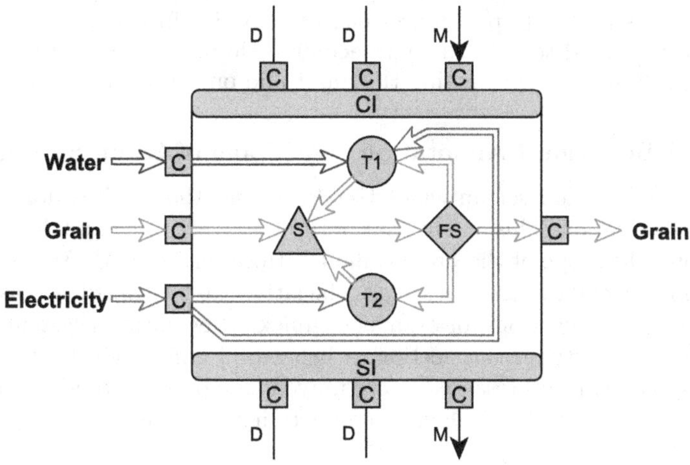

Fig. 4. The UPM diagram for AGROPOL's storage unit component model (Color figure online)

As part of the specific UPM syntax, there is no defined way of presenting information on flowing resources in diagrams, leaving this issue to be decided as part of a specific implementation. From the point of view of readability, the modeler may choose not to indicate on the diagram the type of resource flowing in a given channel, indicate a specific type of resource at the ends of the channel, above or below the channel, or use colors to distinguish different types of channels. The definition of the UPM specific syntax imposes no constraints in this respect. However, it should be remembered that after adopting a specific method for determining the types of flowing resources and channels to be consistent throughout the model.

The storage unit definition consists of the following modules:

S *Storage* module that represents the actual storage unit with properties defined in the Sect. 4.1–Static structures

T1, T2 *Transformation* modules that are used to perform actions of drying or moisturizing the grain

FS *Flow split* module that is responsible to direct the flow of grain according to the rules defined in the Sect. 4.1–Process of Grain Storage Assessment

Apart from the flowing resource connectors, the component definition introduces also the set of connectors, both for data flow and message flow. The principles of their operation are not crucial for the sake of process understanding, thus will not be elaborated here.

The storage unit component definition is used on the higher level of abstraction view presented in the Fig. 3 in a simplified form to model different grain storage units. This view also introduces new components, namely:

TI, SCH *Timer* and *Schedule* that allows the cyclical events to be executed,

P *Procedures* that hold procedures (scripts) specific for the process,
DB *Database* used to hold the data acquired through the process life,
SR *Static Resource* representing the input and output of the grain flow.

4.1 UPM Behavioral Annotation – A Model of Grain Storage Unit

The metastructure and semantics of UPM abstract the facilities for data and behavioral annotation. This means that both process scripting and data processing is beyond the scope of the process model. However, some UPM components favor a definition of database and scrips. The latter can be defined, e.g. as Python scripts, C/C++ applications, or even as complex cyber-physical systems [8]. For the sake of readability, in this section, a behavioral annotation for the process of grain storage will be presented using universal and understandable notation standard, namely UML 2.5.1 [9] class and activity diagrams.

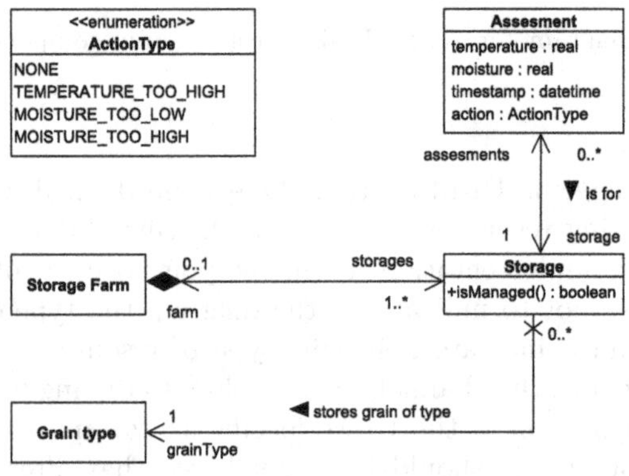

Fig. 5. A UML class diagram for grain storage

Static Structures. The grain storage process operates on the list of structures. Presentation of these structures will help with an understanding of the processes operating on them. In the Fig. 5 we have shown the associations between the following classes:

1. `Storage farm` – a collection of logically and physically connected units of storage, which hold shared *supervisor*.
2. `Storage` – a class representing a single grain storage. It can be managed or non-managed. When it is managed, it is supervised by the farm supervisor; otherwise, its behavior is independent of the other units of storage.
3. `Assessment` – a representation of the situation, when the environmental conditions inside of the storage are assessed, the main observed values consist of temperature and moisture.
4. `ActionType` – enumeration depicting causes for action needed to be taken during non-typical situations in the storage.
5. `Grain type` – a type of grain stored inside a storage.

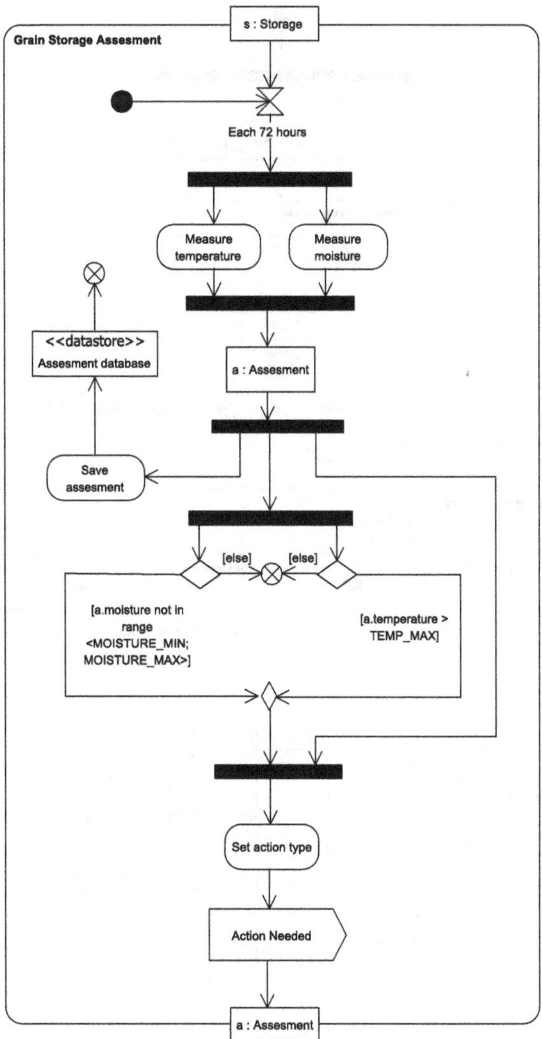

Fig. 6. An UML activity diagram for Grain Storage Assessment process

Process of Grain Storage Assessment. The process of Grain Storage Assessment (GSA) is executed on the single instance of the `Storage` (Fig. 6). Its objective is to monitor storage, thus every 72 h it is checked whether its environmental conditions are in the norm and the assessment instance is created. During the measurement, the temperature and moisture values are acquired. The results are saved to an external database.

If the system should notify about the occurrence of untypical situation, it sets the adequate `ActionType` and sends the signal with the `Assessment` instance attached. The reaction to this signal is not storage-dependent but is linked with the manageability state of storage.

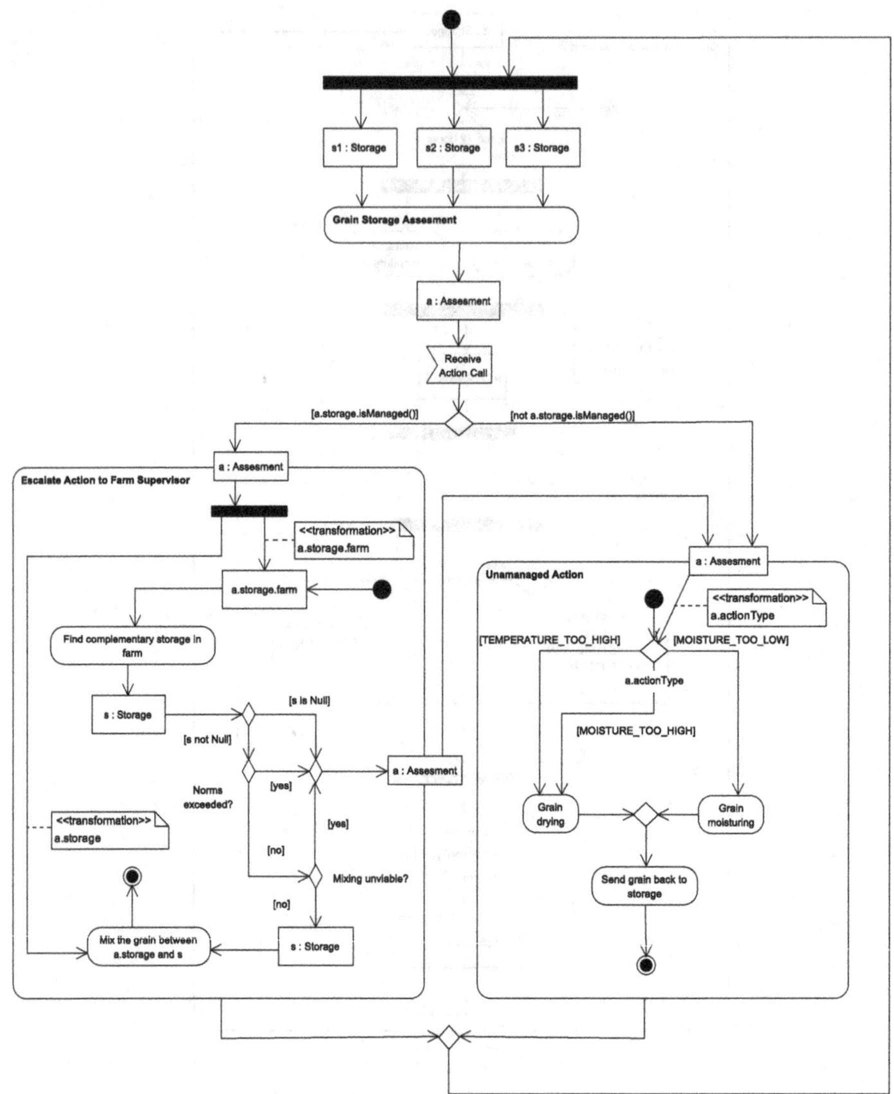

Fig. 7. An UML activity diagram for Storage Farm Process

Model of Storage Farm Process. The model of Storage Farm Process (Fig. 7) is a high-level behavioral description of a `Storage farm`. This means that it is applied to a group of storage units (in the figure there are depicted 3 storage instances: `s1`, `s2`, and `s3`). This behavior involves the dual character of the storage, taking into consideration its manageability. It is supposed to run in an event-like loop while events are delivered by the storage units inside the farm and created in the GSA process. The event handling consists of determination the manageability of the unit and subsequently invokes adequate behavior. The behavior for non-managed storage units is `Unmanaged Action`. In this action, depending on the requested action type from the assessment stage, the grain should be either dried or moisturized and then sent back to the storage. The behavior for managed storage units is `Escalate Action to Farm Supervisor`. In this behavior the supervisor performs several few checks, whether there is a complementary storage within the farm that could be utilized for mixing the grain. If this action is economically unjustifiable or inadequate, the supervisor may delegate the action to the storage, and then the behavior is unmanaged.

5 Summary

This paper presents a case study of modelling real industrial scenario of a Polish grain-trading company in Unified Process Metamodel. UPM is a novel meta-model designed for use in continuous process modelling. The high complexity of the process was described by the UPM model, which allowed for a better process understanding. That, in turn, introduces the possibility of process optimization. Many industrial processes, such as grain storage, are difficult to describe with conventional tools. Token-based approaches cannot easily be applied to the reality, where the infrastructure is static in terms of process instantiation. In such cases, continuous approach results with better-suited models, which could be simulated used specific software tools operating on it. In this article, the following processes were presented:

- storage process for *one* warehouse,
- storage process for *many* warehouses.

In both processes, common elements and requirements were pointed out and analyzed in detail.

The article presents an example that shows the huge possibilities and expressiveness of UPM. The fact that such a complex process could be easily modelled shows many possible UPM applications. In future work, we would like to use UPM in a more nested and complicated process that would show its superiority over other available modelling languages. We would intend to develop simulation software that would use UPM.

References

1. van der Aalst, W.M.P.: Process Mining: Data Science in Action, 2nd edn. Springer, Heidelberg (2016). https://doi.org/10.1007/978-3-662-49851-4
2. Hadjicostas, E.: ISO 9000:2000 Quality Management System, pp. 53–75. Springer, Heidelberg (2004). https://doi.org/10.1007/978-3-662-09621-5_4
3. Kang, H.S., et al.: Smart manufacturing: past research, present findings, and future directions. Int. J. Precis. Eng. Manuf.-Green Technol. **3**(1), 111–128 (2016). https://doi.org/10.1007/s40684-016-0015-5

4. Krótkiewicz, M., Jodłowiec, M., Wojtkiewicz, K., Szwedziak, K.: Unified process management for service and manufacture system—material resources. In: Burduk, R., Jackowski, K., Kurzyński, M., Woźniak, M., Żołnierek, A. (eds.) Proceedings of the 9th International Conference on Computer Recognition Systems CORES 2015. AISC, vol. 403, pp. 681–690. Springer, Cham (2016). https://doi.org/10.1007/978-3-319-26227-7_64
5. Krótkiewicz, M., Jodłowiec, M., Hunek, W.P., Wojtkiewicz, K.: Modeling of unified process metamodel structure in association-oriented database metamodel: flowing resources and transformation. In: 2017 IEEE International Conference on INnovations in Intelligent SysTems and Applications (INISTA), pp. 265–270, July 2017. https://doi.org/10.1109/INISTA.2017.8001168
6. Krótkiewicz, M.: Cyclic value ranges model for specifying flowing resources in unified process metamodel. Enterp. Inf. Syst. **13**(7–8), 1–23 (2018). https://doi.org/10.1080/17517575.2018.1472810
7. Kumar, D., Kalita, P.: Reducing postharvest losses during storage of grain crops to strengthen food security in developing countries. Foods **6**(1), 8 (2017). https://doi.org/10.3390/foods6010008
8. Lee, J., Bagheri, B., Kao, H.A.: A cyber-physical systems architecture for industry 4.0-based manufacturing systems. Manuf. Lett. **3**, 18–23 (2015). https://doi.org/10.1016/j.mfglet.2014.12.001
9. Object Management Group: Omg unified modeling language - version 2.5.1, December 2017. https://www.omg.org/spec/UML/2.5.1
10. Rahman, A., Cho, B.K.: Assessment of seed quality using non-destructive measurement techniques: a review. Seed Sci. Res. **26**(4), 285–305 (2016). https://doi.org/10.1017/s0960258516000234
11. Russell, N.: Foundations of Process-Aware Information Systems. Ph.D. Thesis, Queensland University of Technology (2007)
12. Scheer, A.-W., Nüttgens, M.: ARIS architecture and reference models for business process management. In: van der Aalst, W., Desel, J., Oberweis, A. (eds.) Business Process Management. LNCS, vol. 1806, pp. 376–389. Springer, Heidelberg (2000). https://doi.org/10.1007/3-540-45594-9_24
13. Szwedziak, K.: Artificial neural networks and computer image analysis in the evaluation of selected quality parameters of pea seeds. In: E3S Web Conference, vol. 132, p. 01027 (2019). https://doi.org/10.1051/e3sconf/201913201027
14. Szwedziak, K.: The use of vision techniques for the evaluation of selected quality parameters of maize grain during storage. In: E3S Web Conference, vol. 132, p. 01026 (2019). https://doi.org/10.1051/e3sconf/201913201026
15. Wojtkiewicz, K.: Event condition action approach to process' control layer modeling in unified process metamodel. In: 2018 5th International Conference on Systems and Informatics (ICSAI), pp. 1282–1288, November 2018. https://doi.org/10.1109/ICSAI.2018.8599475
16. Wojtkiewicz, K.: Zunifikowany Metamodel Procesów (eng. Unified Process Metamodel). Ph.D. thesis, Opole University of Technology (2019)

Intelligent Applications of Internet of Things and Data Analysis Technologies

Applications of Fuzzy Inference System on V2V Routing in Vehicular Networks

Yung-Fa Huang[1(✉)], Teguh Indra Bayu[1], Shin-Hong Liu[1], Hui-Yu Huang[2], and Wen Huang[1]

[1] Chaoyang University of Technology, Taichung 413, Taiwan
yfahuang@cyut.edu.tw
[2] National Formosa University, Yunlin 632, Taiwan
hyhuang@nfu.edu.tw

Abstract. The 5th generation (5G) mobile communication technology is an important technology for high throughput, low latency and high reliability. Vehicular Ad-hoc Networks (VANET) provides information switching through Vehicle-to-Vehicle (V2V) wireless network communication technology, where the performance requirements for low latency and high transmission capacity is the most challenging. The multi-hop routing-linking strategies investigated in previous works, First Nearest Vehicle (FNV), Second Nearest Vehicle (SNV) and Third Nearest Vehicle (TNV) can provide different requirements for the transmission delay time and transmission capacity. However, the three routing methods are suffered by the variation situation on the vehicle densities and the transmission range in vehicular network (VN). Therefore, this study explored the application of Fuzzy inference system (FIS) for the V2V routing issues. The proposed fuzzy inference routing (FIR) mechanism was designed to compromise the advantage of the multi-hop routing methods and to reach the requirements of transmission delay time and high reliability. Simulation results show that the proposed fuzzy inference routing-premium (FIR-P) can outperform the multi-hop routing methods and satisfy the 90% transmission delay less than 1 ms for VN.

Keywords: 5G mobile communication · Vehicle-to-Vehicle (V2V) wireless communication · Fuzzy Inference System (FIS) · Low latency

1 Introduction

In 2020, the number of mobile devices and Internet of Things (IOT) devices will reach more than 5 billion. IoT devices such as cars, smart phones, tablets and personal computers will gradually pursue audio and video HD 4K quality and other needs more action bandwidth requirements [1]. For the developing fifth generation communication technology (5G), the major automakers around the world are constantly researching the technology of driverless vehicles, which is making vehicular networks (VNs) a hot topic

N. T. Nguyen et al. (Eds.): ACIIDS 2020, LNAI 12034, pp. 255–265, 2020.
https://doi.org/10.1007/978-3-030-42058-1_21

through vehicle-to-vehicle (V2V) and vehicle-to-infrastructure (V2I), Infrastructure-to-infrastructure (I2I) communication technology. The vehicle-to-everything (V2X) technology provides information exchange between the cars and the cars. The VN is composed of various communication technologies, including Dedicated short-range communication (DSRC) [2], 5G, fourth-generation communication technology (4G), and Wi-Fi, etc. [3]. In the VN environments, there are many situations for V2V communications, such as unbalanced traffic configuration of the multi-path topology and low network resource usage [4]. How to simplify network management and join V2V and V2I communication technology services to build a flexible and programmable architecture will be one of the key requirements of the VN.

The multi-hop routing-linking strategies have been investigated in previous works [5]. The First Nearest Vehicle (FNV), Second Nearest Vehicle (SNV) and Third Nearest Vehicle (TNV) can provide different requirements for the delay time and transmission capacity. However, the three routing methods are suffered by the variation situation on the vehicle densities and the transmission range in the VN [5].

The fuzzy theory was proposed by Professor Zadeh in 1965 [6]. The fuzzy theory is an approximation reasoning mechanism. Therefore, this study will explore the application of Fuzzy inference system (FIS) for the V2V routing issues. The proposed fuzzy inference routing (FIR) mechanism was designed to compromise the advantage of the multi-hop routing methods and reach the requirements of transmission delay time and high reliability. Therefore, in this study an FIS based routing mechanism are proposed to reach the requirements of 90% transmission delay time (90% Td) less than 1 ms for VN.

2 System Models

The channel model includes path loss and shadow fading. According to the characteristics of millimeter wave, the channel model of any pair of V2V links is

$$\mathrm{P}L_{dB}(r_j) = \alpha + \beta \times 10\log_{10}(r_j) + \xi \tag{1}$$

where r_j is the distance between any pair of V2V links (in kilometers); α and β represent the initial offset and path attenuation index, respectively; ξ is the shadow fading effect can be expressed using logarithms normal (Log-Normal) distribution of random variables by N(m, σ^2), where m is the mean value and σ^2 is the variance.

In the V2V communications, there are two channel models with the LOS (Line of sight) and the NLOS (Non-line of sight), which can be expressed by

$$L_{LOS}(r_j)[dB] = 69.6 + 20.9 \times \log_{10}(r_j) + \xi_{LOS}, \tag{2}$$

and

$$L_{NLOS}(r_j)[dB] = 69.6 + 33 \times 10\log_{10}(r_j) + \xi_{NLOS} \tag{3}$$

respectively, where ξ_{LOS} and ξ_{NLOS} are the shadowing fading effect of LOS and NLOS environments.

The simulation environment is shown in Fig. 1; in which it is assumed that the roadside unit (RSU) is located in the fog cells. All the vehicles use the millimeter wave (mmWave) to communicate with each other V2V covered by the fog cell communication range. It is assumed that the vehicle V_a is going to transmit message to the RSU by V2V, where L_a is the transmission length of the vehicle V_a to RSU.

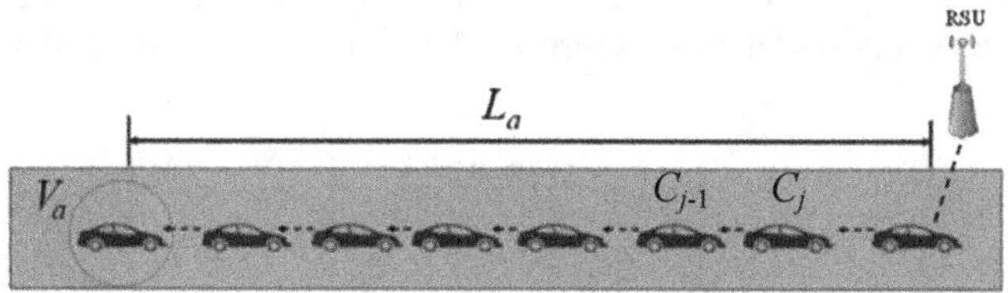

Fig. 1. Multi-hop vehicle routing models [5].

Due to the limited transmission range of the vehicles, it is assumed that there are k relay hops between the vehicle V_a and the RSU. The vehicle V_a selects other vehicles as relays to the RSU, and the total transmission delay time (T_d) of the vehicle V_a is expressed by [7]

$$T_d = (k - 1)T_{pro} + \sum\nolimits_{j=1}^{k} \left(T_{hop_j} + T_{retran_j}\right) \tag{4}$$

where T_{pro} is the data transfer processing time of the vehicle relay node; T_{hopj} is the message transmission delay time between the j-th vehicle relay nodes; $T_{retranj}$ is the V2V retransmission time between the j-th vehicle relay nodes. When the communication transmission distance is too far, the transmission would be probably failed. Then the delay time of waiting and retransmission is required. In this study, it is assumed that it need 10 times of the slot time to retransmitted the message successfully, that is $T_{retran_j} = 10t_{slot}$.

From the channel model Eqs. (2) and (3), if the V2V wireless transmission distance is r_j, the channel gains $h(r_j)$ of LOS and NLOS are equaling to $10^{-L_{LOS}(r_j)}$ and $10^{-L_{NLOS}(r_j)}$, respectively. Then the SNR of the j-th vehicle for the V2V wireless transmission can be expressed by [7]

$$SNR_j = \frac{P_{tx} h_j}{N_0 W_{mmWave}} \tag{5}$$

where P_{tx} is the transmission power plus the antenna gain of the vehicle nodes; h_j is the channel gain between the vehicle relay nodes C_{j-1} and C_j; N_o is the power spectral density of added white Gaussian noise (AWGN)); W_{mmWave} is millimeter wave bandwidth (4 GHz). Then the average transmission capacity of each V2V multi-hop link can be obtained by Shannon theory as

$$C_j = W_{mmWave} \times \log_2(1 + SNR_j) \tag{6}$$

In this study we investigated three V2V transmission modes. In the FNV, the closest vehicle is selected as the relay point for transmission. In the SNV, the second approaching vehicle is selected as the relay point for transmission. In the TNV, it selects the third approaching vehicle as the relay point for transmission [7].

At first, we investigated the transmission delay performance of the three transmission modes FNV, SNV, and TNV for different vehicle density with ρ = 0.01–0.5 vehicle/m. Then we obtained two examples of cumulative distribution function (CDF) of transmission delay in Fig. 2 at the transmission distance $L_a = 300$ m. When the vehicle density is 0.3, as shown in Fig. 2(a), it can be seen that FNV can reach more than 90% of the transmission delay time less than 1 ms threshold. However, from Fig. 2(b), it can be seen

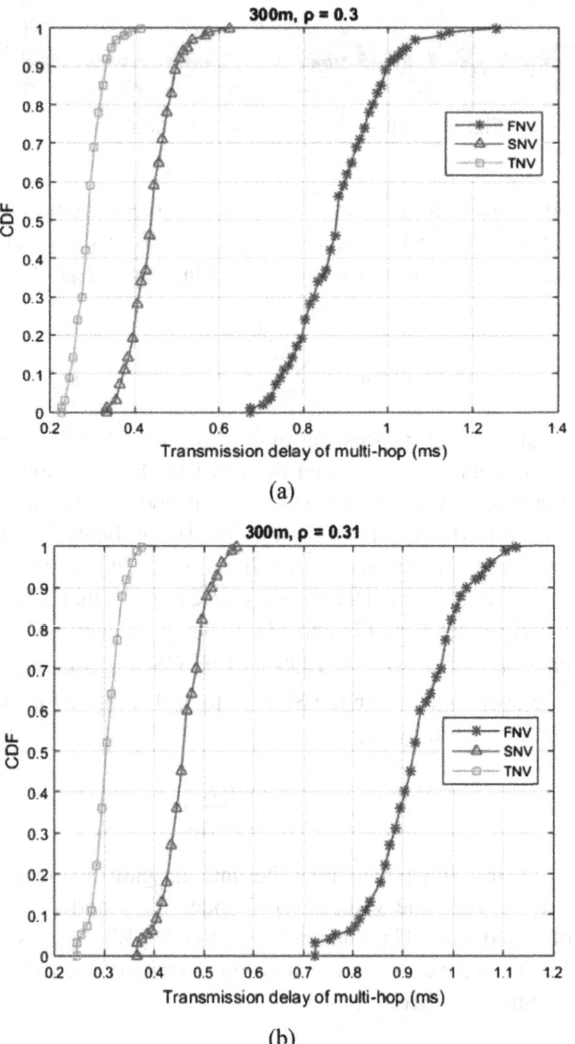

Fig. 2. The CDF of transmission delay for three routing methods in VN with $L_a = 300$ m and (a) $\rho = 0.3$, (b) $\rho = 0.31$ vehicle/m.

that FNV cannot meet the threshold of 90% or more transmission delay time below 1 ms when $\rho = 0.31$. Therefore, the SNV transmission mode needs to be used instead.

3 Fuzzy Inference Routing Mechanism

Fuzzy Inference System (FIS) mainly uses the IF-THEN inferring method as the control rule. The fuzzy inference system can be constituted by four parts. One is the fuzzification on fuzzy input. The value is converted to the membership degree of triggering to the fuzzy linguistic terms. The fuzzy inference engine, according to the fuzzy rule base we designed, calculates the system input trigger to the weight value of each fuzzy rule. At last part, the defuzzification turns the results of fuzzy inference into output values [8].

First of all, we need to establish fuzzy rule bases and membership functions (MBFs) for the input and output variables. We set the membership functions of vehicle density according to the analog transmission distance $L_a = 300$, 400, and 500 m, respectively. Thus, we choose three Triangular MBFs to cover the entire universe of discourse of two inputs, density ρ and distance L_a and one output routing modes P, respectively. The three linguistic terms, Rare (R), Normal (N), and Crowd (C) are chosen to cover its universe of discourse of vehicle density ρ, as shown in Fig. 3(a). The three linguistic terms, short (S), medium (M), and long (L), are chosen to cover its universe of discourse of the transmission distance L_a as shown in Fig. 3(b). The three linguistic terms, First Nearest Vehicle (FNV), Second Nearest Vehicle (SNV), and Third Nearest Vehicle (TNV) are chosen to cover its universe of discourse of routing modes P as shown in Fig. 3(c).

The Triangular MBF of the fuzzy set F_i^l in each interval $[C_i^-, C_i^+]$ of the universe of discourse U can be expressed by

$$\mu_{F_i^l}(x_i) = \begin{cases} 1 - \frac{2|x_i - \bar{x}_i^l|}{w_i}, & \bar{x}_i^l - \frac{w_i}{2} \leq x_i \leq \bar{x}_i^l + \frac{w_i}{2} \\ 0\ , & \text{otherwise} \end{cases}, \tag{7}$$

where $l = 1, 2, 3$, $i = 1, 2, 3$, $x_i \in \left[C_i^-, C_i^+\right]$, and $\bar{x}_i^l$ and w_i are the mean and width of the Triangular MBF, respectively.

The fuzzy control rule is represented by two inputs and one outputs by [8]

$$\begin{aligned} \mathrm{R}_j :\ & \text{IF density}(\rho) \text{ is } F_1^{l_1} \text{ AND } L_a \text{ is } F_2^{l_2} \\ & \text{THEN } P = F_3^{l_3} \end{aligned} \tag{8}$$

where $F_1^{l_1}$, $F_2^{l_2}$ and $F_3^{l_3}$ are the Linguistic Terms, which represent the inputs of vehicle density, transmission distance, and the output of routing methods, respectively, and l_1, l_2, $l_3 = 1, 2, 3$, and the index of rule $j = 1, 2, \ldots, 9$. Then, the FIS for the VN routing can be performed as shown in Fig. 4. The proposed FIS is called as Fuzzy Inference Routing-Preliminary (FIR-P). The fuzzy rules shown in Table 1, including 9 fuzzy IF-THEN rules for FIR-P, can be established heuristically by the experimental results, which infer the relations between the adequate routing methods of vehicle density and distance.

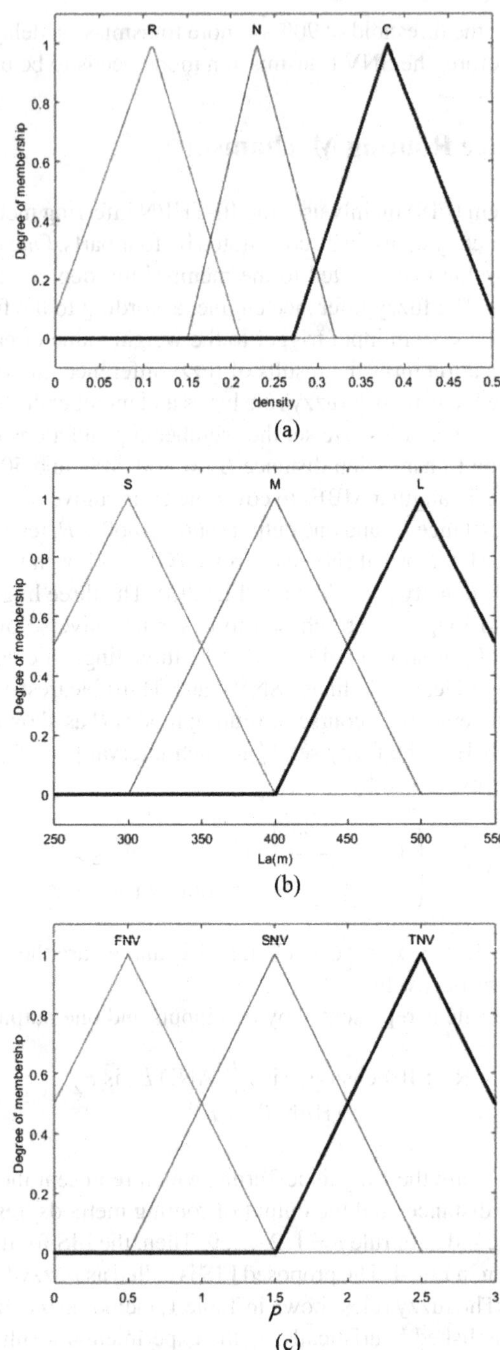

(a)

(b)

(c)

Fig. 3. The membership functions of the input variables (a) density ρ and (b) distance L_a, and output variable (c) routing modes P for the proposed FIR-P scheme.

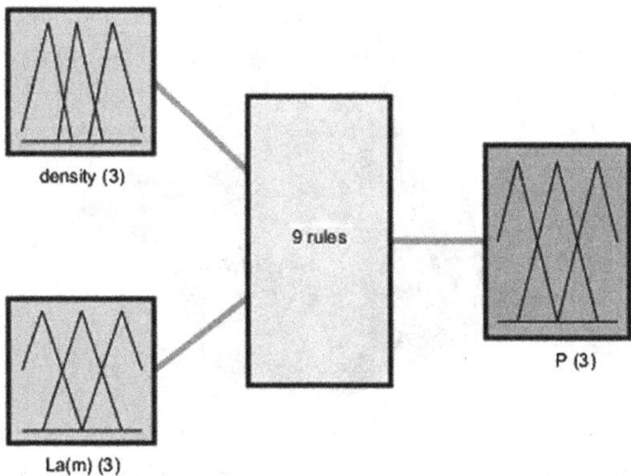

Fig. 4. The fuzzy logic system for multi-hoping routings of VNs with 2 inputs, 1 output and 9 rules.

Table 1. The rule base for the proposed FIR-P

Density (ρ)	L_a		
	S	M	L
R	FNV	FNV	FNV
N	FNV	SNV	SNV
C	SNV	SNV	TNV

There are many defuzzification methods available. However, the following centroid calculation, which returns the center of area under the aggregated MBFs curve, is being employed here [8]:

$$P = \frac{\sum_{i=1}^{n} z_i \times \mu_{F_3^l}(z_i)}{\sum_{i=1}^{n} \mu_{F_3^l}(z_i)} \tag{9}$$

where n is the number of quantization levels of the output area under the aggregated MBFs, z_i is the amount of the inference output at the quantization level i and $\mu_{F_3^l}(z_i)$ is its membership value in the output fuzzy set F_3^l. Then, the inference 3-D results for the routings methods P in VNs can be obtained as shown in Fig. 5. From Fig. 5, the three-dimensional relationship between the FIR-P of the output variable obtained by the fuzzy rules and the vehicle density and L_a.

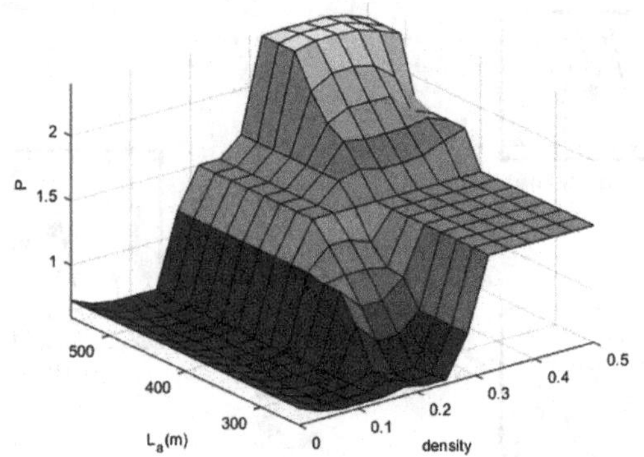

Fig. 5. The 3-D inferring diagram for FIR-P scheme in VN.

4 Simulation Results

In the simulation environments, the transmission distance L_a is set from 250 m to 550 m. The simulation parameters are shown in Table 2. To compromise the different transmission delay of the routing methods, we compare the performance of four methods of FNV, SNV, TNV and FIR-P. In the VN, it requires a transmission delay of less than 1 ms. In the simulation results of Fig. 6 at a transmission distance of 300 m, it can be observed that the 90% transmission delay time of the FNV exceeds 1 ms to 1.4 ms and with only 60% Td less than 1 ms. Moreover, the SNV and TNV can obtain 89% (Td) and 86% (Td) less than 1 ms, respectively. Therefore, the three methods of and the average delay time of FNV, SNV and TNV are all not meet the requirement of low latency and reliability. However, from Fig. 6, it can be seen the proposed FIR-P not only reaches the requirements of 90% (Td) less than 1 ms but also perform the low latency with 90% (Td) less than 0.75 ms and high reliability with 96% (Td) less than 1 ms.

Table 2. Simulation parameters

Total distance (L_a, m)	250–550
Density of Vehicles (ρ, vehicles/meter)	0.01–0.5
Transmission Power of Vehicles (P_{tx}, dBm)	30
Transmission Range of Vehicles (R, m)	70
Power Spectral Density of AWGN (N_0, dBm)	−174
Bandwidth of mm Waves (W_{mmWave}, GHz)	4
SNR Minimum Threshold (θ, dB)	10
Standard Deviation of Shadowing Fading (σ, dB)	LOS: 5, NLOS: 7.6
Processing Time (T_{pro}, μs)	5
Duration of Time Slot (T_{slot}, μs)	5

In addition, we investigate the performance of the strict requirements of 90% (Td) below 1 ms as shown in Fig. 7 for different transmission distance. From Fig. 7, it can be seen that the 90% (Td) of FNV exceeds 1 ms for all distance. the 90% (Td) of SNV and TNV are almost little higher than 1 ms for all distance. However, the 90% (Td) of the proposed FIR-P performed all less than 1 ms.

Furthermore, the performance of system capacity is compared in Fig. 8 for different transmission distances of 250 m to 550 m with the vehicle density of $\rho = 0.01–0.5$ vehicles/m. From Fig. 8, it is observed that due to the FNV performing the nearest distance hops, the capacity of FNV is highest among the schemes. However, the proposed FIR-P outperforms the other two schemes of SNV and TNV.

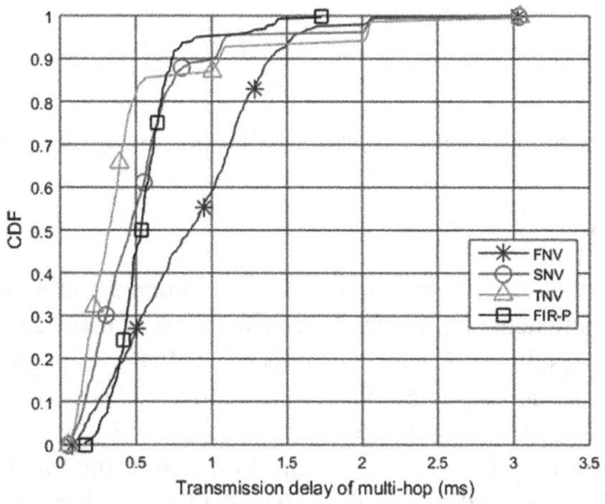

Fig. 6. Comparison of CDF of transmission delay for proposed routing schemes with $L_a = 300$ m.

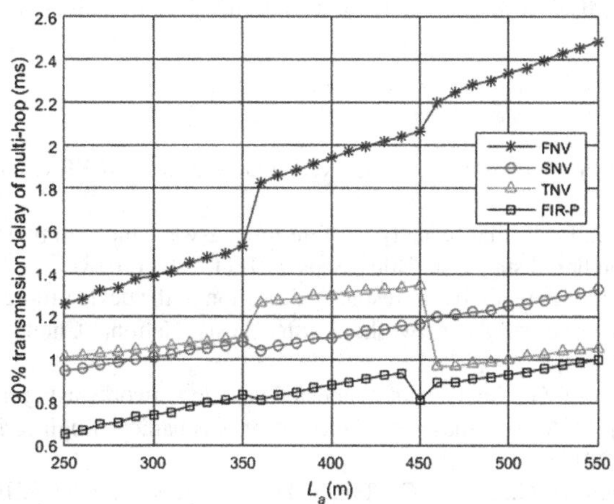

Fig. 7. Comparison of 90% TD for proposed routing schemes.

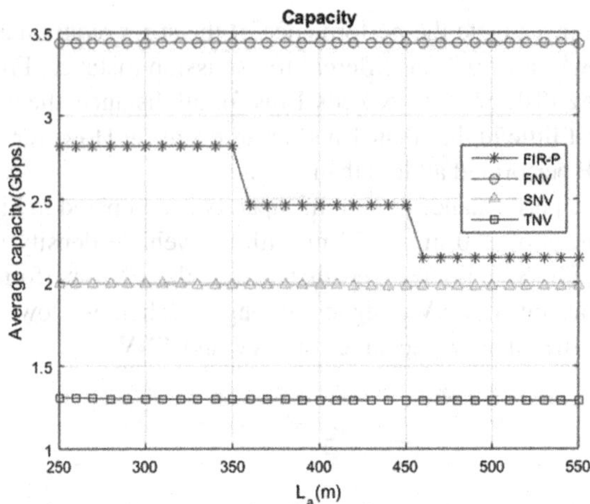

Fig. 8. Comparisons of transmission capacity of the proposed routing schemes in VN.

5 Conclusions

In this study, we applied the FIS to propose a fuzzy inference routing mechanism under different traffic vehicle densities and transmission distances between the road side unit (RSU). Simulation results show that the proposed FIR-P can outperform the multi-hop routing methods and satisfy the 90% transmission delay less than 1 ms for VN under different vehicle densities. Moreover, the proposed FIR-P not only reaches the requirements of 90% (Td) less than 1 ms but also perform the low latency with 90% (Td) less than 0.75 ms and high reliability with 96% (Td) less than 1 ms at transmission distance $L_a = 300$ m.

Acknowledgments. This work was funded in part by Ministry of Science and Technology of Taiwan under Grant MOST 108-2221-E-324-010 and MOST 108-2635-E-150-001.

References

1. Ge, X., Li, Z., Li, S.: 5G software defined vehicular networks. IEEE Commun. Mag. **55**(7), 87–93 (2017)
2. Jung, H., Lee, I.-H.: Performance analysis of millimeter-wave multi-hop machine-to-machine networks based on hop distance statistics. Sensors **18**(1), 204 (2018)
3. Lai, Y.-P.: The grouping methods for resource allocation in the device-to-device uplink underlaying millimeter-wave cellular networks. Master thesis, National Chung Cheng University, Chiayi, Taiwan (2018)
4. Avedisov, S.S., Orosz, G.: Analysis of connected vehicle networks using network-based perturbation techniques. Master thesis, Department of Mechanical Engineering, University of Michigan, USA (2017)
5. Huang, Y.-F., Chen, H.-C., He, B.-C., Tan, T.-H., Huang, S.-C., Chang, F.: Performance of transmission latency in software defined vehicle networks. In: Barolli, L., Leu, F.-Y., Enokido, T., Chen, H.-C. (eds.) BWCCA 2018. LNDECT, vol. 25, pp. 503–512. Springer, Cham (2019). https://doi.org/10.1007/978-3-030-02613-4_45

6. Zadeh, L.A.: Fuzzy sets. Inf. Control **8**, 338–353 (1965)
7. Li, S., Li, Z., Ge, X., Zhang, J., Jo, M.: Multi-hop links quality analysis of 5G enabled vehicular networks. In: Proceedings of 2017 9th International Conference on Wireless Communications and Signal Processing (WCSP), Nanjing, China (2017)
8. Huang, Y.-F.: Performance of adaptive multistage fuzzy-based partial parallel interference canceller for multi-carrier CDMA systems. IEICE Trans. Commun. **E88-B**(1), 134–140 (2005)

A kNN Based Position Prediction Method for SNS Places

Jong-Shin Chen[1], Huai-Yi Huang[1], and Chi-Yueh Hsu[2(✉)]

[1] Department of Information and Communication Engineering, Chaoyang University of Technology, Wufeng, Taichung 41349, Taiwan, R.O.C.
[2] Department of Leisure Services Management, Chaoyang University of Technology, Wufeng, Taichung 41349, Taiwan, R.O.C.
cyhsu@cyut.edu.tw

Abstract. With the growing popularity of Social Network Services (SNS), many researchers put effort into achieving some enhancements for these service. Systems like Facebook (FB), Google Maps, Twitter, Instagram, Foursquare, LinkedIn and so forth are the most acclaimed ones. These services generally contain a large number of geographical places, such as FB check-in places, Google Maps places, Foursquare check-in places. However, it is a very difficult to fast to do place positioning. Notably, place positioning indicates to find the specific geographical area where places are inside to this area. Machine learning (ML) is the scientific study of algorithms and statistical models that computer systems use to perform a specific task without using explicit instructions, relying on patterns and inference instead. With ML, the k-nearest neighbors (kNN) algorithm is a non-parametric method used for classification. Accordingly, in this study, we propose a kNN Based Position Prediction Method for SNS Places.

Keywords: Social Network Services · k nearest neighbors · Position Prediction

1 Introduction

A social networking service (also social networking site or social media) is an online platform which people use to build social networks or social relationship with other people who share similar personal or career interests, activities, backgrounds or real-life connections. With the growing popularity of SNS, many researchers put effort into achieving some enhancements for these service. Systems like Facebook (FB), Google Maps, Twitter, Instagram, Foursquare, LinkedIn and so forth are the most acclaimed ones. These services generally contain big data, such as FB check-in places, Google Maps places, Foursquare check-in places. However, it is very difficult to fast to do positioning for larger number of places.

Machine learning (ML) is the scientific study of algorithms and statistical models that computer systems use to perform a specific task without using explicit instructions, relying on patterns and inference instead. It is seen as a subset of artificial intelligence. Machine learning algorithms build a mathematical model based on sample data, known as "training data", in order to make predictions or decisions without being explicitly

N. T. Nguyen et al. (Eds.): ACIIDS 2020, LNAI 12034, pp. 266–273, 2020.
https://doi.org/10.1007/978-3-030-42058-1_22

programmed to perform the task. ML provides a quick (low time-consuming) positioning solution.

In pattern recognition, the k-nearest neighbors algorithm (kNN) is a non-parametric method used for classification and regression. In both cases, the input consists of the k closest training examples in the feature space. The output depends on whether kNN is used for classification or regression:

1. In kNN classification, the output is a class membership. An object is classified by a plurality vote of its neighbors, with the object being assigned to the class most common among its k nearest neighbors (k is a positive integer, typically small). If $k = 1$, then the object is simply assigned to the class of that single nearest neighbor.
2. In kNN regression, the output is the property value for the object. This value is the average of the values of k nearest neighbors. kNN is a type of instance-based learning, or lazy learning, where the function is only approximated locally and all computation is deferred until classification.

In light of above discussion, we first propose a method of place positioning. However, the time complexity is high. Therefore, we also uses the KNN classification in machine learning to predict the positions of places and study its accuracy.

2 Related Work

Due to the booming SNS, more and more researchers analyze data on social networking services, but geographic type analysis is rarely studied. [1, 5] uses Twitter data to detect hot topics on the SNS by using keyword frequencies. In addition, the analysis of the type of the ground also reflects the path and climate information of the local storm. The article [8, 9] also mentioned how to deal with the corresponding methods of natural disasters.

In addition to using the data on the SNS to detect climate conditions, [2] refers to the time and geographic location associated with a series of tweets, based on observations of people's movement behavior over time and the most recent visits. Position to predict the next position. The methods in [3, 6, 7, 10, 11] also allow users to get better travel tests through the geographic information application on the SNS. From the above examples, it can be found that the application of geographical location information is very extensive. Therefore, in this study, a large number of landmarks on the SNS were also used for positioning. In [12], the extension line is used to locate the landmark but the complexity is relatively high. Therefore, the kNN algorithm in machine learning is used to locate the landmark. Traditional kNN has some drawbacks. One of the problems is that the algorithm assigns equal weight to all neighbors. For this kind of defect, the weighted kNN is improved on the basis of the traditional kNN. For example, the closer the neighbor is, the greater the weight of the neighbor. In [4, 13–15], it is proposed that kNN uses the weighting method to improve the efficiency and accuracy.

3 Proposed Method

A geographical point contains a 2-dimensional coordinate, such as longitude and latitude. For convenience, a geographical point p, is defined as a 2-dimensional coordinate ($x(p)$, y(*p*)), as shown in Definition 1. A geographical area is a polygon in geometry. It is a flat shape consisting of straight, non-intersecting line segments are joined pair-wise to form a closed path. The definition of a polygon ensures the following properties:

1. A polygon encloses a region (called its interior) which always has a measurable area.
2. The line segments that make-up a polygon (called edges) meet only at their endpoints, called vertices.
3. Exactly two edges meet at each vertex.
4. The number of edges always equals the number of vertices.

Vertices of a polygon also are geographical points. In this study, we refer to SNS places as geographical points and also refer to a geographical area as a polygon, where P and A respectively represent the set of SNS places and the set of polygonal vertices. The definitions are shown in Definitions 2 and 3.

Definition 1. A geographical point p is defined as $p = (\mathrm{x}(p), \mathrm{y}(p))$, where $\mathrm{x}(p)$, $\mathrm{y}(p)$ are the values of 2-dimensional coordinate of p.

Definition 2. The set P of m SNS-places is defined as $P = \{p_0, p_1, \ldots, p_{m-1}\}$, where p_i is a geographical point for $i = 0, 1, \ldots, m - 1$.

Definition 3. A geographical area A is polygon with n different points (vertices) is defined as $A = \{a_0, a_1, \ldots, a_{n-1}, a_n\}$, where $a_n = a_0$ and n edges (a_0, a_1), (a_1, a_2),..., (a_{n-1}, a_n) enclose this region of A.

The proposed method includes a place positioning method and a Place Position Predicting method.

3.1 Place Positioning

Given a geographical point set P and a geographical area A, the problem of this study is to find the points of P inside A. In computational geometry, it is the point-in-polygon (PIP) problem. One simple way of finding whether a point p is inside or outside a simple polygon is to test how many times a ray, starting from point p and going in any fixed direction, intersects the edges of the polygon. If point p is on the outside of the polygon the ray will intersect its edge an even number of times. If point p is on the inside of the polygon then it will intersect the edge an odd number of times. Accordingly, the ray of point p is represented as ray(p) that starts from p and go in incremental x-axis direction. Let line l_1 be defined as two distinct points o_1 and o_2 and line l_2 be defined as two distinct points o_3 and o_4. Let ρ be the value, as shown in (1), related to l_1 and l_1. When the two lines are parallel or coincident, ρ is zero.

$$\rho = (x(o_1) - x(o_2)) \cdot (y(o_3) - y(o_4)) - (y(o_1) - y(o_2)) \cdot (x(o_3) - x(o_4)). \quad (1)$$

$$(x, y) = (x(o_1) + \kappa_1 \cdot (x(o_2) - x(o_1)), +\kappa_1 \cdot (y(o_2) - y(o_1))), \tag{2}$$

where $\kappa_1 := ((x(o_1) - x(o_3)) \cdot (y(o_3) - y(o_4)) - (y(o_1) - y(o_3)) \cdot (x(o_3) - x(o_4)))/\rho$;

$$(x, y) = (x(o_3) + \kappa_2 \cdot (x(o_4) - x(o_3)), y(o_3) + \kappa_2 \cdot (y(o_4) - y(o_3))), \tag{3}$$

where $\kappa_2 := ((x(o_1) - x(o_2)) \cdot (y(o_1) - y(o_3)) - (y(o_1) - y(o_2)) \cdot (x(o_1) - x(o_3)))/\rho$;

Otherwise, the intersection point (*x*, *y*) of the lines can be represented as (2) or (3) with the values k_1 or k_2, respectively. The intersection point falls within the first line segment if $0.0 \leq k_1 \leq 1.0$, and the intersection points falls within the second line segment if $0.0 \leq k_2 \leq 1.0$.

As shown in Fig. 1, algorithm EEI is an Edge-Edge Intersection algorithm, where inputs are two edges (o_1, o_2) and (o_3, o_4) and the output is value 0 or 1. If edge (o_1, o_2) and edge (o_3, o_4) are intersected, it results by returning 1. Otherwise, it results by returning 0. As shown in Fig. 2, algorithm PIA is a *P*lace *I*nside *A*rea algorithm, where inputs are a place *p* and an area *A* and output is value 0 or 1. If place *p* is inside to area *A*, it results by returning value 1. Otherwise, it results by returning value 0. Variable *count* counts the amount of intersections for ray(*p*) and edges in *A*. Lines 7 and 8 are executed *n* times. Each time an edge of this area is set as (o_1, o_2). Edge (o_3, o_4) is the segment of ray(*p*) that start from p to point with coordinate (max(x(o_1), x(o_2)), y(*p*)). If edge (o_1, o_2) and edge (o_3, o_4) are intersected, i.e., EEI(o_1, o_2, o_3, o_4) return 1, *count* increments by 1. Therefore, if *count* is odd, i.e., *count* % 2 = 1, place *p* is inside to area *A*. Otherwise, place *p* is non-inside to area *A*.

```
Algorithm EEI(o1, o2, o3, o4)
{
    ρ := (x(o1) – x(o2)) · (y(o3) – y(o4)) – (y(o1) – y(o2)) · (x(o3) – x(o4));
    if ρ = 0 then return 0;
    else
    {
        κ1 := ((x(o1) – x(o3)) · (y(o3) – y(o4)) – (y(o1) – y(o3)) · (x(o3) – x(o4))) / ρ;
        κ2 := – ( ((x(o1) – x(o2)) · (y(o1) – y(o3)) – (y(o1) – y(o2)) · (x(o1) – x(o3))) /ρ;
            if (κ1 ≥ 0 and κ1 ≤ 1) and (κ2 ≥ 0 and κ2 ≤ 1) then return 1;
            else return 0;
    }
}
```

Fig. 1. Edge-edge intersection algorithm

```
Algorithm PIA(p, A)
{
    count := 0;
    for i :=0 to n – 1 do
    {
        o1 := ai; o2 := ai+1;
        o3 := p; o4 :=(max(x(o1),x(o2)), y(p));
        if (EEI(o1, o2, o3, o4) == 1) then count := count + 1;
    }
    if (count % 2 == 1) then return 1;
    else return 0;
}
```

Fig. 2. Place inside area algorithm

3.2 Place Position Predicting

In machine learning and statistics, classification is the problem of identifying to which of a set of categories (sub-populations) a new observation belongs, on the basis of a training set of data containing observations (or instances) whose category membership is known. The kNN algorithm is a non-parametric method used for classification. The input consists of the k closest training examples in the feature space. The output is a class membership. An object is classified by a plurality vote of its neighbors, with the object being assigned to the class most common among its k nearest neighbors (k is a positive integer, typically small).

For each place p with a specific area A, the classification is p is inside A or p is outside A. Therefore, the classification α(p) is defined as (4)

$$\alpha(p) = \begin{cases} 1, & p\,is\,inside\,area\,A \\ -1, & p\,is\,outside\,area\,A \end{cases} \tag{4}$$

The feature p is indicated as 2-dimentional coordinate (x(p), y(p)) and the used distance metric for continuous variables is Euclidean distance. The distance of two points pi and pj is defined as (5).

$$\beta(p_i, p_j) = \sqrt{\left(\mathrm{x}(p_i) - \mathrm{x}(p_j)\right)^2 + \left(\mathrm{y}(p_i) - \mathrm{y}(p_j)\right)^2} \tag{5}$$

Let $p_1, p_2, \ldots, p_k$ be k places nearest to p. Non-weighted ($\gamma(p)$, as shown in (6) and weighted perdition ($\gamma_w(p)$), as shown in (7) methods are represented as, where if the value of $\gamma(p)$ (or $\gamma_w(p)$) is larger than 0, the perdition is p is inside A. Otherwise, p is outside A.

$$\gamma(p) = \sum_{i=1}^{k} \alpha(p_i) \tag{6}$$

$$\gamma_w(p) = \sum_{i=1}^{k} \beta(p, p_i)^{-1} \times \alpha(p_i) \tag{7}$$

4 Experimental Results

Facebook is an online social media and social networking service, which is most popular in the world. Facebook penetration rate in Taiwan is the highest in the world, until July 2015 in Taiwan, the number of daily users reached 13 million for approximately 23 million population. Check-in for places is a location-based function of Facebook. Facebook user can go to some famous scenic spots (Facebook places) or participate some activity and check in on the Facebook to show that the user participate some activities. Taipei officially known as Taipei City, is the capital and a special municipality of Taiwan. The city proper is home to an estimated population of 2,704,810 (2015) and The Taipei city limits cover an area of 271.7997 km^2 (104.9425 sq mi). Therefore, the experiment environment is Taipei city with Facebook check-in places. The geography is located in longitude 121.45° to 121.7° and latitude 24.95° to 25.25°.

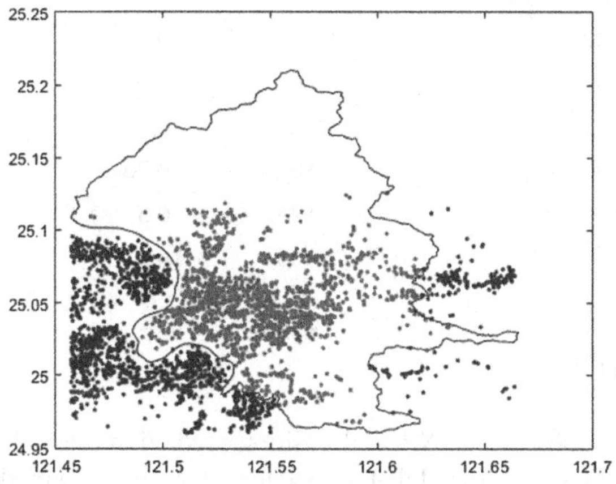

Fig. 3. Place distributions

In this environment, 93549 Facebook check-in places are collected. After positioning, there are 62144 places inside Taipei city and 31405 places outside Taipei city. In a random manner, there are *n* inside places and *n* outside places as the training category. Similarly, there are *n* inside places and *n* outside places as the test category. Moreover, *n* is assigned as 100, 200, 400, 800, and 1600. The place distribution is shown in Fig. 3, with k = 16.

For each prediction, a place *p* in the test category is selected. Then, the nearest *k* places of the training category are evaluated by Euclidean distance, where k is assigned as 3, 5, 7, and 11. The accuracy of the non-weighted method is shown in Table 1. The accuracy of both methods are very high. Moreover, the weighted method presents better accuracy than the non-weighted method.

Table 1. Accuracy of non-weighted and weighted prediction methods

k	*n*	Non-weighted (%)	Weighted (%)	Enhancement (%)
3	100	99.00	100	1.01
	200	99.25	99.75	0.50
	400	99.63	99.75	0.12
	800	99.38	99.44	0.06
	1600	99.63	99.69	0.06
5	100	98.00	99.50	1.53
	200	98.00	99.50	1.53
	400	99.25	99.75	0.50
	800	99.06	99.31	0.25
	1600	99.53	99.63	0.10
7	100	98.00	99.00	1.02
	200	98.00	99.50	1.53
	400	99.25	99.63	0.38
	800	98.88	99.31	0.43
	1600	99.38	99.63	0.25
11	100	98.00	99.00	1.02
	200	98.00	99.25	1.28
	400	98.75	99.50	0.76
	800	98.75	99.25	0.51
	1600	99.00	99.50	0.51

5 Conclusions

In this study, we proposed a kNN based position prediction method for SNS places. Experimental results demonstrated that our strategy is accurate to predict the positions of geographic places. To consider the hot-spot problems are our future work.

Acknowledgment. This research was partially supported by the Ministry Of Science and Technology, Taiwan (ROC), under contract no.: MOST 108-2410-H-324-007.

References

1. Twitter streaming data social topic detection and geographic clustering. In: 2013 IEEE/ACM International Conference on Advances in Social Networks Analysis and Mining (ASONAM), pp. 1215–1220. IEEE (2013)

2. Comito, C.: Where are you going? Next place prediction from Twitter. In: 2017 IEEE International Conference on Data Science and Advanced Analytics (DSAA), pp. 696–705. IEEE, October 2017
3. Rathnayake, W.P.: Google maps based travel planning & analyzing system (TPAS). In: 2018 International Conference on Current Trends towards Converging Technologies (ICCTCT), pp. 1–5. IEEE, March 2018
4. Huang, J., Wei, Y., Yi, J., Liu, M.: An improved kNN based on class contribution and feature weighting. In: 2018 10th International Conference on Measuring Technology and Mechatronics Automation (ICMTMA), pp. 313–316. IEEE, February 2018
5. Funayama, T., et al.: Disaster mitigation support system using web services and SNS information. In: 2015 13th International Conference on ICT and Knowledge Engineering (ICT & Knowledge Engineering 2015), pp. 42–45. IEEE, November 2015
6. Otsuka, T., Suzuki, R., Kawaguchi, S., Ito, T.: An implementation of a geolocation information-sharing system. In: 2011 IEEE International Conference on Service-Oriented Computing and Applications (SOCA), pp. 1–2. IEEE, December 2011
7. Yamamoto, K., Vařacha, P.: Sightseeing navigation system for foreign tourists in Japanese urban area. In: 2019 Smart City Symposium Prague (SCSP), pp. 1–6. IEEE, May 2019
8. Muramoto, T., Nakayama, K., Kobayashi, Y., Maekawa, M.: Resident participating GIS-based tsunami disaster control systems for local communities. In: 2006 International Symposium on Communications and Information Technologies, pp. 701–706. IEEE, October 2006
9. Ma, L., Chen, X., Xu, Y., Gao, Y., Liu, W.: Study on crowdsourcing-compatible disaster information management system based on GIS. In: 2014 International Conference on Information Science, Electronics and Electrical Engineering, vol. 3, pp. 1976–1979. IEEE, April 2014
10. Yamamoto, K., Ikeda, T.: Social recommendation GIS for urban tourist spots. In: 2016 IEEE 18th International Conference on High Performance Computing and Communications; IEEE 14th International Conference on Smart City; IEEE 2nd International Conference on Data Science and Systems (HPCC/SmartCity/DSS), pp. 50–57. IEEE, December 2016
11. Gou, Z., et al.: An interest-based tour planning tool by process mining from Twitter. In: 2016 IEEE 5th Global Conference on Consumer Electronics, pp. 1–4. IEEE, October 2016
12. Chang, S.C., Huang, H.Y.: An efficient geographical place mining strategy for social networking services. In: 2019 IEEE International Conference on Consumer Electronics-Taiwan (ICCE-TW). IEEE, May 2019
13. Xu, S., Luo, Q., Li, H., Zhang, L.: Time series classification based on attributes weighted sample reducing KNN. In: 2009 Second International Symposium on Electronic Commerce and Security, vol. 2, pp. 194–199. IEEE, May 2009
14. Sanei, F., Harifi, A., Golzari, S.: Improving the precision of KNN classifier using nonlinear weighting method based on the spline interpolation. In: 2017 7th International Conference on Computer and Knowledge Engineering (ICCKE), pp. 289–292. IEEE, October 2017
15. Liu, S., Zhu, P., Qin, S.: An improved weighted KNN algorithm for imbalanced data classification. In: 2018 IEEE 4th International Conference on Computer and Communications (ICCC), pp. 1814–1819. IEEE, December 2018
16. Heckbert, P.S.: Graphics Gems IV, pp. 7–15. Morgan Kaufmann (Academic Press), London (1994)
17. Roth, Scott D.: Ray casting for modeling solids. Comput. Graph. Image Process. **18**(2), 109–144 (1982)
18. Antonio, F.: Faster line segment intersection. In: Graphics Gems III (IBM Version), pp. 199–202. Morgan Kaufmann (1992)
19. Altman, N.S.: An introduction to kernel and nearest-neighbor nonparametric regression. Am. Stat. **46**(3), 175–185 (1992)

Performance of Particle Swarm Optimization on Subcarrier Allocation for Device-to-Device Multicasting in Wireless Communication Systems

Yung-Fa Huang[1], Tan-Hsu Tan[2], Chuan-Bi Lin[1], Yung-Hoh Sheu[3](✉), Ching-Mu Chen[4](✉), and Song-Ping Liu[2]

[1] Chaoyang University of Technology, Taichung 413, Taiwan
yfahuang@cyut.edu.tw

[2] National Taipei University of Technology, Taipei 106, Taiwan
thtan@ntut.edu.tw

[3] National Formosa University, Yunlin 632, Taiwan
yhsheu@nfu.edu.tw

[4] Chung Chou University of Science and Technology, Changhua 510, Taiwan
925633@gmail.com

Abstract. In this study, we propose Particle Swarm Optimization (PSO) algorithms for Device-to-Device multicasting (D2DM) subcarrier allocation. The multimedia services provided by 5G communication are required to meet their own Quality of Service (QoS). In addition, it is well known that the Orthogonal Frequency Division Multiple Access (OFDMA) can improve the transmission efficiency. Therefore, this research aims to enhance the effectiveness of D2D multicast resource allocation in OFDMA system by using PSO while maintaining the QoS. Both the QoS threshold and transmission threshold are employed to be the basis for allocation of subcarriers. Simulation results show that the proposed Linear PSO (LPSO) obtains the best performance on subcarrier allocation for the CeUEs with QoS guarantee, and thus achieved higher D2DMs throughput performance than original PSO (OPSO) and traditional heuristic methods.

Keywords: Device-to-Device communications · System throughput · Subcarrier allocation · Particle swarm optimization algorithms · QoS performance

1 Introduction

With the rapid development of wireless broadband networks and mobile communication technologies, various mobile devices have become widespread. Users have become accustomed to handling various aspects of life, such as work, communication, and entertainment through mobile devices. However, data and control signals in the prior art must be transmitted through the core network, which will cause the problem of core network congestion. Direct communication technology (Base Station) through base stations has

N. T. Nguyen et al. (Eds.): ACIIDS 2020, LNAI 12034, pp. 274–284, 2020.
https://doi.org/10.1007/978-3-030-42058-1_23

gradually received attention [1]. In recent years, the IMT-Advanced research plan promoted by the International Telecommunication Union has provided a reference path for future 5G development. Next, the development of 5G technology will continue the 4G issue, and a new generation of device-to-device (D2D) will become a key technology for 3GPP development. D2D is regarded as an important technical standard for mobile communications technology [1], due to that there is no need to rely on Global Positioning System (GPS) and operator's data network.

The challenges of D2D technology development mainly include how to conduct effective proximity device search, inter-device interference control, group communication, mode selection and switching, power allocation and energy management, resource allocation, etc. Two mechanisms are proposed to judge the appropriate timing of D2D switching in the LTE-A environment [2, 3]. The first is to compare the capacity of Cellular and D2D based on the perspective of capacity, which is called Capacity-based. D2D. The second type further adds load to the judgment criterion, and derives Delay-based D2D. Based on the simulation results of Intra-eNB [3], it shows that using D2D communication can reduce the transmission delay compared to the traditional transmission type, and can increase the transmission capacity by 11.7%–13.6%, and also save the resource usage.

Various broadband multimedia services provided by the 5G mobile communications, such as voice, web browsing, video conferencing, etc., must meet their respective Quality of Service (QoS). The multicast communication is a very important and challenging application target in wireless multimedia networks, such as: audio/video editing, mobile TV, interactive games. It contains two key types of traffic, including Unicast Traffics and Multicast Traffics. Among them, multicast communication has more communication transmission efficiency, and Dynamic Resource Allocation is an effective way to improve the system spectral efficiency [4].

Moreover, higher D2DM throughput is realized to further optimize the benefit of subcarrier allocation [3]. The throughput efficiency of Orthogonal Frequency Division Multiple Access (OFDMA) [5] technology need to be improved to become a part of 5G mobile communication systems. While using Particle Swarm Optimization (PSO) [6] to find the best solution to optimize the allocation of subcarriers, thus the throughput of the entire D2DM is further improved.

Many optimization algorithms, including PSO and Genetic Algorithms (GA) [7], have been widely used in different fields, such as maximum power problems [8], personnel. Scheduling problems [9], Job Shop Problem (JSP) [10], and communication system related issues [11, 12], etc. In 1995, Eberhart and Kennedy proposed the original PSO (Original PSO, OPSO) to simulate the behavior of bird foraging [13]. The particles were randomly generated in the solution space. The particles will move according to the best experience of the individual and the best experience of the group to find the best solution. It has the advantages of fast convergence, less parameter setting, and suitable for dynamic environments. Then Eberhart and Shi proposed the Linearly decreasing inertia weight PSO (LPSO) [13] to improve the convergence characteristics and accuracy of PSO. Therefore, this paper proposes the PSO-based subcarrier allocation schemes for the D2D communication to improve the system throughput and the QoS performance of CeUEs.

2 System Models

This paper discusses the subcarrier allocation problem for hybrid systems with D2D and OFDMA systems, as shown in Fig. 1 investigated in previous works [14]. It is assumed that the channel is a Rayleigh distribution, the relationship between path loss and distance can be written as

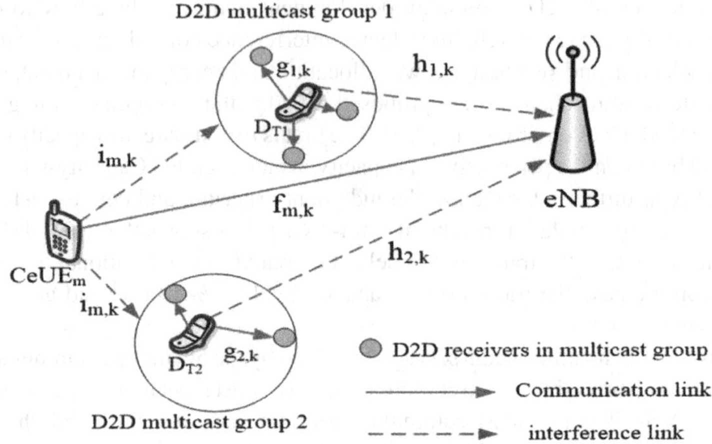

Fig. 1. System model of hybrid systems [14].

$$f_{m,k} = d^{-\alpha}\mu \tag{1}$$

where α is the path-loss exponent, d is the distance between the handset user and the BS, μ is the Independent Unit-Mean Exponential Random Variable, the average value of which is 1.

The ratio of Channel Signal to Interference-Plus-Noise (CSINR) of D2DUEs can be obtained by

$$C_{j(n,k)} = \frac{g_{j(n,k)}}{N_0 B_0 + T_{m,k} i_{m,j(n,k)}} \tag{2}$$

where $g_{j(n,k)}$ is the channel gain of D_T to the jth D_R $(j = 1, 2, \ldots, 30)$ in the nth D2DM group on the k-th subcarrier; $i_{m,j(n,k)}$ is the interference gain on the jth D_R inn the nth D2DM caused by the mth CeUE of the jth cell on the kth subcarrier.

The Sum of Throughput of the k-th subcarrier in the n-th D2DM can be written as

$$R_{n,k} = B_0 X_{j(n,k)} \ln(1 + p_{n,k} C_{j(n,k)}) \tag{3}$$

Then, the maximum total throughput of N D2DMs can be obtained by

$$R_{sum} = B_0 \sum_{k=1}^{K} \sum_{n=1}^{N} X_{j(n,k)} \delta_{n,k} \ln(1 + p_{n,k} C_{j(n,k)}) \tag{4}$$

In order to ensure fairness between D2DMs, we define the minimum throughput threshold of D2DM. Thus, the maximization problem can be expressed by

$$\max B_0 \sum_{k=1}^{K} \sum_{n=1}^{N} \sum_{j=1}^{J} X_{j(n,k)} \delta_{n,k} \ln(1 + p_{n,k} C_{j(n,k)}) \tag{5}$$

where j is the jth D_R, $j = 1, 2, \ldots, J$, of the nth D2DM. Then the constraints for the optimization problems are expressed as the same as the previous works in [14] as

$$s.t \begin{cases} \sum_{k=1}^{K} \delta_{n,k} p_{n,k} = P_{\max}, \ \mathrm{n} \in \mathrm{N} & (6a) \\ p_{n,k} \geq 0, \ \mathrm{n} \in \mathrm{N}, \mathrm{k} \in \kappa & (6b) \\ \frac{f_k T_k}{N_0 B_0 + \sum_{n=1}^{N} \delta_{n,k} p_{n,k} h_{n,k}} \geq \gamma_{th}, \ \ \mathrm{k} \in \kappa & (6c) \\ \sum_{k=1}^{K} \delta_{n,k} = 1, \ \mathrm{n} \in \mathrm{N} & (6d) \\ 0 < X_{j(n,k)} \leq X_n & (6e) \\ \sum_{k=1}^{K} R_{n,k} \delta_{n,k} \geq R_{th} & (6f) \end{cases} \tag{6}$$

3 Performance of PSO for D2DMs

In this section, the PSO algorithms are applied to the subcarrier allocation optimization problem. The K subcarriers of the system spectrum are treated as particles. The particles are randomly and evenly distributed in the solution space of the dimension D_T. The movement of each particle V_i in the particle group The speed is also randomly and evenly distributed. For the number of particle populations, each particle tends to the optimal solution with an iterative procedure. The updating equations of the position and velocity of particles in the OPSO algorithm can be expressed by

$$\mathbf{v}_i^{g+1} = \mathbf{v}_i^g + c_1 \times rand() \times (\mathbf{s}_i^{pbest} - \mathbf{s}_i^g) + c_2 \times rand() \times (\mathbf{s}^{gbest} - \mathbf{s}_i^g) \tag{7}$$

and

$$\mathbf{s}_i^{g+1} = \mathbf{s}_i^g + \mathbf{v}_i^{g+1}, \tag{8}$$

where $\mathbf{v}_i^g = [v_{i,1}^g, v_{i,2}^g, \ldots, v_{i,K}^g]$ and $\mathbf{s}_i^g = [s_{i,1}^g, s_{i,2}^g, \ldots, s_{i,K}^g]$ are the velocity and position of the i-th particle (subcarrier) in the g-th generation, respectively. The best position $\mathbf{s}_i^{pbest} = [s_{i,1}^{pbest}, s_{i,2}^{pbest}, \ldots, s_{i,K}^{pbest}]$ is found for each particle (subcarrier) until now; $\mathbf{s}^{gbest} = [s_1^g, s_2^g, \ldots, s_K^g]$ is the optimal position of the subcarriers of the particles found in the groups for the g-th generation. The learning factors c_1 and c_2 are called acceleration coefficients, and also known as the individual factor and the social factor, respectively. In the study they are set to $c_1 = 2$ and $c_2 = 2$. The random variable $rand()$

is uniformly in [0,1]. $\mathbf{v}_i^{g+1}$ and $\mathbf{s}_i^{g+1}$ are the speed and position of the i-th particle in the (g + 1)th generation after updating. Because the velocity and position of the particle (subcarrier) should be an integer solution, the value $\mathbf{v}_i^{g+1}$ and $\mathbf{s}_i^{g+1}$ will be rounded up.

In order to improve the convergence speed of PSO, Eberhart and Shi add the Inertia Weight (IW) parameter w to the PSO [13] to pursue the balance of global and regional search. The updating equations can be obtained by

$$\mathbf{v}_i^{g+1} = w \times \mathbf{v}_i^g + c_1 \times rand() \times (\mathbf{s}_i^{pbest} - \mathbf{s}_i^g) + c_2 \times rand() \times (\mathbf{s}^{gbest} - \mathbf{s}_i^g) \quad (9)$$

and

$$\mathbf{s}_i^{g+1} = \mathbf{s}_i^g + \mathbf{v}_i^{g+1}, \quad (10)$$

respectively. where the inertia weight value, w can be used to adjust the particle search speed. The following are two ways to set the inertia weight value:

(1) Take the fixed weight value, the experiment shows that the inertia weight value will have a good search ability between 0.9–1.2 [13].
(2) Set the weight to a function that decreases as the number of generations increases. The effect is better than the former. The adaptive weighting [12] is expressed by

$$w = w_{\max} - \frac{w_{\max} - w_{\min}}{G} \times g, \quad (11)$$

where w is the Inertia Weight, $w_{\max}$ is the initial maximum inertia weight value, $w_{\min}$ is the initial minimum inertia weight value, G is the total number of generations, and g is the current iteration number. This method allows the particle to search for the optimal solution extensively. As the number of generations increases, the search range is gradually reduced, and the particle can be avoided from the optimal solution. To avoid finding the best solution, you must develop a particle search and speed range [13]. Assume that the optimization problem searches for the subcarrier range ($\mathbf{s}_{\min} = 1$, $s_{\max} = 32$), and the particle swarm speed range is ($-\mathbf{v}_{\max} = -16$, $\mathbf{v}_{\max} = 16$). If the updated particle position $\mathbf{s}_i^{g+1}$ exceeds the search range ($\mathbf{s}_{\min} = 1$, $s_{\max} = 32$), that is $\mathbf{s}_i^{g+1} \geq \mathbf{s}_{\max}$, then it changes to $\mathbf{s}_i^{g+1} = 32$. And if $\mathbf{s}_i^{g+1} \leq \mathbf{s}_{\min}$, it changes to $\mathbf{s}_{\min} = 1$. On the other hand, if the updated particle speed $\mathbf{V}_i^{g+1}$ exceeds the speed range ($-\mathbf{v}_{\max} = -16$, $\mathbf{v}_{\max} = 16$), that is $\mathbf{v}_i^{g+1} \geq \mathbf{v}_{\max}$, then it is changed to $\mathbf{v}_i^{g+1} = 16$. On the other hand, if $\mathbf{v}_i^{g+1} \leq \mathbf{v}_{\min}$, it changes to $\mathbf{v}_i^{g+1} = -16$

To exhibit the difference between populations, the definition of populations diversity can be defined by

$$Pop\_Div(g) = \frac{\sum_{i=1}^{M} (\mathbf{s}_i^g - \overline{\mathbf{s}^g})^2}{M}, \ \overline{\mathbf{s}^g} = \frac{\sum_{i=1}^{M} \mathbf{s}_i^g}{M}, \quad (12)$$

where g is the number of generations. M is the number of particle populations, $\mathbf{s}_i^g$ is the position of the subcarrier is selected for the ith particle in the gth generation. $\overline{\mathbf{s}^g}$ is the average of the subcarrier positions of the gth generation in M groups. In this study, the reciprocal of the total amount of D2DM throughput is taken as the particle cost function $Cost(\mathbf{s}_i)$ by

$$Cost(\mathbf{s}_i) = \frac{1}{R_{sum}} \tag{13}$$

The higher the diversity, the more scattered the particle distribution of PSO; the more concentrated the particles. This study sets the D2DM user distribution range from a radius of 20 m to a radius of 60 m. The average Sum Throughput for each generation will be used as a performance indicator. In addition, this study explores the characteristics of PSO search and convergence based on changes in populations diversity curves, and compares the advantages and disadvantages of various subcarrier allocation methods.

In the PSO execution process, we gave the initial parameters, including total number of generations G, number of populations M, maximum inertia weight $w_{\max}$, minimum inertia weight $w_{\min}$, learning factors c_1 and c_2, and. The steps are as follows:

(1) In the solution space, the initial position and initial velocity of the particle group are randomly generated.
(2) We set the reciprocal of the system throughput as the cost function, calculated the cost value of all particles, and found the best position of the individual *pbest* and the best position of the group *gbest*.
(3) The values of particles, *pbest* and *gbest* are substituting in (9) and (10). The inertia weight value is substituted into (11). Then the velocity $\mathbf{V}_i^{g+1}$ and position $\mathbf{S}_i^{g+1}$ of the particle in the $(g + 1)$-th generation are obtained.
(4) Calculate the cost of all the particles in the (g + 1)-th generation and find the best position of the individual *pbest* and the best position of the group *gbest* in the (g + 1)-th generation.
(5) Compare the cost values between the g-th generation and the (g + 1)-th generation. If the cost value of the (g + 1)-th generation is better, the best position of the individual *pbest* and the best position of the group *gbest* of the g-th generation are replaced by the values of the (g + 1)-th generation.
(6) If the number of generations reached the total number of generations G, the process is stopped. Else, return to step (2).

4 Simulation Results

In this study, both the OPSO and the LPSO are applied to perform the subcarrier allocation. The simulation parameters listed the Table 1 are performed for PSOs. The system throughput in each generation is used as performance index. The performance of PSOs are investigated based on the diversity and various distribution of D2DM with the radius of [20, 60] m.

At first we let the D2DM radius by 40 m. The performance of OPSO is investigated with different populations M and $G = 100$ generations. We set the number of subcarriers $K = 30$ and the threshold value $\gamma_{th} = 3$ dB for the Rayleigh fading channels. When the

Table 1. Simulation parameters of PSOs.

Parameters	Value
Number of generations (G)	100
Number of populations size of particles, M	5
Maximum of inertia weight, $w_{\max}$ for LPSO	1.2
Minimum of inertia weight, $w_{\min}$ for LPSO	0.9
Inertia weight for OPSO	1
Learning factors, c_1, c_2	2
Number of Dimensions, $D = D_T$	2, 3, 4
Number of particles (subcarriers)	32
Range of subcarriers searching	$\mathbf{s}_{\min} = 1,\ \mathbf{s}_{\max} = 32$
Range of speed of particles	$(-\mathbf{v}_{\max} = -16,\ \mathbf{v}_{\max} = 16)$

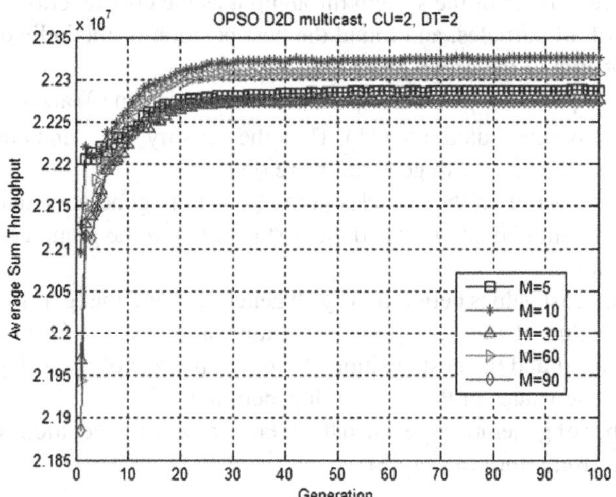

Fig. 2. Total throughput in the generations for OPSO with different M.

searching diversity is higher, the particle distribution of the OPSO is more disordered; on the contrary, the particles are more concentrated. The convergence comparisons are compared as shown in Figs. 2 and 3.

In Figs. 4 and 5, the convergence performance is shown for OPSO and LPSO, respectively, with $M = 5$. The D2DM distribution was set from a radius of 20 m to a radius of 60 m. The characteristics of PSO particle convergence were discussed based on the average Sum Throughput of each generation. From Figs. 4 and 5, it is observed that

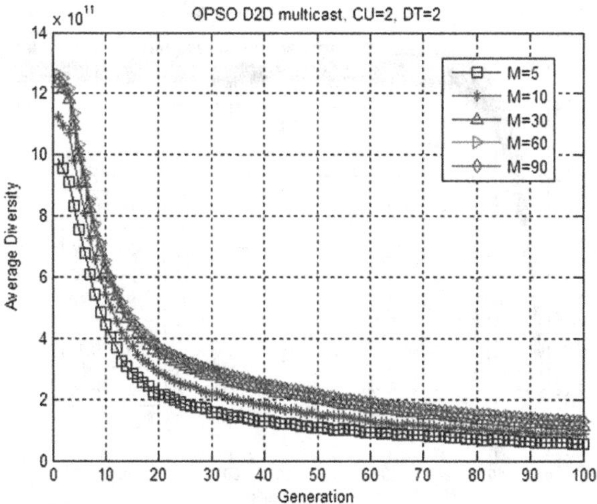

Fig. 3. The diversity of OPSO for number of populations *M*.

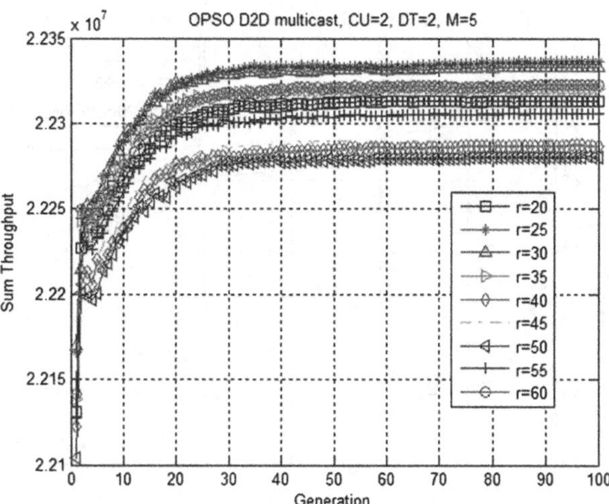

Fig. 4. The convergence curves of OPSO.

OPSO converges around the 30th generation and LPSO particles converge around the 20th generation.

Moreover, we further evaluate the throughput performance for the subcarrier allocation methods with $\gamma_{th} = 3\,\text{dB}$, and $M = 5$. Figure 6 shows the system throughput comparisons for PSO-based and traditional subcarrier allocation method. With QoS threshold $\gamma_{\text{th}} = 3$ dB of the mobile phone user, it is observed that the PSO-based schemes achieve higher throughput than the traditional threshold value $R_{th}= 10^6$ bps. When the number of CeUEs increases, the interference increases from mobile users affects D2DM users and then the throughput drops significantly. Moreover, as compared with the traditional

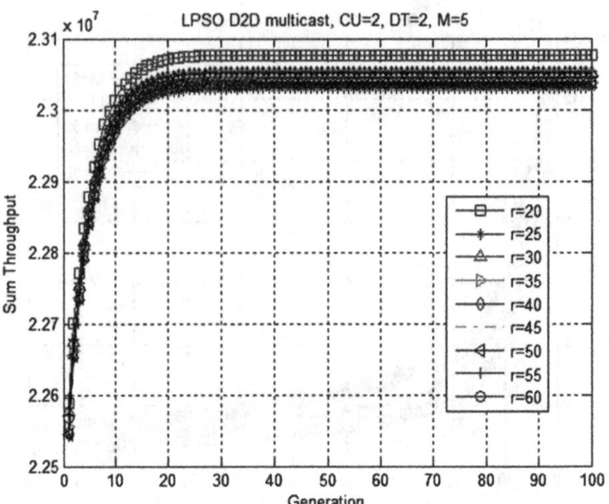

Fig. 5. The convergence curves of LPSO.

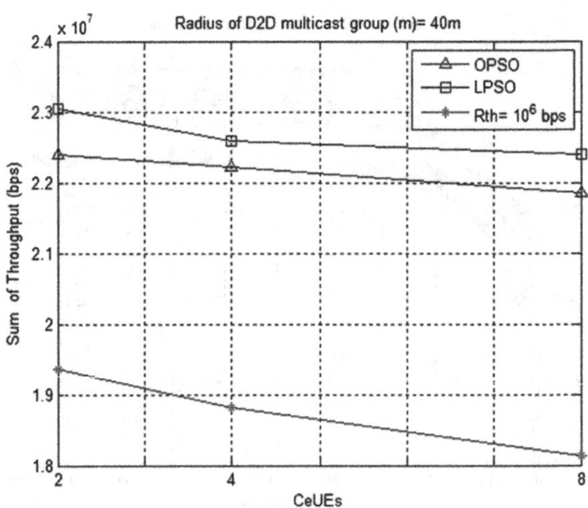

Fig. 6. The system throughput comparisons on PSOs with D2DMs radius of 40 m.

threshold value $R_{th} = 10^6$ bps, the proposed LPSO can select the preferred subcarrier, and obtain the higher throughput.

In the QoS performance, the proposed LPSO allocates subcarriers to set the minimum SINR acceptable to mobile phone users. Then, the throughput under three different thresholds, $\gamma_{\rm th} = 3$ dB, 10 dB and 15 dB are compared as shown in Fig. 7 based on the D2DMs radius of 40 m.

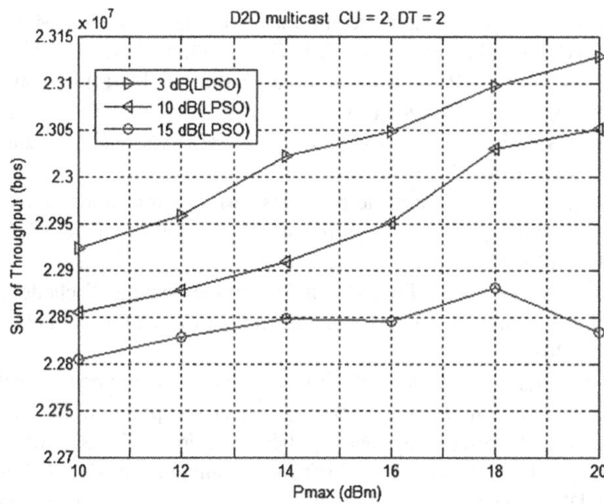

Fig. 7. The performance of different γ_{th} on system throughput with the proposed LPSO.

5 Conclusions

In order to effectively improve the D2DM throughput efficiency and subcarrier usage efficiency, this paper applies the Original Particle Swarm Optimization (OPSO) and the Linearly decreasing inertia weight PSO (LPSO) to the OFDMA system to improve the QoS guarantee. The benefits of D2D multicast resource allocation, and experimentally proved that the system throughput is better than the previous method, and the improvement range is not equal. In addition, the proposed LPSO obtains better subcarrier allocation for the CeUEs with QoS guarantee, and thus achieved higher D2DMs throughput performance than OPSO and traditional heuristic methods.

Acknowledgments. This work was funded in part by Ministry of Science and Technology of Taiwan under Grant MOST 108-2221-E-324-010.

References

1. Camps-Mur, D., Garcia-Saavedra, A., Serrano, P.: Device-to-device communications with wi-fi direct: overview and experimentation. IEEE Wirel. Commun. **20**(3), 96–104 (2013)
2. Liu, J., Chen, W., Cao, Z., Lee, K.B.: Dynamic power and sub-carrier allocation for OFDMA-based wireless multicast systems. In: IEEE International Conference on Communications, pp. 2607–2611, May 2008
3. Wang, D., Wang, X., Zhao, Y.: An interference coordination scheme for device-to-device multicast in cellular networks. In: IEEE Vehicular Technology Conference, pp. 1–5, September 2012
4. Peng, B., Hu, C., Peng, T., Yang, Y., Wang, W.: A resource allocation scheme for D2D multicast with QoS protection. In: OFDMA-based Systems. In: Proceedings of the IEEE 24th International Symposium on, Personal Indoor and Mobile Radio Communications (PIMRC), pp. 2383–2387, September 2013

5. Wei, L., Schlegel, C.: Synchronization requirements for multi-user OFDM on satellite mobile and two-path Rayleigh fading channels. IEEE Trans. Commun. **43**, 887–895 (1995)
6. Eberhart, R.C., Kennedy, J.: Particle swarm optimization. In: IEEE International Conference on Neural Networks, pp. 1942–1948 (1995)
7. Tang, K.S., Man, K.F., Kwong, S., He, Q.: Genetic algorithms and their applications. IEEE Signal Process. Mag. **13**(6), 22–37 (1996)
8. Ishaque, K., Salam, Z.: A deterministic particle swarm optimization maximum power point tracker for photovoltaic system under partial shading condition. IEEE Trans. Ind. Electron. **60**(8), 3195–3206 (2013)
9. Adamuthe, A.C., Bichkar, R.S.: Tabu search for solving personnel scheduling problem. In: IEEE International Conference on Communication, Information Computing Technology, pp. 1–6, October 2012
10. Wang, F., Ma, J., Song, D., Liu, W., Lu, X.: Research on repair operators in the whole space search genetic algorithm of assembly job shop scheduling problem. In: IEEE International Conference on Industrial Electronics and Applications, pp. 1922–1927, July 2012
11. Tan, T.H., Huang, Y.F., Hsu, L.C., Wu, C.H.: Joint channel estimation and multi-user detection for MC-CDMA system using genetic algorithm and simulated annealing. In: IEEE International Conference on Systems, Man, and Cybernetics, pp. 249–256, October 2010
12. Tan, T.H., Huang, Y.F., Hsu, L.C., Lin, K.R.: Carrier frequency offsets estimation for uplink OFDMA systems using enhanced PSO and multiple access interference cancellation. In: IEEE International Conference on Systems, Man, and Cybernetics, pp. 1436–1441, October 2011
13. Shi, Y., Eberhart, R.: A modified particle swarm optimizer. In: IEEE International Conference on Evolutionary Computation, pp. 69–73 (1998)
14. Huang, Y.-F., Tan, T.-H., Liu, S.P., Liu, T.-Y., Chen, C.-M.: Performance of subcarrier allocation of D2D multicasting for wireless communication systems. In: Proceedings of 2018 International Conference on Advanced Computational Intelligence (ICACI 2018), Xiamen, Fujian, China, pp. 193–196, 29–31 March 2018
15. IEEE 802.11, IEEE Wireless LAN Medium Access Control (MAC) and Physical Layer (PHY) Specifications, August 1999

Comparative Analysis of Restricted Boltzmann Machine Models for Image Classification

Christine Dewi[1,2], Rung-Ching Chen[1(✉)], Hendry[2], and Hsiu-Te Hung[1]

[1] Department of Information Management, Chaoyang University of Technology, Taichung, Taiwan, R.O.C.
{s10714904,crching}@cyut.edu.tw, a0970783520@gmail.com

[2] Faculty of Information Technology, Satya Wacana Christian University, Salatiga, Indonesia
hendry@uksw.edu

Abstract. Many applications for Restricted Boltzmann Machines (RBM) have been developed for a large variety of learning problems. Recent developments have demonstrated the capacity of RBM to be powerful generative models, able to extract useful features from input data or construct deep artificial neural networks. In this work, we propose a learning algorithm to find the optimal model complexity for the RBM by improving the hidden layer. We compare the classification performance of regular RBM use *RBM()* function, classification RBM use *stackRBM()* function and Deep Belief Network (DBN) use *DBN()* function with different hidden layer. As a result, Stacking RBM and DBN could improve our classification performance compare to regular RBM.

Keywords: Classification comparison · DBN · RBM · Stack-RBM

1 Introduction

Deep learning has gained its popularity recently as a huge probabilistic models and way of learning complex. Deep neural networks are characterized by the large number of layers of neurons and by using layer-wise unsupervised pre-training to learn a probabilistic model for the data. A deep neural network is typically constructed by stacking multiple Restricted Boltzmann Machines (RBM) so that the hidden layer of one RBM becomes the visible layer of another RBM. Layer-wise pre-training of RBM then facilitates finding a more accurate model for the data. RBM have been particularly successful in classification problems either as feature extractors for text and image data [1] or as a good initial training phase for deep neural network classifiers [2]. However, in both cases, the RBMs are merely the first step of another learning algorithm, either providing a preprocessing of the data or an initialization for the parameters of a neural network.

The main contributions of this work can be summarized as follows: First, we propose a learning algorithm to find the optimal model complexity for the RBM by improving the hidden layer. Second, we will compare the classification performance of regular RBM use *RBM()* function, classification RBM use *stackRBM()* function and Deep Belief Network (DBN) use *DBN()* function with different hidden layer. The rest of the paper is organized

N. T. Nguyen et al. (Eds.): ACIIDS 2020, LNAI 12034, pp. 285–296, 2020.
https://doi.org/10.1007/978-3-030-42058-1_24

as follows. In Sect. 2, we describe brief explanation about the RBM. Section 3 describes the proposed experimental improvement for RBM. In Sect. 4, we present experimental results, and finally, Sect. 5 concludes this paper and suggests a future work.

2 Related Work

2.1 Restricted Boltzmann Machines (RBM)

RBM are undirected graphs and graphical models belonging to the family of Boltzmann machines, they are used as generative data models [3]. RBM can be used for data reduction and can also be adjusted for classification purposes [4]. They consist of only two layers of nodes, namely, a hidden layer with hidden nodes and a visible layer consisting of nodes that represent the data. The discriminate RBM was proposed by Larochelle [4, 5], which uses class information as visible input, so that RBM can provide a self-contained framework for deriving a non-liner classifier. The discriminate RBM model the joint distribution of the inputs and associated target classes, whose graphical model is illustrated in Fig. 1 [5].

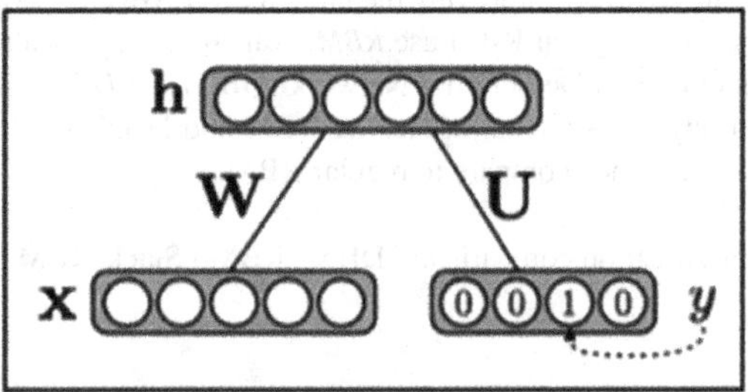

Fig. 1. Discriminative RBM [4, 5]

RBM consists of visible units *v*, binary hidden unit's *h* and symmetric connections between visible units and hidden units. The connections are represented by a weight matrix *W*. RBM uses the energy function for the probabilistic semantics. The energy function is described as follow: [6, 7, 12].

$$E(v, \boldsymbol{h}) = -\sum_i \sum_j v_i w_{ij} h_j - \sum_j b_j h_j - \sum_j c_i v_i \tag{1}$$

where b_j are biases of hidden units and c_i are biases of visible units. This energy function is used to configure a probability model for RBM. *W* is the weight matrix, *v* and *h* represent the visible and hidden layers. a and b are the bias of the visible and hidden layers. When the visible unit state is determined, each hidden element activation state is conditional independent to others. The jth hidden element activation probability is denied as following: [7, 13].

$$P(h_j = 1|v, \theta) = \sigma(b_j + \sum_i v_i w_{ij}) \tag{2}$$

When the hidden element state is determined, the activation state of each visible element is also independent of each other. The probability of ith visible unit is defined as following: [7, 12, 13].

$$P(v_j = 1|h, \theta) = \sigma(a_i + \sum\nolimits_i w_{ij} h_j) \tag{3}$$

2.2 Stack RBM

In general, stacking RBM is only used as a greedy pre-training method for training a Deep Belief Network as the top layers of a stacked RBM have no influence on the lower level model weights. However, this model should still learn more complex features than a regular RBM. We stack some layers of RBM with the stackRBM function, this function calls the RBM function for training each layer and so the arguments are not much different, except for the added layers argument. With the layers' argument we can define how many RBM you want to stack and how many hidden nodes each hidden layer should have. The stack RBM architecture is showed in Fig. 2.

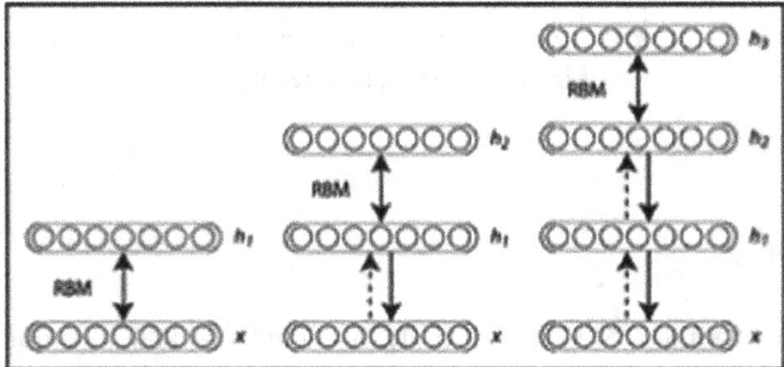

Fig. 2. The stack RBM architecture

2.3 Deep Belief Network (DBN)

Deep Belief Network (DBN), as shown in Fig. 3, is a deep architecture built upon RBM to increase its representation power by increasing depth. In a DBN, two adjacent layers are connected in the same way as in RBM. The network is trained in a greedy, layer-by-layer manner [6], where the bottom layer is trained alone as an RBM, and then fixed to train the next layer. DBN was originally developed by Hinton et al. [8] and was originally trained with the sleep-wake algorithm, without pre-training. However, in 2006 Hinton et al. found a method that is more efficient at training DBNs by first training a stacked RBM and then use these parameters as good starting parameters for training the DBN [9]. The DBN then adds a layer of labels at the end of the model and uses either back propagation or the sleep-wake algorithm to fine tune the system with the labels as the criterion. The *DBN()* function in the RBM package uses the backpropagation algorithm. The backpropagation algorithm works as follows: (1) first a feedforward pass is made through all the hidden layers ending at the output layer (2) then the output is compared to the actual label and (3) the error is used to adjust the weights in all the layers by going back through the whole system. This process is repeated until some stopping criterion is reached, in the *DBN()* function that is the maximum number of epochs but it could also be the prediction error on a validation set.

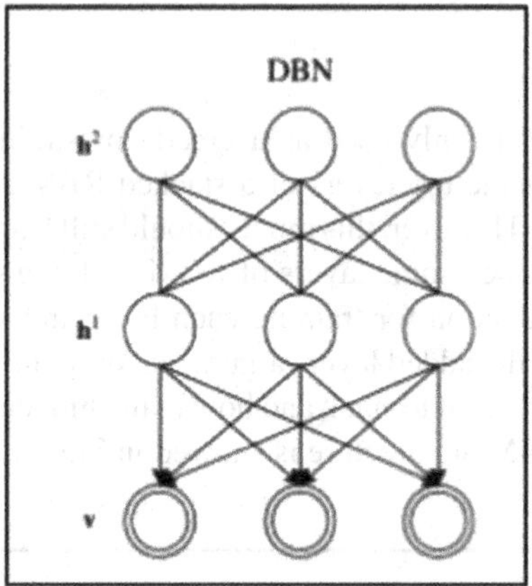

Fig. 3. The DBN architecture.

3 Methodology

This paper use Modified National Institute of Standards and Technology database (MNIST dataset) is a large database of handwritten digits that is commonly used for training various image processing systems. The database is also widely used for training and testing in the field of machine learning [10, 14, 15]. The MNIST database of handwritten digits, has a training set of 60,000 examples, and a test set of 10,000 examples. It is a subset of a larger set available from National Institute of Standards and Technology (NIST). The digits have been size-normalized and centered in a fixed-size image.

In this work we use various type of hidden layer. We raise the nodes in the hidden layer for each model. The configurations of nodes in hidden layer are 50, 100, 150, 200, 250, 300, 350, and 400. We also combine different layer to improve the classification performance of RBM. Moreover, we use 2 and 3 layers for stack RBM and DBN. We will compare the classification performance of regular RBM using RBM function, classification RBM using *stackRBM* function and DBN function with different hidden layer. The *n.hidden* argument defines how many hidden nodes the RBM will have and *size.minibatch* is the number of training samples that will be used at every epoch. For each model we use 1000 as the number of iterations and 10 for the minibatch. The workflow of this research could be seen on Fig. 4.

Furthermore, after training the RBM model, *stackRBM* model and DBN model we can check how well it reconstructs the data with the *ReconstructRBM* function.

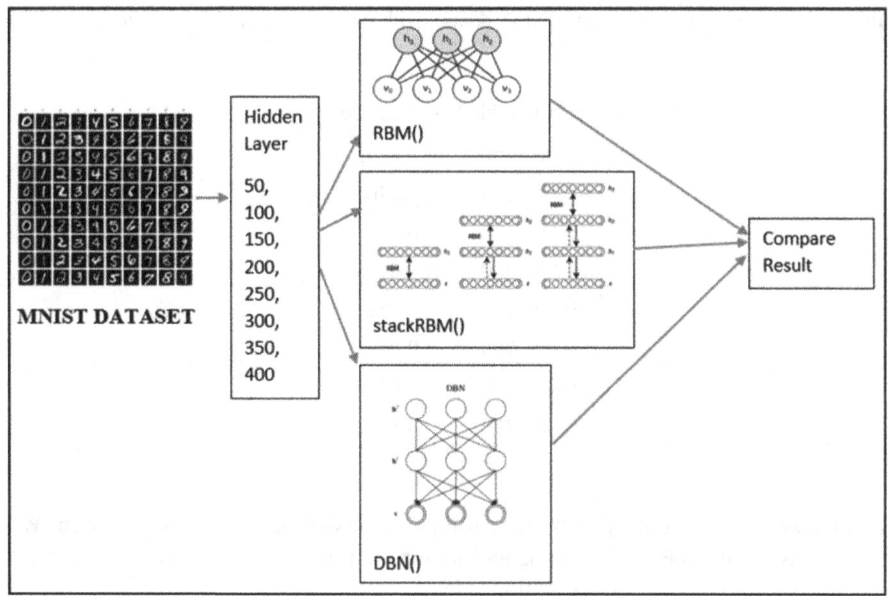

Fig. 4. The workflow of the research.

The function will then output the original image with the reconstructed image next to it. If the model is good, the reconstructed image should look similar or even better than the original. RBM not only good at reconstructing data but can actually make predictions on new data with the classification RBM. So, after we trained our regular RBM, classification RBM and DBN, we can use it to predict the labels on some unseen test data with the *PredictRBM* function. This function will output a confusion matrix and the accuracy score on the test set.

4 Experiment and Result

We evaluate the performance of the proposed learning algorithm using the MNIST dataset [10]. In the classification results, we focused on whether the experiment improvement RBM obtained the best classification accuracy performance. Also, we compared the number of hidden neurons RBM. The classifier used in all the experiment is the Back-Propagation Network (BPN) [11, 15].

Table 1 shows the classification accuracy of MNIST dataset with various type of hidden layer using RBM function. In this experiment we use 50, 100, 150, 200, 250, 300, 350, and 400 nodes in hidden layer. In addition, to train a RBM we need to provide the function with train data, which should be a matrix of the shape *(samples * features)* other parameters have default settings. The number of iterations defines the number of training epochs, at each epoch RBM will sample a new minibatch. When we have enough data it is recommended to set the number of iterations to a high value as this will improve our model and the downside is that the function will also take longer to train. The *n.hidden* argument defines how many hidden nodes the RBM will have and

Table 1. Classification accuracy of MNIST dataset with various type of hidden layer using RBM function.

n.iter	n.hidden	Accuracy
1000	50	0.8245
1000	100	0.846
1000	150	0.851
1000	200	0.8585
1000	250	0.859
1000	300	0.86
1000	350	0.86
1000	400	0.815

size.minibatch is the number of training samples that will be used at every epoch. We use 1000 as the number of iterations and 10 for the minibatch. Moreover, the highest accuracy is 86% with 350 nodes in hidden layer.

After training the RBM model we can check how well it reconstructs the data with the *ReconstructRBM* function. The function will then output the original image with the reconstructed image next to it. If the model is any good the reconstructed image should look similar or even better than the original. The reconstruction model for digit "0" and digit "3" using RBM function could be seen on Figs. 5 and 6. The model reconstruction looks even more like a three and zero than the original image. Furthermore, RBM not only good at reconstructing data but can actually make predictions on new data with the classification RBM. After we trained our classification RBM we can use it to predict the

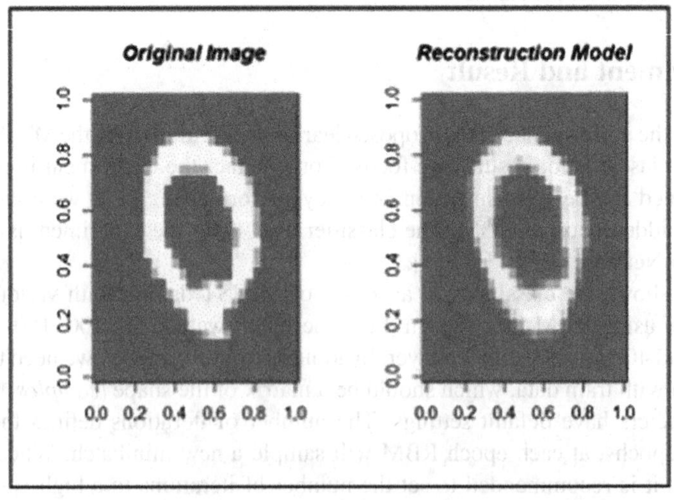

Fig. 5. Reconstruction model digit "0" using RBM model

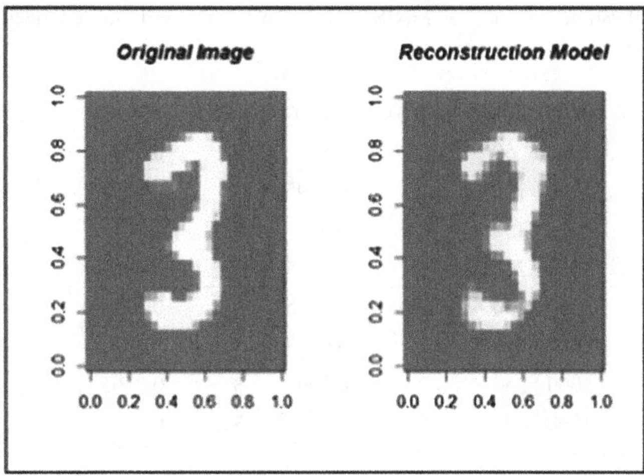

Fig. 6. Reconstruction model digit "3" using RBM model

```
> PredictRBM(test = test, labels = TestY, model = modelClassRBM)
$ConfusionMatrix
    truth
pred   0   1   2   3   4   5   6   7   8   9
   0 189   0   4   2   2   3   2   0   1   0
   1   0 211   0   0   0   1   0   0   3   0
   2   1   0 163   7   3   1   4   4   2   5
   3   0   2   6 161   2  12   0   1   0   5
   4   0   0   2   0 167   0   1   1   1   7
   5   2   1   4   3   6 119   1   3   8   1
   6   2   2   2   3   7   5 206   0   3   0
   7   0   0   4   1   3   0   0 184   0   9
   8   3   8   4  18  14  19   2   1 158  16
   9   0   1   1   3  22   2   0   8   3 162

$Accuracy
[1] 0.86
```

Fig. 7. Confusion matrix MNIST dataset using RBM model

labels on some unseen test data with the *PredictRBM* function. Which should output a confusion matrix and the accuracy score on the test set that could be seen on Fig. 7.

Table 2 shows classification accuracy of MNIST dataset with various type of hidden layer using stackRBM function. In this experiment we use various type of hidden layer consists of 50, 100, 150, 200, 250, 300, 350, and 400 nodes for each layer (2 and 3). In this work the highest accuracy for 2 layers is 90.9% use 350 nodes in hidden layer and for 3 layers is 91.65% use 350 nodes in hidden layers. As we can see on Table 2 stacking RBM use 350 nodes in hidden layer receive 90.9% accuracy and it is higher than on Table 1 normal RBM receive 86% for 350 nodes in hidden layer. We can conclude from this result stacking RBM improves our classification performance. However, stackRBM is not a very elegant method though as each RBM layer is trained on the output of the last layer and all the other RBM weights are frozen. It is a greedy method that will not give us the most optimal results for classification. After training the stackRBM model we can check how well it reconstructs the data with the *ReconstructRBM* function. The

Table 2. Classification accuracy of MNIST dataset with various type of hidden layer using stackRBM Function.

n.iter	n.hidden	Layers	Accuracy	Layers	Accuracy
1000	50	2	0.846	3	0.827
1000	100	2	0.8695	3	0.868
1000	150	2	0.889	3	0.901
1000	200	2	0.8885	3	0.897
1000	250	2	0.9015	3	0.899
1000	300	2	0.9015	3	0.8975
1000	350	2	0.909	3	0.9165
1000	400	2	0.907	3	0.906

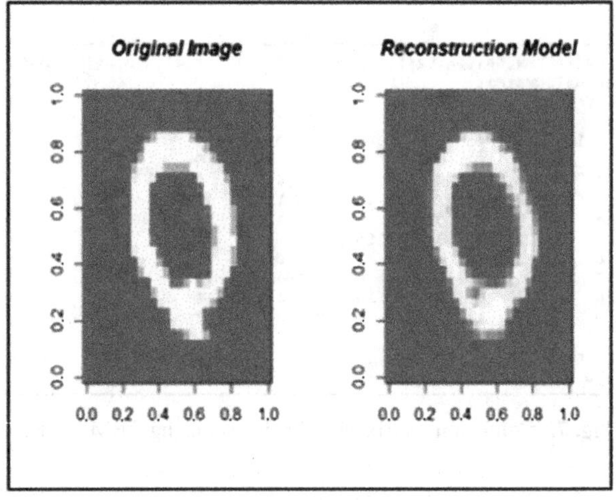

Fig. 8. Reconstruction model digit “0”, 2 layers using stack RBM model

reconstruction model for digit “0” and digit “3” could be seen on Figs. 8 and 9. Figure 10 explain about confusion matrix MNIST dataset using stackRBM function with 350 nodes in hidden layer using 2 hidden layers, which got 90.9% accuracy.

Table 3 shows classification accuracy of MNIST dataset with various type of hidden layer using DBN function. In this experiment we use various type of hidden layer consists of 50, 100, 150, 200, 250, 300, 350, and 400 nodes for each layer (2 and 3). In this work the highest accuracy for 2 layers is 90.15% use 350 hidden layer and for 3 layers is 90% use 400 nodes in hidden layers.

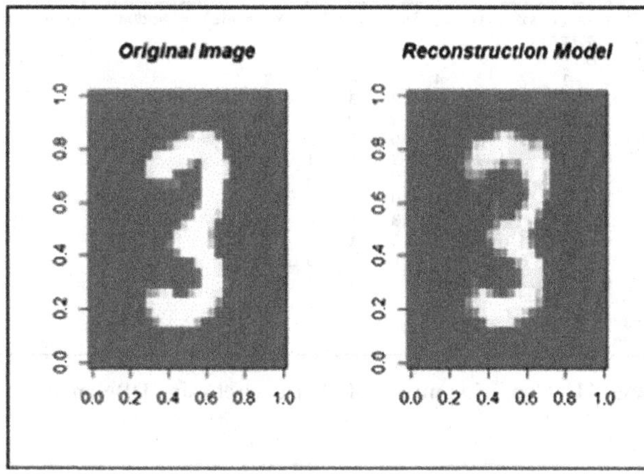

Fig. 9. Reconstruction model digit"3", 2 layers using stack RBM model

```
> PredictRBM(test = test, labels = TestY, model = modStackSup, layers =
 2)
$ConfusionMatrix
     truth
pred   0   1   2   3   4   5   6   7   8   9
   0 189   0   3   0   0   2   0   0   0   2
   1   0 222   0   5   2   3   0   5  10   6
   2   0   0 163   4   0   0   1   2   1   1
   3   0   1   4 168   0   0   0   0   2   2
   4   0   0   2   0 212   2   1   1   2   7
   5   1   0   1  12   0 142   3   0   5   1
   6   3   1   3   4   1   9 211   0   4   0
   7   0   0   5   1   1   1   0 183   2   3
   8   3   1   7   1   0   2   0   0 146   1
   9   1   0   2   3  10   1   0  11   7 182

$Accuracy
[1] 0.909
```

Fig. 10. Confusion matrix MNIST dataset using stack RBM model

Table 3. Classification accuracy of MNIST dataset with various type of hidden layer using DBN Function.

n.iter	n.hidden	Layers	Accuracy	Layers	Accuracy
1000	50	2	0.8485	3	0.8215
1000	100	2	0.8845	3	0.8665
1000	150	2	0.8905	3	0.8865
1000	200	2	0.8895	3	0.899
1000	250	2	0.8675	3	0.889
1000	300	2	0.902	3	0.8985
1000	350	2	0.9015	3	0.8945
1000	400	2	0.89	3	0.90

```
> PredictDBN(test = test, labels = TestY, model = modDBN, layers = 3)
$ConfusionMatrix
     truth
Preds   0   1   2   3   4   5   6   7   8   9
    0 191   0   4   1   1   2   3   0   3   1
    1   0 206   0   0   0   1   0   0   3   0
    2   0   0 169   2   5   0   3   1   3   1
    3   0   1   2 173   0   5   0   1   4   3
    4   0   0   3   0 199   1   1   0   0   5
    5   2   5   4  10   1 140   5   1   4   4
    6   2   1   0   2   2   3 202   0   1   0
    7   0   3   4   3   2   3   1 198   2  24
    8   2   9   3   6   7   7   1   0 159   3
    9   0   0   1   1   9   0   0   1   0 164

$Accuracy
[1] 0.9015
```

Fig. 11. Confusion matrix MNIST dataset using DBN model

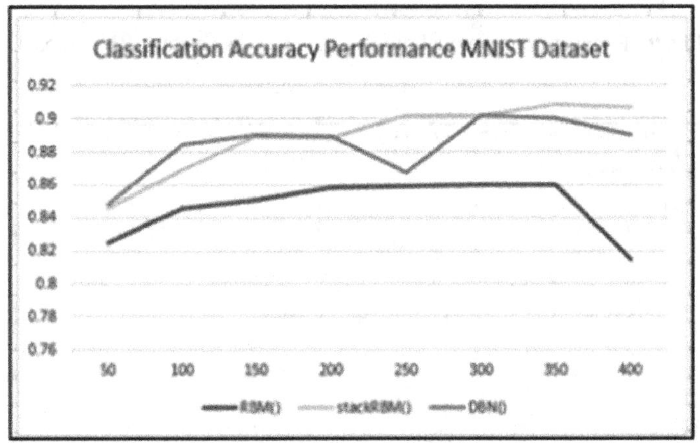

Fig. 12. Classification accuracy performance MNIST dataset with various type of hidden layer.

Figure 11 explain about confusion matrix MNIST dataset using DBN function with 350 nodes in hidden layers, for 2 layers and 1 label layers, which got 90.15% accuracy. Based on experiment result on the Fig. 12 the trends of accuracy increases. When we use more nodes in hidden layer, we can get higher accuracy performance. In this work the highest accuracy is obtained when using 350 nodes in hidden layer. After that the accuracy performance relatively decrease.

5 Conclusions

Based on all experiment result on Tables 1, 2, and 3 the number of hidden units and the key parameter of restricted Boltzmann machine play an important role in the modeling capability. Too many hidden units lead to a large model size and slow convergence speed, even overfitting results in poor generalization ability. And too few hidden units result in low accuracy and bad performance of feature extraction. Stacking RBM and DBN could improve our classification performance compare to regular RBM. Our experiment was focused on comparing the number of hidden neurons use RBM function, stackRBM function, and DBN function. Our future work includes to design a fully automated incremental learning algorithm that can be used in the deep architecture and we will use other advanced types of RBM like Gaussian RBM and Deep Boltzmann Machines.

Acknowledgment. This paper is supported by Ministry of Science and Technology, Taiwan. The Nos are MOST-107-2221-E-324-018-MY2 and MOST-106-2218-E-324-002, Taiwan.

References

1. Gehler, P.V., Holub, A.D., Welling, M.: The rate adapting Poisson model for information retrieval and object recognition. In: Proceedings of the 23rd International Conference on Machine Learning, ser. ICML 2006. ACM, New York, pp. 337–344 (2006). http://doi.acm.org/10.1145/1143844.1143887
2. Hinton, G.E.: To recognize shapes, first learn to generate images. Progress Brain Res. **165**, 535–547 (2007)
3. Hinton, G.E.: A Practical Guide to Training Restricted Boltzmann Machines, pp. 599–619. Springer, Heidelberg (2012). https://doi.org/10.1007/978-3-642-35289-832
4. Larochelle, H., Mandel, M., Pascanu, R., Bengio, Y.: Learning algorithms for the classification restricted Boltzmann machine. J. Mach. Learn. Res. **13**(1), 643–669 (2012)
5. Larochelle, H., Bengio, Y.: Classification using discriminative restricted Boltzmann machines. In: Proceedings of the 25th International Conference on Machine Learning, ser. ICML 2008, pp. 536–543. ACM, New York (2008). http://doi.acm.org/10.1145/1390156.1390224
6. Jongmin, Y., Jeonghwan, G., Sejeong, L., Moongu, J.: An incremental learning approach for restricted Boltzmann machines. In: 2015 International Conference on Control, Automation and Information Sciences (ICCAIS), pp. 113–117 (2015)
7. Jiang, Y., Xiao, J., Liu, X., Hou, J.: A removing redundancy restricted Boltzmann machine. In: 2018 Tenth International Conference on Advanced Computational Intelligence (ICACI), pp. 57–62 (2018)
8. Hinton, G.E., Dayan, P., Frey, B.J., Neal, R.M.: The wake-sleep algorithm for unsupervised neural networks. Science **268**, 1158 (1995)
9. Hinton, G.E., Osindero, S., Teh, Y.W.: A fast learning algorithm for deep belief nets. Neural Comput. **18**(7), 1527–1554 (2006). https://doi.org/10.1162/neco.2006.18.7.1527
10. Lecun, Y., Bottou, L., Bengio, Y., Haffner, P.: Gradient-based learning applied to document recognition. Proceedings IEEE **86**(11), 2278–2324 (1998)
11. Cun, Y.L., et al.: Handwritten digit recognition with a back-propagation network. In: Touretzky, D.S. (ed.) Advances in Neural Information Processing Systems 2, pp. 396–404. Morgan Kaufmann Publishers Inc., San Francisco (1990). http://dl.acm.org/citation.cfm?id=109230.109279
12. Han-Gyu, K., Seung-Ho, H., Ho-Jin, C.: Discriminative restricted Boltzmann machine for emergency detection on healthcare robot. In: 2017 IEEE International Conference on Big Data and Smart Computing (BigComp), pp. 407–409. IEEE (2017)
13. Jinyong, Y., Tao, S., Jiangyun, Z., Xiaolon, B.: Fault prognosis based on restricted Boltzmann machine and data label for switching power amplifiers. In: 2018 12th International Conference on Reliability, Maintainability, and Safety (ICRMS), pp. 287–291. IEEE (2018)

14. Renshu, W., Bin, C., Jingdong, G., Jing, Z.: The image recognition based on restricted Boltzmann machine and deep learning framework. In: 2019 4th International Conference on Control and Robotics Engineering (ICCRE), pp. 161–164. IEEE (2019)
15. Shamma, N., Justine, L.D., Supriyo, B., Amit, R.T.: Low power restricted Boltzmann machine using mixed-mode magneto-tunneling junctions. IEEE Electron. Device Lett. **40**(2), 345–384 (2019)

Intelligent and Contextual Systems

AQM Mechanism with Neuron Tuning Parameters

Jakub Szyguła[1(✉)], Adam Domański[1], Joanna Domańska[2], Tadeusz Czachórski[2], Dariusz Marek[1], and Jerzy Klamka[2]

[1] Institute of Informatics, Silesian University of Technology, Akademicka 16, 44-100 Gliwice, Poland
jakub.szygula@polsl.pl

[2] Institute of Theoretical and Applied Informatics, Polish Academy of Sciences, ul. Bałtycka 5, 44-100 Gliwice, Poland

Abstract. The congestion control is one of the most important questions in modern computer network performance. This article investigates the problem of adaptive neuron based choice of the Active Queue Mechanisms parameters. We evaluate the performance of the AQM mechanism in the presence of self-similar traffic based on the automatic selection of their parameters using the adaptive neuron. We have also proposed an AQM mechanism based on non-integer order PI^{α} controller with neuron tuning parameters and compared it with Adaptive Neuron AQM. The numerical results are obtained using the discrete event simulation model.

Keywords: AQM · Congestion control · PI^{α} controller · Neuron

1 Introduction

The rapid growth of Internet traffic due to the high demand for voice, video and data services has made the problem of congestion hard to be solved. This has led to the recommendation of an active queue management mechanism (AQM) by the Internet Engineering Task Force (IETF) [4]. This mechanism is based on early detection of the possibility of congestion and randomly drop off some packets even though the queue is not yet full. This mechanism prevents the tail dropping of packets when the buffer becomes full, which can drastically degrade the quality of service (QoS).

The AQM mechanisms, when used with the TCP congestion window mechanism in control, the traffic flows in the network enhance the efficiency of network transmission [4,27].

One of the first introduced active queue management algorithms was Random Early Detection (RED), which was proposed in 1993 by Sally Floyd and Van Jacobson [18]. This mechanism is based on the estimation of packet dropping probability, which is a function of the queue length seen by arriving packets. The argument of this function is a weighted mean queue seen by arriving packets. The function is non-decreasing between two thresholds and has the value 0 at the

N. T. Nguyen et al. (Eds.): ACIIDS 2020, LNAI 12034, pp. 299–311, 2020.
https://doi.org/10.1007/978-3-030-42058-1_25

lower threshold (and below) and P_{max} at the upper one. For higher arguments, its value is 1, i.e., all packets are rejected, The other parameter w_q is the weight of the current queue when the moving average is defined [8,18].

The efficiency of the RED mechanism and other AQM algorithms depends on the choice of the parameters P_{max} and w_q. If w_q is too small, the packets may be dropped even if the queue is still small. Also, if w_q is too large, there are high fluctuations of queue size. The articles [18,21] recommended $w_q = 0.001$ or $w_q = 0.002$ and [29] showed the efficiency of the mechanism for $w_q = 0.05$ and $w_q = 0.07$. It was recommended in [3] that the queue size could be analyzed seeing the influence of w_q on waiting time fluctuations, obviously the larger w_q, the higher fluctuations. The value of P_{max} has also a significant influence on the performance of the RED mechanism: if it is too large, the overall throughput is unnecessarily choked and if it's too small the danger of synchronization arises; [17] recommends $P_{max} = 0.1$. The problem of parameter selection was discussed in [5,22].

The numerous propositions of basic algorithms improvements appear [2,9, 20], their comparison may be found e.g. in [19]. Our previous works [6,10], also presented a study of the influence of RED modifications on its performance.

In this paper, we evaluate the performance of AQM mechanisms based on automatic selection of their parameters using the adaptive neuron in the presence of self-similar traffic. The self-similarity of a process means that a change of time scales does not influence the statistical characteristics of the process. The level of self-similarity is characterized by the Hurst parameter H, $H \in (0.5, 1)$ [1,11], the higher H, the higher degree of self-similarity. It results in long-distance autocorrelation and makes possible the occurrence of very long periods of high (or low) traffic intensity. These features have a high impact on network performance [7]. We have also proposed an AQM mechanism based on non-integer order PI^{α} controller with neuron tuning parameters and compared it with Adaptive Neuron AQM [28].

The remainder of the paper is organized as follows: Sect. 2 gives basic notions on AQM mechanisms with neuron tuning parameters and presents briefly theoretical basis for non-integer PI^{α} controller. Section 3 discusses numerical results. Some conclusions are given in Sect. 4.

2 AQM Mechanisms with Neuron Tuning Parameters

This section presents artificial intelligence algorithms used to select the proper AQM parameters. In the presented methods, the neuron's input data are mean queue length and network traffic parameters. The task of the mechanism is a selection of AQM parameters to keep the assumed mean queue length.

2.1 Adaptive Neuron AQM

The article [28] proposes a novel neuron-based AQM scheme. This solution is named Adaptive Neuron AQM (AN-AQM) and uses the single neuron to calculate the probability p of packet dropping. The probability is calculated for each incoming packets and can be presented as follows:

$$p(k) = p(k-1) + \Delta p(k) \tag{1}$$

where $\Delta p(k)$ reflects changes in packet dropping probability and depends on the state of the neuron. The neuron value can be described by the following equation:

$$\Delta p(k) = K \sum_{i=1}^{n} w_i(k) x_i(k) \tag{2}$$

where K is the neuron proportional coefficient and takes positive values, $x_i(k)$ for $i = 1, 2, 3, \ldots n$) are the neuron inputs and $w_i(k)$ is a connection weight of $x_i(k)$. The weights depend on the learning rule. For each incoming packets the algorithms calculates the error $e(k)$. The error is the difference between actual queue length $q(k)$ and the desired queue length Q. Thus:

$$e(k) = q(k) - Q. \tag{3}$$

The algorithm also defines normalized rate error $\gamma(k)$:

$$\gamma(k) = \frac{r(k)}{C} - 1, \tag{4}$$

where $r(k)$ is the input rate of the buffer at the bottleneck link, and C is the capacity of the bottleneck link.

The inputs of the neuron are: $x_1(k) = e(k) - e(k-1)$, $x_2(k) = e(k)$, $x_3(k) = e(k) - 2e(k-1) + e(k-2)$, $x_4(k) = \gamma(k) - \gamma(k-1)$, $x_5(k) = \gamma(k)$ and $x_6(k) = \gamma(k) - 2\gamma(k-1) + \gamma(k-2)$.

The learning rule of a neuron presents formula [25]:

$$w_i(k+1) = w_i(k) + d_i y_i(k) \tag{5}$$

where $d_i > 0$ is the learning rate, and $y_i(k)$ is the learning strategy. The article [25] suggests the following learning strategy:

$$y_i(k) = e(k) p(k) x_i(k). \tag{6}$$

where $e(k)$ is a teacher signal.

Such a strategy implies that an adaptive neuron self-organizes depending on signals $e(k)$ and $\gamma(k)$.

2.2 Non-integer PI^{α} Controller with Neuron Tuning Parameters

Fractional Order Derivatives and Integrals (FOD/FOI) are a natural extension of the well-known integrals and derivatives. Differintegrals of non-integer orders enable better and more precise control of physical processes. A proportional-integral controller (PI^{α} controller) is a traditional mechanism used in feedback control systems. Earlier works show that the non-integer order controllers have better behavior than classic controllers [26].

The articles [12,14,16], describe how to use the response of the PI^{α} (non-integer integral order) to determine the response of the AQM mechanism. This response is based on the probability function and is given by the formula:

$$p_i = max\{0, -(K_P e_k + K_I \Delta^{\alpha} e_k)\} \tag{7}$$

where K_P, K_I are tuning parameters, e_k is the error in current slot $e_k = q_k - q$, i.e. the difference between current queue q_k and desired queue q.

Thus, the dropping probability depends on three parameters: the coefficients for the proportional and integral terms K_p, K_i and integral order α.

In the active queue management, packet drop probabilities are determined at discrete moments of packet arrivals, so the queue model should be considered as a case of discrete systems. There is only one definition of the discreet differ-integrals of non-integer order. This definition is a generalization of the traditional definition of the difference of integer order to the non-integer order and is analogous to a generalization used in Grunwald-Letnikov (GrLET) formula.

For a given sequence $f_0, f_1, ..., f_j, ..., f_k$

$$\Delta^{\alpha} f_k = \sum_{j=0}^{k} (-1)^j \binom{\alpha}{j} f_{k-j} \tag{8}$$

where $\alpha \in R$ is generally a non-integer fractional order, f_k is a differentiated discrete function and $\binom{\alpha}{j}$ is generalized Newton symbol defined as follows:

$$\binom{\alpha}{j} = \begin{cases} 1 & \text{for } j = 0 \\ \dfrac{\alpha(\alpha-1)(\alpha-2)..(\alpha-j+1)}{j!} & \text{for } j = 1, 2, \ldots \end{cases} \tag{9}$$

The neural mechanism of the selection PI^{α} controller parameters for multi-model plants was presented in the articles [23,25]. It presents an adaptation of a proposed solution to the problem of active queue management. The construction of the neuron and the teaching strategy is similar to the AN-AQM algorithm (Sect. 2.1) and the mapping of the neuron response to PI^{α} parameters is as follows [25]:

$$K_P(t) = k_1 \frac{w_1(t) w_4(t)}{\sum_{i+1}^{n} w_i(t)} \tag{10}$$

$$K_I(t) = k_2 \frac{w_2(t) w_5(t)}{\sum_{i+1}^{n} w_i(t)} \tag{11}$$

where k_1 and k_2 are the constant proportional coefficients.

3 Obtained Results

This section presents the influence of the presented mechanism on queue performance. We study the impact of the degree of self-similarity on the examined

AQM mechanisms. The tests analyzed the following parameters of the transmission with AQM: the mean length of the queue, queue waiting times and the number of rejected packets.

All presented results in this article were obtained using the simulation model. The simulations were done using the Simpy Python packet. To accelerate the calculations, a PI^{α} module was written in C language. The input traffic intensity $\lambda = 0.5$ was considered independent of the Hurst parameter. During the tests, we changed the Hurst parameter of the input traffic within the range from 0.5 to 0.90. We use a fast algorithm for generating approximate sample paths for a Fractional Gaussian noise process, first introduced in [24]. We considered different queue loads. The high load was studied for parameter $\mu = 0.25$, where μ is the service intensity (mean number of packets served per time unit). The average load we obtained for $\mu = 0.5$. Small load was considered for parameter $\mu = 0.75$.

The AQM parameters were as follows:

- desired queue length $Q = 100$,
- the neuron proportional coefficient $K = 0,01$,
- the learning rate d_i oscillated between 0.00001 *and* 0.0001,
- PI constant proportional coefficient $K_P, K_I = 0.0001$,
- the non-integer integral order α, depending on the experiment, took the following values: $-0.5, -1.0, -1.8$,
- maximum queue size $Q_{max} = 300$.

The normalized rate error $\gamma(k)$ was calculated as the proportion of input traffic intensity λ and service time μ. In packet-switched networks, service time is the time required to transmit information, and:

$$\gamma(k) = \frac{\lambda(k)}{\mu(k)} - 1, \tag{12}$$

Tables 1, 2 and 3 present the obtained results for Adaptive Neuron AQM (AN-AQM). The results correspond to three types of network traffic. The Table 1 reflects the values for an overloaded network (the packet arrival speed is higher than the queue capacity). The AQM mechanism, independently of the degree of traffic self-similarity (Hurst parameter), maintains the average queue length at the assumed level. The changes in queue size over time are shown in Fig. 1. The queue length, except the initial phase, does not exceed the assumed level. The queue length oscillation increases with the Hurst parameter. Table 2 and Fig. 2 present the results in the case of an average traffic load. The average queue length does not exceed 50 packets (independently of the traffic self-similarity). However, the detailed results suggest that temporarily, the queue exceeds the desired length and the mechanism drops packets. In the case of light traffic (Table 3), the average queue lengths are small, but the number of discarded packets shows that for the higher degree of self-similarity, the queue could exceed the desired size.

In the case of AN-AQM, the neuron controls the packet dropping probability. For considered here PI^{α} controller, the neuron determines the values of the

Table 1. AN-AQM, $\mu = 0.25$, setpoint = 100

Hurst	Avg. queue length	Avg. waiting times	Packet drop by	
			AN-AQM	Queue
0.50	99.2656	39.8558	24945	0
0.70	98.3449	39.1545	24920	0
0.80	96.6901	38.2377	23727	0
0.90	68.5395	29.8168	17410	0

Table 2. AN-AQM, $\mu = 0.50$, setpoint = 100

Hurst	Avg. queue length	Avg. waiting times	Packet drop by	
			AN-AQM	Queue
0.50	47.8359	9.5532	272	0
0.70	48.3206	10.0806	2543	0
0.80	44.8137	9.7713	3585	0
0.90	33.4571	9.3997	5101	0

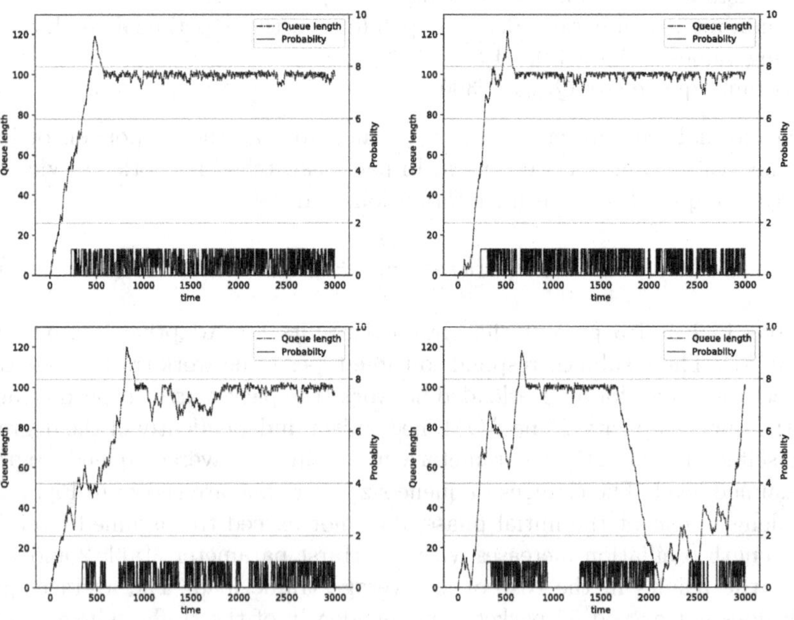

Fig. 1. Queue length and packet dropping probability, AN-AQM, $\mu = 0.25$, top: $H = 0.5$ and 0.7, bottom: $H = 0.8$ and 0.9

Table 3. AN-AQM, $\mu = 0.75$, setpoint = 100

Hurst	Avg. queue length	Avg. waiting times	Packet drop by	
			AN-AQM	Queue
0.50	1.2833	0.1575	0	0
0.70	2.9008	0.4799	0	0
0.80	5.7419	1.0730	35	0
0.90	12.7406	3.0883	360	0

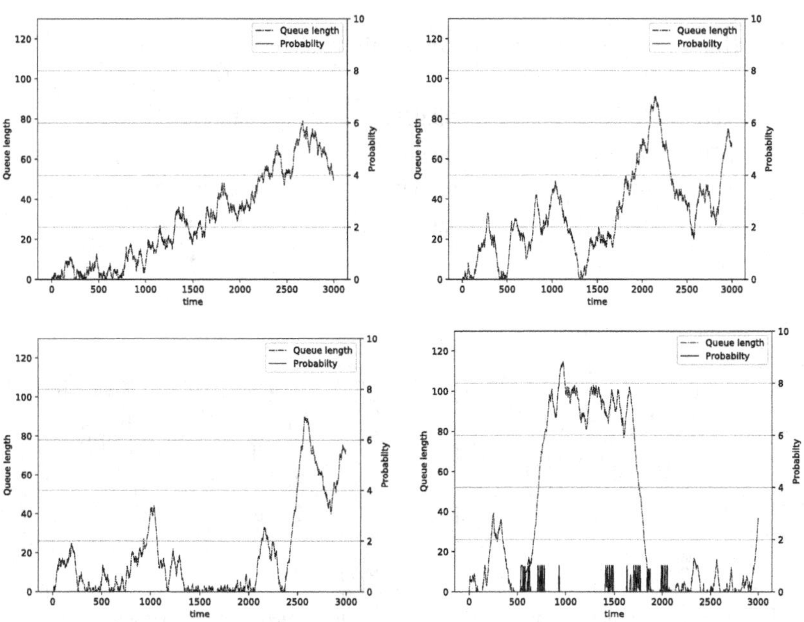

Fig. 2. Queue length and packet dropping probability, AN-AQM, $\mu = 0.50$, top: $H = 0.5$ and 0.7, bottom: $H = 0.8$ and 0.9

controller parameters, and packet dropping probability is obtained as the controller's feedback signal. The results below consider the AQM mechanism when the neuron sets only the K_P and K_I parameters. The value of integral order α is constant. In experiments, we considered different values of α parameter. The results below consider only the situation of an overloaded network ($\mu = 0.25$).

The Table 4 presents the results for standard non fractional PI^{α} controller ($\alpha = -1$). As noticed, the average queue size remains close to the desired level. However, Fig. 3 shows that the queue length varies greatly. This variability increases with the increase of the integration order (see Fig. 4). On the other hand, the oscillations decrease when the integral order decreases (Fig. 4). The comparison of the Tables 4, 5, and 6 allows us to conclude that while the integration order influences the queue evolution, it does not affect the obtained average values (Fig. 5).

The Figs. 6 and 7 display the evolution of the parameters K_I and K_P. It may be seen that regardless of the integral order α, the obtained by the algorithm K_P parameter, is much larger than the K_I. On the other hand, evolution is very similar. As the integration order decreases, the parameter changes become smoother.

Table 4. $\alpha = -1.0$, $\mu = 0.25$, setpoint $= 100$

Hurst	Avg. queue length	Avg. waiting times	Packet drop by	
			PI^{α}	Queue
0.50	100.8211	39.9768	49603	0
0.70	99.5037	39.6021	49838	0
0.80	97.8870	39.0255	49891	0
0.90	76.4802	32.6089	53131	0

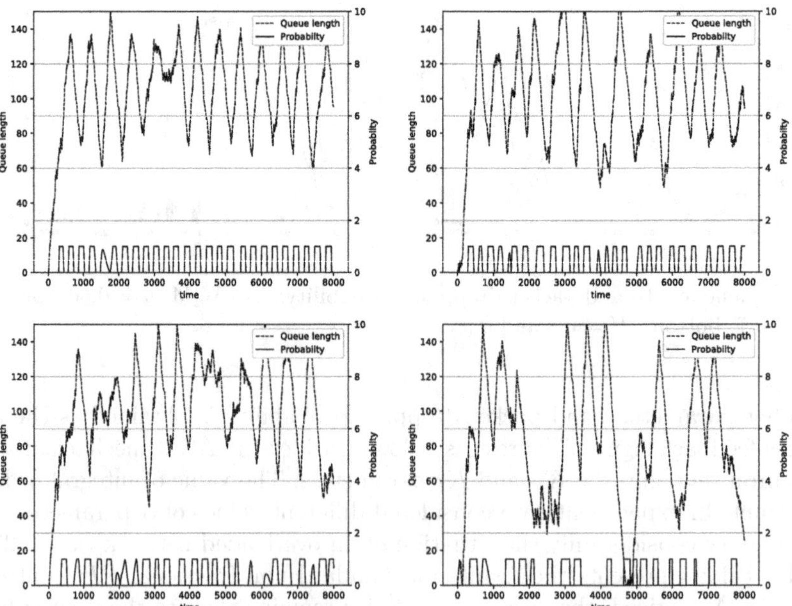

Fig. 3. Queue length and packet dropping probability, $\alpha = -1.0$, $\mu = 0.25$, top: $H = 0.5$ and 0.7, bottom: $H = 0.8$ and 0.9

Table 5. $\alpha = -1.8$, $\mu = 0.25$, setpoint $= 100$

Hurst	Avg. queue length	Avg. waiting times	Packet drop by	
			PI^{α}	Queue
0.50	100.1562	39.8435	49830	0
0.70	98.2007	39.3388	50188	0
0.80	95.7958	38.0802	49762	0
0.90	78.9514	34.0898	53799	0

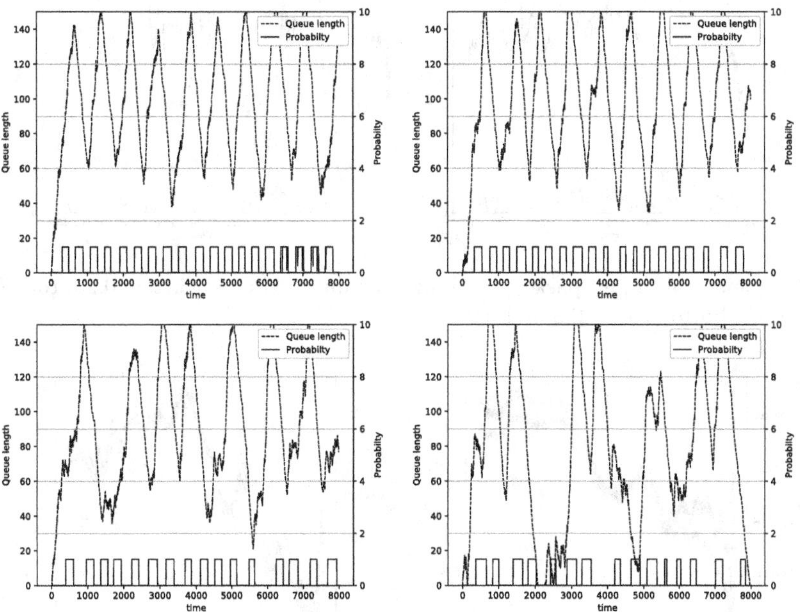

Fig. 4. Queue length and packet dropping probability, $\alpha = -1.8$, $\mu = 0.25$, top: $H = 0.5$ and 0.7, bottom: $H = 0.8$ and 0.9

Table 6. $\alpha = -0.5$, $\mu = 0.25$, setpoint $= 100$

Hurst	Avg. queue length	Avg. waiting times	Packet drop by	
			PI^{α}	Queue
0.50	102.8591	40.8138	49681	0
0.70	102.0629	40.6503	49855	0
0.80	101.2295	40.5846	50184	0
0.90	83.4922	35.14951	52589	0

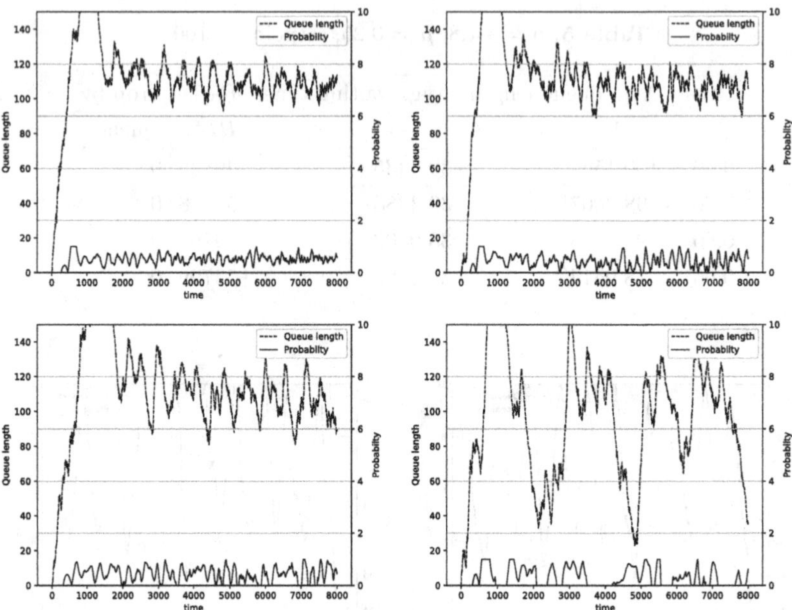

Fig. 5. Queue length and packet dropping probability, $\alpha = -0.5$, $\mu = 0.25$, top: $H = 0.5$ and 0.7, bottom: $H = 0.8$ and 0.9

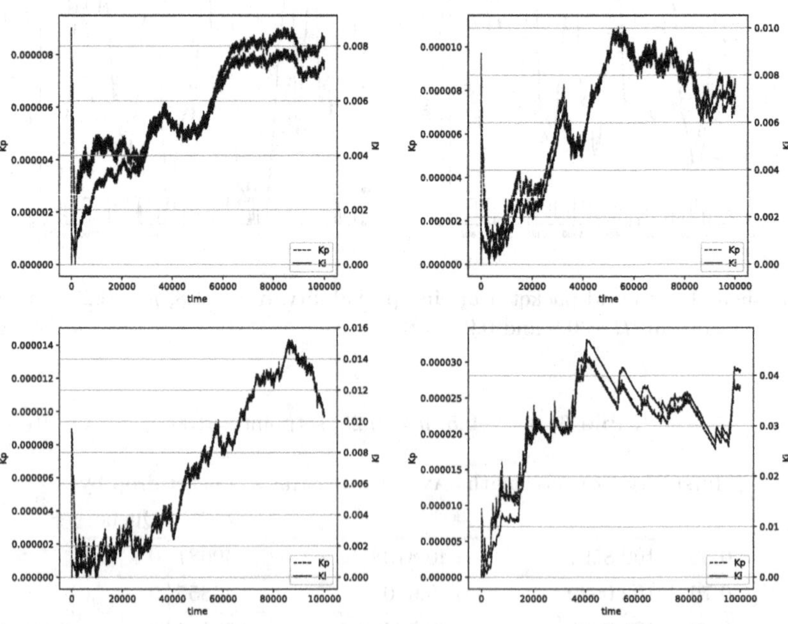

Fig. 6. The parameters K_P and K_I evolution, $\alpha = -1.0$, $\mu = 0.25$, top: $H = 0.5$ and 0.7, bottom: $H = 0.8$ and 0.9

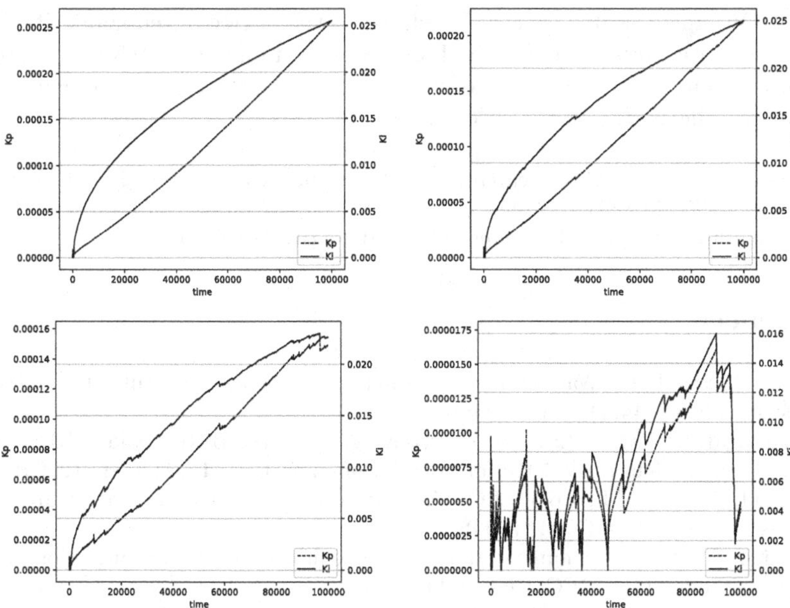

Fig. 7. The parameters K_P and K_I evolution, $\alpha = -0.5$, $\mu = 0.25$, top: $H = 0.5$ and 0.7, bottom: $H = 0.8$ and 0.9

4 Conclusions

The Internet Engineering Task Force (IETF) organization recommends that IP routers should use the active queue management mechanisms (AQMs). The basic algorithm for AQM is the RED algorithm. There are many modifications and improvements to the RED mechanism. One of these improvements is the calculation of the probability of packet loss using the PI^{α} controller. Our previous work has shown the advantages of this solution [12,13].

The efficient operation of the AQM algorithms depends on the proper selection of its parameters. Additionally, the optimal parameters may depend on traffic intensity and degree of traffic self-similarity (expressed in Hurst parameter) [15].

This paper introduces a way of a neuro-intelligent tuning of the AQM controller parameters. In the article, the authors presented two algorithms: Adaptive Neuron AQM and PI^{α} controller with neuron tuning parameters. Both algorithms worked correctly. The AN-AQM maintains the queue size slightly below the assumed level. In the case of PI^{α} the obtained average queue lengths have corresponded to our expectations. However, we observe a large variability of the queue occupancy. These fluctuations can be controlled by proper setting the integral order parameter α.

Our article also presents the impact of the degree of self-similarity (expressed in the Hurst parameter) on the length of the queue and the number of rejected packets. Obtained results are closely related to the degree of self-similarity. The experiments are carried out for the four types of traffic ($H = 0.5, 0.7, 0.8, 0.9$). As can be seen, when the degree of self-similarity increases, the average queue occupancy is below the assumed level.

Acknowledgements. This research was partially financed by National Science Center project no. 2017/27/B/ST6/00145.

This research was partially financed by 02/020/BKM19/0183.

References

1. Abry, P., Veitch, D.: Wavelet analysis of long-range-dependent traffic. IEEE Trans. Inform. Theory **44**(1), 2–15 (1998)
2. Bhatnagar, S., Patro, R.: A proof of convergence of the B-RED and P-RED algorithms for random early detection. IEEE Commun. Lett. **13**, 809–811 (2009)
3. Bonald, T., May, M., Bolot, J.C.: Analytic evaluation of RED performance. In: Proceedings of INFOCOM (2000)
4. Braden, B., et al.: Recommendations on queue management and congestion avoidance in the internet. RFC 2309, IETF (1998)
5. Feng, W.C., Kandlur, D., Saha, D.: Adaptive packet marking for maintaining end to end throughput in a differentiated service internet. IEEE/ACM Trans. Netw. **7**(5), 685–697 (1999)
6. Domańska, J., Augustyn, D.R., Domański, A.: The choice of optimal 3-rd order polynomial packet dropping function for NLRED in the presence of self-similar traffic. Bull. Polish Acad. Sci.: Tech. Sci. **60**(4), 779–786 (2012)
7. Domańska, J., Domański, A.: The influence of traffic self-similarity on QoS mechanisms. In: Proceedings of the International Symposium on Applications and the Internet, Saint, Trento, Italy, pp. 300–303 (2005)
8. Domańska, J., Domański, A., Augustyn, D.R., Klamka, J.: A RED modified weighted moving average for soft real-time application. Int. J. Appl. Math. Comput. Sci. **24**(3), 697–707 (2014)
9. Domańska, J., Domański, A., Czachórski, T.: Fluid flow analysis of RED algorithm with modified weighted moving average. In: Dudin, A., Klimenok, V., Tsarenkov, G., Dudin, S. (eds.) BWWQT 2013. Communications in Computer and Information Science, vol. 356, pp. 50–58. Springer, Heidelberg (2013). https://doi.org/10.1007/978-3-642-35980-4_7
10. Domańska, J., Domańska, A., Czachórski, T.: A few investigations of long-range dependence in network traffic. In: Czachórski, T., Gelenbe, E., Lent, R. (eds.) Information Sciences and Systems 2014, pp. 137–144. Springer, Cham (2014). https://doi.org/10.1007/978-3-319-09465-6_15
11. Domańska, J., Domański, A., Czachórski, T.: Estimating the intensity of long-range dependence in real and synthetic traffic traces. In: Gaj, P., Kwiecień, A., Stera, P. (eds.) CN 2015. CCIS, vol. 522, pp. 11–22. Springer, Cham (2015). https://doi.org/10.1007/978-3-319-19419-6_2
12. Domańska, J., Domański, A., Czachórski, T., Klamka, J.: The use of a non-integer order PI controller with an active queue management mechanism. Int. J. Appl. Math. Comput. Sci. **26**, 777–789 (2016)

13. Domański, A., Domańska, J., Czachórski, T., Klamka, J.: Self-similarity traffic and AQM mechanism based on non-integer order $PI^{\alpha}D^{\beta}$ controller. In: Gaj, P., Kwiecień, A., Sawicki, M. (eds.) CN 2017. CCIS, vol. 718, pp. 336–350. Springer, Cham (2017). https://doi.org/10.1007/978-3-319-59767-6_27
14. Domański, A., Domańska, J., Czachórski, T., Klamka, J., Marek, D., Szyguła, J.: GPU accelerated non-integer order $PI^{\alpha}D^{\beta}$ controller used as AQM mechanism. In: Gaj, P., Sawicki, M., Suchacka, G., Kwiecień, A. (eds.) CN 2018. CCIS, vol. 860, pp. 286–299. Springer, Cham (2018). https://doi.org/10.1007/978-3-319-92459-5_23
15. Domański, A., Domańska, J., Czachórski, T., Klamka, J., Marek, D., Szyguła, J.: The influence of the traffic self-similarity on the choice of the non-integer order PI^{α} controller parameters. In: Czachórski, T., Gelenbe, E., Grochla, K., Lent, R. (eds.) ISCIS 2018. CCIS, vol. 935, pp. 76–83. Springer, Cham (2018). https://doi.org/10.1007/978-3-030-00840-6_9
16. Domański, A., Domańska, J., Czachórski, T., Klamka, J., Szyguła, J.: The AQM dropping packet probability function based on non-integer order $PI^{\alpha}D^{\beta}$ controller. In: Ostalczyk, P., Sankowski, D., Nowakowski, J. (eds.) RRNR 2017. LNEE, vol. 496, pp. 36–48. Springer, Cham (2019). https://doi.org/10.1007/978-3-319-78458-8_4
17. Floyd, S.: Discussions of setting parameters (1997)
18. Floyd, S., Jacobson, V.: Random early detection gateways for congestion avoidance. IEEE/ACM Trans. Netw. **1**(4), 397–413 (1993)
19. Hassan, M., Jain, R.: High Performance TCP/IP Networking - Concepts, Issues and Solutions. Pearson Education Inc., London (2004)
20. Ho, H.-J., Lin, W.-M.: AURED - autonomous random early detection for TCP congestion control. In: 3rd International Conference on Systems and Networks Communications Malta (2008)
21. May, M., Bonald, T., Bolot, J.: Analytic evaluation of RED performance. In: Proceedings of the IEEE Infocom, Tel-Aviv, Izrael (2000)
22. May, M., Diot, C., Lyles, B., Bolot, J.: Influence of active queue management parameters on aggregate traffic performance. Technical report, Institut de Recherche en Informatique et en Automatique (2000)
23. Ning, W., Shuqing, W.: Neuro-intelligent coordination control for a unit power plant. In: IEEE International Conference on Intelligent Processing Systems (Cat. No. 97TH8335), vol. 1, pp. 750–753 (1997)
24. Paxson, V.: Fast, approximate synthesis of fractional Gaussian noise for generating self-similar network traffic. ACM SIGCOMM Comput. Commun. Rev. **27**(5), 5–18 (1997)
25. Ping, Y.D., Wang, N.: A PID controller with neuron tuning parameters for multi-model plants. In: Proceedings of 2004 International Conference on Machine Learning and Cybernetics (IEEE Cat. No. 04EX826), vol. 6, pp. 3408–3411 (2004)
26. Podlubny, I.: Fractional order systems and $PI^{\lambda}D^{\mu}$ controllers. IEEE Trans. Autom. Control **44**(1), 208–214 (1999)
27. Sawicki, M., Kwiecień, A.: Unexpected anomalies of isochronous communication over USB 3.1 Gen 1. Comput. Stand. Interfaces. **49**, 67–70 (2017)
28. Sun, J., Zukerman, M.: An adaptive neuron AQM for a stable internet. In: Akyildiz, I.F., Sivakumar, R., Ekici, E., Oliveira, J.C., McNair, J. (eds.) NETWORKING 2007. LNCS, vol. 4479, pp. 844–854. Springer, Heidelberg (2007). https://doi.org/10.1007/978-3-540-72606-7_72
29. Zheng, B., Atiquzzaman, M.: A framework to determine the optimal weight parameter of RED in next generation internet routers. Technical report, The University of Dayton, Department of Electrical and Computer Engineering (2000)

Deep Reconstruction Error Based Unsupervised Outlier Detection in Time-Series

Tsatsral Amarbayasgalan[1], Heon Gyu Lee[2], Pham Van Huy[3], and Keun Ho Ryu[3,4(✉)]

[1] Database and Bioinformatics Laboratory, School of Electrical and Computer Engineering, Chungbuk National University, Cheongju 28644, Korea
tsatsral@dblab.chungbuk.ac.kr

[2] Big Data Research Center, GAION, Seoul, Korea
hglee@gaion.kr

[3] Faculty of Information Technology, Ton Duc Thang University, Ho Chi Minh City 700000, Vietnam
{phamvanhuy,khryu}@tdtu.edu.vn

[4] Department of Computer Science, College of Electrical and Computer Engineering, Chung-Buk National University, Cheongju 28644, Korea
khryu@chungbuk.ac.kr

Abstract. With all the advanced technology nowadays, the availability of time-series data is being increased. Outlier detection is an identification of abnormal patterns that provide useful information for many kinds of applications such as fraud detection, fault diagnosis, and disease detection. However, it will require an expensive domain and professional knowledge if there is no label which indicates normal and abnormality. Therefore, an unsupervised novelty detection approach will be used. In this paper, we propose a deep learning-based approach. First, it prepares subsequences according to the optimal lag length using Autoregressive (AR) model. The selected lag length for time-series analysis defines the data context in which further analysis is performed. Then, reconstruction errors (RE) of the subsequences on deep convolutional autoencoder (CAE) models are used to estimate the outlier threshold, and density-based clustering is used to identify outliers. We have compared the proposed method with several publicly available state-of-the-art anomaly detection methods on 30 time-series benchmark datasets. These results show that our proposed deep reconstruction error based approach outperforms the compared methods in most of the cases.

Keywords: Outlier · Anomaly · Autoregressive model · Deep convolutional autoencoder · Time-series data

1 Introduction

Outlier detection in time-series is significant over a variety of industries such as internet traffic detection [1, 2], medical diagnosis [3], fraud detection [4], traffic flow forecasting [5] and patient monitoring [6, 7]. Outliers can be created for several reasons, such as data from different classes, natural variation, and data measurement or collection errors [8].

N. T. Nguyen et al. (Eds.): ACIIDS 2020, LNAI 12034, pp. 312–321, 2020.
https://doi.org/10.1007/978-3-030-42058-1_26

Although there may be novelty due to some mistakes, sometimes it is a new, unidentified process [9]. In computer science, an abnormal pattern that is not compatible with most of the data in a dataset is named a novelty, outlier, or anomaly [10]. In practice, human monitoring of these data is difficult and complicated that is a time-consuming task. It can be simplified using automated outlier detection using data mining techniques.

There are three basic ways to detect outlier depending on the availability of data label in data mining [10]. If data is labeled as a normal or outlier, a supervised approach can be used. In this case, we build a classification model from the training dataset to predict normal or outlier of unseen data. However, supervised approaches for outlier detection face to class imbalance problem because the abnormal dataset is smaller than normal [11]. The second method is a semi-supervised; it builds a classification model using only normal dataset. If training dataset includes some outliers, the model cannot detect outliers successfully. Moreover, most of the data have no label in practice, and an unsupervised method, such as the third, is used. In recent years, many unsupervised approaches are suggested. They detect a large number of false positives; it leads to the usefulness of these approaches in practice.

According to time-series characteristics, there are several outlier detection techniques in publicly available. Skyline real-time anomaly detection system [12] was implemented by Etsy Inc. It consists of the Horizon agent that is responsible for collecting, cleaning, and formatting for incoming data points and the Analyzer agent that is responsible for analyzing every metric for anomalies. It has a very simple divide-and-conquer strategy and can be used in a large number of high-resolution time-series.

Extensible Generic Anomaly Detection System (EGADS) is a general time-series anomaly detection framework open-sourced by Yahoo [13]. EGADS consists of two main modules, and each component includes several algorithms. The time-series modeling module (TMM) builds a model on given time-series and produces an expected value. Later the anomaly detection module (ADM) compares the expected value with the actual value and computes anomaly score.

In 2015, Twitter Inc. introduced AnomalyDetection open-source R package. They proposed a Seasonal Hybrid ESD (S-H-ESD) algorithm which is based on generalized extreme studentized deviate test for outliers [14]. Twitter anomaly detection algorithm can be used to detect both global as well as local anomalies from a statistical standpoint.

The Numenta anomaly detection benchmark (NAB) is an open framework that provides real-world time-series benchmark datasets with labeled anomalies and anomaly detectors. Numenta and NumentaTM [15, 16] are Numenta's anomaly detection methods based on Hierarchical Temporal Memory (HTM). At a given time, the HTM model predicts for next time-stamp. After that, the prediction error is calculated on the output of the HTM by compares with actual value. For each time-stamp, anomaly likelihood which defines how anomalous the current state is based on the prediction history of the HTM model and determines whether an anomaly is detected using a threshold.

In this paper, we proposed a novel deep learning-based outlier detection approach in the unsupervised mode for time-series data. First, we find an optimal lag length by the AR model, which is a regression equation for predicting the value of the next time step from observations of previous time steps. Based on lag length, we prepared subsequences from the whole time-series. Then, these subsequences passed into a detector module. This module uses two deep CAE models with Density-based Spatial Clustering of Applications with Noise (DBSCAN) clustering algorithm. CAE is one kind of neural network that tries to learn to represent its input to output as close as possible. First, it compresses its input into a low dimensional representation then reconstructs the input to the output from the compressed representation. RE is a difference between the input and its reconstructed output. We used RE measurement of the CAE model to estimate threshold value for outliers. Based on the threshold, we distinguish clustered data points are normal or an outlier. The proposed method outperformed compared state-of-the-art techniques on publicly available anomaly detection benchmark time-series datasets, in most cases.

The rest of paper is organized as follows: Sect. 2 provides details of the proposed method. Section 3 provides an evaluation of the proposed method and other state-of-the-art anomaly detection approaches. The whole study is concluded in Sect. 4.

2 Materials and Methods

In this section, we describe details of how to create subsequences using AR model and detect outliers by CAE models with density-based clustering. Our proposed approach has two fundamental functions: subsequence creation and outlier detection. The general architecture of the proposed method is shown in Fig. 1.

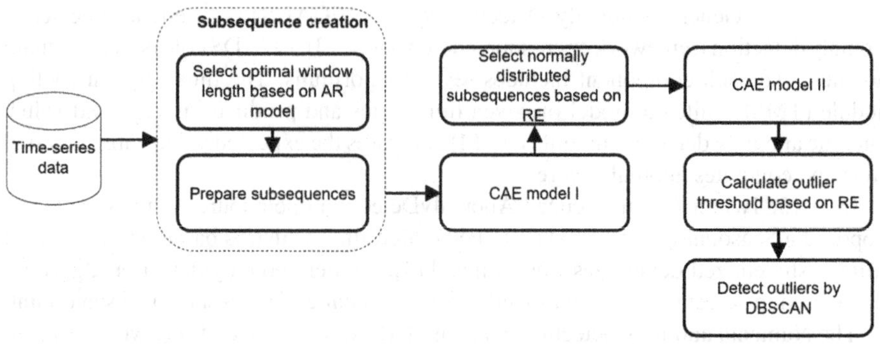

Fig. 1. The general architecture of the proposed method. AR, Autoregressive; RE, Reconstruction error; CAE, Convolutional autoencoder.

2.1 Subsequence Creation

The subsequence creation module of the proposed approach is based on AR model. AR model predicts future value based on its previous values in time-series data. Briefly, a current value of a time series is dependent on its first p values, and an AR model of order p can be written as:

$$y_t = c + \varphi_1 y_{t-1} + \varphi_2 y_{t-2} + \ldots + \varphi_p y_{t-p} + \varepsilon_t \tag{1}$$

Where ε_t is white noise, c is a constant, $\varphi_1 \ldots \varphi_p$ are the model parameters, and $y_{t-1} \ldots y_{t-p}$ are the past series values.

However, the AR process has degrees of uncertainty, the order p (lag length) is always unknown. Therefore, many order selection criteria have been employed to determine the optimal lag length. Akaike's information criterion (AIC) is widely used in order selection [17]. We prepared subsequences based on the selected lag length. Therefore, the presented approach has automated lag selection and can be used for different characteristics of time-series.

2.2 Deep Convolutional Autoencoder Based Outlier Detection

Autoencoder is one kind of artificial neural network, which is the unsupervised algorithm for learning to copy its input (x_1 ... x_n) to its output (y_1 ... y_n) as close ($x_i = y_i$) as possible by reducing the gap between inputs and outputs [18]. Generally, it is used for dimensionality reduction or data denoising. The structure is the same as neural network; it consists of an input layer, hidden layers, and an output layer. First, the input is compressed into a small dimensional space, and then the compressed input is reconstructed back to the output. The difference between input and output is called RE.

CAE is a type of autoencoder which has been widely used; it replaces fully connected layers by convolutional layers. In other words, the Encoder consists of convolutional layers and pooling layers, which reduce the dimension of input. The Decoder consists of convolutional layers and upsampling layers, which provides a reconstruction of the input. The term convolution refers to a mathematical operation on two functions that produces a third function. It merges two sets of information. Convolutional layer produces a feature map by convolution operation on the input data and filter that is learnable weights. A pooling layer converts outputs of the previous convolutional layer to reduced size according to the size of pooling. The upsampling layer reconstructs the pooled feature.

Some data that is different from most data have higher RE than a common dataset. Therefore, we have used the RE of data on CAE model for outlier detection. Like the above mentioned, there are two deep CAE models in the proposed method. The first CAE model trains from the whole dataset and distinguishes normally distributed dataset and highly biased dataset approximately based on reconstruction errors. By this step, the most common subset is selected, and the second CAE model learns from this selected subset. In other words, the second CAE model tries to learn only normal dataset, and it employs an outlier detector. For outlier detection, reconstruction errors of the whole dataset from the second CAE model is calculated.

As shown in Fig. 2, n is the input dimension (selected lag length). The convolutional layers and pooling layers of the proposed deep CAE model reduce the dimension into n/2, and 1, sequentially, and it reconstructed back the output. Each convolutional layer and deconvolution layers use *Relu* activation function, and *Sigmoid* activation function is used to the last deconvolution layer. The parameters of the CAE model was optimized to minimize the Mean Squared Error (MSE), as shown in Eq. 2:

$$\mathrm{MSE} = \frac{1}{n}\sum_{1}^{n}\left(x_i - x_i'\right)^2 \qquad (2)$$

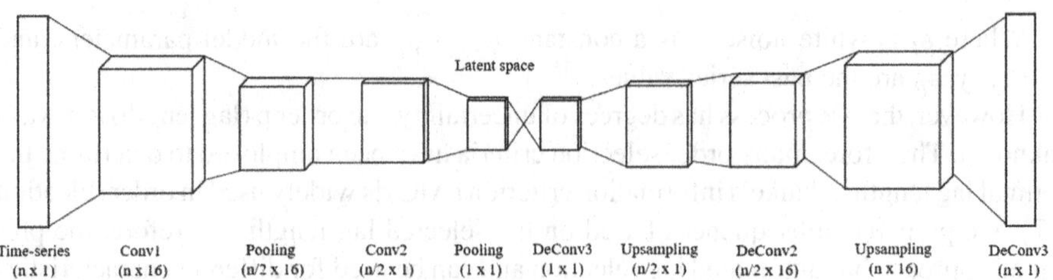

Fig. 2. Structure of the proposed convolutional autoencoder neural network

Where n is the number of subsequences, x is the input, and x' is the output. The reconstruction errors of a dataset are calculated by the squared difference between input and output of the CAE model. To estimate the outlier threshold from the distribution histogram of the reconstruction errors commonly used thresholding technique, the Otsu method [19] is used. The Otsu method calculates the optimum threshold by separating the two classes. Therefore, the desired threshold corresponds to the maximum value of between two class variances.

We used the second CAE model to estimate RE based outlier threshold, and also its latent space and RE are used for dimensionality reduction for density-based clustering. In other words, we reduced the dimension of the input time-series into 2-dimensional space by latent space and RE on the CAE model. DBSCAN [20] is a basic, simple, and effective density-based clustering algorithm; it can find arbitrarily shaped clusters. In this algorithm, a user specifies two parameters, *eps* which determine a maximum radius of the neighborhood and *minPts* which determine a minimum number of points in the *eps* radius. To automatically adjust the value of the *eps* parameter, we calculated the 3rd nearest neighboring space for each point and placed it in ascending order and have chosen the optimal *eps* for the initial maximum change. When grouping the low dimensional data via DBSCAN clustering algorithm, some points are not assigned to any cluster. In the proposed method, we have considered data points which are not clustered as one whole cluster. After grouping time-series data, find out which clusters are the outlier. If the majority of instances of the cluster exceed the outlier threshold, the whole cluster is considered an outlier.

3 Experimental Study

We have evaluated the proposed approach on 30 different datasets and compared with publicly available outlier detection methods.

3.1 Datasets

Numenta Anomaly Benchmark (NAB) [15] is a publicly available streaming anomaly detection benchmark from Numenta [21]. The NAB dataset have a variety of streaming data for anomaly types, e.g., having both true system failures as well as planned shut-downs. Different anomalies display different behavior, and it is important to be able to test for as many as possible. Also, it incorporates a wide variety of metrics, from CPU and network utilization to industrial machine sensors, from web servers to social media activity.

3.2 Evaluation Metrics

The confusion matrix is used to evaluate the performance of a prediction model. It is a summary of prediction results from the number of correct and incorrect predictions. We evaluated the compared and the proposed outlier detection methods among precision, recall, and F-score. The precision and recall show the positive predictive rate and true positive rate, respectively. They can be defined as:

$$\text{Precision} = \frac{\text{number of true positive prediction}}{\text{total number of positive prediction}} \tag{3}$$

$$\text{Recall} = \frac{\text{number of true positive prediction}}{\text{total number of actual positive elements}} \tag{4}$$

F-score combines precision and recall and gives the harmonic mean of them; it can be defined as:

$$\text{F-score} = \frac{2 \times \text{Precision} \times \text{Recall}}{\text{Precision} + \text{Recall}} \tag{5}$$

3.3 Experimental Setups

We have compared the proposed approach with publicly available state-of-the-art unsupervised methods. In the proposed method, we used the AR model based on AIC for the optimal lag selection process. CAE model was trained by the Adam algorithm [22], and the learning rate was 0.001 to minimize MSE. The batch size was 32 and the number of epochs to train model was 1000 with early stopping configuration.

For Twitter anomaly detection method, we used AnomalyDetectionVec function, because the publisher of benchmark datasets replaced time stamps by integers. The AnomalyDetectionVec function detects one or more statistically significant anomalies in a vector of observations. We used all default parameters of this method except *alpha*, *direction*, and *period* parameters. The *alpha* parameter defines the level of statistical significance with which to accept or reject anomalies. We configured it by 0.05 and 0.1. The *direction* parameter defines the directionality of the anomalies to be detected. We used 'both' configuration that finds anomalies in any direction in the dataset. The *period* parameter defines the number of observations in a single period, and we configured this parameter by a maximum of the optimal lags from the AR model on each group of time-series. For example, if a group has n number of time-series files, we will receive n number of optimal lags and choose the maximum lag length for all time-series files in the group.

For Yahoo EGADS, we run SimpleTresholdModel in Anomaly Detection Module (ADM) of the EGADS framework using no time-series model with the automatic static threshold for anomaly detection.

Numenta anomaly detector introduced own NAB score measurement. However, we evaluated results of Numenta by F-score, because NAB score does not show how accurate that algorithm in terms of true and false detections of outliers.

3.4 Results

Table 1 presents the measurements of performance, such as precision and recall of compared algorithms. The recall is a fraction of the true positive predictions over the total amount of positively labeled dataset, while the precision is a fraction of the true positive predictions among the all positive predictions. Therefore, the recall measures what proportion of actual positives was identified correctly, and the precision evaluates the efficiency of true positive predictions.

Table 1. Evaluation of the state-of-the-art algorithms and the proposed method on 30 NAB time-series dataset (%). Precision and recall are described in this table.

	Time-series	Yahoo EGADS SimpleThreshold-Model		Twitter anomaly detection alpha=0.05		Twitter anomaly detection alpha=0.1		Skyline		Numenta		The proposed	
		P.	R.	P.	R.	P.	R.	P.	R.	P.	R.	P.	R.
realAdExhange	exchange-3_cpc	36	5	21	8	21	9	0	0	13	2	61	39
	exchange-3_cpm	11	1	3	1	2	1	0	0	8	1	20	16
	exchange-4_cpm	46	4	15	3	15	3	60	2	33	4	59	12
realAWSCloudwatch	ec2_cpu_utilization_5f5533	50	1	100	0	100	0	100	0	44	2	9	25
	ec2_cpu_utilization_53ea38	37	3	39	3	44	5	0	0	14	1	25	19
	ec2_cpu_utilization_77c1ca	0	0	17	17	17	17	60	2	12	2	4	0
	ec2_cpu_utilization_825cc2	0	0	79	31	79	31	100	9	52	4	70	37
	ec2_cpu_utilization_ac20cd	0	0	45	45	45	45	71	5	27	2	41	44
	ec2_cpu_utilization_fe7f93_	27	3	15	15	15	15	42	6	14	1	9	20
	ec2_disk_write_bytes_c0d644	24	2	18	18	18	18	75	2	30	4	36	19
	ec2_network_in_5abac7	41	3	22	22	21	22	29	2	30	2	36	23
	elb_request_count_8c0756	39	2	32	2	26	3	80	1	17	1	38	18
	iio_us-east-1_i-a2eb1cd9	0	0	32	16	32	17	0	0	0	0	15	3
	rds_cpu_utilization_cc0c53	1	0	33	33	33	33	100	11	38	2	0	0
realk-nownCause	cpu_utilization_asg_misconfiguration	0	0	23	28	23	28	0	0	39	1	52	22
	ec2_request_latency_system_failure	38	5	93	4	93	4	1	1	48	3	34	21
	machine_ temperature_system_failure	1	5	80	41	76	42	97	1	0	0	58	61
	nyc_taxi_labeled_5	89	2	0	0	0	0	0	0	21	1	100	1
realTraffic	occupancy_6005	13	0	17	2	16	2	50	0.4	6	0	19	8
	occupancy_t4013	58	6	79	4	73	4	100	4	26	2	47	10
	speed_6005	18	3	60	1	60	1	100	1.3	17	2	36	9
	speed_7578	89	7	60	28	58	28	83	16	27	6	100	17
	speed_t4013	53	6	59	12	55	12	94	7	35	4	51	17
	TravelTime_387	28	3	25	25	25	25	50	7	15	2	30	13
realTweets	Twitter_volume_AAPL	49	4	40	24	39	25	66	5	47	1	32	14
	Twitter_volume_AMZN	39	3	41	3	38	3	33	1	31	1	36	18
	Twitter_volume_CRM	69	4	54	9	52	1	74	2	50	1	37	25
	Twitter_volume_FB	10	1	11	3	11	3	18	0	26	0	15	3
	Twitter_volume_GOOG	29	5	34	9	33	1	60	3	30	1	45	18
	Twitter_volume_IBM	26	2	18	4	18	4	22	1	42	1	19	4
Average		**37**	**3**	**39**	**14**	**38**	**14**	**55**	**3**	**26**	**2**	**42**	**19**

The proposed method has the highest average recall by 19% on all domains followed by Twitter anomaly detection, Skyline, Yahoo EGADS, and Numenta by 14%, 3%, 3%, and 2%, respectively. From Table 1, we can see that the compared algorithms detect a small number of true outliers. However, the proposed approach detects better than other methods. Also, the precision of the proposed method is higher than compared methods except for Skyline.

Improving the performance of recall typically reduces precision and vice versa. It is difficult to compare models with low precision and high recall or high precision and low recall. Thus, F-score is used to measure recall and precision at the same time, where the highest F-score indicates a good result. Table 2 shows mean F-scores of the compared approaches on the each domain of NAB dataset and each domain has different number of time-series. Best average F-score is highlighted in bold. The CAE based proposed approach outperforms other algorithms. It is 1.2–4.67 times better than best-performing algorithms on each domain of the whole NAB dataset.

Table 2. Evaluation of state-of-the-art algorithms and the proposed method for different domains of NAB time-series dataset (%).

Time-series	**Yahoo EGADS SimpleThresholdModel**	**Twitter anomaly detection alpha=0.05**	**Twitter anomaly detection alpha=0.1**	**Skyline**	**Numenta**	**The proposed**
realAdExchange	5.7	5.8	6.1	1.2	4.1	**28.5**
realAWSCloudwatch	2.3	**20.7**	21	6.1	3.2	20.1
realKnownCause	6	21.8	21.9	1.4	2.1	**29.6**
realTraffic	7.5	16	15.9	10.6	4.6	**19.2**
realTweets	5.6	12.5	12.9	3.6	1.1	**18.4**

4 Conclusions

By this paper, we have proposed a deep learning-based approach to detect outliers by unsupervised mode in time-series and evaluated it on the NAB benchmark datasets. In this method, the AR model is used as an optimal lag selection, and RE of the CAE model is used to estimate outlier threshold. There are two CAE models in our approach. Based on the RE of the first CAE model, the most common subset is selected from the whole dataset. Then, the second CAE model learns from the selected time-series and the final RE based outlier threshold is estimated. In the outlier detection process, we clustered subsequences by DBSCAN algorithm. Then, the RE of each subsequence in each cluster is compared with a previously determined threshold value. If the majority of subsequences of the cluster exceed the threshold, the whole cluster is considered an outlier. Experimental results showed that the proposed method gave higher F-score than all the compared approaches for 22 of 30 files, and the mean F-score outperformed 4 of 5 domains.

Acknowledgement. This research was supported by Basic Science Research Program through the National Research Foundation of Korea (NRF) funded by the Ministry of Science, ICT & Future Planning (No.2017R1A2B4010826), by (NRF-2019K2A9A2A06020672), and by the Private Intelligence Information Service Expansion (No. C0511-18-1001) funded by the NIPA (National IT Industry Promotion Agency).

References

1. Pascoal, C., De Oliveira, M.R., Valadas, R., Filzmoser, P., Salvador, P., Pacheco, A.: Robust feature selection and robust PCA for Internet traffic anomaly detection. In: 2012 Proceedings IEEE INFOCOM, pp. 1755–1763. IEEE. Orlando, Florida USA, March 2012
2. Kim, D.P., Yi, G.M., Lee, D.G., Ryu, K.H.: Time-variant outlier detection method on geosensor networks. In: Proceedings of the International Symposium on Remote Sensing. The Korean Society of Remote Sensing Korea. Daejeon, Korea, October 2008
3. Lyon, A., Minchole, A., Martınez, J.P., Laguna, P., Rodriguez, B.: Computational techniques for ECG analysis and interpretation in light of their contribution to medical advances. J. Royal Soc. Interface **15**(138), 20170821 (2018)
4. Anandakrishnan, A., Kumar, S., Statnikov, A.; Faruquie, T., Xu, D.: Anomaly Detection in Finance: Editors' Introduction. In: Proceedings of the Machine Learning Research. Halifax, Nova Scotia, August 2017
5. Jin, C.H., Park, H.W., Wang, L., Pok, G., Ryu, K.H.: Short-term traffic flow forecasting using unsupervised and supervised learning techniques. In: Proceedings of the 6th International Conference FITAT and 3rd International Symposium ISPM. Cheongju, Korea, September 2013
6. Hauskrecht, M., Batal, I., Valko, M., Visweswaran, S., Cooper, C.F., Clermont, G.: Outlier detection for patient monitoring and alerting. J. Biomed. Inform. **46**(1), 47–55 (2013)
7. Cho, Y.S., Moon, S.C., Ryu, K.S., Ryu, K.H.: A study on clinical and healthcare recommending service based on cardiovascular disease pattern analysis. Int. J. Bio-Sci. Bio-Technol. **8**(2), 287–294 (2016)
8. Tan, P.N., Steinbach, M., Kumar, V.: Introduction to Data Mining. Pearson Education Inc., Boston (2006)
9. Chalapathy, R., Menon, A.K., Chawla, S.: Anomaly Detection using One-Class Neural Networks. arXiv preprint arXiv:1802.06360 (2018)
10. Chandola, V., Banerjee, A., Kumar, V.: Anomaly detection: a survey. ACM Comput. Surv. **41**(3), 15 (2009)
11. Chawla, N.V., Japkowicz, N., Kotcz, A.: Special issue on learning from imbalanced data sets. ACM SIGKDD Explor. Newsl. **6**(1), 1–6 (2004)
12. Skyline. https://github.com/etsy/skyline. Accessed 23 Sept 2019
13. Laptev, N., Amizadeh, S., Flint, I.: Generic and scalable framework for automated time-series anomaly detection. In: Proceedings of the 21st ACM SIGKDD International Conference on Knowledge Discovery and Data Mining, pp. 1939–1947. ACM, Sydney, Australia, August 2015
14. Rosner, B.: Percentage points for a generalized ESD many-outlier procedure. Technometrics **25**(2), 165–172 (1983)
15. Ahmad, S., Lavin, A., Purdy, S., Agha, Z.: Unsupervised real-time anomaly detection for streaming data. Neurocomputing **262**, 134–147 (2017)
16. Lavin, A., Ahmad, S.: Evaluating real-time anomaly detection algorithms–the numenta anomaly benchmark. In: 2015 IEEE 14th International Conference on Machine Learning and Applications, pp. 38–44. IEEE, Miami, Florida, USA, December 2015

17. Akaike, H.: A new look at the statistical model identification. IEEE Trans. Autom. Control **19**(6), 716–723 (1974)
18. Amarbayasgalan, T., Jargalsaikhan, B., Ryu, K.: Unsupervised novelty detection using deep autoencoders with density based clustering. Appl. Sci. **8**(9), 1468 (2018)
19. Otsu, N.: A threshold selection method from gray-level histograms. IEEE Trans. Syst., Man, Cybern. **9**(1), 62–66 (1979)
20. Ester, M., Kriegel, H.P., Sander, J., Xu, X.: A density-based algorithm for discovering clusters in large spatial databases with noise. KDD **96**(34), 226–231 (1996)
21. Numenta's HTM community. https://www.numenta.org/. Accessed 23 Sept 2019
22. Kingma, D.P., Ba, J.: Adam: a method for stochastic optimization. arXiv preprint arXiv:1412. 6980 (2014)

VAR-GRU: A Hybrid Model for Multivariate Financial Time Series Prediction

Lkhagvadorj Munkhdalai[1], Meijing Li[2], Nipon Theera-Umpon[3], Sansanee Auephanwiriyakul[4], and Keun Ho Ryu[5,6](✉)

[1] Database/Bioinformatics Laboratory, School of Electrical and Computer Engineering, Chungbuk National University, Cheongju 28644, Republic of Korea
lhagii@dblab.chungbuk.ac.kr

[2] College of Information Engineering, Shanghai Maritime University, 213, 1550 Haigang Avenue Pudong New Area, Shanghai, China
mjli@shmtu.edu.cn

[3] Department of Electrical Engineering, Faculty of Engineering, Chiang Mai University, Chiang Mai 50200, Thailand
DrNipon@chiangmai.ac.th, nipon@ieee.org

[4] Department of Computer Engineering, Faculty of Engineering, Chiang Mai University, Chiang Mai 50200, Thailand
sansanee@ieee.org

[5] Faculty of Information Technology, Ton Duc Thang University, Ho Chi Minh City 700000, Vietnam
khryu@tdtu.edu.vn

[6] Department of Computer Science, College of Electrical and Computer Engineering, Chungbuk National University, Cheongju 28644, Republic of Korea
khryu@chungbuk.ac.kr

Abstract. A determining the most relevant variables and proper lag length are the most challenging steps in multivariate time series analysis. In this paper, we propose a hybrid Vector Autoregressive and Gated Recurrent Unit (VAR-GRU) model to find the contextual variables and suitable lag length to improve the predictive performance for financial multivariate time series. VAR-GRU approach consists of two layers, the first layer is a VAR model-based variable and lag length selection and in the second layer, the GRU-based multivariate prediction model is trained. In the VAR layer, the Akaike Information Criterion (AIC) is used to select VAR order for finding the optimal lag length. Then, the Granger Causality test with the optimal lag length is utilized to define the causal variables to the second layer GRU model. The experimental results demonstrate that the ability of the proposed hybrid model to improve prediction performance against all base predictors in terms of three evaluation metrics. The model is validated over real-world financial multivariate time series dataset.

Keywords: Multivariate financial time series · Vector Autoregressive · Grange causality · Gated Recurrent Unit

N. T. Nguyen et al. (Eds.): ACIIDS 2020, LNAI 12034, pp. 322–332, 2020.
https://doi.org/10.1007/978-3-030-42058-1_27

1 Introduction

Financial time series prediction is one of the major research topics because the participants in the financial market aim to earn a profit or reduce a loss based on its future prediction. Therefore, the accurate predictions about market fluctuations is the most important aspect of trading in the financial market [1]. In order to accurately predict financial market movements, this paper proposes a hybrid model consisting of Vector Autoregressive and Gated Recurrent Unit (VAR-GRU) for multivariate financial time series prediction. GRU is a powerful variation of Long-short term memory (LSTM) used for processing sequential data such as time series, written natural language, sound, etc. [2]. Vector Autoregression (VAR) is an econometric model which is one of the most commonly used multivariate regression time series analytic techniques [3]. Our proposed approach consists of two main stages. At the first stage, we perform optimal lag length and variable selection based on the Akaike information criterion (AIC) and Granger causality test within VAR framework. In multivariate time series, once each variable depends not only on its historic values but also values of several dependent variables changing with time, it is essential to select contextual variables as input to improve the prediction performance. The first stage of our hybrid model perform this task. At the second stage, we construct a model using GRU deep learning approach for forecasting time series. In recent years, machine learning models, more especially recurrent neural networks, have been widely used for forecasting financial time series [4–9]. However, researchers have not paid much attention to the model for forecasting multivariate financial time series. On the other hands, the multivariate model be able to an extension of the several univariate models, but variable selection procedure is still a challenging step in this analysis [10–13].

In the literature, Zhang [14] proposed a hybrid autoregressive integrated moving average (ARIMA) and artificial neural network model similar to our model for univariate financial time series. The two types of hybrid methods as the multi-layer perceptron, dynamic artificial neural network with GARCH to predict NASDAQ stock market price were proposed in Guresen, et al. [15]. They used the grid search algorithm to define the optimal lag length. For multivariate financial time series analysis, Zhong and Enke [16] applied Principal Component Analysis (PCA) to reduce the dimensionality of the dataset including 60 financial and economic variable over a 10-year period for forecasting daily stock market return. But it is difficult to interpret the features after the PCA transformation. In addition, Atsalakis et al. [17] surveyed more than 100 papers that focused on financial time series prediction. They found that the most commonly inputs used for forecasting the financial market are the opening or closing stock price, market volume, the highest or lowest daily price, etc. In this paper, unlike all previous studies, we collect financial indicators such as 5-year treasury constant maturity rate (DGS5), 5-year forward inflation expectation rate (T5YIFR), Moody's Seasoned Baa Corporate Bond Yield Relative to Yield on 10-Year Treasury Constant Maturity (BAA10Y), etc., from The Federal Reserve Bank of St. Louis to generate multivariate time series dataset as well as three stock market prices, which are S&P 500 index, Dow Jones Industrial (DJI) and Nasdaq, are chosen as the dependent variables.

In the experimental part, VAR-GRU approach is evaluated on our collected dataset and the results show that our proposed hybrid model demonstrates promising performances compared with other baselines.

This paper is organized as follows. Section 2 describes VAR model and GRU neural networks. Section 3 presents our proposed VAR-GRU framework. Then Sect. 4 presents datasets and experimental results. At the end of this study, Sect. 5 summarizes the general findings from this study and discusses possible future research areas.

2 Methodology

2.1 Vector Autoregression

The VAR model is the most successful, flexible, and easy to use model for the analysis of multivariate time series [18]. This model has been extensively used for interpreting the dynamic behavior of economic and financial time series and for forecasting. The VAR is a natural extension of the basic autoregressive (AR) model, which is used to capture the linear interdependencies among multiple time series. The structure of the VAR model is that each variable is a linear function of past lags of itself and past lags of the other variables.

Let $Y_t = (y_{1t}, y_{2t}, \ldots, y_{nt})'$ denote an $(n \times 1)$ vector of time series variables. The basic p-lag vector autoregressive VAR(p) model is defined as:

$$Y_t = c + \sum_{i=1}^{p} \Pi_i Y_{t-i} + \varepsilon_t \tag{1}$$

where t = 1,..., T, Π_i are $n \times n$ coefficient matrices, and ε_t is $n \times 1$ white noise processes that may be contemporaneously correlated. The parameters of VAR model are estimated by minimizing the sum of squared ε_t values using the principle of least squares.

However, there are two important aspects to determine the structure of the VAR model, one of which is to choose the optimal lag length and the other is to find contextual variables for predicting another variable or group of variables.

For choosing suitable lag length, the general approach is to fit VAR(p) models with orders $p = 0, \ldots, p_{max}$ and choose the value of p which minimizes some model selection criteria. Most popular criteria is Akaike Information Criterion (AIC) [19]:

$$AIC(p) = \ln|\Sigma(p)| + \frac{2}{T} pn^2 \tag{2}$$

where $\widetilde{\Sigma}(p) = T^{-1} \sum_{t=1}^{T} \hat{\varepsilon}_t \hat{\varepsilon}_t$ is the residual covariance matrix without a degrees of freedom correction from a VAR(p) model, T is the sample size, and pn^2 is a penalty function which penalizes large VAR(p) models.

For finding the variables, the Granger causality (GS) test is used for determining whether one time series is contextually related or not in forecasting another [20]. The GS test uses the F-test to determine whether independent variables provide statistically significant information about future values of the dependent variable or not.

In a bivariate VAR(p) model for $Y_t = (y_{1t}, y_{2t})'$, if time series y_2 Granger non-causality time series y_1, all of the p VAR coefficient matrices $\Pi_1, \Pi_2, \ldots, \Pi_p$ are equal to zero.

$$\begin{pmatrix} y_{1t} \\ y_{2t} \end{pmatrix} = \begin{pmatrix} c_1 \\ c_2 \end{pmatrix} + \begin{pmatrix} \pi_{11}^1 & \pi_{12}^1 \\ \pi_{21}^1 & \pi_{22}^1 \end{pmatrix} \begin{pmatrix} y_{1t-1} \\ y_{2t-1} \end{pmatrix} + \ldots + \begin{pmatrix} \pi_{11}^p & \pi_{12}^p \\ \pi_{21}^p & \pi_{22}^p \end{pmatrix} \begin{pmatrix} y_{1t-p} \\ y_{2t-p} \end{pmatrix} + \begin{pmatrix} \varepsilon_{1t} \\ \varepsilon_{2t} \end{pmatrix} \tag{3}$$

For equation, test null hypothesis that $\pi_{12}^1 = \ldots = \pi_{12}^p = 0$ using the F-test. Similarly, time series y_1 Granger non-causality time series y_2, if all of the coefficients on lagged values of y_1 are zero in the equation for y_2.

2.2 Gated Recurrent Unit

This study uses GRU networks for constructing prediction model. GRU network is a variation of LSTM neural network. It was proposed by Cho et al. [2] to make each recurrent unit to be able to adaptively capture dependencies of different time scales. GRU has only two gates, a reset gate and update gate, which are reminiscent of forget and input gates of LSTM. Those gates modulate the flow of information inside the unit without having a separate cells. At time step t, the activation h_t^j of the GRU is computed by:

$$h_j^{(t)} = z_j^{(t)} \odot h_j^{(t-1)} + (1 - z_j^{(t)}) \odot h_j^{(t)} \tag{4}$$

where $h_j^{(t-1)}$ and $\widetilde{h}_j^{(t)}$ respectively denote the previous and the candidate activations and $z_j^{(t)}$ denotes the update gate. It decides how much the unit updates its activation. The update gate and candidate activation are computed by:

$$\begin{aligned} z_j^{(t)} &= \sigma(\omega_z x^{(t)} + u_z h^{(t-1)})_j \\ h_j^{(t)} &= \tanh(\omega_r x^{(t)} + u_r h^{(t-1)})_j \end{aligned} \tag{5}$$

where ω_z, u_z, ω_r and u_r denote the weights, x^t denotes input variables at time t, r^t is set of reset gates and $\odot$ is an element-wise multiplication. The reset gate is computed similarly to the update gate:

$$r_j^{(t)} = \sigma(\omega_r x^{(t)} + u_r h^{(t-1)})_j \tag{6}$$

The update and reset mechanisms respectively help the GRU to capture long-term dependencies and to use the model capacity efficiently by allowing it to reset whenever the detected information is not necessary anymore.

3 The Proposed Framework

Our proposed framework of the hybrid model named VAR-GRU is shown in Fig. 1. The VAR- GRU model consists of two main stages. At the first stage as a VAR, it is

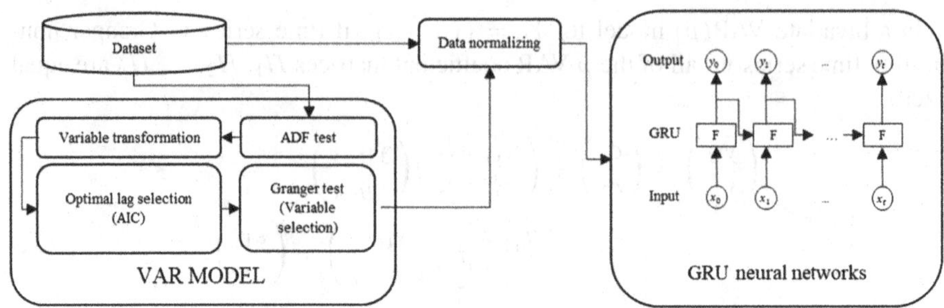

Fig. 1. Proposed VAR-GRU framework.

to determine the optimal lag length and contextual variables using VAR model. The suitable variables and optimal lag selection are the fundamental preprocessing steps for multivariate time series analysis [1]. The variable selection reduces the complexity of the task, potentially increases the model performances and speed up the training in neural networks [21].

At the VAR layer, we used Augmented Dickey-Fuller (ADF) test to assess stationarity or not for each time series. If time series are non-stationarity, those should be changed into stationarity. Consequently, an optimal lag length is determined based on the stationarity time series and the AIC criterion. Finally, statistically significant variables are chosen by GS test with the optimal lag length.

At the second GRU layer, GRU neural network, which is a type of deep neural networks for modeling sequential and time series data, is trained to build a predictive model based on the selected variables and optimal lag length.

4 Experimental Result

4.1 Dataset

Datasets used in this study were retrieved from The Federal Reserve Bank of St. Louis, https://fred.stlouisfed.org and yahoo finance, https://finance.yahoo.com. In order to evaluate the predictive accuracy of VAR-GRU and baselines, datasets are partitioned into three parts; i.e., training (70%), validation (10%), and test sets (20%).

The multivariate financial market dataset consists of S&P 500 index, Dow Jones Industrial (DJI), Nasdaq stock market price, and other seven financial indicators, which are correlated to financial market such as trade weighted U.S. dollar index (DTWEXB), 5-year treasury constant maturity rate (DGS5), 5-year forward inflation expectation rate (T5YIFR), Moody's Seasoned Baa Corporate Bond Yield Relative to Yield on 10-Year Treasury Constant Maturity (BAA10Y), Economic policy uncertainty index for U.S. (EPUI), Wilshire U.S. real estate investment trust price index (WILLREITPR) and Crude oil prices (WTI) (Table 1).

Table 1. Summary of dataset.

№	Variables	Unit	Mean	Max	Min	Sample
1	SP500	U.S dollar	1,444.4	2,453.5	676.5	3,608
2	DJI	U.S dollar	12,946.0	21,529.0	6,547.0	3,608
3	NASDAQ	U.S dollar	2,989.0	6,322.0	1,269.0	3,608
4	DGS5	Percent	2.44	5.23	0.56	3,608
5	DTWEXM	Index	81.3	100.9	68.0	3,608
6	T5YIFR	Percent	2.33	3.05	0.43	3,608
7	BAA10Y	Percent	2.7	6.2	1.5	3,608
8	EPUI	Index	99.4	626.0	3.3	3,608
9	WILLREITPR	Index	203.7	307.8	66.4	3,608
10	WTI	U.S dollar	69.4	145.3	25.3	3,608

4.2 Baselines

We use simple recurrent neural network (RNN) and LSTM as a prediction baseline and Random Forest variable importance and Recursive Feature Elimination algorithm with Support Vector Regression as a variable selection baseline.

A simple RNN was proposed by Elman [22], only RNNs have the feedback connection that gives the ability to memorize and make use of past information. GRU and LSTM are the extensions of simple RNN, which try to solve the vanishing gradient problem and long short-memory term.

LSTM network was proposed by Hochreiter and Schmidhuber [23]. This network helps to solve long-term dependencies by extending their memory cells and utilizing a gating mechanism to control information flow. The memory cell consists of three gates – input, forget and output gate. Those gates decide whether or not to add new input in (input gate), erase the unnecessary information (forget gate) or to add it impact the output at the current time step (output gate).

The baselines for variable selection:

RF-based variable selection uses random forest importance to define the most important variables. We put all variables and their lags as input variables into random forest model. Then we estimate their importance for each variable and lag to choose most important variables.

RFE with SVM (RFE-SVM) refers to Recursive Feature Elimination algorithm with Support Vector Regression. We use SVM to predict target value, then we recursively eliminate the not necessary variables according RFE algorithm.

4.3 Experimental Setup

In RNNs, their hyper-parameters: learning rate, batch size, and epoch number should be pre-defined to train a model. We set the default learning rate to 0.001 in Keras [24], maximum epoch number for training to 2000 and use a mini-batch that is equal to 8 at each iteration as well as an Early Stopping algorithm is used for finding the optimal epoch number using validation set.

For RNN architectures, the size of recurrent layers are chosen from {1, 2} and the number of neurons for each layer are selected from {16, 32, 64}.

Finally, for our dataset, one hidden layer with 32 neurons and linear activation function are utilized to construct RNN models. We set maximum lag length is equal to 12. We added L1 regularization with coefficient 0.0001 to penalize weight parameters and Adam optimizer is used to train the models [25].

In addition, the root mean square error (RMSE), mean absolute error (MAE) and mean absolute percentage error (MAPE) are used to evaluate model performances.

$$RMSE = \sqrt{\sum_{i=1}^{n} \frac{(\hat{y}_i - y_i)^2}{n}} \tag{7}$$

$$MAE = \sum_{i=1}^{n} \frac{|\hat{y}_i - y_i|}{n} \tag{8}$$

$$MAPE = \frac{100\%}{n} \sum_{i=1}^{n} \left| \frac{\hat{y}_i - y_i}{y_i} \right| \tag{9}$$

where $\hat{y}_i$ denotes the *i-th* predicted value, y_i denotes the *i-th* actual value and n is the number of observations.

All experiments in this study were performed by using Python programming language, version 3.6, on a PC with GeForce GTX 1070 Ti GPU with Microsoft Windows 10 operating system. RNN models were implemented by the 'Keras' and 'TensorFlow-gpu' libraries for deep learning [24, 26].

4.4 Prediction Results

Table 2 shows the predictive accuracy of the S&P 500 stock market price. All experiments were repeated 3 times and the average results are reported. The prediction results demonstrated that our proposed model outperforms other baseline models. According to VAR result, the optimal lag length chosen by AIC is equal to 10 and economic policy uncertainty index for U.S. (EPUI), Wilshire U.S. real estate investment trust price index (WILLREITPR) and Crude oil prices (WTI) variables are selected as contextual variables for S&P 500 stock market. In addition, we are able to see that VAR-based variable selection method significantly improved the predictive accuracy in term of the most of prediction models.

For Nasdaq stock price, our proposed model achieves the best performance as well. The results are shown in Table 3. Unlike previous stock markets, 5-year forward inflation expectation rate (T5YIFR), Moody's Seasoned Baa Corporate Bond Yield Relative

Table 2. The predictive performance of the S&P 500 stock market price.

Feature selection	Method	RMSE	MAE	MAPE
SVM	LSTM	712.2 ± 17.1	693.0 ± 17.1	31.39 ± 0.78
	GRU	729.6 ± 102	712.3 ± 111	32.37 ± 5.36
	RNN	773.1 ± 61.8	760.1 ± 68.2	34.64 ± 3.29
RF	LSTM	21.20 ± 0.05	17.61 ± 0.05	0.806 ± 0.00
	GRU	21.45 ± 0.60	17.82 ± 0.59	0.814 ± 0.02
	RNN	18.64 ± 0.59	14.68 ± 0.77	0.680 ± 0.03
VAR	LSTM	133.8 ± 196.	124.8 ± 191.	5.647 ± 8.60
	GRU	**17.24 ± 0.28**	**12.91 ± 0.41**	**0.603 ± 0.01**
	RNN	31.74 ± 2.35	28.36 ± 2.35	1.287 ± 0.10
No feature selection	LSTM	18.90 ± 0.44	14.23 ± 0.77	0.669 ± 0.03
	GRU	17.69 ± 0.39	13.10 ± 0.68	0.616 ± 0.02
	RNN	17.59 ± 0.61	13.10 ± 0.78	0.616 ± 0.03

Table 3. The predictive performance of the Nasdaq stock market price.

Feature selection	Method	RMSE	MAE	MAPE
SVM	LSTM	2293 ± 78.3	2216 ± 78.6	40.99 ± 1.47
	GRU	2218 ± 21.5	2129 ± 23.6	39.28 ± 0.45
	RNN	2299 ± 186	2245 ± 195	41.81 ± 3.77
RF	LSTM	75.86 ± 0.35	64.52 ± 0.35	1.187 ± 0.00
	GRU	75.70 ± 0.17	64.34 ± 0.17	1.184 ± 0.00
	RNN	82.53 ± 3.29	69.16 ± 3.07	1.262 ± 0.05
VAR	LSTM	545.8 ± 860.	492.1 ± 788.	8.942 ± 14.2
	GRU	**49.34 ± 1.72**	**36.15 ± 1.80**	**0.703 ± 0.03**
	RNN	52.28 ± 1.87	41.07 ± 2.18	0.786 ± 0.03
No feature selection	LSTM	61.71 ± 7.69	45.61 ± 6.76	0.884 ± 0.12
	GRU	53.03 ± 2.76	41.25 ± 4.33	0.793 ± 0.07
	RNN	170.3 ± 5.18	156.2 ± 5.26	2.862 ± 0.09

to Yield on 10-Year Treasury Constant Maturity (BAA10Y), Economic policy uncertainty index for U.S. (EPUI), Wilshire U.S. real estate investment trust price index (WILLREITPR) and Crude oil prices (WTI) variable are selected by Granger test.

Regarding DJI stock market price, although VAR-GRU model achieved the best performance in terms of RMSE, VAR-RNN model shows the better performance for the MAE and MAPE as shown in Table 4. Similar to the S&P 500 stock market price, the optimal lag length is equal to 10 and economic policy uncertainty index for U.S. (EPUI), Wilshire U.S. real estate investment trust price index (WILLREITPR) and Crude oil prices (WTI) variables are selected.

Table 4. The predictive performance of the Nasdaq stock market price.

Feature selection	Method	RMSE	MAE	MAPE
SVM	LSTM	5033 ± 283	4717 ± 285	24.37 ± 1.50
	GRU	4618 ± 503	4236 ± 581	21.76 ± 3.15
	RNN	3445 ± 1239	3061 ± 1361	15.71 ± 7.13
RF	LSTM	174.6 ± 2.42	141.6 ± 1.72	0.745 ± 0.00
	GRU	178.8 ± 1.46	145.7 ± 1.30	0.765 ± 0.00
	RNN	184.5 ± 6.49	150.6 ± 6.20	0.790 ± 0.03
VAR	LSTM	179.4 ± 5.42	134.4 ± 9.93	0.730 ± 0.04
	GRU	**144.9 ± 3.25**	106.0 ± 2.84	0.583 ± 0.01
	RNN	145.1 ± 4.76	**104.6 ± 3.62**	**0.576 ± 0.01**
No feature selection	LSTM	205.1 ± 65.7	172.4 ± 66.2	0.909 ± 0.33
	GRU	169.0 ± 5.77	138.2 ± 6.36	0.738 ± 0.03
	RNN	156.3 ± 7.86	117.3 ± 9.33	0.643 ± 0.04

To summarize, for the multivariate financial market prediction, VAR-GRU indicates significantly better performance than other baseline models. In addition, VAR-based variable selection approach can be used with recurrent neural networks for multivariate financial time series prediction.

5 Conclusions

In this paper, we proposed a new predictive framework established by using VAR and GRU approaches for forecasting multivariate financial time series to improve the predictive accuracy. The procedure for developing this framework is as follows: first, the contextual time series variables are chosen by VAR approach using AIC and Granger causality test; second, GRU network is used to construct the prediction model. The predicted values of proposed approach measured using three stock market price are in agreement with the actual values. We used several predictive measurements and compared our proposed model to baseline variable selection methods and prediction models. The results provide evidence as we assumed, our proposed model indicated the better performances than baselines. In addition, it is observed that VAR model-based variable selection method can be used with not only GRU, but also other recurrent neural network approaches for multivariate time series.

Finally, our findings indicate that it is possible to develop a model that is highly efficient and predictable by combining traditional econometric approach with state-of-the-art machine learning algorithm for forecasting multivariate time series.

Acknowledgements. This research was supported by Basic Science Research Program through the National Research Foundation of Korea (NRF) funded by the Ministry of Science, ICT & Future Planning (No. 2017R1A2B4010826) and (No. 2019K2A9A2A06020672) in Republic of Korea, and by the National Natural Science Foundation of China (Grant No. 61702324 and Grant No. 61911540482) in People's Republic of China.

References

1. Beaver, W.H.: Market prices, financial ratios, and the prediction of failure. J. Account. Res. 179–192 (1968). https://doi.org/10.2307/2490233
2. Cho, K., Merrienboer, B., Bahdanau, D., Bengio, Y.: On the properties of neural machine translation: Encoder–Decoder approaches. In: Eighth Workshop on Syntax, Semantics and Structure in Statistical Translation, pp. 103–111. Association for Computational Linguistics, Doha, Qatar (2014). https://doi.org/10.3115/v1/w14-4012
3. Lütkepohl, H.: New Introduction to Multiple Time Series Analysis. Springer, Heidelberg (2005). https://doi.org/10.1007/978-3-540-27752-1
4. Fischer, T., Krauss, C.: Deep learning with long short-term memory networks for financial market predictions. Eur. J. Oper. Res. **270**(2), 654–669 (2018). https://doi.org/10.1016/j.ejor.2017.11.054
5. Ding, X., Zhang, Y., Liu, T., Duan, J.: Deep learning for event-driven stock prediction. In: Twenty-Fourth International Joint Conference on Artificial Intelligence, pp. 2327–2333. IJCAI, Buenos Aires, Argentina (2015)
6. Yang, E., et al.: A simulation-based study on the comparison of statistical and time series forecasting methods for early detection of infectious disease outbreaks. Int. J. Environ Health Res. **15**(5), 966 (2018). https://doi.org/10.3390/ijerph15050966
7. Wang, J., Wang, J., Fang, W., Niu, H.: Financial time series prediction using Elman recurrent random neural networks. Comput. Intell. Neurosci. **2014**, 1–14 (2016). https://doi.org/10.1155/2016/4742515
8. Rather, A.M., Agarwal, A., Sastry, V.N.: Recurrent neural network and a hybrid model for prediction of stock returns. Expert Syst. Appl. **42**(6), 3234–3241 (2015). https://doi.org/10.1016/j.eswa.2014.12.003
9. Munkhdalai, L., et al.: An end-to-end adaptive input selection with dynamic weights for forecasting multivariate time series. IEEE Access **7**, 99099–99114 (2019). https://doi.org/10.1109/ACCESS.2019.2930069
10. Reinsel, G.C.: Elements of Multivariate Time Series Analysis. Springer, New York (2003)
11. Jin, C.H., et al.: A SOM clustering pattern sequence-based next symbol prediction method for day-ahead direct electricity load and price forecasting. Energy Convers. Manag. **90**, 84–92 (2015). https://doi.org/10.1016/j.enconman.2014.11.010
12. Jin, C.H., Pok, G., Park, H.W., Ryu, K.H.: Improved pattern sequence-based forecasting method for electricity load. IEEJ Trans. Electr. Electron. Eng. **9**(6), 670–674 (2014). https://doi.org/10.1002/tee.22024
13. Islam, F., Shahbaz, M., Ahmed, A.U., Alam, M.M.: Financial development and energy consumption nexus in Malaysia: a multivariate time series analysis. Econ. Model. **30**, 435–441 (2013). https://doi.org/10.1016/j.econmod.2012.09.033
14. Zhang, G.P.: Time series forecasting using a hybrid ARIMA and neural network model. Neurocomputing. **50**, 159–175 (2003). https://doi.org/10.1016/S0925-2312(01)00702-0
15. Guresen, E., Kayakutlu, G., Daim, T.U.: Using artificial neural network models in stock market index prediction. Expert Syst. Appl. **38**(8), 10389–10397 (2011). https://doi.org/10.1016/j.eswa.2011.02.068

16. Zhong, X., Enke, D.: Forecasting daily stock market return using dimensionality reduction. Expert Syst. Appl. **67**, 126–139 (2017). https://doi.org/10.1016/j.eswa.2016.09.027
17. Atsalakis, G.S., Valavanis, K.P.: Surveying stock market forecasting techniques–Part II: soft computing methods. Expert Syst. Appl. **36**(3), 5932–5941 (2009). https://doi.org/10.1016/j.eswa.2008.07.006
18. Zivot, E., Wang, J.: Modeling Financial Time Series with S-Plus®. Springer, New York (2007). https://doi.org/10.1007/978-0-387-32348-0
19. Akaike, H.: Fitting autoregressive models for prediction. Ann. Inst. Stat. Math. **21**(1), 243–247 (1969)
20. Granger, C.W.: Investigating causal relations by econometric models and cross-spectral methods. Econometrica 424–438 (1969). https://doi.org/10.2307/1912791
21. Guyon, I., Elisseeff, A.: An introduction to variable and feature selection. J. Mach. Learn. Res. **3**, 1157–1182 (2003). https://doi.org/10.1162/153244303322753616
22. Elman, J.L.: Finding structure in time. Cogn. Sci. **14**(2), 179–211 (1990). https://doi.org/10.1207/s15516709cog1402_1
23. Hochreiter, S., Schmidhuber, J.: Long short-term memory. Neural Comput. **9**(8), 1735–1780 (1997). https://doi.org/10.1162/neco.1997.9.8.1735
24. Gulli, A., Pal, S.: Deep Learning with Keras. Packt Publishing Ltd., Birmingham (2017)
25. Kingma, D.P., Ba, J.: Adam: a method for stochastic optimization. arXiv preprint arXiv:1412.6980 (2014). https://arxiv.org/abs/1412.6980
26. Abadi, M., et al.: Tensorflow: a system for large-scale machine learning. In: 12th Symposium on Operating Systems Design and Implementation. pp. 265–283. USENIX, Savannah (2016)

Contextual Anomaly Detection in Time Series Using Dynamic Bayesian Network

Achyut Mani Tripathi(✉) and Rashmi Dutta Baruah(✉)

Department of Computer Science and Engineering,
Indian Institute of Technology Guwahati, Guwahati 781039, Assam, India
{t.achyut,r.duutabaruah}@iitg.ac.in

Abstract. In this paper, we propose a novel method to identify contextual anomaly in time series using Dynamic Bayesian Networks (DBN). DBN is a powerful machine learning approach that captures temporal characteristics of time series data. In order to detect contextual anomaly we integrate contextual information to the DBN framework, referred to as Contextual DBN (CxDBN). The efficacy of CxDBN is shown using a case study of the identification of contextual anomaly in real-time oil well drilling data.

Keywords: Anomaly detection · Contextual anomaly · Dynamic Bayesian Networks · Oil well drilling

1 Introduction

Increasing demand of sensor-equipped infrastructure for monitoring industrial process requires anomaly detection framework as an essential unit to trigger anomalous situations. At present machine learning techniques are used to learn models from high dimensional data gathered by multiple sensors to identify the transition of dynamical system into a precarious state [22]. Finance [19], video surveillance [23] are among the well-known domains that require extensive use of anomaly detection techniques. Generally, anomaly detection methods can be subdivided based on learning techniques i.e., supervised, semi-supervised, and unsupervised techniques [5]. Various methods have been proposed to detect anomalies in time series data using the aforementioned learning methods. Temporal characteristics, uncertainty, and also high velocity (in case of data streams) of the time series data are major factors that effect the performance of the anomaly detection methods. Though there are several approaches that focus on identifying point anomalies and collective anomalies, not many existing approaches focus on contextual information to identify contextual anomalies in time series data [5]. Contextual anomalies are special kind of anomalies whose identification requires the inclusion of context information with anomaly detection models. For example, temperature sensor used to measure the temperature of a room will show high value during the summer season, however, if it shows the same during

N. T. Nguyen et al. (Eds.): ACIIDS 2020, LNAI 12034, pp. 333–342, 2020.
https://doi.org/10.1007/978-3-030-42058-1_28

winter then this can be alarmed as anomalous while taking into account the weather information as the context information. Contextual anomalies generally occur in areas that require sensors that include sequential, temporal, and spatial characteristics. Remote sensing [11], oil well drilling [4], and healthcare [10] are among the significant areas in which the sensors show contextual and temporal behaviors.

In this paper, our primary focus is on the detection of the contextual anomalies. Contextual information can include behavior attributes and external information like location, structure, sequential, and spatial information. Two approaches have been mainly applied to resolve the problem of contextual anomalies. The first approach requires the transformation of the current problem into point anomaly detection problem, and later the existing point anomaly detection methods are used to detect the anomalies in the given contexts. The second approach is to use the existing model and incorporate the contextual information while learning and inference [6]. The second approach is more feasible as compared to former because former method requires the contextual information a priori to detect the anomalies.

Numerous methods have been proposed to identify point anomalies in time series data. Unsupervised models like Local Outlier Factor (LOF) [3], Incremental Local Outlier Factor (I-LOF) [20] were well applied to detect the anomalies. Supervised predictive models like Artificial Neural Networks (ANN) [21], Long Short Term Memory (LSTM) [14], Recurrent Neural Networks (RNN) [16], Convolutional Neural Networks (CNN) [13] were extensively used to detect a deviation in real values against the predicted values to identify the anomalies in the time series. However, lack of addition of contextual information within the existing framework is the major factor that results in high false alarm rate when applied to detect anomalies in the areas that involve contextual information while making decisions.

Probabilistic models like Hidden Markov model [12], Dynamic Naive Bayesian Classifier [2] and Dynamic Bayesian Network [7] have shown better performance when used to model the temporal characteristics of sequential data like time series data. Belief Network model like the Dynamic Bayesian Network [15] model is capable to model the statistical relationship between multiple time series and is capable of handling heterogeneous sources of information. The DBN encodes these information in terms of directed acyclic graph (DAG), which can be easily modified by adding the contextual behaviour attribute to the DAG. Some of the methods that are used to identify the anomalies in the time series using DBN can be found in [1,17,18,24].

The organization of this paper is as follows: Sect. 2 presents preliminaries of Bayesian Network (BN) and Dynamic Bayesian Network (DBN). Section 3 describes the proposed method to identify contextual anomalies in the time series. Experiments and results are shown in Sect. 4, and finally, Sect. 5 presents the conclusion and future work.

2 Preliminaries

As the proposed method is based on the Dynamic Bayesian Network, we first discuss the basics of Bayesian Network and then the Dynamic Bayesian Network in this section.

2.1 Bayesian Network (BN)

Bayesian Network [8] is a probabilistic Graphical Model that represents relationship between random variables in terms of DAG. Each node in the directed acyclic graph (DAG) shows the random variable and edges represent relationship between the variables. Figure 1 shows a BN with random variables A, B, C, D and E. The edges (A → B), (B → (C, D, E)) show dependency of (B) and (C, D, E) over their parents (A) and (B) respectively.

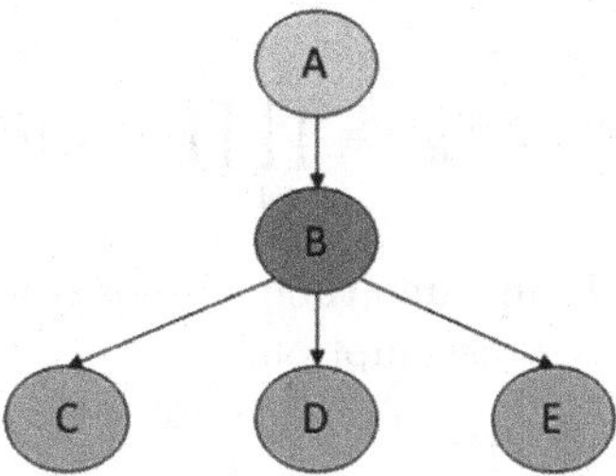

Fig. 1. Bayesian network

The generalized formula to compute the joint probability can be given as

$$P(X_1, X_2, .., X_R) = \prod_{r=1}^{R} P(X_r|Par(X_r)) \tag{1}$$

Where R is the total number of random variables and $Par(X_r)$ is the parent of the attribute X_r. So, the joint probability distribution of the Fig. 1 can be represented as

$$P(A, B, C, D, E) = P(A) * P(B|A) * P(C|B) * P(D|B) * P(E|B) \tag{2}$$

2.2 Dynamic Bayesian Network (DBN)

Dynamic Bayesian Network [9] or Temporal Bayesian Network is an extension of the Bayesian Network that relates variables of the Bayesian Network over the sequenced time stamps. Here the term dynamic denotes modeling of dynamic behavior and structure of network remains same over the time. The DBN is

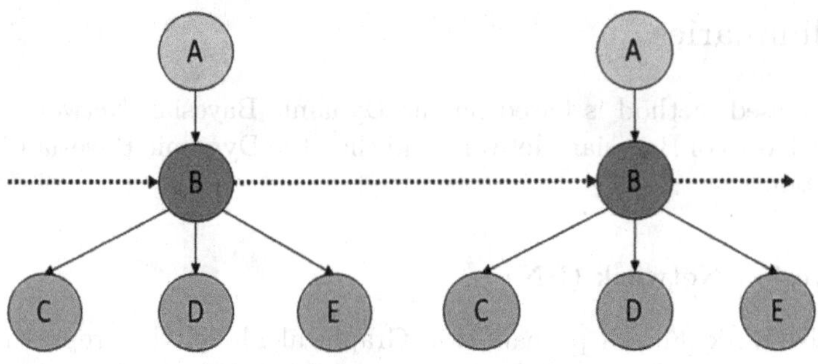

Fig. 2. Two time slice Dynamic Bayesian Network

capable to model the temporal relationship in a multivariate time series. Figure 2 shows a DBN with variable (A, B, C, D, E) that sliced over two time stamps. Joint probability distribution of DBN is given by Eq. (3)

$$P(X_1^{1:t}, X_2^{1:t}, .., X_R^{1:t}) = \prod_{t=1}^{T} \prod_{r=1}^{R} P(X_r^t | Par(X_r^t)) \tag{3}$$

Parents of X_r^t either come from same time slice or from the previous time slice due to the First Order Markov assumption.

3 Methodology

This section presents the methodology used develop the proposed Contextual Dynamic Bayesian Network. Our approach is similar to the work discussed in [6] for multi sensor data fusion.

3.1 Contextual Dynamic Bayesian Network (CxDBN)

CxDBN is a modified version of the DBN. We configure the causal relationship between the state S and the context C. The structure of CxDBN is shown in Fig. 3. Here S denotes the state, C denotes the context information and E denotes the evidence or observation.
The following assumptions are made to perform the inference with CxDBN:

1. Conditional independence between the context attribute and the observation symbols for the given states.
2. Mutual independence between the observation symbols.

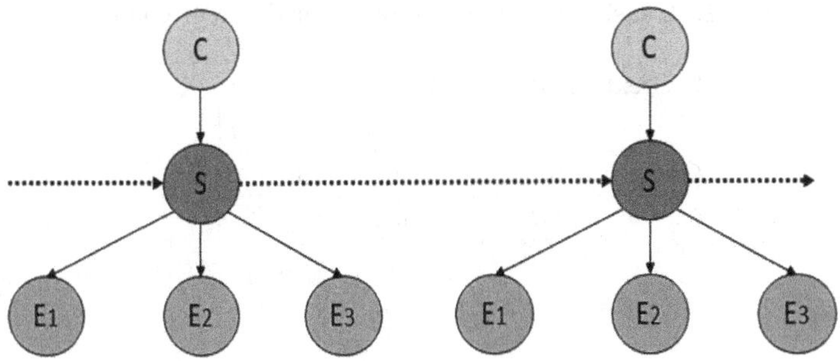

Fig. 3. Two time slice Contextual Dynamic Bayesian Network

3.2 Inference in CxDBN

Inference in CxDBN can be done using the following steps:

1. Equation (4) shows the belief of being in some state at time stamp t for the given observations (Obs) and context (C) seen up to time stamp t.
$$Belief(State(t)) = P(State(t)|Obs^{1:t}, C^{1:t}) \tag{4}$$
2. The Eq. (4) cab rewritten as Eq. (5)
$$Belief(State(t)) = P(State(t)|Obs^{1:t-1}, Obs^{t}, C^{1:t}) \tag{5}$$
3. After further modification Eq. (5) is simplified to Eq. (6)
$$Belief(State(t)) = \alpha * P(Obs^{t}|State(t), Obs^{1:t-1}, C^{1:t}) * P(State(t)|Obs^{1:t-1}, C^{1:t}) \tag{6}$$
Where α is the normalization constant.
4. If mutual independence exits between the observations then first term in the Eq. (6) can be written as
$$P(Obs^{t}|State(t), Obs^{1:t-1}, C^{1:t}) = \prod_{l=1}^{L} P(Obs_{l}^{t}|State(t)) \tag{7}$$
Where l is the number of observations.
5. Second term in the Eq. (6) is a simple one step prediction from the state at t−1 time stamp and can be written as
$$P(State(t)|Obs^{1:t}, C^{1:t}) = \sum_{State(t-1)} P(State(t)|State(t-1), C^{t}) * Belief(State(t-1)) \tag{8}$$
6. Finally the belief at time stamp t can be computed by Eq. (9). Table 1 denotes transition probability table for the different contexts. Here L is number of time series.
$$Belief(State(t)) = \prod_{l=1}^{L} P(Obs_{l}^{t}|State(t)) * P(State(t)|Obs^{1:t}, C^{1:t}) \tag{9}$$

Where a_{ij} is the transition probability from state i to state j, C_i is the context i, T_{C_i} is the transition probability table for the context C_i.

Table 1. Transition probability table for different contexts

Context	$State\,(t-1)$	$State\,(t)$		
		$State_i$	$State_j$	
C_i	$State_i$	a_{ii}	a_{ij}	T_{C_i}
C_i	$State_j$	a_{ji}	a_{jj}	
C_j	$State_i$	a_{ii}	a_{ij}	T_{C_j}
C_j	$State_j$	a_{ji}	a_{jj}	

4 Experiments and Results

This section discusses the experiments and results on the time series data obtained from oil well drilling operation. Supervision of the oil well drilling process is performed by continuous monitoring of various hydraulic and mechanical parameters. The rig contains various sensors equipped to different working units. The measured drilling parameters are stored in a database known as supervisory control and data acquisition (SCADA) system. The variation in different drilling parameters is monitored to identify the anomalies or complications that occurs during the drilling process. The symptoms of these complications change in the presence of different contexts like soil formations. For efficient monitoring of the drilling process, it is required to add the contextual information to the supervision model created to monitor the oil well drilling process.

4.1 Feature Extraction

We used window based method to extract features from the given time series. The extracted features from each window contain two information *trend* and *value.* Table 2 shows <trend, value> pair features extracted from the time series data. Fuzzy sets are created to compute the value features as described in [25] and linear regression is used to estimate the trend feature. Further the computed trend value pair features are used as the observation sequences to train the CxDBN model.

Table 2. <Trend,Value> pair feature for time series. Data

Features	Trend feature	Value feature	<Trend Value Pair>
	Increasing (I)	Very Low (VL)	<VL, I>, <VL, D>, <VL, C>
		Low (L)	<L, I>, <L, D>, <L, C>
	Decreasing (D)	Medium (M)	<M, I>, <M, D>, <M, H>
	Constant (C)	High (H)	<H, I>, <H, D>, <H, C>
		Very High (VH)	<VH, I>, <VH, D>, <VH, C>

4.2 Dataset Description

To test the efficacy of the proposed method we trained our model over the data of oil well drilling process provided by Oil and Natural Gas Corporation Limited (ONGC), India. Hookload parameter is a prime indicator of occurrence of the stuck pipe complication during the drilling process [4]. First, features are extracted from the Hookload parameter, which is in the form of time series, as discussed in Sect. 4.1. These features are extracted for two types of soil formation (context). Table 3 shows the number of the training and test data samples used for the experiment. The training data contains 72072 data samples, including 58.3 % connected with drilling in the first type of the soil and remaining part in the second type of soil. The test data includes 30887 samples out of which 20000 and 10887 vectors belong to the drilling of the first and second type of soil, respectively.

Table 3. Counts of data samples in training and test data

	Training data		Test data	
	Soil-1	Soil-2	Soil-1	Soil-2
	42000	30072	20000	10887
Total	72072		30887	

4.3 Results

The proposed CxDBN is trained over the training data by extracting the trend value pair features with a window size of ten data samples. The same window size is used in the test data to extract the trend value pair features. Figure 4 shows the Hookload values present in the test data. The values of Hookload are plotted by taking the mean of data samples present in each window of the test data. The region covered with a blue rectangular box in Fig. 4 shows the values of the Hookload parameter received for the drilling of the first type of soil. However, the region covered with the green rectangular box denotes the Hookload parameter values collected for the drilling of the second type of soil.

In the Fig. 4, the data samples from range 2380–2440 show anomalies present in the Hookload values in the test data. Figure 5 shows the probabilities values of the normal state computed by the CxDBN. In Fig. 5 the probability of being in normal state decreases for the data range of 2380–2440 by considering the threshold value as 0.5. The drop in the value of the probability at data index 2000 shows the change in the context, i.e., soil formation during the drilling process. It is clear from the Fig. 5 that the model correctly identifies the anomalous values of the Hookload parameter in the presence of different soil formations. The identified anomalous values of the Hookload parameter is shown in Fig. 6.

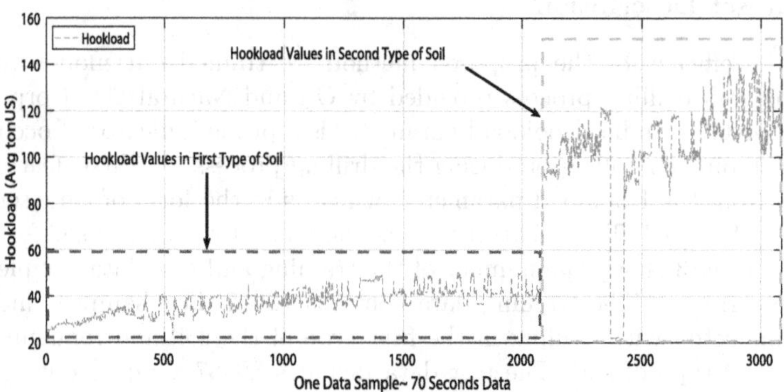

Fig. 4. Hookload values during the drilling of different soils (Color figure online)

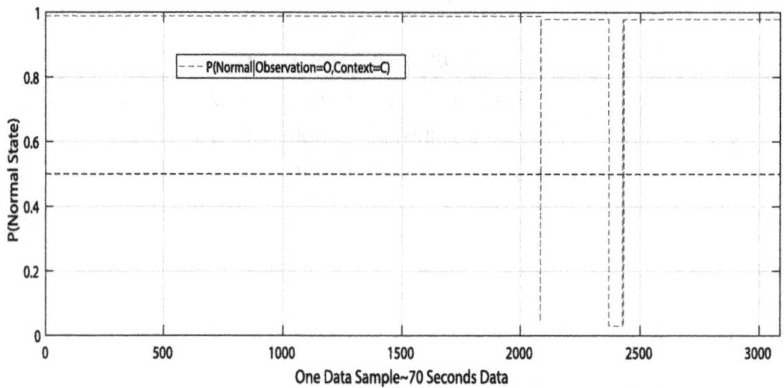

Fig. 5. Probability of normal state computed by CxDBN

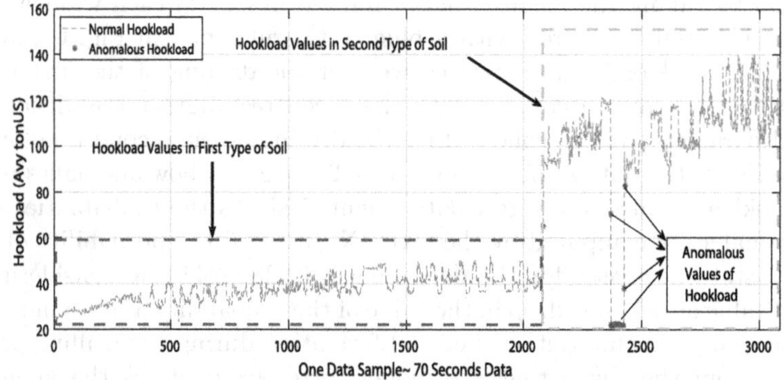

Fig. 6. Contextual anomalies detected by CxDBN (Color figure online)

5 Conclusion

In this paper, we integrated the Contextual information with the Dynamic Bayesian Network to detect contextual anomalies in the time series data. The assumption of the causal relationship between the states and context variables simplifies the computation. The proposed method efficiently identifies the contextual anomalies when applied over the oil well drilling data. The paper presents the initial results. In future, we intend to perform more experiments considering data from different domain where context plays a significant role. Further, a comparison with other contextual anomaly detection method is required. Finally, the method can be extend to identify contextual anomalies in multivariate time series data.

References

1. Abid, N., Kozlow, P., Yanushkevich, S.: Detection of asymmetric abnormalities in gait using depth data and dynamic Bayesian networks. In: 2018 14th IEEE International Conference on Signal Processing (ICSP), pp. 762–767. IEEE (2018)
2. Avilés-Arriaga, H.H., Sucar, L.E., Mendoza, C.E., Vargas, B.: Visual recognition of gestures using dynamic naive Bayesian classifiers. In: Proceedings of the 12th IEEE International Workshop on Robot and Human Interactive Communication, ROMAN 2003, pp. 133–138. IEEE (2003)
3. Breunig, M.M., Kriegel, H.P., Ng, R.T., Sander, J.: LOF: identifying density-based local outliers. In: ACM Sigmod Record, vol. 29, pp. 93–104. ACM (2000)
4. Chamkalani, A., Pordel Shahri, M., Poordad, S., et al.: Support vector machine model: a new methodology for stuck pipe prediction. In: SPE Unconventional Gas Conference and Exhibition. Society of Petroleum Engineers (2013)
5. Chandola, V., Banerjee, A., Kumar, V.: Anomaly detection: a survey. ACM Comput. Surv. (CSUR) **41**(3), 15 (2009)
6. De Paola, A., Ferraro, P., Gaglio, S., Re, G.L., Das, S.K.: An adaptive Bayesian system for context-aware data fusion in smart environments. IEEE Trans. Mob. Comput. **16**(6), 1502–1515 (2016)
7. Ding, N., Gao, H., Bu, H., Ma, H.: RADM: real-time anomaly detection in multivariate time series based on Bayesian network. In: 2018 IEEE International Conference on Smart Internet of Things (SmartIoT), pp. 129–134. IEEE (2018)
8. Friedman, N., Geiger, D., Goldszmidt, M.: Bayesian network classifiers. Mach. Learn. **29**(2–3), 131–163 (1997)
9. Ghahramani, Z.: Learning dynamic Bayesian networks. In: Giles, C.L., Gori, M. (eds.) NN 1997. LNCS, vol. 1387, pp. 168–197. Springer, Heidelberg (1997). https://doi.org/10.1007/BFb0053999
10. Hu, J., Wang, F., Sun, J., Sorrentino, R., Ebadollahi, S.: A healthcare utilization analysis framework for hot spotting and contextual anomaly detection. In: AMIA Annual Symposium Proceedings, vol. 2012, p. 360. American Medical Informatics Association (2012)
11. Liu, Q., et al.: Unsupervised detection of contextual anomaly in remotely sensed data. Remote Sens. Environ. **202**, 75–87 (2017)
12. MacDonald, I.L., Zucchini, W.: Hidden Markov and Other Models for Discrete-valued Time Series, vol. 110. CRC Press, Boca Raton (1997)

13. Malhotra, P., TV, V., Vig, L., Agarwal, P., Shroff, G.: TimeNet: pre-trained deep recurrent neural network for time series classification. arXiv preprint arXiv:1706.08838 (2017)
14. Malhotra, P., Vig, L., Shroff, G., Agarwal, P.: Long short term memory networks for anomaly detection in time series. In: Proceedings, p. 89. Presses Universitaires de Louvain (2015)
15. McAlinn, K., West, M.: Dynamic bayesian predictive synthesis in time series forecasting. J. Econ. **210**(1), 155–169 (2019)
16. Nanduri, A., Sherry, L.: Anomaly detection in aircraft data using recurrent neural networks (RNN). In: 2016 Integrated Communications Navigation and Surveillance (ICNS), pp. 5C2-1. IEEE (2016)
17. Ogbechie, A., Díaz-Rozo, J., Larrañaga, P., Bielza, C.: Dynamic Bayesian network-based anomaly detection for in-process visual inspection of laser surface heat treatment. In: Beyerer, J., Niggemann, O., Kühnert, C. (eds.) Machine Learning for Cyber Physical Systems. Technologien für die intelligente Automation (Technologies for Intelligent Automation), pp. 17–24. Springer, Heidelberg (2017)
18. Pauwels, S., Calders, T.: Extending dynamic Bayesian networks for anomaly detection in complex logs. arXiv preprint arXiv:1805.07107 (2018)
19. Peia, O., Roszbach, K.: Finance and growth: time series evidence on causality. J. Financ. Stab. **19**, 105–118 (2015)
20. Pokrajac, D., Lazarevic, A., Latecki, L.J.: Incremental local outlier detection for data streams. In: 2007 IEEE Symposium on Computational Intelligence and Data Mining, pp. 504–515. IEEE (2007)
21. Pradhan, M., Pradhan, S.K., Sahu, S.K.: Anomaly detection using artificial neural network. Int. J. Eng. Sci. Emerg. Technol. **2**(1), 29–36 (2012)
22. Qiao, H., Wang, T., Wang, P., Qiao, S., Zhang, L.: A time-distributed spatiotemporal feature learning method for machine health monitoring with multi-sensor time series. Sensors **18**(9), 2932 (2018)
23. Xia, H., Li, T., Liu, W., Zhong, X., Yuan, J.: Abnormal event detection method in surveillance video based on temporal CNN and sparse optical flow. In: Proceedings of the 2019 5th International Conference on Computing and Data Engineering, pp. 90–94. ACM (2019)
24. Zhang, H., Zhang, Q., Liu, J., Guo, H.: Fault detection and repairing for intelligent connected vehicles based on dynamic bayesian network model. IEEE Internet Things J. **5**(4), 2431–2440 (2018)
25. Zhou, P.Y., Chan, K.C.: Fuzzy feature extraction for multichannel EEG classification. IEEE Trans. Cogn. Dev. Syst. **10**(2), 267–279 (2016)

Stochastic Optimization of Contextual Neural Networks with RMSprop

Maciej Huk(✉)

Faculty of Computer Science and Management, Wroclaw University of Science and Technology, Wroclaw, Poland
maciej.huk@pwr.edu.pl

Abstract. The paper presents modified version of Generalized Error Backpropagation algorithm (GBP) merged with RMSprop optimizer. This solution is compared with analogous method based on Stochastic Gradient Descent. Both algorithms are used to train MLP and CxNN neural networks solving selected benchmark and real–life classification problems. Results indicate that usage of GBP-RMSprop can be beneficial in terms of increasing classification accuracy as well as decreasing activity of neurons' connections and length of training. This suggests that RMSprop can effectively solve optimization problems of variable dimensionality. In the effect, merging GBP with RMSprop as well as with other optimizers such as Adam and AdaGrad can lead to construction of better algorithms for training of contextual neural networks.

Keywords: Classification · Self-consistency · Aggregation functions

1 Introduction

The popularity of data processing methods based on artificial neural networks is related to their interesting properties and proven usefulness in many different applications from science to business and engineering. They are used in medicine to analyze tissue samples and support diagnostic processes [1] as well as in transport to control autonomous vehicles [2]. Artificial neural networks are used for echo cancelling in telecommunication systems [3] and also are crucial in data acquisition and processing during experiments such as ATLAS of the Large Hadron Collider [4]. They can be found as parts of recommender systems for financial institutions [5] as well as for end customers [6]. And currently they find their place in entertainment serving image enhancement in video games [7, 8].

But to solve different types of tasks different kinds and architectures of artificial neural networks are considered and proposed. Various convolutional neural networks (CNN) are developed and used for image processing systems [9]. Recurrent neural networks including Long Short-Term Memory (LSTM) and Gated Recurrent Units (GRU) networks are very good in recognition of relations within time-series data as well as in analyzing and translation of text [10, 11]. And self organizing maps (SOM) are well known kind of neural systems useful in data clustering for recommender systems and

N. T. Nguyen et al. (Eds.): ACIIDS 2020, LNAI 12034, pp. 343–352, 2020.
https://doi.org/10.1007/978-3-030-42058-1_29

knowledge discovery [12]. Specialized architectures of artificial neural networks such as Generative Adversarial Networks and Variational Autoencoders are used for generation of data of given properties, including images, text and speech [13, 14]. Another example are chaos neural networks can be fast generators of random numbers [15]. Finally, new types of neural networks emerge such as e.g. contextual neural networks (CxNN) with their ability to adjust their internal activity and dimensionality of considered input space to optimize accuracy and cost of processing of given data vectors [16–18].

Contextual neural networks were developed to model situations in which priorities and order of input signals processing can highly influence the results of data analysis [19]. They are using low-level, decentralized selective attention mechanisms in the form of conditional, multi-step aggregation functions [16, 17]. At the same time they can be used easily wherever multilayer perceptron neural networks (MLP) are used. This is because MLP neural networks are just special case of contextual neural networks. Moreover, conditional signals processing allows CxNNs to dynamically limit activity of their internal connections without decrease of data processing accuracy. This makes them a very good replacement for MLP networks in embedded applications with strong limitations of energy and computing power.

Contextual neural networks were used successfully in many practical applications such as e.g. fingerprints classification in crime investigations [20], transmissions prediction in cognitive radio, and classification of cancer gene expression microarray data [18]. Lately contextual neural networks were also implemented in a special version of a very popular H2O machine learning framework. This allows large scale, distributed computation with use of this type of models [21, 22].

In almost all cases mentioned above contextual neural networks were trained with SIF aggregation function [17, 18] and generalized error backpropagation method (GBP) based on stochastic gradient descent approach (SGD, with mini-batch = 1). In this paper we are analyzing properties of contextual neural networks when trained with GBP algorithm modified to use RMSprop [23] stochastic optimization. The tests are performed on three microarray data sets of cancer gene expression, such as: Armstrong (ALL-MLL Leukemia), Golub (ALL-AML Leukemia) and SRBCT (Small Round Blue Cell Tumors) [24–26]. Additionally selected benchmark problems from UCI ML repository were analyzed for comparison with previously reported results [27].

The further parts of the paper are organized as follows. The second section includes description of the GBP algorithm and basic properties of contextual neural networks.

Next, in Sect. 3, the modified GBP method is presented with details related to RMSprop algorithm. Within Sect. 4 the results of experiments with CxNN trained with GBP-SGD are compared with outcomes of GBP-RMSprop. Finally, conclusions are given in Sect. 5 along with planned research.

2 Generalized Error Backpropagation Algorithm

Neurons of contextual neural networks are using multi-step, conditional aggregation functions. Their inputs are clustered in groups of different priorities and in each step of aggregation only one group of inputs is read in and analyzed. The partial activation of the group is calculated and added to the activation of the neuron. This process is repeated for the groups of decreasing priorities till the activation of the neuron is greater than given constant threshold or till all groups are analyzed. Finally, the aggregated activation value is used by activation function (e.g. tanh or linear rectifier) to calculate output value of the neuron. This is generalization of the classical neuron used to build MLP neural networks, in which all inputs belong to only one group.

If the connections which are the most important to solve given problem are assigned to the groups of the highest priorities (one input connection can be assigned only to one group), it can happen that for given input vector not all inputs will be read in and analyzed to calculate the output of the neuron. It can be said that in such case the activity of input connections of the neuron was below 100%. One can also observe that the same neuron with conditional aggregation function can have different activities of inputs for different input vectors. And the lower the activity of input connections of neurons the lower the computational cost of usage of neural network and the lower the influence of low-importance input signals on the output values of the neurons.

It is the role of the training algorithm of neural network to find such grouping of inputs of each neuron to minimize both the output error and average activity of the connections between neurons. To make it possible with gradient-based algorithm and without doubling the number of parameters describing input connection of the neuron, the following assumptions are made:

- the values of number of groups and aggregation threshold parameters are the same for all hidden neurons and are set before the training,
- neurons are using deep coding to store within connections weights both the description of the algorithm of calculation of output values as well as the assignment of inputs to given groups.

In the effect of above assumptions, self-consistence method can be added to the error backpropagation algorithm (BP) to optimize non-continuous, non-differentiable groupings of inputs of neurons by coupling them with continuous and differentiable parameters such as weights of connections [16]. This creates the basic form of the generalized error backpropagation algorithm (GBP) which was shown to effectively train contextual neural networks with different aggregation and activation functions [17, 18] as well as with different schemes of initialization of groupings of neuronal inputs [28]. The schematic block diagram of GBP method is presented at the Fig. 1.

During the training with GBP, weights of connections are updated with usage of generalized delta rule which takes into account the fact that some of input connections for given vectors can have no influence on the output error regardless on the value of the related input signals. At the same time neuron inputs groupings are stored in temporary virtual grouping vectors which are calculated from connections weights vector $\mathbf{w}$ with use of grouping function $\Omega(\mathbf{w})$. Typically, the grouping function Ω forces the relation that for given neuron input connections with higher values of weights belong to groups of higher importance. What is important - update of connections groupings can change the error space of the neural network - and if performed too frequently, can lead to destabilization of the training process.

To overcome this problem, the update of inputs groupings is done only once after each omega training epochs. This controls the coupling between weights vector and grouping

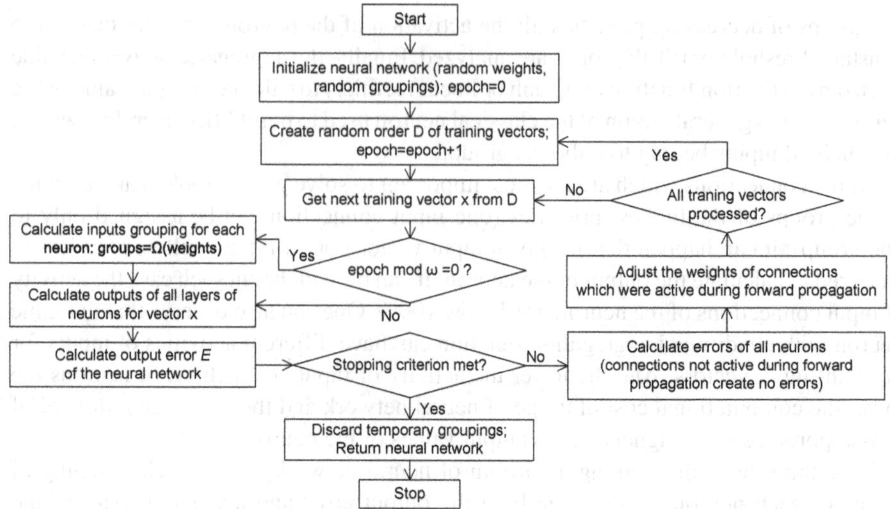

Fig. 1. Generalized error backpropagation algorithm using stochastic gradient descent

vector - and stabilizes the training. Finally, after the training, virtual grouping vectors can be discarded because they can be later calculated from weights of connections of neurons.

3 Generalized Error Backpropagation with RMSprop

In the previous section as well as in applications reported in the literature, the GBP algorithm is considered with the stochastic gradient descent (SGD) method of updating values of connections weights for each training vector. This is because the simplicity of SGD makes it to be faster than batch gradient descent method and allows straightforward derivation of error and weights update rules for contextual neural networks [16, 23]. But SGD method is not good at avoiding local minima of error functions. It also finds it difficult to guide training of neural networks through saddle points of error spaces. In such points the gradient of error is almost zero in all dimensions [29]. This problem can be especially frequent in low–dimensional error spaces, what is typical case in contextual neural networks. In CxNNs the error space evolves from low-dimensional to high-dimensional during conditional aggregation of inputs in neurons. When the model approaches optimum, this evolution often ends after processing of very few first groups of connections. Thus it seems to be worth checking how GBP will function when it will be extended with gradient descent optimization algorithm such as RMSprop [23].

RMSprop, a gradient descent optimization method proposed by Geoff Hinton is a simplified version of AdaDelta method. It can be expressed with the following formula for update of weight w of connection j during the training step t:

$$w_{t+1}^{j} = w_{t}^{j} - \frac{\alpha}{\sqrt{v_{t}^{j} + \varepsilon}} \left(\frac{\partial E_j}{\partial w_j} \right)_t \tag{1}$$

where

$$v_t^j = \gamma\, v_{t-1}^j + (1-\gamma)\left(\frac{\partial E_j}{\partial w_j}\right)_t^2. \tag{2}$$

In such solution the partial derivative of error E over connection weight w is used to adaptively adjust the length of step of the gradient descent. It is partially controlled by the three constant parameters: learning rate α, fraction of gradient decay γ and computability guard ε. Their typical values are: 0.001, 0.9 and 0.0001, respectively. But the most important element of the RMSprop is its partial memory v of the errors caused by given connection for previous training vectors. It allows to speed up the training while sliding from the saddle points and to make it more precise near points close to optimal solutions. This is why for many problems RMSprop outperforms SGD as well as schedule-based gradient descent methods [30].

But in the case of GBP and contextual neural network for given training vector part of connections between neurons can be not active. Thus the generalized formula for error E of j-th neuron in m-th layer of contextual neurons is

$$E_j^m = F'(\phi_j^m)\sum_{i=1}^{n_{m+1}} E_i^{(m+1)} w_{i,j}^{m+1} H(k_i^{*(m+1)} - \theta_{i,j}^{m+1}), \tag{3}$$

where F is the activation function of the neuron, φ is the activation of the neuron and θ is the number of the group to which is assigned connection between j-th neuron in m-th layer and i-th neuron in layer $m+1$. At the same time $k*$ is the maximal number of group which was active during aggregation of signals by i-th neuron in layer $m+1$ and H is the Heaviside function. In the effect, RMSprop formulas to be used with contextual neural network must be rewritten to the following form:

$$w_{t+1}^j = \begin{cases} w_t^j - \frac{\alpha}{\sqrt{v_t^j + \varepsilon}}\left(\frac{\partial E_j}{\partial w_j}\right)_t & : k_j^* \geq \theta_j \\ w_t^j & : k_j^* < \theta_j \end{cases} \tag{4}$$

where

$$v_t^j = \begin{cases} \gamma\, v_{t-1}^j + (1-\gamma)\left(\frac{\partial E_j}{\partial w_j}\right)_t^2 & : k_j^* \geq \theta_j \\ \gamma\, v_{t-1}^j & : k_j^* < \theta_j \end{cases} \tag{5}$$

Thus even if the connection not active for given vector does not contribute to the error of the neuron and its weight is not changed, the related memory of past gradients is modified as given by (5) for $k_j^* < \theta_j$. In the effect, after long inactivity of the connection the actual training step coefficient becomes close to its initial value, what is expected. And finally, by using (3) to calculate right-hand partial derivatives within (4) and (5) one achieves the weights update formula of GBP algorithm combined with RMSprop method.

4 Results of Experiments

To find out how combining GBP with RMSprop modifies the training of contextual neural networks results achieved with GBP-RMSprop algorithm were compared with outcomes of standard version of GBP (further denoted as GBP-SGD). Both methods were used to solve classification problems defined by the three microarray data sets of cancer gene expression, such as: Armstrong (ALL-MLL Leukemia), Golub (ALL-AML Leukemia) and SRBCT (Small Round Blue Cell Tumors) [24–26]. Additionally selected benchmark problems from UCI ML repository were analyzed (Sonar, Heart Cancer and Crx) [27]. During experiments the values of following parameters were examined: number of training epochs, hidden connections activity of resulting models as well as their accuracy of classification for test data. The architectures of trained models were constructed with the use of one-hot encoding of attributes, as well as numbers of neurons and groups of connections as given in the Table 1.

Table 1. Basic properties of data sets and neural networks used within experiments

Data set	Number of classes	Number of attributes	Number of input neurons	Number of hidden neurons	Number of groups	Number of hidden connections	Number of data vectors
Armstrong	2	12582	12582	3	13	33785	72
Golub	2	7129	7129	3	13	21426	72
SRBCT	4	2308	2308	3	13	6963	83
Sonar	2	60	60	30	25	1860	208
Heart C.	5	13	28	10	16	330	303
Crx	2	15	47	20	3	1000	690

Hidden neurons of CxNNs were using SIF aggregation functions and single group initialization of connections grouping [16]. MLP models were trained for reference. In all cases the stopping criterion of training was perfect classification or lack of improvement of accuracy for training data for more than 1200 epochs. For each set of parameters training was performed with the use of repeated 10 times 5–fold cross-validation [31]. During each training models with the lowest test error were stored for analysis.

Values of parameters of considered training methods were: training step $\alpha = 0.01$, mini-batch size = 1, threshold of aggregation function $\varphi^* = 0.6$, groups actualization interval $\omega = 25$, fraction of gradient decay $\gamma = 0.9$, computability guard $\varepsilon = 0.0001$. Activation function of neurons was bipolar sigmoid. Uniform weights initialization was used with range $(-0.2, 0.2)$. All pseudo-random values were generated with the Mersene Twister algorithm (MT19937) [32]. For each type of analyzed training algorithms, aggregation functions and training subsets of data, the same sequences of random values were used during initialization and training of neural networks. Experiments were performed with the use of 3706.6 ± 0.2 MHz 16 core Intel Core i9 9900 K CPU with core-wise separation of simulation processes with appropriate core affinity settings.

The statistical significances of measurements were calculated with the use of two-sample T-Test (confidence above 90%). Shapiro-Wilk normality test was used to analyze the normality of series. Results for MLP neural networks trained with GBP-SGD and GBP-RMSprop are presented in Table 2. Analogous results for CxNNs with SIF aggregation function are given in Table 3.

Table 2. Average results with standard deviations for MLP neural networks trained with GBP-SGD and GBP-RMSprop algorithms. Highlighted values are statistically better.

Data set	Training algorithm	Training epochs [1]	Training time [s]	Test error [%]	Hidden conn. activity [%]
Armstrong	GBP-SGD	**8.2 ± 4.8**	**13.7 ± 8.0**	9.2 ± 8.7	100 ± 0
	GBP-RMSp	56.7 ± 6.1	126.3 ± 148.0	**3.3 ± 4.4**	100 ± 0
Golub	GBP-SGD	**10.8 ± 6.4**	**9.8 ± 5.7**	9.8 ± 8.3	100 ± 0
	GBP-RMSp	30.3 ± 27.2	28.0 ± 25.0	**3.0 ± 4.0**	100 ± 0
SRBCT	GBP-SGD	**11.2 ± 2.9**	**5.6 ± 1.4**	10.9 ± 6.9	100 ± 0
	GBP-RMSp	47.1 ± 21.4	16.2 ± 7.3	**4.2 ± 5.2**	100 ± 0
Sonar	GBP-SGD	280 ± 328	11.1 ± 13.0	13.0 ± 4.5	100 ± 0
	GBP-RMSp	**111 ± 196**	**4.5 ± 7.8**	**10.3 ± 4.1**	100 ± 0
Heart C.	GBP-SGD	528 ± 796	16.4 ± 24.8	12.9 ± 3.8	100 ± 0
	GBP-RMSp	**304 ± 543**	**11.7 ± 20.9**	12.2 ± 3.6	100 ± 0
Crx	GBP-SGD	783 ± 1216	64.9 ± 107.2	11.6 ± 2.2	100 ± 0
	GBP-RMSp	877 ± 928	59.6 ± 63.0	10.9 ± 2.3	100 ± 0

As it can be seen in Table 2 the usage of GBP-RMSprop algorithm allows to generate MLP neural networks which are better than analogous structures trained with usage of GBP-SGD method. The former models for all considered problems have lower average classification error for the test data, and in case of four problems this difference is statistically significant. This is especially evident in the case of neural networks with higher number of connections between neurons (problems Armstrong, Golub, SRBCT). It can be observed that decrease of the average test error in most cases is related with increase of the number of training epochs (except Sonar and Heart Cancer problems). While for all MLP neural networks the activity of hidden connections is by definition equal 100%, changes of number of training epochs are connected with proportional changes of training time. It is also worth to note, that in the case of MLP networks GBP algorithm behaves like standard error backpropagation method (with SGD or RMSprop).

At the same time GBP-RMSprop can train contextual neural networks which produce lower testing error than their counterparts trained with GBP-SGD. And for most considered problems the related change of number of training epochs is lower than for MLP networks. E.g. for Golub data set the number of training epochs increases 3 times when GBP–RMSprop is used for MLP and only by 3% for CxNNs. GBP–RMSprop has also no problems with reduction of activity of hidden connection within CxNNs. For Sonar and Heart Cancer data sets activity of hidden connections is decreased by 20 and 30% points, respectively, while the testing error is lower than results for MLP for both training

Table 3. Average results with standard deviations for CxNNs with SIF aggregation trained with GBP-SGD and GBP-RMSprop algorithms. Highlighted values are statistically better.

Data set	Training algorithm	Training epochs [1]	Training time [s]	Test error [%]	Hidden conn. activity [%]
Armstrong	GBP-SGD	**9.8 ± 5.5**	**14.8 ± 6.6**	6.4 ± 6.6	55.1 ± 24.7
	GBP-RMSp	22.8 ± 17.1	36.3 ± 26.1	**4.3 ± 4.9**	58.7 ± 16.9
Golub	GBP-SGD	15.0 ± 10.8	11.0 ± 9.8	5.9 ± 6.2	50.5 ± 21.6
	GBP-RMSp	15.4 ± 7.5	19.2 ± 9.1	**3.0 ± 4.2**	51.9 ± 13.0
SRBCT	GBP-SGD	**17.1 ± 7.4**	**7.5 ± 2.8**	8.5 ± 8.2	48.3 ± 12.0
	GBP-RMSp	21.4 ± 8.0	10.2 ± 4.6	**6.8 ± 6.3**	72.3 ± 11.3
Sonar	GBP-SGD	1036 ± 755	65.2 ± 47.5	9.7 ± 3.6	46.1 ± 10.3
	GBP-RMSp	**804 ± 836**	**49.5 ± 52.1**	8.9 ± 3.6	**33.1 ± 10.4**
Heart C.	GBP-SGD	1083 ± 1236	71.8 ± 79.9	11.0 ± 3.1	53.6 ± 11.6
	GBP-RMSp	**841 ± 634**	**63.2 ± 48.0**	10.2 ± 3.0	**43.5 ± 10.8**
Crx	GBP-SGD	1150 ± 1407	**116.9 ± 149.1**	10.5 ± 2.2	67.2 ± 7.4
	GBP-RMSp	1168 ± 935	146.5 ± 119.5	10.1 ± 2.3	68.1 ± 6.8

methods. Thus in the case of MLP and CxNNs usage of GBP-RMSprop can be beneficial both in terms of number of training epochs, hidden connections activity and testing error.

5 Conclusions

In this paper a modification of Generalized Error Backpropagation algorithm was presented which includes appropriately adapted RMSprop optimizer. As expected, this allowed statistically significant reduction of test error, activity of hidden connections and number of epochs of training of considered contextual neural networks. Classification results obtained for CxNNs built with GBP-RMSprop in case of four out of six analyzed problems were better than for MLP networks obtained with the same training method. And in all cases models built with GBP-RMSprop were better than those trained with GBP-SGD. This suggests that RMSprop can be effectively adapted and used with contextual optimization models which are operating in error spaces of variable dimensionality which are evolving during processing of given data vectors.

Presented results also open new questions and research directions related with contextual neural networks and optimization algorithms such as GBP-RMSprop. First, it is unknown why for SRBCT data set the measured testing error for models trained with GBP-RMSprop is higher for CxNNs than for MLP. Second, it could be interesting to find out why the usage of GBP-RMSprop decreased number of epochs of training of CxNNs in relation to MLPs in the case of the three biggest of considered neural networks (all for microarray data). Performing analogous analyses for additional benchmark data sets could be very helpful in this task. Finally, presented results of measurements of usage of

GBP–RMSprop indicate that it could be also beneficial to check the results of training of contextual neural networks with other aggregation functions and modifications of GBP method. This would include merging of GBP e.g. with Adam and AdaGrad gradient descent optimizers [23].

References

1. Suleymanova, I., et al.: A deep convolutional neural network approach for astrocyte detection. Sci. Rep. **8**(12878), 1–7 (2018)
2. Chen, S., Zhang, S., Shang, J., Chen, B., Zheng, N.: Brain-inspired cognitive model with attention for self-driving cars. IEEE Trans. Cogn. Dev. Syst. **11**(1), 13–25 (2019)
3. Zhang, S., Zheng, W.X.: Recursive adaptive sparse exponential functional link neural network for nonlinear AEC in impulsive noise environment. IEEE Trans. Neural Netw. Learn. Syst. **29**(9), 4314–4323 (2018)
4. Guest, D., Cranmer, K., Whiteson, D.: Deep learning and its application to LHC physics. Annu. Rev. Nucl. Part. Sci. **68**, 1–22 (2018)
5. Bao, W.N., Yue, J.H., Rao, Y.: A deep learning framework for financial time series using stacked autoencoders and long-short term memory. PloS ONE **12**(7), 1–24 (2017)
6. Tsai, Y.-C., et al.: FineNet: a joint convolutional and recurrent neural network model to forecast and recommend anomalous financial items. In: Proceedings of the 13th ACM Conference on Recommender Systems RecSys 2019, pp. 536–537. ACM, New York (2019)
7. Gao, D., Li, X., Dong, Y., Peers, P., Xu, K., Tong, X.: Deep inverse rendering for high-resolution SVBRDF estimation from an arbitrary number of images. ACM Trans. Graphics (SIGGRAPH) **38**(4), 1–15 (2019). Article no. 134
8. Liu, L., et al.: Automatic skin binding for production characters with deep graph networks. ACM Trans. Graphics (SIGGRAPH) **38**(4), 1–12 (2019). Article no. 114
9. Gong, K., et al.: Iterative PET image reconstruction using convolutional neural network representation. IEEE Trans. Med. Imaging **38**(3), 675–685 (2019)
10. Athiwaratkun, B., Stokes, J.W.: Malware classification with LSTM and GRU language models and a character-level CNN. In: Proceedings of the 2017 IEEE International Conference on Acoustics, Speech and Signal Processing (ICASSP), USA, pp. 2482–2486. IEEE (2017)
11. Huang, X., Tan, H., Lin, G., Tian, Y.: A LSTM-based bidirectional translation model for optimizing rare words and terminologies. In: 2018 IEEE International Conference on Artificial Intelligence and Big Data (ICAIBD), China, pp. 5077–5086. IEEE (2018)
12. Dozono, H., Niina, G., Araki, S.: Convolutional self organizing map. In: 2016 IEEE International Conference on Computational Science and Computational Intelligence (CSCI), pp. 767–771. IEEE (2016)
13. Higgins, I., et al.: beta-VAE: learning basic visual concepts with a constrained variational framework. In: International Conference on Learning Represent, ICLR 2017, vol 2, no. 5, pp. 1–22 (2017)
14. Karras, T., Aila, T., Laine, S., Lehtinen, J.: Progressive growing of GANs for improved quality, stability, and variation. In: International Conference on Learning Representations, ICLR 2018, pp. 1–26 (2018)
15. Alcin, M., Koyuncu, I., Tuna, M., Varan, M., Pehlivan, I.: A novel high speed artificial neural network–based chaotic true random number generator on field programmable gate array. Int. J. Circuit Theory Appl. **47**(3), 365–378 (2019)
16. Huk, M.: Backpropagation generalized delta rule for the selective attention Sigma-if artificial neural network. Int. J. Appl. Math. Comput. Sci. **22**, 449–459 (2012)

17. Huk, M.: Notes on the generalized backpropagation algorithm for contextual neural networks with conditional aggregation functions. J. Intell. Fuzzy Syst. **32**, 1365–1376 (2017)
18. Huk, M.: Training contextual neural networks with rectifier activation functions: role and adoption of sorting methods. J. Intell. Fuzzy Syst. **38**, 1–10 (2019)
19. Huk, M.: Learning distributed selective attention strategies with the Sigma-if neural network. In: Akbar, M., Hussain, D. (eds.) Advances in Computer Science and IT, pp. 209–232. InTech, Vukovar (2009)
20. Szczepanik, M., Jóźwiak, I.: Data management for fingerprint recognition algorithm based on characteristic points' groups. In: Pechenizkiy, M., Wojciechowski, M. (eds.) New Trends in Databases and Information Systems. Advances in Intelligent Systems and Computing, vol. 185, pp. 425–432. Springer, Heidelberg (2013). https://doi.org/10.1007/978-3-642-32518-2_40
21. Janusz, B.J., Wołk, K.: Implementing contextual neural networks in distributed machine learning framework. In: Nguyen, N.T., Hoang, D.H., Hong, T.-P., Pham, H., Trawiński, B. (eds.) ACIIDS 2018. LNCS (LNAI), vol. 10752, pp. 212–223. Springer, Cham (2018). https://doi.org/10.1007/978-3-319-75420-8_20
22. Wołk, K., Burnell, E.: Implementation and analysis of contextual neural networks in H2O framework. In: Nguyen, N.T., Gaol, F.L., Hong, T.-P., Trawiński, B. (eds.) ACIIDS 2019. LNCS (LNAI), vol. 11432, pp. 429–440. Springer, Cham (2019). https://doi.org/10.1007/978-3-030-14802-7_37
23. Ruder, S.: An overview of gradient descent optimization algorithms, pp. 1–14. eprint arXiv:1609.04747v2 (2017)
24. Armstrong, S.A.: MLL translocations specify a distinct gene expression profile that distinguishes a unique leukemia. Nat. Genet. **30**, 41–47 (2002)
25. Golub, T.R., et al.: Molecular classification of cancer: class discovery and class prediction by gene expression monitoring. Science **286**, 531–537 (1999)
26. Khan, J., et al.: Classification and diagnostic prediction of cancers using gene expression profiling and artificial neural networks. Nat. Med. **7**(6), 673–679 (2001)
27. UCI Machine Learning Repository. http://archive.ics.uci.edu/ml
28. Huk, M.: Non-uniform initialization of inputs groupings in contextual neural networks. In: Nguyen, N.T., Gaol, F.L., Hong, T.-P., Trawiński, B. (eds.) ACIIDS 2019. LNCS (LNAI), vol. 11432, pp. 420–428. Springer, Cham (2019). https://doi.org/10.1007/978-3-030-14802-7_36
29. Dauphin Y., Pascanu, R., Gulcehre, C., Cho, K., Ganguli, S., Bengio, Y.: Identifying and attacking the saddle point problem in high dimensional non-convex optimization, pp. 1–14. eprint arXiv:1406.2572 (2014)
30. Darken, C., Chang, J., Moody, J.: Learning rate schedules for faster stochastic gradient search. In: Proceedings of the 1992 IEEE Workshop on Neural Networks for Signal Processing II, September, pp. 1–11 (1992)
31. Bouckaert, R.R., Frank, E.: Evaluating the replicability of significance tests for comparing learning algorithms. In: Dai, H., Srikant, R., Zhang, C. (eds.) PAKDD 2004. LNCS (LNAI), vol. 3056, pp. 3–12. Springer, Heidelberg (2004). https://doi.org/10.1007/978-3-540-24775-3_3
32. Matsumoto, M., Nishimura, T.: Mersenne twister: a 623-dimensionally equidistributed uniform pseudorandom number generator. ACM Trans. Model. Comput. Simul. **8**(3), 3–30 (1998)

Soft Dropout Method in Training of Contextual Neural Networks

Krzysztof Wołk[1], Rafał Palak[2(✉)], and Erik Dawid Burnell[3]

[1] The Wroclaw Institute of Spatial Information and Artificial Intelligence, Wroclaw, Poland
krzysztof.wolk@wizipisi.pl

[2] Wroclaw University of Science and Technology, Wroclaw, Poland
rafal.palak@pwr.edu.pl

[3] Science Applications International Corporation, Reston, USA
erik.d.burnell@gmail.com

Abstract. Various regularization techniques were developed to prevent many adverse effects that may appear during the training of contextual and non-contextual neural networks. The problems include e.g.: overfitting, vanishing of the gradient and too high increase in weight values. A commonly used solution that limits many of those is the dropout. The goal of this paper is to propose and analyze a new type of dropout - Soft Dropout. Unlike traditional dropout regularization, in Soft Dropout neurons are excluded only partially, what is regulated by additional, continuous muting factor. This change can help to generate classification models with lower overfitting. The paper present results suggesting that Soft Dropout can help to generate classification models with lower overfitting than standard dropout technique. Experiments are performed for selected benchmark and real-life datasets with MLP and Contextual Neural Networks.

Keywords: Dropout · Muting factor · Overfitting

1 Introduction

Currently, Artificial Neural Networks are used to perform multiple types of tasks. They serve as valuable tools, e.g. in transport [1], science [2], medicine [3], entertainment [4, 5], and in recommender systems [6]. To do that, various kinds of neural networks were created – from fully connected MLP networks and Self Organizing Maps [7], through Convolutional Neural Networks (CNN) [8] and their recurrent variants using Gated Recurrent Units (GRU) and Long Short-Term Memory (LSTM) [9, 10] to such new solutions as Variational Autoencoders (VAE) and Generative Adversarial Networks (GAN) [11, 12]. Nevertheless, architectures of all mentioned neural networks can be seen as static – after the training, they are always activating all connections between neurons for all input vectors. Thus, in parallel, different types of models: Contextual Neural Networks (CxNN) emerged [13–16], which do not have this limitation.

Contextual neural networks were developed as a generalization of MLP networks [13, 17]. They were used with success in analyses of benchmark and real-life data. Examples include contextual classification of minutia groups in fingerprint detection systems

N. T. Nguyen et al. (Eds.): ACIIDS 2020, LNAI 12034, pp. 353–363, 2020.
https://doi.org/10.1007/978-3-030-42058-1_30

[18, 19], predicting transmissions in cognitive radio [20] and analyses of cancer gene expression microarray data [21–23]. They can also be used in telerehabilitation systems and for context-sensitive text mining [24, 25]. Characteristic properties of CxNNs' are their ability to decrease average activity of connections between neurons, to limit the time cost of calculation of outputs of neural networks, and to increase their accuracy of data processing through contextual filtering of signals [13, 26].

Neurons of contextual neural networks for given data vector aggregate only those input signals that are required to meet the conditions set by their aggregation functions. Unlike the frequently used one-step aggregation functions – such as weighted sum – the operation of conditional aggregation functions is divided into several steps [13, 14]. At each step, signals from one selected subgroup of inputs are aggregated and used to check the aggregation stopping criteria. In effect, the same neuron for different input vectors can set its output by reading in signals from different subsets of inputs. Each subgroup of neuron inputs is defined and prioritized during the training of CxNN. The ordered list of such groups forms a scan-path. One can also notice that each contextual neuron is the realization of Stark's scan-path theory, which describes functional properties of the sensoric system of humans and other primates [27].

Construction of contextual neural networks requires modified training algorithms such as, e.g., generalized error backpropagation algorithm (GBP) [13, 14, 22, 23]. GBP is using self-consistency paradigm to optimize both continuous (weights) and non-continuous parameters (groupings) with gradient-based method [28]. It was shown that GBP can be applied successfully to construct CxNNs with various multi-step conditional aggregation functions, such as CFA, OCFA, RDFA, PRDFA, and SIF (Sigma-if) [14]. The latter was the first analyzed contextual aggregation function and is defined as a direct generalization of the weighted sum [13].

The literature describes research, which shows that the use of Contextual Neural Networks in which the number of groups is greater than 1, for many benchmark datasets can positively affect the accuracy of classification in relation to MLP networks [13, 17, 18, 21]. The reasons for this effect can be two phenomena. Multi-step conditional aggregation of inputs can limit the impact of noise contained in the training data by not aggregating inputs that are less relevant to classification. In addition, scan-paths are modified during the learning process and are activated differently by different training vectors. Due to this in different training epochs and for various data vectors, input signals can be aggregated or not depending on the processed data and actual knowledge accumulated by the neural network. This phenomenon in CxNNs has the features of using dropout regularization [14, 29].

However, both in the case of contextual and non-contextual models, many adverse effects may appear during the training of artificial neural networks, including: overfitting, vanishing of the gradient, and too high increase in weight values. To prevent that, various regularization techniques were developed. One of commonly used solutions is the dropout, which is realized by randomly excluding neurons from training in subsequent training steps [29–31]. This changes the process of training of single neural network into training of an ensemble of subnetworks while subnetworks need to cooperate to give similar results for the same input vectors. At the end of the training, all subnetworks are used together as they would be merged into one neural network of the initial architecture [32, 33]. On the other hand, during the training of CxNNs, not the whole neurons but single connections are independently enabled or disabled during signals processing. Moreover, this type of operation in CxNNs is not random but contextual - driven by processed signals and knowledge aggregated by the whole neural network. Finally, this contextual elimination of connections is done not only during but also after the training leading to a contextual decrease of activity of connections between neurons [13, 14]. The above shows that different types of dropout can exist and be used together in practice.

In this paper, Soft Dropout - a new type of dropout - is proposed and analyzed. The method connects the idea of partial dropout observed in CxNNs with traditional dropout techniques in a form applicable in all types of neural networks that can lead to a considerable reduction of the overfitting effect. Thus the further parts of the paper are organized as follows. The second section describes details of the dropout technique and introduces Soft Dropout method. Within third section results of experiments show the influence of Soft Dropout on the training process of MLP and CxNNs networks while solving selected benchmark problems from UCI Machine Learning Repository [34] and real-life Golub microarray data of ALL-AML leukemia gene expression [35].

2 Soft Dropout in Training of Neural Networks

When training an artificial neural network with the backpropagation algorithm, the network may overfit to the training set. Each weight is changed to minimize the function loss, taking into account what all other neurons do. Because of this, the weights can change so that the neuron corrects errors that appear in other neurons' outputs, which can lead to complex interrelationships between neurons. This, in turn, leads to overfitting because of the numerous relationships between neurons hamper generalization. Dropout regularization limits the formation of interdependencies between neurons by making their presence in the network uncertain [29]. It also means that training of the neural network corresponds to the training of many interdependent neural networks cooperating within the model equivalent to the complex classifier [29, 32]. The use of dropout regularization during the learning process of an artificial neural network may allow searching areas of the error space that are more difficult to search only by the backpropagation algorithm [30, 31]. Results reported in the literature indicate that training neural networks with dropout regularization allows to achieve better classification accuracies than without dropout [31, 33]. It can also be seen that dropout regularization in combination with max-norm regularization allows a significant reduction in classification errors.

Observation of CxNN neural networks can lead to the conclusion that it should not be taken as granted that dropout methods have to enable/disable whole neurons (their all connections). One can also think of partial dropout, which would enable/disable neurons partially. This can be done at least in two ways. First, partial dropout can be realized as it is done by CxNNs, where not all connections need to be disabled/enabled. Second, what is proposed in this paper, partial dropout can be implemented as a partial decrease of neuron output values. The latter allows for smooth control over the strength of the dropout, thus it will be further named Soft Dropout.

In practice, Soft Dropout regularization with smooth regulation of neurons activity is based on dropout regularization. Thus before using each training pattern, neurons dropped out from learning are selected. However, unlike normal dropout regularization, in Soft Dropout neurons are excluded only to some extent determined by the ψ parameter. The output value of each dropped out neuron is the product of its value before dropout and the value $1 - \psi$. In effect, when the value of parameter ψ is equal to 1, soft dropout regularization works as normal dropout regularization. Contrary, when ψ is 0 regularization will not occur at all. Due to this, Soft Dropout can be regarded as a generalization of dropout regularization. But it should also be taken into account, that modification of the neurons operation caused by Soft Dropout need appropriate changes in the training algorithm.

Soft Dropout in the Error Backpropagation Algorithm
When training an artificial neural network using soft dropout regularization, it is necessary to take into account the modification of calculating the output value from the dropped out neuron process. When a neuron is dropped out, its output value is multiplied by $1 - \psi$. When the parameter value is in the range (0;1), the excluded neuron will participate in the forward propagation process only to a certain extent, so its impact on the error at the network output is also partial. During backpropagation, the neuron weights will change proportionally to $1 - \psi$. Thus the formulas for weight change in hidden layers can be calculated and expressed as follows.

Lets' denote the standard neuron aggregation function formula as:

$$net\left(w_k^l, x_p, \theta_k^l\right) = \sum\nolimits_{j=1}^{N} w_{kj}^l * x_{pj} + \theta_k^l \tag{1}$$

where the net function is an aggregation function, w is set of weights of neuron number k for which the aggregation function value is calculated, l is the layer in which the neuron number k is located, x_p is a vector of neuron input values for training pattern p, θ_k^l denotes the bias value of neuron k in layer l, and N is the number of neuron inputs. Then the neuron's output function when Soft Dropout is used can be expressed as:

$$out\left(w_k^l, x_p, \theta_k^l\right) = f\left(net\left(w_k^l, x_p, \theta_k^l\right)\right) * (1 - \psi) \tag{2}$$

where k is the number of neurons in layer l, f is the selected activation function, and ψ is the neuron' muting factor. The error function necessary to calculate the input modification value can be defined as follows:

$$\varrho\left(Y_{pk}, w_k^L, x_p, \theta_k^L\right) = Y_{pk} - out\left(w_k^L, x_p, \theta_k^L\right) \tag{3}$$

where Y_{pk} is the expected value from neuron number k on the last layer L for the pattern p, and $out\left(w_k^L, x_p, \theta_k^L\right)$ is the actual value of the output of neuron number k calculated on the basis of the pattern p on the last layer of the network L. After defining the error function, the definition of the differential loss function is needed:

$$E\left(Y_{pk}, out_{pk}^L\right) = g\left(\varrho\left(Y_{pk}, w_k^L, x_p, \theta_k^L\right)\right) \tag{4}$$

where g is loglos loss function. After completing the calculation of the value of the loss function, the successive weights of all neurons must be modified by the value of Δw_{kj}^l, which is calculated using the formula:

$$\Delta w_{kj}^{l} = \eta * \frac{\partial E_p}{\partial w_{kj}^{l}} \tag{5}$$

where η is the given learning step, and E_p is the value of the loss function for the pattern p. And in the case of Soft Dropout the partial derivative $\frac{\partial E_p}{\partial w_{kj}^{l}}$ of the loss function E_p can be calculated as:

$$\begin{aligned}
\frac{\partial E_p}{\partial w_{kj}^{l}} &= \frac{\partial (g \circ \varrho)}{\partial w_{kj}^{l}} \\
&= \frac{\partial g}{\partial w_{kj}^{l}} * \frac{\partial \varrho}{\partial w_{kj}^{l}} = \frac{\partial g}{\partial w_{kj}^{l}} * \left(\frac{Y_{pk}}{\partial w_{kj}^{l}} - \frac{\partial out}{\partial w_{kj}^{l}} \right) \\
&= \frac{\partial g}{\partial w_{kj}^{l}} * \left(0 - \frac{\partial out}{\partial w_{kj}^{l}} \right) = -\frac{\partial g}{\partial w_{kj}^{l}} * \frac{\partial out}{\partial w_{kj}^{l}} \\
&= \frac{\partial g}{\partial w_{kj}^{l}} * \left(\frac{\partial (f \circ net)}{\partial w_{kj}^{l}} * (1 - \psi) \right) + \\
&+ \left(out\left(w_k^l, i_p, \theta_k^l \right) * \frac{(1 - \psi)}{\partial w_{kj}^{l}} \right) \\
&= -\frac{\partial g}{\partial w_{kj}^{l}} * \left(\frac{\partial (f \circ net)}{\partial w_{kj}^{l}} * (1 - \psi) \right)
\end{aligned} \tag{6}$$

It can be clearly seen that the product of the value of the derivative of the loss function and the derivative of the activation function is regulated by a Soft Dropout neuron fading factor $1 - \psi$. If the neuron is not dropped out, it is equal to 0 and the weights are modified in the same way as in MLP network without dropout regularization; otherwise, the weights are modified in proportion to the value of $1 - \psi$.

3 Results of Experiments

To verify how the proposed Soft Dropout technique can influence the training of neural networks, it was implemented in the H2O environment [36]. Then CxNN and MLP networks were trained with GBP algorithm to solve four selected benchmark problems from the UCI Machine Learning Repository [34] and Golub microarray data of ALL-AML leukemia gene expression [35]. Characteristics of datasets and considered architectures of neural networks are given in Table 1. Values of training parameters were as follows: aggregation functions: SUM/SIF, activation function hard linear rectifier, training step η equal 0.1, number of epochs = 1000, dropout probability = 0.5. All experiments were performed with the use of repeated five times 10-fold cross-validation. Average values and standard deviation of logloss measure for training and testing data were analyzed for muting factor $\psi = \{0.7, 0.9, 1.0, 1.3\}$. Muting factor $\psi = 1.0$ identifies standard dropout.

Table 1. Architectures of neural networks used during the experiments for selected problems.

Dataset	Number of hidden neurons	Number of inputs of neural network	Number of classes (network outputs)	Number of data vectors
Breast cancer	50	9	2	699
Crx	50	60	2	690
Heart Cancer	10	28	2	303
Sonar	30	60	2	208
Golub	20	7130	2	72

Initially, the plan of experiments assumed the analysis of Soft Dropout for values of muting factor ψ below 1.0. However, the initial achieved results were not better than in the case of standard dropout. It was also observed that overfitting decreases with increasing the value of the muting factor. This is why we tried to use also the muting factor $\psi = 1.3$. This can be counterintuitive, but Soft Dropout also allows such situation when $\psi > 1.0$. This represents the behavior in which neurons selected to be "disabled" have an opposite influence on the neurons in the following layers than when they are "enabled." Thus for $\psi = 1.3$, the "disabled" subnetwork partially distracts the "enabled" subnetwork. The results observed during experiments performed on considered MLP networks are shown on Figs. 1 and 2.

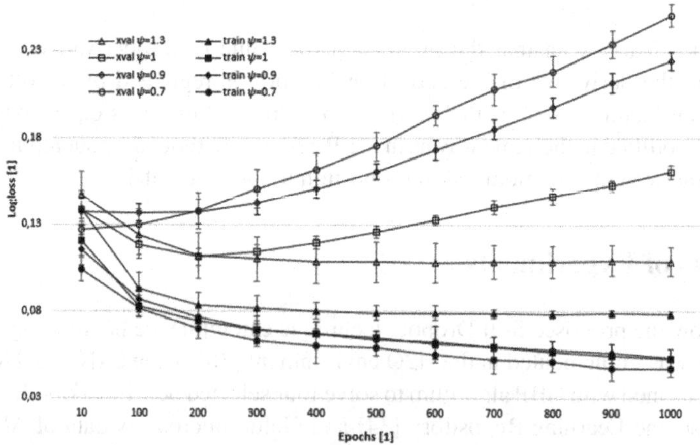

Fig. 1. The average values of logloss measure as a function of the learning epoch for MLP network solving the "Breast cancer" problem with the use of Soft Dropout technique and 5 × 10-fold cv for selected values of muting factor ψ. "xval" indicates results for test data.

The results presented in Figs. 1 and 2 indicate that by selecting the appropriate value of the ψ parameter, it is possible to prevent overfitting of MLP neural networks more effectively with Soft Dropout than when using standard dropout regularization. For both considered datasets, when the value of the muting factor ψ is below 1.0, the observed

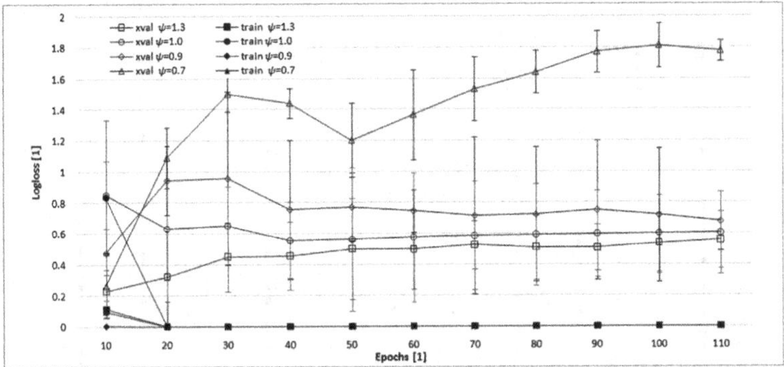

Fig. 2. The average values of logloss measure as a function of the learning epoch for MLP network solving the "Golub" problem with the use of Soft Dropout technique and 5 × 10-fold cross-validation for selected values of muting factor ψ. "xval" indicates results for test data.

overfitting is higher than for standard dropout and is rising as the muting factor decreases. But it can also be observed that when ψ is equal to 1.3, the values of logloss measure show almost no overfitting. In this case, Soft Dropout outperforms standard dropout. For the "Breast Cancer" dataset, it can be noticed that decreasing overfitting causes higher classification errors for training data. However, this is expected and entirely acceptable in practice, as the training errors remain lower than testing errors.

For comparison, it was further checked how Soft Dropout influences training of CxNNs, and results for "Breast Cancer" and "Golub" problems are shown in Figs. 3 and 4, respectively.

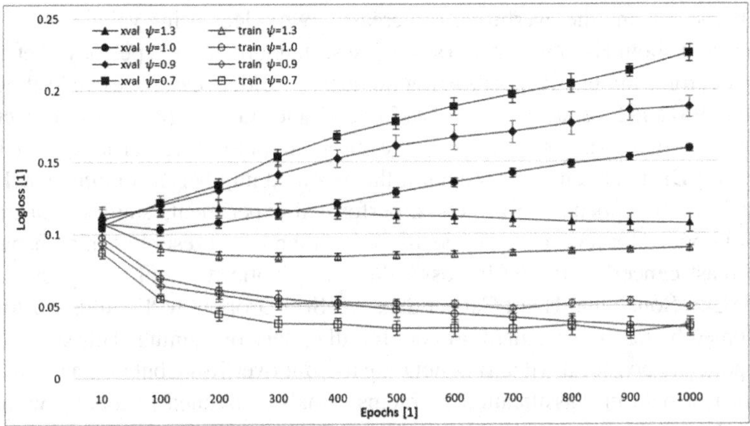

Fig. 3. The average values of logloss measure as a function of the learning epoch for CxNN network solving the "Breast cancer" problem with the use of Soft Dropout technique and 5 × 10-fold cv for selected values of muting factor ψ. "xval" indicates results for test data.

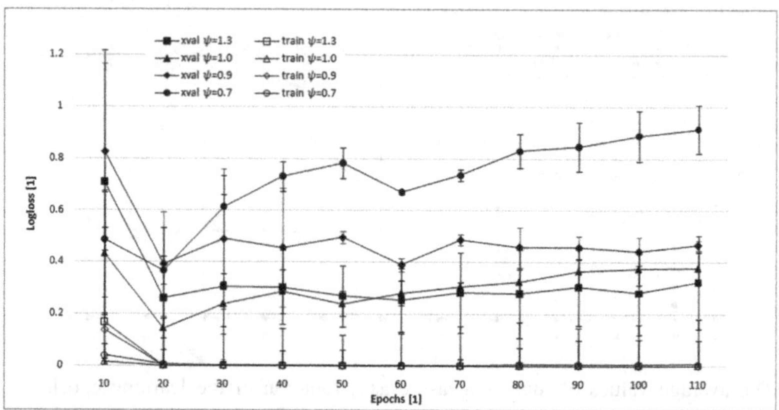

Fig. 4. The average values of logloss measure as a function of the learning epoch for CxNN network solving the "Golub" problem with the use of Soft Dropout technique and 5 × 10-fold cross-validation for selected values of muting factor ψ. "xval" series represent test data results.

The values of parameters used while training of CxNNs were the same as for experiments with MLP networks with the following changes: aggregation function: SIF, GBP update interval = 25, number of groups = 7, and aggregation threshold = 0.6.

Presented results indicate that also for Contextual Neural Networks, the proposed Soft Dropout method with muting factor $\psi = 1.3$ can decrease the overfitting better than a standard dropout. For other considered data sets, i.e., "Crx", "Sonar" and "Heart Cancer", obtained outcomes were analogous as the results displayed in Figs. 1, 2, 3 and 4, both for MLP and CxNN models. In each case, average values of loglos measure for test data are lower for Soft Dropout with muting factor $\psi = 1.3$ than when standard dropout was used, and the overfitting is increasing with decreasing value of ψ.

However, it should be noted that in some cases, standard dropout performs better than Soft Dropout for a number of epochs below some threshold. In cases presented above, the threshold is 300 and 55 epochs for "Breast Cancer" and "Golub" problems, respectively. In the case of "Breast Cancer" this can also be observed for MLP neural networks (Fig. 1). Thus the Soft Dropout can be beneficial if the assumed number of training epochs will be above 200–300 epochs. Nevertheless, performed experiments suggest that training with Soft Dropout for a number of epochs higher than this threshold can be beneficial, as for "Breast cancer" data, the logloss values drop noticeably till the epoch number 1000. For Crx, Sonar, and Heart Cancer data sets Soft Dropout with muting factor $\psi =$ 1.3 performed better than standard dropout for all epochs of training. This suggests that Soft Dropout can be useful – it allows not only to limit overfitting but also achieve better results for test data in classification problems. Finally, obtained results show that the CxNN models perform similarly ("Breast Cancer", "Heart Cancer") or better ("Golub," Crx", "Sonar") than analogous MLP networks – what is expected.

4 Conclusions

The presented results for selected datasets show that the introduction of Soft Dropout regularization can significantly influence the training process, both of the MLP and Contextual Neural networks. Surprisingly Soft Dropout with muting factor $\psi = 1.3$ can lead to lower overfitting than a standard dropout. Thus the usage of "disabled" subnetwork to distract "enabled" subnetwork during dropout can be found as additional, previously not explored, regularization factor. Still, Soft Dropout should be tested with greater number of benchmark problems and for a higher number of values of muting factor ψ above 1.0. The selection of an appropriate value of ψ as well as the number of epochs above which usage of Soft Dropout can be beneficial may be the subject of further research. In addition, studies on the impact of Soft Dropout regularization on the training of Convolutional Neural Networks would also be valuable.

References

1. Chen, S., Zhang, S., Shang, J., Chen, B., Zheng, N.: Brain-inspired cognitive model with attention for self-driving cars. IEEE Trans. Cogn. Dev. Syst. **11**(1), 13–25 (2019)
2. Guest, D., Cranmer, K., Whiteson, D.: Deep learning and its application to LHC physics. Annu. Rev. Nucl. Part. Sci. **68**, 1–22 (2018)
3. Suleymanova, I., et al.: A deep convolutional neural network approach for astrocyte detection. Sci. Rep. **8**(12878), 1–7 (2018)
4. Liu, L., Zheng, Y., Tang, D., Yuan, Y., Fan, C., Zhou, K.: Automatic skin binding for production characters with deep graph networks. ACM Trans. on Graphics (SIGGRAPH) **38**(4), 1–12 (2019). Article 114
5. Gao, D., Li, X., Dong, Y., Peers, P., Xu, K., Tong, X.: Deep inverse rendering for high-resolution SVBRDF estimation from an arbitrary number of images. ACM Trans. Graphics (SIGGRAPH) **38**(4), 1–15 (2019). article 134
6. Tsai, Y.C., et al.: FineNet: a joint convolutional and recurrent neural network model to forecast and recommend anomalous financial items. In: Proceedings of the 13th ACM Conference on Recommender Systems, RecSys 2019, pp. 536–537. ACM, New York (2019)
7. Dozono, H., Niina, G., Araki, S.: Convolutional self organizing map. In: 2016 IEEE International Conference on Computational Science and Computational Intelligence (CSCI), pp. 767–771. IEEE (2016)
8. Gong, K., et al.: Iterative PET image reconstruction using convolutional neural network representation. IEEE Trans. Med. Imaging **38**(3), 675–685 (2019)
9. Huang, X., Tan, H., Lin, G., Tian, Y.: A LSTM-based bidirectional translation model for optimizing rare words and terminologies. In: 2018 IEEE International Conference on Artificial Intelligence and Big Data (ICAIBD), China, pp. 5077–5086. IEEE (2018)
10. Athiwaratkun, B., Stokes, J.W.: Malware classification with LSTM and GRU language models and a character-level CNN. In: Proceedings of the 2017 IEEE International Conference on Acoustics, Speech and Signal Processing (ICASSP), USA, pp. 2482–2486. IEEE (2017)
11. Higgins, I., et al.: β-VAE: learning basic visual concepts with a constrained variational framework. In: International Conference on Learning Representations, ICLR 2017, vol 2, no. 5, pp. 1–22 (2017)
12. Karras, T., Aila, T., Laine, S., Lehtinen, J.: Progressive growing of GANs for improved quality, stability, and variation. In: International Conference on Learning Representations, ICLR 2018, pp. 1–26 (2018)

13. Huk, M.: Backpropagation generalized delta rule for the selective attention Sigma-if artificial neural network. Int. J. App. Math. Comput. Sci. **22**, 449–459 (2012)
14. Huk, M.: Notes on the generalized backpropagation algorithm for contextual neural networks with conditional aggregation functions. J. Intell. Fuzzy Syst. **32**, 1365–1376 (2017)
15. Huk, M.: Learning distributed selective attention strategies with the Sigma-if neural net-work. In: Akbar, M., Hussain, D. (eds.) Advances in Computer Science and IT, pp. 209–232. InTech, Vukovar (2009)
16. Huk, M.: Manifestation of selective attention in Sigma-if neural network. In: 2nd International Symposium Advances in Artificial Intelligence and Applications, International Multiconference on Computer Scientists and Information Technology, IMCSIT/AAIA 2007, vol. 2, pp. 225–236 (2007)
17. Huk, M.: Sigma-if neural network as the use of selective attention technique in classification and knowledge discovery problems solving. Ann. UMCS Sect. AI - Informatica **4**(2), 121–131 (2006)
18. Szczepanik, M., Jóźwiak, I.: Data management for fingerprint recognition algorithm based on characteristic points' groups. Found. Comput. Decis. Sci. **38**(2), 123–130 (2013). New Trends in Databases and Information Systems
19. Szczepanik, M., Jóźwiak, I.: Fingerprint recognition based on minutes groups using directing attention algorithms. In: Rutkowski, L., Korytkowski, M., Scherer, R., Tadeusiewicz, R., Zadeh, L.A., Zurada, J.M. (eds.) ICAISC 2012. LNCS (LNAI), vol. 7268, pp. 347–354. Springer, Heidelberg (2012). https://doi.org/10.1007/978-3-642-29350-4_42
20. Huk, M., Pietraszko, J.: Contextual neural-network based spectrum prediction for cognitive radio. In: 4th International Conference on Future Generation Communication Technology (FGCT 2015), pp. 1–5. IEEE Computer Society, London (2015)
21. Huk, M.: Non-uniform initialization of inputs groupings in contextual neural networks. In: Nguyen, N.T., Gaol, F.L., Hong, T.-P., Trawiński, B. (eds.) ACIIDS 2019. LNCS (LNAI), vol. 11432, pp. 420–428. Springer, Cham (2019). https://doi.org/10.1007/978-3-030-14802-7_36
22. Huk, M.: Training contextual neural networks with rectifier activation functions: role and adoption of sorting methods. J. Intell. Fuzzy Syst. **38**, 1–10 (2019)
23. Huk, M.: Weights ordering during training of contextual neural networks with generalized error backpropagation: importance and selection of sorting algorithms. In: Nguyen, N.T., Hoang, D.H., Hong, T.-P., Pham, H., Trawiński, B. (eds.) ACIIDS 2018. LNCS (LNAI), vol. 10752, pp. 200–211. Springer, Cham (2018). https://doi.org/10.1007/978-3-319-75420-8_19
24. Huk, M.: Context-related data processing with artificial neural networks for higher reliability of telerehabilitation systems. In: 17th International Conference on E-health Networking, Application & Services (HealthCom), pp. 217–221. IEEE Computer Society, Boston (2015)
25. Huk, M., Kwiatkowski, J., Konieczny, D., Kędziora, M., Mizera-Pietraszko, J.: Context-sensitive text mining with fitness leveling genetic algorithm. In: 2015 IEEE 2nd International Conference on Cybernetics (CYBCONF), Gdynia, Poland, 2015, pp. 1–6. Electronic Publication (2015). ISBN 978–1-4799-8321-6
26. Huk, M., Kwaśnicka, H.: The concept and properties of sigma-if neural network. In: Ribeiro, B., Albrecht, R.F., Dobnikar, A., Pearson, D.W., Steele, N.C. (eds.) Adaptive and Natural Computing Algorithms, ICANNGA 2005, pp. 13–17. Springer, Vienna (2005). https://doi.org/10.1007/3-211-27389-1_4. Computer Science
27. Privitera, C.M., Azzariti, M., Stark, L.W.: Locating regions-of-interest for the Mars Rover expedition. Int. J. Remote Sens. **21**, 3327–3347 (2000)
28. Glosser, C., Piermarocchi, C., Shanker, B.: Analysis of dense quantum dot systems using a self-consistent Maxwell-Bloch framework. In: Proceedings of 2016 IEEE International Symposium on Antennas and Propagation (USNC-URSI), Puerto Rico, pp. 1323–1324. IEEE (2016)

29. Srivastava, N., Hinton, G.E., Krizhevsky, A., Sutskever, I., Salakhutdinov, R.R.: Dropout: a simple way to prevent neural networks from overfitting. J. Mach. Learn. Res. **15**, 1929–1958 (2014)
30. Ko, B., Kim, H.G., Choi, H. J.: Controlled dropout: a different dropout for improving training speed on deep neural network. In: Proceedings of the 2017 IEEE International Conference on Systems, Man, and Cybernetics (SMC), Canada. IEEE (2018)
31. ElAdel, A., Ejbali, R., Zaied, M., Ben Amar, C.: Fast deep neural network based on intelligent dropout and layer skipping. In: Proceedings: 2017 International Joint Conference on Neural Networks (IJCNN), Anchorage, USA (2017)
32. Salehinejad, H., Valaee, S.: Ising-dropout: a regularization method for training and compression of deep neural networks. In: Proceedings: 2019 IEEE International Conference on Acoustics, Speech and Signal Processing (ICASSP), Brighton, United Kingdom (2019)
33. Guo, J., Gould, S.: Depth dropout: efficient training of residual convolutional neural networks. In: 2016 International Conference on Digital Image Computing: Techniques and Applications (DICTA), Gold Coast, Australia. IEEE (2016)
34. UCI Machine Learning Repository. http://archive.ics.uci.edu/ml
35. Golub, T.R., et al.: Molecular classification of cancer: class discovery and class prediction by gene expression monitoring. Science **286**, 531–537 (1999)
36. H2O.ai documentation. http://docs.h2o.ai/h2o/latest-stable/h2o-docs/index.html

The Impact of Constant Field of Attention on Properties of Contextual Neural Networks

Erik Dawid Burnell[1], Krzysztof Wołk[2], Krzysztof Waliczek[3](✉), and Rafał Kern[3]

[1] Science Applications International Corporation, Reston, USA
erik.d.burnell@gmail.com

[2] The Wroclaw Institute of Spatial Information and Artificial Intelligence, Wroclaw, Poland
krzysztof.wolk@wizipisi.pl

[3] Faculty of Computer Science and Management, Wroclaw University of Science and Technology, Wroclaw, Poland
238547@student.pwr.edu.pl, rafal.kern@pwr.edu.pl

Abstract. Applications of Artificial Neural Networks are used with success in many fields such as e.g. medicine, economy, science and entertainment. But most of those models processes all input signals without attention directing mechanisms. In this paper contextual neural networks are considered which are using multi-step conditional aggregation functions to direct attention of neurons during data processing. It is verified if aggregation function with constant field of attention (CFA) can help to build classification models of higher accuracy and of lower activity of hidden connections than aggregation function with dynamic field of attention (SIF). Experiments are performed with use of the H2O machine learning framework implementing Generalized Backpropagation Algorithm for selected benchmark problems from UCI ML repository and real-life ALL-AML leukemia gene expression data. Presented analysis of results gives important clues on the most promising directions of research on contextual neural networks and indicates possible improvements of the H2O framework.

Keywords: Contextual processing · Aggregation functions · Selective attention

1 Introduction

Nowadays Artificial Neural Networks form a set of well settled machine learning models and techniques applied in many fields. They are used in financial recommender systems [1] and control units of autonomous vehicles [2], are important elements of medical diagnostic frameworks [3] and setups of important scientific experiments [4] as well as in the newest hardware accelerators for video rendering and images upscaling [5, 6]. This was achieved by development of many types of artificial neural models. The most important are Convolutional Neural Networks (CNN) [7], Variational Autoencoders (VAE) and Generative Adversarial Networks (GAN) [8, 9], recurrent architectures using Long Short-Term Memory (LSTM) and Gated Recurrent Units (GRU) [10, 11]. In most cases mentioned solutions need considerable amounts of memory and time to be trained, thus

N. T. Nguyen et al. (Eds.): ACIIDS 2020, LNAI 12034, pp. 364–375, 2020.
https://doi.org/10.1007/978-3-030-42058-1_31

when it is possible simpler models are still in use: multilayer perceptrons (MLP) [12] and Self Organizing Maps [13]. One can notice that all mentioned types of neural models have a common property – they are always processing signals from all inputs. This simplifies their construction and training but strongly limits their abilities for contextual data processing. But contextual neurons [14–16] and Contextual Neural Networks (CxNN) are also developed providing models with a set of unique properties [17–19].

Contextual Neural Networks can be regarded as generalization of MLP models [20]. Literature reports their successful usage in predicting radio communicates in cognitive radio [21], classification of cancer gene expression microarray data [22–24] as well as in finger-print detection through contextual analyses of minutia groups in crime investigations [25, 26]. CxNNs can be also considered as elements of complex context-sensitive text mining solutions and telemedicine systems [27, 28].

Neurons of CxNNs are realizations of scan-path theory developed by Stark to capture general properties of sensoric system of humans [29]. Conditional, multi-step aggregation functions of such neurons accumulate signals from inputs in groups in a step-by-step manner until given condition is met. The ordered list of groups forms a scan-path which can be encoded directly within weights of connections between neurons. Thus for different data vectors given contextual neuron can calculate its output with use of different subsets of inputs.

In this paper contextual neural networks with different types of aggregation functions are considered. In particular, aggregation functions with static field of attention (CFA, SUM) are compared with SIF aggregation function which implements dynamic field of attention. The presented study tries to answer the question if limited, static field of attention, represented by CFA, can have positive influence on classification properties of CxNNs. Experiments are performed with the use of selected benchmark problems from the UCI Machine Learning Repository [30] as well as with real-life data (Golub cancer gene expression microarray data) [31]. Thus the following parts of the paper are organized as follows. Section 2 contains a description of contextual neural networks, the Generalized Error Backpropagation algorithm. Next, in Sect. 3, CFA aggregation function is presented in opposition to SIF aggregation. Then in Sect. 4 the design and results of experiments are given with discussion of the measured outcomes. Finally, Sect. 5 wraps up with conclusions about the results as well as suggested directions for further research on aggregation functions of contextual neural networks.

2 Contextual Neural Networks and GBP Algorithm

Contextual neural networks (CxNNs) are machine learning models which are using neurons with multistep conditional aggregation functions [23, 32, 33]. Because of such construction, neurons in contextual neural networks aggregate values from their particular inputs under a certain conditions. This is unlike MLP networks, where all inputs are aggregated unconditionally. It was shown that in classification tasks conditional aggregation can reduce the computational cost and the time of the use of obtained CxNN neural networks in relation to MLP [18–20, 34].

The operation of a conditional multi-step aggregation function is carried out in steps. In each step inputs from a single, selected group of inputs, are aggregated. Aggregation of data from successive input groups proceeds according to the input scan-path defined during the training. Once an aggregation stopping condition is met, the aggregation is interrupted, and thus, some inputs might be completely skipped in calculating the neuron output. This type of operation is common for all contextual aggregation functions considered in the literature, including such examples as PRDFA, OCFA and SIF [19, 20]. Figure 1 presents the diagram and formulas used to calculate activation and output of the contextual neuron using SIF aggregation function.

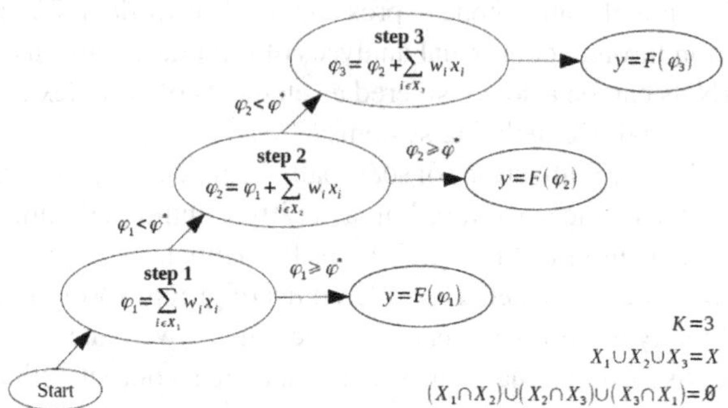

Fig. 1. Diagram of the example SIF aggregation function describing changes of attention field of contextual neuron with number of groups $K = 3$, aggregation threshold φ^*, activation function F and set of inputs X.

In detail, the SIF function aggregates values from subsequent neuron input groups until the total aggregated value of activation does not reach or exceed a specified aggregation threshold φ^*. It is assumed that neuron's inputs are divided into K separate, disjoint groups of similar size. In the case when it is impossible to divide the inputs into groups of equal size, the last group may include lower or higher number of inputs. During the aggregation process in its' each subsequent step $k \leq K$, the values of the k–th group of inputs are aggregated and added to previously accumulated activation of the neuron. If the accumulated activation is higher than threshold $\varphi*$ before all groups are aggregated, the remaining groups (and all their inputs) are skipped and not processed. The number of steps necessary to reach the aggregation threshold depends on the scan-path defined by parameters of the neuron and on the signals from the actually processed input vector.

Training Contextual Neural Networks with GBP Method

One of the method that can be used to train contextual neural networks with neurons using conditional multi-step aggregation functions is Generalized Error Backpropagation algorithm (GBP). It is a version of traditional error backpropagation extended with self-consistency method [19, 20]. The latter is frequently used in physics to solve sets of equations including both differentiable and non-differentiable parameters [35, 36]. In GBP it allows to use a gradient-based optimization to select values of non-continuous and not differentiable parameters describing scan-paths of contextual neurons. The block diagram of GBP is presented at Fig. 2.

The first step of the Generalized error Back Propagation algorithm is to randomly initialize the connection weights and assign all inputs of a each neuron to a single group.

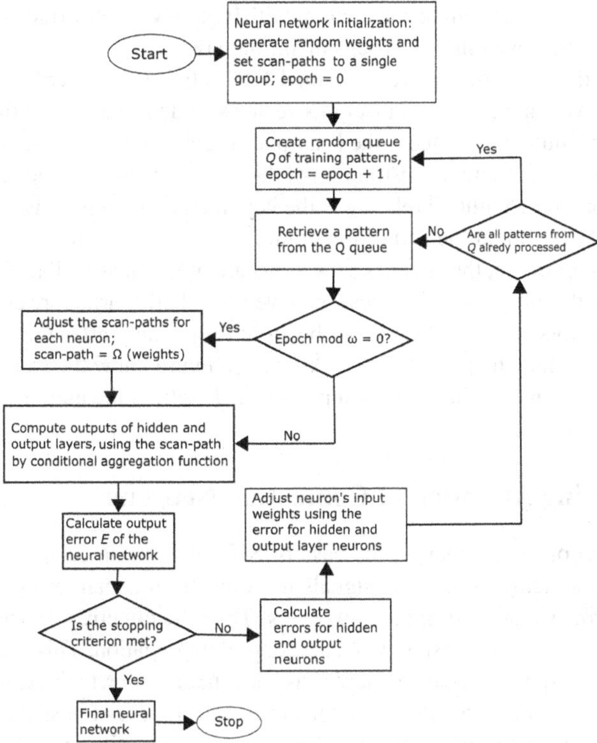

Fig. 2. Diagram of the Generalized error backpropagation algorithm (GBP) for grouping function Ωand scan-paths update interval ω.

In the effect, until groups are updated, the aggregation function operates identically to standard weighted sum aggregation (SUM). Then, during each epoch of training, for each training data vector the outputs of the neural network and batch output error E are calculated. Finally, the error is back-propagated through all the layers of the network and weights of connections active during the forward phase are updated according to stochastic gradient descent method (SGD). This process is repeated till the stopping criterion is not met.

In parallel to the weights optimization process the scan-paths are selected. To do that, after every ω epochs the grouping function $\Omega(w)$is used to calculate actual groupings of inputs of hidden neurons. The groups update interval ω controls the strength of the bound between the weights and scan-paths of neurons, what is important for stability of the self-consistency method. If the bound is too tight (low values of ω) the error space changes too frequently and SGD method can have problems with navigation towards good local optimum. Otherwise (high values of ω) the optimization of weights is much quicker than selection of scan-paths and the latter can be left not optimized. In a special case, when $\omega = \infty$, the GBP method is reduced to standard BP. In typical solutions the value of groups update interval ω is pre-selected from 10 to 25, and grouping function

$\Omega(\boldsymbol{w})$for given neuron is assigning connections with highest weights to the most important groups, dividing the inputs into K groups of similar size.

One can notice, that the above described GBP algorithm describes each hidden neuron with two vectors: connection weights vector $\boldsymbol{w}$ and virtual scan-path vector. Both those vectors are mutually related – the scan-path is calculated as $\Omega(\boldsymbol{w})$and changes of weights $\boldsymbol{w}$ depend on the actual construction of the scan-path used to calculate outputs of the network. This forced mutual relation is the key making the self-consistency method to work within GBP. But at the end of the training process, when the model is near the acceptable local optimum, the changes of weights are minimal as well as the changes of the scan-path. In the effect the virtual scan-path vectors of all neurons are encoded within their weights vectors, and do not need to be stored separately after the training. Thus to use the trained contextual neural network it is enough to know the weights vector $\boldsymbol{w}$, grouping function $\Omega$and single additional parameter K setting the number of the groups in the scan-path.

3 Constraining Attention of Contextual Neurons

It can be noticed on Fig. 2 that, in the case of SIF function, in each subsequent step of aggregation the neuron considers signals not only from actual group of inputs but also from all groups analyzed in previous steps. Thus the effective decision space has dimensionality which is increasing with each step of aggregation. This is referred to as "evolving decision space" an can be viewed as a mechanism in which the neuron at first tries to solve a problem in low-dimensional space, and if it is not possible, it gradually increases the dimensionality of considered decision space until the problem is solved or all available inputs are used. Such operation places SIF in a group of aggregation functions of a dynamic field of attention. It was shown that dynamic field of attention extends properties of neurons. One of the example benefits is that single neuron with SIF aggregation and sigmoidal activation function can solve linearly inseparable problems [18, 20]. Limited field of attention can also lead to considerable decrease of average activity of connections between neurons and in the effect – to decreased computational cost of their usage. And most importantly, adaptation of a field of attention plays role of contextual filtering, what can have effect in increased accuracy of data processing.

Contrary, the weighted SUM aggregation used by neurons of MLP networks is always processing signals from all its inputs. Thus by definition SUM aggregation has constant field of attention. It is interesting that in the case of SUM function the constant field of attention has both positive and negative influences on properties of the neurons. The negatives: neuron models with SUM aggregation have constant, 100% activity of inputs irrespectively on processed data. Even if some of the inputs are not important for classification of given input vector, they will be used. And signals from those inputs can act as a noise decreasing the accuracy of a neuron output values. On the other hand, SUM aggregation gives neuron access to all available data without the need to build appropriate scan-path. This makes neural networks with neurons using SUM aggregation easier to train. Additional benefit is the simplicity of their operation which helps when hardware implementations of NNs need to be considered.

The two above approaches seem to be opposite, but in fact the SIF function is a generalization of weighted sum. When the assumed number of groups of inputs is one,

SIF behaves exactly as the SUM aggregation function. Thus a mixed solution can be also proposed, which has constant size of the field of attention but still allows contextual filtering of input signals. This is why in this work a CFA (Constant Field of Attention) aggregation function is analyzed.

It is presented on Fig. 3. that CFA multi-step aggregation operates in a way very similar to the SIF function but with one important difference: in CFA each step of aggregation forms activation value which is independent from the activations calculated in previous steps. If the information from given group of inputs is not enough to pass the aggregation threshold, it is discarded and information from next group is analyzed. Thus the field of attention of neuron with CFA aggregation has always the width of one group of inputs – it is not growing as in the case of SIF. But the usage of next group is still contextually dependent on the result of processing of the previous groups.

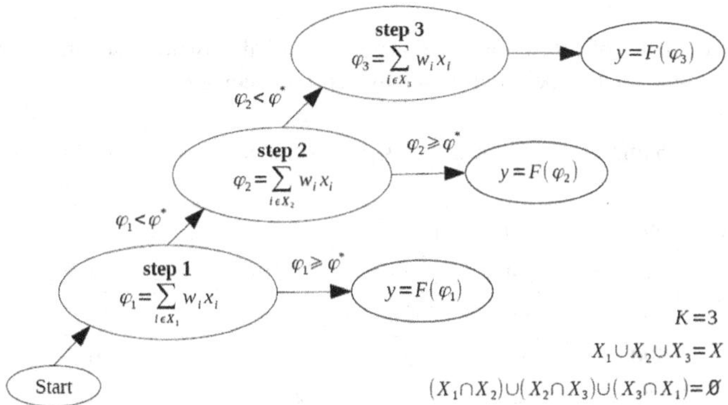

Fig. 3. Diagram of the example CFA aggregation function describing shifts of constant field of attention of contextual neuron with number of groups $K = 3$, aggregation threshold φ^* and set of inputs X.

The potential benefits behind presented construction of CFA aggregation function are:

- limitation of the activity of connections between neurons,
- contextual filtering of processed signals (stronger than in the case of SIF aggregation function),
- higher independence of processing of different groups of inputs, allowing easier parallel computation of output value of the neuron.

Especially the last listed property can be interesting, because dependence of activities of groups within SIF aggregation limits its parallelization and effective usage with modern highly-parallel processors. But the changes introduced in the aggregation process set also additional constraints on the algorithm needed to construct the scan-path of neuron using CFA function. This is why in the following parts of this paper it is verified if GBP method can be successfully used to train CxNNs with CFA aggregation functions, and what are the classification properties of resulting contextual models.

4 Results of Experiments

The goal of the experiments was to study the influence of limited, constant field of attention on properties of contextual neural networks. To perform planned analyses it was needed to compare the classification accuracy as well as the average activity of hidden connections of neural networks utilizing the CFA aggregation function with models using the SIF and weighted sum aggregation methods. The latter is a typical element of MLP and many other non-contextual neural models and was used here as a reference. The experiments were done with selected benchmark problems from the UCI Machine Learning Repository as well as with real-life data (Golub ALL-AML Leukemia gene expression microarray data) [30, 31]. The detailed characteristics of the datasets and values of related parameters of considered neural networks can be found in the Table 1.

Table 1. Characteristics of datasets and parameters of neural networks used in experiments. For MLP neural networks the number of groups is by definition equal one.

Dataset	Samples	Attributes	Classes	Number of groups	Hidden neurons count
Breast cancer	699	10	2	7	35
Iris	150	4	3	2	4
Sonar	208	60	2	9	30
Crx	690	15	2	7	30
Golub	72	7130	2	1000	2
Soybean	307	35	19	10	20
Heart disease	303	75	5	7	10

All presented experiments were conducted with H2O version 3.24.0.3 – an open source framework prepared for creation and testing of machine learning models as well as for statistical analyses of data [37]. It was modified to include GBP algorithm for training of CxNNs [38, 39]. Coding of models outputs was "winner-takes-all" and coding of inputs was "one-hot-encoding with interactions". The training parameters were as follows: activation function = tanh, maximal number of training epochs = 600, number of hidden layers = 1, aggregation threshold = 0.6, groups update interval = 10 epochs, CFA window width = 1. Numbers of hidden neurons and numbers of groups were as given in Table 1. In the case of MLP networks the number of groups by definition was set to one. The measurements were done with use of repeated 10 times 10-fold cross-validation [40].

Results presented in Table 2 show that using the CFA aggregation function with contextual neural networks in some cases allows to achieve models with lower level of classification errors than solutions with the Sum or the SIF aggregation functions (sonar and iris datasets). However, for most considered problems contextual neural networks that are using the CFA function are characterized by statistically higher mean classification errors than CxNNs and MLP models with SIF and Sum aggregation functions, respectively.

Table 2. Probability of classification error for considered aggregation functions and different datasets. The statistically worst results are marked with bold.

Dataset	CFA [1]	SIF [1]	SUM [1]
Breast cancer	**0.045 ± 0.038**	0.032 ± 0.021	0.025 ± 0.007
Crx	**0.142 ± 0.063**	0.138 ± 0.013	0.133 ± 0.009
Golub	**0.087 ± 0**	0.051 ± 0	0.061 ± 0
Heart disease	0.174 ± 0.112	**0.191 ± 0.025**	0.158 ± 0.046
Iris	0.027 ± 0.048	**0.047 ± 0.048**	0.037 ± 0.048
Sonar	**0.152 ± 0.104**	0.137 ± 0.073	0.161 ± 0.028
Soybean	0.051 ± 0.026	0.048 ± 0.043	0.037 ± 0.018

Also for all considered problems, except Crx, the hidden connection activity of CxNNs with CFA aggregation is statistically higher than in analogous models with SIF function (Table 3). In the effect it can be stated that the forced limitation of the field of attention implemented by CFA aggregation has in most cases a negative impact on the basic properties of contextual neural networks. It is probably related to the fact that the CFA function allows neurons to take into account only a given number of their inputs from those that were recognized by the GBP algorithm as the most important for the classification of a given training pattern. Such a strategy does not allow for including the most relevant information together with those that are only valid for some of the analyzed patterns. As a result, it forces decisions based on incomplete data and translates into an increase in the observed level of classification error. At the same time, the lack of accumulation of neuron' signals from subsequent analyzed input groups leads to an increase in the number of analyzed groups and, as a result, to an increase of the mean activity of hidden connections.

Additionally, it can be also noticed that the classification accuracies of contextual networks using SIF aggregation for datasets other than Golub and Sonar are lower than analogous outcomes for MLP networks. These results do not coincide with previous observations reported in the literature [20]. This suggests, that most probably the authors of this and previous works are using different implementations of GBP algorithm. During experiments presented in this work the implementation of CxNNs and GBP algorithm in the H2O environment was used. And the H2O framework implements efficient early-stopping mechanisms that can strongly limit the number of training epochs. But it should be taken into account that the training of the contextual neural networks is a more complex process than training of MLP networks. In the case of a CxNN, the selection of weights of connections must allow not only to meet the requirements set by training patterns, but also to group the inputs of neurons accordingly to their significance for the classification process. The process of CxNN training may therefore require more epochs than for MLP

Table 3. Average activity of hidden connections of considered neural models, aggregation functions and datasets. The statistically worst results for contextual aggregation functions (CFA, SIF) are marked with bold. Activity of MLP networks with SUM aggregation is 100% by definition.

Dataset	CFA [%]	SIF [%]	SUM [%]
Breast cancer	**88.09 ± 8.19**	84.91 ± 0.20	100 ± 0
Crx	76.67 ± 21.06	**83.26 ± 14.9**	100 ± 0
Golub	**19.41 ± 16.32**	7.16 ± 5.59	100 ± 0
Heart disease	**67.61 ± 10.47**	59.99 ± 6.04	100 ± 0
Iris	**100.00 ± 0**	82.69 ± 2.07	100 ± 0
Sonar	**54.18 ± 14.41**	46.35 ± 8.82	100 ± 0
Soybean	**55.09 ± 1.42**	52.45 ± 1.19	100 ± 0

networks. As a result, the mechanism of early-stopping of training currently used by the H2O environment, which is optimized for MLP networks, may not be the best solution for CxNNs.

5 Conclusions

The paper presents the results of studies of contextual neural networks using exemplary aggregation functions with dynamic (SIF) and constant field of attention (CFA). Both their ability to classify and the activity of internal connections were tested. Measurements were made for selected benchmark sets from the UCI ML repository as well as real-life data (Golub cancer gene expression microarray data). During experiments the implementation of contextual neural networks and GBP in the H2O environment was used.

The analysis of obtained results indicates that forced limitation of the field of attention of the aggregation function of contextual neurons can have negative influence on properties of CxNN. Usage of CFA aggregation for many considered data sets decreases the classification accuracy of contextual neural models. It also significantly increases average activity of connections between neurons. This important result suggests that further studies of contextual neural networks should be conducted towards the development of neuron models using aggregation functions with a dynamic field of attention. Such solutions, as presented by SIF aggregation, can build scan-paths which are better suited to efficiently organize inputs processing, than scan-paths of neurons with CFA functions. Also modified versions of CFA can be considered without the limitation that every input of the neuron belongs only to a single group.

Finally, results suggest that adaptation of the early-stopping criterion of the GBP algorithm in the H2O environment can be important element of future research on contextual neural networks.

References

1. Tsai, Y.-C., et al.: FineNet: a joint convolutional and recurrent neural network model to forecast and recommend anomalous financial items. In: Proceedings of the 13th ACM Conference on Recommender Systems RecSys 2019, pp. 536–537. ACM, New York (2019)
2. Chen, S., Zhang, S., Shang, J., Chen, B., Zheng, N.: Brain-inspired cognitive model with attention for self-driving cars. IEEE Trans. Cogn. Dev. Syst. **11**(1), 13–25 (2019)
3. Suleymanova, I., et al.: A deep convolutional neural network approach for astrocyte detection. Sci. Rep. **8**(12878), 1–7 (2018)
4. Guest, D., Cranmer, K., Whiteson, D.: Deep learning and its application to LHC physics. Annu. Rev. Nucl. Part. Sci. **68**, 1–22 (2018)
5. Liu, L., et al.: Automatic skin binding for production characters with deep graph networks. ACM Trans. Graph. (SIGGRAPH) **38**(4), 1–12 (2019). Article 114
6. Gao, D., Li, X., Dong, Y., Peers, P., Xu, K., Tong, X.: Deep inverse rendering for high-resolution SVBRDF estimation from an arbitrary number of images. ACM Trans. Graph. (SIGGRAPH) **38**(4), 1–15 (2019). Article 134
7. Gong, K., et al.: Iterative PET image reconstruction using convolutional neural network representation. IEEE Trans. Med. Imaging **38**(3), 675–685 (2019)
8. Higgins, I., et al.: Beta-VAE: learning basic visual concepts with a constrained variational framework. In: International Conference on Learning Representations, ICLR 2017, vol. 2, no. 5, pp. 1–22 (2017)
9. Karras, T., Aila, T., Laine, S., Lehtinen, J.: Progressive growing of GANs for improved quality, stability, and variation. In: International Conference on Learning Representations, ICLR 2018, pp. 1–26 (2018)
10. Huang, X., Tan, H., Lin, G., Tian, Y.: A LSTM-based bidirectional translation model for optimizing rare words and terminologies. In: 2018 IEEE International Conference on Artificial Intelligence and Big Data (ICAIBD), pp. 5077–5086. IEEE (2018)
11. Athiwaratkun, B., Stokes, J.W.: Malware classification with LSTM and GRU language models and a character-level CNN. In: Proceedings 2017 IEEE International Conference on Acoustics, Speech and Signal Processing (ICASSP), pp. 2482–2486. IEEE (2017)
12. Amato, F., Mazzocca, N., Moscato, F., Vivenzio, E.: Multilayer perceptron: an intelligent model for classification and intrusion detection. In: 31st International Conference on Advanced Information Networking and Applications Workshops (WAINA), pp. 686–691. IEEE, Taipei (2017)
13. Dozono, H., Niina, G., Araki, S.: Convolutional self organizing map. In: 2016 IEEE International Conference on Computational Science and Computational Intelligence (CSCI), pp. 767–771. IEEE (2016)
14. Mel, B.W.: The Clusteron: toward a simple abstraction for a complex neuron. In: Advances in Neural Information Processing Systems, vol. 4, pp. 35–42. Morgan Kaufmann (1992)
15. Spratling, M.W., Hayes, G.: Learning synaptic clusters for nonlinear dendritic processing. Neural Process. Lett. **11**, 17–27 (2000)
16. Gupta, M.: Correlative type higher-order neural units with applications. In: IEEE International Conference on Automation and Logistics, ICAL 2008, pp. 715–718 (2008). Springer Computer Science
17. Vanrullen, R., Koch, C.: Visual selective behavior can be triggered by a feed-forward process. J. Cogn. Neurosci. **15**, 209–217 (2003)
18. Huk, M.: Learning distributed selective attention strategies with the Sigma-if neural network. In: Akbar, M., Hussain, D. (eds.) Advances in Computer Science and IT, pp. 209–232. InTech, Vukovar (2009)

19. Huk, M.: Notes on the generalized backpropagation algorithm for contextual neural networks with conditional aggregation functions. J. Intell. Fuzzy Syst. **32**, 1365–1376 (2017)
20. Huk, M.: Backpropagation generalized delta rule for the selective attention sigma-if artificial neural network. Int. J. App. Math. Comp. Sci. **22**, 449–459 (2012)
21. Huk, M., Pietraszko, J.: Contextual neural network based spectrum prediction for cognitive radio. In: 4th International Conference on Future Generation Communication Technology (FGCT 2015), pp. 1–5. IEEE Computer Society, London (2015)
22. Huk, M.: Non-uniform initialization of inputs groupings in contextual neural networks. In: Nguyen, N.T., Gaol, F.L., Hong, T.-P., Trawiński, B. (eds.) ACIIDS 2019. LNCS (LNAI), vol. 11432, pp. 420–428. Springer, Cham (2019). https://doi.org/10.1007/978-3-030-14802-7_36
23. Huk, M.: Training contextual neural networks with rectifier activation functions: role and adoption of sorting methods. J. Intell. Fuzzy Syst. **38**, 1–10 (2019)
24. Huk, M.: Weights ordering during training of contextual neural networks with generalized error backpropagation: importance and selection of sorting algorithms. In: Nguyen, N.T., Hoang, D.H., Hong, T.-P., Pham, H., Trawiński, B. (eds.) ACIIDS 2018. LNCS (LNAI), vol. 10752, pp. 200–211. Springer, Cham (2018). https://doi.org/10.1007/978-3-319-75420-8_19
25. Szczepanik, M., Jóźwiak, I.: Fingerprint recognition based on minutes groups using directing attention algorithms. In: Rutkowski, L., Korytkowski, M., Scherer, R., Tadeusiewicz, R., Zadeh, L.A., Zurada, J.M. (eds.) ICAISC 2012. LNCS (LNAI), vol. 7268, pp. 347–354. Springer, Heidelberg (2012). https://doi.org/10.1007/978-3-642-29350-4_42
26. Szczepanik, M., Jóźwiak, I.: Data management for fingerprint recognition algorithm based on characteristic points' groups. In: Pechenizkiy, M., Wojciechowski, M. (eds.) New Trends in Databases and Information Systems. Foundations of Computing and Decision Sciences, 38(2), pp. 123–130. Springer, Heidelberg (2013). https://doi.org/10.1007/978-3-642-32518-2_40
27. Huk, M.: Context-related data processing with artificial neural networks for higher reliability of telerehabilitation systems. In: 17th International Conference on E-Health Networking, Application & Services (HealthCom), pp. 217–221. IEEE Computer Society, Boston (2015)
28. Huk, M., Kwiatkowski, J., Konieczny, D., Kędziora, M., Mizera-Pietraszko, J.: Context-sensitive text mining with fitness leveling genetic algorithm. In: 2015 IEEE 2nd International Conference on Cybernetics (CYBCONF), Gdynia, Poland, Electronic Publication, pp. 1–6 (2015). ISBN: 978-1-4799-8321-6
29. Privitera, C.M., Azzariti, M., Stark, L.W.: Locating regions-of-interest for the Mars Rover expedition. Int. J. Remote Sens. **21**, 3327–3347 (2000)
30. Dua, D., Graff, C.: UCI Machine Learning Repository. University of California, School of Information and Computer Science, Irvine, CA (2019). http://archive.ics.uci.edu/ml
31. Golub, T.R., et al.: Molecular classification of cancer: class discovery and class prediction by gene expression monitoring. Science **286**, 531–537 (1999)
32. Huk, M., Kwaśnicka, H.: The concept and properties of Sigma-if neural network. In: Ribeiro, B., Albrecht, R.F., Dobnikar, A., Pearson, D.W., Steele, N.C. (eds.) Adaptive and Natural Computing Algorithms, ICANNGA 2005, pp. 13–17. Springer Computer Science, Vienna (2005). https://doi.org/10.1007/3-211-27389-1_4
33. Huk, M.: Manifestation of selective attention in Sigma-if neural network. In: 2nd International Symposium Advances in Artificial Intelligence and Applications, International Multi-conference on Computer Science and Information Technology, IMCSIT/AAIA 2007, vol. 2, pp. 225–236 (2007)
34. Huk, M.: Sigma-if neural network as the use of selective attention technique in classification and knowledge discovery problems solving. Ann. UMCS Sectio AI Informatica **4**(2), 121–131 (2006)

35. Raczkowski, D., Canning, A., Wang, L.: Thomas Fermi charge mixing for obtaining self-consistency in density functional calculations. Phys. Rev. B **64**(12), 121101–121105 (2001)
36. Glosser, C., Piermarocchi, C., Shanker, B.: Analysis of dense quantum dot systems using a self-consistent Maxwell-Bloch framework. In: Proceedings of 2016 IEEE International Symposium on Antennas and Propagation (USNC-URSI), Puerto Rico, pp. 1323–1324. IEEE (2016)
37. H2O.ai: H2O Version 3.24.0.4, Fast Scalable Machine Learning API For Smarter Applications (2019). http://h2o-release.s3.amazonaws.com/h2o/rel-yates/4/index.html
38. Janusz, B.J., Wołk, K.: Implementing contextual neural networks in distributed machine learning framework. In: Nguyen, N.T., Hoang, D.H., Hong, T.-P., Pham, H., Trawiński, B. (eds.) ACIIDS 2018. LNCS (LNAI), vol. 10752, pp. 212–223. Springer, Cham (2018). https://doi.org/10.1007/978-3-319-75420-8_20
39. Wołk, K., Burnell, E.: Implementation and analysis of contextual neural networks in H2O framework. In: Nguyen, N.T., Gaol, F.L., Hong, T.-P., Trawiński, B. (eds.) ACIIDS 2019. LNCS (LNAI), vol. 11432, pp. 429–440. Springer, Cham (2019). https://doi.org/10.1007/978-3-030-14802-7_37
40. Bouckaert, R.R., Frank, E.: Evaluating the replicability of significance tests for comparing learning algorithms. In: Dai, H., Srikant, R., Zhang, C. (eds.) PAKDD 2004. LNCS (LNAI), vol. 3056, pp. 3–12. Springer, Heidelberg (2004). https://doi.org/10.1007/978-3-540-24775-3_3

Intelligent Systems and Algorithms in Information Sciences

Soft Computing-Based Control System of Intelligent Robot Navigation

Eva Volná[(✉)], Martin Kotyrba, and Vladimir Bradac

Department of Informatics and Computers, University of Ostrava, Ostrava, Czech Republic
{eva.volna,martin.kotyrba,vladimir.bradac}@osu.cz

Abstract. This paper focuses on the study of intelligent navigation techniques which are capable of navigating a mobile robot autonomously in unknown environments in real-time. We primarily focused on a soft computing-based control system of autonomous robot behaviour. The soft computing methods included artificial neural networks and fuzzy logic. Using them, it was possible to control autonomous robot behaviour. Based on defined behaviour, this device was able to deduce a corresponding reaction to an unknown situation. Real robotic equipment was represented by a Lego Mindstorms EV3 robot. The outcomes of all experiments were analysed in the conclusion.

Keywords: Control system · Autonomous robot · Artificial neural network · Fuzzy Logic System

1 The Issue of Autonomous Robot Control

Concerning the issue of autonomy, robotic systems can be divided into controlled, autonomous, and intelligent. Controlled robotic systems are controlled directly by a human operator. They are not able to perform actions themselves. Controlled robotic systems have predefined actions to be performed in a given sequence. On the other hand, autonomous systems have defined procedures based on which they achieve tasks and specified objectives. Intelligent systems are understood as those which can define their own objectives.

This paper focuses on autonomous robots, which can be defined as follows [8]: "A robot is an autonomous system existing in the physical world that can detect the environment and take action to achieve the target."

Autonomy in robotics means that a robot itself makes its own decisions and brings them into practice without any interference of a human operator. Such a system can take action without any human interference in the real world. A system's perception of the surroundings is ensured by various sensors providing data. The control system then provides a tool to establish autonomy of the robotic system using available data and information to plan and perform actions. A typical representation of an environment is a navigational map, which represents part of the real world. Apart from a map representing the environment, the robot can represent objects in the environment and actions to be taken in the environment. Various architectures solve representation of a model of the

N. T. Nguyen et al. (Eds.): ACIIDS 2020, LNAI 12034, pp. 379–390, 2020.
https://doi.org/10.1007/978-3-030-42058-1_32

world in various ways with respect to the requirements on processing, maintenance, construction, and representation of the model. The way the robot represents the surrounding world and how much information will be stored depends on each particular task to be solved. An autonomous robot must be able to process data from sensors, define the state of the surrounding world, plan future actions, take actions, all with limited resources and in real time. These tasks are described by the control architecture defining elements necessary to achieve the required behaviour.

This paper deals with selected methods of soft computing, i.e. artificial neural networks and fuzzy approaches. Using them, control architectures to navigate an autonomous robot in an unknown environment are proposed.

1.1 Controlling a Robot Using Neural Networks

Artificial Neural Networks

Artificial neural networks are structures inspired by biological models. Their primary role is to simulate and implement certain functions of a human brain, primarily adaptation and learning. Each neural network is composed of formal neurons which are interconnected so that the output of one neuron is the input into other neurons, the number of neurons and their interconnection in the network defines the neural network topology. Propagation and processing of information in neural networks are enabled by the change of neuron states. A neural network develops over time, i.e. it changes the states of the neurons in the network and connection weights between them are adapted. Concerning the changes over time, it is suitable to divide the dynamics into three sets and work with three neural networks modes:

- organisational (change of topology)
- adaptive (change of configuration)
- active (change of state)

Various models of neural networks are described in, for example, [13]. Generally, neural network structure can be described by any oriented graph using nodes (neurons) and oriented edges (synapses), where each point (hidden or output neuron) n performs a transfer function f_n in a form (1):

$$output_n = f_n\left[bias_n + \sum\nolimits_{j=1}^{J} \left(w_{nj} \cdot output_j\right)\right] \tag{1}$$

where:

$output_n$ is the output from neuron n
f_n is the activation function of neuron n
J is the number of neurons in the previous layer
$output_j$ is the output from the *j-th* neuron from the previous layer

which is the weighted value on the connection of neuron n and the *j-th* neuron in the previous layer
$bias_n$ is the threshold of neuron n.

Related Works with Controlling a Robot Using Neural Networks

In [15], there was published work on the topic of planning the movement of an autonomous mobile robot using forward neural networks for real-time evaluation of the robot's movement. In that case the training set included 14 expert-defined patterns, the input vector had 5 components and there was one output neuron defining the action: right, left, forward, stop. The autonomous robot was tested in a maze aiming at avoiding obstacles. The work mentions that the use of a neural network improved the robot's performance. In [6] a one-wheeled disk robot was modelled using a one-wheeled disk robot using a neural network. The neural network was used to balance the movement using an additional regulator to generate a compensation output, i.e. error minimisation, of linear regulators. The experiments showed that the ability of neural network generalisation could be a powerful tool. The concept of a navigation system using a self-learning neural network was dealt with in [9]. The proposed robot scanned the environment using a set of sensors. The robot needs a defined target and it then creates an optimal route in real time with a satisfactory result while avoiding obstacles. In [5] a solution deals with of an intelligent autonomous vehicle able to navigate in an unknown environment without hitting obstacles. The navigation was done using two feedforward neural networks, one ensuring reaching the target, the second one avoiding obstacles. The training set was created according to a human operator's control of the robot. The robot achieved satisfactory results. Project OpenWorm [3] mapped the interconnection of a neuron of a nematode (Caenorhabditis elegans) and transferred its perception and behaviour to a robotic body. Although the simulation is not accurate due to programming simplification, the neural network imitated the brain of this worm without the need of learning or programming. The final observations proved similar behaviour of the robot to the worm.

Finally, it can be stated that neural networks provide solution of many problems concerning the control of robot's movement when avoiding obstacles. The used training set and the type of the neural network (node structure, topology, learning algorithm) substantially influences the solved problem. In this paper we propose a training set and a suitable type of a neural network to control the movement of an autonomous robot.

1.2 Use of Fuzzy Logic to Control an Autonomous Robot

Fuzzy Logic Controller

The architecture of a Fuzzy Logic System is depicted in Fig. 1. This system has the following basic parts [4]:

- *Fuzzifier* is used to convert inputs, i.e. crisp numbers into fuzzy sets. Crisp inputs are basically the exact inputs measured by sensors and passed into the control system for processing.
- *Rule base* contains a set of rules and the IF-THEN conditions provided by domain experts to govern the decision making system, on the basis of linguistic information.
- *Inference* determines the matching degree of the current fuzzy input with respect to each rule and decides which rules are to be eliminated according to the input field. Next, the eliminated rules are combined to form the control actions.

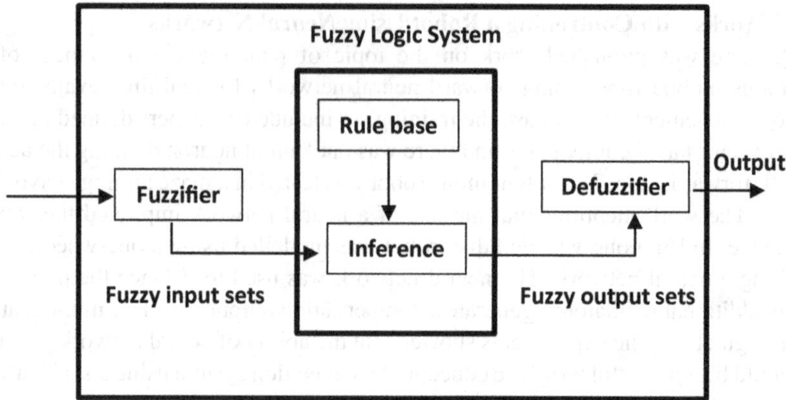

Fig. 1. Fuzzy Logic System

- *Defuzzifier* is used to convert the fuzzy sets obtained by the inference engine into a crisp value.

Related Works with Controlling a Robot Using Fuzzy Logic

The methods of fuzzy logic allow us to build logical-linguistic models of motion control systems, reflecting the entire semantic formulation of the problem, using qualitative representations corresponding to "human" methods of reasoning and decision-making. In [12], a fuzzy logic controller is presented to control the robot's motion along the predefined path. The robot was modelled in Matlab Simulink and the fuzzy logic rules were optimised for the best results possible. In [11], a fuzzy logic algorithm is developed for mobile robot navigation in local environments. A robot perceives its environment through sensors, while the fuzzy logic algorithm performs the main tasks of obstacle avoidance and target seeking. In [18], a behaviour-based hierarchical fuzzy control method for mobile robot navigation in a dynamic environment is proposed. In [10], a behaviour-based fuzzy control method for mobile robot navigation is presented. It is based on a behavioural architecture which can deal with uncertainties in unknown environments and has the ability to accommodate different behaviours. Basic behaviours are controlled by specific fuzzy logic controllers. The proposed approach qualifies for driving a robot to reach a target while avoiding obstacles in the environment. In [1], an optimal Mamdani-type fuzzy logic controller is introduced for trajectory tracking of wheeled mobile robots (WMRs). A dynamic model of a non-holonomic mobile robot was implemented in the Matlab/Simulink environment. The parameters of input and output membership functions, and PID controller coefficients are optimised simultaneously by Random Inertia Weight Particle Swarm Optimization (RNW-PSO). In [14], a mathematical model of a motion control system based on fuzzy logic methods is constructed and analysed. A method for synthesising a fuzzy algorithm of tracked mobile robot motion control is proposed. The input and output fuzzy variables are selected, the forms of their functions and linguistic description are proposed. The applicability of various fuzzy algorithms to solve the problem is analysed. A knowledge base that establishes dependencies between the input and output variables has been developed. The implementation of the mathematical model was carried out in the LabView graphical programming environment.

It can be seent hat fuzzy logic can be successfully used to control the movement of an autonomous robot to avoid obstacles. The subject of this paper is a proposal of Fuzzy Logic architecture in order to control the movement of an autonomous robot.

1.3 Model of the Robot

The proposed model of an autonomous robot is built from the Lego Mindstorm EV3 series. It uses two engines, which provide power to a belt chassis holding a programmable cube. The chassis also carries a sensory head with 4 infrared sensors S_1, S_2, S_3, and S_4. Each infrared sensor broadcasts an infrared signal and detects a reflection from any object which appears in front of the given sensor. The strength of the signal determines the distance from the object. The measurement precision depends on the object size, colour (light objects reflect the signal better than dark ones), material, and other factors. The value ranging 0 to 100 cm does not exactly define the distance from the object, but values with low deviation correspond with values in centimetres. The original Lego Mindstorms firmware of the programmable cube was replaced by LeJOS. It concerns a small Java Virtual Machine, which enables one to program the robot in Java.

2 Proposal of a Control System Based on Artificial Neural Networks

A control neural network will be a three-layer feedforward network. The proposal of the control system was created in the following steps and is based on an approach presented in [2]:

1. *Training set.* Considering that the robot disposes of 4 infrared sensors able to detect the distance from 0 to 100 cm, it is advisable to distribute this interval into five parts, which can be characterised as: very close (0–19 cm), close (20–39 cm), medium-far (40–59 cm), far (60–79 cm), very far (80–100 cm). Each sensor has incorporated all of these terms, which are a binary representing an obstacle within its reach in each interval. Value 1 represents an obstacle in a given interval, value 0 represents an empty space. A set of all terms for a given sensor then forms a binary input vector of the training set for the control neural network. The whole training set for the given configuration includes 1048576 patterns (2^{20} patterns, i.e. a binary representation, 4 sensors * 5 intervals). Defining the required outputs for the whole set would be time-consuming. Thus, we reduced the number of the training set patterns by filtering using a neural network ART1 (Adaptive Resonance Theory) with the vigilance parameter $\rho = 0.5$, which works with a binary input [2]. The resulting filtered set then included 51 patterns and it then incorporated pattern carrying the largest amount of information. The training set was supplied with an additional pattern 52, which defines the default behaviour for forward movement. The output of

the training set form real output vectors with normalised components which represent the speed of engines for forward and backward movement in degrees per second. As two large engines are used, the output vector has 4 components – two for each engine.

2. *Control neural network.* A control neural network is represented as a three-layer feedforward network with a sigmoid function. The topology is 20–20–4 (20 input neurons, 20 hidden neurons, and 4 output neurons). The number of input/output neurons corresponds with the number of component of input/output vector of the training set. The number of neurons in the hidden layer was selected based on the results from previous simulations and testing [17]. It concerned a fully interconnected three-layer neural network with feedforward propagation of the signal. The input-layer neurons were connected by synapses of the hidden-layer neurons and the hidden-layer neurons were connected with the output-layer neurons. The hidden-layer and output-layer neurons are, in addition, affected by a weight value (threshold) of a still active, fictive neuron with the output value equalling zero.
3. *Adaptation rule.* The neural network was adapted by the backpropagation method [13]. The parameters of the backpropagation algorithm were as follows: learning coefficient $\alpha = 0.85$, inertia coefficient $\mu = 0.9$, and the network is a homogeneous logical sinusoid with a transfer function, where steepness coefficient $\lambda = 1$ holds for all neurons. The condition for adaptation termination was selected the value of the maximum error $E \leq 0.0001$. The adaptation was performed in 14 393 cycles.

3 Proposal of a Control Fuzzy Logic System

The proposal of the Control Fuzzy Logic System was done in the following steps and takes basis in [7]:

1. *Fuzzification.* The input quantities of the correction of robot's movement are represented by distances of the robot from an obstacle. They are recorded by individual sensors S_1, S_2, S_3, and S_4. The input variable *Distance* is created by a set of terms {*Small, Medium, Large*}, which are represented by fuzzy sets (Fig. 2), where axis x represents the distance of the robot from an obstacle in the range 0 to 100 cm and μ (y axis) represents the degree of membership of the element to a fuzzy set.
   The inputs are represented by action quantities to control revolutions of individual engines each engine is assigned with one output variable *Engine1* and *Engine2,* created by the following terms: *vqb* (very quick back), *qb* (quick back), *sb* (slow back), *vsb* (very slow back), *stop, vsf* (very slow forward), *sf* (slow forward), *qf* (quick forward), *vqf* (very quick forward). In Fig. 3, axis x represents rpm in the range -500 to 500 and μ (axis y) represents the degree of membership of the element to a fuzzy set.
2. *Rule base* includes fuzzy rules of IF-THEN type, where parts of the antecedent contain the measured infrared values of sensors S_1, S_2, S_3, and S_4. A part of the consequent includes action interferences of the engines (*Engine1* and *Engine2*). These fuzzy rules include all combinations of the input values, which total to 81, i.e.

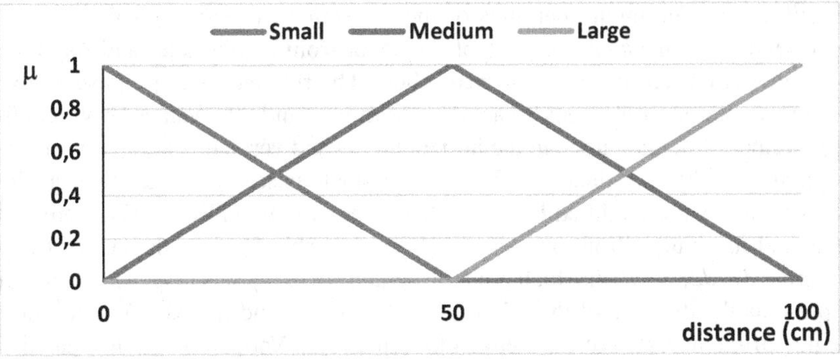

Fig. 2. Fuzzification of the input variable *Distance.*

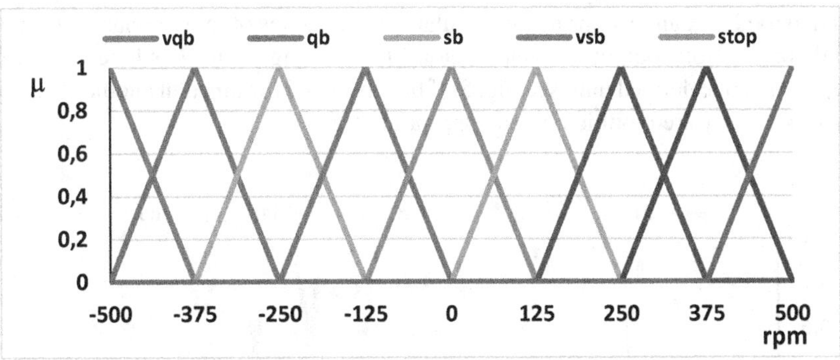

Fig. 3. Fuzzification of output variables *Engine1* and *Engine2.*

3^4 (four sensors, 3 linguistic values). The antecedent and consequent are composed expressions and their individual parts are interconnected by logical conjunctions "AND". The values of the action interferences of individual engines were assigned to each antecedent in an expert way.

3. *Inference and Defuzzification.* The method of defuzzification was the Center of Gravity. This created Fuzzy Logic System suitably adjusts the speed of engines in relation to the distance of the robot from an obstacle. This enables the robot to move without hitting obstacles in the surroundings.

4 Proposal of the Control Fuzzy Logic System - 2D Mapping of the Environment

The proposal of the Control Fuzzy Logic System for an autonomous robot, which includes 2D mapping of the environment was done in the following steps and takes basis on [16]. The model of the autonomous robot includes only one ultrasound sensor in this case.

1. *Fuzzification.* The input quantities of the correction of the robot's movement are represented by *Distance* (distance of the robot from an obstacle) and *Angle* (sum of angles of turned sensor and turned robot). The rule base was proposed only for the first quadrant of the mapped area, i.e. within 0° and 90°. Values [*x*, *y*] represent the number of defined units in the horizontal (x) and vertical (y) direction from the beginning of the coordinate system. If an obstacle is scanned in another quadrant, values *x* and *y* are modified by a sign. The input variable *Distance* is expressed in mm, and its value is from interval ⟨0, 1000⟩. Variable *Distance* is created by a set of terms *d1*, *d2*, …, *d14*, which are represented by fuzzy sets (Fig. 4), where axis *x* represents the distance of the robot from an obstacle and μ (axis y) represents the degree of membership of the element to a fuzzy set. Variable *Angle* is created by a set of terms *a1*, *a2*, …, *a7*, which are represented by fuzzy sets (Fig. 5), where axis *x* represents the size of the angle and μ (axis y) represents the degree of membership of the element to a fuzzy set. Output variables *CoordinateX* and *CoordinateY* are understood as x and y coordinate calculated from values of input variables *Distance* and *Angle*. Both output variables belong to the range 0 to 7, where each value represents one defined unit, see Fig. 5. If the obstacle is scanned in another quadrant, values *x* and *y* are modified by an appropriate sign.

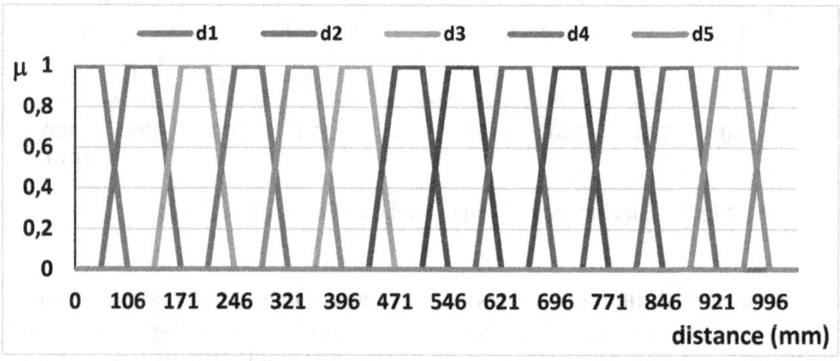

Fig. 4. Fuzzification of input variable *Distance*

2. *Rule base.* If the distance of the robot from an obstacle is between neighbouring circles (input variables *Distance* and *Angle*) while it fully belongs to the whole square area, thus the obstacle lies right in this block. If the distance of the robot from an obstacle overlaps two square areas, the resulting value is considered the nearest overlapped block, see Fig. 5. The Rule base includes a fuzzy rule of IF-THEN type, where the antecedent is formed by input variables *Distance* and *Angle,* which are interconnected by logical conjunction "AND", and the consequent contains values of variables *CoordinateX* and *CoordinateY*. These fuzzy rules include all combinations of the input variables.
3. *Inference and Defuzzification.* The method of defuzzification was the Center of Gravity. This created Fuzzy Logic System suitably adjusts the speed of engines in relation to the distance of the robot from an obstacle. This enables the robot to move without hitting obstacles in the surroundings.

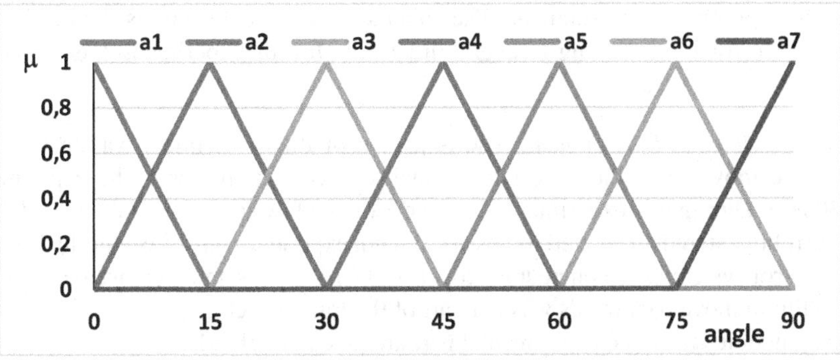

Fig. 5. Fuzzification of input variable *Angle*

5 Comparative Experimental Study of the Proposed Approaches Using Softcomputing Methods to Control an Autonomous Robot

The comparative experimental study of the proposed approaches was done on a robot built on the Lego Mindstorm EV3 series. The robot was placed into the environment demarcated by a square grid. The maximum speed of the robot was set to 500°/s. The objective was the robot to move through the environment without directly hitting obstacles or getting stuck. There were three experimental environments: rectangular, circular, and square, which includes several obstacles placed so that they created "rooms". For each experimental environment, the following data was collected:

- number of primary collisions (direct collisions of the robot with an obstacle),
- number of secondary collisions (collisions of any part of the robot, such as cables, sensors, etc.),
- number of times robot gets stuck,
- time necessary for the experiment,
- robot's trajectory.

Evaluation of the Experiments

Each experimental variant was tested with 30 passages. A problem arose when the robot was turning close to corner parts.

Rectangular Experimental Environment. There were cases when the robot, when turning, hit an obstacle with a sensor edge. It was because the corner appeared in a blind spot of the sensor head or there was incorrect data. In addition, passages sometimes took too long, which was caused by close distances between obstacles so the robot had to operate at very low speeds to find the most suitable position to get to the endpoint. In all cases, the robot reached the endpoint and had no direct collision with an obstacle or got stuck.

Circular Experimental Environment. The objective of this experiment was to test the robot's ability to move under an angle and reach the endpoint without colliding with walls demarcating the environment. The robot's passage took long as there was not enough place for manoeuvring to avoid direct hits. In this experiment, there was no direct nor getting stuck.

Square Experimental Environment. The objective of this experiment was to test the robot's free movement among several obstacles placed as "rooms". The experiment took 30 min. During this experiment, there were only a few secondary collisions. However, the robot entered all created rooms and did not experience any direct hit. In the case of the control system based on neural networks, the robot was able to move backwards. This ability to move backwards was not part of the training set.

The outcomes from the experimental part are given in Table 1.

Table 1. Experimental outcomes.

Experimental environment	Artificial neural network			Fuzzy Logic System		
	Rectangular	Circular	Square	Rectangular	Circular	Square
Max. number of primary collisions	0	0	0	0	0	0
Max. number of secondary collisions	2	0	10	1	5	8
Max. number of getting stuck	0	0	0	0	0	0
Average time of the experiment	24.2 s	47.3 s	30 min.	24.1 s	47.5 s	30 min.

Control Fuzzy Logic System for 2D Mapping of the Environment

The robot was placed into a rectangular environment demarcated by a square grid. The objective of the robot was to pass through the environment without hitting directly an obstacle or getting stuck and to create a 2D map of the environment.

The course of the experiment was as follows. Having initialised the robot and its program, it starts to scan the surroundings and store coordinates into its memory. After scanning, it creates the output image. Placing and turning of the robot at the starting point was arbitrary for each experiment. The resulting map of the scanned environment was compared with the map of the real environment.

In all cases, the robot got through the environment without any direct hit or getting stuck, the created maps of the scanned environments are depicted in Table 2, where black colour represents obstacles, grey colour represents area without obstacles, blue colour represents the robot's starting position, and red colour represents the starting position of the sensor.

Table 2. Experimental environment and the created map of the scanned environment.

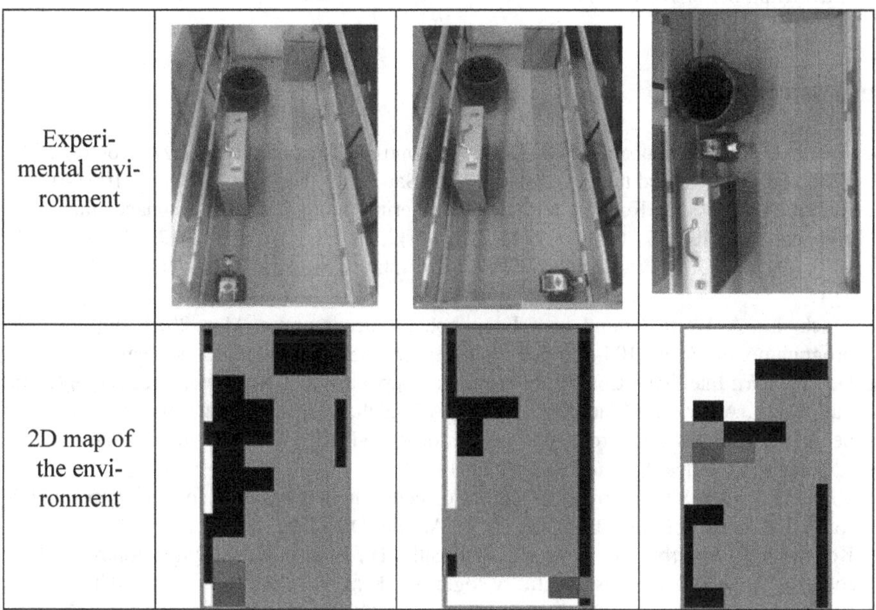

6 Conclusion

The presented paper focuses on the study of intelligent navigation techniques, which are capable of navigating a mobile robot autonomously in unknown environments in real time. We primarily focused on a soft computing-based control system of autonomous robot behaviour. We have proposed a control system based on artificial neural networks and a Control Fuzzy Logic System for the purposes of controlling the movement of an autonomous robot. Both verified control systems were verified on a model of an autonomous robot built on the Lego Mindstorm EV3 series. According to the results of the comparative experimental studies (Table 1), it can be stated that both control systems reached comparable results. In all performed experiments, the robot suitably adjusted its speed and direction based on the given situation. In no experiment, there was any primary collision, i.e. direct hitting an obstacle. However, there occurred secondary collisions, mostly caused by sensors' edges when the robot was turning. The collisions always happened at obstacles corners. It is probable that the infrared signal of the given sensor inappropriately reflected back and thus it resulted in a wrong measurement. In no experiment, the robot got stuck. The biggest limitation of the robot is the inaccuracy of the used infrared sensors. This inaccuracy was partly suppressed by dividing the scanned environment into several intervals.

Our future work will focus on multi-robotic systems, primarily on the coordination of the movement of individual robots in the given environment.

Acknowledgments. The research described here has been financially supported by the University of Ostrava grant SGS05/PRF/2019.

References

1. Abadi, D.N.M., Khooban, M.H.: Design of optimal Mamdani-type fuzzy controller for nonholonomic wheeled mobile robots. J. King Saud Univ. Eng. Sci. **27**(1), 92–100 (2015)
2. Barton, A., Volna, E., Kotyrba, M.: Big data filtering through adaptive resonance theory. In: Nguyen, N.T., Tojo, S., Nguyen, L.M., Trawiński, B. (eds.) ACIIDS 2017. LNCS (LNAI), vol. 10192, pp. 382–391. Springer, Cham (2017). https://doi.org/10.1007/978-3-319-54430-4_37
3. Black, L.: A Worm's mind in a Lego body. I Programmer, May 2019. http://www.i-programmer.info/news/105-artificial-intelligence/7985-a-worms-mind-in-a-lego-body.html
4. De Silva, W.: Intelligent Control: Fuzzy Logic Applications. CRC Press, Boca Raton (2018)
5. Farooq, U., Amar, M., Asad, M.U., Hanif, A., Saleh, S.O.: Design and implementation of neural network based controller for mobile robot navigation in unknown environments. Int. J. Comput. Electrical Eng. **6**(2), 83–89 (2014)
6. Kim, P.K., Jung, S.: Experimental studies of neural network control for one-wheel mobile robot. J. Control Sci. Eng. **2012**, 12 (2012). Article ID 194397
7. Konvicka, J., Kotyrba, M., Volna, E., Habiballa, H., Bradac, V.: Adaptive control of EV3 robot using mobile devices and fuzzy logic. In: Kim, K.J., Baek, N. (eds.) ICISA 2018. LNEE, vol. 514, pp. 389–399. Springer, Singapore (2019). https://doi.org/10.1007/978-981-13-1056-0_40
8. Matarić, M.J., Arkin, R.C.: The Robotics Primer. MIT Press, London (2007)
9. Markoski, A., Vukosavljev, S., Kukolj, D., Pletl, S.: Mobile robot control using self-learning neural network. In: 7th International Symposium on Intelligent Systems and Informatics, pp. 45–48. IEEE (2009)
10. Mo, H., Tang, Q., Meng, L.: Behavior-based fuzzy control for mobile robot navigation. Mathematical Problems in Engineering **2013**, 10 (2013). Article ID 561451
11. Motlagh, O.R.E., Hong, T.S., Ismail, N.: Development of a new minimum avoidance system for a behavior-based mobile robot. Fuzzy Sets Syst. **160**(13), 1929–1946 (2009)
12. Peri, V.M., Simon, D.: Fuzzy logic control for an autonomous robot. In: Annual Meeting of the North American Fuzzy Information Processing Society, pp. 337–342. IEEE (2005)
13. Rojas, R.: Neural Networks: A Systematic Introduction. Springer, Heidelberg (2013). https://doi.org/10.1007/978-3-642-61068-4
14. Tripathi, G.N., Rihani, V.: Motion planning of an autonomous mobile robot using artificial neural network. arXiv preprint arXiv:1207.4931 (2012)
15. Vinogradov, A., Terentev, A., Kochetkov, M., Petrov, V.: Model of fuzzy regulator of mobile robot motion control system. In: 2019 IEEE Conference of Russian Young Researchers in Electrical and Electronic Engineering, pp. 2109–2112. IEEE (2019)
16. Volna, E., Kotyrba, M., Jaluvka, M.: Intelligent robot's behavior based on fuzzy control system. In: International Conference on Industrial Engineering, Management Science and Application, pp. 34 – 38. IEEE (2016)
17. Volna, E., Kotyrba, M., Zacek, M., Bartoň, A.: Emergence of an autonomous robot's behaviour. In: European Conference on Modelling and Simulation, Bulgaria 2015, pp. 462–468 (2015)
18. Wang, D., Zhang, Y., Si, W.: Behavior-based hierarchical fuzzy control for mobile robot navigation in dynamic environment. In: Proceedings of the Chinese Control and Decision Conference, pp. 2419–2424 (2011)

End-to-End Speech Recognition in Agglutinative Languages

Orken Mamyrbayev[1,3](✉), Keylan Alimhan[2], Bagashar Zhumazhanov[1], Tolganay Turdalykyzy[1], and Farida Gusmanova[3]

[1] Institute of Information and Computational Technologies, Almaty 050010, Kazakhstan
morkenj@mail.ru, bagasharj@mail.ru, t_tolganai@inbox.ru
[2] Tokyo Denki University, Tokyo 120-8551, Japan
20787@ms.dendai.ac.jp
[3] Al-Farabi, Kazakh National University, Almaty 050040, Kazakhstan
grfarida77@gmail.com

Abstract. This paper considers end-to-end speech recognition systems based on deep neural networks (DNN). The studies used different types of neural networks, CTC model and attention-based encoder-decoder models. As a result of the study, it was proved that the CTC model works without language models directly for agglutinative languages, but the best is ResNet with 11.52% of CER and 19.57% of WER of using the language model. An experiment with the BLSTM neural network using the attention-based encoder-decoder models showed 8.01% of CER of and 17.91% of WER. Using the experiment, it was proved that without integrating language models, good results can be achieved. The best result showed ResNet.

Keywords: Speech recognition · Agglutinative languages · End-to-End models · Deep learning · CTC

1 Introduction

Speech is a system of human-used audio signals, written signs and symbols to represent, process, store and transmit information. It is also a tool for human-machine interaction [1]. To implement the voice interface requires the participation of a wide range of specialists, namely a computer linguist, DNN programmer, etc. The traditional speech recognition system can be divided into several modules, such as acoustic models, language models and decoding [2]. The modularity design is based on many independent assumptions, and even the traditional acoustic model is trained from frames that depend on the Markov model. In automated speech recognition systems, hidden Markov models (HMM) were popular models to represent the temporal dynamics of speech signals and Gaussian mixture models (GMM) probability density function to represent signal distributions over a stationary short period of time that typically corresponds to a unit of pronunciation. The HMM-GMM model dominated research on automatic speech recognition for several years. Today, the neural network is widely used in the field of speech recognition. Many studies have shown that the use of neural networks at each step of the

N. T. Nguyen et al. (Eds.): ACIIDS 2020, LNAI 12034, pp. 391–401, 2020.
https://doi.org/10.1007/978-3-030-42058-1_33

script of the standard speech recognition system improves the quality of its work. The most popular deep learning approach for automatic speech recognition is the so-called HMM-DNN hybrid architecture, where the HMM structure is preserved and the GMM is replaced by a deep neural network (DNN) to model the dynamic characteristics of speech signals. For example, in studies [3], language models were trained using RNN, in [4] the dictionary was obtained using LSTM networks, in [5] deep neural networks showed high results for constructing acoustic models, in [6] the method of feature extraction using limited Boltzmann machines was presented. Consequently, the idea of using artificial neural networks at all stages of speech recognition appeared.

Deep learning methods using high-performance GPUs in speech recognition have been successfully implemented and this approach has been called the end-to-end method. In the end-to-end approach, when learning a neural network, only one model can produce the desired result without the use of other components and such a model is called an end-to-end. End-to-end networks can be created by adding several convolutional neural network (CNN) and recurrent neural network (RNN) layers, which act as acoustic and language models, and directly correlate speech data at the input with transcription. At the moment, there are several methods of implementing end-to-end models, namely, the connection time classification (CTC) and the attention-based encoder-decoder models, conditional random fields (CRF). In speech recognition problems, special attention is still paid to end-to-end approaches than to traditional methods [7]. Many published works have proved that the success of the results of the end-to-end approach depends on an increase in the amount of data for neural network training. There are applications in the world that work on the basis of an end-to-end approach: Baidu Deep Speech, Google Listen, Attend, Spell, Speech to Translator TTS, Voice to Text Messenger. The main reason for this conclusion is that current end-to-end models are trained based on data. From the above analysis we can see the main problem, it concerns the recognition of a few resource languages, such as Kazakh, Kyrgyz, Azerbaijani, Uighur, Tatar, Turkish, etc. These listed languages are included in the group of agglutinative languages. For agglutinative languages, there are no large corps of training data. Other languages have TIMIT, WSJ, LibriSpeech, AMI and Switchboard which have thousands of hours of training data.

To improve the end-to-end approach in CTC and encoder-decoder models based on the attention mechanism, different variants of networks were introduced. Complex encoders consisting of convolutional neural networks (CNN) were introduced to use local correlations in speech signals. These models take advantage of each submodel and introduce more explicit and strict limitations to the entire model. The above studies in this area significantly improve the performance of end-to-end speech recognition systems. Introducing complex computational layers into a model can use better correlations in both the time and frequency domain, but a model with much more parameters is harder to train. In previous studies, it was determined that deep learning models in different languages are successful, and multitask learning (MTL) is better suited for integrated learning [8, 9].

In the end-to-end speech recognition approach, all signal parameters are determined by calculating gradients that are easily influenced by neural network structures. However, the end-to-end recognition systems of agglutinative languages still do not reach the modern level of research and are not trained. During the analysis, it was determined that the end-to-end models for recognizing agglutinative languages are not sufficiently trained.

In this paper, we propose the recognition of agglutinative languages, which is aimed at solving the problem with a limited speech resource within the framework of the end-to-end architecture. This study is organized as follows. Section 2 describes research in the relevant scientific field. Section 3 describes the principles of the CTC model and the attention-based encoder-decoder models. Below we present our experimental data and describe the equipment for the experiment. Section 4 analyzes the experimental results. The final section provides conclusions.

2 Related Work

Models based on connective time classification (CTC) for speech recognition work without initial alignment of input and output sequences. CTC was designed to decode the language. Hannun et al. [10] and his team used the Baidu model for speech recognition, which implements a parallel network learning algorithm using CTC.

The use of deep recurrent convolutional networks and deep residual networks in conjunction with CTC was proposed in [11]. The best result was obtained with the use of residual networks with batch normalization. Thus, a PER result of 17.33% was obtained on the TIMIT speech corpus.

An alternative to CTC for end-to-end speech recognition is the Sequence to Sequence (Sec2Sec) models with attention [12]. Such models consist of an encoder and a decoder. The encoder compresses audio frame information into a more compact vector representation by reducing the number of neurons from layer to layer, and the decoder recovers a sequence of symbols, phonemes, or even words based on this compressed representation and recurrent neural network.

In [13], a CTC model was proposed using deep convolutional networks instead of recurrent networks. The best model based on convolutional networks had 10 convolutional layers and 3 fully connected layers. The best PER was 18.2%, while the best PER for bidirectional LSTM networks was 18.3%. Tests were conducted on the TIMIT case. It was also concluded that convolutional networks can increase the speed of learning and are more suitable for learning on phoneme sequences.

In a CTC network, the output values of a neural network themselves represent transition probabilities. Bidirectional LSTM networks were chosen as the architecture of the neural network. Three models were compared: the RNN-CTC model, the RNN-CTC model (RNN-WER), the retrained minimized WER, and the basic hybrid model written using Kaldi tools [14].

Soltau et al. [15] performed context-sensitive phoneme recognition by training a CTC-based model in the task of signing a video on YouTube. Sequence-to-sequence models lack recognition by 13–35% compared to base systems.

Graves et al. [16] trained the end-to-end model the CTC criterion without applying frame-level alignment. The Sequence-to-sequence model simplified the problem of automatic speech recognition by training and optimizing the neural network for the acoustic model, pronunciation model, and language model. These models also work as multi dialectic systems, since they are jointly modeled in different dialects.

The RNN-CTC model without a language model showed 30.1%, of WER, although the basic model cannot be trained without LM. But even when using trigram LM, the basic model showed 7.8% of WER, RNN-CTC - 8.7%, and RNN-WER - 8.2%. A combination of RNN-CTC and the base model was also tested, which showed the best result, equal to 6.7%. The Wall Street Journal corpus [17] was used as a speech corpus.

There is a "generalization" of CTC models - the RNN Transducer, which combines two RNNs into a serial Converter system [18]. One of the networks is similar to a CTC network and processes the same moment of time as the input sequence, and the second RNN models the probabilities of the following labels under the condition of the previous one. As in CTC networks, dynamic programming is used for calculations and the forward-backward algorithm, but taking into account the limitations of both RNNs. Unlike CTC networks, the RNN Transducer allows to generate output sequences longer than input ones. RNN Transducers showed good results in recognition of phonemes with 17.7% of PER on the TIMIT corpus.

In [19, 20] three models with CTC were considered: ResNet, BLSTM and combination of LSTM and CNN. A method for combining models similar to ROVER has also been proposed. So, the result was obtained on the WSJ speech data set using ResNet with 8.99% of WER, and using a combination of the three models mentioned above is 7.65%.

Beyond speech recognition, the neural network achieves success in other fields such as natural language processing [21, 22]. In [21], the author presented a deep learning-based model for Kazakh named entity recognition by projecting the word, root and entity label into a vector space. In [22], a neural network for morphological disambiguation was proposed, which learns context embedding from characters by double-layer of BLSTM and compute the similarity score between context and the corresponding morphological analyses. The idea was that the correct analysis should be the most similar to the context.

3 Proposed Automatic Speech Recognition System

The methodology of our work is as follows:

3.1 CTC

To train a neural network, the CTC function is used as a loss function. The output sequence of a neural network can be described as follows: $y = f_w(x)$. The output layer of the neural network contains one block for each symbol of the output sequence and one more for the additional "blank" symbol. Each element of the output sequence is a probability distribution vector for each symbol G' at time t. Thus, the element $\boldsymbol{y}_k^t$ is the probability that at time **t** in the input sequence the symbol **k** from the set G' is pronounced (Fig. 1 (a)).

Let, α be a sequence of blanks indices and symbols of length T, according to x. The probability $P(\alpha|x)$ can be represented as the product of the probabilities of the

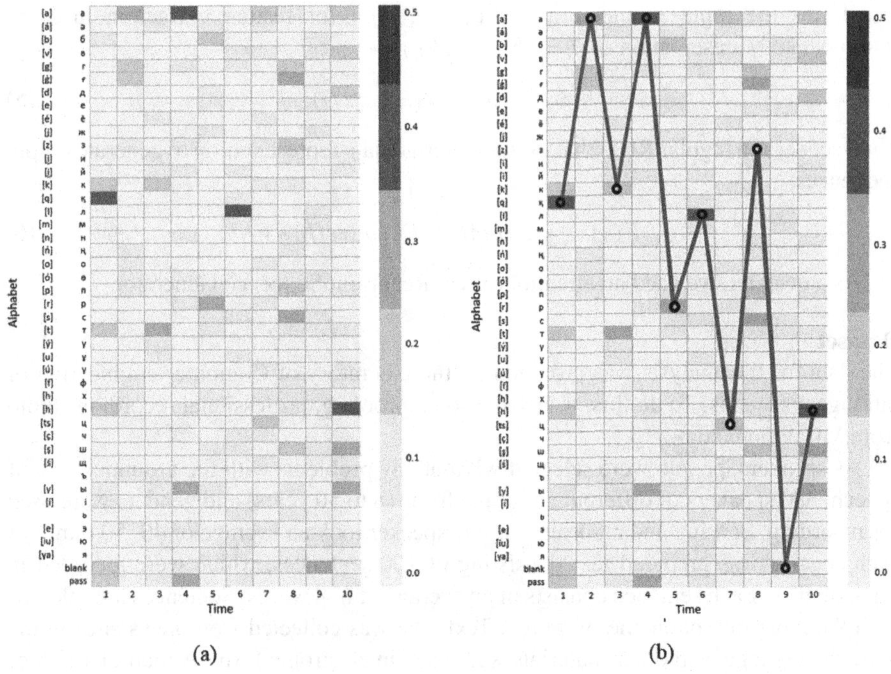

Fig. 1. The matrix predicted by the acoustic model.

appearance of symbols at each moment of time:

$$P(\alpha|x) = \prod\nolimits_{t=1}^{T} y_{\alpha_t}^{t}, \forall \alpha \in G'^{T} \quad (1)$$

Let B be the operator that removes symbol repeats and blanks.

$$P(y|x) = \sum\nolimits_{\alpha \in B^{-1}(y)} P(\alpha|x) \quad (2)$$

The above formula is calculated using dynamic programming, and the neural network will be trained to minimize the CTC function:

$$CTC(x) = -\ln(P(y|x)) \quad (3)$$

Decoding is based on the following assumption:

$$\arg \underset{w}{max}\, P(y|x) \approx B(\alpha^{*}) \quad (4)$$

where $\alpha^{*} = \arg\, max_{\alpha} P(\alpha|x)$. The results of the assumption can be seen in Fig. 1 (b).

3.2 Attention-Based Model

Attention is an Encoder-Decoder mechanism designed to improve RNN performance in speech recognition. Encoder is a neural network, such as: DNN, BLSTM, CNN;

transforms the input sequence $x = (x_1, \ldots, x_{L'})$ for feature extraction in some intermediate representation of $h = (h_1, \ldots, h_L)$.

$$h = Encoder(x_1, \ldots, x_{L'}) \tag{5}$$

Decoder is a regular RNN that uses an intermediate representation to generate output sequences:

$$P(y|x) = AttentionDecoder(h, y) \tag{6}$$

As a decoder, we used an attention-based Recurrent Sequence Generator.

Dataset

The data for the analysis was provided by the laboratory of Computer engineering of intelligent systems. To do this, we used a soundproofing, professional recording studio from Vocalbooth.com.

As speakers, people were selected without any problems with the pronunciation of speech. 380 speakers of different ages (age from 18 to 50 years) and genders were used for recording. Scoring and recording of one speaker took an average of 40–50 min. For each speaker was prepared text consisting of 100 sentences, which were recorded in separate files. Each sentence consists of an average of 6–8 words. Sentences are selected with the most rich phoneme of words. Text data was collected from news sites in the Kazakh language, and other materials were used in electronic form. A total of 123 h of audio data were recorded. During recording, transcriptions were created - a description of each audio file in a text file. The created corpus allows, firstly, to work with large volumes of databases, to check the proposed characteristics of the system and, secondly, to study the impact of database expansion on the recognition speed.

All audio materials have the same characteristics:

- file extension: .wav;
- method of converting to digital form: PCM
- discrete frequency: 44.1 kHz;
- bit capacity: 16 bits;
- number of audio channels: one (mono).

To train the end-to-end recognition system of agglutinative languages, we have chosen 2 corpora:

- Turkish language corpus (9 million words and audio): http://www.tnc.org.tr/
- Tatar language corpus (10 million words and audio): http://www.corpus.antat.ru.

Implementation

End-to-end speech recognition system using CTC function was implemented using TensorFlow. In this system, we used the Eesen toolkit in TensorFlow. This system allows to use language models built in the Kaldi format without additional conversion. We used Tensor2Tensor5 to conduct experiments with attention-based models.

All experiments were carried out using the SuperMicro SYS-7049GP-TRT server. The server configuration has a high-performance NVIDIA TESLA P100 graphics card.

4 Experiments and Results

In the feature extraction experiments, we used the Mel-frequency cepstral coefficients (MFCC) with the first 13 coefficients computed. All training data was divided into training (90%) and cross-validation (10%).

At the second stage of the experiment, we will describe the model results based on the CTC loss function. The results of the corresponding CTC models are presented in Table 1. In our studies, we used several types of neural networks: ResNet, LSTM, MLP, Bidirectional LSTM. Pre-configuration of neural networks without a language model gave us the best results:

Table 1. CTC model results.

Model	CER%	WER%	Decode	Train
	Models that do not use language models			
MLP	48.11	59.26	0.2032	131.2
LSTM	36.43	46.51	0.2152	421.3
Conv+LSTM	34.92	39.31	0.2688	465.2
BLSTM	33.61	37.66	0.2722	491.7
ResNet	32.52	36.57	0.2657	192.6
	Models using language models and MFCC			
MLP	39.11	63.26	0.0192	146.2
LSTM	24.43	46.51	0.0152	521.3
Conv+LSTM	22.92	39.31	0.0088	465.2
BLSTM	13.61	20.66	0.0022	591.7
ResNet	11.52	19.57	0.0051	242.6

- MLP: MLP: there were 6 hidden layers with 1,024 nodes using the ReLU activation function with an initial learning rate of 0.007 and a damping factor of 1.5.
- LSTM: there were 6 layers with 1024 units each with a dropout of 0.5 s, an initial learning rate of 0.001, and a damping factor of 1.5.
- ConvLSTM: one two-dimensional convolutional layer with 8 filters was used, with ReLU activation function. Then it drops out with a retention probability of 0.5.
- BLSTM: used 6 layers with 1,024 units and dropped with a retention probability of 0.5.
- ResNet had 9 residual blocks with batch-normalization.

In the first experiment for the attention-based encoder-decoder models, we used the MFCC algorithm to extract features, and the CTC function was used to train the neural

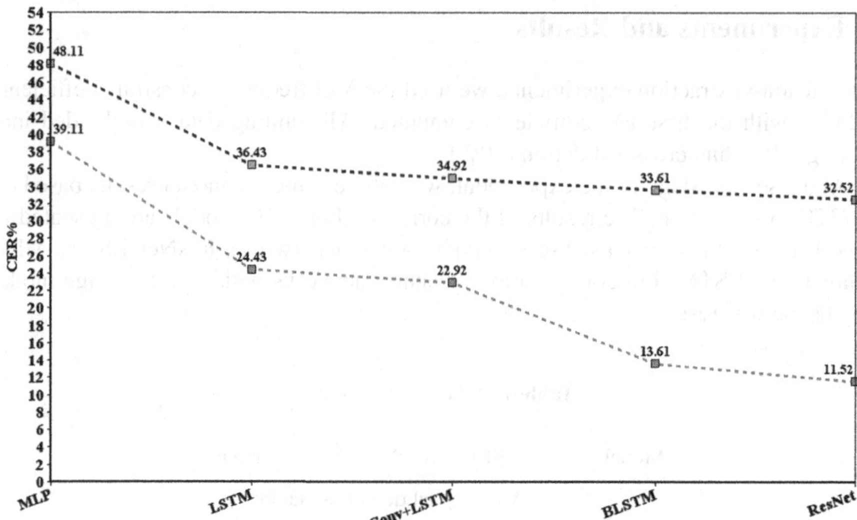

Fig. 2. CTC model results by CER. The blue line is the result of models that do not use language models, as well as the red line - models that use language models and MFCC. (Color figure online)

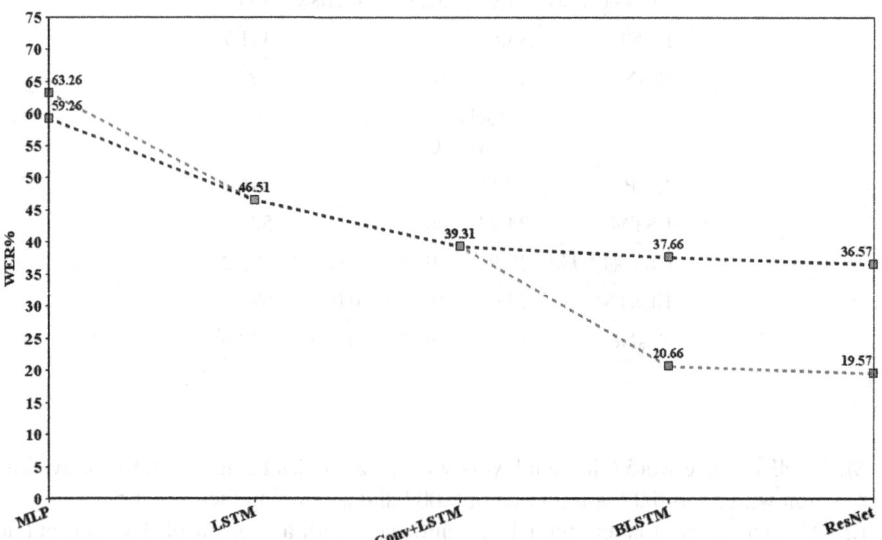

Fig. 3. CTC model results by WER. The blue line is the result of models that do not use language models, and the red line is the result of models that use language models and MFCC. (Color figure online)

network. We did not use language models in this model. In the second experiment, we used MFCC and language models. The results of the experiment can be seen in Figs. 2 and 3.

Table 2. Results of the attention-based encoder-decoder models.

Model	CER%	WER%	Decode	Train
LSTM	8,61	17,58	0,468	476,7
BLSTM	8,01	17,91	0,496	544,3

In the next experiment, we used neural networks LSTM and BLSTM. In our model, 6 layers of 256 units were used with an initial decrease in dropout with a probability of saving in the encoder of 0.7. As a decoder, we used LSTM and an attention-based encoder-decoder models. The results can be seen in Table 2.

Our experiments proved that the CTC model works without language models directly for agglutinative languages, but still the best is ResNet with 11.52% of CER and 19.57% of WER using the language model. Thus, it can be seen that the language model is an important part of speech recognition.

The CTC model makes mistakes in constructing words and sentences from recognized characters, but the resulting phonemic transcription is very similar to the original. But after the experiment, we found that the use of an attention-based encryptor-decoder models for agglutinative languages without integrating language models allows to achieve good results. The BLSTM neural network using the attention-based encoder-decoder models showed 8.01% of CER and 17.91% of WER.

5 Conclusion

In this paper, we consider the problem of recognition of agglutinative languages using an end-to-end approach, such as the CTC model and the attention-based encoder-decoder models. During the experiment we used different types of neural network architectures: MLP, LSTM and their modifications, as well as ResNet. As a result of the experiment, we proved that good results can be achieved without the integration of language models. ResNet showed the best result. In this experiment, good performance was achieved, better than the basic hybrid models.

In the future it is planned to conduct experiments using other types of models for feature extraction and speech recognition. The Conditional Random File model will be applied.

Acknowledgments. This work was supported by the Ministry of Education and Science of the Republic of Kazakhstan. IRN AP05131207 Development of technologies for multilingual automatic speech recognition using deep neural networks.

References

1. Perera, F.P., et al.: Relationship between polycyclic aromatic hydrocarbon–DNA adducts and proximity to the World Trade Center and effects on fetal growth. Environ. Health Perspect. **113**, 1062–1067 (2005)

2. Mamyrbayev, O., Turdalyuly, M., Mekebayev, N., Alimhan, K., Kydyrbekova, A., Turdalykyzy, T.: Automatic recognition of Kazakh speech using deep neural networks. In: Nguyen, N.T., Gaol, F.L., Hong, T.-P., Trawiński, B. (eds.) ACIIDS 2019. LNCS (LNAI), vol. 11432, pp. 465–474. Springer, Cham (2019). https://doi.org/10.1007/978-3-030-14802-7_40
3. Mikolov, T., et al.: Recurrent neural network based language model. Interspeech **2**, 1045–1048 (2010)
4. Rao, K., Peng, F., Sak, H., Beaufays, F.: Grapheme-to-phoneme conversion using long short-term memory recurrent neural networks. In: 2015 IEEE International Conference on Acoustics, Speech and Signal Processing (ICASSP), pp. 4225–4229 (2015)
5. Jaitly, N., Hinton, G.: Learning a better representation of speech soundwaves using restricted boltzmann machines. In: 2011 IEEE International Conference on Acoustics, Speech and Signal Processing (ICASSP), pp. 5884–5887 (2011)
6. Smolensky, P.: Information processing in dynamical systems: foundations of harmony theory. Colorado University at Boulder Department of Computer Science, pp. 194–281 (1986)
7. Vaněk, J., Zelinka, J., Soutner, D., Psutka, J.: A regularization post layer: an additional way how to make deep neural networks robust. In: Camelin, N., Estève, Y., Martín-Vide, C. (eds.) SLSP 2017. LNCS (LNAI), vol. 10583, pp. 204–214. Springer, Cham (2017). https://doi.org/10.1007/978-3-319-68456-7_17
8. Kim, S., Hori, T., Watanabe, S.: Joint CTC-attention based end-to-end speech recognition using multi-task learning. In: IEEE International Conference on Acoustics, Speech and Signal Processing (ICASSP) (2017)
9. Aida-Zade, K., Rustamov, S., Mustafayev, E.: Principles of construction of speech recognition system by the example of Azerbaijan language. In: International Symposium on Innovations in Intelligent Systems and Applications, pp. 378–382 (2009)
10. Hannun, A., et al.: DeepSpeech: scaling up end-to-end speech recognition, arXiv:1412.5567 (2014)
11. Zhang, Z., et al.: Deep recurrent convolutional neural network: improving performance for speech recognition (2016). preprint: arXiv:1611.07174. https://arxiv.org/abs/1611.07174
12. Bahdanau, D., et al.: End-to-end attention-based large vocabulary speech recognition. In: 2016 IEEE International Conference on Acoustics, Speech and Signal Processing (ICASSP), pp. 4945–4949. IEEE (2016)
13. Zhang, Y., et al.: Towards end-to-end speech recognition with deep convolutional neural networks (2017). preprint: arXiv:1701.02720. https://arxiv.org/abs/1701.02720
14. Povey, D., et al.: The Kaldi speech recognition toolkit. In: IEEE 2011 Workshop on Automatic Speech Recognition and Understanding, 4 p. IEEE Signal Processing Society (2011)
15. Soltau, H., Liao, H., Sak, H.: Neural speech recognizer: acoustic-to-word LSTM model for large vocabulary speech recognition. arXiv:1610.09975 (2016)
16. Graves, A., Fernández, S., Gomez, F., Schmidhuber, J.: Connectionist temporal classification: labelling unsegmented sequence data with recurrent neural networks. In: Proceedings of the 23rd International Conference on Machine Learning, pp. 369–376 (2006)
17. Popović, B., Pakoci, E., Pekar, D.: End-to-End large vocabulary speech recognition for the Serbian language. In: Karpov, A., Potapova, R., Mporas, I. (eds.) SPECOM 2017. LNCS (LNAI), vol. 10458, pp. 343–352. Springer, Cham (2017). https://doi.org/10.1007/978-3-319-66429-3_33
18. Boulanger-Lewandowski, N., Bengio, Y., Vincent, P.: High-dimensional sequence transduction. In: 2013 IEEE International Conference on Acoustics, Speech and Signal Processing (ICASSP), pp. 3178–3182 (2013)
19. Wang, Y., Deng, X., Pu, S., Huang, Z.: Residual convolutional CTC networks for automatic speech recognition (2017). preprint: arXiv:1702.07793. https://arxiv.org/abs/1702.07793

20. Rustamov, S., Gasimov, E., Hasanov, R., Jahangirli, S., Mustafayev, E., Usikov, D.: Speech recognition in flight simulator. In: Aegean International Textile and Advanced Engineering Conference. IOP Conference Series: Materials Science and Engineering, vol. 459 (2018)
21. Gulmira, T., Alymzhan, T., Orken, M., Rustam, M.: Neural named entity recognition for Kazakh. In: 20th International Conference on Computational Linguistics and Intelligent Text Processing (CICLing), 7–13 April 2019, La Rochelle, France. Lecture Notes in Computer Science (2019)
22. Toleu, A., Tolegen, G., Makazhanov, A.: Character-aware neural morphological disambiguation. In: Proceedings of the 55th Annual Meeting of the Association for Computational Linguistics (Volume 2: Short Papers), pp. 666–671. Association for Computational Linguistics, Vancouver (2017)

Approach the Interval Type-2 Fuzzy System and PSO Technique in Landcover Classification

Dinh Sinh Mai(✉), Long Thanh Ngo, and Le Hung Trinh

Le Quy Don Technical University, Hanoi, Vietnam
maidinhsinh@gmail.com

Abstract. In fuzzy classification systems, the estimation of the optimal number of clusters and building base-rules are very important and greatly affects the accuracy of the fuzzy system. Base-rules are often built on the experience of experts, but this is not always good and the results are often unstable. Particle swarm optimization (PSO) techniques have many advantages in finding optimal solutions and have been used successfully in many practical problems. This paper proposes a method using the PSO technique to build base-rules for the interval type-2 fuzzy system (IT2FS). Experiments performed on satellite image data for the landcover classification problem have shown that the proposed method works more stably and effectively than the non-PSO technique.

Keywords: Type-2 fuzzy set · Interval type-2 fuzzy system · PSO · Fuzzy system · Landcover · Satellite image

1 Introduction

Type-2 fuzzy set (T2F set) is an extension from the type-1 fuzzy set (T1F set), which is characterized by the fuzzy membership function [1], unlike T1F set, membership function (MF) values are a clear number of [0, 1], MF values of the T2F set are a fuzzy set of [0, 1] [2]. One of the applications of the T2F set is the interval type-2 fuzzy c-means (IT2FCM) clustering algorithm [3], which has been used in many practical problems [4, 5]. However, this is an unsupervised clustering algorithm and they have difficulty in automatic classification [6].

T2FSs are characterized by a three-dimensional fuzzy MF including the footprint of uncertainty (FOU) that can directly model and handle the uncertainty [7]. Once the type-1 MF is selected, all uncertainties will disappear because the type-1 MFs are completely correct [8]. The type-2 fuzzy system (T2FS) based on the T2F set has been used in many practical applications such as predictive problem [9], industrial control [10, 11], data classification [12, 13].

Satellite images with many advantages such as wide coverage, fast update times have been applied in many fields [20]. However, satellite image data also has many uncertainties that use clustering algorithms that are often ineffective [21, 22]. Although T2FS has been widely applied in many areas [14], according to the author's knowledge, applications in problems related to remote sensing data is still very few and mainly based

N. T. Nguyen et al. (Eds.): ACIIDS 2020, LNAI 12034, pp. 402–414, 2020.
https://doi.org/10.1007/978-3-030-42058-1_34

on T1F set, so T2FS-based approach to the problem of remote sensing image processing is a potential research direction. Moreover, due to the complexity of the calculation, it limits T2FS in real applications. One of the case of T2FS is more widely used, which is an interval type-2 fuzzy system (IT2FS) [15]. This study introduces an approach of IT2FS in the remote sensing image landcover classification.

Currently, there are many optimization methods that do not need to use the derivative of objective functions. However, the disadvantage of using the derivative besides complex calculations of derivative formulas, when the UMF or LMF changes the mathematical formula on the specified domain, the calculation will also change; moreover, they are easily stuck at a local extreme [16, 17].

Methods that do not use derivatives are often called evolutionary methods [8] or methods of biological inspiration [19] such as evolutionary programming, genetic algorithms, genetic programming, particle swarm optimization (PSO), quantum particle swarm optimization (QPSO), simulated annealing, differential evolution, ant colony optimization, gravitational search, so on.

These methods tend to be stronger than derivative-based methods because the process of finding a globally optimal solution is repeated many times until convergence. This can be used to optimize the FOU parameter in the IT2 fuzzy system [16]. With so many biological-inspired methods, which method is good to optimize the parameters of the fuzzy system? Each algorithm has advantages and disadvantages, so far no algorithm has been proven to be the best. Which algorithm is used depends on each specific problem and user familiarity [19].

However, these bio-inspired methods often have to use a large number of loops to find the optimal solution, they need a large amount of time to evaluate the objective function for each candidate. If the calculation time is not important, these are very powerful methods. The advantage of PSO algorithm is convergence faster than GA algorithm, which is suitable for large data sets such as satellite image data.

The fuzzy systems generally have difficulty in determining MF values, FOU, the number of rules [8]. When the MF values, the number of inputs and outputs are selected, the IT2FS can be built. Determining these parameters for fuzzy systems is very important and greatly affects the accuracy of the fuzzy system. In the paper, we use the PSO techniques to find the MF parameters for IT2FS.

The paper includes five sections following: Sect. 1 Introduction; Sect. 2 Background; Sect. 3 Method proposed; Sect. 4 Experimental and Sect. 5 Conclusion.

2 Backgrounds

2.1 Interval Type-2 Fuzzy Logic System

The IT2FS is characterized by interval type-2 fuzzy sets (IT2F sets) [8] consisting of 5 main parts as shown in Fig. 1.

There are two types of IT2FSs as shown in Fig. 1 (a and b), one of which is type-reduction and then defuzzification, the second is direct defuzzification. However, in practice the type combines both type-reduction and defuzzification to be more widely used because of lower computational complexity.

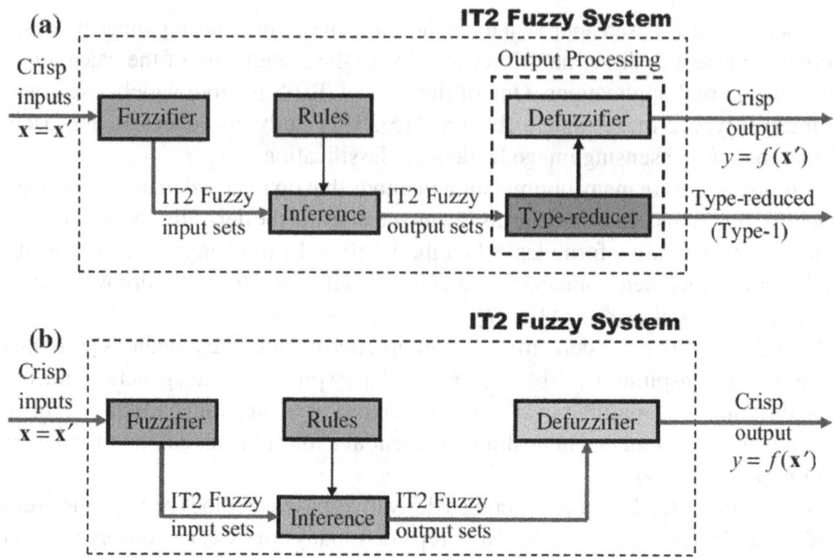

Fig. 1. Interval type-2 fuzzy system architecture [8]

The IT2FS works as follows, the crisp inputs are the attributes of the initial data, which fuzzifier into the input IT2F sets and then activate the inference engine and rule base to maps input IT2F sets into output IT2F sets. These output IT2F sets are then processed by the type-reducer to obtain T1F sets (type reducers). The defuzzifier then defuzzify output T1F sets to create the crisp output.

+ *Fuzzifier*

With T1FS, two types of fuzzifiers are used as singleton and non-singleton, meanwhile with T2FS, there are 3 types of fuzzifiers used including singleton, type-1 non-singleton, and IT2 non-singleton [11]. The fuzzifier maps a crisp input will depend on the choice of the type of fuzzifier. Assuming there are n inputs $X = (x_1, x_2, \ldots, x_n) \in X_1 x X_2 x \ldots x X_n$, $\tilde{A}_x$ is a set of type-2 fuzzy inputs. For example, if $\tilde{A}_x$ is a type-2 fuzzy singleton fuzzifier, then $\mu_{\tilde{A}(x_i)} = 1/1$ when $x_i = x'_i$ and $\mu_{\tilde{A}(x_i)} = 1/0$ when $x_i \neq x'_i$ and $x_i \in X_i$

+ *Rule Base*

Consider the input $x_1 \in X_1, x_2 \in X_2, \ldots, x_n \in X_n$ and c output $y_1 \in Y_1, y_2 \in Y_2, \ldots, y_c \in Y_c$. The rules of T2FS are similar to those of T1FS, only different the antecedents and the consequents instead of T1FS will be replaced by T2FS [11]:

$$R^i : \text{IF } x_1 \text{ is } \tilde{F}_1^i \text{ and} \ldots \text{and } x_n \text{ is } \tilde{F}_n^i \text{ THEN } y_1 \text{ is } \tilde{G}_1^i, \ldots, y_c \text{ is } \tilde{G}_c^i, \quad i = 1, \ldots, M \tag{1}$$

with M is the number of rules in the rule base.

+ *Fuzzy Inference Engine*

The inference engine and the rules that allow the mapping from input T2FS to the output T2FS. Each rule in a fuzzy rule base with M rules having n inputs $x_1 \in X_1, x_2 \in$

$X_2, \ldots, x_n \in X_n$ and output $y_k \in Y_k$, they can be written as follows:

$$R_k^i : \tilde{F}_1^i x \tilde{F}_2^i x \ldots x \tilde{F}_n^i \rightarrow \tilde{G}_k^i = \tilde{A}^i \rightarrow \tilde{G}_k^i \tag{2}$$

Where $\tilde{F}_j^i$ is the jth T2FS, $j = 1, \ldots, n$, which is defined by a lower and upper bound membership function

$$\mu_{\tilde{F}_j^i}(x_j) = [\underline{\mu}_{\tilde{F}_j^i}(x_j), \bar{\mu}_{\tilde{F}_j^i}(x_j)], i = 1, \ldots, M; k = 1, \ldots, c \tag{3}$$

Compute the firing interval of the ith rule, where * denotes the product operation.

$$\underline{f}^i(x) = \underline{\mu}_{\tilde{F}_1^i}(x_1) * \underline{\mu}_{\tilde{F}_2^i}(x_2) * \ldots * \underline{\mu}_{\tilde{F}_n^i}(x_n) \tag{4}$$

$$\bar{f}^i(x) = \bar{\mu}_{\tilde{F}_1^i}(x_1) * \bar{\mu}_{\tilde{F}_2^i}(x_2) * \ldots * \bar{\mu}_{\tilde{F}_n^i}(x_n) \tag{5}$$

+ *Type Reduction*

There are several algorithms used in type reduction, như Karnik-Mendel algorithm (KM) [15, 16], Enhanced Karnik-Mendel algorithm (EKM) [17], iterative algorithm and stopping condition (IASC), enhanced IASC algorithm (EIASC) [11]. In this study, the EIASC algorithm is used because they are easy to setup and the computational complexity is smaller than the remaining algorithms. Compute the output interval of the kth fuzzy rule for the output, which is an interval T1FS $y_k = [y_{kl}, y_{kl}]$, the steps for calculating left most output y_{kl} and right most output y_{kr} using the EIASC algorithm are detailed below.

EIASC algorithm for calculating y_{kl} as follows:

Step 1: Sort $\underline{y}^i$ $(i = 1, \ldots, M)$ by increasing value $\underline{y}^1 \leq \underline{y}^2 \leq \ldots \leq \underline{y}^M$, note is $\underline{f}^i$ will also change the order corresponding to $\underline{y}^i$.
Step 2: Initialization

$$L = 0; a = \sum_{i=1}^{M} \underline{f}^i \underline{y}^i; b = \sum_{i=1}^{M} \underline{f}^i; \underline{y} = a/b \tag{6}$$

Step 3: Calculate:

$$\begin{aligned} L &= L + 1; a = a + \underline{y}^L(\bar{f}^{(L)} - \underline{f}^{(L)}) \\ b &= b + (\bar{f}^{(L)} - \underline{f}^{(L)}); \underline{y} = a/b \end{aligned} \tag{7}$$

Step 4: Stop condition: If $\underline{y} \leq \underline{y}^{(L+1)}$ then stop, otherwise go to Step 3.

Compute the kth left most output:

$$y_{kl} = \frac{\sum_{u=1}^{L} \bar{f}^u \underline{y}^u + \sum_{v=L+1}^{M} \underline{f}^v \underline{y}^u}{\sum_{u=1}^{L} \bar{f}^u + \sum_{v=L+1}^{M} \underline{f}^v} \tag{8}$$

EIASC algorithm for calculating y_{kr} as follows:

Step 1: Sort $\bar{y}^i$ $(i = 1, \ldots, M)$ by increasing value $\bar{y}^1 \le \bar{y}^2 \le \ldots \le \bar{y}^M$, note is $\bar{f}^i$ will also change the order corresponding to $\bar{y}^i$.
Step 2: Initialization

$$R = n; a = \sum_{i=1}^{M} \bar{f}^i \bar{y}^i; b = \sum_{i=1}^{M} \bar{f}^i; \bar{y} = a/b \tag{9}$$

Step 3: Calculate:

$$\begin{aligned} a &= a + \bar{y}^{(R)}(\bar{f}^{(R)} - \underline{f}^{(R)}); b = b + (\bar{f}^{(R)} - \underline{f}^{(R)}) \\ \bar{y} &= a/b; R = R - 1 \end{aligned} \tag{10}$$

Step 4: Stop condition: If $\bar{y} \ge \bar{y}^{(R+1)}$ then stop, otherwise go to Step 3.
Compute the kth right most output:

$$y_{kr} = \frac{\sum_{u=1}^{R} \underline{f}^u \bar{y}^u + \sum_{v=R+1}^{M} \bar{f}^v \bar{y}^u}{\sum_{u=1}^{R} \underline{f}^u + \sum_{v=R+1}^{M} \bar{f}^v} \tag{11}$$

2.5 Defuzzification

The final crisp value of output of the IT2FS model is calculated by combining the corresponding outputs of M rules. For defuzzification solution we calculate the average left most point and right most point, therefore the crisp output for each output is calculated as follows:

$$Y_k(x) = \frac{y_{kl} + y_{kr}}{2} (k = 1, \ldots, c) \tag{12}$$

2.2 Particle Swarm Optimization

The PSO is an adaptive evolution algorithm based on finding the optimal solution for the population, the idea of algorithms comes from the hunting behavior of the birds [19]. Each problem will converge at one or several optimal solutions in the search space, considering each individual is a particle and a set of particles will be a population.

Each state of the population in the search space is considered as a candidate solution, the optimal solution is found by moving particles in the search space according to the position and velocity of the particle as the following formula:

$$\begin{aligned} vt_i^{k+1} &= \omega * vt_i^k + c_1 * r_1 * (P_{ibest} - v_i^k) + c_2 * r_2 * (G_{ibest} - v_i^k) \\ v_i^{k+1} &= v_i^k + vt_i^{k+1} \end{aligned} \tag{13}$$

In which, v_i^k is position of individual i^{th} in k^{th} generation, vt_i^k is velocity of individual i^{th} in k^{th} generation, ω is coefficient of inertia, c_1, c_2 is the acceleration coefficient, with a value of 1.5 to 2.5; r_1, r_2 is the random number, with values in the range [0,1].

In each loop, the optimal position search is performed by updating the velocity and position of the individual. In addition to each loop, the target value of each individual location is determined by an objective function.

3 Proposal Method

The paper developed a method using PSO technique to optimize MFs parameters of IT2FS. Some membership functions are often used in fuzzy systems such as triangular, trapezoidal, Gaussian, Cauchy, Laplace. In this study, we use Gaussian functions to build the MFs for IT2FS (see Fig. 2).

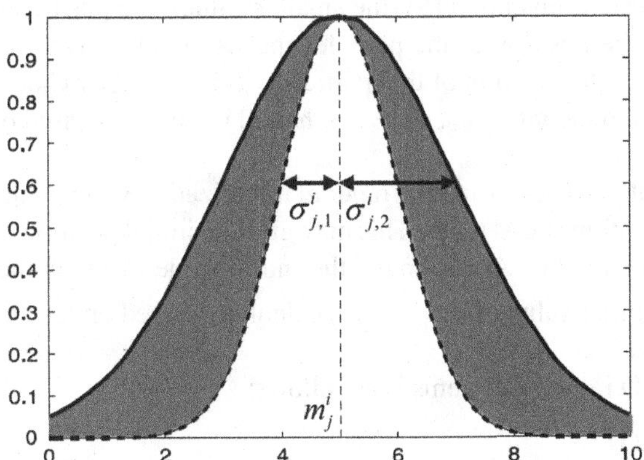

Fig. 2. Example of Gaussian IT2 MF

$$\mu_{\tilde{F}^i_j}(x_j) = \exp\left(-\frac{1}{2}\left(\frac{x_j - m^i_j}{\sigma^i_j}\right)^2\right) = \mathrm{N}(x_j, m^i_j, \sigma^i_j) \text{ with } \sigma^i_j \in \left[\sigma^i_{j,1}, \sigma^i_{j,2}\right] \quad (14)$$

It can be seen that the Gaussian function is characterized by parameters m^i_j, $\sigma^i_{j,1}$, $\sigma^i_{j,2}$. The number of Gaussian functions used will be equal to the number of landcovers. If c is the number of overlays to be classified, then the number of parameters is 3 * c.

The idea of the paper is from labeled data sets, using PSO technique to find the optimal parameters of the Gaussian function for unknown fuzzy systems. Each of the above parameters can be considered as an individual in a population, each individual will include position and velocity moving in the search space. The PSO algorithm will stop if the position of the individual is optimal or when the number of loops is satisfied.

To evaluate the effectiveness of the proposed method, we measure the difference between the actual output and the desired output on the data sets labeled by the following formula:

$$MSE = \frac{1}{n}\sum_{i=1}^{n}(y_i - \bar{y}_i)^2 \quad (15)$$

In PSO, individuals will move through the search space and record optimal locations. PSO consists of 3 main steps: population initialization; evaluate individuals; update position and velocity for individuals (use the formula (13)). Each particle will include position and velocity, which is limited to the search space $v_{\min} \leq v_i^k \leq v_{max}$ and $vt_{\min} \leq vt_i^k \leq vt_{max}$; in case if $v_i^k < v_{\min}(v_i^k > v_{\max})$ then $v_i^k = v_{\min}(v_i^k = v_{\max})$; if $vt_i^k < vt_{\min}(vt_i^k > vt_{max})$ then $vt_i^k = vt_{\min}(vt_i^k = vt_{max})$.

The optimal process is considered by the two parameters P_{best} and G_{best}, P_{best} is the best solution of the individual at the present location and G_{best} is the best solution of the population at the current location. These values will be updated based on the objective function value (15). In each iteration of the PSO algorithm, if a position of the particles optimizes the objective function (15) (the smaller value), the position of the particle is saved by P_{best}; If the position of the particles makes the objective function (15) reach the minimum value, the position of that particle will be saved by G_{best}. The parameters of the PSO algorithm including $c_1, c_2, r_1, r_2, \omega$ will be selected based on the suggestion of the original paper.

The position of particles instead of random initialization will be initialized using the fuzzy c-mean algorithm (FCM) [18]. This may give the initial position of the individuals closer to the optimal value. Accordingly, the initial value of m_j^i is the centroid of the clusters; and the initial value of $\sigma_{j,1}^i, \sigma_{j,2}^i$ is calculated based on the standard deviation of the training data.

The use of PSO in fuzzy systems is as follows:

Algorithm 1: PSO

Step 1: The parameters of each MFs act as individuals in the population.
Step 2: The movement of individuals leads to the optimal solution of individuals and the entire population (considered by the objective function (15)).
Step 3: Once the parameters have been adjusted and it will be used to evaluate the performance of the fuzzy system.

The proposed method includes the following steps:

Algorithm 2: PSO in IT2FLS
Input: $X = (x_1, x_2, \ldots, x_n)$, number of rules M, number of Gaussian functions.

Step 1: Initialize parameters for the Gaussian MFs: Perform the FCM algorithm to initialize m_j^i and using labeled data to initialize the standard deviation.
Step 2: Compute the lower and upper membership function of x_j:

$$\mu_{\tilde{F}_j^i}(x_j) = [\underline{\mu}_{\tilde{F}_j^i}(x_j), \bar{\mu}_{\tilde{F}_j^i}(x_j)]$$

Step 3: Compute the firing interval of the ith rule, where * denotes the product operation.

$$\underline{f}^i(x) = \underline{\mu}_{\tilde{F}_1^i}(x_1) * \underline{\mu}_{\tilde{F}_2^i}(x_2) * \ldots * \underline{\mu}_{\tilde{F}_n^i}(x_n)$$

$$\bar{f}^i(x) = \bar{\mu}_{\tilde{F}_1^i}(x_1) * \bar{\mu}_{\tilde{F}_2^i}(x_2) * \ldots * \bar{\mu}_{\tilde{F}_n^i}(x_n)$$

Step 4: Compute the left most and right most output of the ith fuzzy rule:

$$y^i = [y_l^i, y_r^i]$$

Step 5: The crisp output for each output is calculated as follows: $Y(x) = \frac{y_l + y_r}{2}$
Step 6: Compute MSE: $MSE = \frac{1}{n}\sum_{i=1}^{n}(y_i - \bar{y}_i)^2$
Step 7: If $(MSE < \varepsilon)$ or $(Loop > Loop_{max})$ goto step Output.
Step 8: Implementing Algorithm 1 and goto step 2.

Output: The parameters for IT2FLS.

The complexity of the proposed IT2FLS will also include the complexity of FCM in step 1, the complexity of PSO in step 8. After each implementation of IT2FLS, the parameters of the Gaussian function are adjusted by the PSO algorithm, the process is repeated until the optimal parameter for the fuzzy system is found. After the training is completed, the parameters of the MFs will be used for the fuzzy system.

Because PSO has smaller computational complexity than genetic algorithm, evolutionary computation or deep learning, this method is able to find the optimal parameters faster. However, their disadvantage is that finding the best parameter depends on the selection of the parameters of the PSO algorithm.

4 Experimental Results

Experiment on three data sets downloaded from the UCI Machine Learning Repository [23] including 'Urban land cover data set', 'Crowdsourced Mapping Data Set' and 'Forest type mapping Data Set'. Detailed information about the test data sets are shown in Table 1.

Table 1. Experimental data

Data	Urban landcover	Crowdsourced	Forest type
Attributes	148	29	27
Training	168	10545	198
Testing	507	300	325

The resulting classification performance of the classification is evaluated by determining True Positive Rate (TPR) and False Positive Rate (FPR) defined as follows:

$$TPR = \frac{TP}{TP + FN} \text{ and } FPR = \frac{FP}{TN + FP} \tag{16}$$

Where TP is the number of correctly classified data and FN is the number of incorrectly misclassified data, FP is the number of incorrectly classified data and TN is the number of correctly misclassified data. The better the algorithm is, the higher the TPR value is and the smaller the FTR value is encountered.

4.1 Experiment 1

The experimental data set is 'Urban land cover data set' with the number of samples used for training 168 and used for testing 507.

The landcover classification results are shown in Tables 2 and 3. Accordingly, Table 2 is the accuracy by true positive rate and false positive rate, IT2FS-PSO method gives the highest accuracy with 92.12%, followed by 91.21% for IT2FS-FCM. The lowest accuracy with 84.94% for SVM method. Meanwhile, the lowest false positive rate is 1.18% for IT2FS-PSO, followed by 1.65%, 1.66%, 1.76%, 2.07% and 2.89% for IT2FS-FCM, T1FS-PSO, T1FS-FCM, KNN and SVM.

Table 2. Accuracy by true positive rate and false positive rate

Algorithm	KNN	SVM	T1FS-FCM	T1FS-PSO	IT2FS-FCM	IT2FS-PSO
TPR (%)	85.81	84.94	87.68	89.18	91.21	**92.12**
FPR (%)	2.07	2.89	1.76	1.66	1.65	**1.18**

Table 3. Accuracy according to the landcovers

Class	KNN	SVM	T1FS-FCM	T1FS-PSO	IT2FS-FCM	IT2FS-PSO
Trees	83.41	85.13	86.31	90.34	92.32	**92.34**
Grass	85.72	87.32	88.94	88.01	89.52	**90.58**
Soil	89.14	81.26	84.87	92.32	**96.71**	93.94
Concrete	87.31	83.62	89.72	91.56	93.36	**95.78**
Asphalt	80.52	84.31	87.54	89.19	88.63	**91.49**
Buildings	84.94	88.73	90.09	87.48	91.92	**94.63**
Cars	82.63	**89.52**	86.98	85.83	88.21	86.28
Pools	88.51	80.31	85.71	87.67	90.42	**92.76**
Shadows	90.13	84.23	88.93	90.23	89.83	**91.28**

The overview of Table 3 shows that the IT2FS-PSO method gives the highest accuracy on most landcovers. The overview of Table 3 shows that the IT2FS-PSO method gives the highest accuracy on most coatings. Only the "soil" class has the highest accuracy of 96.71% for IT2FS-FCM and the "cars" class has the highest accuracy of 89.52% for SVM.

4.2 Experiment 2

Crowdsourced data from OpenStreetMap is used to classify satellite images into different land cover classes (impervious, farm, forest, grass, orchard, water) with the number of samples used for training 10545 and used for testing 300.

Table 4 shows the highest accuracy of 90.30% for IT2FS-PSO method, but the lowest false positive rate belongs to T1FS-PSO method with 1.21%. In this experiment, most methods yielded an accuracy of less than 90%. In particular, the lowest accuracy is 87.88% for the KNN method.

Table 4. Accuracy by true positive rate and false positive rate

Algorithm	KNN	SVM	T1FS-FCM	T1FS-PSO	IT2FS-FCM	IT2FS-PSO
TPR (%)	87.88	88.20	88.18	89.12	89.11	**90.30**
FPR (%)	2.12	2.08	1.78	**1.21**	1.34	1.27

Table 5 shows the accuracy by landcovers. The IT2FS-PSO method achieved the highest accuracy for the "impervious", "farm" and "water" classes with 90.84%, 91.87%, and 92.61%, respectively. Meanwhile, T1FS-PSO method has the highest accuracy in the "forest" class with 88.93%; T1FS-FCM method has the highest accuracy in the "orchard" class with 89.03%; KNN method has the highest accuracy in the "grass" class with 90.23%.

Table 5. Accuracy according to the landcovers

Class	KNN	SVM	T1FS-FCM	T1FS-PSO	IT2FS-FCM	IT2FS-PSO
Impervious	87.89	89.21	85.89	87.38	87.38	**90.84**
Farm	90.93	89.49	87.39	90.27	90.99	**91.87**
Forest	85.59	87.65	88.43	**88.93**	88.81	88.38
Grass	**90.23**	86.48	88.21	89.84	88.45	89.45
Orchard	83.41	86.23	**89.03**	87.78	87.92	88.67
Water	89.22	90.12	90.12	90.49	91.08	**92.61**

In this experiment, the T1FS-PSO method gave an accuracy of 89.12% higher than the IT2FS-FCM method with 89.11%. Next to the SVM, T1FS-FCM and KNN methods. The overview can see, the difference in accuracy between the methods is not large.

4.3 Experiment 3

The experimental data set is the 'Mapping Data Set forest type' with the number of samples used for training 198 and used for testing 325.

Table 6 shows that the IT2FS-PSO method has the highest accuracy with 92.98%, while the following classification rate is only 0.98%. While the T1FS-FCM, T1FS-PSO, IT2FS-FCM methods all have accuracy above 90%, the KNN and SVM methods only reach 87.79% and 86.62%.

Table 6. Accuracy by true positive rate and false positive rate

Algorithm	KNN	SVM	T1FS-FCM	T1FS-PSO	IT2FS-FCM	IT2FS-PSO
TPR (%)	87.79	86.62	90.20	91.03	91.72	**92.98**
FPR (%)	2.13	2.38	1.72	1.07	1.19	**0.98**

Table 7. Accuracy according to the landcovers

Class	KNN	SVM	T1FS-FCM	T1FS-PSO	IT2FS-FCM	IT2FS-PSO
'Sugi'	91.34	89.16	90.78	90.32	90.71	**91.92**
'Hinoki'	89.22	87.51	90.82	89.94	91.53	**93.61**
'Mixed deciduous'	83.65	86.34	88.79	93.57	95.28	**95.32**
'Other' non forest land	86.93	83.48	90.41	90.29	89.34	**91.08**

Similarly on Table 7, the IT2FS-PSO method also gives the highest accuracy on all landcovers. The highest accuracy is 95.32% for the 'Mixed deciduous' class.

From the above experiments, it can be seen that the proposed method has the highest accuracy on most landcover. If the T1FS-FCM and IT2FS-FCM methods use the FCM algorithm to initialize parameters for the fuzzy system, the T1FS-PSO and IT2FS-PSO methods use FCM as the initial initialization step before executing the PSO algorithm to find optimal parameters for fuzzy systems. By using the PSO technique to find parameters for fuzzy systems, the landcover classification results achieve higher accuracy.

With the selection of the parameters of fuzzy system based on experience, the selection of parameters based on optimal techniques can help the classification algorithm be more stable, and achieve higher accuracy in most cases.

5 Conclusion

The paper proposed a method using the PSO technique to find the optimal solution for Gaussian MF parameters of IT2FS in landcover classification. Experimental results on 3 data sets from the UCI library show that the proposed method gives significantly higher accuracy than T1FS-FCM, T1FS-PSO, IT2FS-FCM, KNN and SVM methods. Moreover, PSO technology has a low computational complexity which is suitable for large data problems such as satellite image data.

In the next time, we will continue to study other optimization techniques such as evolutionary computation, deep learning for fuzzy classification systems.

Acknowledgment. This research is funded by the Newton Fund, under the NAFOSTED - UK Academies collaboration programme. This work was supported by the Domestic Master/PhD Scholarship Programme of Vingroup Innovation Foundation (VINIF).

References

1. Karnik, N., Mendel, J., Liang, Q.: Type-2 fuzzy logic systems. IEEE Trans. Fuzzy Syst. **7**(6), 643–658 (1999)
2. Mendel, J., John, R.: Type-2 fuzzy sets made simple. IEEE Trans. Fuzzy Syst. **10**(2), 117–127 (2002)
3. Hwang, C., Rhee, F.C.H.: Uncertain fuzzy clustering: interval type-2 fuzzy approach to C-means. IEEE Trans. Fuzzy Syst. **15**(1), 107–120 (2007)
4. Ngo, L.T., Mai, D.S., Nguyen, M.U.: GPU-based acceleration of interval type-2 fuzzy C-means clustering for satellite imagery land-cover classification. In: 12th IEEE International Conference on Intelligent Systems Design and Applications (ISDA), pp. 992–997 (2012)
5. Mai, S.D., Ngo, L.T.: Interval type-2 fuzzy C-means clustering with spatial information for land-cover classification. In: Nguyen, N.T., Trawiński, B., Kosala, R. (eds.) ACIIDS 2015. LNCS (LNAI), vol. 9011, pp. 387–397. Springer, Cham (2015). https://doi.org/10.1007/978-3-319-15702-3_38
6. Ngo, L.T., Mai, D.S., Pedrycz, W.: Semi-supervising interval type-2 fuzzy C-means clustering with spatial information for multi-spectral satellite image classification and change detection. Comput. Geosci. **83**, 1–15 (2015)
7. Mendel, J.M.: Type-2 fuzzy sets and systems: an overview. IEEE Comput. Intell. Mag. **2**(1), 20–29 (2007)
8. Mendel, J.: Uncertain Rule-Based Fuzzy Logic Systems: Introduction and New Directions, 2nd edn. Springer, Cham (2017). https://doi.org/10.1007/978-3-319-51370-6
9. Abbas, K., Saeid, N., Doug, C., Dipti, S.: Interval type-2 fuzzy logic systems for load forecasting: a comparative study. IEEE Trans. Power Syst. **27**(3), 1274–1282 (2012)
10. Das, K., Suresh, S., Sundararajan, N.: A fully tuned sequential interval type-2 fuzzy inference system for motor-imagery task classification. In: IEEE International Conference on Fuzzy Systems, pp. 751–758 (2016)
11. Nguyen, T., Saeid, N.: Modified AHP for gene selection and cancer classification using type-2 fuzzy logic. IEEE Trans. Fuzzy Syst. **24**(2), 273–287 (2016)
12. Mendel, J., Liu, X.: Simplified interval type-2 fuzzy logic systems. IEEE Trans. Fuzzy Syst. **21**(6), 1056–1069 (2013)
13. Liang, Q., Mendel, J.: Interval type-2 fuzzy logic systems: theory and design. IEEE Trans. Fuzzy Syst. **8**(5), 535–550 (2000)
14. Mendel, J.M., John, R.I., Liu, F.: Interval type-2 fuzzy logic systems made simple. EEE Trans. Fuzzy Syst. **14**(6), 808–821 (2006)
15. Karnik, N.N., Mendel, J.M.: Centroid of a type-2 fuzzy set. Inf. Sci. **132**(1–4), 195–220 (2001)
16. Karnik, N.N., Mendel, J.M.: Operations on type-2 fuzzy sets. Fuzzy Sets Syst. **122**(2), 327–348 (2001)
17. Wu, D., Mendel, J.M.: Enhanced Karnik-Mendel algorithms. IEEE Trans. Fuzzy Syst. **17**(4), 923–934 (2009)
18. Bezdek, J.C., Ehrlich, R., Full, W.: FCM: the fuzzy C-means clustering algorithm. Comput. Geosci. **10**(2), 191–203 (1984)
19. Olivas, F., Angulo, L.A., Perez, J., Caraveo, C., Valdez, F., Castillo, O.: Comparative study of type-2 fuzzy particle swarm, Bee Colony and Bat algorithms in optimization of fuzzy controllers. Algorithms **10**, 1–27 (2017)
20. Mai, S.D., Ngo, L.T., Le Trinh, H.: Satellite image classification based spatial-spectral fuzzy clustering algorithm. In: Nguyen, N.T., Hoang, D.H., Hong, T.-P., Pham, H., Trawiński, B. (eds.) ACIIDS 2018. LNCS (LNAI), vol. 10752, pp. 505–518. Springer, Cham (2018). https://doi.org/10.1007/978-3-319-75420-8_48

21. Mai, D.S., Ngo, L.T.: General semi-supervised possibilistic fuzzy c-means clustering for land-cover classification. In: International Conference on Knowledge and Systems Engineering, pp. 1–6 (2019)
22. Dang, T.H., Mai, D.S., Ngo, L.T.: Multiple kernel collaborative fuzzy clustering algorithm with weighted super-pixels for satellite image land-cover classification. Eng. Appl. Artif. Intell. **85**, 85–98 (2019)
23. https://archive.ics.uci.edu/

Intelligent Supply Chains and e-Commerce

Logistics Value and Perceived Customer Loyalty in E-commerce: Hierarchical Linear Modeling Analysis

Arkadiusz Kawa(✉) and Justyna Światowiec-Szczepańska(✉)

Poznan University of Economics and Business, al. Niepodległości 10, 61-875 Poznań, Poland
{arkadiusz.kawa,justyna.swiatowiec-szczepanska}@ue.poznan.pl

Abstract. Value is a fairly capacious concept and there are no clearly defined boundaries. This is due to the unlimited number of all needs, expectations and limitations of potential customers. In e-commerce, however, the most important elements of value that relate to logistics can be distinguished.

The aims of this paper are to identify the components of logistics value for customers, and to present their influence on loyalty in e-commerce. Moreover, we investigated the mediation effect between those two variables using the positions of the value chain members. The hierarchical linear modeling (HLM) method was applied to analyze the data structure and find the dependencies between the variables from different levels.

Keywords: E-commerce · Logistics value · Loyalty · HLM

1 Introduction

In e-commerce, some logistics processes are similar to those in traditional commerce, and some are different. The main difference is access to and collection of goods. In traditional trade, the seller sells a product that is available on the shelf. In e-commerce, on the other hand, s/he sells a promise of order fulfillment, in particular a promise that s/he will deliver the right product, in the right quantity and condition, to the right location, at the expected time, cost and to the right customer. This is in line, then, with the logistic principle of 7R, which refers to seven tips for dealing with the flow of products between individual actors in the supply chain.

The 7Rs refer to the overriding notion of value. The concept of value was introduced in 1954 by Drucker [3]. For more than 60 years it has been defined in many ways. Some of these definitions are characterized by simplicity and are quite general. Most often, value is treated as an evaluation of the usefulness of a product, resulting from the ratio of what was obtained to what was given [16]. Value in management often refers to the client and is therefore referred to as "value for the customer". Kotler defines it as the difference between the total value of the product for the customer and the cost s/he has to bear in acquiring it [6].

The concept of logistics value has a slightly shorter history. It was first described by Novack et al. [11]. Like value itself, logistics value also has many definitions. Most

N. T. Nguyen et al. (Eds.): ACIIDS 2020, LNAI 12034, pp. 417–427, 2020.
https://doi.org/10.1007/978-3-030-42058-1_35

often, it boils down to a combination of quality, price, services provided to the customer, i.e. delivering what he or she wants and when he or she wants. It is also understood as a combination of customer service requirements with simultaneous consideration of minimising supply chain costs and maximising partners' profits [15]. For the purposes of this paper, we assumed that the logistics value for an e-commerce customer is the excess of subjectively perceived benefits over subjectively perceived costs associated with the purchase of a given product via the Internet.

The goals of this article are to identify the elements of logistics value for customers, and to verify their influence on loyalty in e-commerce. We researched the mediation effect between those two variables using the positions of the value chain members. We analyzed 1192 questionnaires from telephone and web interviews (CATI and CAWI). Hierarchical linear modeling method was applied to find the dependencies between variables from different levels.

2 Logistics as a Value

To explain the components of the logistics value, we will use the previously indicated 7R concept. Logistics in e-commerce is to ensure:

- The product in the right condition – it involves safe delivery of the product to the customer. The quality of the product delivered should be exactly the same as if the customer purchased the product himself/herself by means of traditional trade. Appropriate protection and packaging of the product plays an important role. Opening the packaging must be easy and intuitive and should not involve the risk of damage to the goods. In addition, the product is often returned in the same packaging, so it should be durable and designed to be re-shipped. Packaging also has a marketing function. It is often the first element that the customer has physical contact with. The moment of opening the packaging is very emotional for many customers. It can be compared to what happens when a gift is opened. Packaging should therefore encourage people to buy again with its aesthetic appearance and be a kind of advertisement for the online store [1, 10].
- The product in the right place – the possibility to choose the place of delivery or collection of goods makes the customer influence the configuration of his or her value chain. Currently, customers can receive products ordered online in several ways: by courier delivery, delivery to the point of shipment and pickup, pickup from a parcel locker, self-pickup at a stationary store or a different retailer's location, delivery by an online store. The most popular forms of delivery are courier and postal services. Recently, in e-commerce, parcel lockers and pick-up and drop-off (PUDO) points – places to which access is relatively easy, such as press lounges, shopping malls, petrol stations, grocery stores - have gained importance. The models of deliveries to the parcel lockers and PUDO are characterized by flexibility of place and time of delivery. Online retailers who provide varied and convenient ways of collecting and delivering goods can count on greater customer loyalty [7, 8].
- The product at the right time – unlike in traditional trade, the customer does not have immediate access to the product after purchasing it. Therefore, it is important that

the seller specifies the lead time of the order. This time runs from the moment the customer confirms the order to the moment the goods are received by the customer. Several processes affect it, then – completing, packing, preparing for dispatch and delivery. Customers want to be able to choose different delivery dates and change them dynamically, even at the stage of the last mile. Therefore, time and flexibility of delivery are important [7, 8].

- Shipping information – current and accurate information is a very important value factor for the customer in e-commerce. The seller should guarantee information about the availability of the goods, which must be consistent with their current stock status. It is also essential to provide information about the progress of the order fulfillment and the place of delivery or collection of goods (delivery monitoring). This is usually done by sending information about the status of the shipment by e-mail. Another form is a text message or website access. Such information is also provided by external entities connected with the execution of orders placed via the Internet, e.g. by a payment service provider and a logistics operator. With up-to-date information, the customer has more knowledge about the fulfillment of his/her order and a greater sense of security. Therefore, they will be willing to repeat their purchases [1].
- Convenience of return – Internet shopping, in contrast to shopping in stationary shops, is convenient, but at the same time excludes the possibility to check goods before buying them. Customers cannot see or touch them, so if the products fail to meet their expectations, they might want to return them, which is not pleasant for the customers. Returns take extra time, and customers often have to pay for them. Also, the situation may be stressful for some people - especially if they do it for the first time. They do not always know where or how to report a return, how to prepare and pack the goods, how to order a courier or where to bring the shipment. They are not sure if they need to pay for the return, if and when they will be refunded. Returns in e-commerce should, then, be seamless and leave a good experience [9].
- The right cost – the biggest logistical cost in e-commerce is the cost of delivery. The faster the delivery, the higher the cost. Delivery costs are attempted to be reduced by optimizing the last delivery process, the so-called last mile. It consists in offering various forms of shipment self-collection. More and more often, customers are expecting free product delivery and returns, especially when their order exceeds a certain amount of money. Apart from delivery, an important cost factor is the preparation of the order, i.e. product picking and packing. Due to the fact that costs are part of the other value elements mentioned, we did not distinguish them as a separate structure.

All these activities have an impact on the value for the customer and, in turn, on his/her loyalty [2, 12].

The above observations are the basis for the research hypothesis, which is as follows: **H1.** *Logistics value positively affects perceived customer loyalty.*

3 Logistics Value, Closeness to Customer, and Perceived Customer Loyalty

Strategies aimed at creating and delivering the value have led companies to create it together for the final customer and to change business models, which are characterized by cooperation with external specialized partners, co-creating value chains or networks. The joint efforts of companies and buyers can create the desired value.

In terms of entities in e-commerce, the value chain consists of: customers, e-tailers, suppliers, and complementors. The customer is almost any individual or institutional person who has access to the Internet. E-tailers are mostly companies that have their own online shops or sell products on marketplaces, auction platforms, etc. Suppliers offer products sold via the electronic channel and complementors provide services and supporting e-commerce, e.g. logistic services, financial services, IT solutions, comparison shopping websites [5].

These individual links have different customer relationships. Closest to the customers are the e-tailers who sell products directly to them. Then, there are suppliers who deliver the products to the sellers. The most distant ones are complementors. Only residual information reaches them. For this reason, they do not have full knowledge of the customers, especially their expectations and behaviours. They do not know whether customers are satisfied with their purchases or whether they are loyal.

On the basis of the above considerations, we formulated the following research hypothesis: **H2.** *Closeness to the customer in the value chain is negatively related to perceived customer loyalty.*

Because there is a relationship between the logistics value and perceived customer loyalty, it is worthwhile to look for other dependencies. It is interesting to investigate the mediation factor between those two variables. We supposed that the relationship between the logistics value and perceived customer loyalty is stronger in a situation where members of the value chain are further away from the customer.

The presented observations lead to the next research hypothesis: **H3.** *Closeness to the customer in the value chain moderates the relationship between logistics value and perceived customer loyalty.*

4 Methodology

Data Gathering

The research assumed that the respondent was to look at returns through the final customer's "eyes", regardless of their role in e-commerce. The questions addressed to each of these groups were therefore about how customers perceived the issue of logistics value and loyalty. The main reason of this was that the central point of the e-commerce system is the customer who ultimately evaluates the value and converts it into a monetary equivalent for the other network members [5].

CATI (computer-assisted telephone interview) was chosen as the main technique of information collection in the research, preceded by focus group interviews. It was complemented by a CAWI (computer-assisted web interview) method.

The study was conducted from November 2017 to May 2018 by an external agency. A total of 800 correctly filled questionnaires was obtained (200 records in each group – e-tailers, customers, suppliers, and complementors) [5]. In addition, the survey was supplemented by 392 interviews using CAWI. In total, we analyzed 1192 questionnaires.

Measures

Logistics Value. In this study, the logistics value was measured using 23 items. All the items used a 5-point Likert scale ranging from 1 (strongly disagree) to 5 (strongly agree). Exploratory factor analysis (Principal Component Analysis) with varimax rotation was used to explore the factorial structure of the logistics value. This yielded five factors (see Table 1), explaining 48.9% of the total variance. They were named as follows: Convenient packaging, Delivery monitoring, Speed of delivery, Convenient place of delivery, Convenience of return. The results showed that Cronbach's α of these variables was 0.87, 0.81, 0.79, 0.72, 0.74 respectively. All Cronbach's α were above 0.70 and the α of the total scale was 0.79, indicating satisfactory internal consistency of the logistics value variable.

Table 1. Results of exploratory factor analysis of logistics value questionnaire (N = 1192)

Item	Factor 1	Factor 2	Factor 3	Factor 4	Factor 5
	Convenient packaging	Delivery monitoring	Speed of delivery	Convenient place of delivery	Convenience of return
V2.70	**0.768**	0.043	0.177	0.109	0.022
V2.69	**0.752**	0.033	0.142	0.045	0.071
V2.71	**0.720**	0.206	0.104	−0.072	0.047
V2.73	**0.669**	0.031	0.105	0.096	0.238
V2.67	**0.627**	0.263	0.055	0.133	0.077
V2.72	**0.626**	−0.024	0.218	0.240	0.159
V2.62	0.093	**0.741**	0.152	0.069	0.147
V2.63	0.189	**0.714**	0.03	−0.001	0.168
V2.61	0.067	**0.707**	0.063	0.139	0.111
V2.60	0.002	**0.684**	0.199	0.171	0.042
V2.50	0.208	0.084	**0.758**	0.070	0.122
V2.49	0.309	0.014	**0.758**	0.115	0.066
V2.59	0.262	0.320	**0.607**	0.138	0.147
V2.58	0.097	0.415	**0.578**	0.148	0.132
V2.56	0.185	0.034	0.170	**0.756**	0.089
V2.55	0.063	0.151	−0.058	**0.741**	0.153
V2.57	0.238	0.094	0.230	**0.689**	0.044
V2.54	0.065	0.162	0.061	**0.544**	0.034
V2.74	0.142	0.162	0.033	0.177	**0.701**
V2.78	0.070	0.256	0.139	0.015	**0.655**
V2.75	0.093	0.268	−0.054	0.036	**0.616**
V2.77	0.308	0.084	0.337	0.133	**0.582**
V2.76	0.446	0.091	0.256	0.131	**0.509**
Variance	14.48%	11.91%	7.98%	7.40%	7.14%

Customer Loyalty. A 3-item scale was chosen to measure the degree of perceived customer loyalty by members of the value network. A 5-point Likert scale was used in all items. The items and the reliability of each variable are presented in Table 2.

Table 2. Constructs, items and scales

Logistics value. *Cronbach's alpha = 0.79*
Convenient packaging. Cronbach's alpha = 0.87
1. Customers buy from online sellers who use environmentally friendly packaging materials
2. Customers buy from online sellers whose shipments are easy to open
3. Customers buy from online retailers who match the size of the packaging to the size of the product
4. Customers buy from online retailers from which packaging you can easily delete your personal data
5. Customers buy from online retailers whose shipments are aesthetically packed
6. Customers buy from online retailers who offer gift packaging
Delivery monitoring. Cronbach's alpha = 0.81
1. Customers buy from online sellers who cooperate with couriers informing about the time of delivery
2. Customers buy from online retailers cooperating with couriers who are on time
3. Customers buy from online sellers who offer tracking shipments
4. Customers buy from online sellers who inform about the status of the order
Time and flexibility of delivery. Cronbach's alpha = 0.79
1. Customers buy from online retailers who offer delivery of products on the same business day
2. Customers buy from online sellers who offer delivery of products within 2 h
3. Customers buy from online retailers who offer the option of delivery on non-working days
4. Customers buy from online retailers who offer the opportunity to choose delivery times
Convenient place of delivery. Cronbach's alpha = 0.72
1. Customers buy from online sellers who offer deliveries to PUDO (pick up drop off) points (e.g. a traffic kiosk. gas station)
2. Customers buy from online sellers who offer deliveries to self-service terminals (e.g. for a parcel locker)
3. Customers buy from online sellers who offer pickup at their branches
Convenience of return. Cronbach's alpha = 0.74
1. Customers buy from online sellers who offer free return of products
2. Customers buy from online sellers who offer the possibility of returning products over 14 days
3. Customers buy from online sellers who have an easy return procedure

(continued)

Table 2. (*continued*)

4. Customers buy from online sellers who offer return of used products
5. Customers buy from online retailers who offer returnable packaging
(1 = strongly disagree to 5 = strongly agree)
Customer loyalty. *Cronbach's alpha = 0.80*
1. Customers of online retailers will continue to buy with them. even if the products offered by other online retailers are more competitive
2. Customers of online retailers will continue to buy with them. even if the delivery of products offered by other online retailers will be more competitive
3. Customers of online retailers will continue to buy with them. even if payments for products offered by other online retailers are more competitive
(1 = strongly disagree to 5 = strongly agree)

Proximity to the Customer in Value Networks. As mentioned, the members of value networks were divided into four groups: customers, e-tailers, suppliers and complementors. It was assumed that the complementors were furthest from the final customers, the suppliers were a little closer to them, the e-tailers were even closer, while the customers represented the actual level of customer loyalty in the value network. Customer proximity was rated on a scale of 1 to 4, where 1 meant the position of complementors, 2 – suppliers, 3 – e-tailers and 4 – customers, respectively.

Hierarchical Linear Modeling Analysis and Hypothesis Testing

Value networks are multi-level, hierarchical phenomena. Researchers cannot ignore the complex and cross-level nature of value networks when examining the problem of customer value created throughout the network and its consumer loyalty. In this study we analyze two-level hierarchical data structures concerning all members of value networks (level-1 unit) and roles in the value network (network links) (level-2 unit). At the highest level of the hierarchy (level-2) there is a variable related with network links – it is proximity to the customer in the value network. Variables at the individual levels are nested within level-2 groups, these are: logistics value and customer loyalty, which are outcome variables as well.

To test our hypotheses, we used hierarchical linear modeling (HLM), a statistical technique capable of analyzing hierarchical, cross-level data [13]. It simultaneously estimates the relationship within a certain level and between or across hierarchical levels. HLM achieves this process by performing regressions of regressions [4]. To assess the three hypotheses, a sequence of models is required: the null, random-coefficient regression, means-as-outcomes regression, intercept-as-outcomes and intercepts-and-slopes-as-outcomes models. The models and results of HLM are presented in Table 3. The HLM 7 program was used for the hierarchical linear modeling in this study.

Certain prerequisites must be met to perform cross-level analyses. First, there must be systematic within- and between-group variance in the dependent variable. This condition is necessary because it is assumed that the dependent variable (customer loyalty) is significantly related to both the variable at the individual level (logistics value) and the variable at the group level (proximity to the customer in the value network). It tests whether there are any differences at the group level on the outcome variable and confirms that HLM is necessary. This is assessed in HLM using the null model (a one-way analysis of variance). The HLM program creates chi-square statistics to test the significance of variance between groups. The statistically significant chi-square for the dependent variable shows that the variance between the groups is significantly different from zero, thus indicating the differences between the groups. A statistically significant chi-square for "customer loyalty" has been achieved (χ^2 (3) = 143.51; $p < 0.001$); which supports the use of HLM. Additionally, using information estimated in the null model, an intraclass correlation coefficient (ICC) can be computed that represents the percent of the total variance in the dependent variable that is between groups [14]. The following equation was used:

$$ICC = \frac{\tau_{00}}{\tau_{00} + \sigma^2} = \frac{0.13912}{0.13912 + 0.80972} = 0.146$$

This means that about 15% of the variance of "customer loyalty" results from the group level. In other words, 15% of the inter-individual variance of "customer loyalty" is generated only by differences related to the role in the value network.

The essence of model M2 (see Table 3) is the ability to test the relationship between the variable from level 1 and the dependent variable, i.e. between "logistics value" and "customer loyalty". This relationship is positive and statistically significant ($\gamma_{10} = 0.34$. $p < 0.05$). This means that Hypothesis 1 has been supported. The measure of the effect size is the calculation of variance in "customer loyalty", explained by an independent variable from level 1 (logistics value). The result indicates (effect = 0.056; see Table 3) that the logistics value explains 5.6% of the variance in the perceived customer loyalty.

The main purpose of the M3 model is to test the significance and direction of the relationship between the level-2 predictor variable (customer proximity in the value network) and the dependent variable (customer loyalty). The results of the analysis support the fact that proximity to the customer in the value network predicts customer loyalty ($\gamma_{01} = -0.27$; $p < 0.05$). To measure the effect size, the explained variance in the outcome variable by the level-2 predictor variable is calculated (0.767; see Table 3). The results confirm that proximity to the customer in the value network explains 76.7% of the between-measures variance in "customer loyalty".

After determining that there is significant variation between the groups in the level-1 intercept, we can directly test the cross-level hypothesis (Hypothesis 2). The M4 model indicates whether or not the variable at the group level ("customer proximity in the value network") has a significant impact on the dependent variable ("customer loyalty"). The γ_{01} parameter is -0.30; $p < 0.001$. This result confirms Hypothesis 2, proving that the greater closeness to the customer in the value network, the lower the perceived customer loyalty is.

Next, it can be examined whether the variance of the slope coefficient in the groups is significantly related to the independent variable at the group level ("customer proximity"). This is a direct test for a moderator at various levels (Hypothesis 3). Model M5 is a direct test of Hypothesis 3, saying that proximity to the customer in the value network moderates the logistics value-customer loyalty relationship. The HLM results reveal that the interaction is not significant ($\gamma_{11} = 0.15$; $p = 0.39$). It means that there is no cross-level interaction between the level-1 and level-2 predictors. Hypothesis 3 is not supported, then. The interaction is not statistically significant; therefore, it does not confirm that closeness to the customer in the value network moderates the relationship of logistics value and customer loyalty. Thus, we cannot say that the relationship between the logistics value and perceived customer loyalty is stronger in a situation where members of the value network are further away from the customer.

Table 3. Results of HLM analysis

Model	Parameter estimates							
	γ_{00}	γ_{01}	γ_{10}	γ_{11}	σ^2	τ_{00}	τ_{11}	Effect
M1: Null model (One-way ANOVA) L1: $(CL) = \beta_{0j} + r_{ij}$ L2: $\beta_{0j} = \gamma_{00} + u_{0j}$	3.55***				0.809	0.139		
M2: Random Coefficients Regression Model L1: $(CL) = \beta_{0j} + \beta_{1j}(LV) + r_{ij}$ L2: $\beta_{0j} = \gamma_{00} + u_{0j}$ $\beta_{1j} = \gamma_{10} + u_{1j}$	3.50***		0.34*		0.764	0.121	0.080	0.056
M3: Means-as-Outcomes Model L1: $(CL) = \beta_{0j} + r_{ij}$ L2: $\beta_{0j} = \gamma_{00} + \gamma_{01}(PC) + u_{0j}$	4.21***	−0.27**			0.809	0.032		0.767
M4: Intercepts-as-Outcomes Model *L1*: $(CL) = \beta_{0j} + \beta_{1j}(LV) + r_{ij}$ *L2*: $\beta_{0j} = \gamma_{00} + \gamma_{01}(PC) + u_{0j}$ $\beta_{1j} = \gamma_{10} + u_{1j}$	4.25***	−0.30***	0.35*		0.762	0.006	0.073	0.946
M5: Intercepts-and-Slopes-as-Outcomes Model *L1*: $(CL) = \beta_{0j} + \beta_{1j}(LV) + r_{ij}$ *L2*: $\beta_{0j} = \gamma_{00} + \gamma_{01}(PC) + u_{0j}$ $\beta_{1j} = \gamma_{10} + \gamma_{11}(PC) + u_{1j}$	3.49***	−0.27**	0.33	0.15	0.762	0.006	0.074	0.910

Note: ***indicates $p < 0.01$; **indicates $p < 0.05$; *indicates $p < 0.1$;
L1 – Level 1; L2 – Level 2; CL – perceived customer loyalty; PC – proximity to the customer in value networks; LV – logistics value;
γ_{00} = Intercept of Level 2 regression predicting β_{0j}; γ_{01} = Slope of Level 2 regression predicting β_{0j}; γ_{10} = Intercept of Level 2 regression predicting β_{1j} (pooled Level 1 slopes); σ^2 = Variance in Level 1 residual (i.e. variance in r_{ij}); τ_{00} = Variance in Level 2 residual for models predicting β_{0j} (i.e. variance in U_0); τ_{11} = Variance in Level 2 residual for models predicting β_{1j};
Effect = (σ^2 of based model Mn – σ^2 of research model Mn + 1)/σ^2 of based model Mn.

5 Conclusion

Hierarchical linear modeling was used to analyze a data structure where members of value networks (level 1) were nested within groups (level 2) representing the roles in value networks. The relationship between perceived customer loyalty (level-1 outcome variable) and both logistics value (level-1 predictor variable) and proximity to the customer in value networks (level-2 predictor variable) were of special interest. Tests were conducted in 5 phases: the null model, random-coefficient regression model, means-as-outcomes regression model, intercept-as-outcomes model and intercepts-and-slopes-as-outcomes model. The first model revealed the ICC (intraclass correlation coefficient) of 0.146, i.e. 14.6% of the variance in perceived customer loyalty was between members of the value network within a given role in the network. The random-regression coefficient model was tested using the logistics value as the only predictor variable. The result indicated that perceived customer loyalty levels were higher when logistics value levels were also higher. The means-as-outcomes model added proximity to the customer in value networks as a level-2 predictor variable. It turned out that perceived customer loyalty was greater in network links (groups) which were further away from customers in the value network. Finally, intercepts-and-slopes-as-outcomes model was tested with all predictor variables. The aim was to test the presence of interactions between the predictor variables. The cross-level interaction between the logistics value and proximity to the customer in the value networks was not statistically significant, which meant that the degree of distance from the customer in the value network had no influence on the strength of the relationship between the logistics value and the perceived customer loyalty.

Attention should be paid to the limitations of the conducted research procedure – methodological and substantive ones. The former concern the very essence of the model, which simplifies the economic reality and thus reduces the complex factual situation. The latter are the limitations of the research into logistics value. Future research by the authors is to focus on extending the value of e-commerce to include other components.

Acknowledgements. This paper has been written with financial support of the National Center of Science [Narodowe Centrum Nauki] – grant number DEC-2015/19/B/HS4/02287.

References

1. Bansal, H., McDougall, G., Dikolli, S., Sedatole, K.: Relating e-satisfaction to behavioral outcomes: an empirical study. J. Serv. Mark. **18**(4), 290–302 (2004)
2. Chiu, C.-M., Lin, H.-Y., Sun, S.-Y., Hsu, M.-H.: Understanding customers' loyalty intentions towards online shopping: an integration of technology acceptance model and fairness theory. Behav. Inf. Technol. **28**(4), 347–360 (2009)
3. Drucker, P.F.: The Practice of Management: A Study of the Most Important Function in America Society. Harper & Brothers, New York (1954)
4. Hofmann, D.A.: An overview of the logic and rationale of hierarchical linear models. J. Manag. **23**, 723–744 (1997)
5. Kawa, A., Światowiec-Szczepańska, J.: IT value for customer: its influence on satisfaction and loyalty in e-commerce. In: Nguyen, N.T., Gaol, F.L., Hong, T.-P., Trawiński, B. (eds.) ACIIDS 2019. LNCS (LNAI), vol. 11432, pp. 489–498. Springer, Cham (2019). https://doi.org/10.1007/978-3-030-14802-7_42
6. Kotler, P.: Marketing Management: Analysis Planning Implementation and Control. Prentice Hall, Englewood Cliffs (1994)

7. Koufteros, X., Droge, C., Heim, G., Massad, N., Vickery, S.K.: Encounter satisfaction in e-tailing: are the relationships of order fulfillment service quality with its antecedents and consequences moderated by historical satisfaction? Decis. Sci. **45**(1), 5–48 (2014)
8. Lee, G.G., Lin, H.F.: Customer perceptions of e-service quality in online shopping. Int. J. Retail Distrib. Manag. **33**(2), 161–176 (2005)
9. McCollough, M., Berry, L., Yadav, M.: An empirical investigation of customer satisfaction after service failure and recovery. J. Serv. Res. **3**(2), 121–137 (2000)
10. Mentzer, J.T., Flint, D.J., Kent, J.L.: Developing a logistics service quality scale. J. Bus. Logist. **201**, 9–32 (1999)
11. Novack, R.A., Rinehart, L.M.: An internal assessment of logistics value. J. Bus. Logist. **15**, 113–152 (1994)
12. Piyathasanan, B., Mathies, Ch., Wetzels, M., Patterson, P.G., Ruyter, K.: A hierarchical model of virtual experience and its influences on the perceived value and loyalty of customers. Int. J. Electron. Commer. **19**(2), 126–158 (2015)
13. Raudenbush, S.W.: Educational applications of hierarchical linear models: a review. J. Educ. Stat. **13**, 85–116 (1998)
14. Raudenbush, S.W., Bryk, A.S.: Hierarchical Linear Models. Sage, Newbury Park (1992)
15. Rutner, S.M., Langley, C.J.: Logistics value: definition. Process and measurement. Int. J. Logist. Manag. **2**, 73–82 (2000)
16. Zeithaml, V.A.: Consumer perceptions of price, quality, and value: a means-end model and synthesis of evidence. J. Mark. **52**(3), 2–22 (1988)

Technology-Based Value Index to Outline e-Shopper Characteristics

Sergiusz Strykowski and Bartłomiej Pierański(✉)

Poznan University of Economics and Business, al. Niepodleglosci 10, 61-875 Poznan, Poland
{s.strykowski,b.pieranski}@ue.poznan.pl

Abstract. Technological developments shape the way e-commerce has been transforming in recent years. Due to new technologies e-retailers are able to fulfill customer needs. Applications for instant payments, price comparisons, access to product information, personalization of advertisement, chat-bots, and cloud robotics are just some of the areas where retailers are utilizing technology to improve customer performance. On the other hand, the technology-loaded features of on-line commerce could be perceived as too complex for some e-shoppers. As a results this could diminish the value e-retailers try to provide for them. Therefore there is a need to investigate the attitude of e-shoppers towards new technology. Based on this assumption, the Technology-based Value Index (TVI) is proposed in the paper. The index was constructed based on the results of empirical research on e-shopper habits and preferences. The TVI measures the scale of values perceived by the e-shoppers as a result of technology-based features of online shopping services. The index was applied to distinguish three profiles of e-shoppers: the Reluctants, Neutrals, and Enthusiasts, and outline their characteristics.

Keywords: e-Commerce · Value for customer · Technology-based Value

1 Introduction

Introduced in 1954 by P. Drucker, the term of value can be understood in various ways. Broadly speaking, value is defined as an evaluation of product utility understood as a ratio between what has been received and what has been given—value represents a compromise between what can be obtained and what should be given. The notion of value can also refer to customers; in that case it is described as a value "for the customer" [15]. Customers always evaluate a buying effort in terms of investment (e.g., physical labor, financial or opportunity cost, time spent, psychological annoyance) against actual and hedonic benefits achieved from that exchange. In other words, customers while evaluating a purchase, generally attempt to identify the value of the purchase in terms of the effort they put in and the reward they may achieve [21].

For decades, various new technology developments were being incorporated into retail settings in order to provide a greater value for the ultimate customers [25]. Especially within the e-commerce sector, a new technology generally named the Internet allowed to overcome limitations of the brick-and-mortar retailing. In the traditional

N. T. Nguyen et al. (Eds.): ACIIDS 2020, LNAI 12034, pp. 428–436, 2020.
https://doi.org/10.1007/978-3-030-42058-1_36

commerce, transactions must be performed at a specific time and space in order to meet the seller's and buyer's needs [16]. However, e-commerce allows to eliminate those time and space related constraints and restrictions: both the seller and the buyer can start and finish the transaction anytime and anywhere they need [14]. This can be recognized as an example of the value that technology provides to customers.

In this paper, Technology-based Value Index is proposed. The index was constructed on the results of empirical research on e-shopper habits and preferences, and serves to define a measure of the value (benefits over efforts) perceived by the e-shoppers as a result of technology-loaded features of online shopping services. Based on different levels of this value, three profiles of e-shoppers have been distinguished and characterized.

The reminder of this paper is organized as follows. Section 2 provides literature-based theoretical considerations on the technology-resulting value as it is perceived by e-shoppers. Section 3 describes research methodology. In Sect. 4, the proposed Technology-based Value index is presented. Section 5 presents characteristics of e-shopper profiles basing on the proposed index. Finally, Sect. 6 concludes the paper and outlines the directions of possible future research.

2 Technology-Resulting e-Shopper's Value: Theoretical Considerations

The adaptation of a new technology allows online shopping services to provide a number of benefits that can increase the value for e-shoppers. This include: time saving [9], lesser physical [23], psychological [6], or cognitive [2] effort e-shoppers experience and a broad selection of payment methods [4]. What is more, online shopping services due to lower costs can offer better prices than off-line retailers [27]. Internet allows customers to easily compare prices from different retailers [19]. Internet also offers a great variety of products, brands, and shops, and therefore e-shoppers may find niche products more easily [30]. One the other hand, online settings have their drawbacks that could also diminish a value for customers. This could include: the risk of online payment [28] and the risk of sharing personal information [18]; customers may also experience various problems with product shipment [25].

The technology is only expected to deliver value if businesses are able to prioritize actual customer needs, such as more efficient and enjoyable shopping experiences that reduce decision-making uncertainty [8]. For instance, technology has driven the development of applications for instant payments, price comparisons, access to product information, customer opinions and recommendations in social media, recommender systems as well as personalization of advertisement [1]. On the other hand, retailers view technology as a way to decrease costs, especially at the back end; i.e., distribution and supply chain activities, and thus being able to maintain low retail price levels. Innovations in the area of storage-sharing as well as delivery-sharing, chat-bots, and cloud robotics are just some of the areas where retailers are utilizing technology to improve operational and customer performance [17]. Thus, technology has a far-reaching effect on e-retailers, their suppliers, competitors and most importantly, the customers [11]. Furthermore, when a new technology emerge, e-retailers are expected to educate their customers about the benefits of that technology and teach them how to use it [20].

Taking into account that technology seems to be fundamental for e-commerce, surprisingly little research have been carrying on to determine the attitude of e-shoppers towards the technology. In other words, it is very likely that technology can be for e-shoppers sort of motivators for shopping online, for other however technology could be the main inhibitor. Based on that, it seems crucial to pose a question whether technology adopted by e-commerce always contributes to greater value for end-customers. In other words, an important issue is to determine if customers are increasingly interested in and very keen to adopt and use technology for facilitating their routine shopping activities [24]. It is suggested that customers evaluate the value of any purchase based on all the associated complexities and uncertainties. Therefore it can be assumed that technology-based e-commerce transactions can pose some sort of complexities and uncertainties (sometimes risks), at least for some e-shoppers [28].

There is a rich theoretical background for this assumption. This includes for example the theory of situated cognition [22]. Situated cognition implies that customer's experiences seem to be most realistic and the value greater when they integrate information about products in real-time within the immediate decision context (i.e., are embedded), allow for physical interaction with a product or service (i.e., are embodied), and provide opportunities for communication with other customers (i.e., are extended) [12].

Based on this theory, it can be concluded that for customers, a sense of authenticity and realism arises when they can naturally interact with products and services and as a consequence make purchase decisions. Customers quite often find it difficult to imagine how products fit them personally or fit with their environment. That is why physical interaction such as touching, rotating or moving around a product may evoke affective reactions and improve ability to evaluate an offer [3]. The theory of situated cognition allows for the assumption that for some e-shoppers a lack of physical interaction with products can be perceived as the main driver diminishing a value; for instance, by elevating a risk of buying unwanted product [7] or creating distrust toward online vendors and toward the Internet as a shopping channel [10].

3 Research Methodology

The research used in this paper was carried out as a part of the research project aiming to develop a model of value network creation in e-commerce settings. To achieve this goal, the three-stage research was conducted. This includes an in-depth literature review, empirical research, and multi-level modeling. The empirical research was divided into two parts: qualitative and quantitative. This paper is based on the results of the second part. The operational aim of the research was to identify drivers and inhibitors of value for e-shoppers; i.e., customers in e-commerce. It was assumed that the value for customers is created by the value network that is represented by e-shoppers, e-retailers, suppliers and complementors. Nevertheless, the central point of the value network is the customer (e-shopper) who ultimately evaluates the value and converts it into a monetary equivalent for the other network members [13].

The research utilized in this paper was conducted between November 2017 and May 2018. The 200 (N = 200) correctly completed questionnaires were received. All questionnaires were qualified for further analysis, which, assuming the size of the fraction of 50% and the confidence level of 95%, gives an acceptable measurement error of 7%. CATI (computer-assisted telephone interview) was selected as a technique of data collection. The respondents were asked to indicate the extent to which they agreed with a given description. The quality of the results were verified using validity and reliability measures (all Cronbach's alpha coefficients of constructs were higher than 0.75).

4 Technology-Based Value Index

The Technology-based Value Index (TVI) measures the value (the scale of benefits over efforts) perceived by an e-shopper as a result of technology-loaded features of online shopping services. The respondents were asked to evaluate the value they receive if online shopping services offer the following groups of features:

- Ability to design products themselves;
- Modern forms of payment; e.g., PayPal, instant online money transfer, instant mobile payments;
- Personalized and active forms of online advertising and promotion; e.g., blogs with product descriptions and usage tips, social media profiles, websites with customer's feedback and opinions;
- Advanced overall online experience; e.g., web and mobile versions of a shopping service, responsive design across various devices;
- Modern forms of shipment; e.g., self-service terminals, online shipment tracking.

For each feature, the respondents could indicate the perceived value in an ordinal scale ranging from 1 to 5 where one means no value at all, three—moderate value, and five—the great range of values.

Next the Technology-based Value Index was calculated as a mean of all indications. The minimum value for TVI is therefore 1 which indicates e-shoppers perceive no value from technology-loaded features of online shopping services. The maximum value for the index is 5 which denotes the opposite situation: e-shoppers perceive the great range of values from those features.

The frequency distribution of the TVI index values in the surveyed population is close to the normal distribution (Fig. 1); however it demonstrates some negative skewness which means that more respondents received higher TVI ranks. The mean value is 3.383 with quite narrow confidence interval calculated with the robust bootstrapping technique: BCa 95% CI [3.296, 3.472]; standard error of mean: 0.044; standard deviation: 0.606. The very small value of the standard error of mean indicate that the sample is an accurate reflection of the population.

The total range of TVI index values was divided into three interval, defining three e-shopper profiles:

- Reluctant: TVI $\in$ [1.00; 2.78]
- Neutral: TVI $\in$ [2.79; 3.99]
- Enthusiast: TVI $\in$ [4.00; 5.00]

The limit values for particular profiles were determined at the distance of ± the standard deviation to the mean value.

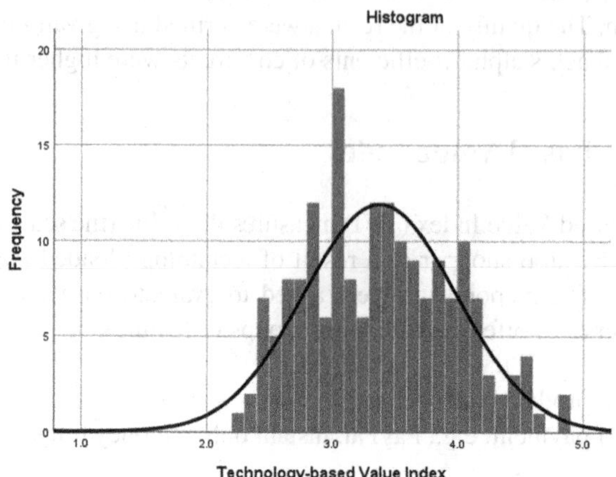

Fig. 1. Distribution of the TVI index value in the surveyed population

The percentage frequencies of the respective TVI profiles in the surveyed population is presented in Fig. 2.

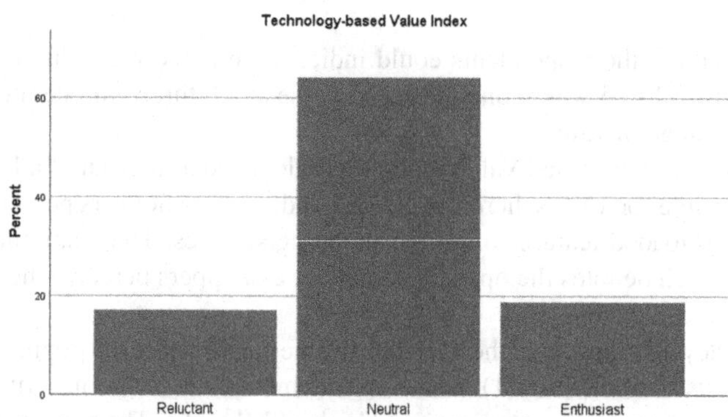

Fig. 2. The sizes of the e-shopper profiles determined according to the TVI index distribution

The Reluctant e-shoppers hardly utilize technology-loaded features of online shopping services and therefore they perceive them as bringing close to nothing value to them.

The Enthusiastic e-shoppers are on the quite opposite side—they are entirely fascinated with technology and while selecting online shopping services they prefer those with the great number of technology-loaded features.

The Neutral e-shoppers demonstrate the balanced attitude. They believe that technology-loaded features of online shopping services in general bring some value to them but this is not a significant argument when deciding which service to select.

5 Technology-Based Value Index: e-Shopper's Profiles

All three profiles of e-shoppers where analyzed against three socio-economical dimensions, namely: age, monthly number of online purchases and net monthly income. When it comes to the age dimension, the Reluctants seem to be the oldest. The group of the oldest technology reluctant e-shoppers (over 50 years) is accounted for 19.4% which is the highest share among all three profiles. The youngest e-shoppers (up to 30 years) and those between 30 and 50 years together constitute 80% of the whole group (22.6% and 58.1% respectively)—the lowest share among all three profiles. On the other hand, the Neutrals and Enthusiast have very similar share of the youngest and middle-aged people (97.4% for the Neutrals and 94.1% for the Enthusiasts). However, within the Enthusiasts, the percentage of the youngest e-shoppers is the highest among all three profiles and accounts to 44.1%. This shows that although there is some relations between age and perceived value resulting from technology-loaded features of online shopping services but the relation is not that strong as it is commonly expected. The age distribution among three profiles of e-shoppers is showed in Fig. 3.

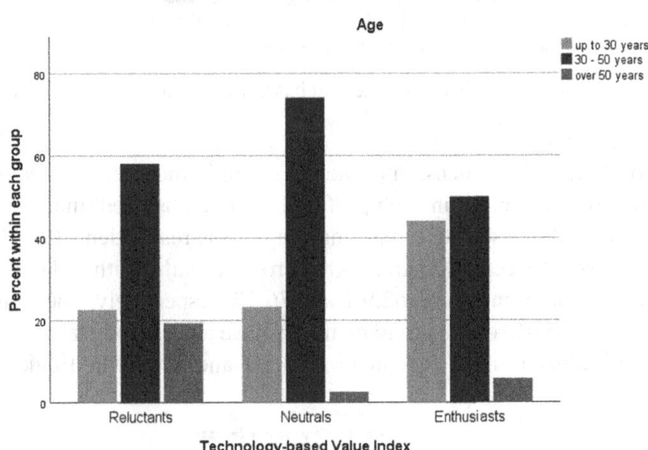

Fig. 3. The age distribution within the e-shopper's profiles

The profiles of e-shoppers are less divers in terms of monthly number of on-line purchases. In all three groups there is a clear domination of purchases with a frequency of 1 to 5 per month (around 80% in each group). Although it could be concluded that the higher the TVI index value the higher the frequency of monthly shopping. Within the Enthusiasts (the highest TVI index level), those who make purchase the most often account to 8.8% of the entire group and those who buy the least often account to 76.5%. Whereas within the Reluctants (the lowest level of TVI index), these values are 6.5% and 87.1% respectively. The share of those who buy with a moderate frequency are identical in the Enthusiasts and the Neutrals (14.7% in both groups), and within the Reluctants equals 6.5%. It seems not to be surprising that general increase of monthly number of on-line purchases aligns with increase of TVI index level. These who perceived technology as the value driver naturally tend to do shopping in the technology-loaded e-commerce environment. On the other hand, those for whom technology brings low or close to nothing value relatively seldom perform on-line shopping. The monthly number of on-line purchases across three profiles of e-shoppers is depicted in Fig. 4.

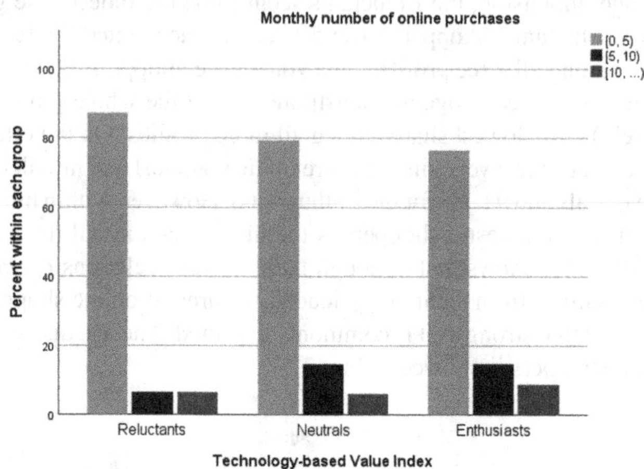

Fig. 4. The monthly number of online purchases within the e-shopper's profiles

Taking into account the dimension of the net monthly income, the income between 2000 and 3000 euros dominates in each profile. It can be observed, that the Reluctants seem to be the most money sensitive. Vast majority of the respondents (90.3%) that fall into this profile earn between 1000 and 3000 euros a month. Within the Neutrals and Enthusiasts this amount of income is 62.9% and 76.5% respectively. The Neutrals seem to be the richest one: 35.4% of the group members have a net income above 3000 euros, while within the Enthusiasts this accounts for 23.5% and within the Reluctants: 9.7%.

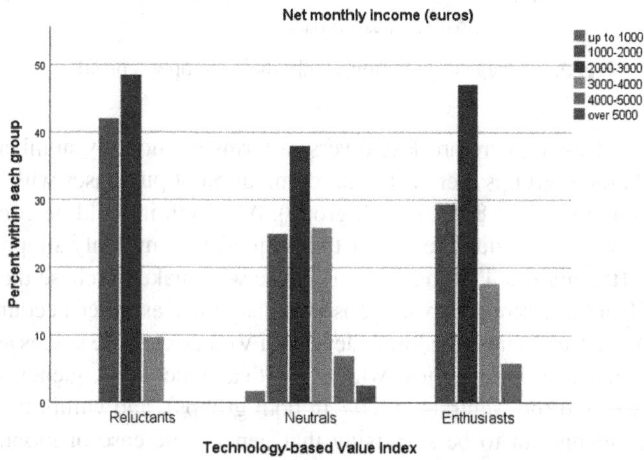

Fig. 5. The net monthly income within the e-shopper's profiles

It also has to be highlighted that the Neutrals are the only ones that compose of members earning more than 5000 euros per month. On the other hand, within the Reluctants, there are no members with net monthly income higher than 4000 euros. The distribution of the net monthly income across all three profiles is depicted in Fig. 5.

6 Conclusions and Future Work

In this paper, Technology-based Value Index has been proposed. The index serves as a measure of the customer's value resulting from technology-loaded features of online shopping services. The distribution of the index was utilized to identified three disparate profiles of e-shoppers: Reluctants, Neutrals, and Enthusiasts. The profiles have been studied and analyzed against socio-economical dimensions, and the results obtained were used to develop their characteristics.

Future work includes to research the profiles against more complex dimensions and develop their further characteristics. The examples of those dimensions could be various aspect of customer's service like product return, shipment, packaging, but also brand and price sensitivity.

Acknowledgment. The work presented in this paper has been supported by the National Science Centre, Poland [Narodowe Centrum Nauki], grant No. DEC-2015/19/B/HS4/02287.

References

1. Borusiak, B., Pierański, B., Romanowski, R., Strykowski, S.: Personalization of advertisement as a marketing innovation. Slovak Sci. J. Manage. Sci. Educ. **3**(1), 18–20 (2014)
2. Bosnjak, M., Galesic, M., Tuten, T.: Personality determinants of online shopping: explaining online purchase intentions using a hierarchical approach. J. Bus. Res. **60**(6), 597–605 (2007)
3. Brasel, S.A., Gips, J.: Tablets, touchscreens, and touchpads: how varying touch interfaces trigger psychological ownership and endowment. J. Consum. Psychol. **24**(2), 226–233 (2014)
4. Brown, M., Pope, N., Voges, K.: Buying or browsing? An exploration of shopping orientations and online purchase intention. Eur. J. Mark. **37**(11–12), 1666–1684 (2003)
5. Chaparro-Peláez, J., Agudo-Peregrina, Á.F., Pascual-Miguel, F.J.: Conjoint analysis of drivers and inhibitors of e-commerce adoption. J. Bus. Res. **69**(4), 1277–1282 (2016)
6. Childers, T.L., Carr, C.L., Peck, J., Carson, S.: Hedonic and utilitarian motivations for online retail shopping behavior. J. Retail. **77**(4), 511–535 (2001)
7. Choi, J., Geistfeld, L.V.: A cross-cultural investigation of consumer e-shopping adoption. J. Econ. Psychol. **25**(6), 821–838 (2004)
8. Dacko, S.G.: Enabling smart retail settings via mobile augmented reality shopping apps. Technol. Forecast. Soc. Chang. **124**, 243–256 (2016)
9. Ganesh, J., Reynolds, K.E., Luckett, M., Pomirleanu, N.: Online shopper motivations and e-store attributes: an examination of online patronage behavior and shopper typologies. J. Retail. **86**(1), 106–115 (2010)
10. Gefen, D.: E-commerce: the role of familiarity and trust. Omega **28**(6), 725–737 (2000)
11. Hagberg, J., Sundstrom, M., Egels-Zandén, N.: The digitalization of retailing: an exploratory framework. Int. J. Retail Distrib. Manage. **44**(7), 694–712 (2016)

12. Hilken, T., Heller, J., Chylinski, M., Keeling, D., Mahr, D., de Ruyter, K.: Making omnichannel an augmented reality: the current and future state of the art. J. Res. Interact. Mark. **12**(4), 509–523 (2018)
13. Kawa, A., Świątowiec-Szczepańska, J.: Value network creation and value appropriation in ecommerce. Przedsiębiorczość i Zarządzanie **19**(6), 9–21 (2018)
14. Khaneghah, E.M., Shadnoush, N., Salem, A.: Artemis time: a mathematical model to calculate maximum acceptable waiting time in B2C e-commerce. Cogent Bus. Manage. **4**(1), 1–30 (2017)
15. Kotler, P.: Marketing Management: Analysis Planning Implementation and Control. Prentice Hall, Upper Saddle River (1994)
16. Lin, X., Li, Y., Wang, X.: Social commerce research: definition, research themes and the trends. Int. J. Inf. Manage. **37**(3), 190–201 (2017)
17. McKenzie, B., Burt, S., Dukeov, I.: Introduction to the special issue: technology in retailing. Baltic J. Manage. **13**(2), 146–151 (2018)
18. McKnight, D.H., Choudhury, V., Kacmar, C.: The impact of initial consumer trust on intentions to transact with a web site: a trust building model. J. Strateg. Inf. Syst. **11**(3), 297–323 (2002)
19. Noble, S.M., Griffith, D.A., Adjei, M.T.: Drivers of local merchant loyalty: understanding the influence of gender and shopping motives. J. Retail. **82**(3), 177–188 (2006)
20. Obal, M., Lancioni, R.A.: Maximizing buyer–supplier relationships in the Digital Era: concept and research agenda. Ind. Mark. Manage. **42**(6), 851–854 (2013)
21. Schloegel, U., Stegmann, S., Maedche, A., van Dick, R.: Age stereotypes in agile software development – an empirical study of performance expectations. Inf. Technol. People **31**(1), 41–62 (2018)
22. Semin, G.R., Smith, E.R.: Socially situated cognition in perspective. Soc. Cogn. **31**(2), 125–146 (2013)
23. Shamdasani, P.N., Yeow, O.G.: An exploratory study of in-home shoppers in a concentrated retail market: the case of Singapore. J. Retail. Consum. Serv. **2**(1), 15–23 (1995)
24. Shareef, M.A., Dwivedi, Y.K., Kumar, V., Davies, G., Rana, N., Baabdullah, A.: Purchase intention in an electronic commerce environment. Inf. Technol. People (2018). https://doi.org/10.1108/itp-05-2018-0241
25. Siregar, Y., Kent, A.: Consumer experience of interactive technology in fashion stores. Int. J. Retail Distrib. Manage. **47**(12), 1318–1335 (2019). https://doi.org/10.1108/IJRDM-09-2018-0189
26. Swinyard, W.R., Smith, S.M.: Why people (don't) shop online: a lifestyle study of the Internet consumer. Psychol. Mark. **20**(7), 567–597 (2003)
27. To, P.L., Liao, C., Lin, T.H.: Shopping motivations on internet: a study based on utilitarian and hedonic value. Technovation **27**(12), 774–787 (2007)
28. Tzavlopoulos, I., Gotzamani, K., Andronikidis, A., Vassiliadis, C.: Determining the impact of e-commerce quality on customers' perceived risk, satisfaction, value and loyalty. Int. J. Qual. Serv. Sci. **11**(4), 576–587 (2019)
29. Vijayasarathy, L.R.: Predicting consumer intentions to use on-line shopping: the case for an augmented technology acceptance model. Inf. Manage. **41**(6), 747–762 (2004)
30. Wolfinbarger, M., Gilly, M.: Consumer motivations for online shopping. In: AMCIS 2000 Proceedings (2000). http://aisel.aisnet.org/amcis2000/112

Visualized Benefit Segmentation Using Supervised Self-organizing Maps: Support Tools for Persona Design and Market Analysis

Fumiaki Saitoh(✉)

Chiba Institute of Technology, Narashino, Chiba, Japan
fumiaki.saitoh@chibakoudai.jp

Abstract. This study provides a visualization technique for market segmentation by benefits using supervised self-organizing maps (SOMs). Recent use of customer personas has attracted attention in the marketing field, and is important in the development of products and services that customers want. Market segmentation tools based on clustering methods, such as SOM, k-means and neural networks, have been widely used in recent years. By associating customer benefits and demographics to classes and attributes, respectively, supervised SOM can be a useful tool to support persona design. Market segmentation is important in order to utilize personas effectively for decision-making in businesses that have accumulated large amounts of customer data. We use a real case of customer data from a hotel in Tokyo, Japan to illustrate our approach for market segmentation and customer analysis.

Keywords: Benefit segmentation · Supervised self-organizing map · Market segmentation · Persona · Customer evaluation · Data visualization

1 Introduction

The purpose of this study is to extract customer benefit and demographic relationships from customer data using a visual-based support tool for creating personas and understanding customers. Persona development is an important tool in the field of product development and marketing. Persona refers to a model of customer characteristics that is identified as a fictitious person and used to target market products and services. It is also used to build empathy between customers and product and service developers, particularly as businesses find it increasingly difficult to differentiate products and services. It is important for businesses to develop products and services that understand customer needs from the customer's point of view. Companies can benefit by increasing their use of accurate personas.

N. T. Nguyen et al. (Eds.): ACIIDS 2020, LNAI 12034, pp. 437–450, 2020.
https://doi.org/10.1007/978-3-030-42058-1_37

With the rapid development of information and communication technology (ICT) in recent years, it has become easier to collect and accumulate customer data. This data can be used to determine typical customer personas objectively. Customer data can be acquired through various routes, such as purchase history in an electronic commerce (EC) site, customer reviews, point-of-sale (POS) data, questionnaires, and so on. Data about customer benefits (such as satisfaction and purpose of use) as well as demographics (such as gender, age, educational background, and occupation) are often available, especially in review sites.

If there is sufficient correlation between demographics and customer needs (or benefits) in the market structure, segmentation information can be obtained by statistical analysis of benefits and demographics. However, with diversification of customer preferences, insufficient data has resulted in inaccurate segmentation based only on demographic information. In persona design, segmentation based solely on demographics is not sufficient for building appropriate market segments. It is important to consider customer benefits in the fields of modern marketing and product development. Benefit segmentation is becoming one of the essential approaches in marketing. Benefits, such as customer purpose, use, and satisfaction with products and services, are important data for determining personas.

Technical data mining tools are becoming more available. In particular, learning models, represented by k-means clustering, self-organizing maps (SOM), artificial neural networks (ANN), and support vector machines (SVM), have been utilized widely in this field, and their effectiveness has been shown in many studies. Since learning models are very powerful for analyzing complex data, highly accurate segmentation is possible based on real customer data, by treating benefit information as teacher signals in supervised learning. However, despite the high identification accuracy of these models, it is difficult to understand the internal structure of these models because many supervised learning models, such as ANN and SVM, are black box models. When using the learning model in marketing, it is important to understanding the learning results.

In this study, we utilize supervised SOM as an effective model to determine correspondences between attributes and classes. By associating customer benefits with class maps from supervised SOM, we can visualize correspondences between demographics and benefits. Marketers, who use this model, are supported in persona design by the results of visualized benefit segmentation. By visualizing the correspondences between demographics and benefits, marketers can more easily understand segmentation personas.

Customer review data published by Rakuten, Inc., a well-known Internet service company in Japan, was used for our analysis. We specifically used data from customer reviews of a major hotel in Tokyo. Our approach was confirmed using this data, which contained demographic and benefit information suitable for knowledge extraction about hotel customers.

2 Background

2.1 Persona Marketing

The concept of persona was initially proposed as an approach for the design and development of software based on user-centered design and a defined virtual user. The persona concept then expanded for use in the marketing field as a customer-centered approach in the design of products and services. In recent years, personas have become an essential idea used in a wide variety of business. The objective of persona use is to develop products and services that are suitable for typical customers, which are represented by virtual personas.

Persona concepts have been applied in various subject areas, and technical methods of persona construction have been widely studied in recent years. The expected outcomes for product developers utilizing persona are as follows: (1) Product designers are able to make decisions based on customer viewpoints. (2) Areas are clarified where developers need to empathize with customers. (3) It is easier to predict the outcome of marketing initiatives and the trends of ideal customer consumption. (4) Information about typical customers is likely to be shared between various parties interested in the product or service development.

The following problems may also arise when utilizing persona for product or service design: (1) Significant time and cost may be associated with the creation of personas. (2) When customer data or understanding is inadequate, designers may create inaccurate personas that do not fit real customers. Resolution of these problems can be expected through quick and objective analysis of adequate large amounts of customer data. Because our analytical approach is based on data mining, we were able to reduce the cost and time required to create accurate and realistic personas.

2.2 Benefit Segmentation

Market segmentation is the identification of subsets of the market, and it is very important in the marketing field. Market targeting is the selection of a market segment that includes current and potential customers. The efficacy of market targeting depends on the quality of market segmentation. In general, market segmentation is based on demographic factors, such as age, gender, earnings, and occupation.

However, market segmentation based only on demographics has been shown to be insufficient to accurately understand customer needs, especially as market needs have become more diversified and complex. Benefit segmentation has been attracting recent attention to address this problem. Benefit segmentation is the concept of market segmentation that focuses on benefits to the customer, such as the utility or purpose of use of products and services [15]. In benefit segmentation, benefits are analyzed in addition to demographic factors.

The benefit to customers is the essential reason for targeting customers in benefit segmentation. Consider the following examples of benefits to restaurant customers: to satisfy appetite quickly and inexpensively, to spend time with friends and family, and to dine in luxury and style. It is clear that customers with diverse demographics (such as age) might share the same benefit, such as wanting to satisfy their appetites quickly and inexpensively. It would therefore be difficult to accurately define the market segments for restaurant customers based only on demographic factors without also looking at customer benefits.

Because customer review data is widely distributed on the Internet, we can easily obtain customer benefit information, such as utility and purpose of use for products and services. Improvements in benefit segmentation can be expected with the effective use of this information. Our goal is the enhancement of persona design support tools with more effective analysis of customer demographics and consideration of customer benefits.

3 Market Segmentation Based on Self-organizing Map

Recently, SOM has been widely used in business applications [10,13] for such things as data mining, optimization and pattern recognition, since it is a robust model with multicollinearity and nonlinearity. SOM can therefore be a powerful tool for visualizing customer data. Clustering is one of the most effective approaches for market segmentation [9]. In general, this statistical approach is typified by the use of the multinomial logit model and hierarchical clustering. In recent years, market segmentation research has actively focused on using mining tools represented by the learning model and neural networks [1]. Prototype-based clustering, such as SOM, k-means, and other mining tools, are widely used for market segmentation [4].

There is widespread use of SOM in various applications that analyze market targets. Models with modified or extended SOM have been widely utilized in this field. For example, hierarchical SOM has been applied to the segmentation of on-demand multi-media markets in Taiwan [5]. Effectiveness of segmentation using SOM and GA has been confirmed in market segmentation of the wireless communication industry market [8]. Hanafizadeh and Mirzazadeh integrated the Fuzzy Delphi method with SOM and produced ADSL service market maps for a communication company in Iran [3]. Kiang et al. extended the SOM model, and applied it to the segmentation of customer data at AT&T [6]. It has been demonstrated that SOM has superiority over typical clustering methods in mature markets. Additionally, k-means clustering, which is similar to SOM, is also widely used in market segmentation.

Research was reviewed that was closely related to the purpose and concepts in our present study. Vosbergen et al. [14] studied a similar persona design method by applying the prototype vector of k-means. Because clustering by SOM and k-means involves data approximation using a set of prototype vectors, it can extract the characteristics of typical customers by focusing on prototype vectors. We focused on previous research related to benefit segmentation. Although the tools utilized in the former study are similar to our tool, that study is significantly different from our research, with respect to benefit segmentation and the subject application. Although the latter study is similar with respect to attention to benefit segmentation, it is significantly different from our research, with respect to persona development, analytical tools and subject application.

SOM is a data mining tool that has been used for a long time and is one of the learning models that has been actively applied to market data analysis in recent years. In addition, it is not limited to market segmentation alone, and advanced development methods have been actively proposed in recent years for its use in the analysis of customer data. Especially, advanced research in recent years have led to improving or expanding of the structure of SOM and its application to large-scale and complex business data. For example, a multilayered SOM with a tree structure has been proposed as a book recommendation tool for the organization of electronic books (e-books) and their corresponding authors [16]. In addition, SOM is used as a component of an inference model that is employed in movie recommendation using the text metadata constructed by movie synopsis, actor list, and user comments.

4 Benefit Segmentation Using a Supervised SOM

Generally, in unsupervised clustering-based segmentation represented by SOM and k-means, only demographic similarity is used as a condition for segment construction. However, customers who belong to the same demographic segment tend to have preferences similar to customers belonging to different segments. Therefore, in order to extract the characteristics of customers with similar demographics, generic clustering tools have been effectively used. However, with recent increased diversification of customer preferences and lifestyles, it is difficult to interpret segments using the conventional approach. Persona designs based on prototype vectors of clusters have a tendency to become ambiguous. It is also difficult to interpret learning results because black box models are typically used (such as ANN and SVM) for segmentation.

We determined that supervised SOM is the best model to use to avoid these problems. This study adopts the supervised SOM as a support tool for persona building and benefit segmentation. Because correspondences between benefits and demographics can be visualized, it is easy to understand the learning results of the model.

4.1 Overview of Supervised SOM

The purpose of this study is to develop a persona design support tool using market segment visualization based on correspondences between customer benefits and demographics. The XY-fused map (or XYF) is one of the supervised self-organizing maps proposed by Melssen et al. [12] for visualization of market data and benefit segmentation. The XY-fused map is a model that extends the SOM in supervised learning and has been utilized in recent years as a data mining tool. Because of its relatively high accuracy similar to other supervised clustering models and its availability as a data visualization tool, we have adopted this model. In this section, we will outline the XY-fused map.

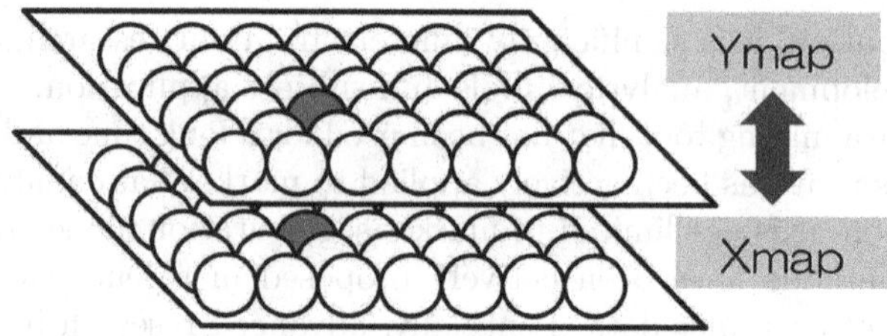

Fig. 1. The schematic diagram of supervised SOM (XY-fused map)

The XY-fused map consists of two maps, the $Xmap$ and $Ymap$. The $Xmap$ is used for learning attributes (explanatory variables), and the $Ymap$ is used for learning classes (objective variables). The schematic diagram of the XYF map is shown in Fig. 1. The $Xmap$ and $Ymap$ must have same structure, such as same map size and same topology. The $Xmap_k$ is the k-th node on the $Xmap$, and it has a weight vector with the same dimension as the attributes. The $Ymap_k$ is k-th node on the $Ymap$, and it has a weight vector with the same dimension as the classes. The i-th data (x_i, y_i) can be represented as a pair, an attribute x_i and class y_i, and the winner node for this data is selected as the same coordinates of the point index on the $Xmap$ and $Ymap$.

The distance between i-th data and the k-th node in this model is defined by Eq. (1), and $Xmap$ and $Ymap$ are associated by this equation. By using the update equations, this model can be learned while maintaining consistency between the two maps.

$$S_{Fused}(i, k) = \alpha(t)S(X, Xmap_k) + (1 - \alpha(t))S(Y_i, Ymap_k)) \tag{1}$$

In this equation, $\alpha(t)(0 < \alpha(t) < 1)$ is a parameter to calibrate weight vectors of $Xmap$ and $Ymap$, and it decreases with an increase in the number of updates t; $S(X_i, Xmap)$ and $S(Y_i, Ymap)$ represent degrees of similarity between i-th data and k-th node. With this update equation, weights between the $Xmap$ and the $Ymap$ are corrected in order to approach the supervised signal.

The parameter $\alpha(t)$ is expressed as a function that decays with increasing learning step number t. The winner node can be calculated by the following equation:

$$c = \arg\min_{k \in K} \{S_{Fused}(i, k)\} \tag{2}$$

where c represents the index corresponding to the winning node for each data, and K shows set of nodes. In this model, the update rules and equations are almost same as normal SOM except for the structure and definition of the distance.

4.2 Support of Persona Design Based on Benefit Segmentation Using Supervised SOM

SOM and k-means are clustering methods that approximate the data distribution by a set of prototype vectors. In this study, analysis was done on customer data composed of customer demographics and benefits. Accordingly, the elements of the prototype vector correspond to customer attribute information and are considered to represent typical customer characteristics. Furthermore, as described in Sect. 3.1, k-means clustering is effective for persona design. Based on the above, a prototype vector-based clustering is believed to be effective for designing personas based on large amounts of customer data.

The goal of this study is the development of a visual-based support tool that shows correspondences between customer demographics and benefits and representative customer personas. We have focused on the use of supervised SOM because it meets the requirements for an effective segmentation tool. In this study, we use XY-focused maps, as discussed earlier, to learn the benefits and demographics of classes and attributes, respectively. The learning results of the *Ymap* correspond to the market segment of benefits, and the learning results of the *Xmap* correspond to the market segment of demographics. By examining nodes with the same *Xmap* and *Ymap* coordinates, relationships can be determined between benefits (e.g., purpose of use, satisfaction) and demographics (e.g., age, gender, accompanying person).

Even though customers may have similar demographics, if benefits are different, customer personas may be significantly different. An example can be seen looking at customer segmentation from the hotel data used in this study. Men in their twenties share an age demographic but segmentation is different when benefits of hotel use are considered. Men in their twenties whose hotel use is for the benefit, or purpose, of leisure have different lifestyles and personas than men in their twenties whose hotel use is for the purpose of business. The former might be students; the latter might be businessmen. The products and services required for these two groups may differ significantly. To determine a more sophisticated persona from the data, benefits must be considered.

The procedural steps used in this study are as follows:

Step 1. XYF learns the data demographics and benefits as attributes and classes.

Step 2. Using the following equation, the variables for each prototype are converted to specified value ranges:

$$W_{jk}^{new} = \frac{W_{jk}^{old} - \min(W_j)}{\max(W_j) - \min(W_j)} \tag{3}$$

where W_{jk} is the weight component of the j-th attribute in the k-th node, and W_j is the set of weight components of the j-th attribute for all the nodes in each map. By fixing a range of values using the transformation, segment determination can be simplified, with color intensity corresponding to strength of the association.

Step 3. The learning results of maps are displayed after Step.2 results. Segments are visualized with a map of association dynamics for each attribute.

Step 4. Features of the desired persona are extracted from the analysis results, and the user can see this result reflected in persona construction.

In the procedure, W^{new} represents the magnitude of reaction for each of the attributes in the set of prototype vectors of the SOM and each of the weight variables are converted to greater than 0 but not less than 1. Through this process, it is possible to display a place where each dummy variable reacts strongly on the map. By fixing the viewpoint to a certain point on the map and by focusing on the variables with strong reactions in that area, it becomes possible for the analyst to determine the persona's image. Through this processing, it becomes easier for the analyst to grasp the displayed segments and their characteristics.

The main contribution in this research is to construct an effective persona building support tool based on data analysis through visualization. By applying the above procedure to market data, personas that reflect the structure of the market segment that is associated with benefits and demographics are constructed. Furthermore, it becomes possible to identify the market segmentation based on the benefits and the corresponding person's image through visualization of the correspondence between the benefit map (Ymap) and the demographics map (Xmap). This procedure provides strong decision-making support in one of the important processes in modern marketing based on market data.

5 Experiments

5.1 Experimental Settings

In our experiments, we confirm the effectiveness of our model with our analysis of real market data. The data used in our analysis is customer review information for a leading hotel in Tokyo, Japan, obtained from Rakuten Inc., a large Japanese Internet company. We were able to obtain a sufficient amount of data in the form of 1,810 samples with no missing values (unanswered items). Our data included customer demographics, type of hotel accommodations, and customer benefits derived from reviewer comments.

Our data specifically included the following demographic information, with accommodation style, used as explanatory (or dummy) variables:

(1) Age group of customer reviewers (i.e., 20s, 30s, 40s, 50s, 60s, and 70s),
(2) Gender of reviewers (male or female),
(3) Presence or absence of a meal
(4) Type of accompanying person(s) (i.e., alone, with family, with lover, with friends, and with colleagues).

The benefit data, used as objective variables, are as follows: (1) Purpose of use (of hotel accommodation) (i.e., for work, leisure, or other), (2) Customer satisfaction rating (i.e., from 1 to 5, where the highest rating is (5). The data attributes are comprised of demographic variables, including accommodation style, and classes are comprised of benefit variables (purpose and satisfaction). The parameters settings used in our study are shown in Table 1.

In order to evaluate the validity of the analysis method, we quantitatively evaluate the time required for learning and the prediction accuracy of the models.

Here, in each evaluation, experiments were conducted for each of the prediction targets that comprise the purpose of use and the degree of satisfaction. Furthermore, in order to investigate the relationship between the number of nodes and the evaluation values, the map size was verified and compared in 20×20, 25×25 and 30×30. Since learning results are affected by initial values etc., we performed 100 enforcements for each setting and calculated the average value and variance.

We calculate the prediction accuracy to evaluate the reliability of the model. Experiments that use purpose as a class are classified into three classes, and classes are made up of "business", "leisure", and "others"; the satisfactory class is constructed with an evaluation value of 4 or more. In either of experimental settings, the average values of the prediction accuracy, evaluated through 5-fold cross validation, are used as the evaluation values.

Table 1. Parameters settings

Parameters used for supervised SOM	Values
Number of iterations	1,000
Initial value of learning rate	0.1
Number of nodes	625 (25 * 25)
Initial value of $\alpha(t)$	0.5
Topological structure of maps	Rectangle
Update lure	Online algorithm

5.2 Experimental Results

Learning Results. The experimental results are shown in Fig. 2. Figure 3 show learning results of class information corresponding to customer purpose of use. In these figures, segments are colored to show state attribute variables close to 1. Since all of the variables are dummy variables, the segments that do not fit the conditions of the variable are displayed in white.

Figure 2 show visualized customer segments based on attribute variables. Here, twelve sub-figures, (a) through (l), correspond to each of the following attributes: gender, 20s age group, 30s age group, 40s age group, 50s age group, 60s age group, meal service, alone, with family, with lover, with friends, and with colleagues, respectively. Since the percentage of customer reviews from the 70s age group was small (about 0.3%), maps of these segments were deleted. For the gender variable, 1 represents male, and 0 represents female. Figure 3 show visualized customer segments based on customer benefit classes. Figure 3 sub-figures, (a) through (c), correspond to each of the following purpose classes: business, leisure, and other, respectively.

The results of the numerical evaluation are shown in Table 2. Table 2 shows the time complexity required for learning the data set. The numerical values in

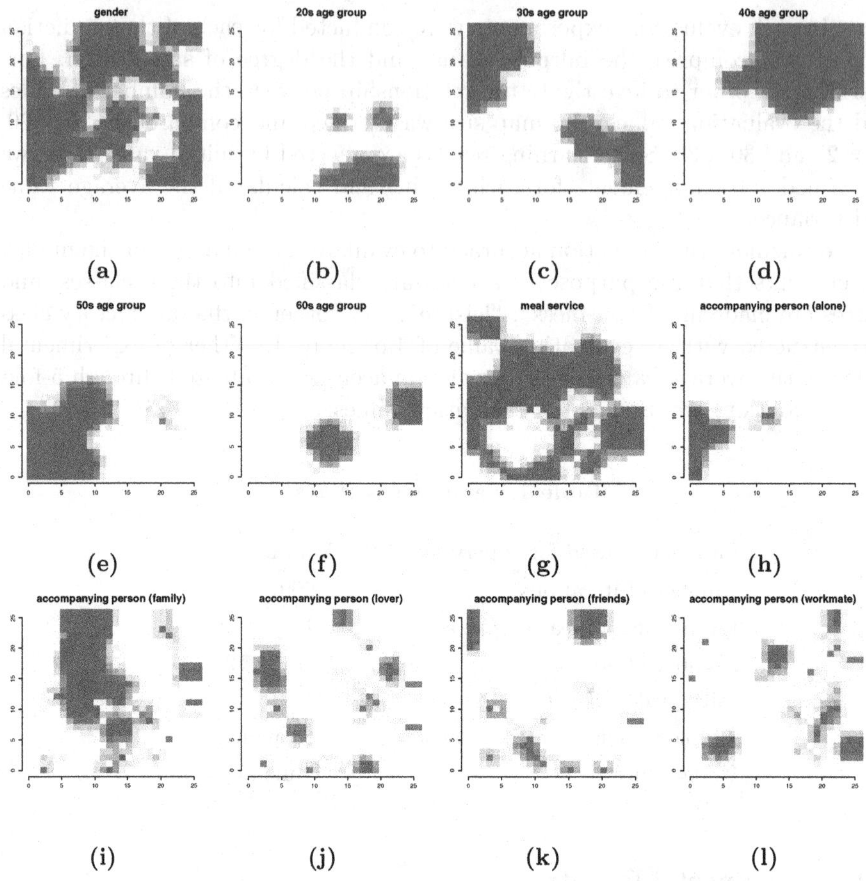

Fig. 2. Market segments that are visualized using *Xmap* based on demographics, including accommodation style. (Color figure online)

the tables are average values of the calculation results for 100 enforcements, and the values in parentheses are the standard deviations for them.

Evaluation and Interpretation of the Learning Results. In this section, we discuss the interpretation of the visualized segment map that is the learning result. By noting the common coordinates on the maps, analysts can interpret segment characteristics. The *Xmap* and *Ymap* are compatible with each other, and nodes positioned in the same coordinates of the *Xmap* and *Ymap* correspond to common segments. Benefits can be identified from the *Ymap* by using features of typical customers that are selected from the *Xmap*. Using an inverse process, we can also identify customer characteristics by satisfaction and by purpose of use.

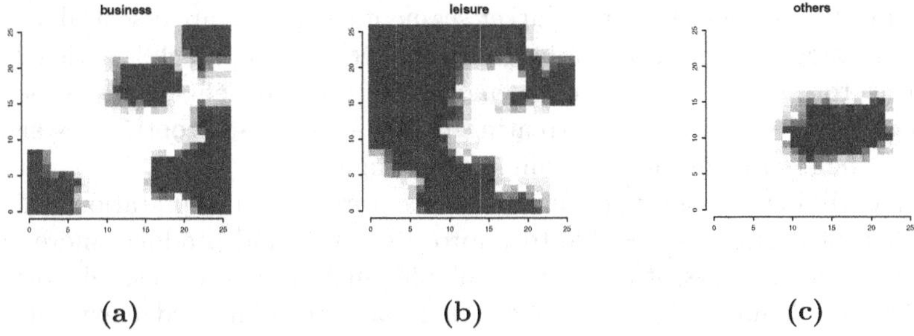

Fig. 3. Market segments that are visualized using *Ymap* based on purpose of use. (Color figure online)

Table 2. Classification accuracy for the experimental data set

Map size	20 × 20	25 × 25	30 × 30
Accuracy for predict purpose	0.702 (0.016)	0.701 (0.017)	0.707 (0.013)

For example, by examining node (2, 25) and its peripheral *Xmap* nodes, shown in Fig. 2, we can identify a segment of 30s women that are taking a meal in the hotel with friends. By examining the *Ymap* in Fig. 3(b), and coordinates (2, 25), the purpose can be determined to be leisure. We can anticipate that 30s women enjoy a meal in a hotel restaurant to spend time with friends. As another example, by examining node (23, 25) and its peripheral *Xmap* nodes, shown in Fig. 2, we can identify a segment of 40s men that are taking a meal in the hotel along with colleagues. By examining the Fig. 3(a) *Ymap*, and coordinates (23, 25), the purpose can be determined to be business. These are considered to businessmen who stayed at the hotel on a Tokyo business trip.

It can be easily imagined that a customer accompanying colleagues is a person using the hotel for the purpose of business. However, when attention is paid to nodes (1, 15) and (3, 21) in the Fig. 2(l), we can identify a segment of 30s people who use the hotel for leisure purposes with colleagues. For such customers, attention to leisure services may lead to increased customer satisfaction, re-purchasing, and improved reputation.

From Table 2, the number of calculations selected by the winner node is proportional to the number of nodes; so the time required for calculation increases as the map size increases. However, even if the map size is set as large, the learning time is within 10 s, and analysis can be performed with a realistic calculation amount. This is sufficient time for a support tool to extract knowledge on persona from data.

5.3 Discussion

The results interpreted in Sect. 5.2 are only examples, and various other segments can be interpreted. This approach is therefore effective as a tool to determine various market segment targets. Market segmentation tools are essential for customer targeting, and data-based decision support. Our approach is a visual segmentation tool based on customer demographics and benefits used to identify personas. Results indicate that treating benefits as classes contributes significantly to the identification of specific persona figures.

By identifying customer personas with accurate data interpretation and segment identification, it is possible to improve services and products more effectively. If large amounts of data are available, such as market use history and more detailed demographics, we believe more advanced knowledge extraction is possible. Additionally, unexpected trends and customer needs may be identified by heuristic analysis through visualization. It is an example of unexpected trends such as the personas contradicting information that is well-known to the analysts or the data deviating from the large clusters of visualized. Such data areas may lead to the development of new customers and markets, so there is a possibility of contributing in measures in marketing. Furthermore, new decision-making support becomes possible after sufficiently validating and examining the obtained unexpected persona with actual data and reality.

We considered the properties of the model and the data. In the framework of pattern recognition, models that treat benefits as classes tend to imbalance data. Customers often have similar benefits, and deviation occurs in the class distribution of the customer data. For example, clusters were distributed because the data used in our study have small evaluation values (e.g., 1 and 2). It is difficult to identify segments in such data. There is also the case where the distribution of an attribute extends to more than one class. In this case, segment features may not be completely identified. To design more sophisticated personas, technical improvements, such as data weighting and attribute incrementing, are required.

6 Conclusion

In this study, we examined benefit segmentation methods with the objective of improving persona design. We proposed a support tool for persona design based on market segmentation using data visualization. This is an important market segmentation concept for improving the usage and utility of products and services. We successfully identify representative customer personas with market segmentation that includes benefits, which is an improvement over segmentation based only on customer demographics. We focus on the use of self-organizing maps (SOM) that are widely used in data mining and market segmentation. We adopted the supervised self-organizing map as a visualization tool. Through a visual-based understanding of the correspondences between benefits and customer attributes, we make it easier to determine personas from supervised learning results. Demographic and benefit data from customer reviews of a Tokyo hotel were used to validate the proposed method.

Areas for future study include the following:

(1) Application of the proposed method to other market data,
(2) Use of review text data as attribute(s),
(3) Optimization of map size and parameter settings.

Further studies are required to address these areas. If we can utilize text data from reviews with additional information on demographics and benefits, we can expect more sophisticated persona development.

Acknowledgements. This work was supported by JSPS KAKENHI Grant-in-Aid for Scientific Research C 19K04887.

References

1. Chattopadhyay, M., Dan, P.K., Majumbar, S., Chakraborty, P.S.: Application of neural network in market segmentation: a review on recent trends. Manag. Sci. Lett. **2**(2), 425–438 (2012)
2. Haines, V., Mitchell, V.: A persona-based approach to domestic energy retrofit. Build. Res. Inf. **42**(4), 462–476 (2014)
3. Hanafizadeh, P., Mirzazadeh, M.: Visualizing market segmentation using self-organizing maps and Fuzzy Delphi method: ADSL market of a telecommunication company. Expert Syst. Appl. **38**, 198–205 (2011)
4. Hong, C.: Using the Taguchi method for effective market segmentation. Expert Syst. Appl. **39**(5), 5451–5459 (2012)
5. Hung, C., Tsai, C.F.: Market segmentation based on hierarchical self-organizing map for markets of multimedia on demand. Expert Syst. Appl. **34**, 780–787 (2008)
6. Kiang, M.Y., Hu, M.Y., Fisher, D.M.: An extended self-organizing map network for market segmentation a telecommunication example. Decis. Support Syst. **42**, 36–47 (2006)
7. Kiang, M.Y., Hu, M.Y., Fisher, D.M.: The effect of sample size on the extended self-organizing map network: a market segmentation application. Comput. Stat. Data Anal. **51**, 5940–5948 (2007)
8. Kuo, R.J., Chang, K., Chien, S.Y.: Integration of self-organizing feature maps and genetic-algorithm-based clustering method for market segmentation. J. Organ. Comput. Electron. Commer. **14**(1), 43–60 (2004)
9. Liu, H., Ong, C.: Variable selection in clustering for marketing segmentation using genetic algorithms. Expert Syst. Appl. **34**, 502–510 (2008)
10. Lopez, M., Valero, S., Senabre, C., Aparicio, J., Gabaldon, A.: Application of SOM neural networks to short-term load forecasting: the Spanish electricity market case study. Electr. Power Syst. Res. **91**, 18–27 (2012)
11. Marshall, R., et al.: Design and evaluation: end users, user datasets and personas. Appl. Ergon. **46**, 311–317 (2015)
12. Melssen, W., Wehrens, R., Buydens, L.: Supervised Kohonen networks for classification problems. Chemom. Intell. Lab. Syst. **83**, 99–113 (2006)
13. Seret, A., Verbraken, T., Versailles, S., Baesens, B.: A new SOM-based method for profile generation: theory and an application in direct marketing. Eur. J. Oper. Res. **220**, 199–209 (2012)

14. Vosbergen, S., et al.: Using personas to tailor educational messages to the preferences of coronary heart disease patients. J. Biomed. Inform. **53**, 100–112 (2006)
15. Yankelovich, D.: New criteria for market segmentation. Harv. Bus. Rev. **42**, 83–90 (1964)
16. Zhang, H., Chow, T.W.S., Wu, Q.M.J.: Organizing books and authors by multilayer SOM. IEEE Trans. Neural Netw. Learn. Syst. **27**(12), 2537–2550 (2016)

Food Safety Network for Detecting Adulteration in Unsealed Food Products Using Topological Ordering

Arpan Barman[1(✉)], Amrita Namtirtha[1], Animesh Dutta[1], and Biswanath Dutta[2]

[1] Department of Computer Science and Engineering, National Institute of Technology Durgapur, Durgapur, India
arpanbarman22@gmail.com, an.15it1110@phd.nitdgp.ac.in, animesh@cse.nitdgp.ac.in

[2] Documentation Research and Training Centre, Indian Statistical Institute, Bangalore, Bangalore, India
bisu@drtc.isibang.ac.in

Abstract. The food supply chains are vast and involve a lot of persons. It is highly probable that any person within the food supply chain can add adulterant in the food product to gain more profit. Therefore, providing safe food through the chain is a big challenge. Though sealed food products can be tracked using various existing technologies such as using IoT devices or by chemical analysis, the traceability of unsealed food products is an unexplored domain. To address this problem, this paper proposes an adulterant check algorithm based on topological ordering. The proposed algorithm checks the adulterant nodes in the unsealed food product supply chain. Additionally, we also monitor the transactions in the unsealed food product supply chain and store the provenance information in the graph database. Thus, we provide a low cost, traceable, adulteration-free unsealed food supply chain.

Keywords: Food adulteration · Unsealed food supply chain · Topological ordering · Graph database

1 Introduction

Recently, buying unadulterated food from market has become a global concern. It is highly probable that one may get foodborne diseases due to food crises and food contamination. Foodborne diseases are more prevalent in developing countries [2,20]. The markets of developing countries are mainly dominated by small shopkeepers, traders on wet markets, and street hawkers [19]. They primarily sell unsealed food products in the market which is very vulnerable to food adulteration. One of the most critical issues is the contamination of look-alike items with unsealed foods [11]. There are many reported cases such as lead chromate in turmeric, brick powder in red chillies, vegetable oil contamination with milk

N. T. Nguyen et al. (Eds.): ACIIDS 2020, LNAI 12034, pp. 451–463, 2020.
https://doi.org/10.1007/978-3-030-42058-1_38

fat [16] and even rotten meat [13]. In the FSSAI report (2016-17) [15], there are 14130 cases of adulterated samples [15]. Consumption of this kind of food might lead to chronic diseases like cirrhosis, cancer and even lead to death [6]. Due to these problems, consumers demand a transparent food supply chain to observe the provenance of the product. Transparent food supply chain means all the transactions involving the flow of food items, from the producers (like farmers, factories) to the consumers, are well documented.

In the literature, many researchers have proposed different methods to identify the adulteration in the sealed food supply chain. Tian [17,18] uses Internet of things (like RFID, GPS, WSN) to collect the relevant data from the food supply chain. However, Accorsi et al. [1] and Tian [17] tell that RFID based solutions are quite costly. Further, Dandage et al. [5] have stated that implementing any costly solutions for the small market traders in the developing countries (such as India, Bangladesh etc.) is itself a problem. Galvez et al. [8] solve the adulteration problem by chemically analyzing the food products. But, Manning et al. [11] have stated that chemically analyzing the food product can detect adulteration if the nature of adulterant is known. Therefore, these approaches fail to monitor the unsealed food supply chain. Moreover, as per our knowledge, it is difficult to find a work in the literature that addresses the traceability issues of the unsealed food products. We propose a system that can be modelled into two stages. Firstly, we store the transactional data into a graph database. Then we run the proposed algorithms based on topological ordering to solve the problem of traceability of unsealed food products. The algorithm finds out the adulterant nodes in the food supply chain. The time taken to execute the algorithm is linearly dependent on the number of persons involved in the supply chain and the number of transactions between them. If we represent the number of persons participating in the supply chain as $|V|$ and the total number of transactions between them as $|E|$, then the time taken by the algorithm is $(|V| + |E|)$. The proposed algorithm is then applied in some random weighted directed acyclic graphs. The experimental result shows that the algorithm identifies the adulterant nodes in the food supply chain.

The rest of the paper is organized as follows: Sect. 2 specifies the problem. In Sect. 3, we make a case study of the existing Indian public distribution system. Sect. 4 models the existing situation into the proposed system. Sect. 5 discusses the proposed algorithms. Sect. 6 analyzes the proposed algorithms with respect to time and space and the necessary proofs. Sect. 7 provides the experiment and result. Finally, the conclusion and future work are summarized in Sect. 8.

2 Problem Statement and Objective

Adulteration in food supply chain often causes health hazards within the society. Unsealed food supply chain is more susceptible to adulteration as the participants in the chain can easily contaminate the food product. Moreover, there is a lack of a proper system that monitors the whole unsealed food supply chain.

So, we propose a system to ensure a transparent unsealed food supply chain. The system will be able to find adulterant nodes within the supply chain. The challenges in this work are as follows:

- Supply chains are vast and involves lot of persons. Every transaction in the supply chain needs to be recorded in the system to ensure food traceability.
- The system must ensure the buyer that the food product is unadulterated.
- Unsealed food supply chain is mostly prevalent in developing countries. So, the system must be cost-efficient and user-friendly.

3 Case Study

To create an unsealed food supply chain, we need to acquire some data. To identify which data elements, we need to store, we have made a case study of the Indian public distribution system (PDS) [12] with respect to the flow of rice. At first, the government buys the raw rice grains from the farmers and bring it to the rice mills. Rice millers process the grains and produce rice. This rice is stored by the government in various godowns of the state. Now, every month, when the government wants to distribute the rice to the customers, they collect the closing balance of the fair price shops. Then, the government generates a delivery order to the wholesalers indicating the amount of rice to be collected from the mentioned godowns and collects the money for it. The government also releases a advice list to the wholesalers indicating the amount of rice, it needs to sell to the mentioned fair price shops. The fair price shops, then buy the rice from the wholesalers and sell it to the customers. In the Fig. 1, we present the detailed data flow diagram of the PDS.

We also show the following cases that may arise in an unsealed food supply chain.

Case 1: A buyer buys a product from a seller.
Solution: The buyer and the seller both have access to the same portal. The seller logs the transaction in the database. The buyer agrees and the transaction is saved in the database. Now, the algorithm runs and checks whether the seller has negative balance after the transaction occurs. If it has negative balance then the product is adulterated.
Case 2: If one seller sells the same product to multiple buyers.
Solution: The buyer and the seller both have access to the same portal. The seller logs the transaction in the database. The buyer agrees and the transaction is saved in the database. Now, the algorithm runs and checks whether the seller has negative balance when all the outgoing transactions are summed up. If yes, the product is adulterated. In Indian PDS, each fair price shop has its own id and each customer has an id in his digital ration card. They both agree to the transaction and log the transaction data. If they try to put different data, the transaction is aborted and the transaction data is not logged.

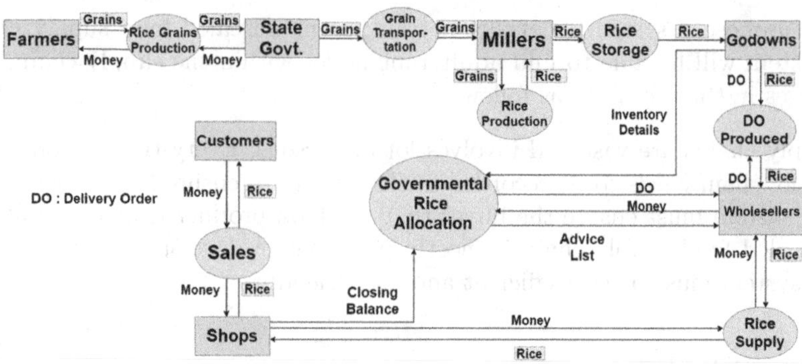

Fig. 1. Data flow diagram of the Indian public distribution system (PDS) involving flow of rice

Case 3: A buyer buys multiple products(with same and different product ids) from different sellers.
Solution: All the products with different product id will create different graphs. If the product ids are same, then the buyer vertex will have multiple incoming edges. In the algorithm, we will deduct the outflow from the sum of the inflow.
Case 4: Can a buyer know the origin and food quality while buying?
Solution: Yes, we store the chronology of the food item along with their timestamps and geo-location. So, a buyer can easily trace back the product using that portal to check its origin. Also, the buyer knows the date when the producer produced the food item. So, he can have an idea of product quality.

From the above case studies, we find that the following data elements are important to store in the database.

1. Unique product id for a product produced by a particular producer at a particular time. Here we have to create a unique id for rice produced at a specific time by a particular producer.
2. Unique participant ids of each person involved in the supply chain that is, farmers, millers, godowns, wholesalers, shops, customers, etc.
3. Weight of the product transferred between any two participants.
4. Timestamp of the transaction.
5. Geo-location to check the proper origin of the product as well as the path it follows before it reaches to the customers.

Besides these, some other data elements that are stored in the database are:

1. Product details such as product name, product type, product dimensions (if any) etc. can be stored for each product having a unique product id.
2. Participant details like participant name, participant address, participant contact details etc can be stored for each participant having a unique participant id.

4 System Model

The proposed system can be modelled in two stages. Firstly, we need to store the data elements, which are discussed in the case study, in a graph database. Graph database provides a more scalable system [3]. Moreover, in [14], the authors have stated that in graph databases (such as Amazon Neptune, Neo4j, etc.) can efficiently retrieve the query result from a huge amount of provenance data. We create separate graphs for different product ids. In the second stage, we model the whole supply chain graph as a weighted directed acyclic graph (DAG) $G = (V, E, W)$ where each vertex ($v \in V$) represents the participants in the supply chain, each edge ($(u, v) \in E$) represents the product flow between vertex u to v where u and $v \in V$ and W is the set of edge weights where $w_{u,v}$ represents the amount of product flow between u and v. A weighted DAG is created because in a food supply chain, food items flow from the producers to the consumers, unidirectionally. Then, we run the proposed algorithms on this graph to check whether there is any adulteration in the unsealed food supply chain graph or not. In Fig. 2, we present the process flow diagram of the proposed system. First part of the diagram shows in brief, an example of a typical food supply chain of unsealed products and the relevant participants. Then, we create a graph database and store the flow of goods. Next, we input the transactions and producers in the proposed algorithm and we get negative or non-negative output based on whether a vertex is an adulterant or not respectively.

Fig. 2. Process flow diagram of the proposed system

5 Proposed Solution

The proposed solution is divided into two algorithms: Algorithms 1 and 2. The purpose of Algorithm 1 is to create the prerequisites of Algorithm 2. Algorithm 1 takes the graph G as input.

- In line 1, we do the initialization operations.
- In line 2 to 4, we input all the edges of the graph and create an adjacency list adj[] because the supply chain graph is generally a sparse graph. In a supply chain, a vertex is only connected to a few predecessors and successors and not to all vertices in the graph. So, adjacency list provides a more efficient representation of sparse graphs, that is, graphs whose $|E|$ is much less than $|V|^2$ [4].

Algorithm 1. PrerequisiteCreation

Input: G = (V, E, W) where G is a weighted directed acyclic graph, V is the set of vertices, E is the set of edges and W is the set of weights where $w_i \in$ W is the weight of edge $e_i \in$ E; producerList[] is a list containing producers and their produced amount respectively. producerList[] $\subseteq$ V.

Output: Adjacency List adj[] of vertices in G; Array balance[] showing balance of all vertices; Array indegree[] containing indegree of all vertices; Queue containing vertices with indegree 0.

```
1  Set balance[] ← 0, indegree[] ← 0, adj[] ← NULL, Queue ← NULL
2  for each item e ∈ E do
3  |   adj[source] ← adj[source] ∪ (destination, weight)
4  end
5  for each item i ∈ producerList[] do
6  |   balance[i.id] ← i.weight
7  end
8  for each item i ∈ V do
9  |   for each item j ∈ adj[ i ] do
10 |   |   indegree[j.destination] ← indegree[j.destination] + 1
11 |   end
12 end
13 for each item v ∈ V do
14 |   if indegree[v] == 0 then
15 |   |   Enqueue v in Queue
16 |   end
17 end
18 return adj[], balance[], indegree[], Queue
```

- In line 5 to 7, we traverse the producerList[] and update the balance of the producers.
- In line 8 to 12, for each vertex v, we traverse its deg(v) neighbours, where deg(v) is the degree of vertex v, and increase the in-degree of the successor by 1. Therefore at the end of line 12, we have the in-degree of each vertex in the graph.
- In line 13 to 17, we traverse each vertex and make a queue of vertices with in-degree zero.
- In line 18, we return the necessary arrays.

Algorithm 2 is used to find the adulterant nodes in graph G. Algorithm 2 uses the help of topological ordering based on Kahn's algorithm [10]. It takes a vertex v with in-degree 0 from Queue. If v has non-negative balance, then in line 6 to 12, we traverse all its successors, that is the vertex in its adjacency list and find the total outflow of that vertex. Also, we decrease the in-degree of the successors by 1. We add the successor vertices, whose in-degree becomes zero to the Queue. Then, if the total outflow of the vertex is more than its current balance, we execute line 13 to 17. There, we make the balance of that vertex and all its successors as -1. Otherwise, we execute 18 to 23 and increase the balance of the successors of v and decrease the balance of v by the edge weight of v and its every immediate successor. If at the beginning, the balance of the vertex v is negative, then we perform the else block, which has a for loop in line 24 to 31. We mark the balance of all the successors of v by -1 and decrease the in-degree of the successor vertex by 1. Now, if the successor vertex has in-degree 0, then we put that vertex in the Queue. In line 33 to 36, we increase the value of count

Algorithm 2. AdulterationCheck

Input: Adjacency List adj[] of vertices in G; Array balance[] showing balance of all vertices; Array indegree[] containing indegree of all vertices; Queue containing vertices with indegree 0; target indicating the vertex id of the seller; V is the number of vertices of graph G.
Output: Balance of the target vertex. Negative balance indicates adulterant vertex and non-negative indicates non adulterant vertex

```
1  Set count ← 0, outflowSum ← 0
2  while Queue ≠ ∅ do
3      v ← Dequeue(Queue)
4      outflowSum ← 0
5      if balance[v] ⩾ 0 then
6          for each successor vertex i ∈ adj[v] do
7              outflowSum ← outflowSum + i.weight
8              indegree[i.destination] ← indegree[i.destination] - 1
9              if indegree[i.destination] = = 0 then
10                 Enqueue i.destination in Queue
11             end
12         end
13         if outflowSum > balance[v] then
14             balance[v] ← -1
15             for each successor vertex i ∈ adj[v] do
16                 balance [i.destination] ← -1
17             end
18         else
19             for each successor vertex i ∈ adj[v] do
20                 balance [v] ← balance[v] - i.weight
21                 balance [i.destination] ← balance [i.destination] + i.weight
22             end
23         end
24     else
25         for each successor vertex i ∈ adj[v] do
26             balance [i.destination] ← -1
27             indegree[i.destination] ← indegree[i.destination] - 1
28             if indegree[i.destination] == 0 then
29                 Enqueue i.destination in Queue
30             end
31         end
32     end
33     count ← count + 1
34     if count > V then
35         return "Cycle exists"
36     end
37 end
38 return balance[target]
```

and check whether the outer loop has been executed at most $|V|$ times or not. If the loop has run for more than $|V|$ times, we break return that their has been a loop which makes the graph invalid. At the end, in line 38, we return the balance of the target vertices. Now, when any person wants to buy any product, he just checks whether the balance of the seller vertex is non-negative, even after the product flow or not. If yes, the food product is unadulterated. Otherwise, it is adulterated.

6 Analysis

6.1 Time Complexity

From Tables 1 and 2, the Algorithm 1 and 2 take $f(n)$ time where $f(n) = c_1 + c_2 * |E| + c_3 * |V| + c_4 * (|V| + |E|) + c_5 * |V| + c_6 + c_7 * (|V| + |E|) + c_8 * (|V| + |E|) + c_9 * (|V| + |E|) + c_{10} * |V|$.
or, $f(n) = c_v * |V| + c_e * |E| + c_x$ where $c_v = c_3 + c_4 + c_5 + c_7 + c_8 + c_9 + c_{10}, c_e = c_2 + c_4 + c_7 + c_8 + c_9, c_x = c_1 + c_6$.
We know $O(g(n)) = f(n)$ such that there exists no positive constants c and n_0 such that $0 \leqslant f(n) \leqslant c * g(n) \forall n \geqslant n_0$ [4]. Here, we take $c = max(c_v, c_e)$ and satisfy the equation i.e. $0 \leqslant c_v * |V| + c_e|E| + c_x \leqslant c * (|V| + |E|)$ where $c > c_v$ and $c > c_e$ and $n_0 = 1$. Thus, we can conclude that the proposed algorithm takes $O(|V| + |E|)$ time. If the graph is fully connected, then the total number of edges $(|E|) = |V| * (|V| - 1) \approxeq V^2$, since all $|V|$ vertices are connected to $|V| - 1$ neighbouring vertices respectively. So, the time complexity at worst case will be $O(V + V^2) = O(V^2)$.

Table 1. Time complexity for Algorithm 1

Line number	Time taken	Remarks
1	c_1	Initialization operations
2–4	$c_2 * \|E\|$	Each input operation takes constant time c_2 and the loop runs $\|E\|$ time
5–7	$c_3 * \|V\|$	Each operation takes a constant time c_3. As producerList[] $\subseteq$ V, the loop will run at most $\|V\|$ times. So, the total time taken in this step will be $c_3 * \|V\|$
8–12	$c_4 * (\|V\| + \|E\|)$	For each vertex v, we traverse its deg(v) successors, where deg(v) is the degree of vertex v, and increase the in-degree of the successor by 1. Each update operation takes time c_4. Thus the inner loop takes c_4* deg(v) time. The outer loop runs $\|V\|$ time. We know, for a directed graph, the sum of degrees of all vertices is $\|E\|$. Thus, the total time taken in the loop is $c_4 * (\|V\| + \|E\|)$
13–17	$c_5 * \|V\|$	If the time taken to traverse one element of the array is c_5, then the time take n to traverse the whole array becomes $c_5 * \|V\|$
18	c_6	The necessary arrays are returned
Total time	$c_1 + c_2 * \|E\| + c_3 * \|V\| + c_4(\|V\|+\|E\|)+c_5*\|V\|+c_6$	

Table 2. Time complexity for Algorithm 2

Line number	Time taken	Remarks
6–12	$c_7 * (\|V\| + \|E\|)$	For each vertex v, we traverse its deg(v) successors, where deg(v) is the degree of vertex v. Each operation takes time c_7. Thus the inner loop takes c_7* deg(v) time. The outer loop runs $\|V\|$ time. We know, for a directed graph, the sum of degrees of all vertices is $\|E\|$. Thus, the total time taken in the loop is $c_7 * (\|V\| + \|E\|)$
13–23	$c_8 * (\|V\| + \|E\|)$	Either the if part in line 13 to 17 executes or the else part in line 18 to 22 executes. In any of the case, the loop runs for deg(v) time which after traversing for $\|V\|$ times adds up to $c_8 * (\|V\| + \|E\|)$
24–31	$c_9 * (\|V\| + \|E\|)$	This loop also runs for deg(v) time which after traversing for $\|V\|$ times adds up to $c_9 * (\|V\| + \|E\|)$
33–36	$c_{10} * \|V\|$	All constant time operations run for $\|V\|$ time. This sums up to $c_{10} * \|V\|$
Total time	$c_7 * (\|V\| + \|E\|)$ $+ c_8 * (\|V\| + \|E\|)$ $+ c_9 * (\|V\| + \|E\|)$ $+ c_{10} * \|V\|$	

6.2 Space Complexity

The proposed Algorithms 1 and 2 store and use the graph in adjacency list representation. Say, a given graph contains $|V|$ vertices and $|E|$ edges. In that case, we need to make an array adj of size $|V|$ indicating all the $|V|$ vertices of the graph, and for each vertex v $\in$ V, adj[v] points to all its outgoing edges in a linked list representation. As the graph is a directed and acyclic one, the total number of outgoing edges will be $|E|$. So, the adjacency matrix representation of the graph, that is adj, takes $O(|V| + |E|)$ space. If the graph is fully connected, then the number of outgoing edges of vertex v $\in$ V will be $|V| - 1$. In that case, total number of edges $(|E|) = |V| * (|V| - 1)$. So, the adj will take $O(V^2)$ space complexity. Other arrays like indegree, balance and Queue are of size $|V|$ respectively, indicating each vertex v of graph G. So, the space taken by them is $3 * |V|$. Thus the total space taken is $O(|V| + |E|)$.

6.3 Proof

Proof of Termination. We are going to prove that the Algorithm 2 terminates when we have traversed all the vertices once. This ensures the proposed algorithm halts after executing. For this, we need to ensure that the Queue will not be empty until all the vertices are traversed. We know that a vertex is inserted in the Queue if it has in-degree zero. So, as we said that the supply chain graph is a directed acyclic graph (DAG), we are going to prove that it must contain at least one vertex of in-degree zero. Let us assume by contradiction that all vertices in graph G = (V, E, W) have at least one incoming edge. Consider an arbitrary vertex v ∈ V. It must have an incoming edge e_1 ∈ E, call it (u, v) where u ∈ V. u also must have an incoming edge e_2 ∈ E, call it (t, u) where t ∈ V. t also must have an incoming edge e_3 ∈ E, call it (s, t) where s ∈ V and so on. We can keep doing this forever, and since the graph is finite, at some point, we must hit the same explored vertex v, which forms a cycle. But a DAG should not have a cycle. This contradicts our assumption. Thus, we can say that there is some element to extract from Queue until all the vertices are traversed and the Algorithm 2 terminates when we have traversed all the vertices once.

Proof of Correctness. We are going to prove that Algorithm 2 produces negative value (−1) as the balance of a vertex if and only if the vertex has transferred more product than its inflow.

- Case 1: For a vertex v ∈ V in G, if the sum of incoming edge weights is greater than equal to that of its outgoing edges, that is, if $\sum$ (incoming edge weights of vertex v ∈ V) $\geqslant \sum$ (outgoing edge weights of vertex v ∈ V), the balance array will have no non-negative value for that vertex v. Hence there has been no adulteration in that vertex v.
- Case 2: Let us assume by contradiction that the proposed Algorithm 2 returns non-negative value for a vertex v ∈ V in G, even if v has transferred more product than his inflow, that is if $\sum$(incoming edge weights of vertex v ∈ V) $\geqslant$ $\sum$(outgoing edge weights of vertex v ∈ V). But the balance array has to have a non-negative value to indicate no adulteration. If any vertex, with initial balance zero, has total outflow more than his balance after it has received all its incoming edge weights from its predecessor, we mark the balance of that vertex v as −1 in line number 14 of Algorithm 2 and also propagate negative values to its successors, in line 26 of Algorithm 2, as a convoy effect. Thus, the negative value of vertex v contradicts the assumption. Therefore, Algorithm 2 will return non-negative value only if someone has sent more production than his inflow.

7 Experiment and Result

In this section, we set some experimental setup to validate the performance of the proposed algorithm in terms of checking of the adulterant node in the network and computational time.

7.1 Experimental Setup

To perform an experiment, we have used the following system configuration: Intel(R) Core(TM) i5-7200U CPU @ 2.50 GHz, 8 GB RAM and 1 TB HDD.

Dataset Description. We create a synthetic dataset using the NetworkX package [9] in Python. We use two different classes of the NetworkX package to generate random graphs having different properties. These graphs serve as the input to test the proposed algorithms. First one is DiGraph which generates a random directed graph with a given number of nodes and edges. We make that graph acyclic by removing the loops. Secondly, we use gnp_random_graph which generates a random graph known as an Erdos-Renyi graph [7] or a binomial graph. The $G_{n,p}$ random graph model takes n: the number of nodes and chooses each of the possible edges with probability p. It is based on Poisson distribution. We randomly assign the edge weights to all edges in both DiGraph and $G_{n,p}$ random graph. The details of the dataset are provided in Table 3.

7.2 Result

We provide those randomly generated graphs as the input in the program. We use the root of these graphs as the producer nodes in the producerList[]. We run the proposed algorithms 100 times per experiment and test the performance of the proposed algorithm. We find the following result in Table 4.

We can see that the proposed algorithm can find the adulterant nodes from the randomly generated graphs. We can also see the time taken by the proposed algorithm to find the adulterant nodes from digraph random graphs 2500, 5000, 7500 and 10000 vertices are 1.67 ms, 3.69 ms, 5.29 ms, 8.1 ms respectively. Similarly, for gnp random graphs with 2500, 5000, 7500 and 10000 vertices, time taken to find all adulterant nodes are 1.55 ms, 3.04 ms, 4.52 ms, 7.04 ms respectively. So, we can see that the time taken to execute the proposed algorithms increase with the number of vertices and edges for different random graphs generated by DiGraph class or Gnp_random_graph class.

Table 3. Dataset table

Class	Number of vertices	Number of edges	Probability of edge creation
DiGraph	2500	6237	–
	5000	25063	–
	7500	55632	–
	10000	99615	–
Gnp_random_graph	2500	6242	0.002
	5000	25092	0.002
	7500	55690	0.002
	10000	99704	0.002

Table shows the classes used and the number of vertices and edges in the generated graph. Also gnp_random_graphs need a probability to create edges which is also shown.

Table 4. Result table

Class	Number of vertices	Number of adulterant vertices	Average time taken
DiGraph	2500	2210	1.67 ms
	5000	4834	3.69 ms
	7500	7331	5.29 ms
	10000	9854	8.1 ms
Gnp_random_graph	2500	1909	1.55 ms
	5000	4507	3.04 ms
	7500	7000	4.52 ms
	10000	9482	7.04 ms

Table shows the experimental result obtained after running 100 iterations in each case

8 Conclusion and Future Work

In this paper, we have proposed a cost-effective solution to the food adulteration problem for unsealed food products which are prevalent in low and medium-income countries like India. It is difficult to use the existing IoT based or chemical analysis based solutions as those are quite a costly solution and might not be economically feasible for a developing country. But the proposed solution is based on storing the transactional data in graph databases. To check whether a given product has been adulterated or not, we check the seller node is an adulterant node or not. We have analyzed the time complexity of the proposed algorithm be linear, that is $O(|V| + |E|)$. We have simulated the proposed algorithm using two type of random graphs and checked the performance of the proposed algorithm in terms of adulterant node identification and the execution time. The experiments on the proposed system have been performed on just two random graphs. However, the real-life networks are vast, complex and consists of diverse network structures. Therefore, it is quite challenging to examine the proposed system in the real-life networks. As the proposed system is mainly reliable on the data, it may happen that some adulterer may not provide data. In our future work, we will explore that domain. Also, our future work will have more detailed comparison of the proposed system with existing algorithms on supply chain adulteration.

References

1. Accorsi, R., Ferrari, E., Gamberi, M., Manzini, R., Regattieri, A.: A closed-loop traceability system to improve logistics decisions in food supply chains: a case study on dairy products. In: Advances in Food Traceability Techniques and Technologies, pp. 337–351. Elsevier (2016)
2. Aung, M.M., Chang, Y.S.: Traceability in a food supply chain: safety and quality perspectives. Food Control **39**, 172–184 (2014)
3. Batra, S., Tyagi, C.: Comparative analysis of relational and graph databases. Int. J. Soft Comput. Eng. (IJSCE) **2**(2), 509–512 (2012)
4. Cormen, T.H., Leiserson, C.E., Rivest, R.L., Stein, C.: Introduction to Algorithms. MIT Press, Cambridge (2009)
5. Dandage, K., Badia-Melis, R., Ruiz-García, L.: Indian perspective in food traceability: a review. Food Control **71**, 217–227 (2017)
6. El Sheikha, A.F.: DNAFoil: novel technology for the rapid detection of food adulteration. Trends Food Sci. Technol. **86**, 544–552 (2019)
7. Erdös, P., Rényi, A.: On random graphs. Publ. Math. Debrecen **6**, 290–297 (1959)
8. Galvez, J.F., Mejuto, J., Simal-Gandara, J.: Future challenges on the use of blockchain for food traceability analysis. TrAC Trends Anal. Chem. **107**, 222–232 (2018)
9. Hagberg, A., Swart, P., Chult, D.S.: Exploring network structure, dynamics, and function using networkX. Technical repoer, Los Alamos National Lab. (LANL), Los Alamos, NM (United States) (2008)
10. Kahn, A.B.: Topological sorting of large networks. Commun. ACM **5**(11), 558–562 (1962)

11. Manning, L., Soon, J.M.: Developing systems to control food adulteration. Food Policy **49**, 23–32 (2014)
12. Nagavarapu, S., Sekhri, S.: Informal monitoring and enforcement mechanisms in public service delivery: evidence from the public distribution system in india. J. Dev. Econ. **121**, 63–78 (2016)
13. NDTV: 20 tonnes of rotten meat meant for supplying to restaurants in Kolkata seized (2018). https://www.ndtv.com/kolkata-news/20-tonnes-of-rotten-meat-meant-for-supplying-to-restaurants-in-kolkata-seized-1843541
14. Rani, A., Goyal, N., Gadia, S.K.: Efficient multi-depth querying on provenance of relational queries using graph database. In: Proceedings of the 9th Annual ACM India Conference, pp. 11–20. ACM (2016)
15. Food Safety and Standards Authority of India: Annual reports (2016-17). Food Safety and Standards Authority of India
16. Shukla, S., Shankar, R., Singh, S.: Food safety regulatory model in India. Food Control **37**, 401–413 (2014)
17. Tian, F.: An agri-food supply chain traceability system for china based on RFID & blockchain technology. In: 2016 13th International Conference on Service Systems and Service Management (ICSSSM), pp. 1–6. IEEE (2016)
18. Tian, F.: A supply chain traceability system for food safety based on HACCP, blockchain & internet of things. In: 2017 International Conference on Service Systems and Service Management, pp. 1–6. IEEE (2017)
19. Trebbin, A.: Linking small farmers to modern retail through producer organizations-experiences with producer companies in india. Food Policy **45**, 35–44 (2014)
20. Uyttendaele, M., Franz, E., Schlüter, O.: Food safety, a global challenge (2015)

Privacy, Security and Trust in Artificial Intelligence

A Proof of Concept to Deceive Humans and Machines at Image Classification with Evolutionary Algorithms

Raluca Chitic(✉), Nicolas Bernard, and Franck Leprévost

University of Luxembourg, House of Numbers, 6, avenue de la Fonte, 4364 Esch-sur-Alzette, Grand Duchy of Luxembourg
{Raluca.Chitic,Nicolas.Bernard,Franck.Leprevost}@uni.lu

Abstract. The range of applications of Neural Networks encompasses image classification. However, Neural Networks are exposed to vulnerabilities, and may misclassify adversarial images, leading to potentially disastrous consequences. We give here a proof of concept of a black-box, targeted, non-parametric attack using evolutionary algorithms to fool both neural networks and humans at the task of image classification. Our feasibility study is performed on VGG-16 trained on CIFAR-10. Given two categories $c_i \neq c_t$ of CIFAR-10, and an original image classified by VGG-16 as belonging to c_i, we evolve this original image to a modified image that will be classified by VGG-16 as belonging to c_t, although a human would still likely classify the modified image as belonging to c_i.

1 Introduction

Neural Network (NN) approaches to image classification have outperformed traditional image processing techniques, becoming today's most effective tool for these tasks [28]. Concretely, a NN is trained on a large set of examples. During this process, the NN is given both the input image and the output category it is expected to associate with the input image. Based on this accumulated "knowledge", the NN can then classify new images into categories with high confidence.

However, the process leading to the classification by a NN still seems to differ significantly from the way humans perform the same task. It is therefore tempting to exploit this difference and cause the NN to make classification errors. Indeed, trying to fool NNs has become an intensive research topic [12,17,18], since the vulnerabilities of NNs used for critical applications can have disastrous consequences. It is hence important to better understand these weaknesses.

Our previous work [3] announced a research program, based on evolutionary algorithms (EA), to deceive machines and humans at image classification. This article presents our first concrete results (limited to one of the scenarii of [3] as explained below) of an on-going work in this direction. More precisely, we choose VGG-16 [24] as NN, and CIFAR-10 [14] as the training image dataset (see Sect. 2.3 for a short presentation). The implementation of a variant of the

N. T. Nguyen et al. (Eds.): ACIIDS 2020, LNAI 12034, pp. 467–480, 2020.
https://doi.org/10.1007/978-3-030-42058-1_39

EA used in [2] leads to the evolutionary algorithm announced in [3]. We run this EA to create adversarial images through the addition of small perturbations to images taken from the CIFAR-10 dataset, that VGG-16 will misclassify.

In our scenario, we choose *a priori* a target category in any of the ten categories of CIFAR-10, and we choose at random a picture of CIFAR's test set belonging to any of the categories but the target one. We evolve this initial picture until VGG-16 classifies it as belonging to the targeted category. In addition, the conception of the fitness function of our EA favours image similarities. In other words, the evolved images remain similar to the original ones for a human eye. Our first results give substantial reasons to believe that humans would not do the same misclassification as VGG-16, and would hence label the modified images as belonging to the original category.

The remainder of this article is organised as follows. We briefly sketch the concepts of Neural Networks and Evolutionary Algorithms in Sect. 1.1. This enables us to position our approach in Sect. 1.2. Then, we describe our strategy and implementation in Sect. 2, before discussing some of the results obtained in Sect. 3. Conclusions and future work directions are sketched in Sect. 4. In particular, we announce our intention (see [4]) to pursue our feasibility study with a statistically relevant study about the human's perception of the modified images, in order to confirm (or infirm) the statement that humans would classify our adversarial images differently than VGG-16 does.

1.1 Background

We give here the very minimum background about Neural Networks (NNs) and Evolutionary Algorithms (EAs) needed to position our contribution (Sect. 2) in the context of some related work (Sect. 1.2).

A Neural Network can be considered here as a black box (see [11] for a detailed description of NNs in the context of Deep Learning), which takes an image p as input and produces the corresponding output, consisting of a vector $\boldsymbol{o}_p$ of length ℓ. Each position $1 \leq i \leq \ell$ in the output vector represents a category c_i of objects defined in a dataset. Since the vector components are probabilities, one has $0 \leq \boldsymbol{o}_p[i] \leq 1$, and $\sum_{i=1}^{\ell} \boldsymbol{o}_p[i] = 1$. A NN is first trained on images of the dataset. Once trained, the NN produces a label for an image p by extracting the category for which the probability is highest, e.g. c_i if $i = \arg\max_{1 \leq j \leq \ell}(\boldsymbol{o}_p[j])$.

Evolutionary Algorithms search for solutions to optimisation problems by imitating natural evolution (see [25, sec. 3] for a quick introduction. For a general introduction to evolutionary algorithms, one can consult [13] and [34]). Whereas NNs are trained on a set of examples, EAs are heuristic and search for solutions in the entire image space. Each generation of the algorithm presents a new, generally improved population g_{i+1}, compared to the previous one g_i. An EA mainly consists of a loop in which (1) an initial population of individuals is created or chosen; (2) each member of the population is evaluated through a fitness function; (3) a new population is created from the existing one, taking into account each individual's fitness.

1.2 Context and Related Work

Consider a sample x, which is classified correctly by M, a machine learning model (MLM), as $M(x) = c_{true}$ (for instance the classification of an image by a NN). To fool this MLM, one can create an adversarial sample x' (this concept was introduced by Szegedy *et al.* [27]) as the result of a tiny (and unnoticed, ideally) perturbation of the original sample x, such that M misclassifies it, i.e. $M(x') \neq c_{true}$. Let us classify these attacks here according to two criteria: white-box vs black-box attacks and targeted vs non-targeted attacks.

White-box and black-box attacks depend on the amount of information about the MLM available to the adversaries. In white-box attacks, the adversaries have full access to the MLM. For instance, if the MLM is a NN, they know the training algorithm, the training set and the parameters of the NN. The adversaries analyse the model constraints and identify the most vulnerable feature space, altering the input accordingly [20,27]. In black-box attacks, the adversaries have no knowledge about the MLM (the NN in our context) and they instead analyse the vulnerabilities by using information about the past input/output pairs. More specifically, the adversaries attack a model by inputting a series of adversarial samples and observing corresponding outputs.

Previous attempts [7,22] to deceive NNs dealt primarily with untargeted attacks. In these, the task is to take some input x, which is correctly labeled as c_{true} by a NN, and to modify it towards a x' with distance $d(x, x') < e$, which is labeled as a different class than c_{true}. For example, a small change imperceptible to humans misleads the network into classifying an image of a tabby cat as guacamole [23].

A targeted attack, while usually harder to perform, is able to produce adversarial samples for a specific target class. Given classes c_a and c_b, with an input x classified as belonging to c_a, a targeted attack produces a similar x', which the NN classifies as belonging to c_b.

Within the category of black-box attacks, which are attempted in this paper, there are multiple methods to fool a NN. Most of these approaches, however, are still reliant on the model and its parameters [21,29]. Bypassing them might be achieved merely by training a NN to recognise the product of these methods. By contrast, EAs are extremely flexible and adaptive. Our contribution positions EA as an alternative approach that has the advantage of being non-parametric (see [16,33] for other non-parametric models for adversarial attacks). Our non-parametric model presents no bias (unless such a bias is programmed in the EA).

For EAs, among the most important aspects are the definition of their target and surrounding landscape, both done through the fitness function. The fitness function also reveals the main difference between our paper and the previous work on adversarial attacks using evolutionary algorithms. While Su *et al.* [25] use an EA to produce misclassifications by only changing one pixel per input image, our method does not limit the number of pixels which can be modified. Limiting the number of modified pixels leads to higher perturbation values for a few pixels, which become very different from their surroundings. However, our focus is on making the modifications less visible by humans, and hence the goal is to limit the overall value change, rather than the pixel count.

The following section describes our targeted black-box attack. Even if it requires no prior knowledge of the NN and its parameters, its feasibility is tested against the concrete example of CIFAR-10 and VGG-16 (see Sect. 2.3).

2 Strategy and Implementation

Our algorithm is an EA that aims to deceive both NNs and human beings. In our scenario (see [3] for others), an EA is initially given an image, the "ancestor" $\mathcal{A}$, labelled by the NN as belonging to $c_{\mathcal{A}}$. We fix a target category $c_t \neq c_{\mathcal{A}}$. The goal of the EA is then to evolve $\mathcal{A}$ to a new image $\mathcal{D}$ (a "descendant") that will be classified into c_t. The constraint is that the evolved adversarial image $\mathcal{D}$ should be very similar to the ancestor $\mathcal{A}$, with the perturbations kept as least visible as possible, in order that a human being should still consider the image as obviously belonging to category $c_{\mathcal{A}}$.

Before describing our EA and the condition in which it ran any further, we start by considering the similarity issue.

2.1 Image Similarity

The difference between two images (of the same size) i and i' can be evaluated in many ways. But only some of them give a hint at the similarity between two images, as a human being would perceive it. We explore here two of them:

- The L_2-distance, which calculates the difference between the initial and modified pixel values:

$$L_2\left(i, i'\right) = \sum_{p_j} \left| i[p_j] - i'[p_j] \right|^2 . \quad (1)$$

  A minimisation of the L_2-norm in the fitness function would lead to a minimisation of the overall value change in the images' pixels.
- The structural similarity (SSIM [32]) method attempts to quantify the perceived change in the structural information of the image, rather than simply the perceived change. The Structural Similarity Index compares pairs of sliding windows (sub-samples of the images) W_x and W_y of size $N \times N$:

$$SSIM_W\left(W_x, W_y\right) = \frac{(2\mu_x\mu_y + c_1)(2\sigma_{xy} + c_2)}{(\mu_x^2 + \mu_y^2 + c_1)(\sigma_x^2 + \sigma_y^2 + c_2)} . \quad (2)$$

  The quantities μ_x and μ_y are the mean pixel intensities of W_x and W_y, σ_x^2 and σ_y^2 the variance of intensities of W_x and W_y, and σ_{xy} their covariance. The purpose of c_1 and c_2 is to ensure that the denominator remains far enough from 0 when both μ_x^2 and μ_y^2 and/or both σ_x^2 and σ_y^2 are small.

The N_W window pairs to consider equals the number of pixels (times the number of colour channels if appropriate) of the picture cropped by a frame that prevents the windows from "getting out" of the picture. With pictures of size $h \times w \times c$ and windows of size $N \times N$, one gets:

$$N_W = \left(h - 2\left\lfloor\frac{N-1}{2}\right\rfloor\right) \times \left(w - 2\left\lfloor\frac{N-1}{2}\right\rfloor\right) \times c. \tag{3}$$

The SSIM value for two images i and i' is the mean average of the values obtained for the N_W window pairs $(i_k,\ i'_k)$:

$$SSIM(i,i') = \frac{1}{N_W}\sum_{k=1}^{N_W} SSIM_W\left(i_k, i'_k\right). \tag{4}$$

Unlike the L_2-norm, the SSIM value ranges from -1 to 1, where 1 indicates perfect similarity.

With both, our EA might thus consider it preferable to slightly modify many pixels, as opposed to changing fewer pixels but in a more apparent way. This contrasts with the approach of Su *et al.* [25], where only one pixel is modified, possibly being assigned a very different colour that can make it stand out.

2.2 Our EA

The EA used here is a descendant—in spirit if not in code—of the one used in [2], and we are only describing here what differs.

Population Initialisation. A population size being fixed, the initial population is set to a number of copies of the ancestor $\mathcal{A}$ equal to the chosen population size.

Evaluation. Our fitness function combines the two following factors. On the one hand, the evolution is directed towards a larger classification of the images as c_t. On the other hand, similarity between the evolved and ancestor images is highly encouraged. The former deals with deceiving machines, the latter with deceiving humans.

Our **fitness function** therefore depends on the type of similarity measure that is used. If using the L_2-norm, it can be written as

$$fit_{L_2}(ind, g_i) = A_{L_2}(g_i)\boldsymbol{o}_{ind}[c_t] - B_{L_2}(g_i)L_2(ind, \mathcal{A}) \geq 0, \tag{5}$$

where *ind* designates a given individual (an image) and g_i the ith generation. $\mathcal{A}$ is the ancestor and d is the measure of the distance between images. Moreover, $\boldsymbol{o}_{ind}[c_t]$ designates the value assigned by the NN to *ind* in the target category c_t. The quantities $A_{L_2}(g_i), B_{L_2}(g_i) \geq 0$ are coefficients to weight the members and balance them. They were created as functions of the current generation of the EA. The purpose of this was to assign different priorities to the different generations. The first generations were thus assigned the task of evolving the image to the target category, while the later generations focused on increasing the similarity to the ancestor, while remaining in c_t.

In the case of structural similarity, a higher value translates into a higher fitness of the individual. Therefore, *mutatis mutandis*, the difference in the fitness function becomes a sum:

$$fit_S(ind, g_i) = A_S(g_i)\mathbf{o}_{ind}[c_t] + B_S(g_i)SSIM(ind, \mathcal{A}) \geq 0. \quad (6)$$

The Evolution and its multiple steps is identical to what was described in [2], but for the replacement of *rectangle intensifying mutations* by *circle* intensifying mutations. In these, a circular area on a image is multiplied by an *intensifying factor* that is chosen with a normal law centred on 1 with a standard deviation of 0.1. The radius and location of the circle are chosen uniformly random.

As in [2], an equivalence is made between the population of the EA and a batch of the NN so that the NN can process the EA population in parallel, as a single batch, using a GPGPU.

2.3 Target Dataset and Neural Network Architecture

The NN architecture chosen here was VGG-16 [24]. VGG-16 is a convolutional neural network (CNN) that takes input images and passes them through 16 layers, to produce a classification output. The dataset used is CIFAR-10 [14]. It contains 50,000 training images and 10,000 test images, of size $32 \times 32 \times 3$. CIFAR-10 has $\ell = 10$ categories. We adjusted VGG-16's input size and trained it on the CIFAR-10 dataset [9] until it reached a validation accuracy of 93.56%. Once the training phase finalised, the optimised parameters were saved and the same model was used throughout the entire EA.

3 Running the EA: Parameters, Results and Discussion

The implementation of the EA was done from scratch, using Python 3.7 with the NumPy [19] library. Keras [5] was used to load and run the VGG-16 [24] model, and Scikit-image [31] to compute SSIM (using Scikit-image's default options).

We ran our experiments on nodes with Nvidia Tesla V100 GPGPUs of the Iris HPC cluster at the University of Luxembourg [30]. The EA was run with a population of size 160, the ancestor images being chosen from the CIFAR-10 [14] test set. The average computation time per generation was 0.034 s. For any source category, a random image was selected from the 1000 test images belonging to that category. This image was then set as the ancestor for the EA. The target category for misclassification was also selected randomly from all labels, excluding the source category.

The goal for the target class probability was set to 0.95, meaning the algorithm would stop when the fittest evolved image reached this value. However, for visualisation purposes, Fig. 2 shows an evolved image classified as the target category with probabilities 0.5, 0.9, 0.95 and 0.99. Depending on the chosen source

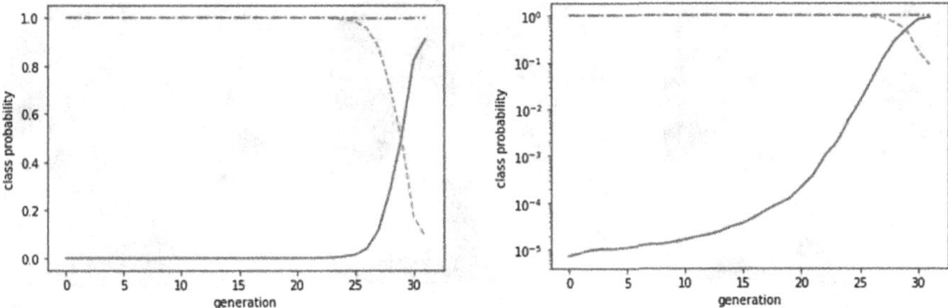

Fig. 1. Linear (left) and logarithmic (right) scale plots of an evolution of source (dashed line) and target (solid line) class probabilities using the L_2 norm, with respect to the generation number for the *dog→horse* combination. Both plots display in addition the sum (dash-dotted line) of the source and target probabilities: The sum remains about constant at a value of 1.0 throughout the evolution process, suggesting that the increase of the target class probability and the decrease of the source class probability take place at the same pace.

and target categories, reaching the probability of 0.95 required a number of generations roughly between 25 and 1500 (the 10 categories of the CIFAR-10 dataset could be divided into two groups: 6 categories are animals and 4 categories are objects. The *animal→animal* and *object→object* evolutions took fewer generations than the *animal→object* and *object→animal* ones. The required number of generations varied significantly not only with the category of an ancestor, but also with the particular image extracted from a given category). An example of the evolution of the source and target class probabilities with the L_2 norm is shown in Fig. 1. The paths of the two probabilities appear to be the inverse of each other, with their sum remaining almost constant at a value of about 1.0 throughout the evolution process. This suggests that the increase of the target class probability and the decrease of the source class probability take place at the same pace.

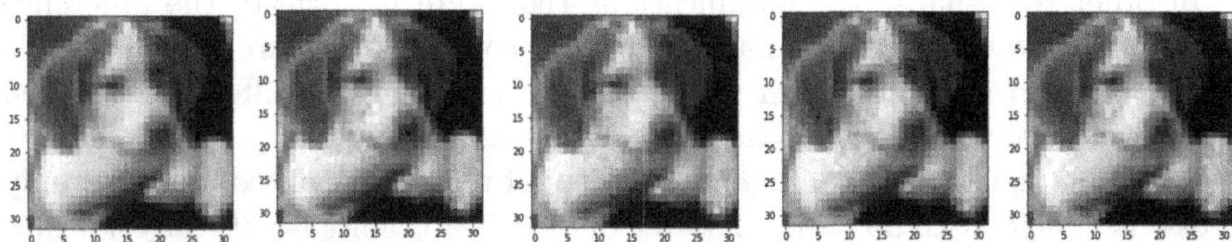

Fig. 2. Comparison of the original and modified (L_2-distance) images at different stages, classified as the target category "horse" with probabilities 6.08×10^{-6}, $0.5, 0.90, 0.95$ and 0.99. These modified images belong to the *dog→horse* combination.

In order to compare between the two similarity measures used in this paper, Fig. 3 displays example results for both the L_2-norm and SSIM. Neither appears

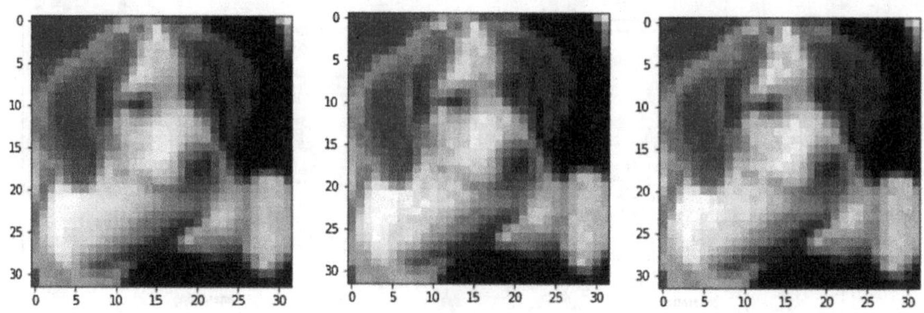

Fig. 3. Comparison of original and modified images for the *dog→horse* combination. From left to right, the images represent the original, the adversarial image created with the L_2-norm and the adversarial image created with the SSIM (both evolved to a confidence ≥ 0.95). No version seems clearly better than the other.

to be clearly better than the other. Both were able to produce misclassification with high confidence, while keeping the adversarial image very close to the original. In some image areas, the similarity to the original is higher with L_2, while in other image areas it is higher with $SSIM$. One should note, however, that the final aspect of the evolved image does not only depend on the used similarity measure. The randomness inherent to the evolution process may indeed contribute to the observed differences.

The image evolution process was repeated to cover all 90 possible combinations of distinct $\{source, target\}$ categories. Yet, since each of source and target can belong to either *animals* or *objects*, we can summarise the results with just four combinations. Figure 4 shows that transitions are achievable for each of these.

These results are encouraging regarding the two goals of our algorithm. The adversarial images are all categorised into the target class with a probability exceeding 0.95. Simultaneously, the evolved images are highly similar to their original versions. However, they are not entirely indistinguishable, the modified image being usually noisier than the unchanged one.

In order to visualise the modifications that were performed, the difference between the evolved and original images was computed. Figure 5 displays this difference, both spatially as an image and as a plot of the difference between original and evolved pixels for the best population individual.

It can be seen that the difference consists of noise distributed almost evenly across the image, with no particular area of focus. This noise is naturally not random, but fine-tuned by our evolutionary algorithm (it has already been proven than random noise is not sufficient to deceive a NN, since they are only vulnerable to targeted noise [8]). The majority of pixels were modified by an absolute value lower than 10. Histograms of the pixel modification values can be seen in Fig. 6, displaying results for both the L_2-norm and SSIM. One observes that both exhibit a bell shape, with the most prominent value being 0, corresponding to no added perturbation. Normalising the two histograms into probability densities,

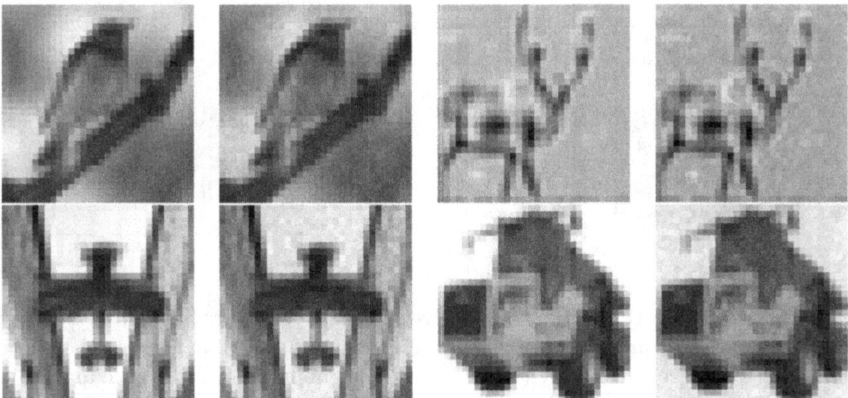

Fig. 4. Comparison of original and modified images. At the top, *animal→animal* and *animal→object* combinations: From left to right, the NN outputs the following class probabilities: 0.999 bird, 0.954 dog; 0.999 deer, 0.951 ship. Below, the *object→object* and *object→animal* combinations: From left to right, 0.998 airplane, 0.951 automobile; 0.999 truck, 0.952 frog. These examples were run with the L_2-norm measure.

the Kullback-Leibler divergence [15] KL between them indicates the proximity of the pattern of the two sampled distributions. Since $KL(p||q) \geq 0$ is not a symmetric function of the probability distributions p and q, one needs to compute two values. It turns out that in our case the Kullback-Leibler divergence values between the probability distribution associated to the L_2 histogram and to the SSIM histogram is 0.015, while the vice-versa value is 0.016, hence providing evidence that the patterns are indeed very similar. However, the SSIM results are more symmetrical than the L_2-norm ones. Although a human being is unlikely to perceive a difference between the modified images obtained either with L_2 or with $SSIM$, the histograms of Fig. 6 indicate that the L_2-based fitness function tends to leave more pixels unchanged.

Interestingly, although this added noise is targeted by the fitness function of the EA, there is no clear spatial pattern that emerges. This leads to the conclusion that the algorithm creates a texture modification, rather than a shape one. This relates closely to the findings of [10] that CNNs have a strong bias towards texture. This would imply that, in order to fool the CNN into outputting a different label, the texture of the original image has to be modified to resemble the texture of the targeted class. However, it is difficult to conclude that this is sufficient to explain why even a small perturbation to the input can produce a misclassification, as we human beings are not able to perceive a "target's texture" when looking at the evolved images.

The explanation might rely in the linearity of NNs. Since inputs are multiplied by weights and then summed up, even a small modification to each pixel can lead to a large difference in the sum. This value is then typically passed through the ReLU activation function ($f(x) = x$ for $x > 0$), which does not

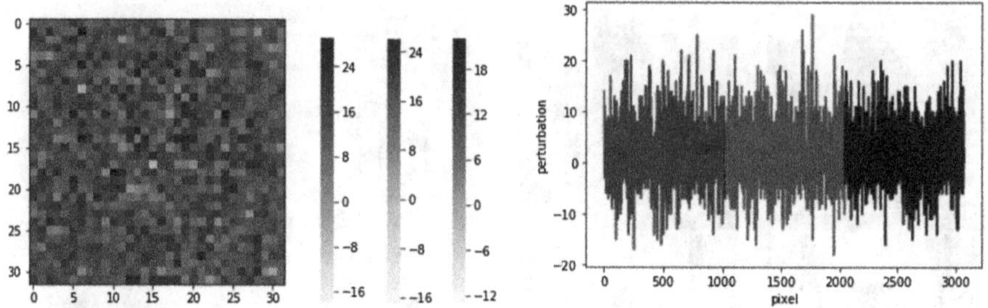

Fig. 5. Difference between original and evolved (using L_2-distance based fitness) images for the *truck→frog* combination. The image on the left gives an idea of the spatial distribution of the changes, each pixel being a combination of three channels, the range of each being given by the scales in the middle. The plot on the right ignores the 2D structure of an image to show more clearly by how much each of the 1024 pixels is changed on the three different channels. Despite the goal oriented nature of the evolution, these changes look like noise, almost evenly distributed across the image. Still, in this example, the range of the blue scale is tighter than the red and green ones, suggesting perhaps that the difference between the truck image and the frog category lies mainly in the red and green channels.

constrain it to a given range, thus leading to very high values [12] and potentially a different predicted class.

Another cause might be that NNs store learned knowledge in a more hybrid way. This hybrid storage may lead to lower distances between different classes and, as a consequence, small modifications may deceive the classifier [35]. This relates closely to recent work by Shamir *et al.* [23], which states that adversarial

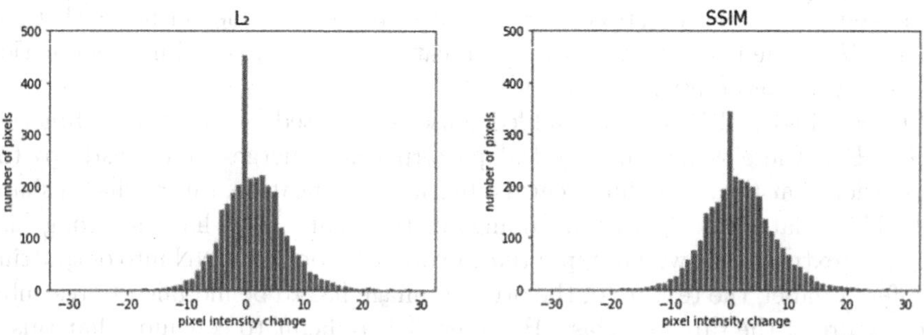

Fig. 6. Changes in pixel intensity through the *dog→horse* evolution (until reaching a confidence ≥ 0.95) of the dog ancestor shown in Fig. 3, with the L_2-norm (left) and SSIM (right). To reduce the possible impact of random fluctuations on the results, the figures are averaged on six evolutions. In both cases, the predominant value is 0, corresponding to a lack of pixel modification. Although both histograms are bell-shaped, SSIM's is more symmetrical. The Kullback-Leibler divergence computed between the L_2-norm and SSIM probability densities is 0.015; Reversing this order leads to 0.016.

examples are a natural consequence of the geometry of $\mathbb{R}^n$ and the intersecting decision boundaries learned by the NN.

4 Conclusions and Future Work

As a first step of the research program announced in [3], we demonstrated an approach using evolutionary algorithms to produce adversarial samples that deceive neural networks performing image recognition, and that are likely to deceive humans as well. The latter means that for a human being the category to which an image belongs will be obvious, yet it will not be the one assigned by the neural network.

Our feasibility study implements an evolutionary algorithm fooling the neural network VGG-16 [24] trained on the dataset CIFAR-10 [14] to label images according to 10 categories. Our algorithm takes an image from the CIFAR-10 dataset, classified by VGG-16 as belonging to one category, and takes as a target category any of the nine remaining categories. It then evolves the original image towards a (modified) adversarial image labeled by VGG-16 as belonging to the target category, while remaining similar to the original one. This second aspect is to ensure that humans would still classify the modified image as belonging to the original category.

Currently, we only have limited confidence that this second aspect is indeed satisfied, in the sense that, while all three authors of this paper classified the modified images as belonging to their original categories, three people represent a small sample. Part of the on-going work [4] is to finalise and send a questionnaire to a significantly larger set of humans, who will be asked to label images (modified or not by our evolutionary algorithm) according to the CIFAR-10 categories. Such a test on sufficiently many unbiased humans is indeed necessary to check whether our evolutionary algorithm does what we intend it to do. We are quite confident that it is the case. Still this requires a statistically relevant confirmation. These aspects, and others, are addressed in [4].

Although our algorithm successfully managed to deceive the neural network, while producing adversarial images similar to the original, a closer comparison of the original and modified images, however, reveals the noisiness of the evolved images. This noisiness is noticeable here as CIFAR-10 images have a low resolution. We observed that although the noise is targeted, it appears to be random and evenly distributed across the image. This leads to two further directions which we plan to explore.

The first one would be attempting to explain why such small perturbations of the input are able to produce misclassifications with high confidence. One interesting idea is to analyse and compare our results to the behaviour of adversarial samples inputted to a Bayesian neural network.

The second further direction could be attempting to produce adversarial images that are almost entirely indistinguishable from the original. Towards this goal, several steps could be taken, such as optimising the EA with different mutations or using other datasets, notably with larger images.

One of the main advantages of our algorithm is the ability to treat the neural network as a black box, with no required knowledge of its architecture or parameters. We hence intend to extend this study to larger training sets (CIFAR-100 [14], ImageNet [6], etc.) with images of larger resolution, and on other NNs (Inception [26], etc.), while measuring the human perception by an ad-hoc questionnaire sent to a statistically relevant group of people [4]. Once these studies performed, we will be armed to conduct a thorough comparison between our EA approach and other adversarial image generation approaches listed for instance in the survey [1].

As pointed out by the referees (allow us to express our thanks for these further directions, and for their observations), one may try to consider a Pareto-based multi-objective fitness function, or a coevolutionary approach to adversarial training.

References

1. Akhtar, N., Mian, A.: Threat of adversarial attacks on deep learning in computer vision: a survey. CoRR abs/1801.00553 (2018). http://arxiv.org/abs/1801.00553
2. Bernard, N., Leprévost, F.: Evolutionary algorithms for convolutional neural network visualisation. In: Meneses, E., Castro, H., Barrios Hernández, C.J., Ramos-Pollan, R. (eds.) CARLA 2018. CCIS, vol. 979, pp. 18–32. Springer, Cham (2019). https://doi.org/10.1007/978-3-030-16205-4_2
3. Bernard, N., Leprévost, F.: How evolutionary algorithms and information hiding deceive machines and humans for image recognition: a research program. In: Proceedings of the OLA 2019 International Conference on Optimization and Learning, Bangkok, Thailand, 29–31 January 2019, pp. 12–15 (2019)
4. Chitic, R., Bernard, N., Leprévost, F.: Experimental evidence of neural networks being fooled by evolved images. Work in Progress (2019–2020)
5. Chollet, F.: Keras. GitHub code repository (2015–2018). https://github.com/fchollet/keras
6. Deng, J., Dong, W., Socher, R., Li, L.J., Li, K., Fei-Fei, L.: The imagenet image database (2009). http://image-net.org
7. Fawzi, A., Fawzi, H., Fawzi, O.: Adversarial vulnerability for any classifier. CoRR abs/1802.08686 (2018). http://arxiv.org/abs/1802.08686
8. Fawzi, A., Moosavi-Dezfooli, S., Frossard, P.: Robustness of classifiers: from adversarial to random noise. In: Lee, D.D., Sugiyama, M., von Luxburg, U., Guyon, I., Garnett, R. (eds.) Advances in Neural Information Processing Systems 29: Annual Conference on Neural Information Processing Systems 2016, Barcelona, Spain, 5–10 December 2016, pp. 1624–1632 (2016). http://papers.nips.cc/paper/6331-robustness-of-classifiers-from-adversarial-to-random-noise
9. Geifman, Y.: cifar-vgg (2018). https://github.com/geifmany/cifar-vgg
10. Geirhos, R., Rubisch, P., Michaelis, C., Bethge, M., Wichmann, F.A., Brendel, W.: Imagenet-trained CNNs are biased towards texture; increasing shape bias improves accuracy and robustness. In: International Conference on Learning Representations (2019). https://openreview.net/forum?id=Bygh9j09KX
11. Goodfellow, I., Bengio, Y., Courville, A.: Deep Learning. MIT Press, Cambridge (2016). http://www.deeplearningbook.org

12. Goodfellow, I.J., Shlens, J., Szegedy, C.: Explaining and harnessing adversarial examples (2014). arXiv preprint arXiv:1412.6572
13. De Jong, K.A.: Evolutionary Computation: A Unified Approach. A Bradford Book. MIT Press, Cambridge (2006)
14. Krizhevsky, A., Nair, V., Hinton, G.: The CIFAR datasets (2009). https://www.cs.toronto.edu/~kriz/cifar.html
15. Kullback, S., Leibler, R.: On information and sufficiency. Ann. Math. Stat. **22**, 79–86 (1951)
16. Li, X., Chen, Y., He, Y., Xue, H.: AdvKnn: adversarial attacks on *k*-nearest neighbor classifiers with approximate gradients. CoRR abs/1906.06591 (2019). http://arxiv.org/abs/1906.06591
17. Moosavi Dezfooli, S.M., Alhussein, F., Pascal, F.: DeepFool: a simple and accurate method to fool deep neural networks. In: Proceedings of the 2016 IEEE Conference on Computer Vision and Pattern Recognition, pp. 2574–2582. IEEE (2016)
18. Nguyen, A., Yosinski, J., Clune, J.: Deep neural networks are easily fooled: high confidence predictions for unrecognizable images. In: Proceedings of the 2015 IEEE Conference on Computer Vision and Pattern Recognition (CVPR), pp. 427–436 (2015)
19. Oliphant, T.E.: A Guide to NumPy. Trelgol Publishing, New York (2006)
20. Papernot, N., McDaniel, P., Wu, X., Jha, S., Swami, A.: Distillation as a defense to adversarial perturbations against deep neural networks. In: Proceedings of the 2016 IEEE Symposium on Security and Privacy (SP), pp. 582–597. IEEE (2016)
21. Papernot, N., McDaniel, P., Goodfellow, I., Jha, S., Celik, Z.B., Swami, A.: Practical black-box attacks against machine learning. In: Proceedings of the 2017 ACM on Asia Conference on Computer and Communications Security, pp. 506–519. ACM (2017). https://doi.org/10.1145/3052973.3053009
22. Shafahi, A., Huang, W.R., Studer, C., Feizi, S., Goldstein, T.: Are adversarial examples inevitable? CoRR abs/1809.02104 (2018). http://arxiv.org/abs/1809.02104
23. Shamir, A., Safran, I., Ronen, E., Dunkelman, O.: A simple explanation for the existence of adversarial examples with small hamming distance. CoRR abs/1901.10861 (2019). http://arxiv.org/abs/1901.10861
24. Simonyan, K., Zisserman, A.: Very deep convolutional networks for large-scale image recognition. CoRR abs/1409.1556 (2014). http://arxiv.org/abs/1409.1556
25. Su, J., Vargas, D.V., Sakurai, K.: One pixel attack for fooling deep neural networks. CoRR abs/1710.08864 (2017)
26. Szegedy, C., et al.: Going deeper with convolutions. arXiv 1409.4842 (2014). https://arxiv.org/pdf/1409.4842.pdf
27. Szegedy, C., et al.: Intriguing properties of neural networks (2013). arXiv:1312.6199
28. Taigman, Y., Yang, M., Ranzato, M., Wolf, L.: Deepface: closing the gap to human-level performance in face verification. In: Proceedings of the 2014 IEEE Conference on Computer Vision and Pattern Recognition (CVPR), pp. 1701–1708. IEEE (2014)
29. Tramèr, F., Zhang, F., Juels, A., Reiter, M.K., Ristenpart, T.: Stealing machine learning models via prediction APIS. In: Proceedings of the 25th USENIX Security Symposium Austin, TX, USA, 10–12 August 2016, pp. 601–618. USENIX (2016)
30. Varrette, S., Bouvry, P., Cartiaux, H., Georgatos, F.: Management of an academic HPC cluster: the UL experience. In: Proceedings of the 2014 International Conference on High Performance Computing & Simulation (HPCS 2014), pp. 959–967. IEEE, Bologna, July 2014. https://hpc.uni.lu

31. van der Walt, S., et al.: scikit-image: image processing in Python. PeerJ **2**, e453 (2014). https://doi.org/10.7717/peerj.453
32. Wang, Z., Bovik, A.C., Sheikh, H.R., Simoncelli, E.P.: Image quality assessment: from error visibility to structural similarity. IEEE Trans. Image Process. **13**(4), 600–612 (2004)
33. Yang, Y., Rashtchian, C., Wang, Y., Chaudhuri, K.: Adversarial examples for non-parametric methods: attacks, defenses and large sample limits. CoRR abs/1906.03310 (2019). http://arxiv.org/abs/1906.03310
34. Yu, X., Gen, M.: Introduction to Evolutionary Algorithms. Springer, Heidelberg (2010). https://doi.org/10.1007/978-1-84996-129-5
35. Zhou, X., Ma, T., Zhang, H.: Explaining adversarial examples with knowledge representation (2019). https://openreview.net/forum?id=BylRVjC9K7

Scenario Generation for Validating Artificial Intelligence Based Autonomous Vehicles

Christopher Medrano-Berumen[1] and Mustafa İlhan Akbaş[2(✉)]

[1] Florida Polytechnic University, Lakeland, FL 33805, USA
cmedranoberumen2844@floridapoly.edu
[2] Embry-Riddle Aeronautical University, Daytona Beach, FL 32114, USA
akbasm@erau.edu

Abstract. The progress in the development of artificial intelligence engines has been driving the autonomous vehicle technology, which is projected to be a significant market disruptor for various industries. For the public acceptance though, the autonomous vehicles must be proven to be reliable and their functionalities must be thoroughly validated. This is essential for improving the public trust for these vehicles and creating a communication medium between the manufacturers and the regulation authorities. Existing testing methods fall short of this goal and provide no clear certification scheme for autonomous vehicles. In this paper, we present a simulation scenario generation methodology with pseudo-random test generation and edge scenario discovery capabilities for testing autonomous vehicles. The validation framework separates the validation concerns and divides the testing scheme into several phases accordingly. The method uses a semantic language to generate scenarios with a particular focus on the validation of autonomous vehicle decisions, independent of environmental factors.

Keywords: Autonomous vehicles · Validation · Simulation · Testing

1 Introduction

Automated driving system (ADS) technology and ADS-enabled vehicles, commonly referred to as autonomous vehicles (AVs), have the potential to affect the world as significantly as the internal combustion engine [2]. Successful ADS technologies could fundamentally transform the automotive industry, transportation system, energy sector and more. Rapid progress is being made in artificial intelligence (AI), which sits at the core of ADS platforms. Consequently, autonomous capabilities such as those afforded by advanced driver assistance systems (ADAS) and other automation solutions are increasingly becoming available in the marketplace.

Testing and validation will help to ensure the safety and reliability of ADS platforms. Therefore, to achieve highly or fully autonomous capabilities, a major

N. T. Nguyen et al. (Eds.): ACIIDS 2020, LNAI 12034, pp. 481–492, 2020.
https://doi.org/10.1007/978-3-030-42058-1_40

leap forward in the validation of the ADS technologies is required. The key challenges for ADS validation are the comprehension of the unwritten rules of traffic, development of object models for perception and the creation of a progressively evolving safety measure for AI-based decision-making systems. Even though the current ADS validation methods offer potential improvements, they fall short in addressing these key challenges. The AVs must be proven to be at least as safe as human-driven vehicles before they are accepted as a new mode of transportation. Current methods of AV validation and verification such as shadow driving or annotated images-based testing are costly, slow, and resource intensive [2]. Hence, modeling and simulation is an indispensable asset to achieve validation goals for AVs.

This paper focuses on the simulation development for the validation of AV decision making. The proposed approach follows a framework that is built according to the 'separation of concerns' principle [4,10] and separates the roles of low-fidelity simulation testing, high fidelity simulation testing, hardware in the loop testing, sensor testing, proving-ground testing, and real-world testing. In this paper, the simulation model development for the low-fidelity testing is described.

In our solution, a semantic language is designed to describe the scenarios. Details used to describe roads are parameterized, allowing us to both randomly and deterministically generate roads and furthermore street networks. The geometry of the roads are paired with different road rules that can affect the tested vehicle's interaction with the road. The parameterization is done with the behavior of actors in a driving scenario as well, giving us a full description of the scenario that the vehicle under test perceives. This will give us the means by which test scenarios can be generated with entirely randomizable features. By designing the language to describe all possible scenarios, complete coverage of the AV's complex domain can be virtually achieved.

Focusing on the decision-making at AVs enables the utilization of low-fidelity simulation. The efficiency of this approach can be used with validation features such as assertions to classify scenarios and detect edge cases. We use a module built in MATLAB to take the scenario definition language as input and generate a test scenario in a way that our simulator can run. The module generates the road layout piece by piece, validating itself at each new piece to ensure the generated scenario is still legal. The scenario generation capability allows positioning the AV under test within this simulation by Hardware or Software in the Loop using any existing systems modeling environment, where the vehicle perceives its virtual environment and acts at each step. Data for further analysis is collected during and after the run as part of this framework.

2 Related Work

As the introduction of AVs becomes more inevitable, questions of its safety and viability also become more apparent. To gain the trust of the public in adopting the technology and state and federal entities in allowing them on their streets, a way to validate their competence needs to be established. Many companies are already testing their AVs in various ways, but most of these methods fall short by some measure.

The most common form of AV testing is known as 'Shadow Driving,' where a driver is ready to take over in case the AV decides to disengage or to preemptively prevent an accident [7]. To verify that AVs will be at least as safe as humans using this method will take at least 275 million miles [8]. Some if not all of those miles would also need to be repeated if there were to be any update to the AV under test. Using test tracks is also common for real world testing, as this allows companies to test specific, extreme scenarios. Hardware-in-the-loop (HiL) testing is another option that allows connecting the AV's brain into a simulation test and testing interaction with specific hardware components simultaneously [12].

While HiL or vehicle-in-the-loop forms of testing function in real time, Software-in-the-Loop (SiL) allows the test to run in simulation time. In these systems, the scene is generated within the simulation software and sent from the perspective of where the vehicle under test would be to either simulated sensors for SiL or the sensors for HiL. This view is then sent to the decision making component, either in the real world or simulated world based on what type of testing is being done. From there, once a decision is made, it is sent through any physics component of the vehicle, such as the tire or axis properties, which can also be done in the real or simulated world. Similar approaches are have been employed in simulation of the unmanned aerial vehicle systems [3,5].

Simulators used for AV testing vary by great degree in what their approach and focus are. Some of the testing approaches focus on using simulation to verify that newly learned maneuvers (e.g. u-turns, merging) are tested until they can be performed at a satisfactory rate. Others use simulations based on scenarios that their vehicles in the real world encountered [6]. Each important scenario from real life can then be fuzzed into generating many more scenarios based on the original to strengthen the coverage of that test. There are also several recent initiatives aiming for the standardization of AV validation by integrating different techniques. Intelligent Testing Framework [9] and PEGASUS [14] are two important examples of such approaches. Vehicle-in-the-Loop simulators place a physical representation of the vehicle under test (often an actual vehicle) on a platform surrounded by projected screens to create the full driving experience from the driver's perspective. These simulators often have higher cost and are mainly used to study the human computer interaction aspect of driving.

Several companies and research groups have been developing their own framework for testing AVs, but there is yet to be one adopted as a standard. For instance, Waymo tests each of their primary subsystems, the vehicle, their hardware, and their software, individually as well as together using simulation, closed-course testing, and real world driving [1]. The scenarios they focus on are based on training the vehicle behavioral competencies as designed by Berkeley's PATH, such as detecting bikes and pedestrians or responding to emergency vehicles [11], as well as behavioral competencies designed by Waymo themselves [1]. PEGASUS, a joint effort between multiple groups from science and industry in Germany, aims to develop a full toolchain for AV validation, looking at traditional vehicle validation and new innovations in the field for inspiration. A common trait with these is that they are still in progress of being designed and provide no hard definition for when an AV can be demonstrated to be sufficiently competent.

The majority of the current testing approaches target testing of the full vehicle stack, from scene perception and understanding to making a decision for action in that scene. In our approach, we focus mainly on the decision step. In other words, our approach is designed to test the decision making of the AV under test with the assumption that perception (sensors and sensor fusion) and action (vehicle dynamics) are functioning perfectly.

This focus point of testing the decision making aspect played into deciding the requirements of the simulation platform as well. The framework requires bare bones modeling of the physical world without details on environmental conditions. For that reason, we developed a method in several candidate simulation tools for the definition of scenes in our previous work [10] and MATLAB is chosen as the simulation platform. The MATLAB AD Toolbox represents actors in the scenario as cuboid boxes [13]. We chose MATLAB as our simulation platform, as it provides a full toolchain testing platform, the Automated Driving (AD) Toolbox, without focusing on the vehicle dynamics or realistic visuals for sensor testing, two things that we are not looking at this point of the validation process. By rendering all objects in the scenario, from the vehicles to the pedestrians as 3d boxes, we can get the level of abstraction necessary to make the testing vehicle agnostic. MATLAB's native implementations of matrix math functions make implementation of the model of computation simpler as well.

3 Scenario Generation Approach

The current methods used for AV validation and testing fall short in addressing the AV validation challenges. Our validation method makes use of hardware validation methods since the validation of complex digital hardware such as a modern central processing unit chip has similar challenges [2,4] such as complex underlying behavior, high rate of scenarios and high cost of failure. In the hardware verification, there are three key methods to solve these challenges: Abstraction and Separation of Concerns, Constrained Pseudo-Random Test Generation, Assertions, and Coverage Analysis. Within the hardware design and validation process, there is a strong use of abstraction to separate concerns and decompose the process of validation. Since the top-level validation has been a much more difficult challenge, in addition to conventional "real world" simulation of program streams, hardware validation has made heavy use of the Constrained Pseudo-Random Test Generation, Coverage Driven Testing, and Functional Assertions. Hence, we also use the separation of concerns principle.

One of the most important parts of the proposed method in this paper is verifying the decision making component of AVs. This is where, given an understanding of the environment, the AV decides on what it must do to safely get to its desired destination. The decision making part of the framework utilizes a model of computation to verify the AV's decision made in simulation [4]. The simulation model provides functionalities for the implementation of validation features. This makes it possible for the full space of random input to create all permutations for driving scenarios, with logical constraints keeping them in the space of realistic scenarios. A deep understanding of road design standards is also required to maximize the coverage. Then, the scenario generation model can be used to detect edge scenarios. Additionally our approach allows implementation of approaches from software verification for assertions and exception handling.

The model must be capable of creating scenarios to reflect the set of all possible situations. We developed a matrix-based semantic language for breaking down the factors that define a scenario including roads, actors, and traffic logic. Therefore, the first phase in our approach is the creation of test scenarios in our matrix based system to generalize scenario characteristics and create an efficient system of labeling and sorting. The numerical matrix is read as input where each row is a different assertion describing a single road piece or actor that can then be parsed to generate the scenario. To do this, road network and actors are reduced to their most basic elements in terms of Newtonian physics such as center of mass and dimensions. These elements are then parameterized according to real life conditions. It's important to note that the model contains no environmental factors. The overall scenario generation flowchart is given in Fig. 1.

The road pieces that make up the scenarios are made to be as general as possible in order to maximize coverage with the goal of eventually being able to create all possible roads. Each new piece will bring with it new possible roads making the definition of these roads a large ongoing effort. With new pieces, new parameters will have to be defined and old ones redefined, meaning the current parameters may not be totally representative of later iterations. It should also be noted that certain parameters may have different meanings in different road pieces, as each piece will have needs specific to them.

4 Implementation Study

For the implementation of custom scenarios, we create simulation matrices to define the roads and actors. Within these matrices, each row defines a new road piece or actor, and each column describes a certain trait of said piece or actor, from curvature to speed limit and from vehicle type to moving behavior.

4.1 Road Pieces

Geometric centers are used for placing the roads in simulation models. The width of the road at each of those centers, the banking angles, and the details of the lanes. Then, generating the roads is a matter of taking an input and converting it into a series of points that the road follows within the context of the scene. The first value in the input defines what type of road piece will be

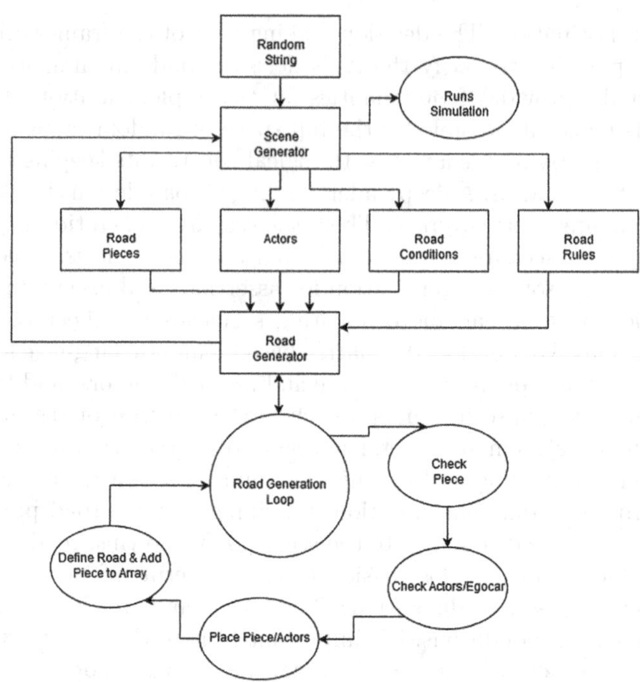

Fig. 1. The flowchart for creating a test scenario

created. By limiting the roads to specific road pieces, the generated scenario can be constrained to realistic road networks and situations. As each road piece is generated, it is stitched to the previous piece by rotating it so that the tangent line along the first point is lined up with the tangent line of the last point of the previous piece and shifting it to that coordinate in the driving scenario. Between two consecutive pieces, there is also an intermediary piece to smooth transitions between two pieces with different number of lanes. The Fig. 2 shows a randomly created street network with multiple road types.

The road pieces can also be created for specific street networks to demonstrate the methodology of breaking down an environment into its characteristic driving components (road pieces and parametrization) as a means of verifying an AV for driving in said environment (Level 4 Autonomous Driving - environment specific). The current parameters are Road Type, Length, Lanes, Bidirectional, Mid-Lane, Speed Limit, Intersection Pattern, Curvature (two types), Pedestrian Pathways, Outlets and Show Markers. In the following subsections, we present several road piece examples and show how the parameters are used to generate these pieces.

Multi-lane Road. The multi-lane road piece has variable length and width that can take on virtually all geometries with G2 continuity, which is a desired

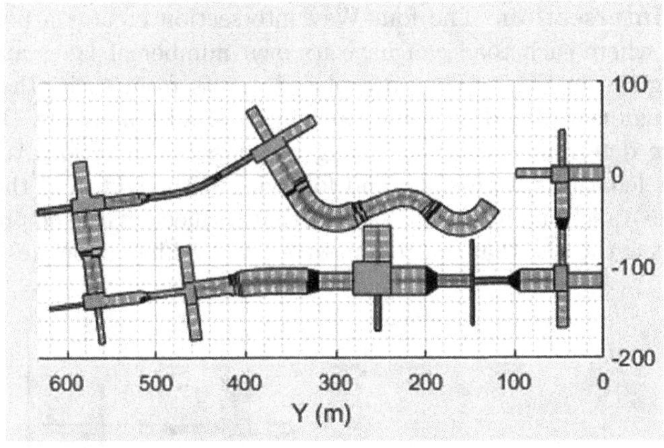

Fig. 2. Example of a street network with multiple road types.

property when designing real roads. It achieves this by making permutations of three geometric primitives: lines, clothoids, and arcs, using the clothoid to transition between different curvatures at a constant rate. The parameters determining the geometry of the road are curvature1 and curvature2. If either of them are zero, the road is made of a line to a clothoid to an arc, where the arc is the non-zero value. If both are zero, a straight line is made. And if both are non-zero values, then either a Clothoid-Arc-Clothoid (0 to curvature1 to curvature2) or an Arc-Clothoid-Arc (curvature1 to curvature2) is generated, the current deciding factor between the two being random. By composing the roads using these three primitives, the permutations should make up most all possible roads. The length of each of the three parts is one third of the total length given from the determining matrix row. The Fig. 3 shows a multilane road example.

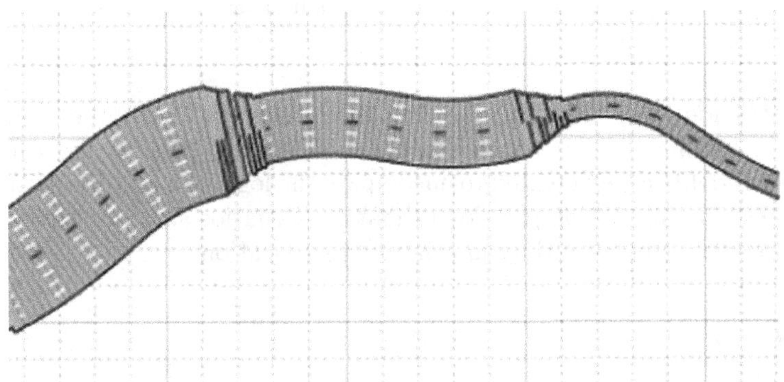

Fig. 3. Multilane road example.

Four-Way Intersection. The four-Way intersection creates a perpendicular intersection, where each road can have its own number of lanes and direction of travel as given in Fig. 4. They are placed across from each other around a central rectangle calculated by using the widths of all four roads. Once placed, the paths are determined by distributing the possible directions to each lane, starting with left turns if possible, and following with right, and then forward. This is for the direction on the intersection that connects with the previous road piece, as that is the one that matters most for the vehicle under test.

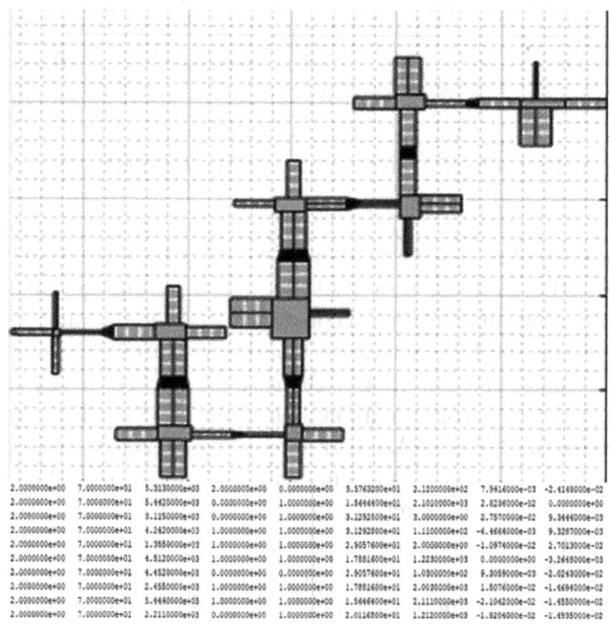

Fig. 4. Four-way intersection example.

Side Entrance. The Side Entrance is a road piece that splits the road with a median. It also has a side entrance in the middle to a space as given in Fig. 5. This piece is deliberately created to encompass the logic of a real-life path. A side entrance can be highly complex for validation scenarios since multiple vehicles and actors may interact with each other at this location.

4.2 Actors

The actors in the scenarios can be randomly generated for both testing and demonstration purposes. However, defining a language to encompass all potential actor behavior is an ongoing task in our project. Currently, actors can follow a straight path along the road with the option to have variability in several

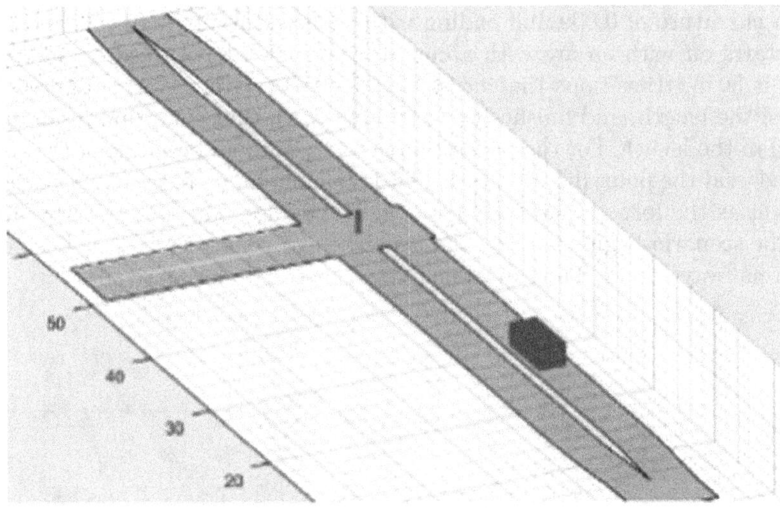

Fig. 5. Side lot entry road piece example.

actions. To generate actors in the scenarios provide the variability, we use various parameters such as type, speed and dimensions. The main two actor types are given as examples in the following subsections.

Vehicle. The vehicle actor is generated along the path (*StartLocation* parameter) and moves forward along a lane or in the opposite direction given the *Forward* parameter and the availability of an opposite direction (the road is bidirectional). Its size is determined based on its vehicle type (*Car*, *Truck*, *Motorcycle*) and varies according to its dimensions. In MATLAB, this sets the x, y, and z dimensions of the actor as they are all represented by boxes. Its path can be set to follow the lanes exactly or offset to varying degrees from the lane which also uses an offset parameter. It moves at the road's speed limit, set in the road pieces parameters, with variation according to the *MoveSpeed* parameter.

Pedestrian. Pedestrians are set on either side of the road somewhere along the scenario based on their initial location and their directions. Then the actors move across at an average walking speed varied by using a speed parameter. Two path options are walking straight across and walking across with a pause at some point in the middle.

4.3 Example Scenario

The Fig. 6 gives an example scenario created in our system. Here, a curving multi-lane road with a median is constructed as well as a pedestrian actor that crosses said road. The road's geometry follows an Arc-Clothoid-Arc pattern starting with a curvature of 0.009 and ending with a curvature of −0.014, meaning the road starts off with an arc with a curvature of 0.009 for a third of the length, which is 50 m, transitions the curvature to −0.014 with a clothoid curve for a third of the length, and finishes by continuing with that curvature in an arc for a third of the length. For the pedestrian actor, a random point along the road is selected, and the point directly to the left is calculated, its path moving across to the right as the forward parameter was given as false. This actor starts moving once the scenario begins, and by the time the ego vehicle arrives, it has almost gotten halfway across. The ego vehicle was generated automatically.

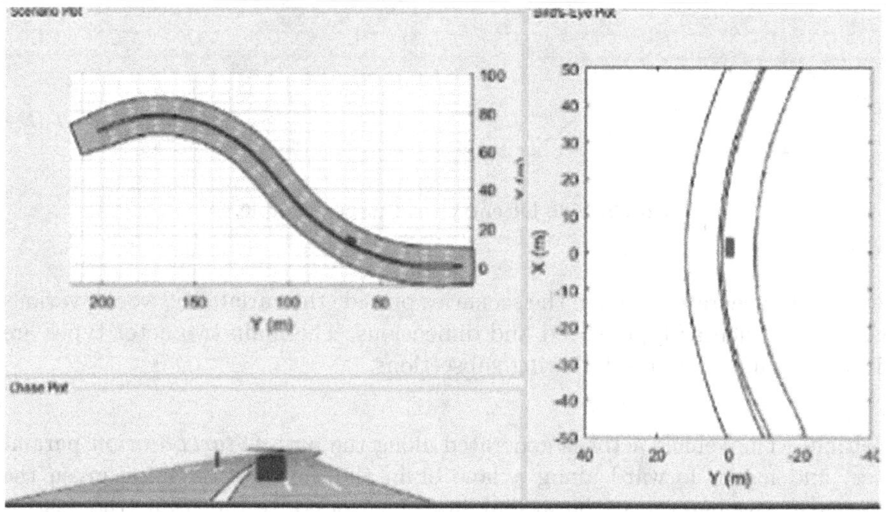

Fig. 6. Sample scenario including a multilane road with a median and a pedestrian.

The simulation data is collected during and after each run. The current collected data points include safety and legality of each decision and the scenario's definition. Since the scenarios are generated by a certain input, they are available to be recreated whenever it is required.

For the actors, in order to minimize their box representation unrealistically occupying the same space, roads store information regarding the actors and at what point in the scene they are going over that lane. If two actors are at the same lane at the same point in the simulation, a different lane is used or the actor is made to slow down for a random but small period of time before moving forward. This does not apply to pedestrians as how they interact with each other is less critical in simulations.

5 Conclusion

The validation is critical for AV technology as for other complex AI-based systems. Testing and certification of AVs will be crucial for their deployment in the transportation system. In this paper, we present the details of a test scenario generation method for AV simulation testing. We create a semantic language and propose a road network generation mechanism for random and constrained scenario generation. The generated scenarios are also automatically checked as the new road pieces are added. The current system is capable of taking a particular street network and producing the testing scheme for it.

As the future work, we plan to enable scenario export and import features to exchange scenarios automatically with other available scenario generation tools.

References

1. Waymo safety report: On the road to fully self-driving (2018)
2. Akbaş, M.İ., et al.: Unsettled technology areas in autonomous vehicle test and validation. Society of Automotive Engineers (SAE) International Journal of Connected and Automated Vehicles, June 2019
3. Akbaş, M.İ., Solmaz, G., Turgut, D.: Actor positioning based on molecular geometry in aerial sensor networks. In: IEEE International Conference on Communications (ICC), pp. 508–512 (2012)
4. Alnaser, A.J., Akbaş, M.İ., Sargolzaei, A., Razdan, R.: Autonomous vehicles scenario testing framework and model of computation. Society of Automotive Engineers (SAE) International Journal of Connected and Automated Vehicles, December 2019
5. Brust, M.R., Akbaş, M.İ., Turgut, D.: VBCA: a virtual forces clustering algorithm for autonomous aerial drone systems. In: IEEE Systems Conference (SysCon), pp. 1–6. IEEE (2016)
6. Dolgov, D.: Google self-driving car project-monthly report-September 2016-on the road. Technical report, Google (2016)
7. Favarò, F.M., Nader, N., Eurich, S.O., Tripp, M., Varadaraju, N.: Examining accident reports involving autonomous vehicles in California. PLoS ONE **12**(9), e0184952 (2017)
8. Kalra, N., Paddock, S.M.: Driving to safety: how many miles of driving would it take to demonstrate autonomous vehicle reliability? Transp. Res. Part A: Policy Pract. **94**, 182–193 (2016)
9. Li, L., Huang, W.L., Liu, Y., Zheng, N.N., Wang, F.Y.: Intelligence testing for autonomous vehicles: a new approach. IEEE Trans. Intell. Veh. **1**(2), 158–166 (2016)
10. Medrano-Berumen, C., Akbaş, M.İ.: Abstract simulation scenario generation for autonomous vehicle verification. In: Proceedings of the IEEE SoutheastCon. IEEE, April 2019
11. Nowakowski, C., Shladover, S., Chan, C.: Behavioral competency requirements methodology project background regulatory issues and potential regulatory strategies for highly automated vehicles (AVs) how to ensure safety prior to deployment? In: Automated Vehicles Symposium. AVS, Ann Arbor (2015)

12. Sarhadi, P., Yousefpour, S.: State of the art: hardware in the loop modeling and simulation with its applications in design, development and implementation of system and control software. Int. J. Dyn. Control **3**(4), 470–479 (2015)
13. The MathWorks Inc.: MATLAB and Automated Driving System Toolbox (Release R2018b), Natick, Massachusetts, United States
14. Winner, H., Lemmer, K., Form, T., Mazzega, J.: PEGASUS—first steps for the safe introduction of automated driving. In: Meyer, G., Beiker, S. (eds.) Road Vehicle Automation 5. LNM, pp. 185–195. Springer, Cham (2019). https://doi.org/10.1007/978-3-319-94896-6_16

Privacy Protection in Location Based Service by Secure Computation

Masato Ishikawa and Noriaki Yoshiura(✉)

Department of Information and Computer Sciences, Saitama University, 255, Shimo-ookubo, Sakura-ku, Saitama, Japan
{mishikawa,yoshiura}@fmx.ics.saitama-u.ac.jp

Abstract. Mobile devices with Global Positioning System enable various kinds of location information searches using location based services (LBS). Location information and query contents of searchers are privacy information. Privacy protection requires to hide location information and query contents of searchers from not only third parties such as eavesdroppers but also location database servers which answer queries. This paper focuses on privacy protection in nearest neighbor search. There are several researches of secure computation in nearest neighbor search for other than LBS. This paper proposes a new method of privacy protection in LBS. The method is based on Asymmetric Scalar-product Preserving Encryption. This paper also shows that the proposed method is faster and more secure than conventional privacy protection methods in LBS.

1 Introduction

Mobile devices with Global Positioning System (GPS) have been widely spread. The mobile devises with GPS enable to easily use location based services (LBS). An example of LBS is neighborhood search. "Where is the nearest hospital from here?" is a typical query of neighborhood search. This kind of search is implemented in a client server model. In using LBS, searchers provide the location information of searchers and query contents to location database servers which answer queries. This information is private and therefore sharing it without consent may violate the privacy of the searchers. Therefore, privacy protection requires to hide location information of searchers and query contents from not only eavesdroppers but also location information servers which answer queries.

Several methods have been proposed to resolve this problem. One of them uses secure computation. This method is as follows; a searcher encrypts the location information of the searcher and a query content. A database server also encrypts all data. The searcher sends the encrypted query and location information to a cloud server and the database server also sends the encrypted data to the cloud server. The cloud server computes on the encrypted data and query to obtain the encrypted data of the nearest neighbor, but does not know what the nearest neighbor is. The cloud server sends the encrypted data of the nearest neighbor to the searcher. The searcher decrypts the encrypted data to obtain

N. T. Nguyen et al. (Eds.): ACIIDS 2020, LNAI 12034, pp. 493–504, 2020.
https://doi.org/10.1007/978-3-030-42058-1_41

the nearest neighbor. A disadvantage of the method is that the computation on encrypted data, encryption and decryption require much computation power and that the much encrypted data is transmitted among the database and cloud servers and the searcher. In order to resolve the disadvantage, Efficient Privacy-preserving Location-Based Service (ePriLBS) [3] has been proposed. ePriLBS uses K-anonymity and homomorphic encryption. ePriLBS has, however, several disadvantages; one of them is that location information of searchers is transmitted to the cloud servers.

Secure computation is used in the fields different from LBS. Asymmetric Scalar-product Preserving Encryption (ASPE) is a secure computation [12,13]. The original aim of ASPE is secure computation on multi-dimensional data such as human biological data. ASPE uses the characteristics of the matrix to compute the similarity between a query and the data contained in databases by using encrypted data. This paper proposes a new method of privacy protection in LBS by applying ASPE. This paper also shows that the proposed method is faster and more secure then conventional LBS methods.

This paper is organized as follows; Sect. 2 explains related works. Section 3 explains ASPE. Section 4 proposes the proposed method. Section 5 shows the performance comparison between the conventional LBS method and the proposed method. Section 6 concludes this paper.

2 Related Works

This section discusses related works. The researches related to privacy protection in LBS include the methods using "Trusted Third Party" (TTP), the method using no TTP, the method using untrusted intermediate party and so on.

2.1 Methods with TTP

TTP relays queries from searchers to database servers and hides privacy of searchers. "Mix zone" method was proposed by Beresford in 2004 [2]. This method uses a range called "mix zone", which is managed by TTP. Users in the same mix zone exchange user IDs with other users to hide the location information of searchers. The search results are not the best but available if the size of mix zone is appropriate. Recently, a method called "MobiMix" has been proposed [10]. MobiMix extends mix zone onto road networks.

K-anonymity is known as an information hiding method and is unrelated with LBS [6]. Gruteser and Grunwald first used K-anonymity for privacy protection in LBS. They use an anonymizer which is a trusted third party between servers and users. Searchers send their location information and queries to the anonymizer. The anonymizer obscures query contents by sending the location information of $k-1$ users near the searcher and the query to the servers. The search results from the servers are the nearest neighbors which match query contents for $k-1$ users and the searcher. The anonymizer selects and sends the nearest neighbor to the searcher.

Using K-anonymity has several disadvantages. First, some users must be near the searcher. Second, the servers can guess query contents or location information of searchers if the servers obtain several additional information. Third, searchers cannot use LBS continuously. In order to resolve disadvantages, several methods have been proposed. Some of them use l-diversity [1] and t-approximation [8], which redeem disadvantages of K-anonymity. A method "Anonymity of Motion Vectors" (AMV) [11], which supports continuous searches, has also been proposed. However, K-anonymity has a disadvantage that compromise of the anonymizers exposes location information of searchers and query contents.

2.2 Methods Without TTP

There are several advantages in methods using TTP; TTP gathers query contents and location information of searchers. TTP is also easily targeted by attackers. Therefore, methods without TTP have been proposed. One of the methods protects privacy of the searchers by sending location information of searchers to the servers with much dummy location information [7]. The servers send all search results for the much location information. The searcher obtains an appropriate result among all the search results. Generating dummy information omits TTP, but it is impractical to generate much dummy information in order to preclude obtaining the location information of searchers.

"Private Information Retrieval" (PIR) [4] is known as a method of information hiding. Ghinita first used PIR to protect location information in 2008 [5]. PIR enables searchers to receive information from the servers without the servers knowing what information the searchers receive. A disadvantage of PIR is that the amount of communication and the amount of computation are enormous; both the amount of communication and the amount of calculation are $O(N^m)$ where N is the number of POI and m is the number of bits that represent POI location information, where POI (Point of Interest) is a search target place. For example, "Where is the nearest hospital from here?" is a query and POI is a location of a hospital.

2.3 Methods with Untrusted Intermediate Party

Preparing trusted TTP is difficult and methods using untrusted intermediate party in LBS have been proposed. "efficient Privacy-preserving Location-Based Service" (ePriLBS) consists of a searcher, a location database server, and an intermediate proxy server. Searchers send a query to the proxy server, which generates dummy queries and sends them to the database server. The dummy queries hide location information of searchers and query contents. The communications between the searchers and the database servers are encrypted with a shared key and the proxy server does not obtain the search results. Furthermore, in ePriLBS, using a homomorphic encryption reduces computation cost in the devices of searchers and communication cost between the servers and the searchers. A disadvantage of ePriLBS is that proxy servers obtain the location information of searchers.

Another proposed method is based on DPT (Distance Preserved Transformation) [9]. DPT encrypts the location information of searchers and POI with maintaining the distance between searchers and POI. Secure computation on encrypted data on the database servers or cloud servers obtains the nearest neighbors with hiding the location information of searchers and query contents. However, if correct location information of several POIs is exposed, the location information of searchers and query contents can be guessed.

3 Asymmetric Scalar-Product Preserving Encryption

Asymmetric Scalar-product Preserving Encryption (ASPE) has been proposed to improve the disadvantage of DPT. ASPE maintains only the magnitude relationships of the distances between elements in databases. The distances are not obtained from the magnitude relationships. Because of this feature, ASPE is strong against even attacks using the correspondence between encryptions and decryptions of some elements on databases. The original aim of ASPE is secure computation on multi-dimensional data such as human biological data. This paper proposes the method using ASPE in LBS.

ASPE consists of database servers, cloud servers and searchers. Cloud servers have huge data capacity and huge computation resources. Data servers have m d-dimensional data $p_1, p_2, \cdots, p_m \in \mathbb{R}^d$ where $\mathbb{R}$ is a set of real numbers. Suppose that a searcher searches the nearest neighbor of the query point $q \in \mathbb{R}^d$. The searcher sends the encryption of $q \in \mathbb{R}^d$ to the cloud servers. The database servers send form $D' = \{E_T(p_1),\ E_T(p_2), \cdots, E_T(p_m)\}$ to the cloud servers where $E_T(p_i)$ is a encryption of p_i. The cloud servers obtain the encryption of the k nearest neighbors by computations on D' and the encryption of q without decryption. The cloud servers send the encryption of the k nearest neighbors to the searcher. The searcher decrypts the encryption of the k nearest neighbors.

The following terminology is used in k-NN search based on ASPE.

- $D = \{p_1, \cdots, p_m\}$ is a set of data.
- $D' = \{p'_1, \cdots, p'_m\}$ is a set of encrypted data where p'_i is an encryption of p_i.
- q is a query location.
- q' is an encryption of q.
- d is a dimension of elements in D.
- $d(p_i, q)$ is a distance between p_i and q where $d(p_i, q) = \sum_{j=1}^{d}(p_{ij} - q_j)^2$.
- M is a matrix.
- $M_{i,j}$ is an element of i-th row and j-th column in M.
- I_q is a set of k nearest neighbors of query location q.
- Key is a encryption key that is generated by the database servers.

ASPE consists of the following procedures. The method proposed in the paper uses these procedures.

KeyGen. The database servers randomly generate an invertible matrix $M \in \mathbb{R}^{(2d+2)\times(2d+2)}$, a positive real number $\alpha \in \mathbb{R}^+$ and two $(d+1)$-dimensional vectors $r = (r_1, \cdots, r_{d+1})$ and $s = (s_1, \cdots, s_{d+1}) \in \mathbb{R}^{d+1}$. The encryption key Key is defined as $\{\alpha, r, s, M, M^{-1}\}$.

TupleEnc. The database servers encrypt the data and send them to the cloud servers as follows.

1. The database servers generate p'_i by encrypting each data $p_i \in D$. $p'_i = \dot{p}_i M^{-1}$, where
$\dot{p}_i = \left(r_1 - 2\alpha p_{i1}, s_1, \cdots, r_d - 2\alpha p_{id}, s_d, r_{d+1} + \alpha \sum_{j=1}^{d} p_{ij}^2, s_{d+1}\right)$
2. The database servers send $D' = \{p'_1, \cdots, p'_m\}$ to the cloud servers.

QueryEnc. The searcher sends a query of k-NN search to the cloud servers as follows.

1. The searcher generates a random d-dimension vector $t = (t_1, t_2, \cdots, t_d) \in \mathbb{R}^d$, and sets $\dot{q}_l = q_l + t_l$ $(1 \leq l \leq d)$. The searcher sends $\dot{q}_1, \dot{q}_2, \cdots, \dot{q}_d$ to the database servers.
2. The database servers generate a random real number $u \in \mathbb{R}$ and a positive real number $\beta_p \in \mathbb{R}^+$. Moreover, the database servers $(2d+2)$-dimension vector X and $(2d+2)$ by d matrix Y as follows;
$X_j = \beta_q\left(M_{j,2d+1} + uM_{j,2d+2} + \sum_{l=1}^{d} \dot{q}_l(2M_{j,2l-1} - M_{j,2l})\right),$
$Y_{jl} = \beta_q(M_{j,2l} - M_{j,2l-1}) \quad 1 \leq j \leq 2d+2, \;\; 1 \leq l \leq d$
where X_i is i-th element of X and Y_{ij} is an element of i-th row and j-th column in Y. The database servers send X and Y to the searcher.
3. The searcher calculates $q' = (q'_1, q'_2, \cdots, q'_{2d+2})$ where $q'_j = X_j + \sum_{l=1}^{d} Y_{jl}$ $(q_l + 2t_l)$ $(1 \leq j \leq 2d+2)$. The searcher sends q' to the cloud servers.

OutKNN. The cloud servers receive D' and q' and calculate $p'_i(q')^T$ for all i. $p'_i(q')^T$ preserves magnitude relation; if the distance between p_i and q is longer than that between p_j and q, $p'_i(q')^T$ is greater than $p'_j(q')^T$. Therefore, the cloud servers can select the k nearest neighbor elements of D' for the query location q. The cloud servers send the indexes of the k elements of D' to the searcher, which obtains the k nearest neighbors of the query location.

4 Proposed Method

This paper proposes the method using ASPE in LBS. ASPE provides higher secureness than conventional ePriLBS because ePriLBS must provide the query location information and query contents to the cloud servers by using K-anonymity. On the other hand, ASPE does not require to provide query location information or query contents. However, the original aim of ASPE is not privacy protection in LBS. LBS uses query location and query contents. Query location is a vector of xy coordinates, but query contents are not a scaler or a vector. Suppose "Where is a hospital near here?" is a query. "hospital" is a type of POI, but is not a scaler or a vector. The database servers have the POI information $(x, y, kind)$ where x and y are xy coordinates and $kind$ is a type of POI. One of the problems in using ASPE in LBS is handing types of POI.

The other problem is the method of sending the search results. As shown in the previous section, ASPE obtains k-NN results only in the form of database indexes. In LBS, the searchers must obtain the location information of the k-NN results. The proposed method uses AES encryption for the search results. AES is

Algorithm 1. ASPE1

```
○KeyGen
Server :
key ← generateKey(r)
○TupleEnc
Server :
P'← EncPOI(P,key.s,key.r.key.α,key.M^{-1})
○QueryEnc
User :
q̇ ← q + t
aeskey ← generateAESKey()
Send q̇, kind, aeskey to Server
Server :
for all p_i ∈ P do
    if p_i.kind == kind then
        p_kAES ← EncPOIwithAES(p_i, aeskey)
        P_AES ← p_kAES
        P'_q ← p'_i
    end if
end for
Send P_AES, P'_q to Cloud
X ← generateX(q̇, key.M, u, β_q)
Y ← generateY(q̇, key.M, u)
Send X, Y to User
User :
q' = EncQuery(q, X, Y, t)
Send q' to Cloud
○OutKNN
Cloud :
I_q ← KNN(P'_q, q')
EncResult_q ← choicePOI(I_q, P_AES)
Send EncResult_q to User
User :
Result_q ← DecResultwithAES(EncResult_q, aeskey)
```

a shared key encryption method. In the proposed method, a searcher generates a shared key of AES and sends it to database servers. The database servers encrypt the database by AES and send it to the cloud servers. The searchers receive the search results and AES encrypted database from the cloud servers. Therefore, the cloud servers and the database servers do not obtain the search results, but the searchers obtain them.

The second problem is resolved by AES encryption, but the first problem still remains unresolved. From the viewpoint of the first problem, this paper proposes the three patterns of methods to handle types of POI.

- Pattern 1 (Algorithm 1)
  A method informs database servers what types of POI searchers search.
- Pattern 2 (Algorithm 2)
  A method does not inform database servers what types of POI searchers search. The cloud servers find all nearest neighbors for each type of POI and send them to searchers. The searcher selects the nearest neighbors for the POI type which the searchers search.

Algorithm 2. ASPE2

○KeyGen
Server :
key ← generateKey(r)
○TupleEnc
Server :
P'← EncPOI(P,key.s,key.r.key.α,key.M^{-1})
○QueryEnc
User :
$\dot{q} \leftarrow q + t$, aeskey ← generateAESKey()
Send $\dot{q}, aeskey$ to Server
Server :
for all $p_i \in P$ **do**
    p_{kAES} ← EncPOIwithAES($p_i, aeskey$)
    switch $p_i.kind$
        case j : $P_{jAES} \leftarrow p_{kAES}$, $P'_{jq} \leftarrow p'_i$
    end switch
end for
Send P_{jAES}, P'_{jq}(all j) to Cloud
X ← generateX($\dot{q}, key.M, u, \beta_q$), Y ← generateY($\dot{q}, key.M, u$)
Send X, Y to User
User :
q' = EncQuery(q, X, Y, t)
Send q' to Cloud
○OutKNN
Cloud :
for all j **do**
    $I_j q$ ← KNN(P'_{jq}, q'), $EncResult_j q$ ← choicePOI($I_j q, P_{jAES}$)
end for
Send $EncResult_{jq}$ (all j) to User
User :
for all j **do**
    $Result_{jq}$ ← DecResultwithAES($EncResult_{jq}, aeskey$)
    if ($q_i.kind \in Result_{jq}$) == a kind User want **then**
        $Result_q \leftarrow Result_{jq}$
    end if
end for

- Pattern 3 (Algorithm 3)
  A method encrypts what types of POI searchers search and sends it to cloud or database servers.

Table 1 shows secureness of ePriLBS and the proposed methods. Algorithm 1 is an pseudo code of Pattern 1. **KeyGen** generates cryptographic key for ASPE using a random number r and **TupleEnc** encrypts POI by EncPOI. In **QueryEnc**, the searcher randomly generates a vector t and a shared key $aeskey$ of AES and sends, to the database server, $\dot{q}$, $aeskey$ and what type of POI is searched. The database server encrypts, by AES, the same type of POI as the POI type which the searcher searches. The database server sends the data encrypted by AES and the data encrypted by TupleEnc to the cloud server. The database server also sends key X and Y to the searcher. The searcher encrypts the query location by X and Y by **QueryEnc** and sends it to the cloud server. In **OutKNN**, the cloud server obtains k nearest neighbors by computation on the encrypted data from the database server and the searcher. The

Algorithm 3. ASPE3

○KeyGen
Server :
key ← generateKey(1^r)
○TupleEnc
Server :
P'← EncPOI(P,key.s,key.r.key.α,key.M^{-1})
○QueryEnc
User :
$\dot{q} \leftarrow q + t$
aeskey ← generateAESKey()
Send $\dot{q}, aeskey$ to Server
Server :
for all $p_i \in P$ **do**
  $p_{kAES} \leftarrow$ EncPOIwithAES$(p_i, aeskey)$, $P_{AES} \leftarrow p_{kAES}$, $P'_q \leftarrow p'_i$
end for
Send P_{jAES}, P'_{jq}(all j) to Cloud
$X \leftarrow$ generateX$(\dot{q}, key.M, u, \beta_q)$, $Y \leftarrow$ generateY$(\dot{q}, key.M, u)$
Send X, Y to User
User :
$q' =$ EncQuery(q, X, Y, t)
Send q' to Cloud
○OutKNN
Cloud :
$I_q \leftarrow$ KNN(P'_q, q'), $EncResult_q \leftarrow$ choicePOI(I_q, P_{AES})
Send $EncResult_q$ to User
User :
$Result_q \leftarrow$ DecResultwithAES$(EncResult_q, aeskey)$

Table 1. Secureness of ePriLBS and the proposed methods

	ePriLBS		The proposed methods	
	Proxy servers	Database servers	Cloud servers	Database servers
Query location	Known	K-anonymity	Unknown	Unknown
Query region	Unknown	K-anonymity	Unknown	Unknown
Query type	Unknown	K-anonymity	Unknown	Unknown (pattern 2, 3) Known (pattern 1)
Search result	Unknown	K-anonymity	Unknown	Unknown

cloud server sends the AES encrypted data k nearest neighbors. The searcher decrypts the AES encrypted data and obtains k nearest neighbors. In Algorithm 1, the database server knows the what type of POI the searcher searches. This is privacy violation of the searcher and a disadvantage of Algorithm 1.

Algorithm 2 is a pseudo codes of Pattern 2. Difference between Algorithms 1 and 2 is that the searcher does not inform the database server what type of POI the searcher searches. In **EncPOIwithAES**, the database server encrypts the location information and type of POI by AES. The database server sends all encrypted POI data to the cloud server. The searcher sends an encrypted query location to the cloud server. In **OutKNN**, the cloud server obtains k nearest neighbors for all types of POI by computation on the encrypted data. The cloud server sends all the search results to the searcher. The search results include what the searcher does not search and the searcher selects, among all the search results, k nearest neighbors of type of POI which the searcher searches. In Algorithm 2, the database server and the cloud server do not know what type of POI the searcher searches or the query location. However, the cloud server computes on encrypted data for all pairs of query location and all POI locations, and sends k nearest neighbors for each type of POI. The amount of computation and communication are enormous and are disadvantages of Algorithm 2.

Algorithm 3 is a pseudo code of Pattern 3. Like Algorithm 2, the searcher does not inform the database server what type of POI the searcher searches. Algorithm 3 assigns integers to types of POI and handles POI data as three dimensional vector. The database server encrypts three dimensional vectors of POI data and sends them to the cloud server. The searcher also encrypts a query including query location and type of POI and sends the encrypted query to the cloud server. In **OutKNN**, the cloud server computes on the encrypted data from the searcher and the database server to obtain the encryption of k nearest neighbors. The cloud server sends AES encrypted data of the k nearest neighbors to the searcher. The searcher obtains the k nearest neighbors by AES decryption. In Algorithm 3, the database and cloud servers do not know the query location or what type of POI the searcher searches. Moreover, Algorithm 3 reduces a mount of computation and communication as compared with Algorithm 2. The key point of Algorithm 3 is assignment of integers to types of POI. Integers assigned to a type of POI is much different from each other. Much difference between the integers of types of POI enables to distinguish types of POI on secure computation.

5 Experiment

This paper experiments the proposed methods. The environment of experiment is Windows 10 PC with Intel i5-6500 CPU 3.2 GHz and 16 GB memory. Java is used for the implementation of the proposed methods. The experiment uses a road map of Saitama City in Japan to create a data set of location information. The map data has 16284 segments and 24914 nodes. The experiment generates POI datasets on the map; POI are randomly assigned to a node on the map and ten types of POI datasets are generated with the same data density. Query location is also randomly generated.

The experiment compares the processing times of the proposed methods using ASPE and the conventional ePriLBS, and communication traffic among the cloud and database servers and the searchers. The three algorithms in the previous section are represented by ASPE1, ASPE2, and ASPE3. The processing time is defined to be the time from sending queries to receiving the search results. In ASPE1 and ASPE3, the processing time is the time required to obtain the top 10 POIs, and in ASPE2 the processing time is the time required to obtain the top 10 POIs for all POI types. For both ePriLBS and ASPE, the AES encryption key is 128 bits and the RSA encryption key is 512 bits. Each element in a matrix is 256 bits for ASPE. The number of POIs is varied from 160 to 8100.

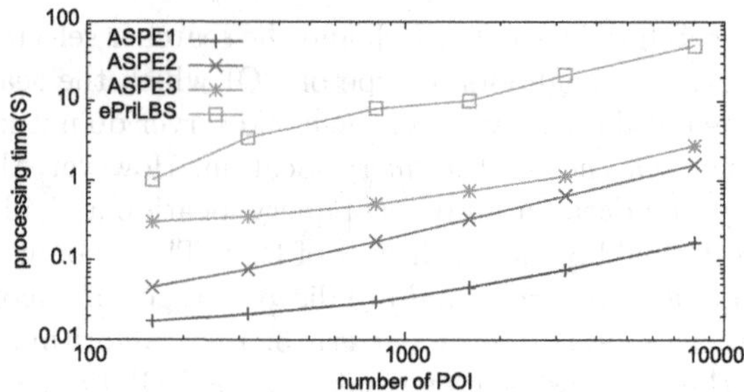

Fig. 1. Processing times versus the number of POI

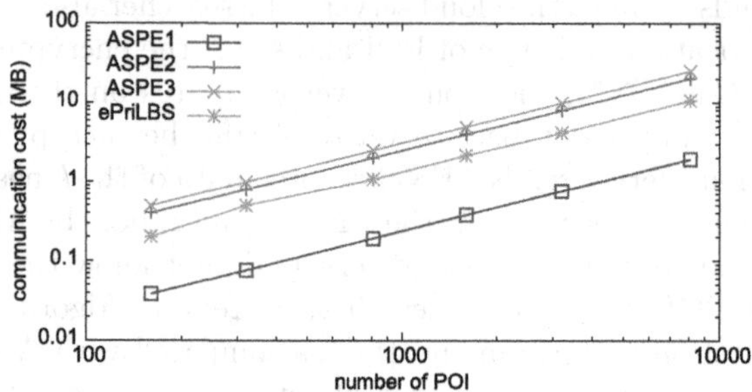

Fig. 2. Data traffic from database servers to cloud servers versus the number of POI

For each number of POI, twenty queries are performed and the graphs in this section show the averages.

Figure 1 shows processing times of ASPE and ePriLBS where the vertical axis is the logarithm of processing time and the horizontal axis is the logarithm of the number of POIs. ASPE is always faster than ePriLBS. As the number of POIs increases, the difference in processing time between ASPE and ePriLBS increases. The fastest method is ASPE1. Figure 2 is a graph of the data traffic sent from database servers to cloud servers in ASPE and ePriLBS, where the vertical axis is the logarithm of data traffic and the horizontal axis is the logarithm of the number of POIs. ASPE1 is able to search with the least traffic. The data traffic of ASPE1 is about one tenth less than other methods. The data traffic of ASPE3 is about 1.1 times more than that of ePriLBS. Figure 3 is a graph of the data traffic sent from cloud servers to searchers in ASPE and ePriLBS, where the vertical axis is the data traffic and the horizontal axis is the logarithm of the number of POIs. ASPE1 and ASPE3 always have the same data traffic and the traffic data of ASPE2 exceeds 1600. The data traffic of all ASPE methods are constant.

Figure 1 shows that the methods of ASPE are always faster than ePriLBS. ePriLBS uses Paillier cryptography and ASPE uses a matrix for secure computation. ASPE takes less time than ePriLBS for one comparison of POIs. ASPE2 and ASPE3 are slower than ASPE1 because ASPE1 encrypts and decrypts

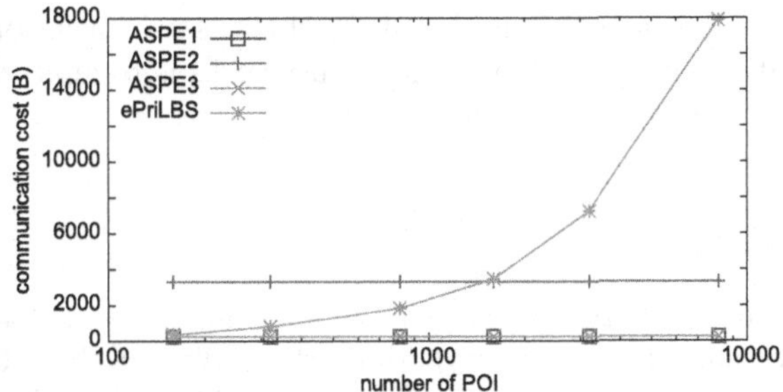

Fig. 3. Data traffic from cloud servers to searchers versus the number of POI

location information of POI without the types of POI. Increase of POIs reduces the difference in processing time between ASPE2 and ASPE3. ASPE2 encrypts only location information and ASPE3 encrypts location information and types of POI. It follows that ASPE3 takes more processing time than ASPE2 even if POIs increase.

Figure 2 shows the relations of data traffic. Since ASPE1 sends the data for only one type of POI to the cloud servers, and ASPE2 and ASPE3 send the data for all types of POIs to the cloud servers, the traffic data of ASPE1 is less than those of ASPE2 and ASPE3. ePriLBS uses K-anonymity and encrypts less data than ASPE2 and ASPE3, and thus ePriLBS sends less data traffic than ASPE2 and ASPE3. In order to reduce the data traffic of ASPE methods, the methods must decrease secureness of queries. One of the ways to reduce the data traffic is that searchers inform database servers of the search area. Suppose that the searcher is at "Shinjuku" in "Tokyo". The searcher would be afraid that the database servers know that the searcher is at "Shinjuku", but not that the database servers know that the searcher is in "Tokyo". The searcher informs the database servers that the searcher is in "Tokyo" and reduces the computation time of encryption and data traffic.

Figure 3 shows that the data traffic of the methods of ASPE are constant regardless of the number of POIs. ePriLBS increases the data traffic as the number of POIs increases. This is because ASPE methods uses k-NN search while ePriLBS uses a range search. Therefore, as POIs increase, data traffic also increases in ePriLBS. The methods of ASPE send only the data of k nearest neighbors and the traffic data is constant when the number of types of POIs is constant.

To summarize the experiment, the proposed methods are faster than conventional ePriLBS, but the proposed methods should be improved regarding with the traffic data.

6 Conclusion

This paper proposed the methods that protect the privacy of searchers. The methods are based on ASPE, secure computation using matrix calculation. The proposed methods are faster than ePriLBS, which is a conventional method of location information services. One of the future works is to analyze the strength of encryption in the proposed methods.

References

1. Bamba, B., Liu, L., Pesti, P., Wang, T.: Supporting anonymous location queries in mobile environments with privacygrid. In: Proceedings of the 17th International Conference on World Wide Web, WWW 2008, pp. 237–246 (2008)
2. Beresford, A.R., Stajano, F.: Mix zones: user privacy in location-aware services. In: 2004 Proceedings of the Second IEEE Annual Conference on Pervasive Computing and Communications Workshops, pp. 127–131 (2004)
3. Chen, J., He, K., Yuan, Q., Chen, M., Du, R., Xiang, Y.: Blind filtering at third parties: an efficient privacy-preserving framework for location-based services. IEEE Trans. Mob. Comput. **17**(11), 2524–2535 (2018)
4. Chor, B., Goldreich, O., Kushilevitz, E., Sudan, M.: Private information retrieval. In: Proceedings of the 36th Annual Symposium on Foundations of Computer Science, FOCS '95, pp. 41–50 (1995)
5. Ghinita, G., Kalnis, P., Khoshgozaran, A., Shahabi, C., Tan, K.L.: Private queries in location based services: anonymizers are not necessary. In: Proceedings of the 2008 ACM SIGMOD International Conference on Management of Data, SIGMOD 2008, pp. 121–132 (2008)
6. Gruteser, M., Grunwald, D.: Anonymous usage of location-based services through spatial and temporal cloaking. In: Proceedings of the 1st International Conference on Mobile Systems, Applications and Services, MobiSys 2003, pp. 31–42 (2003)
7. Kido, H., Yanagisawa, Y., Satoh, T.: An anonymous communication technique using dummies for location-based services. In: ICPS 2005, Proceedings of the International Conference on Pervasive Services, pp. 88–97 (2005)
8. Li, N., Li, T., Venkatasubramanian, S.: t-closeness: privacy beyond k-anonymity and l-diversity. In: 2007 IEEE 23rd International Conference on Data Engineering, pp. 106–115 (2007)
9. Oliveira, S.R.M., Zaïane, O.R.: Privacy preserving clustering by data transformation. JIDM **1**(1), 37–52 (2010)
10. Palanisamy, B., Liu, L.: Mobimix: protecting location privacy with mix-zones over road networks. In: 2011 IEEE 27th International Conference on Data Engineering, pp. 494–505 (2011)
11. Song, D., Sim, J., Park, K., Song, M.: A privacy-preserving continuous location monitoring system for location-based services. Int. J. Distrib. Sen. Netw. **11**(8), 14:14–14:14 (2015)
12. Wong, W.K., Cheung, D.W.l., Kao, B., Mamoulis, N.: Secure KNN computation on encrypted databases. In: Proceedings of the 2009 ACM SIGMOD International Conference on Management of Data, SIGMOD 2009, pp. 139–152 (2009)
13. Zhu, Y., Xu, R., Takagi, T.: Secure k-NN query on encrypted cloud database without key-sharing. Int. J. Electron. Secur. Digit. Forensic **5**(3/4), 201–217 (2013)

Interactive Analysis of Image, Video and Motion Data in Life Sciences

Exploration of Explainable AI in Context of Human-Machine Interface for the Assistive Driving System

Zenon Chaczko[1,3], Marek Kulbacki[2,3](✉), Grzegorz Gudzbeler[4], Mohammad Alsawwaf[1,6], Ilya Thai-Chyzhykau[1], and Peter Wajs-Chaczko[5]

[1] Faculty of Engineering and IT, University of Technology Sydney, Ultimo, NSW, Australia
{zenon.chaczko,mohammad.alsawwaf,ilya.thai-chyzhykau}@uts.edu.au

[2] Polish-Japanese Academy of Information Technology, R&D Center, Warsaw, Poland
mk@pja.edu.pl

[3] DIVE IN AI, Wroclaw, Poland

[4] Faculty of Political Science and International Studies University of Warsaw, Warsaw, Poland
grzegorz.gudzbeler@gmail.com

[5] Macquarie University, Sydney, NSW, Australia
peter.Wajs-Chaczko@students.mq.edu.au

[6] Imam Abdulrahman Bin Faisal University, Dammam, Saudi Arabia

Abstract. This paper presents the application and issues related to explainable AI in context of a driving assistive system. One of the key functions of the assistive system is to signal potential risks or hazards to the driver in order to allow for prompt actions and timely attention to possible problems occurring on the road. The decision making of an AI component needs to be explainable in order to minimise the time it takes for a driver to decide on whether any action is necessary to avoid the risk of collision or crash. In the explored cases, the autonomous system does not act as a "replacement" for the human driver, instead, its role is to assist the driver to respond to challenging driving situations, possibly difficult manoeuvres or complex road scenarios. The proposed solution validates the XAI approach for the design of a safety and security system that is able to identify and highlight potential risk in autonomous vehicles.

Keywords: Explainable AI · HMI · Convolutional Neural Network · Assistive system for vehicles

1 Background

The Artificial Intelligence (AI) component of many modern systems is often perceived as a black box [2], where data goes in, and a prediction goes out.

N. T. Nguyen et al. (Eds.): ACIIDS 2020, LNAI 12034, pp. 507–516, 2020.
https://doi.org/10.1007/978-3-030-42058-1_42

The inner mechanisms of the module are often not well understood even by the designers of the solution. This, however, poses a serious problem in the real-world solutions, where a human is unable to qualify the prediction accurately without additional and time-consuming investigation. As the AI mechanism (module) is represents an enclosed "black box" subsystem, it is not easy to determine the real factors that influenced the computational component to provide result A over result B. Looking at how users perceive an explanation, it can be seen that an explanation is influenced by the cognitive bias, contrasting events that occurred, social beliefs and often may refer to cause rather than a statistical measure [4,6,9]. In most implemented algorithms, the statistical measure is often the qualifying value to choose event A over B, regardless of the "cause" of the event. This occurs since computer algorithm machine is unable to process the entire context of an event, and produce an explanation based on the cause. Rather, the algorithm looks for patterns that it has "learned" and provides an output based on the possible results set by humans. Hence, if the computer program is unable to provide "reasoning" to create a satisfactory explanation, then how can an Explainable Artificial Intelligence (XAI) [2,7,10] exist? The underlying research project investigates an approach into how an intelligent system can, not only make a sound prediction [3], but also explain or reason the outcome to a human user. In doing so, enhancing the Human-Machine Interaction [1] and promoting trust [14]. The investigation looked into boundary (edge) values case scenarios of autonomous vehicles, taking into account the time constraints of human users to react, and the fact that an explanation has to be easy to understand, provide a sound level of explanation and can be processed by the human operator as quickly as possible [12,13]. These factors influence the overall design. The main goal of the designed prototype is to have a system that would detect, analyse and provide explanation of the predicted risks within the edge scene of an autonomous vehicle [3]. This approach provides a sufficiently wide range of base cases, where the system could use the AI module for risk detection and provide a satisfactory level of reasoning. The design of the project was separated into two primary parts:

- Training a Convolutional Neural Network (CNN)
- Create a system that delivers CNN output and compatible reasoning

1.1 Implemented Convolutional Neural Network

Given the "real time" constraints of the project, CNN by itself is unable to provide quick reasoning for the prediction it makes. The predictions made by the CNN are based on the patterns the model was trained on [8]. Considering the need for explanation, a design decision was made to train a model to detect specific entities, referred to as Vulnerable Road Users (VRUs). This approach added an extra layer of explainability to the overall system, as the CNN's primary goal was to detect certain entities within the scene, but not do the actual risk analysis. Chosen architecture was also able to localize the detected object, thus providing further reasoning for the decision made.

The chosen CNN was based on the Singe Shot Multi-Box Detector (SSD) architecture (Fig. 1), which scans the image once, and utilizes bounding box proposals to detect entities with an image [11]. VGG16 was adopted as the backbone feature extractor for SSD given its good performance and good accuracy in non-high-performance equipment.

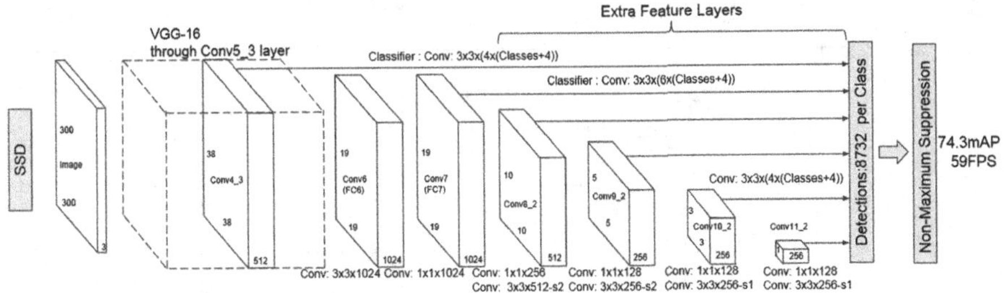

Fig. 1. SSD network architecture [11]

CNN was trained on manually collected dataset comprised of people talking and looking at mobile phones (otherwise referred to as VRUs), cyclists and general pedestrians (Table 1).

Table 1. Dataset distribution

Label name	Total	Training/Test	Validation
VRU_Pedestrians	234	180	54
Cyclist	150	135	15
People	200	150	50

Low volume of dataset posed a great challenge for training of CNN. However, through training techniques such as transfer learning and further hyperparameters tuning, the final accuracy of model was around 73 points. For an initial prototype and given the time and resource constraints of the project, this result was deemed "good enough".

2 System Design

As the accuracy of the CNN model was low, the second part of the project, was designed to integrate precautionary measures to further analyse the risk and actions required. The below diagram (Fig. 2) depicts relations between entities integrated in the system, as well as the communication sequence between the modules. Unlike early attempts of out-of-the-loop automatic control and performance evaluation of such systems [5], various functions of the proposed Assistive Driving System (ADS) operate within a continuous loop process that includes the input video streams, identifies objects, performs analysis of risks, and generates actions according to the identified risks in context of HMI [1].

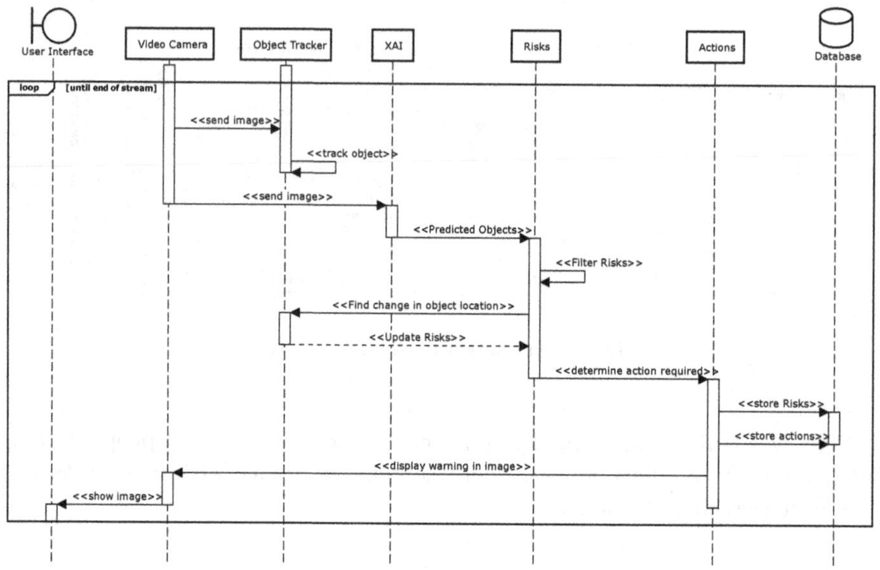

Fig. 2. Communication sequences, modules and entities of the system

The intention behind "Risks" module was to analyse the identified object in the context of risk of collision. Each risk was calculated within the context of a single entity, however, further risk analysis is possible where all entities are considered as a group, thus provisioning a way to calculate contextual overall environment risk. This further calculation and analysis was not within the scope of the project. Risk of collision was calculated based on factors including:

- Total speed (subject + vehicle speed)
- Risk factor based on subject location
- Distance to subject (calculated using "similar triangles" property and defined by equation below)

$$Distance\ to\ object = \frac{Real\ Object\ Height * Focal\ length}{Object\ Height\ in\ Image} \tag{1}$$

"Actions" primary focus is to advise the vehicle operator of identified risks using a number of visual cues and messages. Following the underlying project theme of "explainability", a big design focus was on figuring out how to warn the vehicle operator of the upcoming risk in the most efficient way. An optimal solution was to use a col-our indicator based on the analysed risk. Risk classification (see Table 2) in the previous module would provide details of the consequence and the likelihood of a collision event. The colour coding match those used on road (i.e., Safety Signs), and thus are easy to understand the meaning of. The localized box around the potential risk would utilise the colour coding, as well as limited meta-data constrained to:

- Warning icon (representative of the identified risk factor)
- Distance to object in meters
- Likelihood of collision value

Table 2. Risk classification

Consequence						
Likelihood		Insignificant	Minor	Moderate	Major	Severe
	Rare	Low	Low	Low	Medium	High
	Unlikely	Low	Low	Medium	High	High
	Possible	Low	Medium	Medium	High	Extreme
	Likely	Medium	Medium	High	Extreme	Extreme
	Almost certain	Medium	High	High	Extreme	Extreme

3 Results and Evaluation

The results were split based on the two distinct project phases: CNN design and overall system design. CNN model testing was evaluated based on the accuracy, loss, validation loss and validation accuracy metrics. During the tests, there were issues raised if the model was over-fitting to a rather limited dataset available, as this was a primary risk of the CNN design. As seen in the table below, a number of hyper-parameters were turned at each test. Each test design was done iteratively, as the main aim was to identify the parameters that had the best outcome. The overall system was tested using two approaches. This is mainly due to the fact that "explanation", by its nature, is qualitative and cannot be quantitatively judged, as what determines it to be a "good" explanation. What humans consider to be "good" explanation is usually based on the reasoning, personal bias (desires/intentions) and social belief. Hence, the system's performance was judged on (1) quantitative values of proximity to real time (based on FPS) and the accuracy of prediction; and (2) qualitative values of ease of use (usability) and level of explanation evaluated by human users.

3.1 Quantitative Results

The quantitative tests (Table 3) and real-time proximity benchmarking was done using two different machines, with the calculated average FPS results (Table 4).

Using a better performance model of GPU, the system was able to obtain a good level of "Frames Per Second" (FPS). This can be seen as an acceptable result as the output of the video was very close to real time, with minimum lag. However, the accuracy of the model did not perform as well as anticipated. At times, the model picked up "normal pedestrians" as potential risks. Further analysis and evaluation indicted that the AI module required a much larger training data set to achieve an acceptable accuracy for the live system implementation. Also, the test results indicated that the solution had some performance issues evaluating poor environmental conditions scenarios, as in such scenarios it was much harder to identify subjects. To address the described performance issues further experiments will involve a wider range of various difficult environmental conditions and a much larger training dataset.

Table 3. Quantitative tests

TEST #	weight	Labels	Optimizer	Batch size	Scheduler
0	vgg16_2_classes (COCO)	3 (risk)	SGD	8	Ir_schedule
1	vgg16_2_classes (COCO)	4 (risk)	SGD	8	Ir_schedule
2	vgg16_2_classes (COCO)	5 (risk)	SGD	8	Ir_schedule(updated to 6)
3	vgg16_2_classes (COCO)	6 (risk)	SGD	8	Ir_schedule(updated to 5)
4	vgg16_2_classes (COCO)	7 (risk)	SGD	8	PolynomialDecay(maxEpochs=20, initAlpha=1e-3, power=5)
5	vgg16_2_classes (COCO)	8 (risk)	SGD	8	PolynomialDecay(maxEpochs=20, initAlpha=1e-3, power=2)
6	vgg16_2_classes (COCO)	9 (risk)	SGD	8	PolynomialDecay(maxEpochs=20, initAlpha=1e-3, power=1)
7	vgg16_2_classes (COCO)	10 (risk)	SGD	8	PolynomialDecay(maxEpochs=20, initAlpha=1e-3, power=3)
8	vgg16_2_classes (COCO)	11 (risk)	Adam (lr=0.001, be)	8	PolynomialDecay(maxEpochs=20, initAlpha=1e-3, power=3)
9	vgg16_2_classes (COCO)	12 (risk)	Adam (lr=0.003, be)	8	PolynomialDecay(maxEpochs=20, initAlpha=1e-3, power=3)
10	vgg16_2_classes (COCO)	13 (risk)	SGD	8	PolynomialDecay(maxEpochs=20, initAlpha=1e-3, power=3)
11	vgg16_2_classes (COCO)	14 (risk)	SGD	8	PolynomialDecay(maxEpochs=20, initAlpha=1e-3, power=2)

TEST #	Epochs	Steps/epoch	Acc	Loss	val_loss	val_acc
0	20	120	0.67	2.681	3.2149	0.65
1	20	30	0.7359	2.9011	2.8885	0.68
2	20	30	0.7475	2.92	2.8929	0.7008
3	20	24	0.7480	3.0476	2.9868	0.7083
4	20	24	0.7194	3.0663	3.0066	0.6515
5	20	24	0.7198	2.7959	2.7637	0.6705
6	20	24	0.7248	2.8759	2.7143	0.6847
7	20	24	0.7249	2.8508	2.8967	0.6843
8	20	24	0.2277	6.1883	5.9808	0.2253
9	20	24	0.3674	6.0277	5.8737	0.3428
10	20	24	0.6611	nan	invalid loss	
11	20	24	0.7367	2.93	2.8858	0.6874

3.2 Qualitative Results

The qualitative evaluation and results was based on the feedback provided by a small group of users who were given a video on their mobile phones showing the streets with pedestrians and the assistive system evaluating risk of each pedestrian scenarios (Fig. 3). Prediction and Accuracy tests (Table 5) indicate cases where not all pedestrians at risk were detected and predicted as being

Table 4. Real time proximity tests

System spec	Average FPS	Note
Desktop, Intel i5 CPU Nvidia GTX 960 (4GB GPU)	14–18	The FPS saw a significant drop using a mobile phone as the input device, however, a potential cause of that is the poor WIFI available
Laptop, Intel i7 CPU Nvidia GTX 1050 TI (4GB GPU)	18–22	The laptop contained an in built camera, and seemed to be processing image stream a faster rate due to better Graphical Unit

at risk, as well as, cases of pedestrians evaluated by users at being at no risk, were indicted and predicted by the system as being a potential risk (Table 5). User feedback approach was used to evaluate and compare risk values based on a positive detection indicted by the system with risk values based on user perceptions (Table 6).

Fig. 3. CNN model testing using video captures

Table 5. Prediction and accuracy tests.

Scene	Description	Expected detections	Positive detections	False positive detections	Note
1	Single pedestrian on mobile	1	1	0	
2	Multiple pedestrians on mobile	3	3	0	
3	Multiple pedestrians on a phone call	5	4	0	1 pedestrian not detected
4	Multiple pedestrians with mobile and normal pedestrians	3-mobile 1-normal	3	1	Normal pedestrian detected as risk
5	Normal pedestrians	3	0	2	2 pedestrians detected risks
6	Single cyclist	1	0	0	Cyclist not detected
7	Multiple cyclists	4	3	0	1 cyclist not detected
8	Cyclists and pedestrians on mobile	2-cyclists 6-mobile	2-cyclists 4-mobile	0	2 pedestrians on mobile not detected
9	No subjects in screen	0	0	0	
10	Multiple subjects on screen (lower visibility)	1-cyclist 2-normal pedestrians 1-mobile	0-cyclists 0-mobile	1	1 normal pedestrian detected as "with mobile"

Table 6. Risk values based on positive detections

Scene	Positive detections	Correct risk value	Incorrect risk value
1	1	1	
2	3	3	
3	4	2	2 identified as high risk (low risk)
4	3	3	
5	0	0	
6	0	0	
7	3	2	1 identified as low risk (mid-high risk)
8	2 cyclists 4 mobile	3	1 (mobile pedestrian) identified as high risk (low-mid risk)
9	0	0	
10	0 cyclist 0 mobile	0	

4 Conclusion

Discussion on explainability of threats and risks detected by the Assistive Driving System conducted with a group of users, indicated their preference for an explainable AI system. The current implementation of the system is still lacking the required accuracy, as the users often saw a higher risk produced by the system, then they would otherwise visually and cognitively classify themselves. The user group feedback validated the preference of having short and high-quality information (fiducial markers), over long texts. The test users mentioned that in "edge case" situations, where quick decisions are necessary, they would prefer not to read a large amount of text, but rather, have the system identify and locate the risks on the screen. They also noted that too much "information" on the screen could potentially occlude objects and reduce the vision of the driver. This could create a potentially even more dangerous situation.

Acknowledgements. This study was supported by the Scientific Research from Technical University of Technology Sydney, School of Electrical and Data Engineering, Macquarie University and DIVE IN AI.

References

1. Abdul, A.M., Vermeulen, J., Wang, D., Lim, B.Y., Kankanhalli, M.S.: Trends and trajectories for explainable, accountable and intelligible systems: an HCI research agenda. In: Proceedings of the 2018 CHI Conference on Human Factors in Computing Systems, CHI 2018, Montreal, QC, Canada, 21–26 April 2018, p. 582 (2018). https://doi.org/10.1145/3173574.3174156
2. Adadi, A., Berrada, M.: Peeking inside the black-box: a survey on explainable artificial intelligence (XAI). IEEE Access **6**, 52138–52160 (2018). https://doi.org/10.1109/ACCESS.2018.2870052
3. Badue, C., et al.: Self-driving cars: a survey. CoRR abs/1901.04407 (2019)
4. Doshi-Velez, F., Kim, B.: Towards a rigorous science of interpretable machine learning (2017)
5. Endsley, M.R., Kiris, E.O.: The out-of-the-loop performance problem and level of control in automation. Hum. Factors **37**(2), 381–394 (1995). https://doi.org/10.1518/001872095779064555
6. Flint, A., Nourian, A., Koister, J.: xAI toolkit: practical, explainable machine learning (2019). https://www.fico.com/en/latest-thinking/white-paper/xai-toolkit-practical-explainable-machine-learning. Accessed 06 Oct 2019
7. Goebel, R., et al.: Explainable AI: the new 42? In: Holzinger, A., Kieseberg, P., Tjoa, A.M., Weippl, E. (eds.) CD-MAKE 2018. LNCS, vol. 11015, pp. 295–303. Springer, Cham (2018). https://doi.org/10.1007/978-3-319-99740-7_21
8. Gu, J., et al.: Recent advances in convolutional neural networks. CoRR abs/1512.07108 (2015)
9. Holzinger, A., Kieseberg, P., Tjoa, A.M., Weippl, E. (eds.): CD-MAKE 2018. LNCS, vol. 11015. Springer, Cham (2018). https://doi.org/10.1007/978-3-319-99740-7

10. van Lent, M., Fisher, W., Mancuso, M.: An explainable artificial intelligence system for small-unit tactical behavior. In: Proceedings of the Nineteenth National Conference on Artificial Intelligence, Sixteenth Conference on Innovative Applications of Artificial Intelligence, San Jose, California, USA, 25–29 July 2004, pp. 900–907 (2004)
11. Liu, W., et al.: SSD: single shot MultiBox detector. In: Leibe, B., Matas, J., Sebe, N., Welling, M. (eds.) ECCV 2016. LNCS, vol. 9905, pp. 21–37. Springer, Cham (2016). https://doi.org/10.1007/978-3-319-46448-0_2
12. Miller, T.: Explanation in artificial intelligence: insights from the social sciences. Artif. Intell. **267**, 1–38 (2019). https://doi.org/10.1016/j.artint.2018.07.007
13. Redmon, J., Divvala, S.K., Girshick, R.B., Farhadi, A.: You only look once: unified, real-time object detection. In: 2016 IEEE Conference on Computer Vision and Pattern Recognition, CVPR 2016, Las Vegas, NV, USA, 27–30 June 2016, pp. 779–788 (2016). https://doi.org/10.1109/CVPR.2016.91
14. Ribeiro, M.T., Singh, S., Guestrin, C.: "Why should I trust you?": explaining the predictions of any classifier. In: Proceedings of the 22nd ACM SIGKDD International Conference on Knowledge Discovery and Data Mining, San Francisco, CA, USA, 13–17 August 2016, pp. 1135–1144 (2016). https://doi.org/10.1145/2939672.2939778

Combining Results of Different Oculometric Tests Improved Prediction of Parkinson's Disease Development

Albert Śledzianowski[1(✉)], Artur Szymanski[1], Aldona Drabik[1], Stanisław Szlufik[2], Dariusz M. Koziorowski[2], and Andrzej W.Przybyszewski[1]

[1] Polish-Japanese Academy of Information Technology, 02-008 Warsaw, Poland
albert.sledzianowski@gmail.com

[2] Neurology, Faculty of Health Science, Medical University of Warsaw, 03-242 Warsaw, Poland

Abstract. In this text we compare the measurement results of reflexive saccades and antisaccades of patients with Parkinson's Disease (PD), trying to determine the best settings to predict the Unified Parkinson's Disease Rating Scale (UPDRS) results. After Alzheimer's disease, PD statistically is the second one and until today, no effective therapy has been found. Luckily, PD develops very slowly and early detection can be very important in slowing its progression. In this experiment we examined the reflective saccades (RS) and antisaccades (AS) of 11 PD patients who performed eye-tracking tests in controlled conditions. We correlated neurological measurements of patient's abilities described by the Unified Parkinson's Disease Rating Scale (UPDRS) scale with parameters of RS and AS. We used tools implemented in the Scikit-Learn for data preprocessing and predictions of the UPDRS scoring groups [1]. By experimenting with different datasets we achieved best results by combining means of RS and AS parameters into computed attributes. We also showed, that the accuracy of the prediction increases with the number of such derived attributes. We achieved 89% accuracy of predictions and showed that computed attributes have 50% higher results in the feature importance scoring than source parameters. The eye-tracking tests described in this text are relatively easy to carry out and could support the PD diagnosis.

Keywords: Parkinson's Disease · Reflexive saccades · Antisaccades · Eye tracking · Data mining · Machine learning · UPDRS

1 Introduction

The eyes make projectile movements called saccades when watching surroundings and fixing on important elements. The reason of this behaviour is in the specifics of the fovea which is the part of retina responsible for extracting details from the available view. The fovea is able to receive information only from a small field,

N. T. Nguyen et al. (Eds.): ACIIDS 2020, LNAI 12034, pp. 517–526, 2020.
https://doi.org/10.1007/978-3-030-42058-1_43

so eyes must be often moved from between different locations to scan the visible area [5]. Saccades occur both involuntarily and intentionally and are varied in terms of range, as we make small saccades when we read and longer when we look around ourselves. Saccades are common in most of our daily activities and its mechanism seems to be complex as many brain areas take part in the analysis of visual information. Therefore saccades seem to be sensitive for neurodegenerative changes and it is sensible perform tests based on this kind of eve move during diagnostic connected to diseases like the PD. Several types of tests to study saccades have been developed, usually consisting calculation of the parameters while subject moves his eyes from the fixation point to the peripheral target. One of them is a test for the RS, based on simply move from the fixation point towards the stimuli and the AS test, based on the reverse action. The AS is an eye move in the opposite direction of appearing stimulus. The idea is in suppressing reflexive transfer of the focus to the emerging target and forcing eyes to look into to the mirror location of the target. As the reflexive eye movement have to be inhibited, this test is generally more difficult and usually takes more time than the RS [8]. In terms of the PD disease, various studies have shown that patients have impaired executive function, including deficits in attention, movement initiation, motor planning and decision making [10]. It usually leads to impairments in control of suppression of involuntary behavior, as variety of neurological diseases result in dysfunctions and errors in this mechanism [13]. It was found that the degree of advancement of the PD significantly increases mean latency and error rate in the AS tasks and significantly decreases the velocity [9,11].

In this experiment we used results of both RS and AS tests of PD patients, combining them with their results of neurological tests expressed in the UPDRS, which is most common rating scale of the PD progression. Th UPDRS consists clinician-scored and monitored personal behavior, mood, mental activity, activities of daily living, motor evaluation and the evaluation of complications during the therapy [7]. Unfortunately, most of the patients diverge in their combinations of symptoms, which leads to constant need of development in methods of evaluation of the disease progression.

In this experiment we researched different datasets containing results of RS and AS of PD and tried to find the best features system to predict clinically-scored attribute, the UPDRS Total. The UPDRS Total represents all sections of the UPDRS evaluation, therefore it is considered as a good generalization of the PD progression. We are investigating methods of predictions that could extend information about patients, could be automated and be available for everyone on personal devices like PC or Smart-Phone.

2 Methods of the Experiment

We performed tests with 11 patients in one of neurosurgical clinic. Patients differed in terms of the disease treatment. Also our data distinguished patients who were treated with pharmaceuticals (BMT - Best Medical Treatment) and patients who underwent an electrode implantation. (DBS - Deep Brain Stimulation). Implanted electrode stimulates patient's Subthalamic Nucleus (STN), an area of the brain that has a strong effect on the dopaminergic system, damaged as a result of the PD. Patients qualified for the neurosurgery are mostly characterized by the low sensitivity to the stabilizing effects of the L-Dopa [14]. Variants of patients session are presented in the list below:

- Sesion type 0 : BMT off, DBS off
- Sesion type 1 : BMT off, DBS on
- Sesion type 2 : BMT on, DBS off
- Sesion type 3 : BMT on, DBS on

We performed test with a head-mounted eye-tracker, the JAZZ-Novo. This device works in frequency of 1000 Hz, thus provides very high spatial and temporal resolution needed to carry out this kind of experiments. The location of the sensor on the subject's forehead allowed for compensation of a head and body movements, which is very important during eye-tracking measurements of the PD patients because of involuntary quivering movements (the tremor). Experimental attempts consisted of tracking the light marker, moving in horizontal directions behind patient's eyes. Both the RS and the AS attempts started from fixating in the initial position (0°). When fixation point disappeared at the same moment targets of the RS (green eclipses) or the AS (red ellipses) revealed to the patient 10° to the left or right. Patient task was to move eyesight on the appearing target (RS) or in the opposite direction (AS) with the highest precision and shortest latency. Next attempt started after 100 ms without any break ("no-gap" model introduced by Saslow) [15]. Schema on Fig. 1 presents both models of the RS and the AS trials.

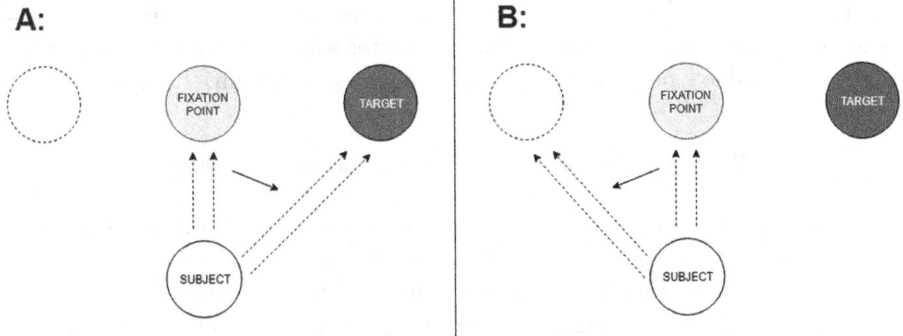

Fig. 1. The models of oculometric tests: A: The reflexive saccade, B: The antisaccade.

All tests were conducted in the same room with the same lighting. From the eye-tracking data (time series of eye positions) we algorithmically calculated different parameters of the RS and the AS - mean delay, mean duration and maximum speed. We used windows as a search space of the eye moves with the beginning and the end designated by appearance and disappearance of the peripheral target. Depending on the test type, algorithm expected straight move between fixation point and the target (RS) or to the opposite direction (AS) - both below the latency of 500 ms. All records not passing the requirements of move direction and latency, including missed records (blinks, head movements, etc.) have not been qualified for further analysis.

3 Computational Basis

As mentioned before, the subject of the calculation from the eye tracker data were mean latency, mean duration and max speed (because we think that max speed is better for overviewing the eye performance than the average speed). We calculated the latency as a difference in time between showing the target and starting the eye movement and the beginning of a duration from the starting point when eyes began to follow the target and when simultaneously the eye speed started to rise. The end of the duration was determined by the eye movement inhibition. The means of those two parameters were calculated arithmetically for each patient. The max speed was counted as the maximum value determined in the period of the eye move duration.

After performing oculometric tests and calculations we created the dataset from parameters of the RS and AS data performed in different sessions. As previously mentioned, the neurological data was expressed in the attribute UPDRS TOTAL. The dataset contained 52 records (26 RS/26 AS) collected from 11 patients. Patients differed between methods of the treatment, so it was not possible to collect data of the same patient in each of the session type, this why.

The oculometric parameters were calculated for each eye separately. The dataset contained additional features: standard deviations of the mean parameters (latency and duration) and attributes representing the number of frames on the basis of which given oculometric parameter was calculated (for particular record). The Table 1 presents the examples of rows of the initial dataset.

Table 1. The example of dataset before preprocessing

id	updrs	sess	type	mtre	stre	mspre	mrdur	stdredur	delay_r_cn	ms_r_cnt	dur_r_cnt
1	55	0	RS	0.265	0.043	5.068	1.951	1.224	2	8	4
1	55	0	AS	0.296	0.055	5.427	4.372	0.785	5	30	15
2	52	1	RS	0.422	0.073	5.240	4.014	1.703	3	17	8
2	52	1	AS	0.376	0.071	5.641	2.965	1.760	7	19	9

Where columns: "id" is a patient number, "sess", "type" describes types of the session and test, "mtre", "stre" - mean and standard variance for latency, "mrdur", "stdredur" - mean and standard variance for duration, mspre, max speed and delay_r_cn, ms_r_cnt dur_r_cnt are frame counters corresponding to the parameters. Presented eye move parameters were calculated separately for each eye

In the next phase, the dataset has been preprocessed by several methods implemented in the scikit-learn [1]. First we obtained matrix ("X") containing all the feature values of all instances in the dataset, excluding the target attribute UPDRS. The target attribute data ("y") was discretized using pandas "cut" method into 5 bins (classes) [2]. Then in the matrix "X" we aligned the scales of attribute values using scikit-learn's "Standard Scaler" which standardizes features by removing the mean and scaling to unit variance [1]. As discretized "y" class bins differed in number of records (from 4 to 17 items), we used Synthetic Minority Over-sampling Technique (SMOTE) to oversample the dataset with number of neighbours depending on number of items in the smallest group [1]. After the oversampling the "X" contained an equal (17) number of records for each of the five classes of discretized UPDRS. The number of records has increased by 38% from 52 to 85. Next, because we wanted to compare different combinations of dataset in terms of predictions accuracy, we generated 2 additional matrices separating the RS records from the AS. We also created 2 copies of initial "X" matrix in order to add new features. We wanted to check the accuracy of prediction when we average the results of particular patient - regardless the session and the test type. In order to do that, we calculated and combined means for particular patients on the basis of the eye move parameters and added it to first copy of the matrix in number of 4, and to the second copy in number of 8 new features. We also unified eye move parameters for both eyes in those two datasets. Table 2 presents the differences in the features between different datasets. In the next stage we split each of the 5 versions of the dataset into training and test/validation sets using Stratified Shuffle Split (SSS). The SSS returns stratified and randomized folds with preserved percentage of samples for each class. In stratified sampling the data is divided into homogeneous subgroups ("stratas") and the right number of instances is sampled from each stratum to guarantee that the test and training sets are representative [1].

Table 2. The comparison of the features for different datasets

RS	AS	RS + AS	RS + AS + 4	RS + AS + 8	Features
N	N	Y	Y	Y	Contains oculometric test type parameter
Y	Y	Y	N	N	Separated parameters for left/right eye
N	N	N	Y	Y	Unified parameters for both eyes
N	N	N	Y	Y	Calculated attributes from source parameters

Where RS is the dataset containing only saccade and AS only antisaccade data, RS + AS, +4/+8 contain both types of test and respectively 4 and 8 new attributes computed from means of parameters for the patient, regardless oculometric test type and session.

4 Results

With prepared 5 versions of datasets we started supervised learning using different classifiers implemented in the scikit-learn, as we decided to check predictions using different methods:

- Nearest Neighbors
- Support Vector Machine
- Decision Tree
- Random Forest
- Multi-Layer Perceptron
- Naive Bayes
- Quadratic Discriminant Analysis
- Gaussian Process

For each of listed classifiers we used the GridSearchCV (GSCV), the feature implemented in scikit-learn for searching possible best values of hyper-parameters. The GSCV performs exhaustive search over parameters values optimized by the cross-validation returning various types of scoring and also best values of parameters [1]. For ranking different input variants we used GSCV "best score" which is a calculation of cross-validated mean score of the best estimation achieved for particular classifier [1]. Table 3 shows score ranking for different datasets and classifiers.

Table 3. The prediction score ranking for different datasets

Dataset type	Best classifier	Score
Only RS	QDA	0.68
Only AS	QDA	0.75
RS + AS	Decision tree	0.61
RS + AS + 4 combined features	Decision tree	0.77
RS + AS + 8 combined features	Decision tree	0.89

As we can see in Fig. 3 the Quadratic Discriminant Analysis (QDA) performed the highest accuracy predictions for smaller and simpler datasets containing only the RS or only the AS. Discriminant analysis is research method where the criterion (or the dependent variable) is divided into categories and the predictor (or the independent variable) is an interval in nature. The quadratic decision boundary determining the division criterion is set by Bayes rule which fits conditional densities of a class to the data. The QDA have no assumption that the covariance of each of the classes is identical, also doesn't require any hyper-parameter tuning. For larger datasets consisting both the RS and the AS, the best predictions were obtained by the Decision Tree. As can bee seen in the Table 3, its accuracy rises linearly, dependently on number of added features computed on basis of the oculometric parameters (described in the previous section). Figure 2 is showing the confusion matrix of this dataset, obtained with the Decision Tree classifier [1].

We also examined feature importance between different datasets. Figure 3 shows 5 most important features for the Decision Tree (the best ranked classifier)

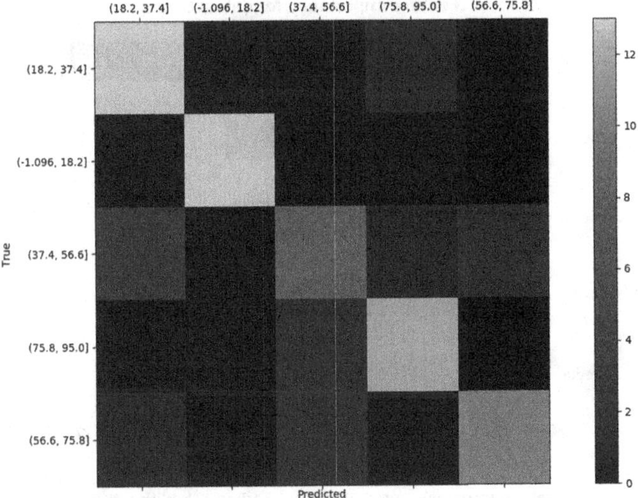

Fig. 2. The confusion matrix of the total UPDRS.

in 3 different variants of large dataset (containing both the RS and the AS data). As can be seen on the charts, computed features occupy a high places in the ranking of top 5 important features and their position and number increases with the number of such parameters added into the dataset.

5 Discussion and Conclusions

In the results we can see that computed attributes specifying joined results of the patient are more sensitive in predicting scoring group of the total UPDRS, than standard parameters (Fig. 3). What can be noticed, with the number of computed attributes added, the accuracy of the prediction increases, as well as the importance of those attributes. Attributes "redur_join" being the average of the patient's results in terms of eye move duration and "stre_join" - the average of standard deviations of patient's latency results, played the most important role in the predictions in terms of the most accurate results.

Also importance scoring for those attributes seems to be much higher than for standard ones. Attribute "sess" describing type of the session, previously the most important decision attribute, in the final shape of the dataset plays only the third role in the importance ranking, with a 50% worse scoring result than the best computed attribute. The results obtained using final dataset suggests that we can increase the sensitivity of the UPDRS groups prediction by increasing the dataset with combined attributes, representing the averages of oculometric parameters from different tests. We think that it can be hypothesized that combined attributes create an union between same features from different tests which helps classifiers in prediction and creates bridge between those separated

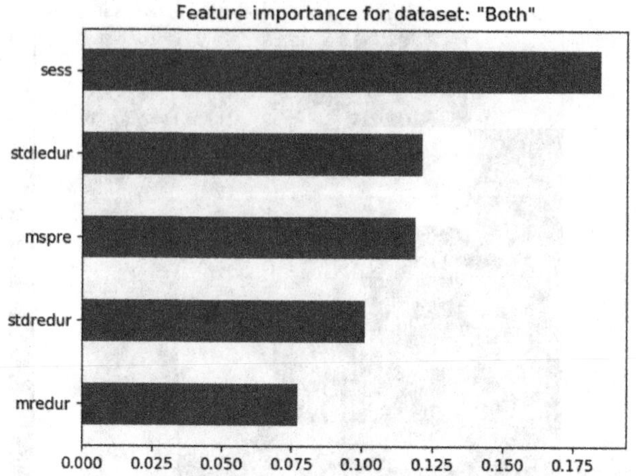

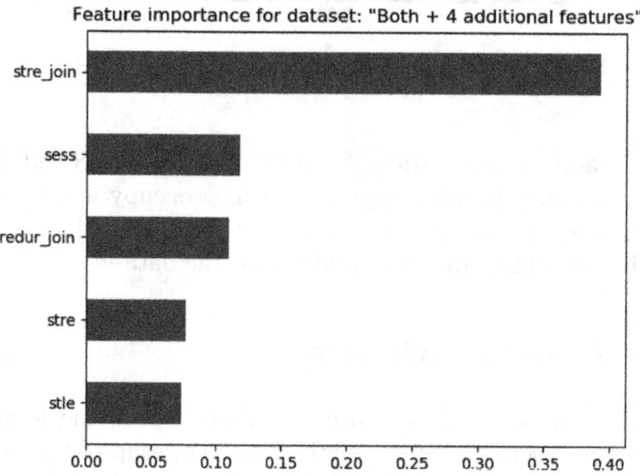

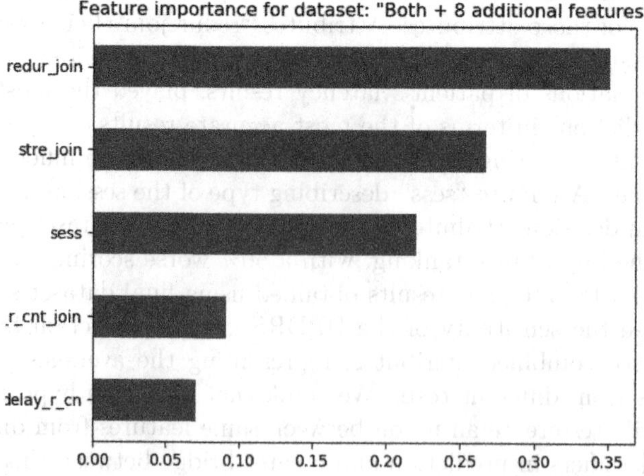

Fig. 3. The decision tree feature importance comparison for different datasets.

results of a patient. Additional attributes have also extended the size of the dataset and larger sizes of samples can improve the accuracy [16].

We found presented results as very indicating with so small group of records and the fact that we were able to increase prediction accuracy for our dataset over 21%. We think that the approach presented in this article can improve the quality of the UPDRS predictions based on eye movement testing. We also think that the eye-move performance tests are good indicators of the motor skills of the PD patients, revealing the scales of disease progression. Our results also proves the sense of further investigations of correlations between parameters of different eye movements tests, computed attributes and the UPDRS evaluations. We hope that development of methods like the one described in this text may help in improving the PD diagnosis. We believe that in the upcoming future, oculometric tests combined with the machine-learning pipelines could improve detection and progression evaluation of the neurodegenerative diseases. If such tests could be available on personal devices, patients could perform tests under different conditions providing more information about the disease.

Ethic Statement. This study was carried out in accordance with the recommendations of Bioethics Committee of Warsaw Medical University with written informed consent from all subjects. All subjects gave written informed consent in accordance with the Declaration of Helsinki. The protocol was approved by the Bioethics Committee of Warsaw Medical University.

References

1. Pedregosa, F.: Scikit-learn: machine learning in python. J. Mach. Learn. Res. **12**, 2825–2830 (2011)
2. McKinney, W., et al.: Data structures for statistical computing in python. In: Proceedings of the 9th Python in Science Conference, vol. 445, pp. 51–56 (2010)
3. Hallett, P.E., Adams, B.D.: The predictability of saccadic latency in a novel voluntary oculomotor task. Vision. Res. **20**(4), 329–339 (1980)
4. Purves, D., Augustine, G.J., Fitzpatrick, D., et al. (eds.): Types of Eye Movements and Their Functions. Neuroscience, 2nd edn. Sinauer Associates Inc, Sunderland (2001)
5. Pratt, J., Abrams, R., Chasteen, A.: Initiation and inhibition of saccadic eye movements in younger and older adults an analysis of the gap effect. J. Gerontol.: Psychol. Sci. **52B**(2), 1O3–1O7 (1997)
6. Robinson, D.A.: Control of eye movements. In: Handbook of Physiology, Section I: The Nervous System. American Physiological Society, Bethesda (1981)
7. Bhidayasiri, R., Martinez-Martin, P.: Clinical assessments in Parkinson's disease: scales and monitoring. In: International Review of Neurobiology, vol. 132, pp. 129–182 (2017)
8. Olk, B., Kingstone, A.: Why are antisaccades slower than prosaccades? A novel finding using a new paradigm. Cogn. Neurosci. Neuropsychol. **14**(1), 151–155 (2003)
9. Kitagawa, M., Fukushima, J., Tashiro, K.: Relationship between antisaccades and the clinical symptoms in Parkinson's disease. Neurology **44**(12), 2285 (1994)

10. Hood, A.J., Amador, S.C., Cain, A.E., et al.: Levodopa slows prosaccades and improves antisaccades, an eye movement study in Parkinson's disease. J. Neurol. Neurosurg. Psychiatry **78**, 565–570 (2007)
11. Everling, S., DeSouza, J.F.X.: Rule-dependent activity for prosaccades and antisaccades in the primate prefrontal cortex. J. Cogn. Neurosci. **17**(9), 1483–1496 (2005)
12. Nakamura, T., Funakubo, T., Kanayama, R., Aoyagi, M.: Reflex saccades, remembered saccades and antisaccades in Parkinson's disease. J. Vestib. Res. **4**(6), S32 (1996). (Suppl. 1)
13. Everling, S., Fischer, B.: The antisaccade: a review of basic research and clinical studies. Neuropsychologia **36**(9), 885–899 (1998)
14. Przybyszewski, A., Szlufik, S., Szymanski, A., Habela, P., Koziorowski, D.: Multimodal learning and intelligent prediction of symptom development in individual Parkinson's patients. Sensors **16**(9), 1498 (2016)
15. Saslow, M.G.: Effects of components of displacement-step stimuli upon latency for saccadic eye movement. J. Opt. Soc. Am. **57**(8), 1024–1029 (1967)
16. Chu, C., Hsu, A.-L., Chou, K.-H., Bandettini, P., Lin, C.: Does feature selection improve classification accuracy? Impact of sample size and feature selection on classification using anatomical magnetic resonance images. NeuroImage **60**(1), 59–70 (2012)

In Your Face: Person Identification Through Ratios of Distances Between Facial Features

Mohammad Alsawwaf[1,3](✉), Zenon Chaczko[1,2], and Marek Kulbacki[2,4]

[1] School of Electrical and Data Engineering, University of Technology Sydney, Ultimo, Australia
{mohammad.alsawwaf,zenon.chaczko}@uts.edu.au
[2] DIVE IN AI, Wroclaw, Poland
[3] Imam Abdulrahman Bin Faisal University, Dammam, Saudi Arabia
[4] Polish-Japanese Academy of Information Technology, R&D Center, Warsaw, Poland
mk@pja.edu.pl

Abstract. These days identification of a person is an integral part of many computer-based solutions. It is a key characteristic for access control, customized services, and a proof of identity. Over the last couple of decades, many new techniques were introduced for how to identify human faces. The purpose of this paper is to introduce yet another innovative approach for face recognition. The human face consists of multiple features that when considered together produces a unique signature that identifies a single person. Building upon this premise, we are studying the identification of faces by producing ratios from the distances between the different features on the face and their locations in an explainable algorithm with the possibility of future inclusion of multiple spectrum and 3D images for data processing and analysis.

Keywords: Person identification · Human face recognition · Biometrics · Facial features · HMI

1 Introduction

1.1 Background

Person identification based on facial features is becoming a widely used authentication and authorization technique. Nowadays, the technique is seen in smart phones, tablets, in proof of identity applications, and even in access to critical data, like bank accounts, or critical locations or resources, like research facilities, labs and pharmacies [1].

Facial recognition is a process that includes multiple steps [7]. First, a face detection is applied to decide whether a face exists in a given video or image. Then, features' extraction of the detected face. Finally, a comparison algorithm

N. T. Nguyen et al. (Eds.): ACIIDS 2020, LNAI 12034, pp. 527–536, 2020.
https://doi.org/10.1007/978-3-030-42058-1_44

can be applied to obtain results for the purpose of identification or verification (Fig. 1). Moreover, there are multiple methods to conduct facial identification and authentication. Some methods are based on features while others are image based (Fig. 2). Deep convolutional neural networks (CNNs) is one of the most successful image-based techniques due to its significantly improved results over past methods [2] which is based on deep learning. It is also considered the current state-of-the-art in terms of accuracy and bench-marking.

Just like many computer-based system, facial recognition has its issues and challenges. Current methods struggle with unusual expressions, obstacles that block the view, illumination, and even with difficult backgrounds [9]. In addition to that, morphing attacks are also one of the issues that these systems need to consider [12].

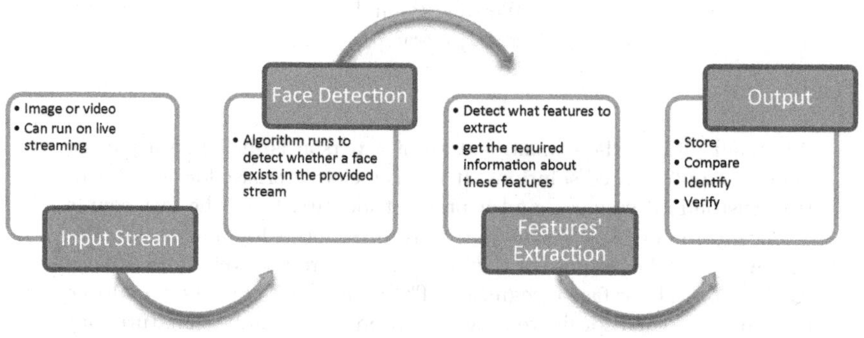

Fig. 1. Generic face recognition steps

Among these disadvantages, some currently used methods for recognition lack attention to: (1) privacy by storing peoples' images on databases, (2) storage space consumption due to storing these face images and comparison templates [10], (3) fast results due to requiring image comparison, and most importantly, in many cases, (4) we do not know exactly how the identification process happens and what are the basis of the identification. The algorithm is a black box and we are to take what it offers for granted.

The aim of this paper is to present an innovative and reliable method for facial identification from digital images or a video live stream. It introduces the identification of a person rather than identifying gender only. It focuses on gathering the distances between the different facial features and use width and length ratios to calculate a number identifier for each person. The rest of the paper is organized as follows. Section 1.2 gives an overview of related works. Overview of the methodology of the new potential approach is discussed in Sect. 2. In Sect. 3, case studies in progress. Finally, conclusion and future work are in Sect. 4.

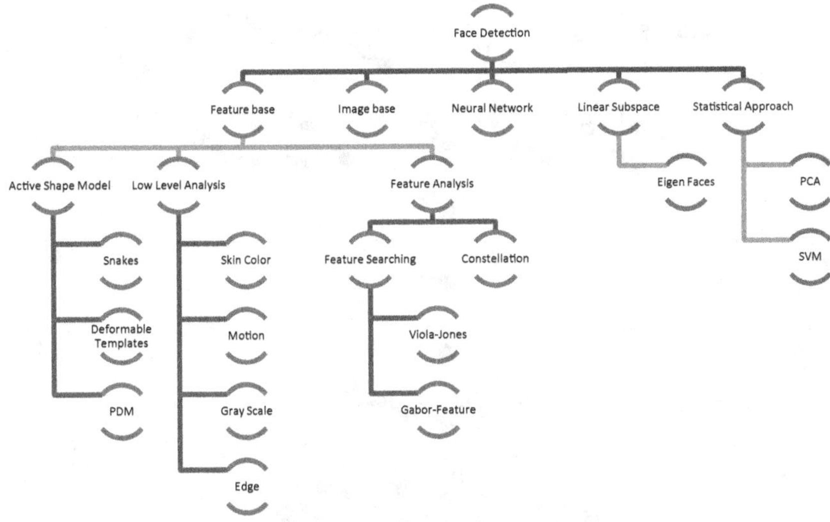

Fig. 2. A taxonomy of face detection techniques [12]

1.2 Related Work

In this approach we refer to python's open source face recognition library by [4] which is based on dlib [8]. Dlib is a toolkit written in C++ with support and examples in python. It has machine learning algorithms for a wide variety of purposes; One of which is face detection. Even though this library offers a complete face recognition solution, this approach only references the detection part of the approach because the identification procedures are based on unknown calculations. The term unknown calculations here refers to the deep learning aspect of the calculations. For any deep learning approach, the users don't know why and how the algorithm reached the conclusion whether it was positive or negative. It is based on similarity factors that the algorithm uses to reach the conclusion. In this python library, the face detection feature offers the pixel coordinates of faces from the input stream. It prints the top, right, bottom and left coordinates of each detected face in pixels. It finds 68 points on the face (see Fig. 3).

This approach used the open source library in python to detect a face passing in front of the camera or in an input image. Then the developed algorithm was used to calculate the distance between (1) the near ends of the eyes which correspond to points 39 and 42 on Fig. 3, (2) Ear-to-ear which correspond to points 0 and 16, (3) Nose-top to bottom of nose which are points 27 and 33, and (4) Nose-top to chin which are 27 and 8 on the same figure. These calculations produced the face width and length which were used to produce the face ratios. Based on the calculated distances and ratios, a unique identifier was issued per person. This identifier was the base advantage of the proposed method and what potentially can be used as a new technique for the purpose of identification and authenticating.

Fig. 3. Face points template extracted from dlib [4]

2 The Algorithm of the Proposed Method

2.1 Overview and Setup

The idea of using a unique identifier results in multiple advantages. It saves huge amounts of storage space in the central database. It also eliminates the need for com-paring new and stored images from a single person for identification. Instead, only numbers are stored and compared each time an identification is requested. Thus, making the comparison process much faster. Moreover, there is an elimination of the fact that face recognition invades privacy. This is done by storing only the unique identifier and not a picture of the face.

Pairing this approach up with multi-spectral cameras including infrared and ultraviolet will add to the reliability of this system. By doing this, ambient light will play much less significance in the acceptance and usefulness of an obtained picture. On the other hand, UV offer a much deeper analysis of the face in terms of livelihood, skin scars and skin tone differences that are not always visible with the visual spectrum. This proposed method assumes the following:

- Frontal face pictures are used.
- Pictures a re lighted enough around the face to show the features.
- Full faces show in the pictures.

Table 1. Distances calculated on facial features and their corresponding points from Fig. 3

No.	Description	Corresponding points in Fig. 2
1	Near end of eyes	$39(x_1, y_1)$–$42(x_1, y_1)$
2	Ear to ear	0–16
3	Nose length	27–33
4	Nose to chin	27–8

Based on Table 1, the approach locates the coordinates of each given point, for example 39. The approach Obtains the x and y coordinates of the point and do the same with all other points. Then, use the following formula to find the distance:

$$Distance = \sqrt{(X_i - X_{i-1})^2 + (Y_i - Y_{i-1})^2} \tag{1}$$

where X_i is the x-axis coordinate of the second point of the facial feature (left eye), X_{i-1} is the x coordinate of the facial feature (right eye). We then use the following formula for the calculation of the ratio:

$$Ratio_width = \frac{D_{Between\ eyes}}{D_{Between\ ears}} \tag{2}$$

$$Ratio_length = \frac{D_{Nose\ tip\ to\ bottom\ of\ nose}}{D_{Nose\ tip\ to\ chin}} \tag{3}$$

The software is supplied with an image, batch of images or a live video stream as an input. It can be observed that the software is capable to detect faces and apply the facial template on them based on the adapted open source library components. The test set has two siblings' pictures. The settings used are identical for the photo shoot. In both cases the camera was 120 cm from the subjects and the camera was placed at line-of-sight in terms of height. The pictures were taken at the same time ensuring same ambient light. The two siblings were asked to make three different stances, smiling, mouth-open, and neutral Fig. 4. After detecting the faces and applying the face features template, the specific algorithms from the development approach are applied on the test

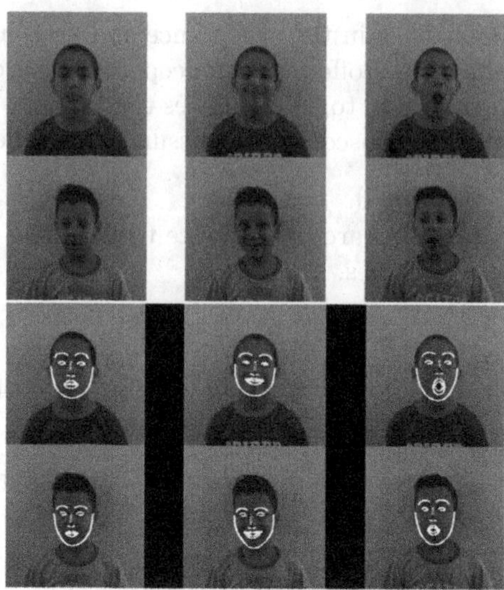

Fig. 4. Two siblings' photos taken with identical situations - Row 1 Subject(1) with three stances, Row 2 Subject(2) with three stances - Owned Pictures

set of the two siblings to produce the identifiers for each of these images as per Table 2 below.

Table 2. Identifiers for test set of siblings

Subject	Stance	Width ratio	Length ratio	Identifier
1	Neutral	0.2707018	0.4184043	**0.2707 : 0.4184**
2	Neutral	0.2516501	0.4317872	**0.2483 : 0.4317**
1	Open mouth	0.2677019	0.3428050	**0.2677 : 0.3428**
2	Open mouth	0.2361776	0.3632493	**0.2361 : 0.3632**
1	Smiling	0.2815420	0.4134484	**0.2815 : 0.4134**
2	Smiling	0.2528079	0.4283028	**0.2528 : 0.4283**

These results can be very valuable to the study as these are siblings with significant similarities in look and very close in age but not twins. These results show enough difference to distinguish between these two siblings when conducting the authentication. To test this, we take a second photo of subject 2 with eyes not looking straight at the camera and neutral stance (Fig. 5).

2.2 Population and Sampling

For the purpose of this experiment, the Labeled Faces in the Wild (LFW) library has been used with 13233 labeled pictures for 5749 subject 1680 of them with 2

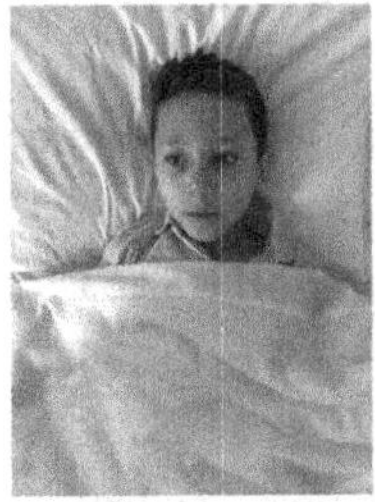

Fig. 5. Subject 2 authentication - owned picture

or more images. This library is used because it was designed for the purpose of "studying the problem of unconstrained face recognition" and developed by the University of Massachusetts, Amherst [6]. The size of this library allows for it to be used in research without worrying about storage space or using it with high performance computing power. Moreover, it is used widely for bench-marking by many literature making it ideal for this experiment.

Since this approach is divided into two parts, mainly detection and identification, the 10-fold cross validation was used to benchmark the detection part of the approach along with other state-of-the-art approaches. K-fold cross validation is a widely used algorithm evaluation technique by many researches. Using this technique eliminates the "too optimistic" results and makes the results more realistic [14]. On the other hand, the identification part was bench-marked using the Equal Error Rate due to the statistical nature of the identification algorithm [13].

2.3 Results

Using the 10-fold cross validation for the detection of faces resulted in accuracy of 99.38% (See Table 3). This means that out of the 13233 images, faces were correctly detected and landmarked in 13150 of them [4].

Table 3. Statistics of publicly available academic benchmark datasets.

Dataset	Purpose	No. of images	Accuracy
LFW [6]	Testing	13233	99.38%

Out of these detected images, the identification algorithm populated ratios for all subjects with more than one image for a total of 1680 subjects. However, after registering all the of these ratios for the 1680 images, the identification ran on 100 subjects only and received the results in the following Table 4.

3 Case Studies: In Progress

In this section, we discuss some of the applications of the proposed approach in real scenarios with related significance for each study. The light nature of the

Table 4. Performance comparison on LFW data set.

Algorithm	Accuracy	EER
HoG + Sub-SML	88.34	11.45
LBP + Sub-SML	83.54	16.00
FV + Sub-SML	91.30	8.85
Deep features	99.38	0.62
This approach	93.30	6.85

proposed approach in terms of size and ability to run on low power computing devices like Raspberry-Pi allows for the use of this identification technique in places like vehicles. Furthermore, no continuous internet connection is required as the device could potentially store relevant information and conduct comparisons locally for specified users until a connection is restored when all information could be synced.

3.1 Case Study - Patient Identification System in Hospitals: Verification of Identity Before Medical Procedure

Using this technique in settings like pre-operation verification could contribute to any identification conflicts that might occur in patients scheduled to go through a procedure or in medicine administration [5]. From the system point of view, a specific operation with details like time and date and patient identifier could be entered into the system. When the time comes for the operation, the system will assure that the patient is the one matching the one in the system. Currently, this verification procedure is done manually and by checking the wristband tag. This could result in many errors that could potentially be harmful. Thus, using a computer-based system for this matter can increase the safety of the patients, decrease errors, and enhance the verification process.

3.2 Case Study - Assisted Driving System: Verification of Drivers' Identities Before Authorizing Them to Use or Drive Special Vehicles

Special vehicles in this context include police cars, ambulances, firetrucks, any construction vehicles, or heavy machinery with restricted access. Misuse of these vehicles have multiplied circumstances over these vehicles. Nevertheless, it can be used in any car to help against theft, collecting and analyzing valuable information, and determining driver behavior [3,11]. Using the authorization technique discussed in these vehicles will contribute to limiting misuse occurrence. It can be used by users, government agencies and insurance companies for the use case examples above. There are more applications where implementing this technique can be very useful specially with extended features like including location services, fuel consumption, and mileage analysis.

4 Conclusion and Future Work

We are in the process of validating our approach and obtaining benchmarking results against different state-of-the-art methods using industry standards. There are many advantages to the proposed approach including maintaining privacy using a unique identifier per person instead of storing images. This also saves storage space. In addition, faster computations due to comparing simple numbers. Finally, this approach is done in an explainable matter using a glass-box approach rather than black-box in which it is unknown how the comparison results are obtained. We presented some of the case studies or application for our approach where it will be helpful to use including protecting assets, enhancing security authentication, and increasing safety specially in patients. Future extensions of this approach will look at expanding the basis to include multiple spectrum like IR and UV including 3D images as an input.

Acknowledgements. This study was supported by the Scientific Research from Technical University of Technology Sydney, School of Electrical and Data Engineering and DIVE IN AI.

References

1. Bud, A.: Facing the future: the impact of apple FaceID. Biometric Technol. Today **2018**(1), 5–7 (2018). https://doi.org/10.1016/s0969-4765(18)30010-9
2. Cheng, G., Han, J., Zhou, P., Xu, D.: Learning rotation-invariant and fisher discriminative convolutional neural networks for object detection. IEEE Trans. Image Process. **28**(1), 265–278 (2019). https://doi.org/10.1109/TIP.2018.2867198
3. Ezzini, S., Berrada, I., Ghogho, M.: Who is behind the wheel? Driver identification and fingerprinting. J. Big Data **5**, 9 (2018). https://doi.org/10.1186/s40537-018-0118-7
4. Geitgey, A.: Machine learning is fun! part 4: modern face recognition with deep learning (2016). https://medium.com/@ageitgey/machine-learning-is-fun-part-4-modern-face-recognition-with-deep-learning-c3cffc121d78. Accessed 6 Oct 2019
5. Härkänen, M., Tiainen, M., Haatainen, K.: Wrong-patient incidents during medication administrations. J. Clin. Nurs. **27**(3–4), 715–724 (2017). https://doi.org/10.1111/jocn.14021
6. Huang, G.B., Ramesh, M., Berg, T., Learned-Miller, E.: Labeled faces in the wild: a database for studying face recognition in unconstrained environments. Technical report, 07-49, University of Massachusetts, Amherst, October 2007
7. Iqbal, M., Sameem, M.S.I., Naqvi, N., Kanwal, S., Ye, Z.: A deep learning approach for face recognition based on angularly discriminative features. Pattern Recogn. Lett. **128**, 414–419 (2019). https://doi.org/10.1016/j.patrec.2019.10.002
8. King, D.E.: Dlib-ml: a machine learning toolkit. J. Mach. Learn. Res. **10**, 1755–1758 (2009). https://dl.acm.org/citation.cfm?id=1755843
9. Kumar, A., Kaur, A., Kumar, M.: Face detection techniques: a review. Artif. Intell. Rev. **52**(2), 927–948 (2019). https://doi.org/10.1007/s10462-018-9650-2
10. Liu, J., Liu, W., Ma, S., Wang, M., Li, L., Chen, G.: Image-set based face recognition using K-SVD dictionary learning. Int. J. Mach. Learn. Cybern. **10**(5), 1051–1064 (2019). https://doi.org/10.1007/s13042-017-0782-5

11. Martinelli, F., Mercaldo, F., Nardone, V., Orlando, A., Santone, A.: Who's driving my car? A machine learning based approach to driver identification. In: Proceedings of the 4th International Conference on Information Systems Security and Privacy, ICISSP 2018, Funchal, Madeira - Portugal, 22–24 January 2018, pp. 367–372 (2018). https://doi.org/10.5220/0006633403670372
12. Scherhag, U., Rathgeb, C., Merkle, J., Breithaupt, R., Busch, C.: Face recognition systems under morphing attacks: a survey. IEEE Access **7**, 23012–23026 (2019). https://doi.org/10.1109/ACCESS.2019.2899367
13. Sengupta, S., Chen, J.C., Castillo, C., Patel, V.M., Chellappa, R., Jacobs, D.W.: Frontal to profile face verification in the wild. In: 2016 IEEE Winter Conference on Applications of Computer Vision (WACV), pp. 1–9. IEEE (2016)
14. Wong, T.T.: Performance evaluation of classification algorithms by k-fold and leave-one-out cross validation. Pattern Recogn. **48**(9), 2839–2846 (2015)

Eye-Tracking and Machine Learning Significance in Parkinson's Disease Symptoms Prediction

Artur Chudzik[1](✉), Artur Szymański[1], Jerzy Paweł Nowacki[1], and Andrzej W. Przybyszewski[1,2]

[1] Polish-Japanese Academy of Information Technology, Koszykowa 86 Street, 02-008 Warsaw, Poland
{artur.chudzik,artur.szymanski,nowacki,przy}@pjwstk.edu.pl
[2] Department of Neurology, University of Massachusetts Medical School, 65 Lake Avenue, Worcester, MA 01655, USA
andrzej.przybyszewski@umassmed.edu

Abstract. Parkinson's disease (PD) is a progressive, neurodegenerative disorder characterized by resting tremor, rigidity, bradykinesia, and postural instability. The standard measure of the PD progression is Unified Parkinson's Disease Rating (UPDRS). Our goal was to predict patients' UPDRS development based on the various groups of patients in the different stages of the disease. We used standard neurological and neuropsychological tests, aligned with eye movements on a dedicated computer system. For predictions, we have applied various machine learning models with different parameters embedded in our dedicated data science framework written in Python and based on the Scikit Learn and Pandas libraries. Models proposed by us reached 75% and 70% of accuracy while predicting subclasses of UPDRS for patients in advanced stages of the disease who respond to treatment, with a global 57% accuracy score for all classes. We have demonstrated that it is possible to use eye movements as a biomarker for the assessment of symptom progression in PD.

Keywords: Eye-tracking · Saccades · Parkinson's Disease · Machine learning

1 Introduction

In Parkinson's disease, we can distinguish multiple therapies, which could be combined. The gold-standard treatment for PD is a pharmacological treatment with Levo-dopa (L-dopa) [1, 2]. Nevertheless, L-dopa associates with long-term disturbances. We can distinguish hypo-hyperkinetic phenomena and psychosis as examples of motor and mood side effects. [3]. The practical and safe procedure, which is lacking these effects and remains the preferred surgical treatment for advanced Parkinson's disease, is Deep Brain Stimulation of the subthalamic nucleus (STN) [4]. However, there is still a necessity of the support for the neurologists in the field of optimum treatment parameters, because due to the huge diversity of cases, even the most experienced doctors could not be sure how the therapy would influent on the patient.

N. T. Nguyen et al. (Eds.): ACIIDS 2020, LNAI 12034, pp. 537–547, 2020.
https://doi.org/10.1007/978-3-030-42058-1_45

2 Methods

2.1 The Subject of the Study

The research group was composed of 62 patients who have Parkinson's disease and who are under the supervision of the Warsaw Medical University (Warsaw, Poland) neurologists. We differentiate the patients into three groups. The first one, we named the Best Medical Treatment (BMT). In this group, we placed patients who were treated only by the medication. The second one, the Deep Brain Stimulation (DBS), was a group where patients, who had implanted electrodes in the STN during our study. The last group, which was named Post-Operative Patients (POP) aggregates individuals who had had surgery earlier (before the beginning of our research). Every PD patient had three visits, which were underdone approximately every six months. Every patient from the DBS group had his/her visit before surgery.

In order to obtain the countable value of the disease, it is crucial to provide a precise neurological tool which could objectively measure all of the symptoms and determine a score. For the metric of Parkinson's disease advancement, there are two common neurological standards: the Hoehn and Yahr scale and the Unified Parkinson's Disease Rating (UPDRS). The UPD rating scale is the most commonly used in the clinical study of Parkinson's disease [5], and we also decided to adopt it in this study. Altogether with UPDRS, every patients' disease metric was combined with the disease duration, the result of Parkinson's Disease Questionnaire PDQ39 (which is a disease-specific health-related quality-of-life outline), the result of Epworth Sleepiness Scale (which is intended to measure daytime sleepiness), and the parameters of saccadic eye movements, described further.

The mean age of patients was 51.1 ± 10.2 (standard deviation) years. The mean duration of the disease was 11.6 ± 4.3 years. The mean of UPDRS score (for all symptoms) was 33.8 ± 19.4. The mean of PDQ39 score was 50.5 ± 26.0, and the mean of Epworth score was 8.7 ± 4.6.

2.2 Eye Movements

The often diagnosed impairment of automatic behavioral responses accompanies the slowness of initiation of voluntary movements in individuals with PD [6]. An individual set of behavioral tasks may provide insight into the neural control of response suppression with the usage of motor impairments analysis based on saccadic eye movements [7, 8]. Saccades are a quick, simultaneous movement of both eyes between two or more phases of fixation in the same direction, and can be measured quickly and precisely.

We choose this marker because of the considerable understanding of the neural circuitry controlling the planning and execution of saccadic eye movements [9].

During this study, we have used a head-mounted saccadometer JAZZ-pursuit (ober-consulting.com), which was able to measure the reflexive saccades (RS) in the high frequency (1000 Hz). We have chosen this device because it is optimized for easy set-up and provides minimal intrusiveness while can keep stable 1 kHz frequency of measurement. During the experiment, we created a task of the horizontal reflexive saccades analysis. Used hardware allowed us to obtain high accuracy and precision in eye tracking and the compensation of possible subjects' head movements relative to the monitor. Thus subjects did not need to be positioned in an unnatural chinrest, which has a positive influence on the ergonomy of the experiment. However, we asked patients to use a headrest in order to minimize the head motion because they could have a significant influence on the accuracy of the high-frequency measurements. Each patient was seated in front of the monitor at a distance of 60–70 cm.

The patients' task was to fix their eyes on the spot placed in the middle of the screen (0°). Then, the spot changed color into one of the possible variations and shift horizontally to one of the possible directions: 10° to the left, or 10° to the right, after arbitrary time ranging between 0.5–1.5 s. During that task, we measured the fast eye movements of the patient, according to the spot color transition.

When the transition was from white to green, it was a signal for the execution of RS. We also prepared an additional protocol for antisaccades (AS) measurement. In this task, the individual was asked to make a saccade in the direction away from the stimulus. A signal for that was when the spot changed color from white to red. After that, the central marker was hidden, and one of the two peripheral targets was shown. The selection was made randomly, with the same probability.

According to the task, each patient looked at the spots and followed them as they moved (in the RS task) or made opposite direction eye movement (in the AS task). After that, the target remained still for 0.1 s before the next experiment initialization.

In each test, the subject had to perform twenty saccades and antisaccades in a row-twice. The recording of the first session (marked as S1) was with the patient who has temporarily disabled treatment (without medicine/with the disabled neurostimulator). In the next session (marked as S3), the patient took medication and had a break for one half to one hour, and then the same experiments were performed (with L-dopa/with enabled neurostimulator).

In this experiment, we have investigated only RS data using the following population parameters averaged for both eyes: mean latency (*RSLat* $\pm$ SD), mean amplitude (*RSAmp* $\pm$ SD), mean of the maximum velocity (*RSPVel* $\pm$ SD), and mean duration of the saccades (*RSDur* $\pm$ SD).

2.3 Dataset

We implemented a dedicated database for measurements storage that was designed and maintained by Polish-Japanese Academy of Information Technology (Warsaw, Poland). For operations described further, we flattened the data, placing every experiment in each row, with the results and metadata in the separated columns. Therefore it could be represented by a single table represented by comma-separated-values, which is a universal format for computational engines. The basic structure of the dataset contains 374 observations, and each has 13 variables, which are: *Duration* - the duration of the disease; *UPDRS, PDQ39, Epworth* - the score for each test; *RSLat, RSDur, RSAmp, RSPVel* - the parameters of recorded saccades; *Session, Visit* - the indexes of the experiment; and *BMT, DBS, POP* - boolean variables which describe the kind of patients' therapy.

2.4 Computational Learning Theory

Our data contains a set of N training samples of the form $(x_1, y_1), \ldots, (x_n, y_n)$ such that x_i is the feature vector of the $i-th$ sample with the class denoted by y_i. Thus, it is possible to use a supervised learning algorithm which seeks for a function $g : X \to Y$, where the X is the input space, and the Y is the output space. The g function is an element of some space of possible functions G, known as the hypothesis space.

The task itself could is a multiclass classification. Our goal is to predict a level of the disease measured as UPDRS value binned into intervals, for different groups of patients. For the predictions, we have created a dedicated machine learning framework, written in Python and based on two libraries: Scikit Learn [11] and Pandas, which are high-quality, well-documented collection of canonical tools for data processing and machine learning. We have chosen well-known models that implement different multiclass strategies, such as K Neighbors Classifier, Support Vector Classifier, Decision Tree Classifier, Random Forest Classifier, Gradient Boosting Classifier. The framework allowed us to find an optimal solution by the examination of multiple algorithms with different parameters.

2.5 Hypothesis

Our goal was to predict Parkinson's disease progression in the advanced stage, based on the data obtained from the patients in the different treatment and stage of this disease. This task is non-trivial because there are significant differences between symptom developments and the effects of different treatments in individual PD patients.

As a training dataset, we used patients from the BMT group (3rd visit), DBS (3rd visit), and POP (1st visit). The independent test set consisted of the POP group from the second visit.

3 Results

Our framework was responsible for every step of data processing in order to evaluate the best model based on given data. Therefore, we implemented procedures which were responsible for the creation of the Profiling Report, Correlation Matrices, Missing Data Imputation, Data Discretization, One-Hot Encoding or Data Normalisation of selected variables and Machine Learning Algorithms Evaluation for various parameters.

3.1 Profiling

Our dataset consists of 374 observations, where each has 13 variables. First, we generated a Pearson correlation coefficient matrix, where each cell in the table shows the correlation between two variables. It is a useful tool that proves if a correlation between the parameters from measurements and the symptoms suggest that some close relations exist (Fig. 1).

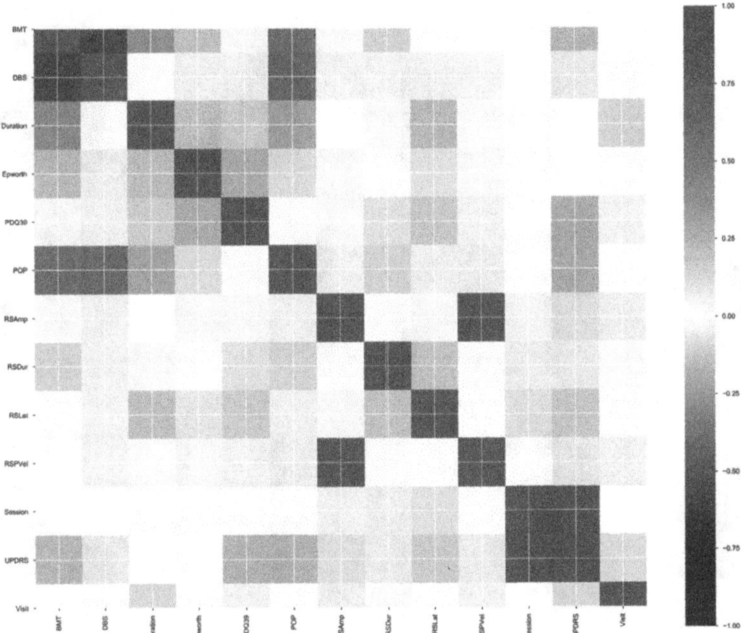

Fig. 1. Correlation matrix. The similarity between all paired parameters is correlated with the color of the cell in the matrix (red shade represents high similarity; blue stands contrariwise). (Color figure online)

3.2 Pre-processing

In the report, we noticed that some of the values are empty. *Epworth* column had 5 (1.3%) zeros; *RSAmp*, *RSDur*, and *RSLat* columns had 6 (1.6%) zeros. Because of the Epworth scale ranges between 0–24, we decided not to apply data imputation on that column. When we analyzed records related to a single patient, we noticed, that on an early stage of the disease (duration about 7 ± 0.3 years) two of them indeed reported no problems with the daytime sleepiness, yet visits in the following years revealed a linear increase in the results. For missing parameters of the saccades, we applied the imputation transformer for completing missing values which replaced missing values using the mean along each column.

For neurological and neuropsychological tests results (Fig. 2), we applied k-bins discretization, which provides a way to partition continuous features into discrete values. Thus those features are a one-hot encoded allowing the model to be more expressive while maintaining interpretability.

For neurological and neuropsychological tests results, we applied k-bins discretization, which provides a way to partition continuous features into discrete values. Thus those features are a one-hot encoded allowing the model to be more expressive while maintaining interpretability.

$$Epworth = (-\infty, 6.00), [6.00, 11.00), [11.00, +\infty)$$

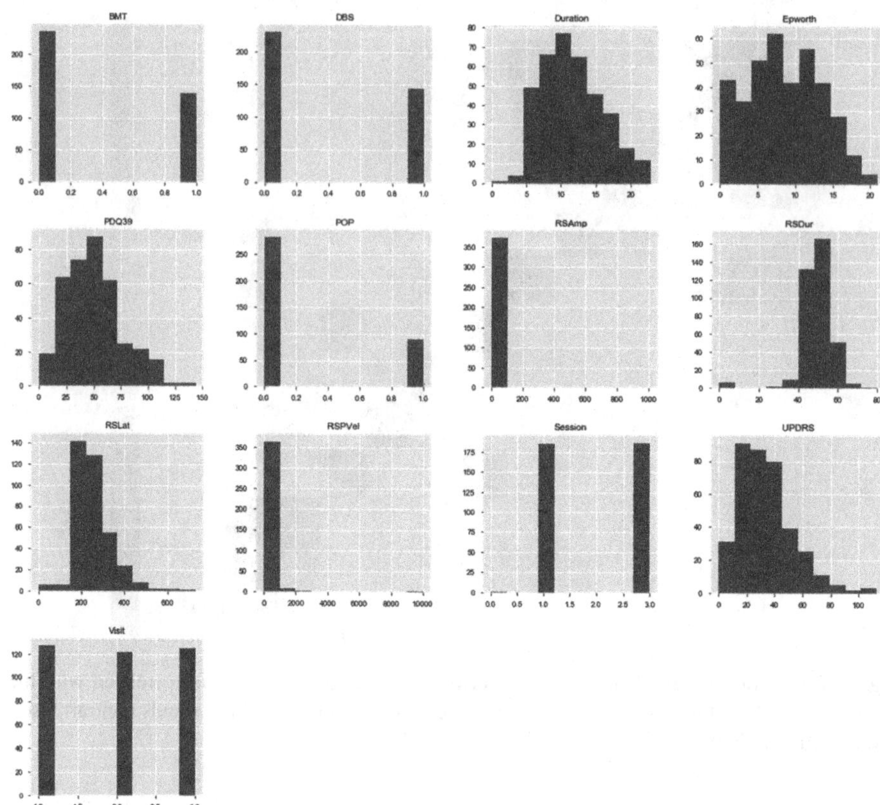

Fig. 2. A histogram, which is a representation of the distribution of data for every measurable property in the data set. Y-Axis represents a number of samples. The X-Axis presents a parameter value.

$$PDQ39 = (-\infty, 25.0), [25.0, 38.0), [38.0, 47.5), [47.5, 58.0), [58.0, 74.0), [74.0, +\infty)$$

Other columns (*Duration, RSLat, RSDur, RSAmp, RSPVel*) were standardize by removing the mean and scaling to unit variance to keep the subtle representation of the eye movement signal. The calculation of the standard score of X sample is defined by:

$$z = \frac{x - u}{s}$$

where u is the mean of the training samples, and s is the standard deviation of the training samples.

Target value, the *UPDRS* score was optimally divided into four ranges

$$UPDRS = (-\infty, 19.25), [19.25, 30.50), [30.50, 44.00), [44.00, +\infty)$$

This split ensured approximately the same data set size for each class.

3.3 Machine Learning

We decided to evaluate multiple algorithms in order to determine the best approach in the meaning of accuracy of the predictions. In the previous research [10], Random Forrest Classifier achieved the high prediction level, being second to Rough Set. This article covers the evaluation of a few machine learning models which were not used in the previous research, such as K Neighbors Classifier, Support Vector Classifier, Decision Tree Classifier, Random Forest Classifier, Gradient Boosting Classifier. The following sections present each algorithm with evaluated parameters and scores obtained from our machine learning framework.

K Neighbors Classifier

Classifier implementing the k-nearest neighbors' vote with uniform weights achieved maximum accuracy score (0.50) with a parameter defining the number of neighbors on the level of 12 and 13 (Fig. 3).

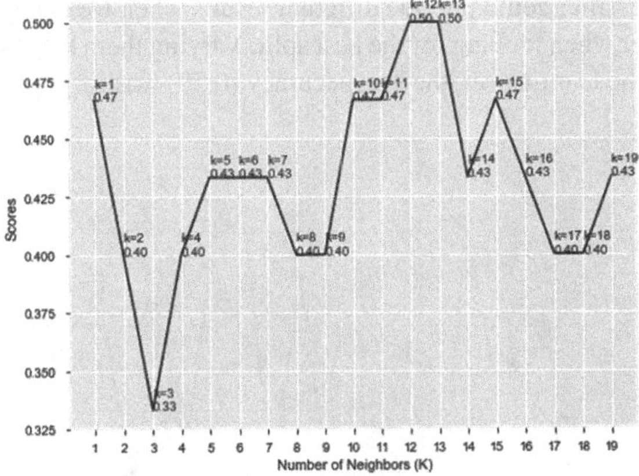

Fig. 3. K Neighbors Classifier accuracy scores for different K values.

Support Vector Classifier

We evaluated the C-Support Vector Classification against different kernels (linear, polynomial, radial-basis, and sigmoid). The "linear" kernel achieved the best accuracy score (0.40) under the penalty parameter $C = 0.3$ (Fig. 4).

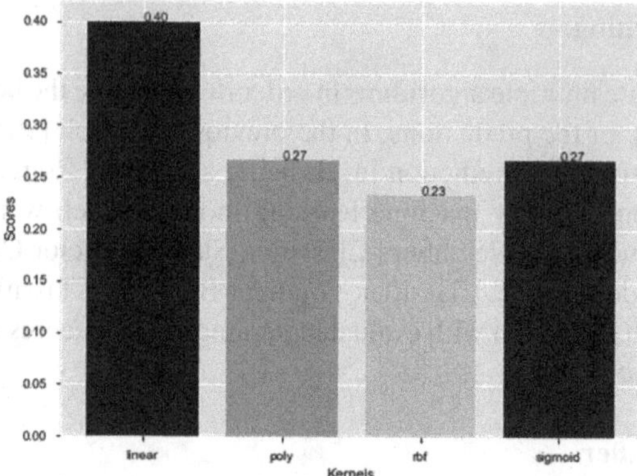

Fig. 4. Support Vector Classifier accuracy scores for different kernels.

Decision Tree Classifier

A decision tree classifier could provide different results when we change the number of features to consider when looking for the best split. Varying them between 1 and the size of all columns of the learning set, the best accuracy (0.50) was for 6, 14, and 15 features (Fig. 5).

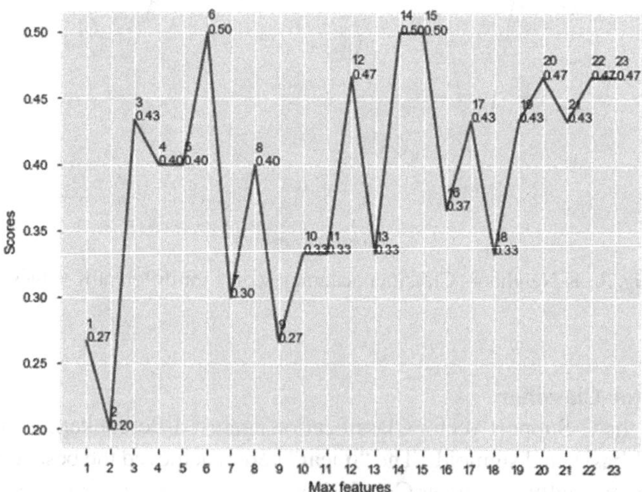

Fig. 5. Decision Tree Classifier accuracy scores for different number of maximum features.

Gradient Boosting Classifier

Gradient Boosting it allows for the optimization of arbitrary differentiable loss functions. We challenged a different number of boosting stages to perform. Gradient boosting is relatively robust to over-fitting, so a large number usually results in better performance.

The scope was between 10 and 1000 estimators, and the highest score (0.43) reveals in the 200 boosting stages (Fig. 6).

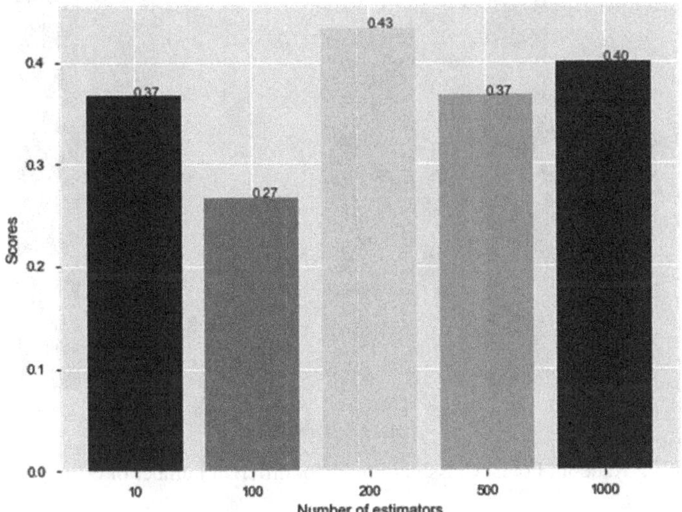

Fig. 6. Gradient Boosting Classifier accuracy scores for different number of estimators.

Random Forest Classifier

In this classifier, we evaluated a parameter which defines the number of trees in the forest. We used Gini impurity as a criterion of the quality of a split. Random Forest Classifier achieved the highest overall accuracy score (0.57) among other machine learning algorithms when the number of estimators exceeded 120 (Table 1, Fig. 7).

Table 1. Confusion matrix based on Random Forest Classifier result.

		Predicted label				
		(−Inf, 19.25)	[19.25, 30.5)	[30.5, 44.)	[44., +Inf)	Accuracy
True label	(−Inf, 19.25)	6	1	1	0	**0.75**
	[19.25, 30.5)	1	7	2	0	**0.70**
	[30.5, 44.)	3	2	3	0	**0.38**
	[44., +Inf)	1	1	1	1	**0.25**

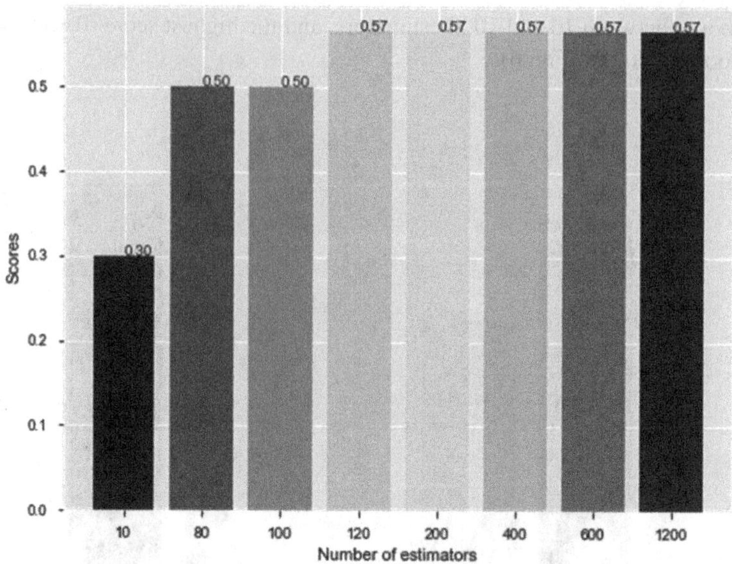

Fig. 7. Random Forest Classifier scores for different number of estimators.

4 Discussion

Examination of the variation of the saccades is a non-invasive method of evaluating the neural networks involved in the control of eye movements. The examples above demonstrated that it is possible to use eye movements as a biomarker for the assessment of symptom progression in PD.

We live in the "age of implementation" of the machine learning models. Commercial companies are acquiring data on a large scale in order to present the content which is best-fitted to the end-user, which influences the businesses. However, there is still an emerging necessity of medical data extension because it is a crucial factor in the context of machine learning. Paradoxically, we have access to much fewer data of the medical records (even about ourselves) than most of the managers of the commercial websites who aggregates about our shopping behavior. The modeled data sample, presented in this article, is relatively abundant in the neuroscience scale. Based on the examinations conducted by neurologists, we were able to create predictions based on the different patient population with different treatments for the most advanced stages of the disease. This results supported with further extended and in-depth research could lead to a new approach in the development of a follow-up tool for PD symptoms. As an outcome, this automated mechanism could provide to a doctor an objective opinion about applied therapy symptoms. Hence we can conclude that when the patient is doing significantly worse than others, their treatment is not optimal and should be changed. However, there is still a large field for the model improvements that should lead to more accurate results. The proposed protocol allowed us to evaluate multiple machine learning models in a relatively agile process of data aggregation. Consequently, more complex variations of saccadic tasks can give insight into higher-order eye movement control [12]. Our work is an evaluation of well-known models that implement different multiclass strategies, such as K Neighbors Classifier, Support Vector Classifier, Decision Tree Classifier, Random Forest Classifier, Gradient Boosting Classifier in the context of saccades research. In this trial, Random Forest Classifier achieved the highest overall accuracy score, which could lead to a direction in further discoveries in the field of bioinformatics.

5 Conclusions

We believe that the multidisciplinary cooperation between neurologists and information engineering is essential to achieve significant results in the open-science approach. We attempted to predict the longitudinal symptom developments during different treatments based on the neuropsychological data aligned with the parameters of the eye movements. As a training dataset, we used patients from the BMT group (3rd visit), DBS (3rd visit), and POP (1st visit). The independent test set consisted of the POP group from the second visit. For predictions, we used a machine learning framework, written in Python. The best classifier - Random Forest - reached 75% and 70% of accuracy while predicting subclasses of UPDRS for patients in advanced stages of the disease who respond to treatment, with a global 57% accuracy score for all classes. Thanks to collaborative research, we have presented a comparison of different machine learning models that could be useful in the context of bioinformatics. Our direction is to create a new research ecosystem, that would significantly increase (by a factor of 10) the number of attributes and measurements in order to implement deep learning methods.

References

1. Connolly, B.S., Lang, A.E.: Pharmacological treatment of Parkinson disease: a review. JAMA **311**(16), 1670–1683 (2014)
2. Goldenberg, M.M.: Medical management of Parkinson's disease. Pharm. Therapeutics **33**(10), 590 (2008)
3. Thanvi, B.R., Lo, T.C.N.: Long term motor complications of levodopa: clinical features, mechanisms, and management strategies. Postgrad. Med. J. **80**(946), 452–458 (2004)
4. Benabid, A.L., et al.: Deep brain stimulation of the subthalamic nucleus for the treatment of Parkinson's disease. Lancet Neurol. **8**(1), 67–81 (2009)
5. Ramaker, C., et al.: Systematic evaluation of rating scales for impairment and disability in Parkinson's disease. Mov. Disord. Off. J. Mov. Disord. Soc. **17**(5), 867–876 (2002)
6. Henik, A., et al.: Disinhibition of automatic word reading in Parkinson's disease. Cortex **29**(4), 589–599 (1993)
7. Jones, G.M., DeJong, J.D.: Dynamic characteristics of saccadic eye movements in Parkinson's disease. Exp. Neurol. **31**(1), 17–31 (1971)
8. White, O.B., et al.: Ocular motor deficits in Parkinson's disease: II. Control of the saccadic and smooth pursuit systems. Brain **106**(3), 571–587 (1983)
9. Chan, F., et al.: Deficits in saccadic eye-movement control in Parkinson's disease. Neuropsychologia **43**(5), 784–796 (2005)
10. Przybyszewski, A., et al.: Multimodal learning and intelligent prediction of symptom development in individual Parkinson's patients. Sensors **16**(9), 1498 (2016)
11. Pedregosa, F., et al.: Scikit-learn: machine learning in Python. J. Mach. Learn. Res. **12**, 2825–2830 (2011)
12. Nij Bijvank, J.A., et al.: A standardized protocol for quantification of saccadic eye movements: DEMoNS. PLoS ONE **13**(7), e0200695 (2018)

IGrC: Cognitive and Motor Changes During Symptoms Development in Parkinson's Disease Patients

Andrzej W. Przybyszewski[1](✉), Jerzy Paweł Nowacki[1], Aldona Drabik[1], Stanislaw Szlufik[2], and Dariusz M. Koziorowski[2]

[1] Polish-Japanese Academy of Information Technology, 02-008 Warsaw, Poland
{przy,jerzy.nowacki,adrabik}@pjwstk.edu.pl

[2] Department of Neurology, Faculty of Health Science, Medical University of Warsaw, Warsaw, Poland
stanislaw.szlufik@gmail.com, dkoziorowski@esculap.pl

Abstract. Cognitive symptoms are characteristic for neurodegenerative disease: there are dominating in the Alzheimer's, but secondary in Parkinson's disease (PD). However, in PD motor symptoms (MS) are dominating and their characteristic helps neurologist to recognize the disease. There are a large number of data mining publications that analyzed MS in PD. Present study is related to the question if development of cognitive symptoms is related to motor symptoms or if they are two independent processes? We have responded to this problem with help of IGrC (intelligent granular computing) approach. We have put together eye movement, neurological and neuropsychological tests. Our study was dedicated to 47 Parkinson's disease patients in two sessions: S#1 - without medications (Med_{OFF}) and S#2 after taking medications (Med_{ON}). There were two groups of patients: Gr1 (23 patients) less advanced and Gr2 more advanced PD. We have measured Gr1 in three visits every 6 months: $Gr1_{VIS1}$, $Gr1_{VIS2}$, $Gr1_{VIS3}$. Gr2 (24 patients) has only one visit (no visit number). With rough set theory (RST) that belongs to IGrC we have found from Gr2 three different sets of rules: a) general rules (G_{RUL}) determined by all attributes; b) motor related rules (M_{RUL}) – motor attributes; c) cognitive rules (C_{RUL}) determined by cognitive attributes. By applying these different sets of rules to different Gr1 visits we have found different set of symptoms developments. With G_{RUL} we have found for $Gr1_{VIS1}$ accuracy = 0.682, for $Gr1_{VIS2}$ acc. = 0.857, for $Gr1_{VIS3}$ acc. = 0.875. With M_{RUL} we have found for $Gr1_{VIS1}$ acc. = 0.80, for $Gr1_{VIS2}$ acc. = 0.933, for $Gr1_{VIS3}$ acc. = 1.0. With C_{RUL} we have found for $Gr1_{VIS1}$ acc. = 0.50, for $Gr1_{VIS2}$ acc. = 0.60, for $Gr1_{VIS3}$ acc. = 0.636. Cognitive changes are independent from the motor symptoms development.

Keywords: Neurodegeneration · Rough set theory · Intelligent granular computing

N. T. Nguyen et al. (Eds.): ACIIDS 2020, LNAI 12034, pp. 548–559, 2020.
https://doi.org/10.1007/978-3-030-42058-1_46

1 Introduction

Cognitive changes are leading in the most common neurodegenerative disease (ND) Alzheimer's disease (AD), but in Parkinson's disease (PD) they are secondary to dominating motor symptoms. In the most cases of AD neurodegeneration starts from the hippocampus and frontal cortex and it related to memory and orientation problems. With the disease progression other brain regions become also affected. In PD neurodegeneration starts from basal ganglia (substantia nigra) and is related to the lack of dopamine. Dopamine (Dopa) controls adaptation of movements to the environment. Therefore PD patients have primary motor symptoms but some of them may have also cognitive changes [1].

As SN (substantia nigra) neurodegeneration causes depletion the Dopa, in addition to the movements' problems there are also potentially emotional and cognitive decays in some PD. As and individual patient has not only a unique neurodegeneration development but also distinctive compensatory processes then as result symptoms might be various and therefore finding optimal treatment is an art for an experienced neurologist.

We have estimated disease progression in different sets of attributes in order to find if motor and cognitive symptoms have similar or dissimilar developments.

This study is an enlargement of our earlier works by using additional to our IGrC (RST) a new attributes – Trail Making Test (TMT) Part A and Part B (see below).

2 Methods

We have evaluated data from Parkinson Disease (PD) patients separated into two main groups with different disease duration:

- **Gr1** contained of 23 patients that received only appropriate medication (L-Dopa). As mentioned above, because PD starts with major neurodegeneration in substantia nigra that regulates the level of the dopamine, the major medications are related to the dopamine precursors or inhibitions of the dopamine reuptake.
- **Gr2** involved 24 patients also as Gr1 only on medications, but they were in the more advanced disease stage. They had longer mean disease duration, as well as their UPDRS was higher than that in Gr1 group. Also these patients went later to the surgery of DBS (deep brain stimulation) that was related to placing stimulating electrode in the basal ganglia – here in the STN (subthalamic nucleus).

All subjects were tested in two related to medications meetings: session 1 (Ses#1) was for patients that stopped taking their medications one day before our tests; session 2 was performed when patients were on their normal medications.

All PD patients had the following measurements: neuropsychological - related to the quality of life (PDQ39), sleep problems (Epworth test), depression (Beck test), and two additional TMT A&B tests; disease duration; UPDRS (Unified Parkinson's Disease Rating Scale) as basic neurological test for PD and fast eye movement tests. TMT tests present the same motor and perceptual demands, namely drawing lines to connect randomly arranged circles. TMT A has only circles with numbers (motor task). In part

B there are numbers and letters that additionally measures divided attention and mental flexibility (TMT B – cognitive task). All patients were tested in the Dept. of Neurology, Brodno Hospital, Faculty of Health Science, and Medical University of Warsaw, Poland. In this study, we have registered reflexive saccadic (RS) eye movements as validated in our earlier articles [1, 2]. In short, each subject was placed *vis-à-vis* computer monitor paying attention to screen before him/her. The experiment began as subject fixated on the spot light in the center of the computer display. The session normally started from the spot light slow movements with increasing speed and continued with random in directions (ten degrees to the right or ten degrees to the left) light spot springing that patient's eyes should followed it. This test took about 1.5 min.

We have registered instantaneous light spot and eyes positions by the clinically proven head-mounted saccadometer (JAZZ novo, Ober Consulting, Poland). In the current work, we have only analyzed fast, saccadic responses of both eyes. We have compared light spot and eyes positions and calculated the following attributes of the fast saccades (RS): the delay (*RSL*) measured as the differentiation between the initiation of the light spot and eyes movements; the amplitude of the saccadic - *RSAm*, its duration (*RSD*) and the mean speed (*RSV*) of both eyes during the saccade.

All procedures and measurements were repeated for each session also in each session patient has to perform 10 RS and means values of above-mentioned parameters were used for the analyses.

2.1 Theoretical Basis

The intelligent granular computing (IGrC) analysis was implemented in RST (rough set theory proposed by Zdzislaw Pawlak [3]) and recently extended by Andrzej Jankowski [4].

In the standard RST procedure all our results were adapted into the decision table with rows showing actual attributes' values for the dissimilar or the same subject and columns were related to different attributes. Following [3] an information system is a pair $S = (U, A)$, where U, A are nonempty finite sets. The set U is the universe of objects, and A is the set of attributes. If $a \in A$ and $u \in U$, the value $a(u)$ is a unique element of V (where V is a value set). The *indiscernibility relation IND(B)* of any subset B of A is defined after [3]: $(x, y) \in IND(B)$ *iff* $a(x) = a(y)$ for every $a \in B$ where the value of $a(x) \in V$. This relation divides A into *elementary granules* and it is the basis of RST. In the information system S set $B \subset A$ is a reduct if $IND(B) = IND(A)$ and it cannot be further reduced. Other important RST properties such as *lower approximation* and *upper approximation* were defined and discussed in [3, 5] and illustrated in [6].

In this work we have used different AI: machine learning methods (RSES 2.2) such as: exhaustive, genetic [7], covering, or LEM2 algorithms [8].

An extension of the information system is the decision table as a triplet: $S = (U, C, D)$ where attributes A are divided into C and D as condition, and decision attributes [9]. As in a single row there are many condition attributes and one decision attribute related to a particular measurement of the individual subject we can interpret it as unique rule *if*$(condition\text{-}attribute_1 = value_1 \& \ldots) => (decision\text{-}attribute_n = value_n)$. If we gather all such rules (measurements) for a single patient they might be basis for so-called *precision (personalized) medicine* with one condition that they are not contradictory. As in our decision table we have measurements of different patients in several conditions our main purpose is to find *universal rules*. This is a possible thanks to IGrC implemented in RST that generalize all specific rules into the knowledge with proposals that are always true as it is related to the *lower approximation*, and others are only partly true that is related to the *upper approximation*. Notice that the decision attribute (in our case UPDRS) is the result of tests performed by the neurologist specialized in the Parkinson's disease. Therefore, *the supervised machine learning with a doctor as the teacher determines our knowledge*.

Even for very experienced neurologist finding optimal treatment is difficult and it is related to differences between patients and dissimilar effects of similar treatments. It is connected to differences between individual patients, as the neurodegenerative processes begin many years earlier than first symptoms in PD and during this time plastic compensatory processes in each brain are distinctive.

Our algorithms are related to IGrC that takes into account individual patients differences, but thanks to RST we can obtain abstraction and generalization of our rules that might simulate intuitions of an experienced neurologist. Also we would like to mention that our IGrC mimics such advanced processes in the brain as for example complex objects recognition [10]. We have recently demonstrated that RST can describe processes in the visual brain that are related to different objects classification [10]. Therefore, we might assume that our IGrC rules are sufficiently flexible, abstract and universal (like a visual brain) to resolve Parkinson's disease progressions issues related to different disease stages and various treatments.

We have used as IGrC the RSES 2.2 [11] that generalizes rules from decision table to treat diverse patients. In our earlier work, we have demonstrated that the RST application provides enhanced results than proposed by others AI algorithms [1].

3 Results

As explained above in the Methods section we had two groups of patients: Gr1 of less advanced PD (23 patients) and more advanced group (24 patients). Both group were tested in two sessions: Ses#1 – without medications and in Ses#2 on medications.

Statistical Results

Patients from group Gr1 had three visits: $Gr1_{VIS1}$, $Gr1_{VIS2}$, $Gr1_{VIS3}$ every 6 months. Their mean age was 58 ± 13 (SD) years with mean disease duration of 7.1 ± 3.5 years.

For $Gr1_{VIS1}$ mean total UPDRS in Ses#1 was 48.3 ± 17.9, in Ses#2 was 23.6 ± 10.3 (statistically different with $p < 0.0001$).

For $Gr1_{VIS2}$ mean total UPDRS in Ses#1 was 57.3 ± 16.8, in Ses#2 was 27.8 ± 10.8 (statistically different with $p < 0.0005$).

For $Gr1_{VIS3}$ mean total UPDRS in Ses#1 was 62.2 ± 18.2, in Ses#2 was 25 ± 11.6 (statistically different with $p < 0.0001$).

For Gr2 patients their mean age was 53.7 ± 9.3 years, and they have 10.25 ± 3.9 years disease duration. Their total UPDRS in Ses#1 was 62.1 ± 16.1, and in Ses#2 29.9 ± 13.3.

3.1 IGrC for Reference Gr2 Group

We have placed Gr2 data in the following information table (Table 1).

Table 1. Part of the decision table for three Gr2 patients

P#	dur	S#	PDQ39	Epw.	RSL	RSD	RSAm	RSV	Beck	TrA	TrB	UPDRS
45	13.3	1	56	10	212	46	9.5	407	19	49	90	76
45	13.3	2	56	10	284	47	9.69	402	19	49	90	42
46	7.3	1	48	0	202	55	4.7	168	19	49	60	53
46	7.3	2	48	0	386	49	9.8	367	19	49	60	18
47	9.3	1	94	6	360	60	10.1	353	37	63	333	70
47	9.3	2	94	6	206	50	9.6	337	37	63	333	33

The complete Table 1 has 48 rows (24 PD each in two sessions). There are the following condition attributes: P# - number given to each patients, S# - session number (1 or 2), dur –duration of the disease, PDQ39 – quality of life test result, Epworth test (quality of sleep) results, RS parameters (as describe in the Methods section): RSL – delay. RSD – duration, RSAm – amplitude, RSV – velocity. The last attribute (called decision attribute) was UPDRS (total UPDRS) as described above.

Table 2. Discretized-table extract for above (Table 1) Gr2 patients

P#	dur	S#	PDQ39	Epw.	RSL	RSD	RSAm	RSV	Beck	TrA	TrB	UPDRS
45	"(8.5,Inf)"	1	"(-Inf,58.5)"	*	"(-Inf,219)"	*	*	*	"(12.5,Inf)"	*	*	"(54,Inf)"
45	"(8.5,Inf)"	2	"(-Inf,58.5)"	*	"(219,Inf)"	*	*	*	"(12.5,Inf)"	*	*	"(18.5,43)"
46	"(-Inf,8.5)"	1	"(-Inf,58.5)"	*	"(-Inf,219)"	*	*	*	"(12.5,Inf)"	*	*	"(43.0,54)"
46	"(-Inf,8.5)"	2	"(-Inf,58.5)"	*	"(219,Inf)"	*	*	*	"(12.5,Inf)"	*	*	"(-Inf,18.5)"
47	"(8.5,Inf)"	1	"(58.5,Inf)"	*	"(219,Inf)"	*	*	*	"(12.5,Inf)"	*	*	"(54,Inf)"
47	"(8.5,Inf)"	2	"(58.5,Inf)"	*	"(-Inf,219)"	*	*	*	"(12.5,Inf)"	*	*	"(18.5,43)"

Table 2 is a discretized table for three patients: 45, 46, and 47 in two sessions: S#1 (session 1), and S#2 (session 2). Significant parameters were: disease duration (dur), session number, PDQ39, RSL (saccade delay), Beck depression test results, and UPDRS. Not significant were: Epworth test results, RSD, RSAm, RSV – saccades parameters, and Trail A and B results.

We have used RSES 2.2 for an automatic discretization of the Gr2 measurements. We have found that UPDRS has 4 ranges: "(-Inf, 18.5)" , "(18.5, 43.0)", "(43.0, 54.0)", "(54.0, Inf)".

We had obtained 70 rules for Gr2 patients, and after filtering reduced them to 8 rules. As an example we present below 4 rules filled by the most cases:

(dur="(8.5,Inf)")&(Ses=1)&(Beck="(12.5,Inf)")=>(UPDRS ="(54.0,Inf)"[8]) 8 (1)

(Ses=1)&(RSLat="(219,Inf)")&(Beck="(12.5,Inf)")=>(UPDRS ="(54.0,Inf)"[6]) 6 (2)

(dur="(8.5,Inf)")&(Ses=1)&(PDQ39="(58.5,Inf)")=>(UPDRS ="(54.0,Inf)"[5]) 5 (3)

(Ses=2)&(PDQ39="(58.5,Inf)")&(Beck="(12.5,Inf)")=>(UPDRS ="(18.5,43)"[5]) 5 (4)

We can interpret rules as following: Eq. 1 claims for 8 cases that if disease duration was longer than 8.5 years and patients were without medications with Beck test above 12.5 (indicating depression) then his/her total UPDRS was above 54. Previously by addition the depression results we have obtained higher accuracy than without Beck depression attribute [12], but in this study we do not see very strong influence of the depression (compare Eqs. 2 and 3). In Eq. 4 patients were on medication with not good quality of life and depressive but his/her UPDRS was medium (between 18.5 and 43).

3.2 IGrC for Estimation of General Disease Progression for Gr1 Group

We have used above general rules from Gr2 to predict UPDRS of Gr1 (Table 3).

Table 3. Confusion matrix for UPDRS of $Gr1_{VIS1}$ group by rules obtained from Gr2-group

		Predicted				
		"(54.0, Inf)"	"(18.5, 43.0)"	"(43.0, 54.0)"	"(-Inf, 18.5)"	ACC
Actual	"(54.0, Inf)"	7. 0	0.0	0.0	0.0	1.0
	"(18.5, 43.0)"	2.0	8.0	0.0	0.0	0.8
	"(43.0, 54.0)"	3.0	1.0	0.0	0.0	0.0
	"(-Inf, 18.5)"	0.0	1.0	0.0	0.0	0.0
	TPR	0.6	0.8	0.0	0.0	

TPR: True positive rates for decision classes; ACC: Accuracy for decision classes: the global coverage was 0.48 and the global accuracy was 0.68, the coverage for decision classes was 0.8, 0.45, 0.57, 0.125.

Table 4. Confusion matrix for UPDRS of $Gr1_{VIS2}$ group by rules obtained from Gr2-group

		Predicted				
		"(54.0, Inf)"	"(18.5, 43.0)"	"(43.0, 54.0)"	"(-Inf, 18.5)"	ACC
Actual	"(54.0, Inf)"	6.0	0.0	0.0	1.0	1.0
	"(18.5, 43.0)"	1.0	7.0	0.0	0.0	0.875
	"(43.0, 54.0)"	2.0	2.0	0.0	0.0	0.0
	"(-Inf, 18.5)"	0.0	0.0	0.0	0.0	0.0
	TPR	0.7	0.8	0.0	0.0	

TPR: True positive rates for decision classes; ACC: Accuracy for decision classes: the global coverage was 0.39 and the global accuracy was 0.72, the coverage for decision classes was 0.46, 0.5, 0.4, 0.0.

Table 5. Confusion matrix for UPDRS of $Gr1_{VIS3}$ group by rules obtained from Gr2-group

		Predicted				
		"(54.0, Inf)"	"(18.5, 43.0)"	"(43.0, 54.0)"	"(-Inf, 18.5)"	ACC
Actual	"(54.0, Inf)"	7.0	0.0	0.0	0.0	1.0
	"(18.5, 43.0)"	0.0	7.0	0.0	0.0	1.0
	"(43.0, 54.0)"	1.0	1.0	0.0	0.0	0.0
	"(-Inf, 18.5)"	0.0	1.0	0.0	0.0	0.0
	TPR	0.9	0.8	0.0	0.0	

TPR: True positive rates for decision classes; ACC: Accuracy for decision classes: the global coverage was 0.37 and the global accuracy was 0.82, the coverage for decision classes was 0.5, 0.438, 0.29, 0.11.

Results from Table 3 to 5 we interpret that with the disease development in time patients from Gr1 become more similar to patients from Gr2 as predictions of their symptoms become more precise from 0.68 to 0.82.

3.3 IGrC for Estimation of Motor Disease Progression for Gr1 Group

We have used the following attributes in order to predict UPDRS of Gr1 group on the basis of motor attributes such as Trail A rests (time of number connecting), eye movements results (RL – latency of saccades) and in addition disease duration and quality of sleep (Epworth test results). In this case, we have obtained from Gr2 33 rules, as examples:

(Ses=1)&(TrailA="(42.0,Inf)")&(dur="(5.695,Inf)")&(Epworth="(-Inf,14.0)")
&(RSLat= "(264.0,Inf)") =>(UPDRS ="(63.0,Inf)"[4]) 4 (5)

(dur="(5.695,Inf)")&(Ses=2)&(RSLat="(-Inf,264.0)")&(Epworth="(-Inf,14.0)")&
(TrailA= "(-Inf,42.0)")=>(UPDRS ="(-Inf,33.5)"[4]) 4 (6)

(dur="(5.695,Inf)")&(Ses=2)&(TrailA="(42.0,Inf)")&(RSLat="(-Inf,264.0)") &
(Epworth="(14.0,Inf)")=>(UPDRS ="(-Inf,33.5)"[3]) 3 (7)

Notice that these rules Eqs. 5–7 describe UPDRS in two ranges: below 33.5 or above 63. In the consequence in Tables 6, 7 and 8 (below) there is no prediction for other ranges of UPDRS. Equation 5 says that for patients without medication, with disease duration above 5.7 years with long delay of saccades and slow Trail A then their UPDRS was above 63. Even with similar disease duration as in Eq. 6 if patients were on medications and had a short saccadic delay then their UPDRS was below 33.5 (Eq. 7).

Table 6. Confusion matrix for UPDRS of $Gr1_{VIS1}$ group by motor rules obtained from Gr2-group

		Predicted				
		"(63.0, Inf)"	"(33.5, 43.0)"	"(43.0, 63.0)"	"(-Inf, 33.5)"	ACC
Actual	"(63.0, Inf)"	2.0	0.0	0.0	0.0	1.0
	"(33.5, 43.0)"	0.0	0.0	0.0	0.0	0.0
	"(43.0, 63.0)"	1.0	0.0	0.0	0.0	0.0
	"(-Inf, 33.5)"	0.0	0.0	0.0	2.0	1.0
	TPR	0.7	0.0	0.0	1.0	

TPR: True positive rates for decision classes; ACC: Accuracy for decision classes: the global coverage was 0.22 and the global accuracy was 0.5, the coverage for decision classes was 0.7, 0.0, 0.08, 0.08.

Confusion Tables 6, 7 and 8 shows that the motor symptoms give very accurate estimation of the UPDRS with accuracy from 0.8 to 1. There is only problem that coverage is small between 0.11 and 0.32. On the basis of motor attributes, we have very precise predictions but only for a small percentage of the population subjects.

Table 7. Confusion matrix for UPDRS of $Gr1_{VIS2}$ group by motor rules obtained from Gr2-group

		Predicted				
		"(63.0, Inf)"	"(33.5, 43.0)"	"(43.0, 63.0)"	"(-Inf, 33.5)"	ACC
Actual	"(63.0, Inf)"	3. 0	0.0	0.0	0.0	1.0
	"(33.5, 43.0)"	0.0	0.0	0.0	0.0	0.0
	"(43.0, 63.0)"	1.0	0.0	0.0	0.0	0.0
	"(-Inf, 33.5)"	0.0	0.0	0.0	11.0	1.0
	TPR	0.8	0.0	0.0	1.0	

TPR: True positive rates for decision classes; ACC: Accuracy for decision classes: the global coverage was 0.33 and the global accuracy was 0.6, the coverage for decision classes was 0.86, 0.25, 0.33, 0.0.

Table 8. Confusion matrix for UPDRS of $Gr1_{VIS3}$ group by motor rules obtained from Gr2-group

		Predicted				
		"(63.0, Inf)"	"(33.5, 43.0)"	"(43.0, 63.0)"	"(-Inf, 33.5)"	ACC
Actual	"(63.0, Inf)"	1.0	0.0	0.0	0.0	1.0
	"(33.5, 43.0)"	0.0	0.0	0.0	0.0	0.0
	"(43.0, 63.0)"	0.0	0.0	0.0	0.0	0.0
	"(-Inf, 33.5)"	0.0	0.0	0.0	8.0	1.0
	TPR	1.0	0.0	0.0	1.0	

TPR: True positive rates for decision classes; ACC: Accuracy for decision classes: the global coverage was 0.24 and the global accuracy was 0.64, the coverage for decision classes was 0.45, 0.13, 0.3, 0.11

3.4 IGrC for Estimation of Cognitive Disease Progression for Gr1 Group

We have used the following attributes in order to predict UPDRS_T of Gr1 group on the basis of motor attributes such as Trail B rests (time of number and letters connecting), eye movements results (RL – latency of saccades) and in addition session number and quality of sleep (Epworth test results). In this case, we have obtained from Gr2 rules, as examples are:

(TrailB="(127.5,Inf)")&(Ses=1)&(Epworth="(-Inf,7.5)")=>(UPDRS="(63.0,Inf)" [4]) 4 (8)

(Ses=1)&(RSLat="(244.5,Inf)")&(Epworth="(-Inf,7.5)")=>(UPDRS="(63.0,Inf)" [3]) 3 (9)

(TrailB="(127.5,Inf)")&(Ses=2)&(RSLat="(-Inf,244.5)")=>(UPDRS="(18.5, 43.0)" [3]) 3 (10)

(TrailB="(-Inf,52.0)")&(Ses=1)=>(UPDRS ="(43.0,63.0)"[2]) 2 (11)

The Eq. 8 described the rule for 4 cases when patients were without medications (Ses = 1) and cognitively slow (Trail B time longer than 127.5 s) and without sleep problems (Epworth below 7.5) then UPDRS was above 63 (Eq. 8). In the next Eq. 9 there were similar condition attributes (no medications, no sleep problems), but slow cognitive was replaced but slowness in the reflexive saccades (long latency) that determined an advanced (above 62) UPDRS.

Equation 10 demonstrated that if patient was on medication and had a short saccadic latency, even if his/her Trail B was slow, the UPDRS was in medium range. But without medications even if Trail B was good his/her UPDRS was high (between 43 and 63) (Eq. 11).

Table 9. Confusion matrix for UPDRS of $Gr1_{VIS1}$ group by cognitive rules obtained from Gr2-group

		Predicted				
		"(63.0, Inf)"	"(18.5, 43.0)"	"(43.0, 63.0)"	"(-Inf, 18.5)"	ACC
Actual	"(63.0, Inf)"	2.0	0.0	0.0	0.0	1.0
	"(18.5, 43.0)"	3.0	1.0	0.0	0.0	0.3
	"(43.0, 63.0)"	2.0	0.0	2.0	0.0	0.5
	"(-Inf, 18.5)"	0.0	0.0	0.0	0.0	0.0
	TPR	0.3	1.0	1.0	0.0	

TPR: True positive rates for decision classes; ACC: Accuracy for decision classes: the global coverage was 0.22 and the global accuracy was 0.5, the coverage for decision classes was 0.7, 0.0, 0.08, 0.08.

Table 10. Confusion matrix for UPDRS of Gr1$_{VIS2}$ group by cognitive rules obtained from Gr2-group

		Predicted				
		"(63.0, Inf)"	"(18.5, 43.0)"	"(43.0, 63.0)"	"(-Inf, 18.5)"	ACC
Actual	"(63.0, Inf)"	5.0	0.0	0.0	0.0	0.8
	"(18.5, 43.0)"	1.0	3.0	0.0	0.0	0.8
	"(43.0, 63.0)"	4.0	0.0	1.0	0.0	0.2
	"(-Inf, 18.5)"	0.0	0.0	0.0	0.0	0.0
	TPR	0.5	1.0	0.5	0.0	

TPR: True positive rates for decision classes; ACC: Accuracy for decision classes: the global coverage was 0.33 and the global accuracy was 0.6, the coverage for decision classes was 0.86, 0.25, 0.33, 0.0.

Tables 9, 10 and 11 demonstrated that cognitive changes were not changing so fast with disease development as general or motor changes. They are more stable than other changes and not in the main stream of the neurodegeneration processes in PD. It is related to the fact that accuracy of cognitive changes (from 0.5 to 0.64) was significantly lower in comparison to general (from 0.68 to 0.88) or to motor (from 0.8 to 1.0) symptoms development during disease progression.

Table 11. Confusion matrix for UPDRS of Gr1$_{VIS3}$ group by cognitive rules obtained from Gr2-group

		Predicted				
		"(63.0, Inf)"	"(18.5, 43.0)"	"(43.0, 63.0)"	"(-Inf, 18.5)"	ACC
Actual	"(63.0, Inf)"	4.0	0.0	1.0	0.0	0.8
	"(18.5, 43.0)"	0.0	2.0	0.0	0.0	1.0
	"(43.0, 63.0)"	2.0	0.0	1.0	0.0	0.3
	"(-Inf, 18.5)"	0.0	1.0	0.0	0.0	0.0
	TPR	0.7	0.7	0.5	0.0	

TPR: True positive rates for decision classes; ACC: Accuracy for decision classes: the global coverage was 0.24 and the global accuracy was 0.64, the coverage for decision classes was 0.45, 0.13, 0.3, 0.11.

4 Discussion

We have used IGrC for evaluation of disease development in our longitudinal study of patients with Parkinson's disease (PD). We applied IGrC (intelligent granular computing with RST [3] that looks into "crisp" granules and estimates objects by upper and lower approximations that determine precision of the description as dependent from properties of granules [3]. In order to follow time changes we have used similarities measured by estimating accuracy between more advance group of patients (Gr2) and another group Gr1 measured three times every 6 months. General and motor symptoms are following disease progression in contrast to cognitive changes that have lower accuracies that might suggest that there different mechanisms of neurodegeneration even if there are some similarities between Parkinson's and Alzheimer's diseases. The next interesting step might be to look into emotional developments during disease progression if they progress like motor symptoms or are relatively low like cognitive changes? In our previous work [13] we have proposed that in analog to the Model of different objects learned in the visual brain we can think about the Model of the Parkinson's disease in the advanced stage of the disease. Therefore, we can take the group Gr2 as a Model and look into similarities between disease time developments of Gr1. For general and motor symptoms similarity (accuracy of rules) increases to near 1 with disease development. But for the cognitive change it is not exactly the case. Therefore Gr2 is not good model for cognitive developments in time for Gr1. It might be related to large interindividual variability in cognitive decay between PD patients. However, the most data mining studies related to Parkinson's disease concentrate on motor symptoms such as gait [14] or speech articulation [15].

References

1. Przybyszewski, A.W., Kon, M., Szlufik, S., Szymanski, A., Koziorowski, D.M.: Multimodal learning and intelligent prediction of symptom development in individual Parkinson's patients. Sensors **16**(9), 1498 (2016)
2. Przybyszewski, A.W.: Fuzzy RST and RST rules can predict effects of different therapies in Parkinson's disease patients. In: Ceci, M., Japkowicz, N., Liu, J., Papadopoulos, G.A., Raś, Z.W. (eds.) ISMIS 2018. LNCS (LNAI), vol. 11177, pp. 409–416. Springer, Cham (2018). https://doi.org/10.1007/978-3-030-01851-1_39
3. Pawlak, Z.: Rough Sets: Theoretical Aspects of Reasoning About Data. Springer, Dordrecht (1991). https://doi.org/10.1007/978-94-011-3534-4
4. Jankowski, A.: Interactive Granular Computations in Networks and Systems Engineering: A Practical Perspective. LNNS, vol. 17. Springer, Cham (2017). https://doi.org/10.1007/978-3-319-57627-5
5. Bazan, J., Nguyen, S.H., et al.: Desion rules synthesis for object classification. In: Orłowska, E. (ed.) Incomplete Information: Rough Set Analysis, Physica – Verlag, Heidelberg, pp. 23–57 (1998)
6. Przybyszewski, A.W.: Applying data mining and machine learning algorithms to predict symptom development in Parkinson's disease. Ann. Acad. Med. Siles. **68**(5), 332–349 (2014)
7. Bazan, J., Nguyen, H.S., Nguyen, S.H., et al.: Rough set algorithms in classification problem. In: Polkowski, L., Tsumoto, S., Lin, T. (eds.) Rough Set Methods and Applications, pp. 49–88. Physica-Verlag, Heidelberg (2000). https://doi.org/10.1007/978-3-7908-1840-6_3
8. Grzymała-Busse, J.: A New Version of the Rule Induction System LERS. Fundamenta Informaticae **31**(1), 27–39 (1997)
9. Bazan, J.G., Szczuka, M.: The rough set exploration system. In: Peters, J.F., Skowron, A. (eds.) Transactions on Rough Sets III. LNCS, vol. 3400, pp. 37–56. Springer, Heidelberg (2005). https://doi.org/10.1007/11427834_2

10. Przybyszewski, A.W.: The neurophysiological bases of cognitive computation using rough set theory. In: Peters, J.F., Skowron, A., Rybiński, H. (eds.) Transactions on Rough Sets IX. LNCS, vol. 5390, pp. 287–317. Springer, Heidelberg (2008). https://doi.org/10.1007/978-3-540-89876-4_16
11. Bazan, J.G., Szczuka, M.: RSES and RSESlib - a collection of tools for rough set computations. In: Ziarko, W., Yao, Y. (eds.) RSCTC 2000. LNCS (LNAI), vol. 2005, pp. 106–113. Springer, Heidelberg (2001). https://doi.org/10.1007/3-540-45554-X_12
12. Przybyszewski, A.W., Nowacki, J.P., Drabik, A., Szlufik, S., Habela, P., Koziorowski, D.M.: Granular computing (GC) demonstrates interactions between depression and symptoms development in Parkinson's disease patients. In: Nguyen, N.T., Gaol, F.L., Hong, T.-P., Trawiński, B. (eds.) ACIIDS 2019. LNCS (LNAI), vol. 11432, pp. 591–601. Springer, Cham (2019). https://doi.org/10.1007/978-3-030-14802-7_51
13. Przybyszewski, A.W.: Parkinson's Disease Development Prediction by C-Granule Computing. In: Nguyen, N.T., Chbeir, R., Exposito, E., Aniorté, P., Trawiński, B. (eds.) ICCCI 2019. LNCS (LNAI), vol. 11683, pp. 296–306. Springer, Cham (2019). https://doi.org/10.1007/978-3-030-28377-3_24
14. Khoury, N., Attal, F., Amirat, Y., Oukhellou, L., Mohammed, S.: Data-driven based approach to aid Parkinson's disease diagnosis. Sensors (Basel) **19**(2), 242 (2019). https://doi.org/10.3390/s19020242
15. Yadav, G., Kumar, Y., Sahoo, G.: Predication of Parkinson's disease using data mining methods: a comparative analysis of tree, statistical, and support vector machine classifiers. Indian J. Med. Sci. **65**(6), 231–242 (2011)

A New Facial Expression Recognition Scheme Based on Parallel Double Channel Convolutional Neural Network

D. T. Li[1,2], F. Jiang[1,2(✉)], and Y. B. Qin[1]

[1] Guangxi Key Lab of Multi-source Information Mining and Security, College of Electronic Engineering, Guangxi Normal University, Guilin 541004, China
Franknsw2011@gmail.com

[2] Faculty of Engineering and IT, University of Technology Sydney, Ultimo, NSW 2007, Australia

Abstract. The conventional deep convolutional neural networks in facial expression recognition are confronted with the training inefficiency due to many layered structure with a large number of parameters, in order to cope with this challenge, in this work, an improved convolutional neural network—with parallel double channels, termed as PDC-CNN, is proposed. Within this model, we can get two different sized feature maps for the input image, and the two different sized feature maps are combined for the final recognition and judgment. In addition, in order to prevent over-fitting, we replace the traditional RuLU activation function with the Maxout model in the fully connected layer to optimize the performance of the network. We have trained and tested the new model on JAFFE dataset. Experimental results show that the proposed method can achieve 83% recognition rate, in comparison with the linear SVM, AlexNet and LeNet-5, the recognition rate of this method is improved by 14%–28%.

Keywords: Facial expression recognition · Parallel · Double channel · Convolutional neural network

1 Introduction

Facial expression recognition is the process of facial expression image acquisition, facial expression image preprocessing, facial expression feature extraction and expression classification by computer. It learns the information contained in the expression content obtained by computer analysis, infers the psychological state of human beings, and finally realizes the intelligent interaction between human beings and computers [1].

A complete facial expression recognition system includes three parts: face detection, feature extraction, classification and recognition [2]. Among them, feature extraction is particularly important. The efficiency of facial expression recognition system depends on the quality of features. Research on feature extraction is also emerging in endlessly. The related algorithms can be summarized into three categories: Extract texture features, extract collection features, and extract mixed features [3]. In contrast, the recognition

N. T. Nguyen et al. (Eds.): ACIIDS 2020, LNAI 12034, pp. 560–569, 2020.
https://doi.org/10.1007/978-3-030-42058-1_47

effect of hybrid feature method is better than that of the former two methods. The reason is that hybrid feature can make many features complement each other, thus avoiding the disadvantage of single feature.

Expression recognition technology is one of the important contents in the field of emotional computer research. At present, facial expression recognition has been applied to human-computer interaction [4], security detection [5], operator fatigue monitoring [6], emotional music [7], Clinical medicine [8] and so on. Because of its wide application, facial expression recognition has attracted a large number of scholars to study [9–12]. In the literature [13–15], the convolutional neural network is used to identify the expression, and good results are obtained. The accuracy and reliability of the convolutional neural network to identify the expression are verified.

The work in the literature is introduced in Sect. 2, and then a new model called PDC-CNN is introduced further in Sect. 3. In order to test the superiority of the model, we compare the linear SVM, AlexNet, LeNet-5 with the proposed method in this paper and report the experimental results in Sect. 4.

2 Related Work

Research on expression began in the field of biology. In the mid-19th century, biologist Darwin studied human facial expressions and compared them with animal expressions, and found facial expressions to be consistent across different races, genders, and regions. This conclusion became the theoretical basis for the subsequent study of facial expressions.

The turning point that plays an important role in facial expression research comes from the field of psychology. In 1978, two American psychologists, Paul Ekman and Friesen, studied facial muscles. They proposed the Facial Action Coding System (FACS) [16]. They defined six basic expressions in the field of facial expression recognition and divided facial muscles into 44 action units, these unit combinations can intuitively represent nearly 10,000 kinds of expressions, and have the advantages of being objective, concise, and easy to understand.

After the 20th century, the rapid development of computer technology and information technology has brought breakthroughs to the research of facial expression recognition. The popular research direction of expression recognition has attracted many universities and research institutions to invest in this research field. Carnegie Mellon University has developed a facial image analysis system for facial expression recognition [17]. Advanced Telecommunication Research Institute released a highly used JAFFE data set in Japan. In 2015, Google released an online expression recognition system API, which can detect faces and recognize expressions in static images, and analyze different emotional components in expressions.

In recent years, the use of convolutional neural networks for image data feature extraction has become more and more popular. From low-level image processing (such as super-resolution) to high-level image understanding (such as image classification, object monitoring and scene annotation), the success of CNN is mainly attributed to its ability to represent complex natural images at different levels.

Andre Teixeira Lopes [18] proposed a CNN-based expression recognition method for expression recognition. This method is divided into three steps. The first step is to standardize the image space and align the eyes of the face with the horizontal axis of the image by rotation. The second step is to expand the data set by using a 2D Gaussian distribution to introduce random noise in the locations of the center of the eyes. The third step is to send the cropped, down-sampled and normalized images to CNN for facial expression recognition.

Al-Shabi [19] proposed the Hybrid CNN-SIFT Aggregator for facial expression recognition. In this method, feature extraction is divided into two channels. The first channel is a six-layer convolution layer neural network. The second channel uses Scale Invariant Feature Transform to extract image features. And the feature maps obtained from the two channels are then fused using the fully connected layer and classified.

The above two methods both focus on the convolutional neural network to resolve the specific problem in face recognition, and these problems have strong connections with our problem in this paper, hence, it appears the improved CNN networks is a promising approach to achieve good results, the next section of the paper will illustrate the proposed parallel double channel convolutional neural network model termed as PDC-CNN.

3 Refining PDC-CNN

3.1 CNN

In a neural network, the input-output relationship of a neuron is given by (1), where a_i^l represents the input of the i-th neuron in the j-th layer, w_{ij}^l represents the connection weights between the i-th neuron of the l-th layer and the j-th neuron of the $(l-1)$-th layer, b_i^l indicates the bias, $f(\cdot)$ represents the activation function.

$$a_i^l = f\left(\sum\nolimits_j w_{ij}^i a_j^{i-1} + b_i^l\right) \tag{1}$$

CNN is a network model of deep learning, which originates from multi-layer perceptron. In the convolutional layer, the connection of neurons is achieved by convolution operations. The relationship between input and output is given by (2).

$$A_i^l = f\left(\sum\nolimits_j K_{ij}^l * A_j^{l-1} + b_i^l\right) \tag{2}$$

Where A_i^l represents the i-th feature map of the l-th layer, the symbol "*" indicates convolution operation, K_{ij}^l denotes a convolution kernel of a convolution map between the i-th feature map of the l-th layer and the j-th feature map of the $(l-1)$-th layer.

The convolutional neural network can directly input the original image when identifying the image, so the complicated image preprocessing process and the feature extraction algorithm design process in the traditional image recognition method can be omitted. Inspired by convolution neural network, the PDC-CNN model is proposed in this paper.

3.2 Maxout

Because the number of JAFFE datasets used in this paper is small, the model may be over-fitting when training the model, so the Maxout [20] model is applied.

Maxout is a recently emerging network model. Unlike other models, its activation function is Maxout. Since the output of the Maxout model is non-sparse, this article adds Maxout after the fully connected layer and uses it in conjunction with the Droupout technique to reduce over-fitting of the network. Given the input $x \in R^d$, the calculation formula for each neuron in the Maxout model are given by (3) and (4), and Fig. 1 shows the calculation process of Maxout.

$$h_i(x) = max\, z_{ij}\ j \in [1, k] \tag{3}$$

$$z_{ij} = x^T W_{ij} + b_{ij}\ W \in R^{d \times m \times k},\ b \in R^{m \times k} \tag{4}$$

Where W and b represent the weight and bias of the Maxout model, W is a three-dimensional matrix with size (d, m, k), d denotes the number of input nodes and K denotes the number of nodes in the Maxout model.

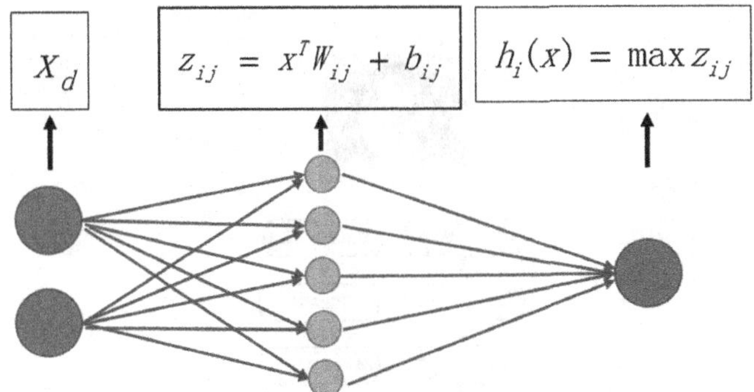

Fig. 1. Calculation process of Maxout

3.3 Feature Fusion

The feature maps extracted from two sub-channels, we first convert them into pixel matrices. The matrix is then stretched and spliced into a one-dimensional array as input to fully connected layer. In this way, we can fuse the extracted two different sized features maps as the basis for the training. The process of feature fusion is shown in the following Fig. 2.

3.4 PDC-CNN

The PDC-CNN model proposed in this paper contains 9 layers. The most important part is two parallel sub-convolutional neural networks. The structure is shown in Fig. 3.

The first sub-convolutional neural network consists of C1, P1 and C2, its convolution kernel and stride are both smaller than the second one, and the size of the output feature map is 32*32, which is larger than the second one. The second sub-convolutional neural network consists of C3, P2 and C4, and the size of the feature map is 8*8.

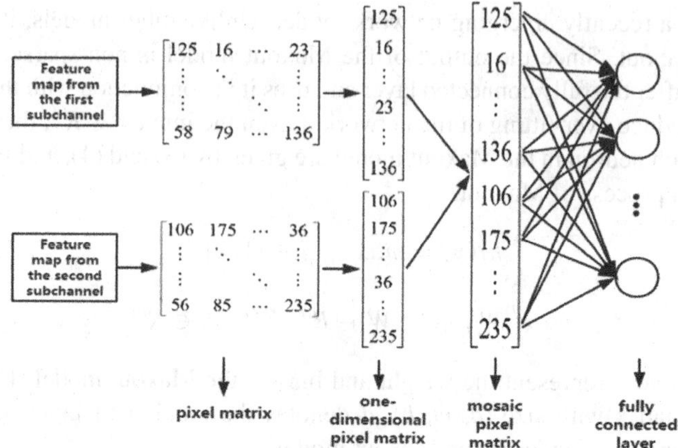

Fig. 2. Process of feature fusion

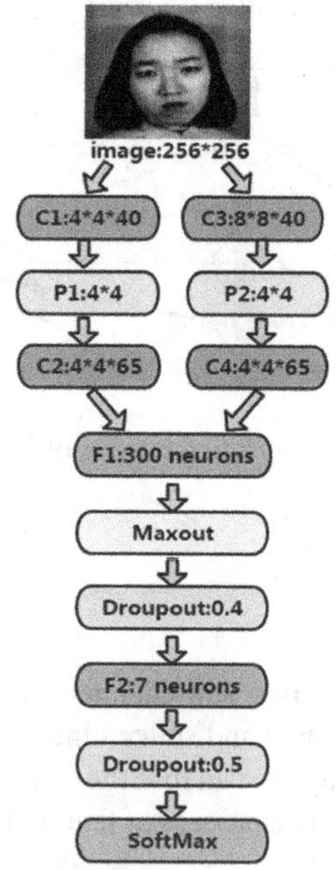

Fig. 3. PDC-CNN structure

The input of the first fully connected layer F1 is a one-dimensional matrix after feature fusion. In this layer, we use Maxout instead of ReLU function, and then Droupout, which can prevent the model from over-fitting. The second fully connected layer F2 has seven neurons, which correspond to the classification of seven expressions, namely, angry, happy, surprised, fear, disgust, sad and neutral, and it only performs Droupout operations. The parameters and function of each layer are shown in Table 1.

In Table 1, column 1 indicates the names of the layers, column 2 is the type of each layer, and column 3 is the size and number of the feature maps. Column 4 and 5 represent the size of the convolution kernel and pooling area of each layer, respectively. Column 6 shows the stride size of each layer and column 7 shows the activation function applied by each layer. Among them, the convolution kernel size, the pooling area size and the step size of each layer are obtained after many experiments and debugging. Only by using these debugged layer parameters can we get the best recognition effect.

Table 1. Parameters and function of each layer

Layer	Type	Feature map	Convolution kernel	Pooling area	Stride	Function
C1	Convolutional	128*128*40	4*4	–	2	ReLU
P1	Max Pooling	64*64*40	–	4*4	2	ReLU
C2	Convolutional	32*32*65	4*4	–	2	ReLU
C3	Convolutional	64*64*40	8*8	–	4	ReLU
P2	Max Pooling	16*16*40	–	4*4	4	ReLU
C4	Convolutional	4*4*65	4*4	–	4	ReLU
F1	Fully Connected Layer	300	–	–	–	Droupout, Maxout
F2	Fully Connected Layer	7	–	–	–	Softmax

4 Experimental Results and Analysis

4.1 Experimental Data Preprocessing

We trained our iterative refinement model on JAFFE data set, The data set has a total of 213 Japanese female expression pictures, each of which consists of a grayscale image with a fixed size of 256*256. The data set is divided into seven kinds of expressions: angry, disgust, fear, happy, sad, surprised and neutral, which correspond to the digital labels 0–6.

Because the number of data set is too small, it is easy to fall into over-fitting when training the model, so we expand the data set.we randomly select 5 angles to rotate each image in the range of 0° to 45°. In this way, five rotated images can be obtained from an original image, which enlarges the number of original data set by five times. The number of expanded data set becomes 1065, an example of a rotated picture is shown in Fig. 4.

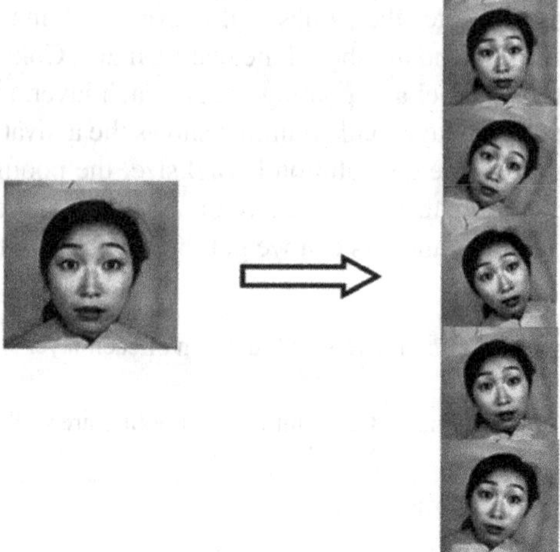

Fig. 4. An example of a rotated picture

4.2 Experimental Results

We implemented our model on Jupyternotebook and use the 1065 rotated images as the data set. we randomly selected 850 pictures as the training set, and the remaining 215 pictures as the test set. The classical approaches for expression classifications include linear SVM, AlexNet and LeNet-5, they are widely adopted in the literature and industry, therefore, we compare our proposed model with these benchmarked models under the same experimental environment and data set in this paper. Table 2 lists the accuracy of each model on the data set, Table 3 shows the time cost of different model when the loss function decrease to 0.05 on training set, Fig. 5 shows the variation of each model's loss function as the number of iterations increases on training set. We can get the following conclusions:

1. As can be seen from Table 2, in the test set, the model proposed in this paper achieves 83% recognition rate, compared with the other three models, the recognition rate is increased by 14%–28%, the model is more resistant to over-fitting than the other three models.
2. From Fig. 5 and Table 3, it can be concluded that PDC-CNN's convergence speed of the loss function is faster than the other three models when the model is iterated on the training set.

In the above results, we can see that these four models on the test set do not perform as well as on the training set, the reason is that facial expressions have certain similarities. In real life, the human eye is hard to distinguish. We printed two graphs on the test set that were misidentified by the model. The expression shown in Fig. 6 is angry and the Fig. 7 is neutral, the similarity between the two is extremely high.

Table 2. Accuracy of different models on data set

Model	Train accuracy	Test accuracy
PDC-CNN	99%	83%
Linear SVM	99%	69%
Alexnet	96%	67%
LeNet-5	94%	55%

Table 3. Time cost of different models

Model	Time cost (s)
PDC-CNN	41.16
Linear SVM	156.12
Alexnet	53.29
LeNet-5	47.34

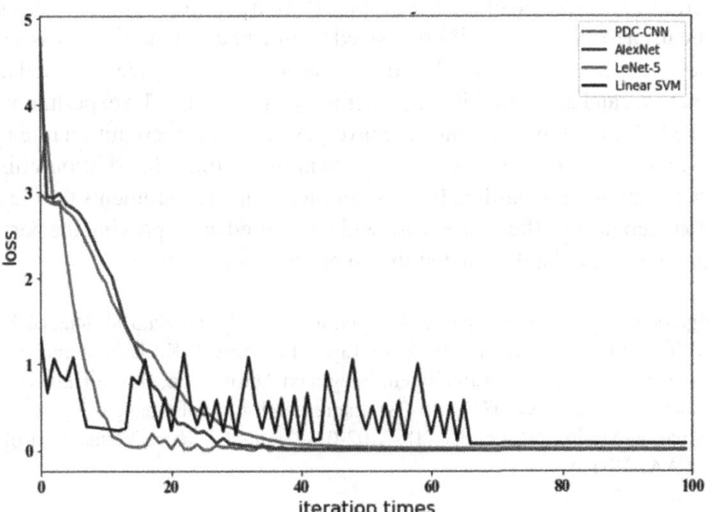

Fig. 5. Loss-iteration graph of each model

Fig. 6. Angry

Fig. 7. Neutral

5 Conclusion

In this paper, a new PDC-CNN model is proposed. Two parallel sub-convolutional neural networks are used to extract two different sized feature maps of facial expression images, and then they are merged for the final recognition and judgments. We validated the model on JAFFE data set and compared it with other network models. The experimental results show that the loss function of the model converges faster, the recognition rate is high on the test set, and the model can effectively prevent over-fitting. In addition, this method has no extra computational burden. In consideration of the requirements for the accuracy of expression recognition, the future work will be focused on improving the performance which does not require the designated image pre-processing stage.

Acknowledgements. This work is partially supported by National Natural Science Foundation of China (61762018), the Guangxi 100 Youth Talent Program (F-KA16016) and the Colleges and Universities Key Laboratory of Intelligent Integrated Automation, Guilin University of Electronic Technology, China (GXZDSY2016-03), the research fund of Guangxi Key Lab of Multi-source Information Mining & Security (18-A-02-02), Natural Science Foundation of Guangxi (2018GXNSFAA281310).

References

1. Takalkar, M., Xu, M., Wu, Q., Chaczko, Z.: A survey: facial micro-expression recognition. Multimedia Tools Appl. **77**(15), 19301–19325 (2018)
2. Zhang, L., Verma, B., Tjondronegoro, D., Chandran, V.: Facial expression analysis under partial occlusion: a survey. ACM Comput. Surv. **51**(2), 1557–7341 (2018)
3. Kumar, N., Bhargava, D.: A scheme of features fusion for facial expression analysis: a facial action recognition. J. Stat. Manage. Syst. **20**(4), 693–701 (2017)

4. Bartlett, M.S., Littlewort, G., Fasel, I., Movellan, J.R.: Real time face detection and facial expression recognition: development and applications to human computer interaction. In: 2003 Conference on Computer Vision and Pattern Recognition Workshop. IEEE (2008)
5. Kim, J.O., Seo, K.S., Chung, C.H., Hwang, J., Lee, W.: On facial expression recognition using the virtual image masking for a security system. Lect. Notes Comput. Sci. **3043**, 655–662 (2004)
6. Ji, Q., Zhu, Z., Lan, P.: Real-time nonintrusive monitoring and prediction of driver fatigue. IEEE Trans. Veh. Technol. **53**(4), 1052–1068 (2004)
7. Dureha, A., Dureha, A.: An accurate algorithm for generating a music playlist based on facial expressions. Int. J. Comput. Appl. **100**(9), 33–39 (2014)
8. Collin, L., Bindra, J., Raju, M., Gillberg, C., Minnis, H.: Facial emotion recognition in child psychiatry: a systematic review. Res. Dev. Disabil. **34**(5), 1505–1520 (2013)
9. Shan, C., Gong, S., Mcowan, P.W.: Facial expression recognition based on local binary patterns: a comprehensive study. Image Vis. Comput. **27**(6), 803–816 (2009)
10. Palestra, G., Pettinicchio, A., Coco, M.D., Carcagnì, P., Leo, M., Distante, C.: Improved performance in facial expression recognition using 32 geometric features. Image Anal. Process. **9280**, 518–528 (2015)
11. Liu, W., Wen, Y., Yu, Z., Li, M., Raj, B., Song, L.: Sphereface: deep hypersphere embedding for face recognition. arXiv, 9 (2017)
12. Shan, C., Gong, S., Mcowan, P.W.: Robust facial expression recognition using local binary patterns. In: IEEE International Conference on Image Processing. DBLP (2005)
13. Li, Y., Zeng, J., Shan, S., Chen, X.: Occlusion aware facial expression recognition using CNN with attention mechanism. IEEE Trans. Image Process. **28**(5), 2439–2450 (2019)
14. Pramerdorfer, C., Kampel, M.: Facial expression recognition using convolutional neural networks: state of the art. arXiv, 6 (2016)
15. Lu, G., He, J., Yan, J., Li, H.: Convolutional neural network for facial expression recognition. J. Nanjing Univ. Posts Telecommun. **36**, 16–22 (2016)
16. Ekman, P., Friesen, W.V.: Facial action coding system (facs): a technique for the measurement of facial action. Rivista di Psichiatria **47**(2), 126–138 (1978)
17. Torre, F.D.L., Campoy, J., Ambadar, Z., Cohn, J. F.: Temporal segmentation of facial behavior. In: IEEE International Conference on Computer Vision (2007)
18. Lopes, A.T., Aguiar, E.D., Oliveira-Santos, T.: A Facial expression recognition system using convolutional networks. In: 28th Sibgrapi Conference on Graphics, Patterns and Images, pp. 273–280 (2015)
19. Al-Shabi, M., Cheah, W.P., Connie, T.: Facial expression recognition using a hybrid CNN-SIFT aggregator. Multi disc. Trends Artif. Intell. **10607**, 139–149 (2017)
20. Goodfellow, I.J., Warde-Farley, D., Mirza, M., Courville, A., Bengio, Y.: Maxout networks. Comput. Sci. **28**, 1319–1327 (2013)

Reversible Data Hiding Based on Prediction Error Histogram Shifting and Pixel-Based PVO

Nguyen Kim Sao[1(✉)], Cao Thi Luyen[1(✉)], Le Kinh Tai[2(✉)], and Pham Van At[1(✉)]

[1] University of Transport and Communications, Ha Noi, Vietnam
{saonkoliver,luyenct}@utc.edu.vn, phamvanat83@gmail.com
[2] Trade Union University, Ha Noi, Vietnam
tailk@dhcd.edu.vn

Abstract. As already known, prediction error histogram shifting (PEHS) is a remarkable reversible data hiding method in which a sequence E of prediction errors of pixels is constructed and data is embedded by histogram shifting (HS) on E. In this paper, we propose a method that is a combination of PEHS and PPVO method. In this scheme, after dividing the host image into equal-sized blocks, the block is considered based on its flat property. If it is flat, it embeds a data bit based on the PESH method. In remaining case, it embeds a data bit based on PPVO. Experimental results show that the proposed method outperforms some state-of-the-art methods in terms of embedding capacity and stego image quality.

Keywords: Reversible data hiding · Histogram shifting · Prediction error · PVO · Pixel value ordering

1 Introduction

Reversible data hiding (RDH) has received much attention from scientists because of his usefulness in the authentication and protection of multimedia products.

A RDH method was first introduced in 2000 [6] where Macq and his partners used modulo 256 to do the inverse. It means that pixel values and embedded bits are added modulo with 256, which solved the pixel domain overflow problem but the stego image quality remained low.

Shortly after that, Fidrich et al. [2] introduced a RDH method based on lossless compression, in which the lower bits of the pixels are compressed to create empty space for inserting hidden bits. This method successfully solved the overflow and underflow as well as the quality of stego images, but the capacity is very poor. In 2002, Tian [15] introduced a RDH technique based on difference expansion. In this method, the difference of a selected pair of two neighboring pixels is extended for embedding a data bit. The difference expansion methods

N. T. Nguyen et al. (Eds.): ACIIDS 2020, LNAI 12034, pp. 570–581, 2020.
https://doi.org/10.1007/978-3-030-42058-1_48

have a very high capacity, however, stego image quality depends on magnitude of the differences.

In 2006, Ni et al. [7] proposed a method based on the histogram shift. The histogram of the image is the frequency of each pixel. The histogram shift is based on the highest frequency point called the peak point, and pixels that are greater than the peak point will be incremented by 1 to create empty space. The data bits are concealed at the peak point by rule that the pixels equal the peak point are used to embed bits 0 and unchanged, on the opposite, ones to embed bits 1 and increased to 1. In [12], we proposed a method by using two adjacent peaks in order to increase capacity.

Because a difference expansion method has very high capacity and stego image quality depends only on considered differences, scientists always try to decrease the value of the differences. In 2004, Thodi and his colleagues [14] proposed the prediction of each pixel to reduce the value of the differences, this not only increases the embedding capacity but also increases the quality of stego images.

In prediction methods, each pixel x is evaluated by a value $y = P(x)$. $P(x)$ is determined based on a neighborhood of x and as close to x as possible. Then the difference $e = x - y$ is used to embed a data bit. It shows that if the prediction method is good, then stego image quality is also significantly improved.

In order to improve the embedding capacity, Li et al. [5] compute the difference of each pair of two adjacent pixels, then use the HS on these differences. This method uses one pixel for predicting another pixel, so the prediction is not good and the capacity is not high.

Recently, the prediction method based on Pixel Value Ordering (PVO), which was first proposed by Li et al. in 2013 [4]. The image is divided into blocks. In each block, pixels are sorted in ascending order. The PVO-based method is performed by shifting the histogram of the prediction errors, which has very high stego image quality. However, these methods accept block constraints, which means that every block can only embed maximum two bits, therefore its embedding capacity is limited.

Next, there are some generalizations of PVO such as IPVO [10], PVO-K [8], and GePVO-K [3], that allow embedding data in largest or smallest pixels of each block. However, all these methods also accept block constrains, so their embedding capacity is still limited.

Peng et al. [10] proposed IPVO method, is an improved method of PVO. In IPVO, instead of embedding in the prediction errors that has value 1, the predicted errors are selected to embed are 0 and 1. This change get a hight capacity for this method.

Another improved method of PVO is PVO-K, this method is proposed by Ou et al. [8] in 2014. In this method, each block could embed a bit into k largest or smallest values. Addition, this method also uses absolute flat blocks for embedding to improve the capacity.

Li et al. [3] give another enhance of PVO-K method called GePVO-K. The brilliant point of this method is that they can embed k bit in each side of block, where k is the number of the largest/smallest values.

Qu et al. proposed the PPVO method [11] (Pixel -based pixel value ordering). This method uses a prediction context for each pixel and maximum or minimum value of context is used as the prediction of considered pixel.

This method is a most effective among PVO-used methods. It has embedding capacity larger as well as stego image quality better than other PVO-based methods.

In this paper, we propose a combination of PESH and PPVO methods. Theoretical analysis in Sect. 3.1 proves that embedding capacity of proposed method is higher than one of PPVO. Moreover, by experiment results, it shows that proposed method has embedding capacity larger and stego image quality better than some recent state-of-the-art methods such as method of Li et al. (Li method) [5], PVO [4], IPVO [10], PVO-K [8], PPVO [11], GePVO-K [3], and TRES [9].

The rest of the paper is organized as follows: Sect. 2 reviews the related notions and works. Section 3 presents proposed method. Section 4 provides experimental results and evaluates the performance of the methods. Finally, some conclusions are introduced in Sect. 5.

2 Related Notions and Works

2.1 PEHS-MED Method

Histogram. Let $X = (x_1, x_2, \cdots, x_K)$ be a given sequence of integers, the histogram of X is a function, denoted h, that is defined as follows:

$$h(x) = \#\Omega\{1 \leq i \leq K | x_i = x\}, \forall x \in Z,$$

where $\#\Omega$ means the cardinal number of set Ω and Z is the set of integers. In other words, $h(x)$ is the number of times that x occurs (frequency) in X. For example, for $X = (1, 3, 4, 5, 3, 3, 4, 3, 4)$:

$$h(x) = \begin{cases} 1, if\ x = 1 \\ 4, if\ x = 3 \\ 3, if\ x = 4 \\ 1, if\ x = 5 \\ 0, otherwise. \end{cases}$$

A point x (*or bin* x) with $h(x) = 0$ is called zero-point or empty-point. The points at that h obtains the highest values are called peak-points. The HS methods often use two pairs: (LPP, LZP) and (RPP, RZP) where LPP and RPP are left peak-point and right peak-point, LZP is zero-point smaller (left) and closest to the LPP, RZP is zero-point bigger (right) and closest to the RPP. For the above example, we have: $LPP = 3, LZP = 2, RPP = 4, RZP = 6$.

Histogram Shifting. The points in segment $[LZP+1, LPP-1]$ are shifted to the left by 1 to make point $LPP-1$ to be empty-point. Similarity, points in the segment $[RPP+1, RZP-1]$ are shifted to the right by 1 to make point $RPP+1$ to be empty-point. In other word, the elements x_i in $[LZP+1, LPP-1]$ are subtracted from 1, while x_i in $[RPP+1, RZP-1]$ are added by 1. For the above example, after shifting histogram, X will become: $X = (1, 3, 4, 6, 3, 3, 4, 3, 4)$.

Embedding and Extracting Algorithm. Now, we can embed C bits (with $C = h(LPP) + h(RPP)$) at most in elements x_i that have value equal LPP or RPP as follows. Scan X on left-to-right direction, when finding $x_i \in \{LPP, RPP\}$ then x_i is modified to x_i' for embedding a bit b by the formula:

$$x_i' = \begin{cases} LPP, & if\ x_i = LPP\ and\ b = 0 \\ LPP-1, & if\ x_i = LPP\ and\ b = 1 \\ RPP, & if\ x_i = RPP\ and\ b = 0 \\ RPP+1, & if\ x_i = RPP\ and\ b = 1. \end{cases}$$

Thus, the maximal embedding capacity of this method is $C = h(LPP) + h(RPP)$.

In this case, $C = 4 + 3 = 7$. For illustrating the embedding algorithm, we take a sequence of 7 data bits $B = (1001101)$ and use above X. Then obtained X' will be: $X' = (1, 2, 4, 6, 3, 2, 5, 3, 5)$.

For extracting B and restoring X from X', we only do inverses to the above embedding procedure. Namely, firstly we scan X' for extracting B and restoring elements x_i having values equal LPP or RPP. Then we restore elements x_i belong to $[LZP+1, LPP-1]$ and $[RPP+1, RZP-1]$ by histogram shifting in the opposite direction to Sect. 2.1.

The above presented method can be considered as a general schema of HS methods. The different methods differ mainly in the ways to construct a sequence of integers X from a host image.

Finally, it is noted that can select two arbitrary non-empty bins, denote as LSP and RSP (left selected point and right selected point) for replacing LPP and RPP, respectively. Then the maximal embedding capacity will be $h(LSP) + h(RSP)$.

PESH-MED Method

Suppose $X = (x_1, \ldots, x_K)$ is a sequence of pixels. For each pixel x_i, create a three-pixel prediction context as in Fig. 2. Then, compute the prediction value $\tilde{x}_i$ according to predictor MED:

$$\tilde{x}_i = \begin{cases} min\left(c_1^i, c_2^i\right), & if c_3^i \geq max\left(c_1^i, c_2^i\right) \\ max\left(c_1^i, c_2^i\right), & if c_3^i \leq min\left(c_1^i, c_2^i\right) \\ c_1^i + c_2^i - c_3^i, & otherwise. \end{cases} \quad (1)$$

and prediction error $\tilde{e}_i$ as follows:

$$\tilde{e}_i = \tilde{x}_i - x_i \quad (2)$$

Using histogram shift method in error sequence

$$\tilde{E} = (\tilde{e_1}, \ldots, \tilde{e_K})$$

to embed data bits is called as the PESH-MED method in X.

2.2 PPVO Method

This method uses two parameters m and n for constructing a prediction context for each pixel. Suppose the host image I has the size of $M \times N$. Firstly, scan in $(M - m + 1)$ first rows and $(N - n + 1)$ first columns of I in the order left to right and top to bottom to obtain the sequence of pixel values:

$$X = (x_1, x_2, \cdots, x_K),$$

where $K = (M - m + 1) \times (N - n + 1)$. For every $x_i \in X$, Qu et al. created an image block of size $m \times n$ such that x_i is the element at left-upper corner. The following Fig. 1 illustrates a block with $m = 3, n = 4$.

The sequence $C_i = \left(c_1^i, \cdots, c_{m\times n-1}^i\right)$ is used as a prediction context for x_i. The prediction value $\widehat{x_i}$ for x_i is calculated from C_i as follows:

$$\widehat{x_i} = \begin{cases} min\left(C_i\right), \; if \; x_i \le min\left(C_i\right) \\ max\left(C_i\right), if \; x_i \ge max\left(C_i\right) > min\left(C_i\right). \end{cases} \tag{3}$$

and the prediction error is computed by formula:

$$e_i = x_i - \widehat{x_i}. \tag{4}$$

Then the histogram shift method is applied in sequence

$$E = \{e_1, e_2, ..., e_K\}.$$

However, Qu et al. used a predefined bin for shifting and embedding, namely: $LSP = 0$ (Left Selected Point) for the case $x_i \le min\left(C_i\right)$ and $RSP = 0$ (Right Selected Point) for the case $x_i \ge max\left(C_i\right) > min\left(C_i\right)$.

Therefore, do not need to record the values e_i as well as to construct the histogram $h(e)$. Shifting and embedding can be performed whenever having a value e_i as follows: For every $x_i, 1 \le i \le K$, first create the context C_i, then calculate prediction error e_i and finally modify x_i to x_i' by formulas:

$$x_i' = \begin{cases} x_i - 1, \; if \; e_i < 0 \; - shift \; to \; the \; left \\ x_i - b, \; if \; e_i = 0 \; - embed \; bit \; b \\ x_i + 1, \; if \; e_i > 0 \; \& \; max\left(C_i\right) > min\left(C_i\right) \; - shift \; to \; the \; right \\ x_i + b, \; if \; e_i = 0 \; \& \; max\left(C_i\right) > min\left(C_i\right) \; - embed \; bit \; b. \end{cases}$$

For extracting data bits and restoring the host image, only perform in reverse order to embedding procedure. It is noted that using a predefined bin for embedding also makes decreasing the capacity. In addition, the number of shifted points is increased as without using zero points.

$$\begin{matrix} x_i & c_1^i & c_2^i & c_3^i \\ c_4^i & c_5^i & c_6^i & c_7^i \\ c_8^i & c_9^i & c_{10}^i & c_{11}^i \end{matrix}$$

Fig. 1. PPVO prediction context

$$\begin{matrix} x_i & c_1^i \\ c_2^i & c_3^i \end{matrix}$$

Fig. 2. MED prediction context

3 Proposed Method

In this section, we propose a method that is a combination of PPVO method and PEHS-MED method.

The detail of proposed method is presented below:

3.1 Motivation

In this section, a combination of PPVO method and PEHS-MED method is proposed. First, we construct the sequence of pixels from $M-1$ rows and $N-1$ columns of a original image ($M \times N$ is the size of image):

$X = (x_1, \ldots, x_K)$, where $K = (M-1) \times (N-1)$

Then using an integer threshold $t \geq 0$ for dividing the sequence X, $\tilde{E}$ (defined in (2)) and E (defined in (4)) into pairs of two separated subsequences:

$$X_t^- = \{x_i \in X \mid FL_i \leq t\};\ X_t^+ = \{x_i \in X \mid FL_i > t\},$$

$$E_t^- = \{e_i \in E \mid FL_i \leq t\}; E_t^+ = \{e_i \in E \mid FL_i > t\},$$

$$\tilde{E_t^-} = \{\tilde{e_i} \in \tilde{E} \mid FL_i \leq t\}; \tilde{E_t^+} = \{\tilde{e_i} \in \tilde{E} \mid FL_i > t\},$$

where $FL_i = max(C_i) - min(C_i), C_i$ are defined in Sect. 2.2. Then we have

$$Q(E) = Q(E_t^-) + Q(E_t^+),$$

where $Q(E), Q(E_t^-), Q(E_t^+)$ denote the capacities of PPVO method on X, X_t^-, X_t^+, respectively. Now a combination of PEHS-MED and PPVO methods is proposed as follows: Applying PEHS-MED method in X_t^- and PPVO method in X_t^+. Then the capacity of combination method, denoted $C(t)$, will be:

$$C(t) = S(\tilde{E_t^-}) + Q(E_t^+),$$

where $S(\tilde{E_t^-})$ denotes the capacity of the PEHS-MED method on X_t^-. In the case t = 0, it can be proved $S(\tilde{E_0^-}) > Q(E_0^-)$. Indeed, if h^0 denotes histogram of subsequence E_0^-, then from Sect. 2.2, we have

$$Q(E_0^-) = h^0(0).$$

On the other hand, it follows from Sect. 2.1 that:

$$S(\tilde{E_0^-}) = \tilde{h^0}(LPP^0) + \tilde{h^0}(RPP^0),$$

where $\tilde{h^0}$ denotes the histogram of subsequence $\tilde{E_0^-}$, LPP^0 and RPP^0 are bins at that $\tilde{h^0}$ reaches highest values. Moreover, it is easily that $\tilde{h^0} = h^0$, thus $S(\tilde{E_0^-})$ is about twice compared $Q(E_0^-)$. From this, it follows $C(0) > Q(E)$. To increase embedding capacity, we need find value t at which $C(t)$ reaches the highest value. Thus, the combination method is performed as follows: First, determine the optimal value T such that

$$T = arg\ max\{C(t)|0 \leq t \leq 255\}.$$

(Values of T are different for test images - see Table 1). Then performing PEHS-MED method on X_T^- and the PPVO method on X_T^+.

3.2 Location Map

In histogram shifting, every pixel can be modified at most by one. Therefore, if there are pixels 0 in the host image, these pixels must be added by 1 to avoid underflow. To distinguish between origin pixels 1 and obtained ones, a left location map, denote by ML, is constructed. The map is a binary sequence in which zero-bits corresponding to origin values and one-bits with obtained values. When $h(0) = 0$ (h is the histogram of the host image) then ML is empty. It is noted that the size of ML and the number of shifting operations (distortion), denoted as DL, can be defined as:

$$DL = h(0),$$

$$size\,(ML) = \begin{cases} 0, & if\ h(0) = 0 \\ h(0) + h(1), & otherwise. \end{cases}$$

The right location map, denoted MR, is also constructed in the same way for avoiding overflow. $Size\,(MR)$ and corresponding distortion DR can be also defined as:

$$DR = h(255),$$

$$size\,(MR) = \begin{cases} 0, & if\ h\,(255) = 0 \\ h\,(255) + h\,(254)\,, & otherwise. \end{cases}$$

The location map LM is defined as $LM = ML \bigoplus MR$, where $\bigoplus$ is the concatenation between two sequences.

3.3 Determining the Auxiliary Information

With a host image I, size $M \times N$, we perform following steps:

Step 1: Partition the host image into overlapping blocks with the same size 2×2 similar as in Fig. 2.

Step 2: Define $T, \widetilde{E_T^-}, E_T^+$ as in Sect. 3.1

Step 3: Construct the histogram $\tilde{h}$ of $\widetilde{E_T^-}$ and determine the values LPP, RPP, LZP, RZP as in Sect. 2.1 for $\tilde{h}$.

Step 4: Determine the binary sequence, denoted B, that consists data, denote as D, location Map is called LM, and $A = LPP \bigoplus RPP \bigoplus LZP \bigoplus RZP \bigoplus T \bigoplus END$, where x_{END} is last embedded pixel (END is the index). Thus, B is defined as:

$$B = D \bigoplus LM \bigoplus A.$$

If denote $size(A)$ as SA, then SA can be considered as a constant that is known to both sides: embedding and extraction. For a host eight-bit grayscale image with size $M \times N$, length of SA (in bit) can be selected as:

$$SA = 2 \times 4 + 2 \times 8 + 8 + \lceil log_2(M \times N) \rceil.$$

3.4 Embedding Procedure

Let I be the host image of size $M \times N$ and D be the sequence of embedded data bits. We must embed D in I to get stego image I'. First add 1 to pixels 0, subtract 1 from pixels 255, construct location map as in Sect. 3.2, determine auxiliary information $E, LPP, RPP, LZP, RZP, LM$. as in Sect. 3.3. Then we do the following steps:

Step 1: Implement same as Sect. 3.1 to get $T, \widetilde{E_T^-}$ and E_T^+.

Step 2:

- With E_T^+ do as follow: Each pixel x_i corresponding to a prediction error e_i in E_T^+, is transformed to x_i' as follows:

$$x_i' = \begin{cases} x_i - 1, \ if \ e_i < 0 \\ x_i - b, \ if \ e_i = 0 \\ x_i + 1, \ if \ e_i > 0 \ \& \ max\,(C_i) > min\,(C_i) \\ x_i + b, \ if \ e_i = 0 \ \& \ max\,(C_i) > min\,(C_i)\,. \end{cases}$$

  where $max(C_i)$, $min(C_i)$ and C_i are defined in Sect. 2.2, b is a embedded bit.
- With $\widetilde{E_T^-}$ do as follow:

  Each pixel x_i corresponding to a prediction error $\tilde{e_i}$ in $\widetilde{E_T^-}$, is transformed to x_i' as follows:

$$x_i' = \begin{cases} x_i - 1, if \ LZP < \tilde{e_i} < LPP \\ x_i + 1, if \ RPP < \tilde{e_i} < RZP \\ x_i, \quad\ if \ \ (\tilde{e_i} = LPP \ or \ \tilde{e_i} = RPP) \ \ and \ b = 0 \\ x_i - 1, if \ \tilde{e_i} = LPP \ and \ b = 1 \\ x_i + 1, if \ \tilde{e_i} = RPP \ and \ b = 1 \\ x_i, \quad\ otherwise. \end{cases}$$

  Where b is a bit picked out from the binary sequence $D \bigoplus LM$.

Step 3: After embedding all bits of $D \bigoplus LM$ is completed, we extract the LSB (Least Significant Bit) of x'_i with $i = 1, \cdots .SA$.
Step 4: Continue embedding SA bits extracted at step 3 into x_i as in step 2.
Step 5: Denote the last embedded pixel as x_{END}, where END is the index. After having index END, we create the binary sequence A consisting of $LPP, RPP, LZP, RZP, T, END$. Then insert A in LSB of $x'_i, i = 1, 2, \cdots, SA$.

As shown in Sect. 3.3, SA is known to both sides: embedding and extracting.

3.5 Extraction Algorithm

Let I' be the stego image, we must extract data D and restore host image I. For this objective, we do the following steps:

Step 1: Extracting LSB of the first SA pixels x'_i of I' to obtain $LPP, RPP, LZP, RZP, T, END$.
Step 2: Browse the stego image in reverse order to embedding algorithm, begin at x'_{END}. For each x'_i, we extract embedded bit b (if x'_i carries a bit) and restore x_i, as follows:
 2.1: Compute the $FL_i = max(C_i) - min(C_i)$, where C_i are defined in Sect. 2.2 for block sized 2×2. Then, compute $\tilde{x}_i$ as in formula (1) if $FL_i \leq T$ or $\widehat{x}_i$ as in formula (3) if $FL_i > T$ and e'_i as follows:

$$e'_i = \begin{cases} x'_i - \tilde{x}_i, & if\ FL_i \leq T \\ x'_i - \widehat{x}_i, & if\ FL_i > T. \end{cases}$$

 It is noted that, by executing in reverse order from x'_{END} to x'_1, so here, prediction context is still as the same as in embedding.
 2.2. Extract embedded bit and restore host pixel:
 - If $FL_i \leq T$ do as follow:
 Extract b:

$$b = \begin{cases} 0, & if\ e'_i = LPP\ or\ e'_i = RPP \\ 1, & if\ e'_i = LPP - 1\ or\ e'_i = RPP + 1 \end{cases}$$

 Restore x_i:

$$x_i = \begin{cases} x'_i - 1, if\ RPP + 1 \leq e'_i \leq RZP, \\ x'_i + 1, if\ LZP \leq e'_i \leq LPP - 1, \\ x'_i, \quad otherwise. \end{cases}$$

 - If $FL_i > T$ do as follow:
 Extract b:

$$b = \begin{cases} 0, & if\ e'_i = 0 \\ 1, & if\ |e'_i| = 1 \end{cases}$$

 Restore x_i:

$$x_i = \begin{cases} x'_i, & if\ e'_i = 0 \\ x'_i + 1, & if\ e'_i \leq -1 \\ x'_i - 1, & if\ e'_i \geq 1 \end{cases}$$

Step 3: After extracting SA bits from step 2, we insert these bits into LSB of $x'_{SA}, \cdots, x'_1$.

Step 4: Repeat step 2 until to reach x'_1. At this time, we get hiding data D and location map LM.

Step 5: from LM, restore pixels having values 0 or 255.

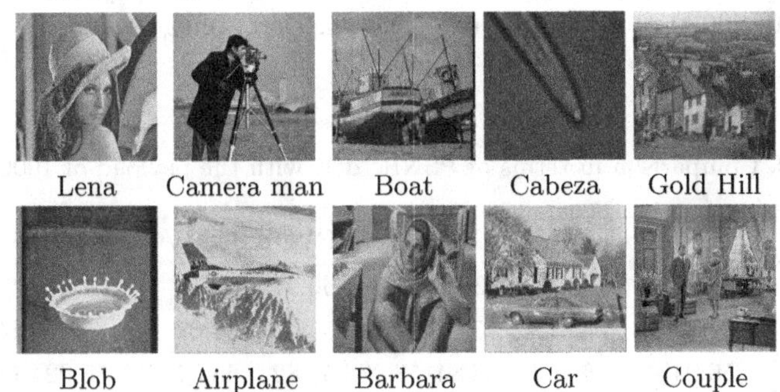

Fig. 3. Experimental images.

Table 1. Comparison in terms of embedding capacity (number of bits)

Images	Li	PPVO	PVO	IPVO	PVOK	GePVOK	TRES	Proposed (Threshold: T)
Camera man	100276	95826	43458	71097	63378	69772	74621	132551 (120)
Couple	37602	41159	21599	29950	25127	26869	32461	44174 (138)
Boat	39493	41446	27392	35159	32784	33321	36477	59432 (108)
Gold hill	35868	31350	25955	28662	29919	31210	31899	39970 (112)
Lena	44565	45865	33115	40176	40373	43198	46968	55540 (129)
Cabeza	90618	85712	54263	71946	73417	79199	77475	98413 (63)
Barbara	54152	60072	29490	48066	40203	41008	47099	85790 (160)
Blob	68052	66083	41495	53326	52774	56473	59182	81200 (168)
Car	69107	65212	31378	47327	41594	44644	53097	78508 (187)
Airplane	71856	68305	38454	53055	51505	56305	58200	81594 (138)
Average	61159	60103	34660	47876	45107	48200	51748	**86356**

4 Experimental Results

To illustrate the results of theory analysis, we perform experiments on the sample image set in [1] with some types of images as common images, texture images, images having many flat regions, etc as shown in Fig. 3. Images have size

512 × 512. The hidden data is a random bit sequence. Programs are written in the language Matlab and ran on IdeaPad S410p Lenovo computer.

Proposed method is compared with Li method [5], PPVO method [11], PVO method [4], IPVO method [10], PVOK method [8], GePVOK method [3].

Capacity comparison results between other methods applied on sample images, are given in Table 1. Meanwhile Tables 2, 3 present the stego image quality comparison results of the methods with the same payloads having the size of 10000, 20000 random bits, respectively. Here, image quality is computed in term of PSNR defined in [13].

Table 2. Comparison in terms of PSNR (dB) with the payload of 10000 bits

Images	Li method	PPVO	PVO	IPVO	PVOK	GePVOK	TRES	Proposed
Camera man	60.69	63.14	63.07	62.47	60.27	52.94	61.47	63.41
Couple	54.77	59.80	55.83	56.77	55.09	52.82	55.86	59.84
Boat	56.24	60.53	58.84	58.93	57.82	53.57	55.92	60.66
Gold hill	57.74	61.41	58.05	58.62	58.16	54.75	58.78	61.56
Lena	55.95	60.83	58.38	58.77	57.56	54.09	58.01	60.93
Cabeza	58.43	61.84	61.62	61.02	59.80	54.00	59.93	62.00
Barbara	56.11	60.65	57.76	58.78	57.03	52.29	57.55	60.78
Blob	56.07	60.02	58.60	58.43	57.72	52.85	60.33	60.10
Car	61.48	63.66	58.27	62.50	59.41	54.68	61.87	64.27
Airplane	56.91	60.48	59.60	59.18	58.48	52.78	58.26	60.58
Average	57.44	61.24	59.00	59.55	58.13	53.48	58.80	**61.41**

Table 3. Comparison in terms of PSNR (dB) with the payload of 20000 bits

Images	Li method	PPVO	PVO	IPVO	PVOK	GePVOK	TRES	Proposed
Camera man	57.66	60.10	59.5	59.52	57.33	51.72	58.44	60.34
Couple	50.86	56.49	52.64	53.41	51.9	50.04	53.04	56.54
Boat	52.54	57.45	54.05	55.17	53.85	50.74	52.54	57.55
Gold hill	52.16	56.12	53.32	53.44	53.21	50.91	53.10	56.21
Lena	52.42	56.79	54.97	55.13	54.34	51.4	53.56	56.86
Cabeza	55.31	58.78	58.62	57.98	56.81	52.02	56.86	58.91
Barbara	53.38	57.56	54.92	56.04	54.37	50.75	55.69	57.66
Blob	53.34	57.29	56.15	55.87	55.08	51.03	55.64	57.37
Car	58.04	60.50	54.4	56.88	54.15	51.17	57.59	61.07
Airplane	54.31	57.81	56.82	56.7	55.81	51.18	55.64	57.92
Average	54.00	57.89	55.54	56.01	54.69	51.10	55.21	**58.04**

From above experimental results, we can deduce the remarks as follows: The proposed method have better stego image quality and higher embedding capacity.

5 Conclusions

This paper proposes a new reversible data hiding which is combination of prediction error histogram shifting and PPVO method.

By theoretical analysis and experimental results, it is proved that proposed method outperform some recent state-of-the-art methods in terms of embedding capacity and stego image quality.

References

1. Images data (2017). http://decsai.ugr.es/cvg/dbimagenes. http://www.imageprocessingplace.com
2. Fridrich, J., Goljan, M., Du, R.: Invertible authentication. In: Security and Watermarking of Multimedia Contents III, vol. 4314, pp. 197–208. International Society for Optics and Photonics (2001)
3. Li, J., Wu, Y.H., Lee, C.F., Chang, C.C.: Generalized PVO-K embedding technique for reversible data hiding. IJ Netw. Secur. **20**(1), 65–77 (2018)
4. Li, X., Li, J., Li, B., Yang, B.: High-fidelity reversible data hiding scheme based on pixel-value-ordering and prediction-error expansion. Sig. Process. **93**(1), 198–205 (2013)
5. Li, Y.C., Yeh, C.M., Chang, C.C.: Data hiding based on the similarity between neighboring pixels with reversibility. Digit. Signal Proc. **20**(4), 1116–1128 (2010)
6. Macq, B.: Lossless multiresolution transform for image authenticating watermarking. In: 2000 10th European Signal Processing Conference, pp. 1–4. IEEE (2000)
7. Ni, Z., Shi, Y.Q., Ansari, N., Su, W.: Reversible data hiding. IEEE Trans. Circuits Syst. Video Technol. **16**(3), 354–362 (2006)
8. Ou, B., Li, X., Zhao, Y., Ni, R.: Reversible data hiding using invariant pixel-value-ordering and prediction-error expansion. Sig. Process. Image Commun. **29**(7), 760–772 (2014)
9. Pan, Z., Gao, E.: Reversible data hiding based on novel embedding structure PVO and adaptive block-merging strategy. Multimedia Tools Appl. **78**, 1–25 (2019)
10. Peng, F., Li, X., Yang, B.: Improved pvo-based reversible data hiding. Digit. Signal Proc. **25**, 255–265 (2014)
11. Qu, X., Kim, H.J.: Pixel-based pixel value ordering predictor for high-fidelity reversible data hiding. Sig. Process. **111**, 249–260 (2015)
12. Sao Nguyen, K., Le, Q.H., Pham, V.A.: A new reversible watermarking method based on histogram shifting. Appl. Math. Sci. **11**(10), 445–460 (2017)
13. Taubman, D., Marcellin, M.: JPEG2000 Image Compression Fundamentals, Standards and Practice: Image Compression Fundamentals, Standards and Practice, vol. 642. Springer, Heidelberg (2012)
14. Thodi, D.M., Rodríguez, J.J.: Reversible watermarking by prediction-error expansion. In: 6th IEEE Southwest Symposium on Image Analysis and Interpretation, pp. 21–25. IEEE (2004)
15. Tian, J.: Reversible watermarking by difference expansion. In: Proceedings of Workshop on Multimedia and Security, vol. 19 (2002)

An Improved Reversible Watermarking Based on Pixel Value Ordering and Prediction Error Expansion

Le Quang Hoa[1(✉)], Cao Thi Luyen[2(✉)], Nguyen Kim Sao[2(✉)], and Pham Van At[2(✉)]

[1] Institute of Applied Mathematics and Informatics, Hanoi University of Science and Technology, Hanoi, Vietnam
hoa.lequang1@hust.edu.vn

[2] Faculty of Information Technology, University of Transport and Communications, Hanoi, Vietnam
{luyenct,saonkoliver}@utc.edu.vn, phamvanat83@gmail.com

Abstract. In this paper, a new pixel value ordering (PVO) and prediction error Expansion (PEE) based reversible watermarking method is proposed. PVO based method firstly introduced by Li et al.'s has been concerned and expanded because it can achieve high embedding capacity without losing the implication of the cover. We improve related work in two points: the first is the way to build location map, the second is technique of embedding secret data. Only one bit is used to create location map while related scheme utilizes two bits. So that we get the location map with shorter size. The less size location map the more embedding capacity because we have more space to embed data, the embedding capacity of the proposed scheme is thus increased. Theory analysis and experimental results showed that the proposed scheme is clearly better in terms of embedding capacity compared with state-of-the-art works. The watermarked image that is applied our technique is good imperceptibility.

Keywords: Revesible watemarking · PVO · PVOK · GePVOK · PEE

1 Introduction

Today, the storage and transmission of data is made popular through the Internet environment. The problem of digital product infringement negatively affects the economy, politics, society... On the other hand, the digital product is transformed before reaching the recipient, if the received data has been modified previously, then which area is changed. That is the problem of digital data authentication. Finding effective ways to solve copyright protection and content validation is especially important in economics, politics, security and defense, it receives special attention not only for scientists but also for society. Watermarking has been considered as a efficient solution to the problem of digital content copyright protection and authentication in a networked invironment.

Digital watermarking is the art of hiding useful information (watermark) in digital contents such as images, audio clips and videos, etc. Watermark may contain confidential messages that need to be exchanged or information about copyright, multimedia

N. T. Nguyen et al. (Eds.): ACIIDS 2020, LNAI 12034, pp. 582–592, 2020.
https://doi.org/10.1007/978-3-030-42058-1_49

products. Watermark can then be restored as evidence to prove legal rights as well as to test digital products which are in integrity or not or there is unauthorized access before reaching the user [12]. Watermarking technique includes embedding process and recovering process. In embedding process, the watermark W is hidden into the cover I to get watermarked image I'. The technique of embedding ensures that it is hard to distinguish I and I' by human eyes. The exchange of watermarked images I' on the Internet can be transformed into an image I^*. To verify the copyright or the integrity of the image I^* we apply recovering process. W^* is gotten by using recovering procedure. The extracted watermark W^* in the tested image $I^*$is extracted and compared to the original data W. If both match, we consider that the image I^* has not been tampered with; otherwise, we conclude the image I^* was fake.

In some applications such as health, military, security - defense, the host image needs to be restored intact besides restoring the original data is requied. This kind of watermarking is called the reversible data watermarking.

Current watermarking studies aim to successfully develop a watermarking system that best satisfies the basic requirements of a watermarking system such as embedding capacity, watermarking image quality, computational complexity, reversible.... This work focuses on reversible watermarking schemes to enhance embedding capacity of the watermarking scheme applying to solve digital authentication problem.

Many valuable reversible watermarking methods have been researched in the literature. Some primary approaches are proposed such as difference expansion (DE) [3, 5 13, 26], lossless compression [11, 16, 21], histogram shifting (HS) [2, 7, 22, 27], prediction error expansion (PEE) [1, 6, 9, 17, 18] and pixel value ordering (PVO) [1, 4, 8, 12, 23, 25]. A lot of schemes are created by incorporating methods together in order to gain higher capacity and a lower distortion [1, 24]. For example, DE technique is combined with other methods such as histogram shifting, prediction error expansion, PEE is merged HS; Histogram shifting is put together with prediction error expansion, PVO, PEE, etc....; In this paper, we research the works related to PVO based method, which has been seen as the most currently interested technique because it achieves impressive performance for moderate pay load size.

PVO method was firstly proposed by Li et al.'s. [24] has two advanges: The prediction utilized from sorted pixel values is more exacting than the previous one such as MED (median edge detector) [1, 27], GAP (gradient adjusted prediction) [15, 21] and the mean value predictor [17]; Larger sized blocks can be used to better image quality for low capacity case. In [24], the cover image is divided into blocks with size $s = m \times n$. Each bock is scanned in zigzag order to obtain $X = (x_1, x_2, \cdots, x_s)$. Sorting $X = (x_1, x_2, \cdots, x_s)$ in ascending manner to get array $H = \left(x_{\pi(1)}, x_{\pi(2)}, \cdots, x_{\pi(s)}\right)$. Then, the largest value $dmax = x_{\pi(s)} - x_{\pi(s)-1}$ is embedded one bit if $x_{\pi(s)} - x_{\pi(s)-1} = 1$. Besides, by also considering the smallest value $dmin = x_{\pi(2)} - x_{\pi(1)}$, so maximum two bits are embedded into each block with n pixels if $dmax = 1$ or $dmin = -1$. In [24], the blocks that $dmax = 0$ and $dmin = 0$ are abandoned. In fact that is the peak of the histogram. Embedding into which leads watermarked image quality better. Especially, there are a lot of prediction errors 0 in natural image. Peng et al. [10] improved PVO to get higher capacity and better image quality called IPVO method. But IPVO still skips the blocks which have largest and second largest pixel values equal. [4] improved that disadvange by considering pixels with largest (or smallest) value of a block as a unit to embed data. Assume that there are K pixels with largest (or smallest) pixel values, that method is called PVOK. In this way, more blocks can be used to embed data, that means the capacity is increased. But, in [4], all K pixels are modified together to embed only one bit. [12] proposed by Jian Li et al. is call GePVOK scheme expanded PVOK method in order to embed K bits into K largest (smallest) valued pixels. As a result, embedding capacity of the GePVOK scheme is greater than the PVOK method. In [12], location map is created to handle overflow (or underflow) as well to server for recovering phrase. Jian Li et al.'s uses two bits to built location map. That map will be compressed to get PM and PM is embedded into the original image. The longer the PM is, the less embedding capacity is. In this paper, we only use one bit to built location map therefore the length of location map as well the compression map is less than the GePVOK scheme. So that the capacity of the proposed scheme is larger than the GePVOK. Moreover, there are a lot of blocks are used to embed by our algorithm whereas the related work are not. Therefore, the proposed algorithm is more efficient than the related algorithm. The detail explanations are presented in Sect. 3.

The rest of the paper has the structure as below. In Sect. 2 we briefly review the related scheme, called GePVOK. We describe the proposed reversible watermarking method based on PVO in Sect. 3. The experimental results are shown in Sect. 4. Lastly, the conclusion of the paper is presented in Sect. 5.

2 GePVOK

GePVO scheme proposed by Li et al. [12] can embed K bits into K largest (smallest) valued pixels so that the capacity of this method is high. The detailed steps of the embedding and extraction procedures in the largest pixel values of a block are introduced as belows. The embedding on the K smallest valued pixels is performed the same.

2.1 Embedding Process

Firstly, the cover image is divided into blocks with size $s = m \times n$. Each bock is scanned in zigzag order to obtain $X = (x_1, x_2, \cdots, x_s)$. Next, a location map is built based on each blocks as belows:

$$L(X) = \begin{cases} 2\, if \exists x_i \in \{0, 1, 254, 255\}, i = 1 \cdots s \\ 1 \qquad\qquad if\ x_i = x_j, i \neq j \\ 0 \qquad\qquad otherwise \end{cases}$$

Then, the secret data is embedded into the largest pixel values depending on each block is described as following. Embedding into the smallest pixel values is performed the same.

If L(X) = 2 then the block X is skipped, that means we do nothing.

If L(X) = 1 then the secret data $B = (b_1, b_2, \cdots, b_{s-1})$ are hided into block X by using the Eq. 1.

$$x_i' = \begin{cases} x_1 & if\ i = 1 \\ x_i + b_{i-1} & if\ i = 2, 3, \cdots s \end{cases} \tag{1}$$

If LM(X) = 0, we have to perform following steps:

Step 1: Sorting X.

The block X is sorted in ascending order to get $H = \left(x_{\pi(1)}, x_{\pi(2)}, \cdots, x_{\pi(s)}\right)$. Assum that, array H has K largest pixel values and L second-largest pixel values. It is clearly H is displayed as belows:

$$x_{\pi(1)} \leq x_{\pi(2)} \leq \cdots \leq x_{\pi(s-K-L)} < x_{\pi(s-K-L+1)} = \ldots = x_{\pi(s-K)} < x_{\pi(s-K+1)} = \cdots = x_{\pi(s)}$$

Step 2: Calculating the maximum prediction error by applying Eq. 2.

$$dmax = x_{\pi(s-K+1)} - x_{\pi(s-K)} \tag{2}$$

Case 2.1: if $dmax > 1$, secret data is not embedded but all K largest valued pixels are increased by one according to formular 3.

$$x'_{\pi(i)} = x_{\pi(i)} + 1, i = s - K + 1, \cdots s \tag{3}$$

Case 2.2: if $dmax = 1$, the secret data B $= b_1, \cdots, b_K$ is embedbed into block X by applying in turn Eqs. 4 and 5.

$$x'_{\pi(i)} = x_{\pi(i)} + 1, i = s - K - L + 1, \cdots s \tag{4}$$

$$x'_{\pi(j)} = x_{\pi(j)} + b_j, j = s - K + 1, \cdots s \tag{5}$$

That the way we get the watermarked image.

2.2 Extraction Process

The restoring secret data and the original image based on the location map is restored as belows:

– Firstly, the watermarked image I' is devided into the block X' with same size in the embedding process $s = m \times n$. Next, the location map is computed based on remainding order of PVO method as belows:

If $L\left(X'\right) = 2$ then we do nothing, the original block X is the same as X' and this block has not secret data.

If $L\left(X'\right) = 1$ then d'_i is computed according to Eq. (6).

$$d'_i = x'_i - x_1, i = 2 \cdots s \tag{6}$$

And the secret data and the original pixel are restored by using Eq. 7.

$$x_1 = x'_1; x_i = \begin{cases} x'_i, b_{i-1} = 0 \, if \, d'_i = 0 \\ x'_i - 1, b_{i-1} = 1 \, if \, d'_i = 1 \end{cases}, i = 2 \cdots s \tag{7}$$

If $L\left(X'\right) = 0$ then the block X' is sorted in ascending order to get array denoted by $H' = \left(x'_{\pi(1)}, x'_{\pi(2)}, \cdots, x'_{\pi(s)}\right)$. Assum that, array H' has R largest pixel values (m_1),

S second-largest pixel values (m_2) and T third-largest pixel values (m_3) where $R + S + T = K + L$. It is clearly H is displayed as belows:

$$x'_{\pi(1)} \le x'_{\pi(2)} \le \cdots x'_{\pi(s-R-S-T)} < x'_{\pi(s-R-S-T+1)} = \cdots$$
$$= x'_{\pi(s-R-S+1)} < x'_{\pi(s-R-S)} < x'_{\pi(s-R-S+1)} = \cdots$$
$$= x'_{\pi(s-R)} < x'_{\pi(s-R+1)} = \cdots = x'_{\pi(s)}$$

Two prediction errors d_1, d_2 are calculated by applying Eq. 8.

$$d_1 = x'_{\pi(s-R+1)} - x'_{\pi(s-R)};\ d_2 = x'_{\pi(s-R-S+1)} - x'_{\pi(s-R-S)} \quad (8)$$

If T does not equal to 0, it is clear that the prediction errors $d_1 \ge 1$ and $d_2 \ge 1$ else there are not the third-largest pixel values, only d_1 exits and $d_1 \ge 1$.

If $d_1 > 2$ respectively case 2.1 in embedding process then no data is embedded so no data is restored and the original pixels are recorved according to the Eq. 9.

$$x_{\pi(i)} = x'_{\pi(i)} - 1, i = s - R + 1 \cdots s \quad (9)$$

If $d_1 \le 2$ and $d_2 = 1$ then first, we compute $dmax'$:

$$d'_i = x'_{\pi(i)} - x'_{\pi(s-R-S)}, i = s - R - S + 1 \cdots s. \quad (10)$$

Next, the value of pixels in the m1, m2 and m3 zones are decreased by one:

$$x'_{\pi(i)} = x'_{\pi(i)} - 1, i = s - R - S - T + 1 \cdots s \quad (11)$$

Finally the original pixels and secret data are restored by using Eq. 12.

$$x_{\pi(i)} = \begin{cases} x'_{\pi(i)} - 1, b_i = 0\, if\, d'_i = 1 \\ x'_{\pi(i)} - 2, b_i = 1\, if\, d'_i = 2 \end{cases}, i = s - R - S + 1 \cdots s \quad (12)$$

If $d_1 \le 2$ and ($d_2 \ge 2\, or\, T = 0$) then we compute d'_i by applying Eq. 13:

$$d'_i = x'_{\pi(i)} - x'_{\pi(s-R)}, i = s - R + 2 \cdots s. \quad (13)$$

Next, the original pixels and secret data are restored by using Eq. 14.

$$x'_{\pi(j)} = x'_{\pi(j)} - 1;\ x_{\pi(i)} = \begin{cases} x'_{\pi(i)}, b_i = 0\, if\, d'_i = 1 \\ x'_{\pi(i)} - 1, b_i = 1\, if\, d'_i = 2 \end{cases} \quad (14)$$

where $j = s - R - S + 1 \cdots n; i = s - R + 1 \cdots s.$

Experimental results show that the GePVOK scheme consistently enhances the performance of PVO embedding too. There is still space for improvement GePVOK. To restore the embedded secret data and the host image, location map must be compressed and hided into the host image together secret data. The capacity of each image is limit, the more space spending on hiding location map the less one for embedding secret data. GePVO uses two bits to create the location map is disadvantage. Morever, the block that has one of set $\{0, 1, 254, 255\}$ will be ignored. That is the cause to descrease the embedding capacity and imperceptibility. This paper will solve those problems by improving the location map and proposing the novel embedding technique.

3 The Proposed Scheme

3.1 The Embedding Procedure

Embedding secret data into the largest pixel values of a block are discribled as belows.

Firstly, the cover image I is divided into subblocks with size $s = m \times n$. Each bock is scanned in zigzag order denoted by $I_j = (x_1, x_2, \cdots, x_s)$. Next, a location map is built based on each blocks as belows:

$$L(I_j) = \begin{cases} 1\, if\ I_j\ is\ not\ smooth\ and\ \exists x_i \in \{0, 1, 245, 255\},\ i = 1 \cdots s \\ 0 \qquad\qquad\qquad\qquad\qquad\qquad\qquad otherwise \end{cases}$$

If $L(I_j) = 1$ then the block X is skipped,

If $L(I_j) = 0$ then

Case 1: I_j is absolutely smooth block.

In this situation, *s-1* bits are embedded into *s-1* pixels where the first pixel of block is unchanged as equation belows.

$$x'_{\pi(1)} = x_{\pi(1)}$$

if $x_{\pi(1)} = 255$ then $x'_{\pi(i)} = x_{\pi(i)} - b_i$ else $x'_{\pi(i)} = x_{\pi(i)} + b_i, i = 2, \cdots s$

It is easy to find that $dmax'$ equals to 0 or 1 but the pixel in this block is divided into two groups, that means $dmax'_2$ does not exit.

Case 2: if I_j is not smooth, we execute as following steps:

Step 1: Sorting I_j to get array denoted by $H = \left(x_{\pi(1)}, x_{\pi(2)}, \cdots, x_{\pi(s)}\right)$. Assume that, array H has K largest values pixels.

Step 2: Calculating the maximum prediction error by applying Eq. 8.

Case 2-1: if *dmax* > 1, secret data is not embedded but all K largest valued pixels are increased by two.

$$x'_{\pi(i)} = x_{\pi(i)} + 2, i = s - K + 1, \cdots s \tag{15}$$

So that $dmax'$ is greater than or equal to 4.

Case 2-2: if *dmax* = 1

Case 2-2-1: if $b_i \equiv 1, i = 1 \cdots K$ then $x'_{\pi(i)} = x_{\pi(i)} + 2,$

$$i = s - K + 1, \cdots s \tag{16}$$

In this case, $Dmax'$ equals to 3.

Case 2-2-2: if $b_i \equiv 0, i = 1 \cdots K$ then

$$x'_{\pi(i)} = x_{\pi(i)} + 1, i = s - K + 1, \cdots s \tag{17}$$

Modified Pixel by applying Eq. 18 leads $Dmax'$ changes to 2.

Case 2-2-3: if $b_i = 0|1, i = 1 \cdots K$ then

$$x'_{\pi(i)} = x_{\pi(i)} + b_i, i = s - K + 1, \cdots s \tag{18}$$

Using formular 18 make I_j changing into I'_j where $dmax' = 0$ or $dmax' = 1$ and $dmax2' = 1$.

We combine all the subblocks to get the watermarked image I'.

3.2 The Extracting Procedure

Firstly, the watermarked image I' is divided into subblocks with size $s = m \times n$. Each bock is scanned in zigzag order to get $I_j^{'} = \left(x_1^{'}, x_2^{'}, \cdots, x_s^{'}\right)$. Next, the location map is restored. Then the secret data and the host image are recorved as belows:

If $L\left(I_j^{'}\right) = 1$, that means, this block is not used to hide data and $x_i = x_i^{'}$.

If $L\left(I_j^{'}\right) = 0$ then at first we sort I_j to get array denoted by $H' = \left(x_{\pi(1)}^{'}, x_{\pi(2)}^{'}, \cdots, x_{\pi(s)}^{'}\right)$. Assume that, array H' has K largest pixel values and S second largest pixel values. Next, we compute $dmax_1^{'}$ and $dmax_2^{'}$ as formula 19.

$$\begin{aligned} dmax_1^{'} &= x_{\pi(s-K+1)}^{'} - x_{\pi(s-K)}^{'}; \\ d' &= x_{\pi(s-R-S+1)}^{'} - x_{\pi(s-R-S)}^{'} \end{aligned}; \tag{19}$$

Case 1: $dmax_1^{'} \geq 4$ (corresponding case 2.1 in embedding process) that means the secret data wasn't embedded into this block and the host pixel is restored by applying fomular 20.

$$x_{\pi(i)} = x_{\pi(i)}^{'} - 2, i = 1, \cdots s \tag{20}$$

Case 2: $dmax_1^{'} = 3$ (corresponding case 2.2.1 in embedding process), the secret data and the host pixel are restored by applying fomula 21.

$$b_i \equiv 1, x_{\pi(i)}^{'} = x_{\pi(i)} - 2, i = s - K + 1, \cdots s \tag{21}$$

Case 3: $dmax_1^{'} = 2$ (corresponding case 2.2.2 in embedding process), the secret data and the host pixel are restored by applying fomula 22.

$$b_i \equiv 0, x_{\pi(i)}^{'} = x_{\pi(i)} - 2, i = s - K + 1, \cdots s \tag{22}$$

Case 4: $dmax_1^{'} = 1$

If there is not $dmax_2^{'}$ (corresponding case 1 in embedding process) then the secret data and the host pixel are restored by applying fomula 23.

$$b_i = x_{\pi(i)}^{'} - x_{\pi(1)}^{'}, x_{\pi(i)} = x_{\pi(i)}^{'} - b_i, i = 2, \cdots s. \tag{23}$$

If $dmax_2^{'}$ equals to 1 (corresponding case 2.2.3 in embedding process) then the K bits of secret data and the host pixel are restored by applying Eq. 24 from block $I_J^{'}$.

$$\text{If}\, x_i^{'} = \begin{cases} x_{\pi(s)}^{'}\, then\, b_j = 1, x_i = x_i^{'} - 1 \\ x_{\pi(s-K)}^{'}, then\, b_j = 0, x_i = x_i^{'} \end{cases}, i = 1, \cdots s,\ j = 1, \cdots K \tag{24}$$

Finally, the original pixels and secret data are restored by combining all restored subblocks image as well sub-secret data arrays.

3.3 Analysis

Given subblock belows:

4 6 4
5 5 6

This block is scanned in zigzag order and sorted to get array X = (4, 4, 5, 5, 6, 6). Applying embedding process of GePVOK as well the proposed method, embedded array is shown in following Table 1.

Table 1. Example of embedding capacity of GePVOK scheme and the proposed method

Input	Secret data	GePVOK method						The proposed method					
4 4 5 5 6 6	1 1	4	4	6	6	8	8	4	4	5	5	8	8
	0 0	4	4	6	6	6	6	4	4	5	5	7	7
	1 0|0 1	4	4	6	6	6	7	4	4	5	5	6	7

Data only can not be continuosly embedded into the smallest pixel values in GePVOK scheme because of $dmin = -2$ contrariwise all cases of the proposed scheme absolutedly are hided in the smallest pixel values since $dmin = -1$.

Morever, the GePVOK scheme skips all smooth blocks containing 0, 1, 254, 255 while the proposed scheme has done. On the other hand, the space to save location map of the proposed scheme is less than GePVOK method. Thus, the proposed scheme always gives higher embedding capability.

4 Experimental Results

In this section, several experiments were performed in order to verify the compressed location map size, embedding capacity and the quality image of those watermarking schemes. Six standard uncompressed images of size 512 × 512 pixels are used as host images (Fig. 1). The host image is classified into blocks with size 2 × 2. Watermark data is randomly created with different length depend on the embedding capacity of each image.

Firstly, we survey the length of compressed map of each host image. The shorter location map size, the better watermarking technique is. The experiment results are displayed in Table 2.

PVO uses one bit to build location map while two bits are used in GePVOK and the proposed scheme. The location map size of PVO method is the shortest length. The location size of the proposed scheme is shorter than GePVOK.

Next, we evaluate the embedding capacity of watermarking schemes: PVO, GePVOK and the proposed scheme. The experiment results are displayed in Table 3.

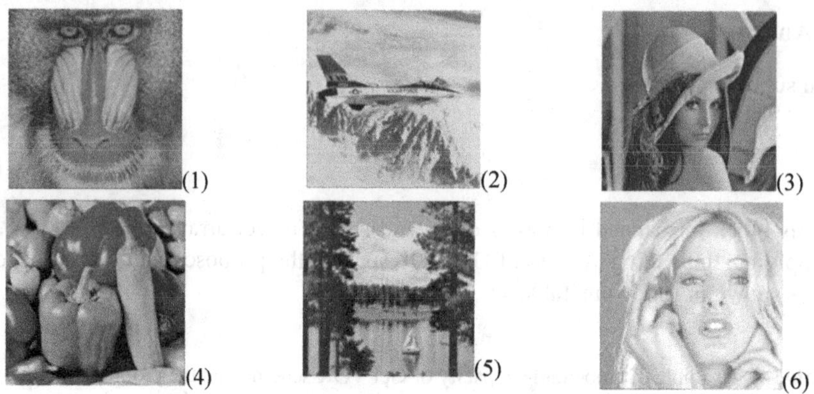

Fig. 1. Six standard test images

Table 2. The length of compressed map of the watermark schemes

Image	PVO	GePVOK	The proposed method
1	65	448	360
2	0	8714	0
3	0	2428	0
4	865	32575	7111
5	0	681	51
6	1461	10196	5089

As reported in Table 3, the embedding capacity of the proposed method is highest. That means the proposed method gets better performance than the other schemes.

Last, we apply the embedding procedure of those schemes to test the imperceptibility of the proposed scheme and several widely used watermarking schemes of PVO, and GePVOK. The peak signal-to-noise ratio *PSNR* between the original images and corresponding watermarked images is calculated by using Eq. 25. The higer *PSNR*, the better quality image. The experimental results are shown in Table 4.

$$PSNR = 20log_{10}\left(\frac{255}{\sqrt{MSE}}\right) \tag{25}$$

where *MSE (Mean Square Error)* is computed as following formular:

$$MSE = \frac{1}{M \times N}\sum_{i=1}^{M}\sum_{j=1}^{N}(I(i,j) - I'(i,j))^2$$

Watermarked image is concerned good quality image if *PSNR* is gretter than 35 dB.

Table 4 shows that, PSNR of the proposed method is less than the other schemes that means the quatily image of the proposed scheme is not as good as the related arts.

Table 3. Embedding capacity of the watermark schemes

Image	PVO	GePVOK	The proposed method
1	13211	14994	15239
2	38963	65472	70131
3	32108	43845.67	46795.33
4	852	2261.333	41456.67
5	11187	13753	13853
6	35554	50967	54769

Table 4. PSNR of the watermarked images

Image	PVO	GePVOK	The proposed method
1	51.618	50.3823	45.6328
2	54.461	51.2719	49.2702
3	53.055	50.143	47.1799
4	61.753	64.5754	69.981
5	51.870	51.013	46.206
6	53.673	50.4732	47.660

However, *PSNR* of the proposed scheme is high (greater than 45 dB) so the diffrence between the original and the watermaked image is almost not distinguishable. That means, the proposed scheme works well in terms of imperceptibility.

5 Conclusion

Reversible digital watermarking has been proposed as a effective solution to the problem of digital content copyright protection and authentication in a networked environment. This paper improves the reversible watermarking based on GePVOK method in terms of both embedding capacity and security. We used only one bits to create location map so there are more spaces to embed data. Our work selected the smooth subbocks included pixel values 0, 1, 254, 255 to hide secret data so the security of propsed scheme is better. Morever, the embedding technique was improved to get higher capacity. Theory analysis and experimental results demonstrate that the proposed algorithm is much higher capacity than the related works. In the our next works we will research the way to get better quatily image.

References

1. Hong, W., Chen, T.S., Shiu, C.W.: Reversible data hiding for high quality images using modification of prediction errors. J. Syst. Softw. **82**(11), 1833–1842 (2009)
2. Tsai, Y.Y., Tsai, D.S., Liu, C.L.: Reversible data hiding scheme based on neighboring pixel differences. Digit. Signal Process. **23**(3), 919–927 (2013)

3. Alattar, A.M.: Reversible watermarking using the difference expansion of a generalized integer. IEEE Trans. Image Process. **13**, 1147–1156 (2004)
4. Ou, B., Li, X., Zhao, Y., Ni, R.: Reversible data hiding using invariant pixel-value ordering and prediction-error expansion. Signal Process. Image Commun. **29**(7), 760–772 (2014)
5. Luyen, C.T., At, P.V.: An efficient reversible watermarking method and its application in public key fragile watermarking. Appl. Math. Sci. **11**(30), 1481–1494 (2017)
6. Coltuc, D.: Improved embedding for prediction-based reversible watermarking. IEEE Trans. Inf. Forensic Secur. **6**(3), 873–882 (2011)
7. Thodi, D.M., Rodriguez, J.J.: Expansion embedding techniques for Reversible watermarking. IEEE Trans. Image Process. **16**(3), 721–730 (2007)
8. Peng, F., Li, X., Yang, B.: Improved pvo-based reversible data hiding. Digit. Signal Proc. **25**, 255–265 (2014)
9. Wu, H.T., Huang, J.: Reversible image watermarking on prediction errors efficient histogram modification. Signal Process. **92**(12), 3000–3009 (2012)
10. Cox, I.J., Miller, M.L., Bloom, J.A., Fridrich, J., Kalker, T.: Digital Watermarking and Steganography, 2nd edn. The Morgan Kaufmann Series in Computer Security (2007)
11. Fridrich, J., Goljan, M., Du, R.: Invertible authentication in Security and Watermarking of Multimedia Contents III. SPIE **4314**, 197–208 (2001)
12. Li, J.J., Wu, Y.H., Lee, C.F., Chang, C.C.: Generalized PVO-K embedding technique for reversible data hiding. Int. J. Network Secur. **20**(1), 65–77 (2018)
13. Tian, J.: Reversible data embedding using a difference expansion. IEEE Trans. Circ. Syst. Video Technol. **13**, 890–896 (2003)
14. Luo, L., Chen, Z., Chen, M., Zeng, X., Xiong, Z.: Reversible image watermarking using interpolation technique. IEEE Trans. Forensic Secure **5**(1), 187–193 (2010)
15. Fallahpour, M.: Reversible image data hiding based on gradient. IEICE **5**, 870–876 (2008)
16. Celik, M.U., Sharma, G., Tekalp, A.M.: Lossless watermarking for image authentication, a new framework and an implementation. IEEE Trans. Image Process **15**(4), 1042–1049 (2006)
17. Sachnev, V., Kim, H.J., Nam, J., Suresh, S., Shi, Y.Q.: Reversible watermarking algorithm using sorting and prediction. IEEE Trans. Circuits Syst. Video Technol. **19**(7), 989–999 (2009)
18. Hong, W.: An efficient prediction-and-shifting embedding technique for high quality reversible data hiding. EURASIPJ. Adv. Signal Process (2010)
19. Weng, S., Shi, Y., Hong, W., Yao, Y.: Dynamic improved pixel value ordering reversible data hiding. Inf. Sci. **489**, 136–154 (2019). Elsevier
20. Zhang, W., Hu, X., Li, X., Yu, N.: Recursive histogram modification establishing equivalency between reversible data hiding and lossless data compression. IEEE Trans. Image Process. **22**(7), 2775–2785 (2013)
21. Gao, X., An, L., Yuan, Y., Tao, D., Li, X.: Lossless data embedding using Generalized statistical quantity histogram. IEEE Trans. Circuits Syst. Video Technol. **21**(8), 1061–1070 (2011)
22. Li, X., Li, B., Yang, B., Zeng, T.: General framework to histogram shifting based reversible data hiding. IEEE Trans. Image Process. **22**(6), 2181–2191 (2013)
23. Li, X., Yang, B., Zeng, T.: Efficient reversible watermarking based on Adaptive prediction-error expansion and pixel selection. IEEE Trans. Image Process. **20**(12), 3524–3533 (2011)
24. Li, X., Li, J., Li, B., Yang, B.: High fidelity reversible data hiding scheme based on pixel value ordering and prediction error expansion. Signal Process **93**(1), 198–205 (2013)
25. Qu, X., Kim, H.J.: Pixel-based pixel value ordering predictor for high-fidelity reversible data hiding. Signal Process. **111**, 249–260 (2011)
26. Hu, Y., Lee, H.K., Li, J.: DE-based reversible data hiding with improved Overflow location map. IEEE Trans. Circuits Syst. Video Technol. **19**(2), 250–260 (2009)
27. Ni, Z., Shi, Y.Q., Ansari, N., Su, W.: Reversible data hiding. IEEE Trans. Circuits Syst. Video Technol. **16**(3), 354–362 (2007)

Early Detection of Coronary Artery Diseases Using Endocrine Markers

Felician Jonathan More[1], Zenon Chaczko[1,2], and Julita Kulbacka[2,3](✉)

[1] School of Electrical and Data Engineering, University of Technology Sydney, Ultimo, Australia
felician.j.more@student.uts.edu.au, zenon.chaczko@uts.edu.au
[2] DIVE IN AI, Wroclaw, Poland
[3] Faculty of Pharmacy, Department of Molecular and Cellular Biology, Wroclaw Medical University, Wroclaw, Poland
julita.kulbacka@umed.wroc.pl

Abstract. Cardiovascular diseases including coronary artery disease is the leading cause of death in the well developed and developing countries of the 21st century and has a higher rate of mortality and morbidity. Dysfunction of the pituitary, thyroid, and parathyroid glands caused cardio/cardiovascular diseases including changes in blood pressure, contractility of myocardium - systolic and diastolic myocardial functions, endothelial and dyslipidemia. Dysfunction of thyroid, parathyroid and adrenocorticotropic hormones caused imbalance of endocrine system such as hyper and hypo function, effects on pathophysiology of the cardiovascular system.

Keywords: Cardiovascular disease · Thyroid dysfunction · Endocrine markers

1 Background

Cardio/cardiovascular diseases were the prime cause of death in the world in the 21st century and had a higher rate of morbidity and mortality. Statistical analysis reveal that 12–14% of men and 10–12% of women between the age of 65 and 84 years have been treated for angina. In the United States of America, one in three individuals around 92.1 million have cardiovascular diseases including 10 million adults with angina pectoris, and also it is stated that there were 836,546 deaths with one in three deaths in USA; of which 46% were males and 47.7% of women in non-Hispanic black adults had died in the USA in some form of cardiovascular diseases, and 2,300 people died each day. In USA 43.8% of deaths were the cause of coronary artery diseases and accounts for one in 7 deaths. In Australia, CAD is the main cause of cardiovascular disease. Australian Bureau of Statistics of National Health Survey in 2014–2015 report shows, that 645,000 people aged 18 and over had CAD and affecting one in six individuals aged 75 and over. In 2015 hospital based on data revealed that 61,600 individuals

N. T. Nguyen et al. (Eds.): ACIIDS 2020, LNAI 12034, pp. 593–601, 2020.
https://doi.org/10.1007/978-3-030-42058-1_50

aged 25 and over had CAD which accounts 170 events each day. Statistical data shows that 19,100 deaths in Australia including 40% of CAD deaths from heart attack which is 8000 deaths, 43% of cardiovascular deaths and it signifies 12% of all deaths; 78% death of individuals aged 55 and over had CVD and 52% were women. Due to advanced technologies and diagnostic methods, and treatment results older age groups death rate declined and improved from 0.5 to 4.9% for aged 85 and over, and aged 75–84 to 2.2 to 6.3%. Death rate fell by 86% that is 204 to 28 deaths per 100,000 population of 5.8 per year in myocardial infarction aged over 35 years. In year 2015, 17.9 million deaths represent cardio-vascular diseases in globally and expected to raise by 23.6 million in year 2030 [4].

2 Endocrine Markers and Cardiovascular Systems

In 2001 NIHC (National Institute of Health Consortium) presented a definition of a biomarker as a "characteristic that is objectively measured and evaluated as an indicator of normal biological processes, pathogenic processes, or pharmacologic responses to a therapeutic intervention" [1]. Since the endocrine molecules were detected to be released by cardiomyocytes and vascular endothelial cardiac cells, then endocrine markers seems to be promising in cardiovascular system monitoring [2]. The following endocrine markers can be used to detect cardio/cardiovascular diseases in early stages as a result of effects on the cardiovascular system by disordered function of thyroid, parathyroid and adrenocorticotropic endocrine gland functions.

2.1 Thyroid Hormone

Physiology and Function. Thyroid hormone function is regulated by feedback loop mechanism, and is coordinated by the hypothalamus-pituitary-thyroid axis. Hypothalamus secretes the hormone and stimulates the anterior pituitary to produce hormones called the thyroid stimulating hormone and it releases thyroid hormone from the thyroid glands. The production of thyroid stimulating hormone (TSH) and release of thyroid hormone levels are regulated by the thyrotropin releasing hormone (TRH). Larger changes in thyroid stimulating hormone can be observed by small changes in thyroid hormone concentrations because of the log linger relationship with T4 (thyroxine) and TSH. Two main iodinated thyroid hormones are thyroxine (T4), and triiodothyronine (T3) and it is considered as active hormones and both have pathophysiological/biological effects. Thyroid endocrine negative feedback mechanism can be disordered by illnesses such as heart failure or acute myocardial infarction and by a reduction in serum thyroid hormone, without a rise in levels of TSH. The concentrations of circulating thyroid hormone changes of TSH is highly sensitive and is able to detect changes in thyroid function which lead to thyroid diseases [6]. The thyroid cellular mechanisms for the release of thyroxine, triiodothyronine and iodine transport into the blood is depicted in Fig. 1.

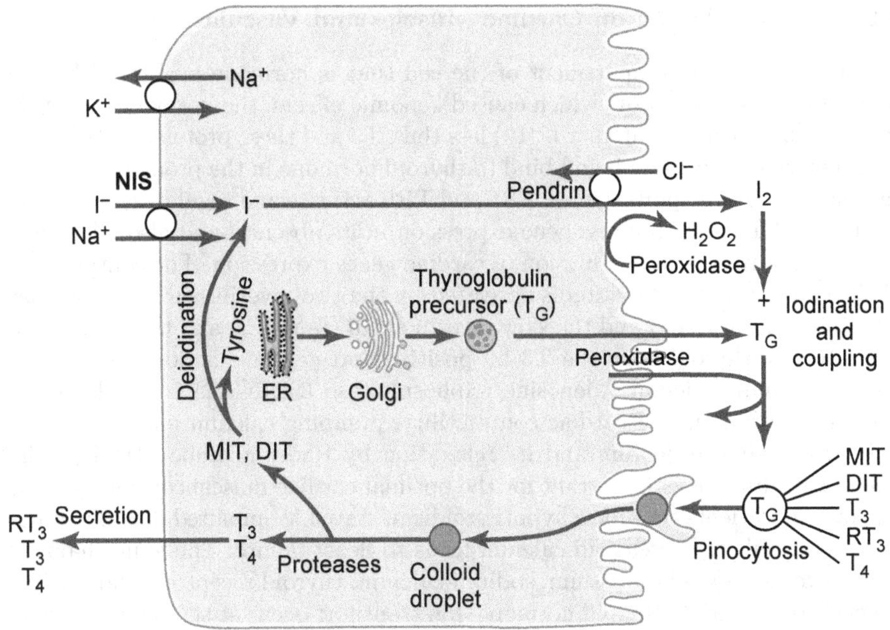

Fig. 1. Thyroid cellular mechanisms [3]

Genomics. The architecture of thyroid hormone genomics has a polygenic function and is stated as Mendelian disorders which has been linked with thyroid function as single-nucleotide variants (SNVs). The polygenic basis of thyroid dysfunctions addressed also as 37 genomic loci has been found to increase the risk of thyroid hormones as a result of that effect on the parameters of thyroid functions. Thyroid hormone metabolism can be directly involved by the gene encoding enzymes such as e.g. DIO1 or transport, FOXE1 - growth factors, MCT8 - transcription factors for thyroid development and related to proteins signaling hypothalamus-pituitary-thyroid axis. Studies show that DIO1 and DIO2 loci can cause phenotypes such as neurocognitive, and bone disorders and independent of thyroid hormones. There was proved that hypothyroidism caused by autoimmune disorders and several genes are the HLA class1 region, PTPN22, SH2B3 or VAV3 and hyperthyroidism including SLAMF6 or the 6q27, Tg, GPR174-ITM2A, the C1QTNF6-RAC2 locus and 14q32.2 loci. Degradation of cAMP involved in the loci encoding a phosphodiesterase of PDE8B is important for the TSH loci and thyroid functions. Therefore the polygenic basis of thyroid disorders can be addressed as the variation of FT4 and TSH throughout the general population and changes within each patients. Additionally, the available data indicate that genomic effect on TH are also mediated by TH nuclear receptors. Authors observed that TRα1 (TH receptor) isoform has been shown to play an important role in the regulation of cardiac genes [5].

2.2 Thyroid Action on Cardiac Muscles and Vasculature

In the intracellular compartment of the cell that is hormone facilitated by the thyroid hormone receptors which caused genomic effects, the hormone receptors (protein) bind with T4 affinity (<10) less than T3 and these proteins exist in two isoforms which are a and b and bind to thyroid hormone in the promoter areas of thyroid hormone response genes. TRa and TRb activate and regulate genes in the existence of T3 and suppress gene expression in its absence, and also TRa1 isoform plays a role in the regulation of cardiac genes expression. The contractility of the cardiac muscle function is activated by the two myosin heavy chains: fast myosin chain (a-MHC) and the slow myosin chain (b-MHC), and the actions are regulated by thyroid hormone T3 by positive and negative actions. Sarcoplasmic Reticulum Calcium Adenosine triphosphatase (SERCA2) is regulated by T3, and the function of cardiac contractility; pumping calcium ions Ca2+ into the sarcoplasmic reticulum and its relaxation by Phospholamban (PLB). The role of calcium ions is important for the optimal cardiac muscle contraction and relaxation. Patients who have hypothyroidism, stated as impaired diastolic function as a result of decrease in calcium leads to heart failure. The ionic channels such as voltage gated potassium, sodium-calcium, thyroid receptors, and cardiac gene are regulated by thyroid hormone. Ion transport occur at the cell membrane and is regulated by the nongenomic effects of the receptor independently and the rate of action can be identified as high. Thyroid hormones can be regulated by actions of several ionic channels and post translationally by the mechanisms of nongenomic effects. Also the heart rate can be increased by the actions of both mechanisms of the atrial myocyte. T3 is transported into the cell cytoplasm of the responsive cell. The numerous families of thyroid hormone transporters are highly iodothyronines which are the heterodimeric L-type amino acid, the Naþ-independent organic anion transporting polypeptides, the Naþ-taurocholate co-transporting polypeptides and the monocarboxylate transporters (MCTs) 8 and 10. The functions of the MCT8 in thyroid hormone transport in cardiac myocytes tissues by mutations are the cause of heart rate abnormalities, specific thyroid and neurological defects. However, MCT10 has a less affinity to T4 than T3 and MCT8 has a greater capacity to transport T3. Also T4 is transported across the cardiac myocytes sarcolemma nor deiodinated into T3. If the serum levels of T3 falls and the levels of T4 and TSH are considered to be lower or normal then the heart action will express a hypothyroidism. Genomic and nongenomic effects on vasculature occur at the endothelial cell and vascular smooth muscle. Ion channels activations such as sodium, potassium, and calcium channels and regulation of signal pathways are the indirect effects of the non-genomic mechanism of the thyroid hormone. Research studies show that endothelial nitric oxide production and vascular tone can be regulated by thyroid hormones. The patients with chronic heart failure and undergoing coronary artery bypass grafting with similar vasodilatory effects can produce thyroid hormone T3 and affects the systemic vasculature than the pulmonary vasculature [6] (Fig. 2).

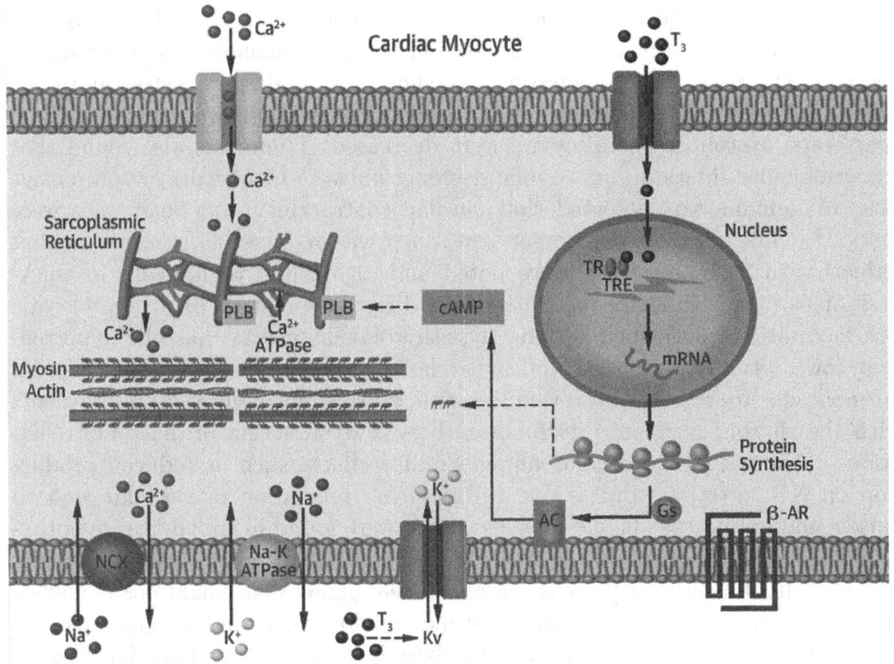

Fig. 2. Thyroid action on cardiac myocyte [6]

2.3 Thyroid Hormones Disorders

Hyperthyroidism. The disordered of thyroid hormones caused cardiac/ cardiovascular diseases and leads to heart failure result either of hypothyroidism or hyperthyroidism. The hyper-dynamic conditions can be found such as atrial fibrillation with patients who have hyperthyroidism. The study shows that an increased preload, contractility, and reduced vascular resistance with respect to increasing cardiac output [9]. Untreated patients with hyperthyroidism may undergo arrhythmias, cardiac hypertrophy, increased preload and caused heart failure. High cardiovascular mortality rate has been identified with the individuals who have hyperthyroidism and did not undergo treatment such as hormone therapy. The study shows that an increased carotid intima media thickness with patients who has hyperthyroidism. Aortic stiffness can be found in patients with dysfunction of thyroid hormones such as high levels of free thyroxine T4. The population base study shows that with an absence of hypertension as noted in hyperthyroidism patients with changes in pulse or blood pressure, have a higher risk of cardiovascular diseases such as coronary artery diseases, atrial fibrillation (AF) and heart failure (HF) events noted with the low serum thyroid stimulating hormone as a result of an increase in plasma fibrinogen [8]. Coronary artery diseases are present in patients with hyperthyroidism, hence, thyroid hormones can be used to detect cardio/cardiovascular diseases in early stage.

Hypothyroidism. The dysfunction of thyroid hormone is caused of hypothyroidism as a result of cardio/cardiovascular diseases including coronary artery diseases. The research conducted in rats, shows an increase in systole and diastole of left ventricle diameters; while an ejection fraction, wall thickness, and heart rate; systolic blood pressure were decreased. There was also found that left ventricular dilation and systolic dysfunction were observed in hypertensive rats. In humans was detected that cardiac contractility and heart rate were reduced, while peripheral vascular resistance was reduced as a result of that reduction in cardiac output were noted and also found to increase in thickness of carotid intima media. There is a direct association of atherosclerotic risk factors with patients who have hypothyroidism such as diastolic hypertension, lower production of NO and hypercholesterolemia. However, that thyroid hormone disordered can be reversible with thyroid hormone therapy. Patients with the thyroid hormones dysfunction have a dysfunction of diastolic, relaxation and impaired ventricular filling. Similar effects such as reduced production on NO, arterial stiffness due to impairing relaxation of vascular smooth muscle and an increase in vascular resistance were found in subclinical hypothyroidism and can be reversible by treating with thyroid hormone therapy. Patients who had subclinical hypothyroidism, also reveal higher concentrations of cholesterol, and increased in systolic and diastolic blood pressure and increased carotid intima media thickness. The population based study showed that higher levels of TSH caused increased arterial stiffness in women. Meta-analysis study shows that an increased risk of cardiovascular events include coronary artery diseases and death with high concentrations of TSH levels (TSH levels $> 10\,\mathrm{mU/l}$). Subclinical hypothyroidic patients undergo levothyroxine hormone therapy showed promising effects in cardiovascular functions and also had lower death rate in cardiovascular events. Levothyroxine hormone treatment occurred effective in the diseases involving mitochondrial function in cardiac muscles [8]. Therefore the thyroid hormone dysfunction led to cardio/cardiovascular diseases and it is a vital marker for early detection of coronary artery diseases.

2.4 Parathyroid Hormone

The calcium and phosphorus homeostasis can be maintained by the parathyroid hormone and plays a vital role in human endocrine system. As a result of that para-thyroid hormone affects major organs such as bone, intestinal mucosa and kidney. Parathyroid hormone is an important marker for early detection of cardio/cardiovascular diseases. Statistical study shows that 21.6 per 100,000 annually of primary hyperparathyroidism and incidence in females reaching 63.2 per 100,000 annually at the age of 65–67. The dysfunction of parathyroid hormone such as inappropriately elevated levels of calcium concentrations led to hyperparathyroidism or hypoparathyroidism, causing cardio/cardiovascular diseases including coronary artery diseases. Due to the imbalance of parathyroid hormone, it may disturb the endocrine homeostasis and cause parathyroid gland hyperplasia, autonomous adenoma, vitamin D deficiency and secondary hyperparathyroidism. 85% of the patients with primary hyperparathyroidism are symptomatic or asymptomatic with elevated calcium levels. Parathyroid hormones caused cardiovascular diseases such as hypertension, arterial stiffness by increased calcium deposition, stimulation of renin aldosterone system, endothelial and increased sympathetic activity. Patients with parathyroid hormone disorders such as hyperparathyroidism have higher carotid intima media thickness with vessel stiffness. Research studies shows that primary hyperparathyroidism patients have left ventricle hypertrophy with hypertension. Whereas animal studies showed that severe effects on cardiac muscles led to hypertrophy due to parathyroid hormone. Diastolic dysfunction has been noted in patients with primary hyperparathyroidism with decreased E/A ratio and isovolumetric relaxation time. Diastolic dysfunction also had an impact on the aortic and mitral valve, and myocardium calcifications with significant hypercalcemia. If calcium concentrations levels were higher than 12 mg/dL, the reduction of the ventricular cardiac contractility action and the effective refractory period due to hypercalcemia was observed. In hypercalcemia can be detected an increased QRS complex, shortened intervals of QT and QTc, biphasic T waves and shortened ST segment in ECG. Cardiac conduction abnormalities can be also identified by serum calcium concentrations levels [7].

Hypoparathyroidism can be defined by low concentrations levels of calcium "hypocalcemia" or low parathyroid hormone as a result of damage to the parathyroid glands. Diagnostic assessment can be included in serum measurements such as calcium, albumin, magnesium, 25-hydroxyvutamin D, phosphorus, and creatinine. It can be reversible by calcium supplement replacement treatment. Patients with hypocalcemia or hypoparathyroidism caused heart failure due to impaired contraction-relaxation mechanism-decreased myocardial action and dilated cardiomyopathy. Calcium channels actions can be regulated by the rate of intake in extracellular calcium levels. This led to change in QT interval, T wave flattening or T wave inversion or ST segment elevation due to coronary artery spasm noted in ECG [7]. Therefore that parathyroid hormone is a vital marker for early detection of coronary artery diseases including cardiac diseases.

2.5 Adrenocorticotropic Hormone

Corticotroph cells of the anterior pituitary gland are responsible for the synthesis and secretion of adrenocorticotropic hormone or corticotropin, which regulates the secretion of adrenal cortisol. The excess production of adrenocorticotropic hormone can cause pituitary corticotroph adenoma such as lung cancer, medullary thyroid cancer, Cushing's syndrome or hypercortisolism. The excess secretion of adrenocorticotropic hormone caused 80% of dependent cases, and 20% of independent cases which include dysplasia, bilateral adrenal hyperplasia, and unilateral adrenal adenomas. The statistical results show that endogenous Cushing's syndrome is detected in 2.3 cases per million annually, and can be demonstrated by an increased levels of cortisol. The treatment of Cushing's syndrome can be monitored and controlled by the levels of cortisol for long term, and as a result of weight gain, muscle weakness, hypertension, obesity, hirsutism, menstrual disorders for women and fatigue. Hypercortisolism can be treated by transsphenoidal surgery, radiotherapy and uni-/bilateral adrenalectomy. The available research shows that hypercortisolism is directly associated with hypertension, insulin resistance, obesity, platelet function and dyslipidemia. 80% of adult patients who have endogenous Cushing's syndrome present hypertension and results vasodilation, changes in regulation of plasma volume, systemic vascular resistance, increased lipoprotein, triglycerides and decreased HDLc. The treatment of Cushing's syndrome with hormone therapy can improve hypertension. Coagulation factors can be controlled by increased cortisol levels as a result of Willebrand, and concomitantly increasing factor. There is a direct association between the cardio/cardiovascular diseases such as diastolic dysfunction, left ventricle systolic dysfunction and Cushing's syndrome. Patients who have Cushing's syndrome have increased posterior wall thickness, increased left ventricle mass index, interventricular septum and increased relative wall thickness. Diastolic dysfunction can be determined by myocardial relaxation impairment with early left ventricle relaxation, and isovolumetric relaxation times. Therefore, the dysfunction of adrenocorticotropic hormone can cause cardio/cardiovascular diseases, and it is an important marker for early detection of coronary artery diseases [7].

3 Conclusions

The anterior pituitary gland that synthesize and secrete major hormones such as thyroid hormone, parathyroid and adrenocorticotropic hormones are vital for homeostasis of the body. Dysfunction of major hormones can cause cardiac disorders as was described in the above sections of this paper. Therefore, thyroid, parathyroid and adrenocorticotropic hormones with subtle changes can lead to cardio/cardiovascular diseases including coronary artery diseases. The restoration of normal endocrine function based on hormonal therapy can be a suitable method for treating cardiovascular diseases. Theses hormones are the important pathophysiological markers for early detection of cardio/cardiovascular diseases including coronary artery diseases. Moreover, the hormones detection is a relatively not complicated, thus endocrine hormones seem to be more reliable indicators in prognosis, pharmacodynamic and predictive cardiovascular studies.

Acknowledgements. This study was supported by the Scientific Research from Technical University of Technology Sydney, School of Electrical and Data Engineering and DIVE IN AI.

References

1. Dhingra, R., Vasan, R.S.: Biomarkers in cardiovascular disease: statistical assessment and section on key novel heart failure biomarkers. Trends Cardiovasc. Med. **27**(2), 123–133 (2017)
2. Fioranelli, M., Bottaccioli, A.G., Bottaccioli, F., Bianchi, M., Rovesti, M., Roccia, M.G.: Stress and inflammation in coronary artery disease: a review psychoneuroendocrineimmunology-based. Front. Immunol. **9** (2018). https://doi.org/10.3389/fimmu.2018.02031

3. Hall, J.E.: Pocket Companion to Guyton & Hall Textbook of Medical Physiology E-Book. Elsevier Health Sciences (2015)
4. More, F.J., Chaczko, Z.: Non-invasive methods in the detection of coronary artery disease. In: 2018 26th International Conference on Systems Engineering (ICSEng), pp. 1–5. IEEE (2018)
5. Pietzner, M., Kacprowski, T., Friedrich, N.: Empowering thyroid hormone research in human subjects using OMICs technologies. J. Endocrinol. **238**(1), R13–R29 (2018). https://doi.org/10.1530/joe-18-0117
6. Razvi, S., et al.: Thyroid hormones and cardiovascular function and diseases. J. Am. Coll. Cardiol. **71**(16), 1781–1796 (2018). https://doi.org/10.1016/j.jacc.2018.02.045
7. Rhee, S.S., Pearce, E.N.: The endocrine system and the heart: a review. Revista Española de Cardiología (Engl. Ed.) **64**(3), 220–231 (2011). https://doi.org/10.1016/j.rec.2010.10.016
8. Vale, C., Neves, J.S., von Hafe, M., Borges-Canha, M., Leite-Moreira, A.: The role of thyroid hormones in heart failure. Cardiovasc. Drugs Ther. **33**(2), 179–188 (2019). https://doi.org/10.1007/s10557-019-06870-4
9. Vincent, J.L.: Understanding cardiac output. Crit. Care **12**(4), 174 (2008)

Author Index